AF566666

HOMOGENEOUS AND HETEROGENEOUS CATALYSIS

HOMOGENEOUS AND HETEROGENEOUS CATALYSIS

Proceedings of the Fifth International Symposium on Relations between Homogeneous and Heterogeneous Catalysis

Novosibirsk, 15-19 July 1986

Editors

Yu. Yermakov
V. Likholobov

Utrecht, The Netherlands

VNU Science Press BV
P.O. Box 2073
3500 GB Utrecht
The Netherlands

© 1986 VNU Science Press BV

First published 1986

ISBN 90-6764-073-5

All rights reserved. No part of this publication may be reproduced, stored in a retrieval system, or transmitted, in any form or by any means, electronic, mechanical, photocopying, recording, or otherwise, without the prior permission of the copyright owner.

Printed in Great Britain

Contents

*The invited lectures

FORWORD. WHAT CONNECTS HOMOGENEOUS AND HETEROGENEOUS CATALYSIS?

Yu.I. Yermakov
Institute of Catalysis, Novosibirsk 630090, USSR

The 5th, by account, symposium is a kind of jubilee volte-face in the organization of international meeting of scientists, who are interested in the search of a connection between homogeneous and heterogeneous catalysis. Therefore, it is suitable to recollect some facts from yet a not-very-long history of the organization of these meetings.

The 1st symposium was organized on Prof. Delmon's initiative, and held in Brussels from 23 to 25 of October, 1974. So marked Prof. Delmon in his Introduction to the proceedings of this symposium (Ref. 1), the main task of the organization was laid upon the shoulders of Prof. Janess. Nasty weather with the autumn rains in Brussels, could not ruin the good mood of the participants of the symposium, roused by the interesting programme and by the excellent organization of the symposium. In his lecture, Prof. Delmon noted that the symposium had been devoted to the consideration of the common features of the phenomena of homogeneous and heterogeneous catalysis, and to the consideration of similarities of catalytic reactions, occuring in the presence of homogeneous and heterogeneous catalysts. The organization of the 1st symposium on the consideration of these problems in 1974, proved to be proper time. By this time the comprehension of that the many reactions on heterogeneous catalysts proceed through the formation of the same intermediate species, that are in the reactions in the presence of soluble metal complexes as catalysts began to spread widely.

Since 1960's began to develop a new trend of preparing catalysts by the immobilization of metal complexes on the surface of carriers or in the different matrices. By the same time, several endeavours have been made to consider the clusters of metals as the analogs of dispersed metal particles and their use in catalysis.

A great interest in the occurrences laid on the junction between the earlier developing separately spheres of homogeneous and heterogeneous catalysis, estimated attention to the first international symposium: in its list of participants we find 230 people from 24 countries. The International catalysis community appreciated the successes of the symposium and its timeliness.

There were such enthusiasts, thanks to whose efforts were held the next symposia in Lyons (November 1977), Groningen (May, 1981), Asilomar (October, 1983) (the detailed information about these symposia look further in this volume).

By the way, precisely during the symposium held in Groningen, during the discussion by the International advisory committee on the organization of symposium, of the question of where should the next symposia take place, Prof. Kaesz occupied himself in the organization of symposium in Asilomar and then the idea of holding the symposium in Novosibirsk, where the Institute of Catalysis of the Siberian Branch was conducting works on the study of catalysis by anchored organometallic compounds.

Let's recollect that the specialists on catalysis long ago realized the necessity of organizing regular meetings. The 1st International Congress on Catalysis was organized in 1956 in Philadelphia in the city, where (in 1949) the first club on catalysis was organized. Since that time these congresses are held regularly every four years, having become a formal organization with a clear succession of organizers. Now this congress is in fact a congress of heterogeneous catalysis. The specialists on homogeneous catalysis began to gather on their own meetings since 1978, when Prof. Tsutsui organized the 1st International Symposium. Since then the Symposium on homogeneous catalysis is held every two years. Hardly anybody has any doubt in the distinct difference in the areas of heterogeneous and homogeneous catalysis, the specialists of every one of which regularly assemble on their own international meetings. But a question might occur on how wide and definite is the intermediate area between homogeneous and heterogeneous catalysis, and is it worth holding a special symposium, devoted to this field.

To the last part of the question a positive answer gives the tradition - all the symposia held until now were of great interest. One of the main tasks of these symposia - to create an opportunity of meeting of the specialists from different areas of catalysis for exchange of ideas and redistribution of experience of some areas of research into others. Precisely on the junction of different areas of catalysis new approaches could be found to the explanation of catalytic phenomena and the search for new catalysts.

But just like any other interdisciplinary area the space between the phenomena of homogeneous and heterogeneous catalysis represents a fuzzy interface the position of which undoubtedly depends on the view of whoever establishes it.

I should say, a few main topics of symposia on relations between homogeneous and heterogeneous catalysis remain unrevised:

- heterogenization of metal complexes catalytically active in solutions,
- the synthesis of active centers (including those that do not have analogues among soluble metal complexes) by different types of attaching of the metal complexes on different surfaces, matrices or in cavities,
- the discussion of the action of heterogeneous catalytic systems with the use of data on stoichiometric reactions of organometallic compounds and experience of homogeneous catalysis.

On the scheme an endeavour has been given to present the divisions of chemistry which could be regarded as parts of a bridge, connecting classical banks of homogeneous and heterogeneous catalysis. All the space between the banks could get into the subjects of symposia on relations between homogeneous and heterogeneous catalysis. In the centre of the bridge is a part corresponding to catalysis by immobilized metal complexes, equidistant from both banks. Area of homogeneous catalysis sometimes is considered as occupying part of interstitial space and comes up to this part (on this part the subjects of symposia on homogeneous and heterogeneous catalysis overlap). Lately, classical realm of the heterogeneous catalysis begins to spread up to the area of immobilized metal complexes. On the one hand, this event is connected with that in the region of catalyst preparation more and more methods of synthesis of the individual complexes as precursors of the catalytic active centres are applied. On the other hand, in explaining the phenomena of heterogeneous catalysis, at present, the conception of active centres is used more and more frequently.

By the active centre here one should comprehend a surface compound, which includes such a number of atoms that is sufficient for the catalytic cycle to occur. The adjacent to the active centre atoms of the surface or the volume of a catalyst render an electronic or a "steric" influence, which plays the same role as the action of ligands under the catalysis by metal complexes in a solution.

The problem of the role of active centres in catalysis is not new. It was Taylor, who proposed that catalysis occurs on special sites of the surface, whereas the main surface area is inactive. In catalysts, in which reaction takes place on separate active centres, but not on the entire surface, catalytic activity, referred to the unit of surface (specific catalytic activity, SCA) does not necessarily have to be constant. Besides, SCA in this case could depend upon the method of preparing the catalyst.Catalysts with such features, may be called catalysts corresponding to Taylor's conception.

G.K. Boreskov proposed another conception - the whole surface of the catalyst is responsible for catalytic activity (in case of the supported catalysts - the surface of the supported component) (Ref. 7). SCA is determined by the strictly chemical composition of the catalyst and is independent of the method of catalyst preparation. According to G.K. Boreskov's opinion the effect of the reactive environment levels the differences in the initial state of the catalyst surface. In catalysis, corresponding to Boreskov's conceptions, the proportionality between activity and the surface of working catalyst must be observed.

There are no doubts that the two points of view might be just for extreme cases. Catalysts of Taylor's type are first of all the systems functioning under the relatively low temperatures. In such cases coordinatively unsaturated sites on the surface could remain for a long time, playing the role of active centres. Typical catalysts of Taylor's type are the catalysts of polymerization on the base of $TiCl_3$ in which the concentration of the active centres located not on the entire surface, but only on the lateral planes of crystals, depends, for instance, upon the catalyst's degree of grinding. In this case the correlation between the activity per one gram of catalyst and the BET surface of catalyst might not occur.

Lately, the data have been accumulating that in case of many classical catalysts active centres are being localized just on the one of the planes of the catalyst crystallities and are mononuclear. For instance, such a conception proposed for the explanation of the action of sulphide catalysts, where the active centres are, for example, Co ions fixed in the matrix of MoS_2 (Ref. 9), or Ni ions on the side plane of WS_2 (Ref. 10).

And for metal catalysts the explanations are being put forward, that mononuclear active centres (the atoms of metal in metallic matrix) are their active sites (Ref. 11).

Probably, in the future as well, the development of the methods of preparation of heterogeneous catalysts by the purposeful synthesis of the surface complexes, and the identification of discrete sites in catalysts, prepared by traditional methods, would make the boundary between classical heterogeneous catalyst and the intermediate region examined here more vague.

Still the central part of this region - the catalysis by immobilized complexes on the surfaces and in matrices remains a sufficiently clear link of the bridge between homogeneous and heterogeneous catalysis. Up to the present moment many approaches have been made to the preparation of such catalysts with the use of the oxide carriers, polymer matrices, mixed polymer-inorganic supports, etc. For a while many researches took great interest in "heterogenization" of homogeneous catalysts and revealed new and new facts and prooofs, that the complexes on the surface function similarly as their soluble analogues of the close composition. But now it seems possible to say that the straightforward heterogenization does not prove its value as a method of getting interesting practical catalysts, because with such a method all attempts to achieve systems stable in reactive environment were unsuccessful.

Generally, the total results of the practical use of catalysts prepared through fixation of metal complexes may rouse disappointment. Evidently, the only catalyst of such a type which is used in practice in a large scale process (the author would be thankful if he'll be pointed out other examples) is a system prepared by an interaction of biscyclopentadienylchromium with SiO_2. This catalyst is used for obtaining some types of polyethylene in the Union Carbide in gas-phase process (Refs. 12,13). However, other catalysts for olefin polymerization, whatever the method of preparing, contain as active centres surface organometallic compounds. There is no problem of long-time stability of the catalyst in processes of olefin polymerization - active centre, before its death, has time to conduct such a number of cycles, that the catalyst remains as a small admixture (less than 0.1%) in the polymer.

Undoubtedly, the use of catalysts, obtained through fixing metal complexes pays off in such a case if they possess qualitatively different properties than the catalysts obtained through traditional methods. For instance, it would have been tempting to create catalysts providing selectivity close to one hundred percent for the necessary product. There are reports, that the catalysts, obtained with the use of organometallic compounds can display special properties. Thus, supporting carbonyl complexes of rhodium (Ref. 14) and cobalt (Ref. 15) on the oxides, such catalysts have been prepared, on which ethanol is produced from syn-gas.

However, in this case as well, such a level of selectivity and stability of catalysts has not been reached, which would justify their practical use. As before elaboration of the methods of purposeful synthesis of the active centres on the surface is still a major problem of the area, which is intermediate between homogeneous and heterogeneous catalysis. Research directed on its solution is important; during this research data will be accumulated on the synthesis of precursors of the active centres, their reactivity and peculiarities of catalysis with their participation.
Researches of such a kind give fresh impulse to the working up of the new catalysts and appearing of new conceptions for explaining the mechanism of their action. One can hope, that the 5th International symposium on relations between homogeneous and heterogeneous catalysis shall make its own contribution to the development of this direction. However, at the moment of writing this introduction (March 1986) I do not think it possible to value this contribution.

In conclusion I would like to express gratitude to those, who applied great efforts to the preparation of the Proceedings of the Symposium - to L.Ya. Startseva (technical editing of the works in the English language), T.B. Anufrienko (technical editing of the works in the Russian language), A.D. Gindina (translation into English), E.N. Myakisheva (preparation of the camera-ready typescript in English).

REFERENCES

1. B. Delmon, in Catalysis, Heterogeneous and Homogeneous, B. Delmon, G. Jannes (Edts.), Elsevier (1975).
2. B. Delmon and G. Jannes, ibid, p. 25.
3. H. Heinemann, ACS Symp. Ser., 222, 519-521 (1983).
4. M. Tsutsui (Ed.), Fundamental Research in Homogeneous Catalysis, vol. 3, Corpus Christi, Texas, 1978, Plenum Press, N.Y., London (1979).

5. H.S. Taylor, Z. Electrochem. Ang. Phys. Chem., 35 542 (1929).
6. J. Turkevich, ASC Symp. Ser., 222, 463-490 (1983).
7. G.K. Boreskov, in Mechanism of Action of Solid Catalysts in "Heterogeneous Catalysis in Chemical Industry" (in Russian), Goskhimizdat, Moscow (1954).
8. Yu.I. Yermakov and V.A. Zakharov, in J.C.W. Chien (Ed.), Coordination Polymerization, Academic Press, Inc., N.Y., San Francisko, London (1975).
9. H. Topsøe, R. Candia, N.-Yu. Topsøe and B.S. Clausen, Bull. Soc. Chim. Belg., 93, 783-806 (1984).
10. Yu.I. Yermakov, A.N. Startsev, V.A. Burmistrov, O.N. Shumilo and N.N. Bulgakov, Appl. Catal., 18, 33-46 (1985).
11. J.F. Rooney, J. Mol. Catal., 31, 147-159 (1985).
12. F.J. Carol, G.L. Karapinka, C. Wu, A.W. Dow, R.N. Johnson, W.L. Carrick, J. Polymer. Sci., p. A-1, 10, 2621-2637 (1972).
13. Chem. Eng., 104 (1974).
14. M. Ichikawa, CHEMTECH, 674-680 (1980).
15. A.S. Lisitsyn, A.V. Golovin, V.L. Kuznetsov and Yu.I. Yermakov, J. Catal., 95, 527-538 (1985).

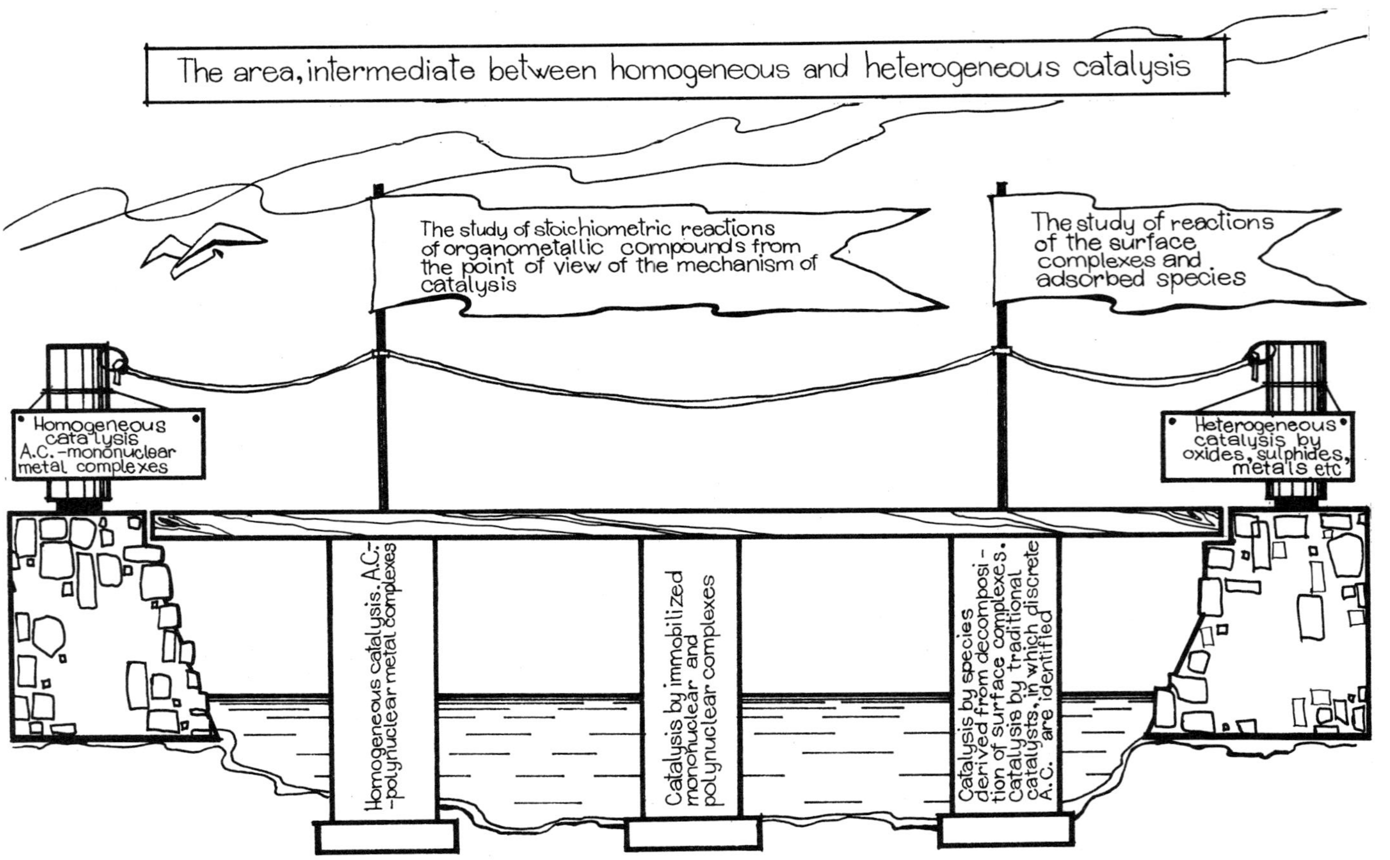
The area, intermediate between homogeneous and heterogeneous catalysis
The study of stoichiometric reactions of organometallic compounds from the point of view of the mechanism of catalysis
The study of reactions of the surface complexes and adsorbed species
Homogeneous catalysis A.C.-mononuclear metal complexes
Heterogeneous catalysis by oxides, sulphides, metals etc
Homogeneous catalysis. A.C.-polynuclear metal complexes
Catalysis by immobilized mononuclear and polynuclear complexes
Catalysis by species derived from decomposition of surface complexes. Catalysis by traditional catalysts, in which discrete A.C. are identified

I N F O R M A T I O N

ON INTERNATIONAL SYMPOSIA ON RELATIONS BETWEEN HOMOGENEOUS AND HETEROGENEOUS CATALYSIS

The First International Symposium on Relations between Homogeneous and Heterogeneous Catalysis

October 23-25, 1974
Brussels, Belgium

Topics of the Symposium:

- interpretation of phenomena observed in heterogeneous catalysis thanks to fundamental approaches pertaining to homogeneous catalysis;
- heterogenization of homogeneous catalysts.

Proceedings of the Symposium were published:

- as a volume (280 pages) of preprints distributed between the participants of the Symposium. The volume comprised the part of the Symposium materials - the texts of 25 lectures and communications presented at the Symposium.
- as a book (547 pages), which contains the whole materials of the Symposium (38 lectures and communications with discussions): "Catalysis. Heterogeneous and Homogeneous". B. Delmon, G. Jannes (Editors), Elsevier Scientific Publishing Company, Amsterdam-Oxford-New York, 1975.

The list of the invited reviews and papers:

1. I. de Agirre, B. Duque
   Les Modificateurs de Sélectivité en Catalyse Hétérogène.

2. D.G.H. Ballard
   Use of Homogeneous Metal-Alkyl Compounds as Models for the Behaviour of Heterogeneous Catalysts for Olefin Reactions.
3. J.M. Basset
   Emploi de "Coordinate Marqueurs" en Catalyse sur Métaux Supportés. Analogie avec la Chimie de Coordination.
4. P.A. Gosselain
   Activité et Selectivité de Catalyseurs Organométalliques d'Hydrogénation Transposés sur Matrice Solides.
5. G. Jannes
   Transition of Catalytic Coordination Complexes into Heterogeneous Systems.
6. J. Manassen
   Intermediate Regions between the Usual Catalyst Classification.
7. G. Martino
   Comparaison de la Catalyse Homogène et la Catalyse Hétérogène en Hydrogénation.
8. P.R. Rony, J.I.Roth
   Supported Metal Complex Catalysts.
9. S. Siegel, D.W. Ohrt
   Stereochemical Probes into the Mechanism of Homogeneous and Heterogeneous Catalytic Hydrogenation.
10. Ph. Teyssie
   The Importance of Coordination Aggregates in Homogeneous Catalysis by Polynuclear Metal Complexes.

The Second
International Symposium on Relations between
Homogeneous and Heterogeneous Catalysis

November 3-6, 1977
Lyons, France

Topics of the Symposium:

- catalysis by homogeneous catalytic species that are in heterophase by virtue of being attached to an insoluble material such as cross lined polymers, silica, etc.;
- study of cluster complexes as border line between homogeneous and heterogeneous catalysis;
- understanding of heterogeneous catalytic reactions based on the knowledge of coordination chemistry.

Proceedings of the Symposium were published:

- in a special issue of the Journal of Molecular Catalysis, vol. 3, No 1-3, November-December 1977 (220 p., texts of 23 communications). This volume was distributed between Symposium participants.
- in a separate volume "Relations between Homogeneous and Heterogeneous Catalysis" published by Centre National de la Recherche Scientifique, Paris, 1978 (323 p.). This volume contains the full texts of 4 plenary lectures and 7 communications, abstracts of 22 communications and discussions for all lectures and communications.

The list of the plenary lectures:

1. J. Halpern
   Homogeneous and Heterogeneous Catalysis: Kinetic and Mechanistic Aspects.
2. B.F.G. Johnson
   Metal Clusters of Osmium and their Potential in Homogeneous Catalysis.
3. C.U. Pittman
   Catalytic Reactions Promoted by Homogeneous Catalysts Attached to Polystyrene Matrices.
4. R. Ugo
   Oxidation Reactions of Olefins in the Homogeneous and Heterogeneous Phase: Mechanisms of Dioxygen Activation and Selective Transfer.

The Third International Symposium on Relations between Homogeneous and Heterogeneous Catalysis

May 25-27, 1981

Groningen, The Netherlands

Scope of Symposium was formulated in the 1st Circular as follows:

"This Symposium will continue the tradition established by the previous meetings in Lyon (1977) and Brussels (1974). These proved the value, to both academic and industrial research workers, of a forum for the dissemination and exchange of results relevant to both traditional major catalytic disciplines. It is increasingly recognized that new departures in science and technology often originate at the boundaries between separate but related areas of inquiry.

With this in mind, the Scientific Committee invites contributions which reflect the interdisciplinary character of the Symposium. Some examples of suitable "interfacial" areas are metal cluster catalysis, immobilized homogeneous catalysts, photocatalysis and synthetic metalloenzymes. The emphasis is to be upon the study of catalytic reactions, their stoichiometry, stereochemistry, kinetics and mechanism."

Proceedings of the Symposium were published in a special issue of the Journal of Molecular Catalysis, vol. 11, No 2/3, May-June 1983 (407 pages).

Guest editors: J.W.E. Coenen, R.S. Downing and J.J.F. Scholten.

This volume contains the full texts of 27 communications presented at the Symposium. Besides, the booklet of abstracts of 16 poster presentations was issued and distributed between participants.

The list of the plenary lectures:

1. B.C. Gates

   Catalysis by Supported Metal Clusters.

2. F. Ciardelli

   Polymer-Supported Transition-Metal Catalysts: Established Results, Limitations and Potential Developments.

3. J.M. Lehn

   Homogeneous and Heterogeneous Catalysis in the Photoinduced Decomposition of Water.

The Fourth

International Symposium on Relations between

Homogeneous and Heterogeneous Catalysis

October 2-5, 1983

Asilomar, California, U.S.A.

Scope of Symposium:

- immobilized catalysts, including mononuclear, dinuclear and metal cluster complexes on polymers, surfaces and in zeolites;
- chemical modification of surfaces and/or electrodes, photodriven surface reactions, photoelectrochemical processes at interfaces;

- organized enzyme assemblies;
- organometallic chemistry relevant to surface reactions;
- theoretical studies in any of the above.

The programme of the Symposium included 28 plenary and invited lectures and 17 poster presentations.

Symposium proceedings were published in a special triple issue of the Journal of Molecular Catalysis, vol. 21, October 1983 (430 pages).

Guest editor: B.C. Gates.

Plenary and invited lectures included into the programme:

1. J.M. Basset

   Surface Organometallic Chemistry: A New Approach to Heterogeneous Catalysis.

2. P. Biloen

   Transient Kinetic Methods.

3. C.P. Casey

   Reactions of a Cationic Bridging Methylidene Iron Complex with Carbon Monoxide and with Alkenes.

4. G. Challa

   Polymer Chain Effects in Polymeric Catalysis.

5. M. Che

   Dimerization of Olefins with Nickel Surface Complexes in X-Type Zeolite or on Silica.

6. A.J. Deeming

   Binding and Rearrangement of Small Oxy-Organics at Metal Clusters.

7. W. Drenth

   Zeolite H-ZSM-5 as a Pseudo-Transition Metal. Calculations on the Bonding of Carbene to a Zeolite Model.

8. R.G. Finke

   Trisubstituted Heteropolytungstates as Soluble Metal Oxide Analogs: Their Synthesis, Characterization,

^{183}W NMR and Supported Organometallic Derivatives.

9. W.A. Goddard
   Relationships of Mechanisms for Homogeneous and Heterogeneous Reactions.
10. A.T. Hubbard
    Electrochemical Oxidation of Oriented Absorbed Aromatic Molecules.
11. L.E. Iton
    A New Approach to the Generation of Metal-Bearing, Medium-Pore Shape-Selective Zeolites for Fischer-Tropsch Catalysis; Spectroscopic Studies of Zeolites.
12. P.A. Jacobs
    Influence of Geometry and Structure of Zeolite Pores on the Product Selectivity of Hydrocarbon Conversion Reactions.
13. V.B. Kazansky
    Preparation of Monovalent Nickel Surface Complexes by Selective Photoreduction of Supported Ni(II) Ions and their Activity in Acetylene Cyclotrimerization and Ethylene Oligomerization Reactions*.
14. K.J. Klabunde
    The Bilding of a Catalytic Metal Particle One Atom at a Time. Solvated Metal Atom Dispersed Catalysts.
15. R.M. Laine
    Modelling Heterogeneous Catalysts with Homogeneous Catalysts: Modelling the HDN Reaction.
16. R.J. Madix
    Adsorption and Reaction of Cyanogen on Clean and on Oxygen Dosed Copper Surfaces.
17. H. Nijs
    Structural Aspects in the Photooxidation and Photoreduction of Water in Clay Minerals Suspension.
18. G. Ozin
    Metal Atoms in Solution: Versatile Reagents for Pre-

paring Metal Cluster Zeolite Catalysts; Application to the Selective Reduction of CO to Butene.

19. G. Pez
Selective Hydrogenation of Benzene to Cyclohexene with Zirconium Hydride Catalysts.
20. R. Pierantozzi
Support and Cluster Effects on Synthesis Gas Conversion with Supported Ruthenium Clusters.
21. T.J. Pinnavaia
Clay Intercalation Compounds for Selective Triphase Catalysis. Reaction of Alkyl Bromides with NaCl.
22. D.F. Shriver
Intermediates in the Proton-Induced Reduction of CO in Iron Clusters.
23. J.K. Stille
Asymmetric Hydroformylation of Styrene on Polymer-Supported Catalyst-Bound Chiral Phosphines.
24. R. Ugo
Surface Supported Metal Cluster Carbonyls: an Investigation of the Behaviour of $Os_3(CO)_{12}$ Supported on Highly Decarbonated Magnesia.
25. H. Vahrenkamp
Cluster Reactions Related to Catalysis: Isomerization of $RuCo_2(CO)_9(HCCR)$.
26. M.E. Volpin
Transition Metal Complexes of Graphite and their Catalytic Properties; The New Bridge between Homogeneous and Heterogeneous Catalysis.*
27. H. Yamazaki
Organic Reactions with Rhodium Carbonyl Cluster Catalysts.
28. Yu.I. Yermakov
Organometallic Compounds in the Preparation of Supported Catalysts*.

* The lectures were not delivered at the Symposium.

The Fifth
International Symposium on Relations between Homogeneous and Heterogeneous Catalysis

July 15-19, 1986
Novosibirsk, U.S.S.R.

Topics of the Symposium:

- catalytic properties of metal complexes attached to various interfaces (organic and inorganic supports and membranes) or bonded to soluble or swollen polymers;
- interaction of metal complexes and organometallic compounds with support surfaces as a stage of catalyst preparation, reactivity of surface organometallic species;
- interrelation between homogeneous and heterogeneous acid-base catalysis;
- interrelation between homogeneous and heterogeneous photocatalytic systems;
- elucidation of the role of organometallic intermediates in the mechanism of reactions on heterogeneous catalysts;
- catalysis by polynuclear complexes, in particular, by heteronuclear clusters, on the surface and in solution;
- application of immobilized complexes for enantioselective catalysis.

The programme of the Symposium includes 39 invited lectures and 267 poster communications.

The full texts of lectures and some of poster communications are published in this book.

The titles of the lectures are given in the content of this book.

Part 1

Common Features between Homogeneous and Heterogeneous Catalysis

ACTIVE SITES AND THE ROLE OF THE MEDIUM IN HOMOGENEOUS, HETEROGENEOUS AND ENZYMATIC CATALYSIS: SIMILARITIES AND DIFFERENCES

K.I. Zamaraev and G.M. Zhidomirov
Institute of Catalysis, Novosibirsk 630090, USSR

ABSTRACT - Similarities and differences in behaviour of analogous active sites (ASs) in homogeneous, heterogeneous and enzymatic catalysis are discussed using Brönsted and Lewis acid sites, acid-base pairs, metal complexes, metal clusters and small metal particles as examples. Far-going analogies in the structures and the mechanisms of catalytic action of these types of active sites in homogeneous, heterogeneous and enzymatic systems are demonstrated. The reasons for these analogies arise from similar chemical properties of analogous ASs in various systems as well as from qualitatively the same character of the influence of the reaction medium on the mechanisms and the rates of the key steps of catalysis. For ASs of heterogeneous catalysts the role of the medium can be played by functional groups of the lattice. At the same time differences in local environment may have dramatic effects on the mechanisms of action of similar ASs in catalysts of different nature.

1. INTRODUCTION

Nowdays it is widely recognized that the key role in the mechanisms of both chemical and biological catalysis is played by intermediate chemical interactions of the reacting molecules with certain groups of atoms of catalysts. Such groups of atoms are usually called active sites (ASs) and their number is often rather small as compared to the total number of atoms in the catalyst (Ref. 1). Detection of these ASs, the study of their structure and chemical properties become these days one of the main objectives of the fundamental research in catalysis since such data seem to give a key both for the search for novel catalysts and catalytic reactions and for the control of their course.

The chemical nature of ASs is rather diverse. These can be Brönsted and Lewis acid sites (BAS and LAS), acid-base pairs, transition metal ions with low coordination numbers; atomic groups of various structural defects on solid surfaces such as adatoms and steps on the surface of metals; structural defects in the regions where dislocations come out onto the surface of solids. Some ASs can be inherent in the catalyst; others can be formed upon precoordination or prechemisorption of various substances (reagents, additives or solvents) or during destruction of initial portions of reagents. Sometimes several types of ASs are necessary to provide the catalytic transformation.

The properties of ASs of catalysts are determined primarily by those their fragments (atomic groups) which are involved in direct chemical interactions with the molecules of reaction mixtures either in intermediate complexes or in transition states. In the following discussion these atomic groups will be referred to as active fragments of the AS. Large effect on the properties of ASs is also exerted by neighbouring functional groups of the catalyst that are chemically bound to active fragments. Though less significant, but still quite important can be the effect of more remote functional groups which serve as the medium for the active fragments. An impressive example of the effect of the reaction medium on catalysis is provided by stabilization via solvation of ionic intermediates, e.g. carbonium ions, in solutions.

The above approach to the interpretation of the nature of ASs and of the effect of media on catalytic properties follows as a logical consequence from the idea of the chemical nature of catalysis substantiated by Prof. G.K. Boreskov (Ref. 2).

Chemical interactions of reaction mixture components with the same AS must be similar regardless of whether this AS belongs to a homogeneous or a heterogeneous catalyst, or to an enzyme. For this reason, one can expect certain mechanistic similarities in the action of all the three types of catalysts provided that their ASs are of the same chemical nature. But at the same time differences between local environments certainly can have a marked effect on activities and selectivities of similar ASs in different catalysts.

In this lecture we shall discuss similarities and differences in behaviour of similar ASs in homogeneous, heterogeneous and enzymatic catalysis using Brönsted and Lewis acid centers, acid-base pairs as well as transition metal complexes, metal clusters and small metal particles as examples. Along with experimental data, quantum-chemical calculations made in the framework of the so-called cluster approximation (Refs. 3-5) play an important part in our discussion. The applicability of the quantum-chemical methods used in this paper to the description of the properties of ASs we are interested in has been confirmed previously by consistency of the results of the calculations with the experimental data (Ref. 3).

2. BRÖNSTED ACID SITES

2.1. Bridged Hydroxyl Groups. Solvation of Acid Residues by Lewis Acid Sites

A typical example of BAS on the surface of a heterogeneous catalyst is a bridged hydroxyl group of silica-

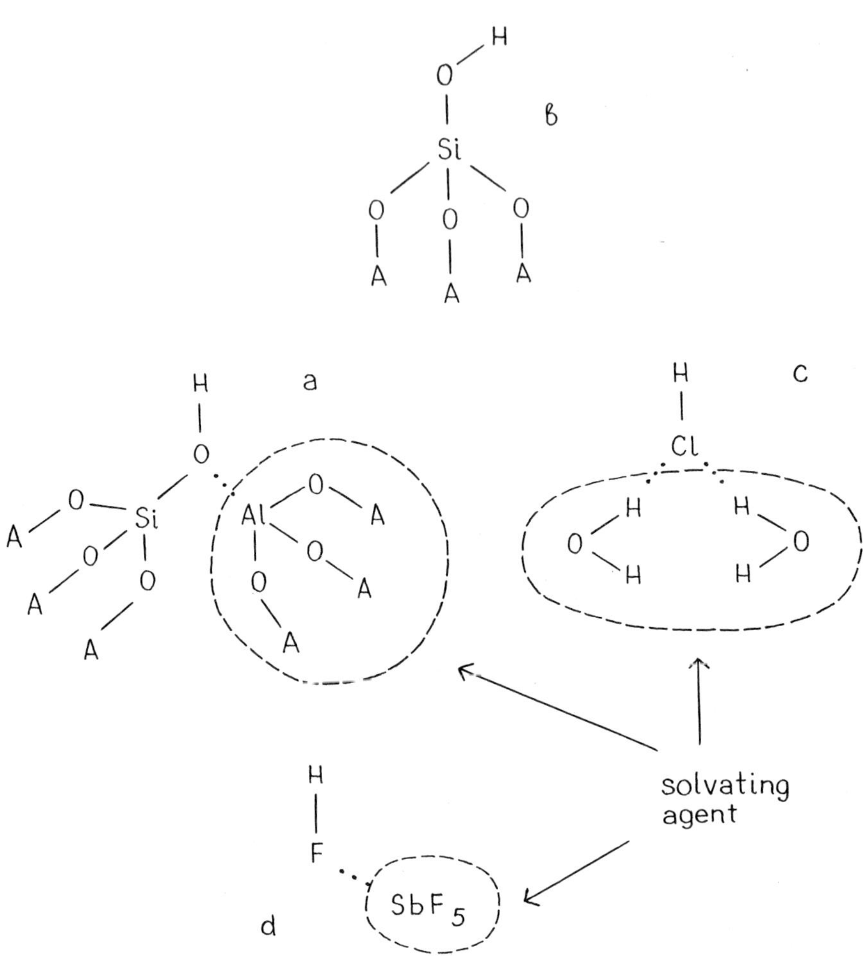

Figure 1. Cluster structures modeling BAS: a - bridged hydroxyl group in silica-alumina; b - terminal hydroxyl group in silicates; c - BAS in polar solvent; d - BAS in superacid. In quantum-chemical calculations of the properties of BASs, for saturation of terminal valent bonds in clusters a and b univalent atoms A with appropriately selected parameters of electronic structure (pseudoatoms) are used; this allows account of the otherwise neglected effect of the rest part of the solid lattice (Refs. 3,6,7).

alumina (Fig. 1a). Higher acidity of this group in comparison with that of the terminal OH-group of e.g. silicates (Fig. 1b) is explained in monographs on heterogeneous catalysis by a higher electronegativity of the ⩾Si-O-Al⩽ group in comparison with the ⩾Si-O group. Or, using slightly different terminology which is more accepted in homogeneous catalysis, the greater strength of BAS (Fig. 1a) is due to solvation of the $(\geqslant Si\text{-}O)^-$ anion by LAS Al⩽.

With this approach, the following similarity in the origin of strong acid properties of bridged hydroxyl surface groups, acid molecules in aqueous solutions (Fig. 1c) and superacids in organic solvents (Fig. 1d) becomes immediately obvious. The great strength of all these sites is caused to a large extent by solvation effects.

Whatever the nature of catalysts which involve BAS of the type $T_1\text{-}O^{\diagup H}\cdots T_2$ (homogeneous, heterogeneous or enzymatic), the following general conclusions about the tendencies of variations of their acidity can be made.

First, the proton-donor strength of these BASs has to increase with increasing coordination strength of LASs T_2. For instance, according to quantum-chemical calculations, the heat effect Q of coordination of water molecules to LAS Al⩽ is 30 kJ/mol larger than that for LAS B⩽ . Here and below, unless otherwise stated, the calculations by MINDO/3 method within a cluster approximation (Ref. 7) are reported. This difference in coordination strength induces a serious difference in deprotonation energy, which is 60 kJ/mol lower for BAS $\geqslant Si\text{-}O^{\diagup H}\ldots Al\leqslant$ than for BAS $\geqslant S\text{-}O^{\diagup H}\ldots B\leqslant$. The result obtained provides a reasonable explanation for a weaker Brönsted acidity of boralites in comparison with zeolites.

The relative coordination strength of various LASs can be characterized by the shift in the stretching vibrational frequency, $\Delta\nu$ pr., of a molecule-probe (e.g. CO) upon its coordination to the LAS. The bigger $\Delta\nu$ pr. the greater is the heat effect Q of the coordination. The proton-donor ability of the hydroxyl group of the BAS can be characterized by the shift in OH group stretching vibration frequencies, $\Delta\nu_{OH}$, on adsorption or coordination of molecules of some base B (e.g. CO) via formation of the hydrogen bond such as B... H-O$\genfrac{}{}{0pt}{}{T_1}{T_2}$. By way of example see in Figure 2 a correlation between the proton-donor ability of BAS of the type T_1-O^{H}... T_2 and the coordination strength of LAS T_2 for the series of heterogeneous zeolite catalysts (T_1 = =Si; T_2 = B, Be, Fe, Ga, Al) (Ref. 8). Here PA^c is the proton affinity of the acid residue of the BASs of the

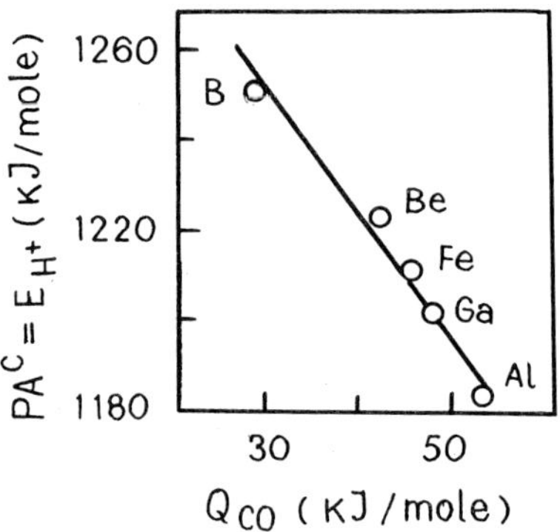

Figure 2. Comparison of the strength of Brönsted, PA^c, and Lewis, Q_{CO}, acid sites of high-silica zeolites

catalyst, and Q_{CO} is the heat of coordination of a CO molecule to the LAS T_2. It can be seen that the energy of proton abstraction from the OH group, $E_{H^+} = PA^c$, monotonously falls as Q_{CO} increases. The values of E_{H^+} and Q_{CO} for the catalysts indicated have been found from the correlations between E_{H^+}, $\Delta\nu_{OH}$ and Q_{CO},

$\Delta\nu_{CO}$ (Ref. 8).

The second important factor that affects the proton-donor ability of the BASs is the change in acidity of the T_1-OH fragments. As found by quantum-chemical calculations (Ref. 9) for aluminophosphates, the energy of proton abstraction from fragment ≥P-OH is 125 kJ/mole lower than from ≥Si-OH. Correspondingly, for the BAS ≥P-O$^{\diagup H}$... Al≤ , deprotonation energy is 25 kJ/mole lower than that for BAS ≥Si-O$^{\diagup H}$... Al≤ . Hence, if the bridging groups were in sufficient concentrations in aluminophosphates, the latter catalysts would have far stronger Brönsted acidity than silicaaluminas. However, the concentration of the bridged BASs in aluminophosphates seems to be very low. In fact, the quantum-chemical calculation (Ref. 9) indicates that centers I are energetically more favorable than centers II (Fig. 3).

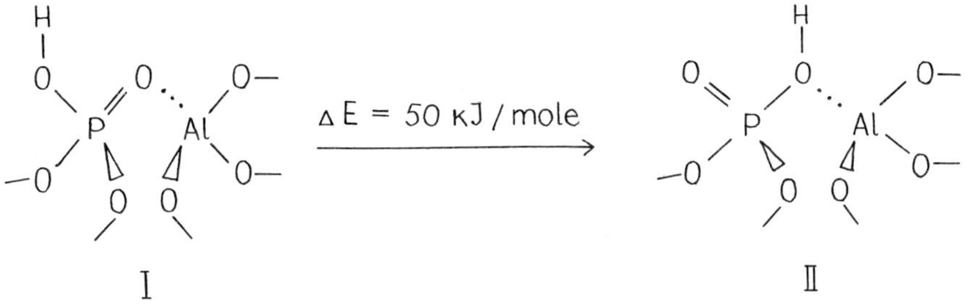

Figure 3. Cluster structures corresponding to the two types of BASs on the surfaces of aluminophosphate catalysts

The energy, ΔE, consumed in transition I → II is about 50 kJ/mole. An experimental density (concentration) of centers I, n_I, is about 10^{14} cm^{-2}. At a typical reaction temperature T = 573 K the concentration of centers II is thus expected to be considerably smaller: $n_{II} \simeq$

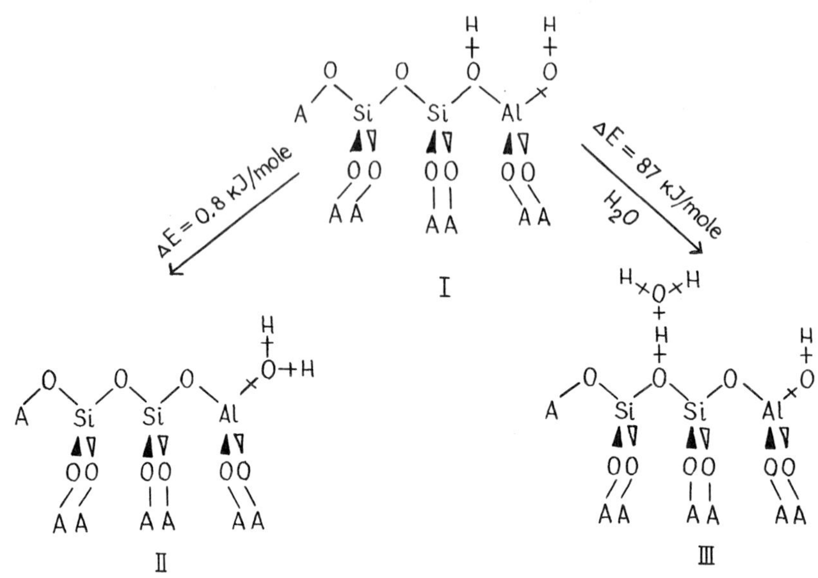

Figure 4. Cluster structures used in quantum-chemical calculations of the relative strength and stability of the various BASs on silica-alumina, ΔE is the energy consumed in the reactions indicated by the arrows

$\simeq n_I \exp\left(-\frac{\Delta E}{RT}\right) \simeq 10^9$ cm^{-2}. Indeed, centers II have not been detected by any spectroscopic technique, presumably just because of their very low concentration.

Similar estimates for possible types of BASs in silica-alumina (Fig. 4) indicate that $n_I \sim n_{II}$, $n_{III} \simeq 10^{-8}$ n_I (Ref. 10). In this situation, the catalytic activity is expected to be determined primarily by BASs I and II; their experimental surface concentration is about 10^{14} cm^{-2}. Despite that centers III are stronger than centers I and II, the concentration of BASs III seems to be insufficient to make them of practical interest for catalysis (Ref. 10).

From all of the above, it follows that the strength of the BASs of heterogeneous catalysts is determined by the same factors as that of homogeneous catalysts, that is, by acidity of proton-containing fragments and by solvating strength of the LASs present in the system. The former value is characterized quantitatively by the energy, E_{H^+}, of proton abstraction from the acid fragment and the latter one by the heat effect, Q, of LAS addition to the acid residue. The nature of the LASs and, thus, their solvating strength with respect to the same acid residue, can significantly change when passing from one catalytic system to another. In aqueous solutions, the role of solvating LASs is often played by water molecules; in nonaqueous solutions by molecules of polar organic substances or by coordinatively unsaturated compounds, such as SbF_5, and on the surface of heterogeneous catalysts by the LASs neighbouring to hydroxyl groups.

2.2. Heteropolyacids. Intramolecular Solvation of Acid Residues by Lewis Sites

Interesting data for discussing the similarity between the actions of Brönsted acid sites in homogeneous and heterogeneous catalysis is provided by the studies on catalytic properties of heteropolyacids (HPAs) conducted at our Institute (Refs. 11,12) and by Japanese scientists (Ref. 13). HPAs with the Keggin structure, $M_{8-n}[XM_{12}O_{40}]$, where X is the central atom (P^V, Si^{IV}, etc.), n is its oxidation state, and M stands for W^{VI} or Mo^{VI} atoms, seem to be most interesting for catalysis. These HPAs are soluble in many organic ligands; as regard to Brönsted acidity, they are more acidic than strong mineral acids and traditional heterogeneous acid catalysts. Figure 5 displays one of the four triads of atoms M that compose an exterior structural fragment of the HPA. The most probable position of an acid proton is also indicated. The strong Brönsted acidity of such a fragment seems to be accounted for by a strong solvating

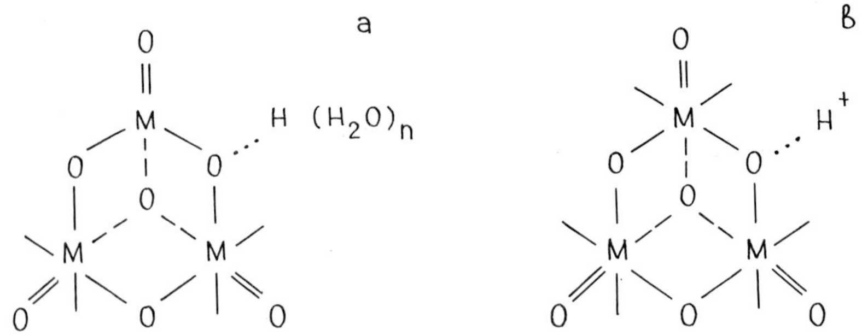

Figure 5. Schematic representation of the structure of BAS in HPA a) solvent, b) dispersed HPA on the support.

ability of the cationic centers M = W^{VI}, Mo^{VI} with respect to acid residues. The strength of the HPA in solution and in a crystalline state changes in a similar manner depending on the composition of HPA in the order:

$$H_3PW_{12}O_{40} > H_4SiW_{12}O_{40} \simeq H_3PMo_{12}O_{40} > H_4SiMo_{12}O_{40}.$$

This observation is in agreement with the above suggestion concerning the determining effect of intramolecular solvation of the acid residue by the Lewis centers of HPAs on the strength of these latter not only in a crystalline state but in solutions as well. Further support for this conclusion comes also from the quantum-chemical calculations (Ref. 14).

As found, the above order of the decrease in the HPA activities is valid for both homogeneous (decomposition of cumene hydroperoxide into phenol and acetone, addition of carboxylic acids to olefins, etc. (Ref. 12)) and heterogeneous acid reactions (isopropanol dehydration (Ref. 13)).

Thus, the HPAs, indeed, provide a good example of the similarity between mechanisms of the catalytic action of the BASs of homogeneous and heterogeneous catalysts. The strong acidity of the HPAs in solutions seems to be ensured by solvation of the acid residue by neighbouring Lewis acid centers of the HPA molecule, i.e. by the mechanism typical for heterogeneous catalysis.

2.3. Contribution of Solvation to Proton Transfer

It is known that proton transfer in a liquid phase is always accompanied by solvation of the protonated substrate by a counter-ion (formation of ion pairs) and/or by solvent molecules. Both these effects favour the proton transfer. On the surface of heterogeneous catalysts proton transfer is also assisted by the formation of ion pairs between the protonated substrate and the acid residue. However, solvation of the protonated substrate by reaction mixture molecules usually is not expected to occur since in the typical reaction conditions the coverage of the surface of heterogeneous catalysts with physically adsorbed molecules is rather low. Still its contribution can, to some extent, be substituted by effective "solvation" of the protonated substrate by basic centers of the catalyst lattice. To illustrate this statement, compare the results of quantum-chemical calculations by nonempirical method for the reaction of NH_4^+ formation from NH_3 in aqueous solution of hydrochloric acid (supermolecule approximation) and on the surface of silica-alumina (cluster approximation) involving saturation of broken boundary bonds with H atoms). Note, first, that according to the quantum-chemical calculations, in both cases the interaction in the acid agent + NH_3 system leads only to the formation of hydrogen bonds, but not to proton transfer (Fig. 6a). Experimental studies of the systems silica + NH_3 (Ref. 15)

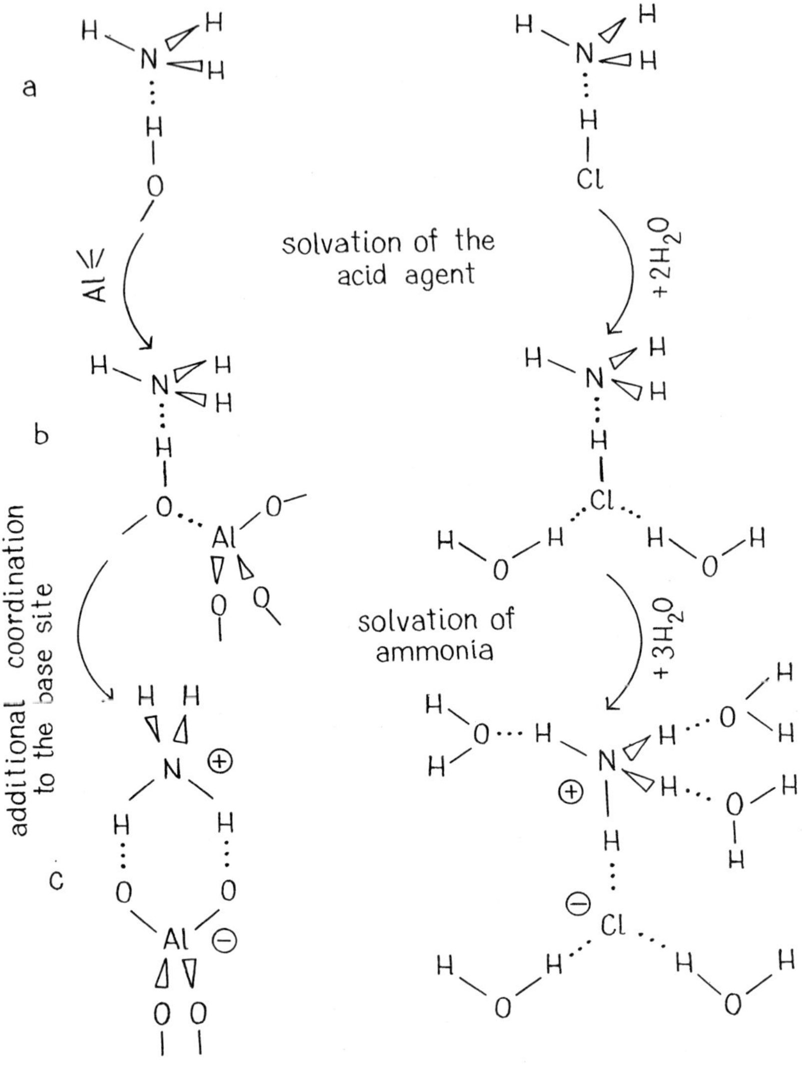

Figure 6. Scheme illustrating the sequence of factors in the interaction of ammonia with the BAS in liquid (right-hand side) and on the surface of the oxide catalyst (left-hand side).

and HCl + NH_3 (inert matrix) (Ref. 16) have indicated that in the absence of solvation NH_4^+ particles indeed are not formed. Solvation of acid agents (Fig. 6b) en-

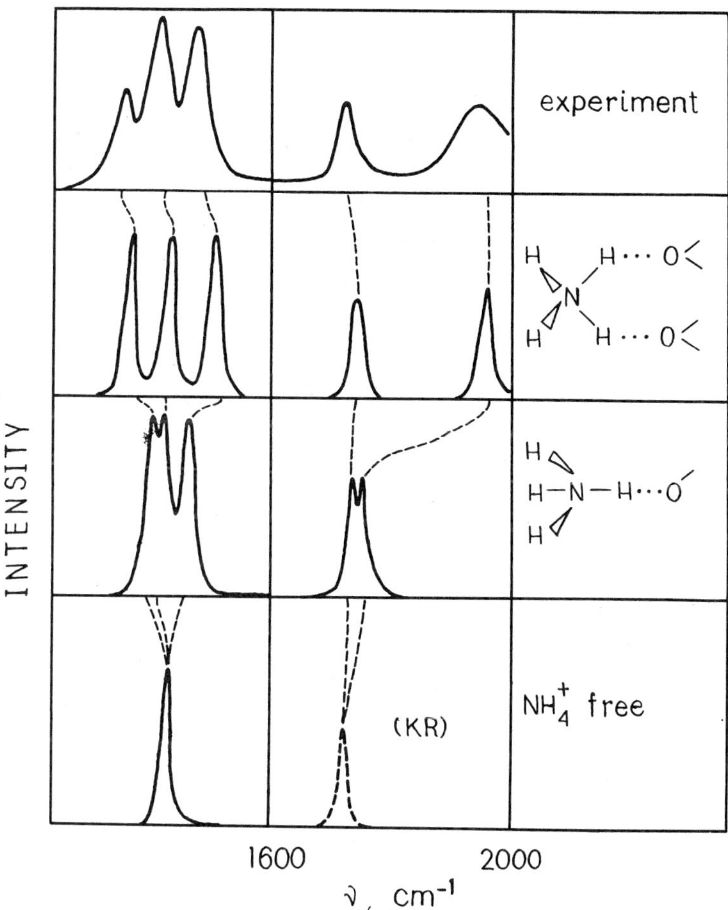

Figure 7. Comparison of the experimental IR spectrum of NH_4^+ on the HNaY zeolite with that calculated for different types of chemisorption structures

hances their proton-donor ability which results in the strengthening of the hydrogen bond; but proton transfer still remains energetically unfavourable. Only when solvation of the protonated substrate, in other words, the interaction between NH_4^+ and basic centers (Fig. 6c) is considered, the formation of NH_4^+ begins to be energetically favourable. In case of the surface these are lattice oxygen atoms that serve as solvating agents providing a two-centered coordination of the NH_4^+ cation (Ref. 17).

The idea of a two-centered coordination of NH_4^+ on the surface of silica-aluminas is experimentally supported by IR data. Figure 7 shows a comparison of an IR spectrum from NH_4^+ on the HNaY zeolite registered at 100 K with theoretical IR spectra from free NH_4^+ ions and from NH_4^+ ions that are bound to the surface via one- and two-centered interactions.

It can be seen that the experimental spectrum corresponds best to the spectrum calculated for the use of the two-centered coordination (adsorption).

Following the two-centered model, suggested for NH_4^+ coordination by the surface on the ground of quantum-chemical and IR data, one might expect larger heat effects for adsorption producing protonated states in the case of amines capable of forming two hydrogen bonds with the surface (ammonia, butylamines) than in the case of amines that can form only one hydrogen bond (pyridine and its derivatives). Experimental evidence obtained favours this prediction.

The dependence between the heats Q_{H^+} of the formation of protonated amines on HNaY (calorimetric measurements (Ref. 18)) and the proton affinity PA^r for the same amines in the gas phase (ion-cyclotron measurements) is plotted in Figure 8. By inspecting the figure one can see that for butylamines and ammonia a correlation line is indeed shifted upward in comparison with the corresponding line for pyridine and its derivatives. From this shift it is possible to find the average energy of the hydrogen bond between the cations of the protonated amines and the oxygen anions of the zeolite lattice: Q_{HB} = 30 kJ/mole.

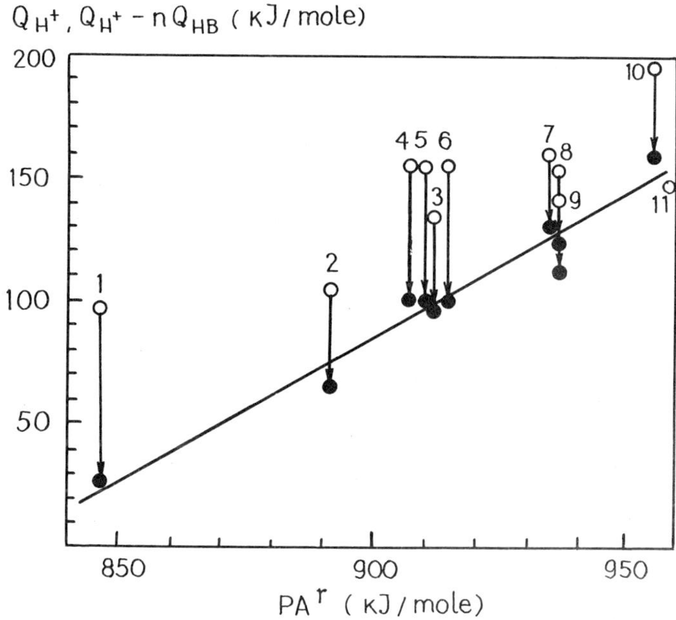

Figure 8. The heats of protonation of amines, Q_{H^+}, (light circles) upon their adsorption on HNaY zeolites characterized by a high uniformity of BASs, vs proton affinity, PA^r of these amines. 1 - ammonia; 2 - chloropyridine; 3 - pyridine; 4 - n-butylamine; 5 - secondary butylamine; 6 - tertbutylamine; 7 - 3,5-dimethylpyridine; 8 - 2,3-dimethylpyridine; 9 - 2,5-dimethylpyridine; 10 - 2,4,6-trimethylpyridine, 11 - triethylamine. Black circles indicate the value $Q_{H^+} - nQ_{HB}$, Q_{HB} is the energy of one hydrogen bond, n = 2 for ammonia and amines and n = 1 for pyridine and its derivatives; $Q_{HB} = 1.3\,(3400 - \nu_{NH})^{1/2}$ kJ/mole, ν_{NH} in cm^{-1} (Ref. 18).

2.4. Do Carbonium Cations Exist on the Surface of Heterogeneous Catalysts?

It follows from the previous section that on the surface of heterogeneous catalysts, BASs, even those of moderate strength (e.g., bridged OH-groups of silica-aluminas), can form stable protonated structures upon the adsorption of such strong enough bases as amines. A further important question is whether on the surface

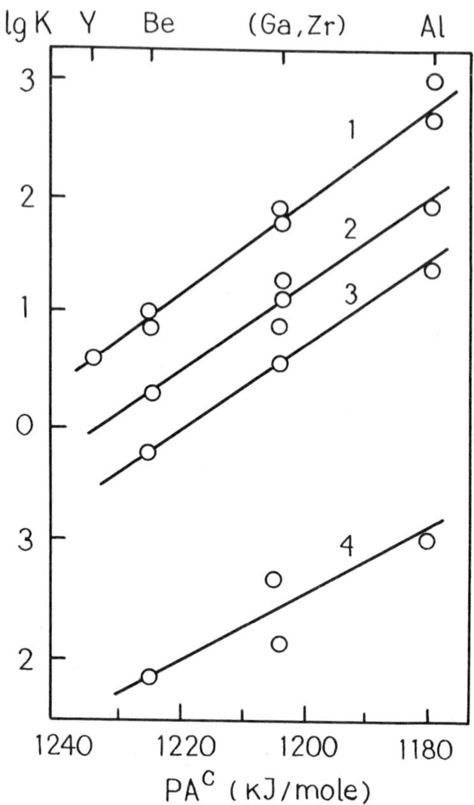

Figure 9. Plot of reaction rate constants (K, s^{-1}) at 573 K vs the strength of BASs, PA^c, for a series of amorphous binary oxide catalysts M_nO_m-SiO_2 (M = Al, Ga, Zr, Be, Y): 1 - dehydration of secondary butyl alcohol, 2 - dehydration of isobutyl alcohol, 3 - dehydration of n-butyl alcohol, 4 - isomerization of n-butene-1 to n-butene-2 (Refs. 19, 20).

of traditional acid heterogeneous catalysts the protonated states of considerably weaker bases (e.g. of hydrocarbons) can be formed as intermediates of catalytic reactions. Such a possibility has been postulated in a number of works on organic acid catalysis. In particular, it is traditionally assumed that double bond shift and dehydration of alcohols occur on heterogeneous acid catalysts via the formation of carbonium ions as intermediates. Support for this mechanism comes from the data on skeleton isomerization of the hydrocarbon chain in these systems. Modern theory of organic reactions in solutions usually associates skeleton isomerization with the formation of carbonium cations. Kinetic data such as a correlation between the reaction rate and the strength of the surface BASs (Fig. 9) lend further support for the carbonium ion mechanism.

In order to get more direct evidence for the formation of carbonium ions during the reactions under consideration, numerous attempts have been made to detect these ions on surfaces of heterogeneous catalysts by spectroscopic methods. However, in spite of numerous studies initiated in many laboratories the success here seems to be as yet rather modest. More or less reliable evidence for the formation of carbonium cations on the surface of traditional silica-alumina catalysts seems to be obtained by UV method only for such extremely basic aromatic hydrocarbons as triphenylmethane and diphenylethylene (Ref. 21). As regard to acyclic hydrocarbons, to our knowledge, nobody has succeeded so far in observing spectroscopically their protonated states RH^+ on the surface of heterogeneous catalysts. On the other hand, these protonated structures have reliably been identified in solutions (Ref. 22).

Methods of quantum chemistry have been employed in (Ref. 23) to analyze whether carbonium cations of acyclic hydrocarbons can exist as real chemical compounds on the surfaces of silica aluminas. Among semiempirical quantum-chemical methods, MINDO/3 is considered nowdays as most reliable for calculating energy characteristics of hydrocarbons and of their protonated derivatives. Unfortunately, this method, in its original version, is not appropriate for description of hydrogen bonds, while these latter play quite an important role in che-

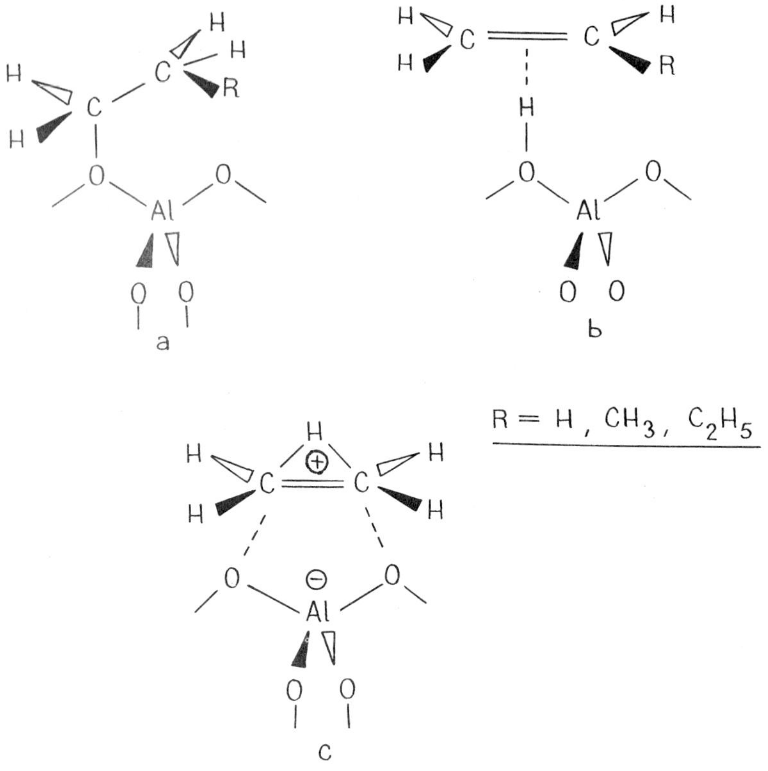

R = H, CH_3, C_2H_5

Figure 10. Cluster structures for various types of olefin chemisorption on BASs of silica-alumina: a) alcoxide structure, b) complex with a hydrogen bond, c) carbocationic form.

misorption processes on the surfaces of oxides. For this reason, a new MINDO/3-HB method has been developed (Ref. 23). The method preserves (with minimum corrections) the original parametrization of MINDO/3 for covalent bonds and also makes it possible to calculate complexes with hydrogen bonds with the same accuracy as a nonempirical method in an extended STO-4-31 G basis (Ref. 24).

Calculations made by MINDO/3-HB in a cluster approximation for the interaction of a bridged hydroxyl group of silica-alumina with hydrocarbons R = C_2H_4, C_3H_6 and C_4H_8 have indicated that alcoxide structures are energetically most stable (Fig. 10a). Next in the stability come complexes having hydrogen bonds (Fig. 10b). All the possible carbocationic forms occured to have still higher energy. Of a wide set of the protonated structures studied, bridged carbonium cations were found to have the lowest energy (Fig. 10c). From the standpoint of experimental observation of surface carbonium cations of principle importance is the question whether these structures correspond to a local minimum of the potential energy surface. Calculations by MINDO/3-HB method suggest that in the systems studied the carbonium cationic structures, RH^+, are in fact unstable, that is, they do not correspond to such a minimum. At the same time, the calculated energies of carbonium cations formation from the noninteracting pair BAS plus an olefin molecule are within 20 kJ/mole, in other words, are not high. Even if one considers a possible calculation error as well as the fact that in reality the initial reaction state may be a complex with a hydrogen bond (Fig. 10b), rather than a noninteracting pair of the BAS + the olefin (the latter fact is expected to increase further by some 20-30 kJ/mole the energy, needed for the formation of the surface RH^+), the resulting formation energies of the carbonium cationic structures still occur close to typical activation energies for acid-type reactions on heterogeneous catalysts (vide infra Fig. 14).

The above experimental data in conjunction with the quantum-chemical calculations allow us to suggest the following hypothesis concerning the role of carbonium cations of acyclic hydrocarbons in organic heterogeneous acid catalysis. Most probably, on the surfaces of traditional heterogeneous acid catalysts these carbonium cations play the role of transient states through which chemical transformations occur rather than exist in the form of real intermediate compounds. With this hypothesis it is possible simultaneously to account for the dependence of the catalytic activity on proton affinity of catalyst acid residues and on that of reagents (see, e.g., Fig. 9) as well as to explain repeated failures to detect carbonium cations of acyclic hydrocarbons on the surface of heterogeneous catalysts by spectroscopic methods.

The question naturally arises: can the situation qualitatively change when passing to the surfaces of solid superacids? To obtain an argumentative answer, further systematic experimental and theoretical investigations are needed. However, at this stage, we may say that preliminary attempts toward this direction suggest that the answer perhaps may be negative. Quantum-chemical calculations of thc system of tertbutylchloride + dehydroxylated surface of alumina have been made in a cluster approximation (see Fig. 11, left-hand side structure). Regarding the energy of the formation of the carbonium cationic structures, this system is similar to the system R·HCl·Al$\lessgtr$, where R is an olefin and HCl·Al$\lessgtr$ is a superacid. Like in the previous case, no local minimum on the potential energy surface has been found in this system for the carbonium ion structure RH^+. The complex with an adsorption energy, E_{ads}^{theor}, of 105 kJ/mole shown in Figure 11 was found to be energetically most favorable. This complex has a strongly polarized C^+-Cl^- bond; however, it cannot be regarded as a carbonium cation, as ascertained by the analysis of the electronic structure of its fragments. Additional evidence for this is also only a small change $\Delta R_{C-Cl} \simeq 0.004$ nm in the length of the C-Cl bond upon adsorption.

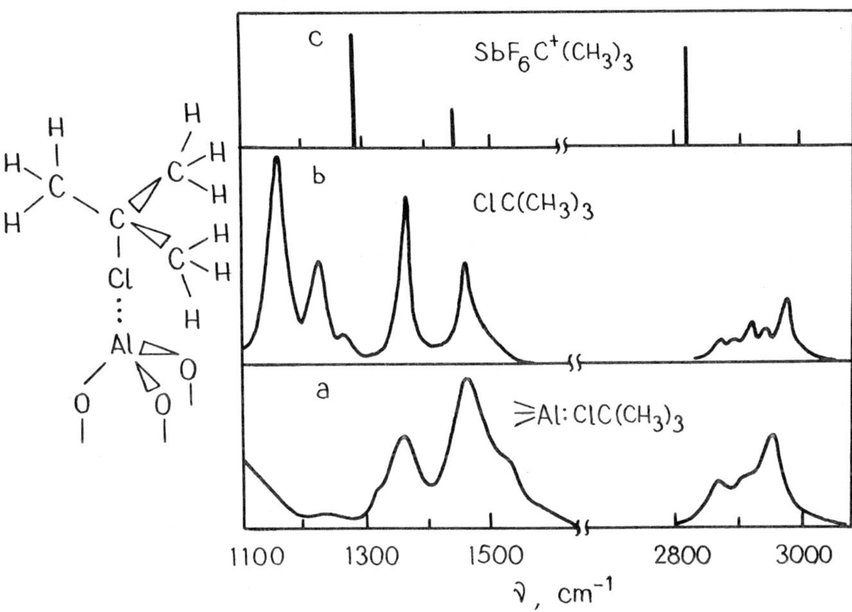

Figure 11. Comparison of the IR spectrum at 210 K of adsorbed tertbutylchloride on fluorinated alumina at a stoichiometric ratio $[ClC(CH_3)_3]$: LAS = 1 : 1 (a) with the spectrum of tertbutylchloride in solution (b) with the spectrum of carbocation $C^+(CH_3)_3$ (c) (Ref. 25). A cluster structure of the chemisorbed complex is shown in the left-hand side

Experimental data on adsorption of tertbutylchloride on fluorinated alumina are in good agreement with quantum-chemical calculations. According to calorimetric data heat of adsorption, E^{exp}_{ads} , is 84 ÷ 101 kJ/mole. An IR spectrum of the adsorbed tertbutylchloride is shown in Figure 11a. As is clearly seen, it is significantly different from the spectrum of $C^+(CH_3)_3$ carbonium cation (Ref. 25) (Fig. 11b); in particular, it has no 1290 cm^{-1} band characteristic of $C^+(CH_3)_3$. Thus, the adsorbed tertbutylchloride structure cannot be interpreted as a carbonium cation.

The proposed hypothesis on the role of carbonium cations of acyclic hydrocarbons in heterogeneous acid catalysis has the following evident consequences. If the carbonium cation is a real and sufficiently stable intermediate it can isomerize and also take part in secondary reactions with reagents and products. The both possibilities are often postulated under the assumption of the reaction mechanism involving carbonium cations. When the carbonium cation is a transition state, these possibilities are precluded.

One might attempt to use the difference mentioned above to choose between the two forms of existence of the carbonium cations on the surface, i.e. as real intermediate compounds or as transient states. Unfortunately, in practice the use of these criteria is met with certain difficulties. Regarding the first criterion, it is often assumed in related literature that the life-time of the carbonium cation is short because of its high reactivity and may be insufficient for the cation to participate in isomerization processes (Ref. 26). As regard to secondary reactions, since the initial reaction state appears to be a complex having a hydrogen bond, the same products distribution as in the case of the carbonium cationic intermediate could result from the attack on this complex with the hydrogen bond of either reagent or of product molecules resulting in the formation of a transient state of the carbonium cationic type.

The above difficulties can be illustrated by the reactions of dehydration of butyl alcohols, which have been examined experimentally at our Institute. An isomeric composition of the butenes produced was found to depend strongly on the reagent isomeric structures (Fig. 12). Obviously, this result is not consistent with the assumption of a fairly long life-time of the carbonium cationic intermediate (would this be the case one should expect isomerization of primary carbonium

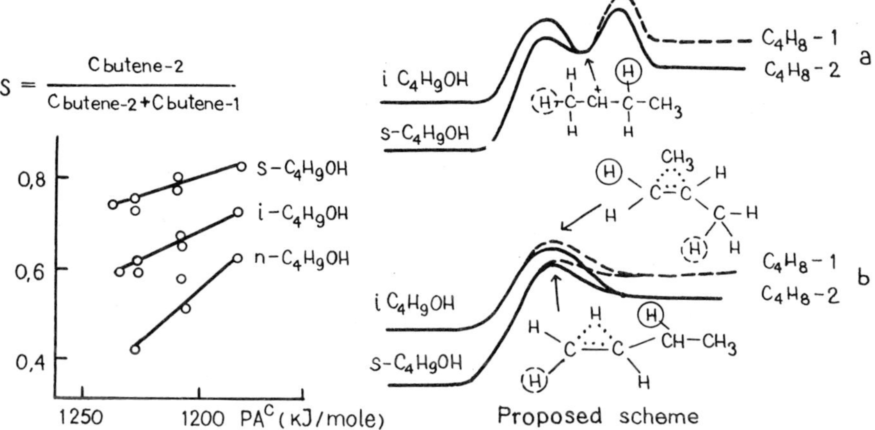

Figure 12. Plot of the selectivity (S) for the dehydration of various butyl alcohols vs. the strength of BASs of the catalysts. The reactions were studied at 573 K in a flow-circulation system at low conversion and zero order with respect to the reagent. On the right-hand side are energy diagrams for dehydration reactions: a) classical scheme with a carbocationic intermediate, b) proposed scheme with the carbocations as transient states. The diagram shows transformations to n-butenes only, the difference in cis- and trans-butenes-2 being neglected. Two different reaction routes indicated by solid and dotted lines correspond to the abstraction from the carbocation of hydrogen atoms enclosed in solid line and dotted line circles, respectively

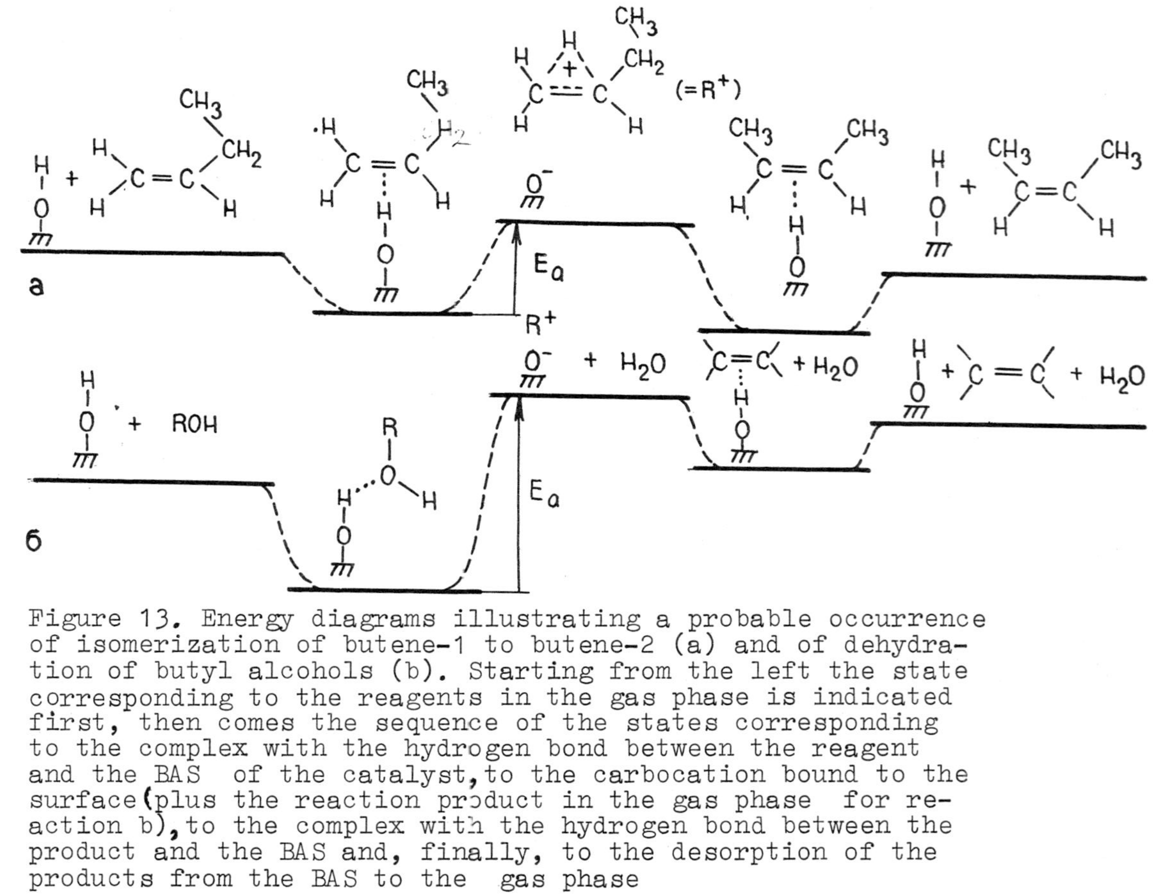

Figure 13. Energy diagrams illustrating a probable occurrence of isomerization of butene-1 to butene-2 (a) and of dehydration of butyl alcohols (b). Starting from the left the state corresponding to the reagents in the gas phase is indicated first, then comes the sequence of the states corresponding to the complex with the hydrogen bond between the reagent and the BAS of the catalyst, to the carbocation bound to the surface (plus the reaction product in the gas phase for reaction b), to the complex with the hydrogen bond between the product and the BAS and, finally, to the desorption of the products from the BAS to the gas phase

cations to give some more stable secondary butylcarbonium cation (Fig. 12a) and as a result, similar compositions of the butenes formed). However, these experimental results can be equally satisfactorily explained by assuming either that there exists a transient state of the carbonium cationic type (Fig. 12b) or that the reaction proceeds via carbonium cationic intermediates but is in a kinetically controlled regime when these intermediates live too shortly and have no time to isomerize. One may propose the following simple method for estimating activation energies of isomerization and decomposition reactions that occur on BASs of heterogeneous catalysts through transient states of carbonium cation nature (see Fig. 13). Apparently, the heat of protonation of the reagent molecule can be described as follows

$$Q_{H^+} = PA^r - PA^c + Q_s, \qquad (1)$$

where PA^r and PA^c are proton affinities of the reagent and of the acid residue of the catalyst BAS, and Q_s is the energy of the reaction of the acid residue with the carbonium cation ("solvation" energy).

If one assumes that the hydrogen bonds between the cation and basic centers of the lattice can make a contribution to Q_s (see, e.g., section 2.3), the latter one may be approximated as a sum of two contributions

$$Q_s = nQ_{HB} + \Delta, \qquad (2)$$

where n is the number of hydrogen bonds, Q_{HB} is the average energy of a hydrogen bond, while the main contribution to Δ is, most probably, made by the Coulomb interactions. Further, if we assume that Δ is constant (or that Δ linearly depends on $PA^r - PA^c$), a linear correlation between Q_{H^+} and PA^r for amines and pyridine derivatives (Fig. 8) will readily be understood.

Moreover, if we propose that the above initial assumptions are also valid for carbonium cations of acyclic hydrocarbons, and use the linear dependence displayed in Fig. 8, the heat Q_{H^+} of olefin protonation on various catalysts can be estimated provided that PA^c values for these catalysts are known. These dependences can be obtained for any catalyst by shifting the straight line in Fig. 8 along the abscissa by the value taken equal to the difference between PA^c for the catalyst in question and the HNaY zeolite.

The activation energies of acid reactions of the above mentioned type may be described as (it is assumed that $E_a > 0$):

$$E_a = Q^r_{HB} - Q_{H^+} - Q_r , \qquad (3)$$

where Q^r_{HB} is the heat of formation of the hydrogen-bonded complex of the reagent with the BAS of the catalyst which is the initial state of the reaction, see Fig. 13, Q_r is the reaction heat (in the gas phase). The values of E_a^{theor} calculated by this method, have been compared to the corresponding experimental values, E_a^{exp}, for reactions of isomerization of butene and dehydration of butyl alcohols on heterogeneous acid catalysts of different nature (Fig. 14).

On the whole, E_a^{exp} and E_a^{theor} are in good agreement. Thus, in terms of the proposed model of carbocations being transient states one can predict quite successfully activation energies of the catalytic reactions under study proceeding from the adsorption heat of reagents on BASs, proton affinity of the reagents in the gas phase, proton affinity of the surface acid residue and the reaction heat.

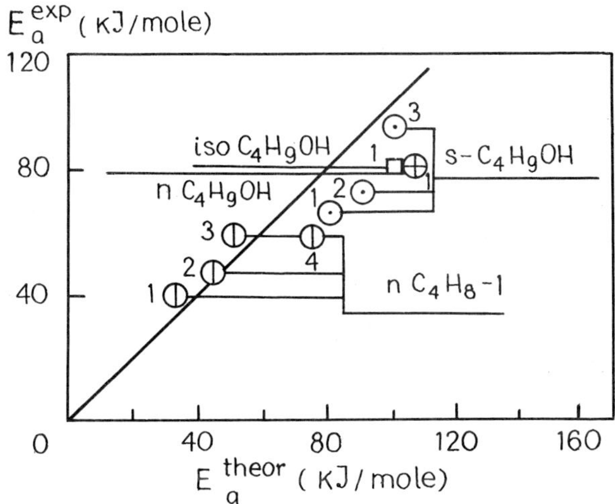

Figure 14. Comparison of E_a^{exp} and E_a^{theor} for dehydration of secondary butyl (o), n-butyl (⊗) and isobutyl (□) alcohols; (o), isomerization of butene-1 to butene-2 (Φ) on amorphous catalysts: 1 - (Al_2O_3-SiO_2); 2 - (Ga_2O_3-SiO_2); 3 - (BeO-SiO_2); 4 - (P_2O_5-ZrO_2). The values of E_a^{exp} were determined in the temperature range from 437 to 673 K at a zero reaction order and small conversion in a flow system in the conditions of ideal mixing.

3. LEWIS ACID SITES

By Lewis acid sites are usually meant such centers which have a vacant, localized orbital with a low enough energy and which thus tend to a partial capture of an unshared electron pair of bases. One of the chemical structures showing LAS properties, which is most widely discussed in the literature on heterogeneous catalysis, is Al(III) site with vacancies in its coordination sphere. Simple cluster models of this center in the oxide lattice which are typically used in quantum-chemical calculations are illustrated in Fig. 15a. An adsorption interaction of this center with a base molecule is shown in Fig. 15b, and an example of a similar

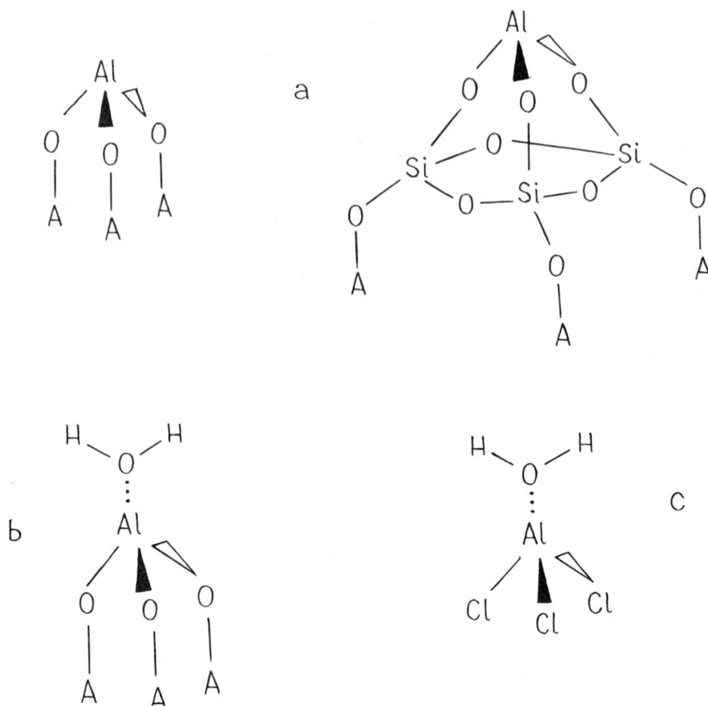

Figure 15. Simple cluster models of a LAS (a) and chemisorption complexes (b) on the surfaces of heterogeneous catalysts and of coordination of the base to a LAS in liquid (c)

LAS in solution in Fig. 15c.

Quantum-chemical calculations of the electronic structure and of the coordination strength of these centers indicate that they do correspond to the above mentioned qualitative ideas concerning their electronic properties, that is, they indeed have a vacant localized orbital with low enough energy and have the tendency to the capture of an electron pair from bases (Ref. 28).

However, quantum-chemical calculations unexpectedly predict very strong adsorption properties for such LASs. Theoretical heats of both molecular and dissociative adsorption of a number of molecules (H_2O, NH_3, alcohols)

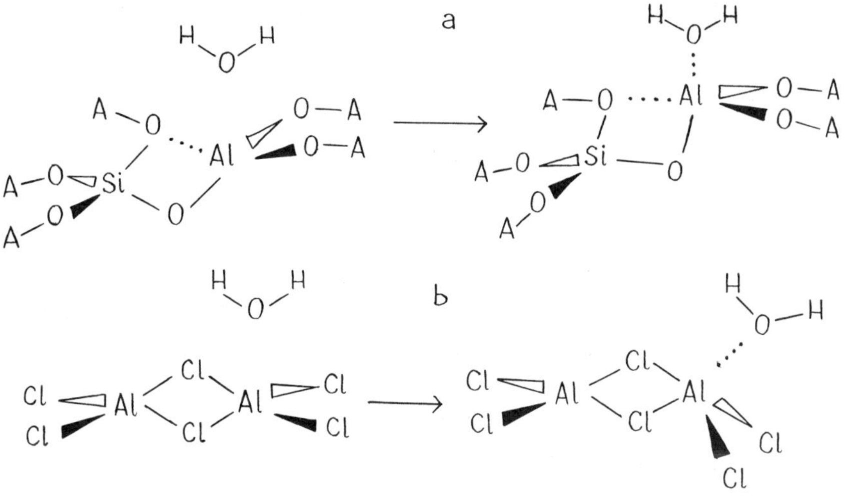

Figure 16. Cluster models accounting for partial shielding of LASs by anions on the surface of heterogeneous catalyst (a) and via formation of dimeric structures in solution (b)

were found to be 80-120 kJ/mole higher than average experimental heats. A possible explanation of this discrepancy is that the LAS models in Fig. 15 are too idealized, in a sense that they neglect a possible shielding of the LASs by the basic fragments of the medium both for heterogeneous and homogeneous catalysts. According to the quantum-chemical calculations in (Ref. 29), anions of the oxide lattice can, indeed, react with LASs and thus shield them. The extent of the shielding, characterized, e.g., by the decrease in heat effect of addition of H_2O molecules to LASs, is expected to depend strongly on the mutual disposition of the LASs and of the shielding anions (Fig. 16a). Therefore, Lewis acid sites widely different in strength can be anticipated to exist on the surfaces of real heterogeneous catalysts.

An extreme situation of a strong shielding where an

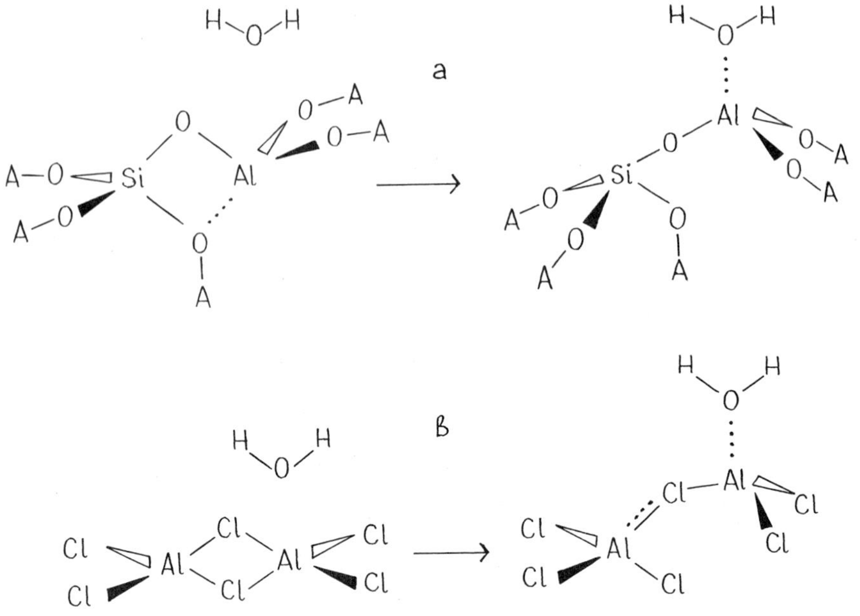

Figure 17. Cluster models of LASs which become available for chemisorption or coordination only after a substantial reconstruction: a - LAS of a heterogeneous catalyst, b - LAS in a solution

aluminium atom becomes in fact almost inserted into the subsurface layer of the lattice (Fig. 17a), will result in that the adsorption of a base molecule will be activated. In fact, the coordination of the adsorbed molecule must be preceded by the rupture of the bond between an aluminium atom and a surface oxygen atom and by the emergence of the aluminium atom on the catalyst surface. Special experiments to test this assumption seem to be quite expedient.

The effect of shielding of the LASs is also observed in the liquid phase. For example, the shielding of the LAS of homogeneous catalysts owing to the addition of the molecules of polar solvents or to dimerization (oligomerization) of the LAS is well known. A typical example is the shielding of the LAS of $AlCl_3$ molecules in solutions through dimerization (or oligomerization). Figures 16b and 17b demonstrate possible versions of the coordination of the base to the LAS having a dimeric structure. It can be seen that the coordination requires distortion of the initial flat structure of the center. This distortion is associated with the loss in energy; as a result, the heat effect of the base addition is expected to decrease.

A particular case of shielding is the addition of the LASs to oxygen atoms of OH groups on the surfaces of heterogeneous catalysts. As we have already mentioned, this process considerably enhances the proton-donor properties of the OH groups. A thermal stability of the strong BASs so formed is largely determined by the energy of the coordination interaction between the LAS and the OH group. To clarify this idea, consider now the quantum-chemical calculations of the BAS stability in zeolites and boralites (Ref. 30). The stability of this center is seriously influenced by the strength of the coordination interaction of the $Al(OA)_3$ and $B(OA)_3$ LASs with the lattice fragment, $SiOH(OA)_3$ (Fig. 18). It is essential that minimum energies of isolated $Al(OA)_3$ and $B(OA)_3$/ or $Al(OH)_3$ and $B(OH)_3$ centers correspond to their flat structures. Insertion of these centers into the lattice requires distortion of their flat structures producing almost tetrahedral structures.

The corresponding loss in energy is greater for $B(OA)_3$; besides the coordination interaction, in this case, is weaker. The resulting estimates are shown in Figure 18, from which it follows that the BASs of interest must be far more stable in zeolites. This conclusion is consistent with the experimental data.

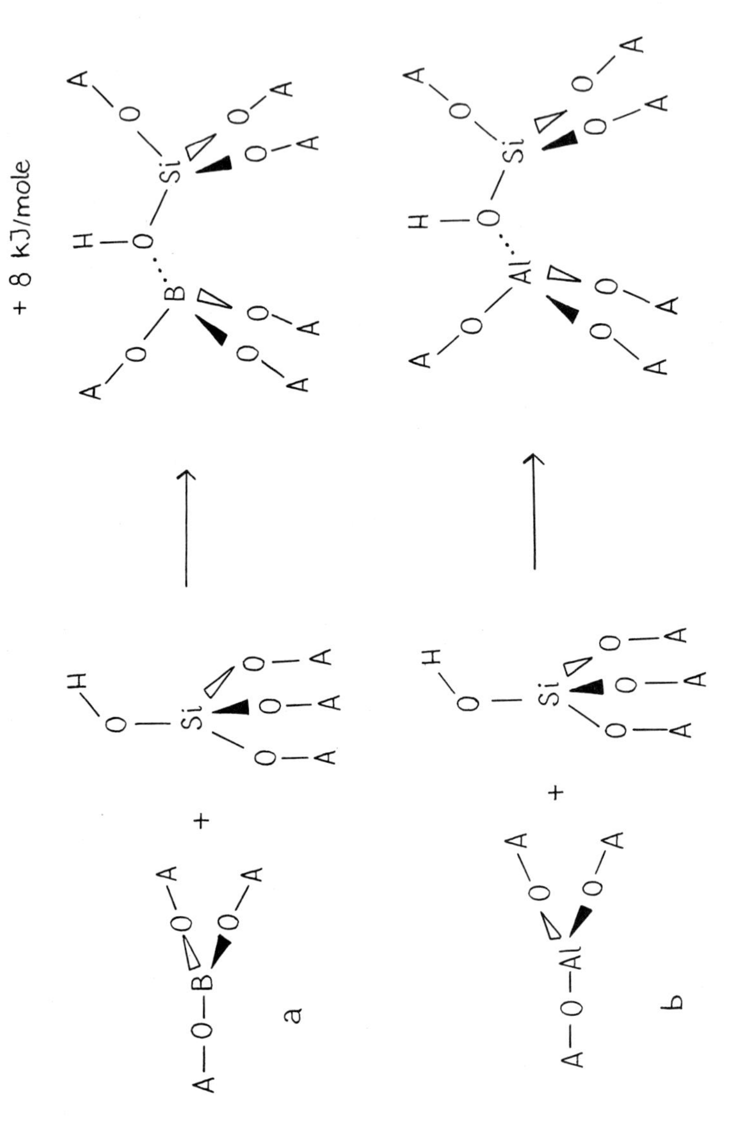

Figure 18. Cluster models used in the quantum-chemical calculation of the relative stability of BASs in boralites (a) and zeolites (b)

4. ACID-BASE PAIRS. CONCERTED MECHANISMS OF CATALYSIS

A conjugated action of acid and base centers is assumed for many catalytic transformations. When these centers are sufficiently close to each other and have appropriated mutual orientation, possibilities for concerted mechanisms of acid-base catalysis arise. Characteristic features of these mechanisms are usually a comparatively low activation energy and a high selectivity of catalytic action.

Concerted mechanisms appear to be widely spread in enzymatic acid-base catalysis. According to available data acid and base centers of enzymes are inferior in strength to inorganic superacids and superbases. Nevertheless, owing to concerted mechanisms, complex acid-base transformations in the presence of enzymes occur at remarkably fast rates and with very high selectivities. In the past 15-20 years, with the aid of X-ray analysis impressive success has been attained in identifying the structure of active sites and in determining the location of substrate molecules relative to active fragments of ASs in intermediate enzyme-substrate complexes for various acid-base enzymatic reactions.

As a typical example consider the mechanism of hydrolysis of a glycoside bond of polysaccharides by the enzyme lysozyme. The active center of this enzyme (Fig. 19) is an acid-base pair that consists of a carboxyl group of glutamine acid (Glu-35) and a carboxylate-anion of aspartic acid (Asp-52) (see (Ref. 31) and Refs. therein). As suggested by X-ray data, the substrate site subjected to hydrolysis is located in the intermediate enzyme-substrate complex in a fashion which is shown schematically in Figure 19a. As can be seen, the hydrolysis of the glucoside bond follows a concerted mechanism involving proton transfer from acid

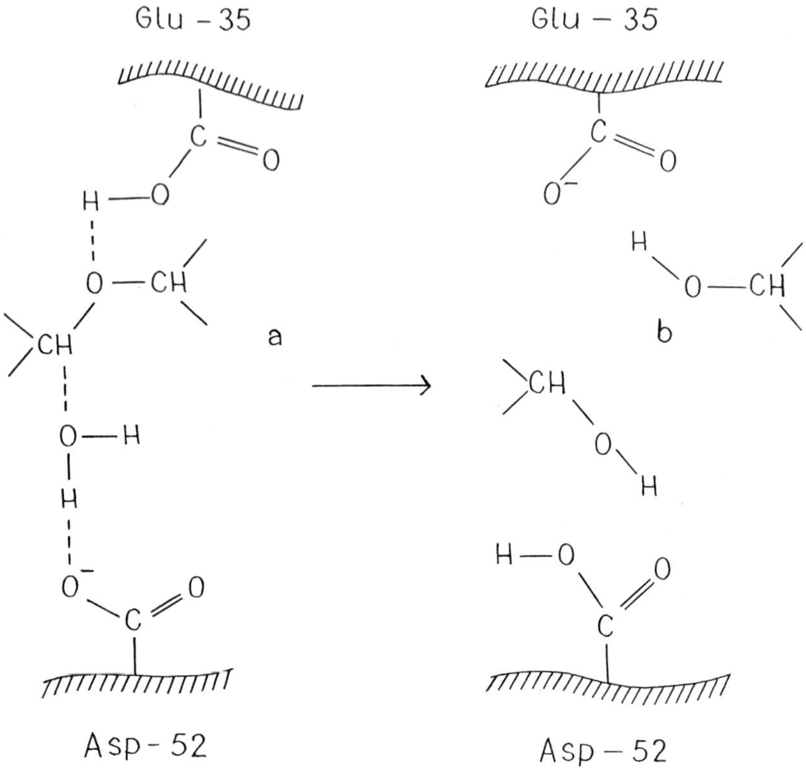

Figure 19. Concerted mechanism of the hydrolysis of a glycoside bond of a polysaccharide by an acid-base pair of the enzyme lysozyme: a - the intermediate complex of the substrates with the enzyme; b - the result of the concerted transformation

to base group of the AS of the enzyme via a chain of chemical and hydrogen bonds. In this case, the molecules of both substrates - a polysaccharide and water serve as links of the chain for proton transport.

Examples of nonenzymatic homogeneous catalytic systems are also known, in which reactions proceed via concerted acid-base mechanisms. Their significance seems to increase in nonpolar media at low temperatures, that is, under conditions favouring the association of reagents (Ref. 32). A classical example is mutarotation of tetramethylglucose catalyzed by phenols and amines in a benzene solution. Most effective are α-oxypyridine and carboxylic acids, which are bifunctional catalysts with acid-base pairs as ASs (Ref. 33). The mechanism of action of acid-base pairs of carboxylic acids can be represented as follows:

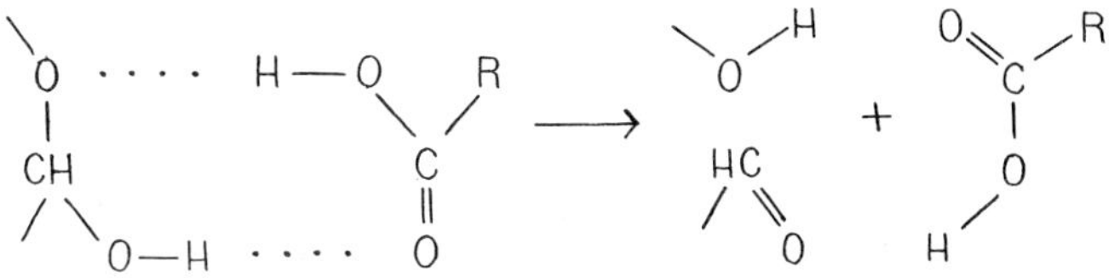

Recently, at our Institute concerted mechanisms have been found for various acid-base transformations of organic molecules proceeding in the coordination sphere of metal complexes. A typical example is the formation of Schiff bases in the coordination sphere of Cu(II), Zn(II) and Ni(II) complexes (Ref. 34). The mechanism of this reaction has been studied by NMR, EPR and optical spectroscopies. The reaction was found to occur due to the outer-sphere attack of a free (nonbound to metal) aniline molecule on a salicylaldehydate-ion which is coordinated to a metal atom. This attack leads to the formation of the carbinolamine in the coordination sphere of the metal with a subsequent dehydration of this products producing the Schiff base:

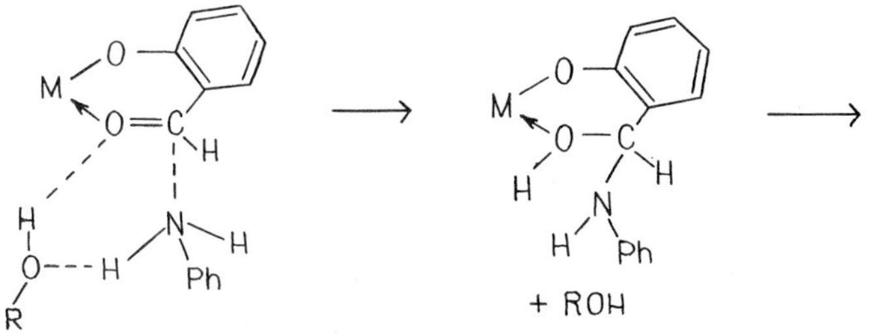

+ ROH

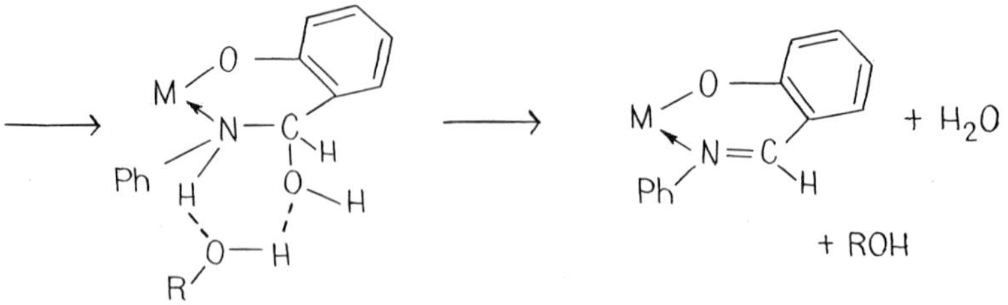

As found, alcohols and amines catalyze this reaction. This effect is accounted for by a synchronous transfer of protons in cyclic intermediate structures, where the second molecule of amine, or the molecule of alcohol, serves as a bifunctional acid-base catalyst (see the scheme of the reaction mechanism given above).

The second example refers to the formation of urethan from isocyanate and alcohol in the coordination sphere of Fe(III) acetylacetonate (Ref. 35). An analysis of the kinetic data, obtained by Prof. S.G. Entelis et al. at the Institute of Chemical Physics, and of the NMR studies on reagent coordination to the catalyst carried out at the Institute of Catalysis, has indicated that the role of the active form $Fe(AA)_2OR$ of the catalyst (AA and OR are acetylacetonato-anion and alcoholate anion) is that of providing an optimal for the reaction mutual orientation of the molecules of the isocyanate and of the alcohol.

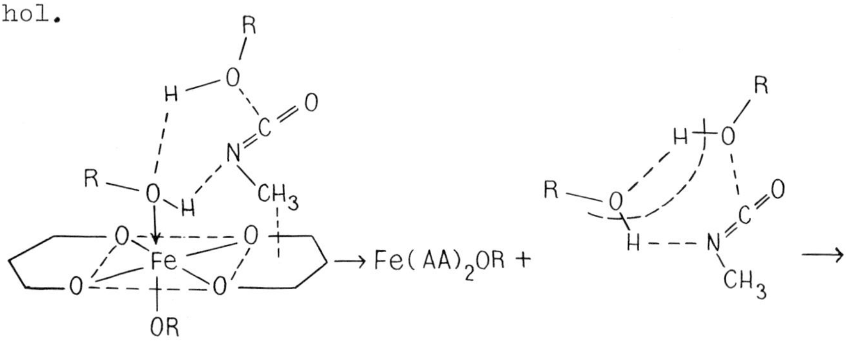

$$\longrightarrow Fe(AA)_2OR + ROH + \underset{H_3C}{\overset{H}{\diagdown}}\!\!N{-}C\!\!\begin{smallmatrix}\nearrow OR\\ \searrow O\end{smallmatrix}$$

It was found that in the intermediate complex with the catalyst, the alcohol molecule is most probably coordinated to the Fe(III) atom as ligand, while the isocyanate molecule is coordinated into the outer sphere of $Fe(AA)_2OR$ at the expense of the interaction between its methyl group and the π-system of the chelate AA ligand. This complex seems to be additionally stabilized by the second molecule of the alcohol, which bridges via hydrogen bonds the orbitals of the coordinated molecules of the alcohol and of the isocyanate. The process of urethan formation seems to proceed via the synchronous proton transfer in a cyclic intermediate located in the coordination sphere of Fe(III) (see the scheme of the reaction mechanism given above). It seems that the synchronous proton transfer through the chain of chemical and hydrogen bonds in cyclic intermediates can be important for quite a broad class of organic reactions proceeding in the coordination sphere of metal complexes. It should be noted that a structurally organizing role in these processes often seems to be played by rather weak outer-sphere interactions (Ref. 36), owing to which reacting molecules are located in the coordination sphere of metal complex at the sites providing formation of the chains of bonds through which proton transfer is accomplished. Moreover, like in enzymatic catalysis, an important role in proton transfer through such chains is provided by hydrogen bonds.

Concerted mechanisms seem to be possible in the case of heterogeneous catalysis as well. For example, the concerted mechanism, has been discussed in the literature for a double bond shift in olefins on bifunctional

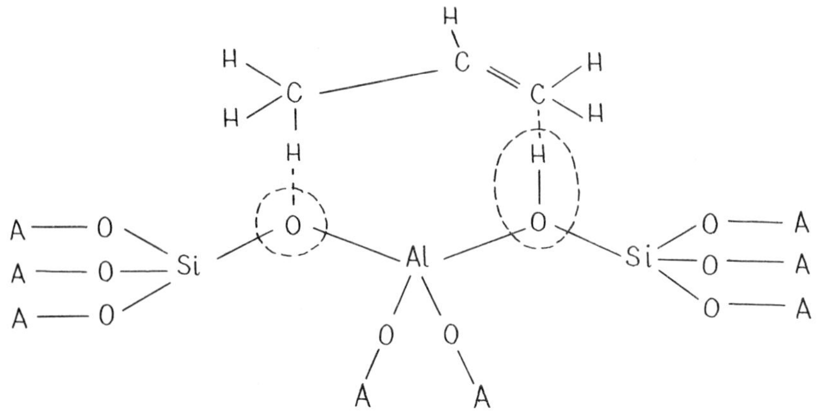

Figure 20. Cluster model of the olefin chemisorption on the acid-base pair of the silica-alumina catalyst

ASs of silica-aluminas which have a bridged hydroxyl group and a base anion of the lattice oxygen (Fig. 20) (Ref. 37). For this center a definite geometric correspondence between the structure of olefins and the position of the active fragments of ASs is necessary for a concerted mechanism to occur. However, at the moment we cannot state with certainty what is a real function of the structures such as in Figure 20 in heterogeneous catalysis; this should be a subject of further special investigations.

Thus, a comparison of the structure and of the chemical properties of the active sites for heterogeneous, homogeneous and enzymatic acid-base catalysts, indeed, demonstrates quite far-going similarities in the mechanisms of their catalytic action. However, in enzymatic catalysis, concerted mechanisms that provide both a lower activation energy and a higher reaction selectivity seem to occur much more often. This phenomenon seems to result from certain favourable for catalysis position of AS fragments and molecules of the reactants and of the media in intermediate enzyme-substrate complexes. The above data suggest that, in principle, such favourable arrangements of the active sites are possible for homogeneous and heterogeneous catalysts as well. One can, therefore expect that an intentional design of bifunctional acid-base ASs, capable of providing catalytic transformations by concerted mechanisms similar to those used by enzymes, will make it possible to substantially improve both homogeneous and heterogeneous acid-base catalysts.

5. ANCHORED COMPLEXES

Far-going similarities are known between the catalytic properties of metal complexes in homogeneous solutions and metal complexes anchored on surfaces of organic polymers or inorganic supports in various reactions (Ref. 38). These similarities are known to exist both in the key steps of catalytic cycles and in the nature of ASs of catalysts. However, anchored complexes often reveal also their specific properties in catalysis which are caused by peculiarities of their interactions with the medium. These peculiarities may noticeably affect both the concentration and the reactivity of ASs of the same type. Numerous illustrations of these statements have been presented at this Symposium.

Typical similarities and differences between the reactivities of metal complexes in liquids and on solid surfaces can be examplified by two reactions of individual alkyl and hydride complexes of Ti, Zr and Hf in solutions and on oxide supports (SiO_2, Al_2O_3) studied at the Institute of Catalysis in the laboratory of Prof. V.A. Zakharov:

1. The reaction of carbon monoxide with these alkyl and hydride complexes, which is known to be an important step of many catalytic processes.

2. The catalytic reaction of ethylene polymerization in the presence of these alkyl and hydride complexes.

For numerous alkyl and hydride complexes of Ti, Zr and Hf it has been shown that CO can insert into the M-C and M-H bonds (Refs. 39-41). This reaction occurs through the preliminary step of CO coordination to a metal ion. When these complexes are supported on SiO_2 or Al_2O_3, CO also can easily be inserted into the M-C and M-H bonds, the reaction of CO with a hydride proceeding even at 77 K (Ref. 42). Like for homogeneous systems, the reaction of CO insertion obeys a two-step mechanism involving preliminary coordination of CO molecule to a metal ion. The main differences between the behaviour of homogeneous and anchored complexes are associated with a rigid fixation of the latter on supports, which suppresses further transformations of acyl and formyl complexes formed. For example, while for surface hydrides of Zr the insertion of CO ends up in the formation of surface formyl complexes

$$L_xZr\text{-}H \xrightarrow{CO} L_xZr\text{<}(CH{=}O) ,$$

for homogeneous hydrides of Zr formyl complexes further interact with Zr hydrides to give more complex dimeric structures of the type $L_xZr\text{-}CH_2\text{-}O\text{-}ZrL_x$ (Ref. 43).

Still more pronounced differences between the behavior of organometallic complexes in solutions and on surfaces are observed for the catalytic polymerization of ethylene. Individual complexes with composition MR_4 or Cp_2MR_2 are inactive in this reaction. A very low but still non-zero activity of MR_4 compounds in ethylene polymerization, observed in a number of works (Refs. 43,44) seems to be caused by the formation of small amounts of side products due to the decomposition of these compounds (for instance, via interaction with water microimpurities) (Refs. 45, 46). Only when supported on SiO_2 or Al_2O_3, complexes MR_4 become catalytically active in ethylene polymerization. For supported systems prepared under optimal conditions, the fraction of metal atoms that are active in catalysis achieves 30-50% of the total amount. The transformation of inactive MR_4 complexes into active centers seems to result from the change in their coordination state upon anchoring, to be more precise, from formation of surface coordinatively unsaturated octahedral complexes, which can effectively activate olefin via its coordination to a metal ion (Ref. 47). In a similar way one may explain the formation of ASs in homogeneous systems when adding special co-catalysts to organometallic compounds. For example, homogeneous systems containing Cp_2MR_2 (M = Ti, Zr) and a co-catalyst methylalumoxane $[-Al(CH_3)-O-Al(CH_3)-O-]_n$ (Ref. 48), where n = 3-10, are known to be very active and stable. Most probably, in these systems, the rolc of the co-catalyst consists of transforming an inert coordinatively saturated tetrahedral complex into a coordinatively unsaturated octahedral active center able to effectively coordinate olefins. In this case a maximum activity is attained at a very low concentration of a transition metal complex (ca. 10^{-7} mole/l of Cp_2MR_2) and at a large excess of alumoxane (ca. 10^{-2} mole/l). In view of these facts methylalumoxane may perhaps be considered to be a kind of a "support" that stabilizes organometallic complexes in their active forms and excludes their deactivation via dimerization or oligomerization.

Thus there do exist far-going similarities on a molecular level between the factors determining the catalytic properties of the complexes under consideration in solutions and on the surface of oxides.

6. METAL COMPLEXES AND ACTIVE SITES OF METAL CATALYSTS

It might appear first that an anology between the nature of the catalytic actions of metal-complex and metal catalysts is arguable because of a considerable difference in their electron states. It is well known that, in contrast to metal complexes, metals demonstrate pronounced collective effects in their electron properties. However, numerous experimental data and quantum-chemical calculations evidence that for transition metals the d-zones are typically very narrow. For this reason the influence of collective effects on chemisorption interactions typically must be rather weak. In practice this means that when describing catalytic properties the interaction of reactants with metal surfaces can be interpreted in terms of the formation of localized chemical bonds between the atoms of the reactant and a small number of selected metal surface atoms. Relative to these selected metal atoms the remainder metal atoms function as ligands and as a reaction medium. With such approach the drawing of some analogies between ASs of metal catalysts and of metal complexes seems quite reasonable.

A good example for the comparison of the catalytic action of ASs of metal complexes and of metals is furnished by reactions of carbon monoxide and olefin hydrogenation. As a rule, atoms on metal surfaces demonstrate a higher activity and a lower selectivity with respect to these reactions than complexes of the same metals. A reason for this seems to be the dispersion of surface ASs as regard to their structure and properties. In fact, the surface may have both relatively smooth areas at which metal atoms have comparatively high coordination numbers, and the defects (steps, for example) at which the coordination number may be as low as 4 or 3.

In addition, during the reaction, various forms of both one-centered and multi-centered adsorption of reacting molecules can be formed on the surface (Ref. 49).

As distinct from metal surface, transition metal complexes typically have much more uniform ASs, which provides their higher selectivity. Moreover, it is possible to control the selectivity by varying metal environment (ligands and solvents). In the case of CO hydrogenation examples of such selective reactions are the formation of ethylene glycol in the presence of the $Rh(CO)_2(acac)$ complex (Ref. 50) and of $i\text{-}C_4H_{10}$ paraffins in the presence of the cluster complex $Ir_4(CO)_{12}$ (Ref. 51). However, the activities of these catalysts are low, and the processes occur very slowly, often only at high pressures. Thus, one of the most important problems in catalysis science is the development of systems, which would combine a high selectivity of homogeneous metal complex catalysts and a high activity of heterogeneous metal catalysts.

As examples of such systems one may consider supported rhodium catalysts for synthesis of hydrocarbons and alcohols from CO and H_2, obtained by Prof. Yu.I. Yermakov et al. at the Institute of Catalysis (Ref. 52) and hydrogenation catalysts based on the giant palladium cluster complexes obtained by Prof. I.I. Moiseev et al. at the Institute of General and Inorganic Chemistry (Ref. 53). Structures of these catalysts have been studied by EXAFS and electron microscopy methods (Refs. 52,53). As found, the main structural units of these catalysts are metallic species of 1.2-3 nm in size. In rhodium systems this cluster is attached to oxide supports SiO_2, $\gamma\text{-}Al_2O_3$, La_2O_3, whereas in palladium systems it is a nucleus of the giant complex with phenanthroline (or bipyridine) ligands and anionic solvating shell.

Structurally, these systems can be related with classical heterogeneous catalysts in view of the existence of a large number of metal atoms directly bound with each other. On the other hand, they resemble metal complexes because of a strong dependence of the structure of a cluster and of its chemical properties on the nature of the environment. For example, Rh species on La_2O_3 are considerably flatter than on SiO_2 and form clusters with a thickness of only 2-3 atomic layers.

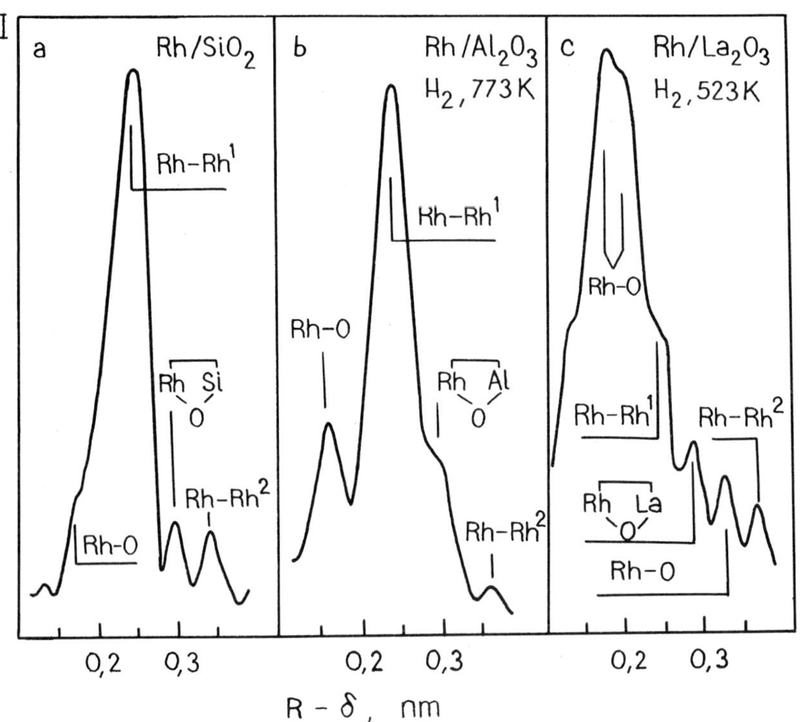

Figure 21. Radial atomic distribution curves obtained by Fourrier transform EXAFS spectra of the reduced rhodium catalysts: a - Rh/SiO_2, T_{red}=773 K; b - Rh/Al_2O_3, T_{red}=773 K; c - Rh/La_2O_3, T_{red}= 523 K. The distance R from the Rh atom to the other atoms of the catalyst is shown on the abscissa in the R-δ scale, where δ is the phase shift, I is the relative intensity

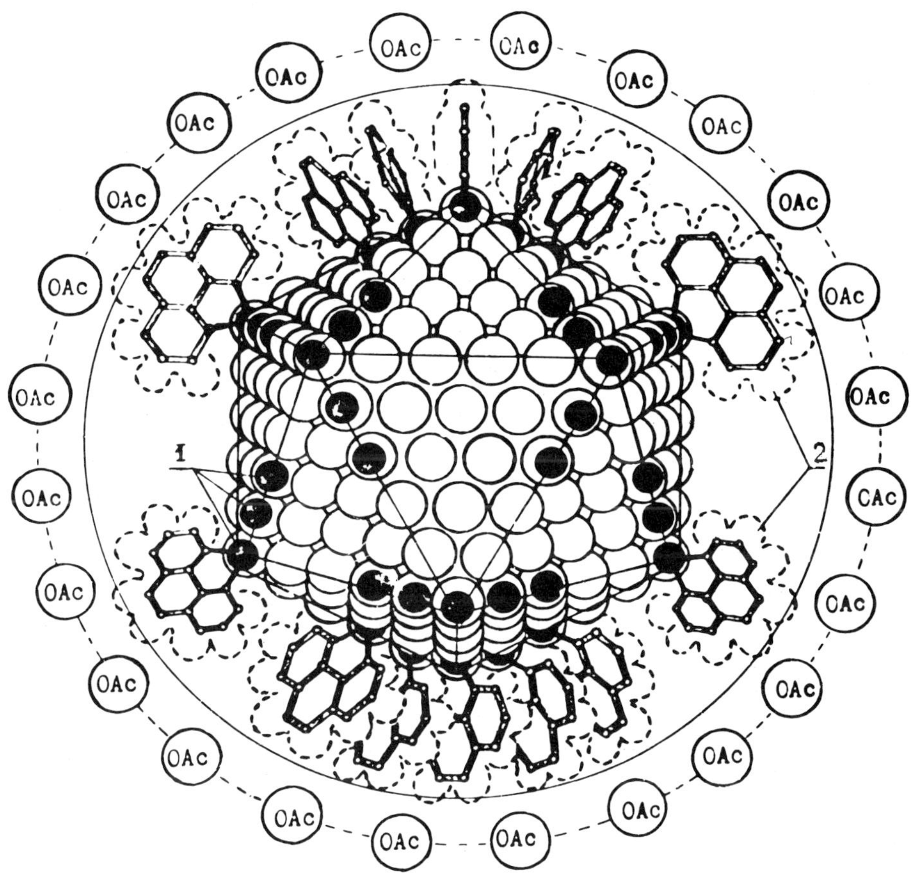

Figure 22. The most probable structure of the icosahedral palladium cluster complex $Pd_{561}Phen_{60}OAc_{180}$. 1 - Pd atoms on cluster edges; 2 - phenanthroline ligands chemisorbed on the boundary Pd atoms in position 1 (according to Ref. 53)

The dependence of the structure of rhodium clusters on the nature of the support can be elucidated if one compares the intensities of various peaks of radial atomic distribution curves obtained from EXAFS spectra of supported rhodium catalysts (Fig. 21). It can readily be seen in the figure that for the catalyst Rh/La_2O_3 the number of Rh-O bonds considerably exceeds the number of Rh-Rh bonds (and vice versa for Rh/SiO_2 and Rh/Al_2O_3 catalysts).

Similarly, a giant palladium cluster with the solvation shell consisting of the acetate anions has an icosahedral structure (Fig. 22), while substitution of acetate anions for PF_6^- anions leads to a dramatic rearrangement of this structure into a cubooctahedral one (Ref. 54). These Pd and Rh cluster catalysts are similar to classical metal heterogeneous ones in their relatively high catalytic activity, which is higher than that typical of metal complexes. At the same time, in their selectivity, which can vary as a function of the environment, these clusters certainly resemble more homogeneous catalysts. For example, CO hydrogenation over Rh/SiO_2 and Rh/Al_2O_3 species results mainly in the formation of paraffins, whereas CO hydrogenation over Rh/La_2O_3 clusters leads to the formation of alcohols.

In a similar way, a palladium cluster catalyzes a selective hydrogenation of dienes to monoolefins.

The above-mentioned peculiarities of the catalytic action of the clusters may be caused, on the one hand, by that these clusters have a large enough number of directly bound with each other and coordinatively unsaturated metal atoms. This relates them with classical heterogeneous catalysts. On the other hand, under the influence of the environment of the support or of the ligands, and also because of the finite number of metal atoms in the clusters, the dispersion of their ASs as regards to their structure and properties, decreases. As a result, most of the ASs of these clusters are of similar structure (like in the case of metal complexes) which provides a higher selectivity of their catalytic action.

Experimental data of Ref. 52 seem to indicate that

in supported metal catalysts clusters of about 30 Å in size, in which a sufficiently large fraction of metal atoms interacts with the environment, tend to catalyze reaction routes characteristic of catalysis by metal complexes, whereas larger clusters tend to catalyze the routes characteristic of traditional heterogeneous metal catalysts. Thus, metal clusters, anchored on supports or stabilized by ligands, seem to be a sort of structures bridging homogeneous and heterogeneous catalysts. By varying the size of clusters and their environment, it seems possible to shift their properties to a certain extent deliberately towards those of either homogeneous or heterogeneous catalysts.

The difference in the properties of ASs of metal and metal-complex catalysts may also be caused by different oxidation states of metal atoms in these two types of compounds. One may attempt to control the oxidation state of atoms on surfaces of metal catalysts by introducing promoting additives. It is known, for example, that additives of alkaline metals to transition metals facilitate dissociation on the surface of the latter of such molecules as CO and H_2O (Refs. 55,56). Additives of halide atoms have an opposite (poisoning) effect on this process. These effects can easily be explained on the ground of the quantum-chemical analysis of the changes in electronic structure and in adsorption properties of local metal sites at which adsorption of dissociating molecules takes place. In calculations, these sites can be imitated by clusters composed of a limited number of metal atoms. In (Ref. 57) a dissociative adsorption of CO on Ni_4, Ni_3K and Ni_3Cl clusters has been calculated by ASED method. As found, the presence of K or Cl atoms decreases or increases, respectively, the activation barrier for CO dissociation (the corresponding estimates for Ni_4, Ni_3K and Ni_3Cl clusters are 2.2, 1.2 and 2.5 eV). A qualitative explanation of this phenomenon is as follows. The dissociation of CO during its adsorption is initiated by the weakening of the C-O bond upon electron transfer from the cluster into the 2 π^*-antibonding vacant molecular orbital of CO. This transfer is promoted by an electron-donor atom K and is inhibited by an electron-acceptor chlorine atom. The above approach to the analysis of the dissociative adsorption of CO clearly resembles those used to analyze the properties of homogeneous metal complex catalysts both as regard to the type of the theoretical model used and to the nature of the effect of additives on CO dissociation.

7. CONCLUSIONS

The results reported in this paper indicate close similarities in the structures and in the mechanisms of operation of active sites in homogeneous, heterogeneous and enzymatic catalysis. These similarities arise from the same chemical properties of the ASs of the same type in these three types of catalysts as well as from a similar character of the effects of the medium on the mechanisms and the rates of the key steps of homogeneous, heterogeneous and enzymatic catalysis. For ASs of heterogeneous catalysts these are functional groups of their lattice as well as adsorbed molecules that can serve the role of the medium.

From the above consideration one may conclude, for example, that depending on structural peculiarities of ASs and on their local environment, it seems likely that for each of the three types of catalysts all known mechanisms of catalysis, including those which are traditionally attributed to catalysts of an alternative type, are possible. Thus, one may hope to substantially improve both homogeneous and heterogeneous acid-base catalysts by deliberate designing bifunctional acid-base active sites capable of providing catalytic transformations via concerted mechanisms typical for enzymes. Also, it looks quite possible to develop organometallic catalysts that would combine the high activity typical for heterogeneous metal catalysts and the high selectivity typical for homogeneous metal-complex catalysts.

ACKNOWLEDGEMENT. The authors are grateful to their co-workers from the Institute of Catalysis Prof. V.A. Zakharov, Dr. A.G. Pelmenschikov, Dr. E.A. Paukshtis, Prof. I.V. Kozhevnikov, Dr. O.V. Gritsenko, Dr. G.A. Nesterov, Dr. D.I. Kochubey and Prof. V.M. Nekipelov for stimulating discussions and for their most valuable help in the preparation of this paper.

REFERENCES

1. H.S. Taylor, Proc. Roy. Soc., A, 108, 105 (1925).
2. G.K. Boreskov, Zh. Vses. Khim. Obschestva, XXII, 495 (1977).
3. G.M. Zhidomirov and I.D. Mikheikin, in: Annual Review by All-Union Institute of Sci.-Tech. Information, ser. Structure of Molecules and Chemical Bonds (in Russian), v. 9, Moskva (1984).
4. M. Simonetta and A. Gravezzotti, Adv. Quant. Chem., 12, No 1, 103 (1980).
5. R.P. Messmer, Surf. Aci., 106, No 1/3 225 (1981).
6. I.D. Mikheikin, I.B. Abronin, G.M. Zhidomirov and V.B. Kazansky, J. Molec. Catal., No 3, 435 (1977/78).
7. A.G. Pelmenscikov, I.D. Mikheikin and G.M. Zhidomirov, Kinet. Katal., 22, 1427 (1981).
8. E.A. Paukshtis and E.N. Yurchenko, Usp. Khim., 52, No 3, 426 (1983).
9. A.G. Pelmenschikov and G.M. Zhidomirov, React. Kinet. Catal. Lett., 23, No 3-4, 295 (1983).
10. A.G. Pelmenschikov, G.M. Zhidomirov and K.I. Zamaraev, React. Kinet. Catal. Lett., 21, No 1-2, 115 (1982).
11. I.V. Kozhevnikov and K.I. Matveev, Usp. Khim., 51, 1875 (1982).
12. I.V. Kozhevnikov and S.M. Kulikov, Kinet. Katal., 22, 956 (1981).
13. M. Mesono, K. Sakata and Y. Yonoda, 7th Intern. Congr. on Catalysis, Preprints, Tokyo, B27 (1980).
14. J.B. Moffat, in Catalysis by Acids and Bases, B. Imelik (Ed.) et al., Elsevier Sci. Publ., (1985).
15. A.V. Kiselev and V.I. Lygin, IR Spectra of Surface Compounds (in Russian), Nauka, Moskva (1972).
16. B.S. Ault and G.C. Pimentel, J. Phys. Chem., 77,1649 (1973).

17. G.M. Zhidomirov and A.G. Pelmenschikov, Proc. Int. Symp. Electrodynamics and Quantum Phenomena at Interfaces, p. 136, Telavi, 1984.
18. Yu.D. Pankratiev, E.A. Paukshtis, V.M. Turkov and E.N. Yurchenko, Proc. Int. Symp. Zeolites Catalysis, Hungary, Shiofok, p. 55 (1985).
19. L.V. Malysheva, N.S. Kotsarenko and E.A. Paukshtis, Kinet. Katal., 21, No 2, 536 (1980).
20. L.V. Malysheva, N.S. Kotsarenko and E.A. Paukshtis, React. Kinet. Catal. Lett., 16, No 4, 365 (1981).
21. Kh.F. Leftin, M.K. Khobson, Usp. Khim., 35, 938 (1966).
22. G.A. Olah, Usp. Khim., XLIV, 793 (1975).
23. G.M. Zhidomirov, Proc. 9th All-Union Meeting on Quantum Chemistry, p. 34, part I, Ivanovo, 1985, Chernogolovka, 1985.
24. A. Pullman, A.M. Armbruster, Int. J. Quant. Chem., S8, 169 (1974).
25. G.A. Olah, E.B. Baker, T.C. Evans, W.S. Tolgyesi, J.S. McIntyre and I.J. Bastien, J. Am. Chem. Soc., 86, 1360 (1964).
26. V.A. Koptyug, Arenonium Ions. Structure and Reactivity (in Russian), Nauka, Novosibirsk (1983).
27. N.S. Kotsarenko and L.V. Malysheva, Kinet. Katal., 24, No 4, 877 (1983).
28. I.N. Senchenya, I.D. Mikheikin, G.M. Zhidomirov and A.I. Trokhimets, ibid, 24, 35 (1983).
29. A.G. Pelmenschikov, I.N. Senchenya, G.M. Zhidomirov and V.B. Kazansky, ibid, 24, 233 (1983).
30. A.G. Pelmenschikov, G.M. Zhidomirov, D.V. Kuroshvili and G.V. Tsitsishvili, in Structure and Reactivity of Modified Zeolites, P.A. Jacobs (Ed) et al., Elsevier, Amsterdam (1984).
31. O.M. Poltorak, in Proc. All-Union Conf. on Mechanisms of Heterogeneous Catalytic Reactions, p. 94, Chernogolovka, 1971, O.V. Krylov (Ed.), Nauka, Moskva.
32. R.P. Bell, The Proton in Chemistry, Chapman and Hall, London (1973).
33. P.R. Rony, J. Am. Chem. Soc., 91, 6090 (1969).
34. E.P. Talsi, V.M. Nekipelov, K.I. Zamaraev, Zh. Fiz. Khim., 58, No 2, 273 (1984).
35. O.V. Nesterov, V.M. Nekipelov, Yu.P. Chirkov, A.N. Kitaigorodskii, S.G.Entelis, Kinet. Katal., 21, No 5, 1238 (1980).
36. V.M. Nekipelov and K.I. Zamaraev, Coord. Chem. Rev., 61, 185 (1985).
37. A.I. Lumpov, I.D. Mikheikin, G.M. Zhidomirov and V.B. Kazansky, Kinet. Katal., 20, 811 (1979).
38. Yu.I. Yermakov, V.A. Zakharov and B.N. Kuznetsov, Supported Complexes in Catalysis (in Russian), Nauka, Novosibirsk (1980).
39. C. Masters, Adv. Organom. Catal., 17, 61 (1975).
40. J. Manziger, D. McAlister, R. Sanner and J. Bercaw, J. Am. Chem. Soc., 100, 2716 (1978).

41. J. Marsella, C. Cartis, J. Bercaw and K. Calton, J. Am. Chem. Soc., 102, 7244 (1980).
42. G.A. Nesterov, E.A. Paukshtis and V.A. Zakharov, React. Kinet. Catal. Lett., 26, 357 (1984).
43. V. Giaunini, V. Zucchini and E. Albizzati, J. Polym. Sci., Polym. Lett., 8, 405 (1970).
44. D.G.H. Ballard, Adv. Catal., 23, 263 (1973).
45. E.A. Demin, V.A. Zakharov, Yu.I. Yermakov, O.S. Ovchinskii, Kinet. Katal., 13, 248 (1972).
46. P. Dumas, D. Martineau and P. Sigwalt, Proc. Int. Symp. Macromol., vol. 2, p. 71, Florence, Italy (1980).
47. I.I. Zakharov and V.A. Zakharov, React. Kinet. Catal. Lett., 23, 61 (1983).
48. W. Kaminsky, M. Miti, H. Sinn and R. Woldt, Macromol. Chem. Rapid Comm., 4, 417 (1983).
49. O.V. Krylov and V.F. Kiselev, Adsorption and Catalysis on the Transition Metals and their Oxides, Khimia, Moskva (1981).
50. W. Keim, M. Berger, J. Schlupp, J. Catal., 61, No 2, 359 (1980).
51. G.C. Demitras and E.L. Muetterties, J. Am. Chem. Soc., 99, No 8, 2796 (1977).
52. V.L. Kuznetsov, A.V. Romanenko, I.L. Mudrakovsky, V.M. Mastikhin, V.A. Shmachkov, Yu. I. Yermakov, 8th Int. Congr. on Catalysis, Berlin, 1984, Verlag Chemie, v. 5, p. 3, 1984.
53. M.N. Vargaftik, V.P. Zagorodnikov, I.P. Stolyarov, I.I. Moiseev, V.A. Likholobov, D.I. Kochubei, A.L. Chuvilin, V.I. Zaikovsky, K.I. Zamaraev and G.I. Timofeeva, J. Chem. Soc., Chem. Commun., 937 (1985).
54. V.P. Zagorodnikov, M.N. Vargaftik, D.I. Kochubei, A.L. Chuvilin, S.G. Sakharov and M.A. Maifat, Izv. Akad. Nauk. SSSR, ser. khim., No 1, 253 (1986).
55. M.P. Kiskinova, Surf. Sci., 111, 584 (1981).
56. M.P. Kiskinova, G. Pirug and H.P. Bonzel, Surf. Sci., 150, 319 (1985).
57. D. Tomanek and K.H. Benneman, Surf. Sci., 127, LIII (1983).

AN APPROACH TO ELUCIDATION OF REACTION MECHANISMS OVER INORGANIC OXIDE-ATTACHED METAL CATALYSTS

Y.Iwasawa* and K.Asakura

Department of Chemistry, Faculty of Science, The University of Tokyo, Hongo, Bunkyo-ku, Tokyo 113, Japan

ABSTRACT. Elucidation of reaction mechanisms on solid metal catalysts should include the characterization of structures of active sites and the in situ observation of their dynamic behaviors during catalysis, and also correlate catalytic phenomena with the properties of active structures. The surface behaviors and catalysis of Mo-dimers attached on SiO_2, Rh hydride-monomers attached on TiO_2, and Rh metal-clusters attached on TiO_2 are reported in relation to molecular-level surface designs and reaction mechanisms.

1. INTRODUCTION

Recent preparations of catalysts have been directed to provide well defined and high-quality catalysts on a molecular level(Ref.1). Developments of unique and specific surfaces may lead to the finding of new catalytic reaction systems(Ref.2). It may be the time when we should recognize a high possibility of surface design in chemical (molecular-scale) sense on the basis of recent surface-analysis spectroscopies. Usually reaction sites dynamically change their structures and electronic properties during catalysis. In order to understand reaction mechanisms of solid catalysts the dynamic behaviors of active sites may be necessary to monitor directly. However, traditional supported metal catalysts have ill defined and heterogeneous properties of catalyst surfaces including different types of active sites and hence it may be acutely difficult to obtain a definite conclusion on the reaction sites and reaction mechanisms. It is impossible to observe atoms or molecules on catalyst surfaces, but EXAFS(extended x-ray absorption fine structure) spectroscopy can determine the local structure of a particular metal atom included in reaction sites and also follow the change in structures of active sites during a catalytic reaction, that is, it can analyze the behaviors of molecular structures of catalyst, which is equivalent to a direct watch of catalyst surface on a molecular scale. In the present article the surface designs of molybdenum-dimers, rhodium-monomers and rhodium-clusters on SiO_2 or TiO_2, the the direct observation of dynamic behaviors of active sites of molybdenum-dimers attached on SiO_2 are reported in relation to potential preparations of a new class of catalysts with ultimate activities and selectivities and elucidation of reaction mechanisms.

2. TiO_2-ATTACHED Rh CATALYSTS

2.1. Preparations and Transformations of Rh Species on TiO_2

$Rh(\eta^3\text{-}C_3H_5)_3$ was synthesized from $[RhCl(C_3H_5)_2]_2$ and C_3H_5MgCl according to the previous report(Ref.3). The attachment of $Rh(C_3H_5)_3$ onto TiO_2 surface was conducted in a Pyrex-glass reactor equipped with several U-shaped glass tubes with breakseals. TiO_2P-25, decarbonated) was treated at 473 K in the reactor in situ before the Rh-attaching. The attaching reaction at 273 K was completed within 20 min to give the TiO_2-attached Rh catalyst(1) with Rh/TiO_2=2.0 wt%. The incipient surface Rh complexes were converted to Rh hydride species and Rh metallic clusters with H_2. The attachment and transformations are illustrated in Scheme 1(Ref.4).

2.2. Structures of Surface Rh Species

The temperature programmed decomposition products of the surface

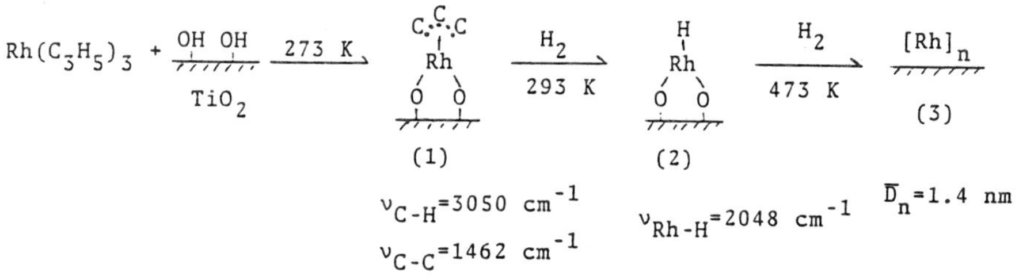

$$Rh(C_3H_5)_3 + \underset{TiO_2}{\overline{OH\ OH}} \xrightarrow{273\ K} \underset{(1)}{\text{C}\cdot\overset{\text{C}}{\cdot}\cdot\text{C}-\text{Rh}(\text{O})(\text{O})} \xrightarrow[293\ K]{H_2} \underset{(2)}{\text{H}-\text{Rh}(\text{O})(\text{O})} \xrightarrow[473\ K]{H_2} \underset{(3)}{[Rh]_n}$$

(1): $\nu_{C-H}=3050\ cm^{-1}$, $\nu_{C-C}=1462\ cm^{-1}$

(2): $\nu_{Rh-H}=2048\ cm^{-1}$

(3): $\bar{D}_n=1.4$ nm

Scheme 1. Attachment and Transformations of Rh Species on TiO_2

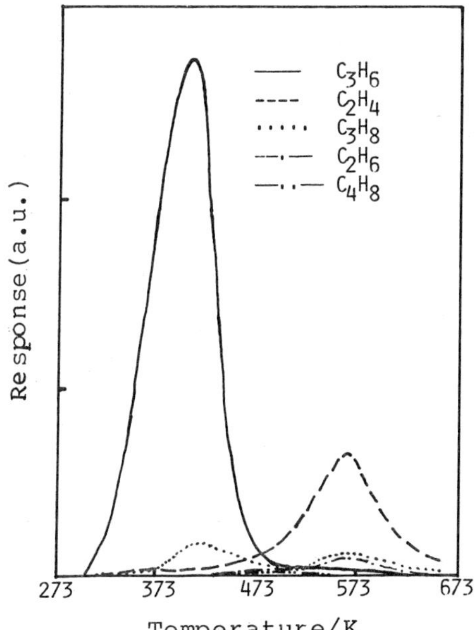

Figure 1. Temperature Programmed Decomposition Chromatogram of the Rh-allyl Species(1) Attached to TiO_2 Surface; heating rate: 4 K min^{-1}.

complex(1) were mainly propene, exhibiting a single peak(see Fig.1). This implies that the Rh allyl complexes were almost uniformly distributed on TiO_2. Besides propene peak, a single peak for ethene evolution was observed around 573 K. The feature of the thermal decomposition almost unchanged by decreasing the Rh loadings to 1.0 wt%, and the decomposition to ethene was also observed with platinum allyl complexes attached on TiO_2. Hence the higher-temperature peak may be derived from the coordinated ligands to Ti(bridged). Two C_3H_5 groups of $Rh(C_3H_5)_3$ were lost by the attaching reaction with two surface OH groups of TiO_2. The characteristic absorption bands of (1), 3050 and 1462 cm^{-1} for ν_{C-H} and ν_{C-C}, respectively, indicate that the C_3H_5 ligand of the surface complex is still of π-allyl character. The species(1) reacted with H_2 at 293 K for 1 h to form the Rh-hydride(2) with concomitant evolution of propane. The IR bands for the allyl groups disappeared to be replaced by a new peak at 2048 cm^{-1}. The peak shifted to 1495 cm^{-1} by excahnge with D_2. The 2048 cm^{-1} peak also disappeared with contact with ethene: two peaks at 2970 and 2916 cm^{-1} with shoulder which are assigned to the C-H stretching vibrations of C_2H_5 groups were observed. Reversely, the ν_{C-H} peak was replaced by the 2048 cm^{-1} peak by introducing H_2 again. Thus the peak of 2048 cm^{-1} was assigned to the mononuclear Rh-monohydride sepcies(2).

In order to clarify the structures of the Rh species(1) and (2) on TiO_2 we performed the EXAFS analyses(Ref.5). The Fourier transform of the EXAFS spectrum of (2) showed only one peak assignable to Rh-O bond and no peak for Rh-Rh bond was observed in the range 0 - 5 Å. The bond distance of Rh-O was determined to be 2.15 Å by a curve-fitting technique. The bond number of the Rh-O was 2. These results suggest that the Rh atoms in the Rh-hydride catalyst are monoatomically attached to the TiO_2 surface in a bidentate form(see Scheme 2). The three Rh-C bond lengths between η^3-allyl ligand and Rh atom in $(DBM)Rh(C_{12}H_{16})$ are known to be 2.11, 2.18 and 2.21 Å, resepctively. The close values to the Rh-O bond length would make it difficult to distinguich Rh-C and Rh-O bonds in the species(1). The cond distances of Rh-O and Rh-C(allyl) determined by one-shell

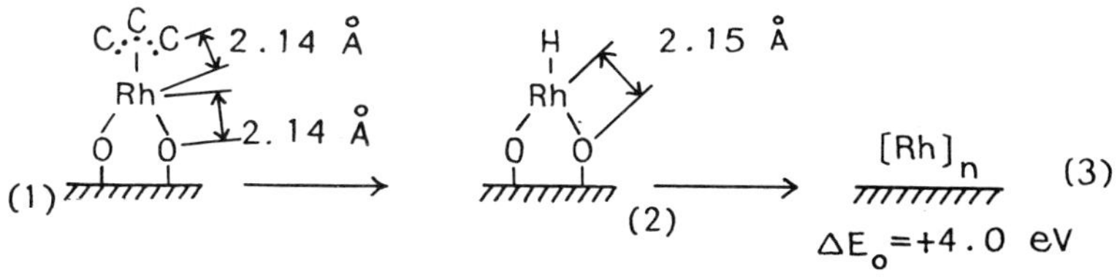

Scheme 2. Transformations of the Attached Rh Species on TiO_2

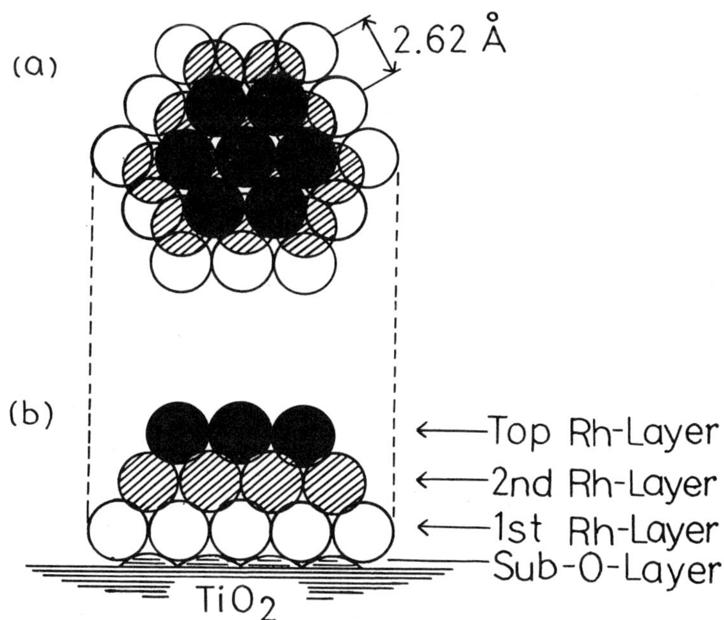

Figure 2. A Proposed Structure of the Attached Rh Cluster(3) on TiO_2 ; (a) Top View, (b) Side View

EXAFS analysis are illustrated in Scheme 2.

The species(2) was reduced with H_2 at 473 K or 773 K to form Rh particles. The transmission electron microscope showed relatively narrow particle distributions; the mean diameter was 14 Å and 15 Å for 473 K-reduced and 773 K-reduced Rh samples. An impregnation Rh/TiO_2 catalyst depicted a much broader distribution of Rh particles with the mean diameter of 34 Å. A peak around 2.5 Å assignable to Rh-Rh bond was observed in the EXAFS Fourier transform. The peak was analyzed by a curve-fitting method. The Rh-Rh bond distance was determined to be 2.62 Å(see Fig.2) which is shorter than 2.66 Å for Rh metal and the impregnated Rh/TiO_2 catalyst. $\overline{N}_j/N_j$ values($\overline{N}_j$ is the average coordination number determined from EXAFS and N_j is the bulk coordination number)(Ref.6) was calculated to be 0.64. The value together with the particle size of 14 Å suggests that the Rh particles exist as raft-like particles consisting of 2 or 3-atomic Rh layers(see Fig.2).

For comparison, a SiO_2-attached Rh catalyst was similarly obtained. The $\overline{N}_j/N_j$ for the SiO_2-attached Rh particles was 0.42 from EXAFS and the Rh particlc size was 13 Å from TEM, which suggests a monolayer-raft Rh structure(see Fig.3). The one-atomic layer structure of Rh assembly on SiO_2 well explains the 100% dispersion in H_2 adsorption measurements(see Fig.4).

The Rh K-absorption edge of the Rh catalyst(3) shifted towards the higher energy by 4.0 eV compared to that of Rh metal. In contrast, the impregnation Rh/TiO_2 catalyst showed only 0.4 eV shift and the SiO_2-attached Rh catalyst showed no edge shift. The K-edge energy of the SiO_2-attached Rh catalyst was the same as that of Rh metal although the Rh particle size(13 Å) was similar to that of the attached Rh/TiO_2 catalyst(3)(14 Å). The positive edge shift may be owing to the positive charge of Th particles in the catalyst(3) rahther than the reduction of the relaxation energy. The species(3) were prepared form the species(2) which have the Rh-O-Ti bonds. These chemical bonds to the surface appear to retain partially during the reduction of (2) with H_2, by which electron transfer from TiO_2 to Rh particles occurs to induce a positive charge on the

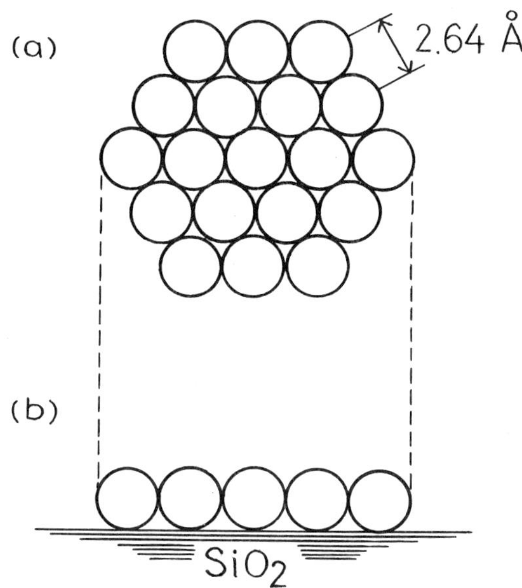

Figure 3. A Proposed Structure of the Rh Cluster Attached onto SiO_2 Surface ; (a) Top View, (b) Side View .

raft-like small Rh particles(see Fig.2). The contraction of the Rh-Rh bond distance in the attached catalyst(3) is ascribed to the positive charge which would reduce effective Rh-metallic redius and the surface-lattice effect by the formation of Rh-O(TiO_2) bonds. Thus Rh particles can be called to be attached Rh metal clusters in the sense that the Rh assembly exists under strong interaction with the surface oxygen atoms(Ref.5).

2.3. Chemisorption and Catalysis over Rh/TiO_2 Catalysts

Conventional Rh/TiO_2 catalysts prepared by an impregnation method

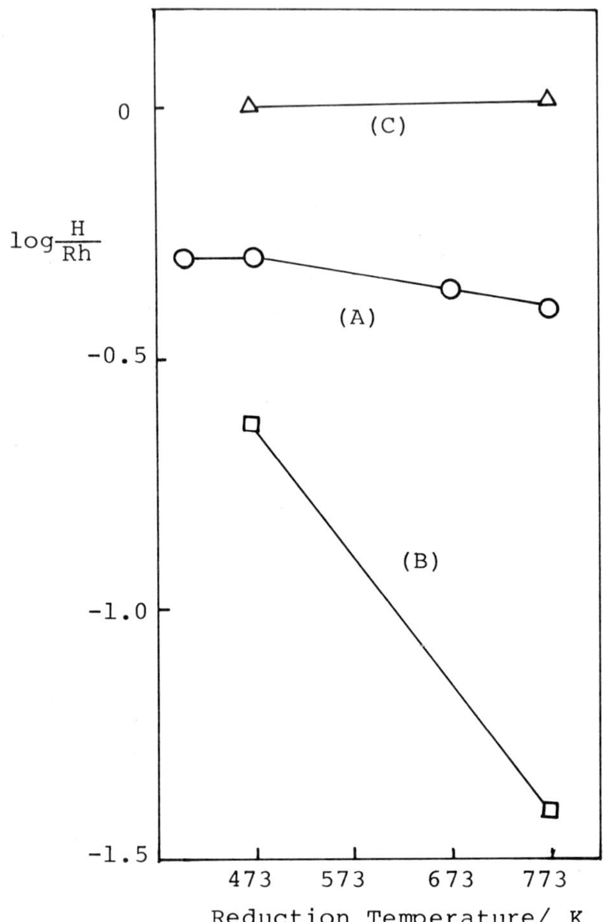

Figure 4. H_2 Adsorption on TiO_2-Attached Catalyst(A), TiO_2-Impregnated Catalyst(B), and SiO_2-Attached Catalyst Prereduced at Different Temperatures;
Rh loadings: 2.0 wt%.

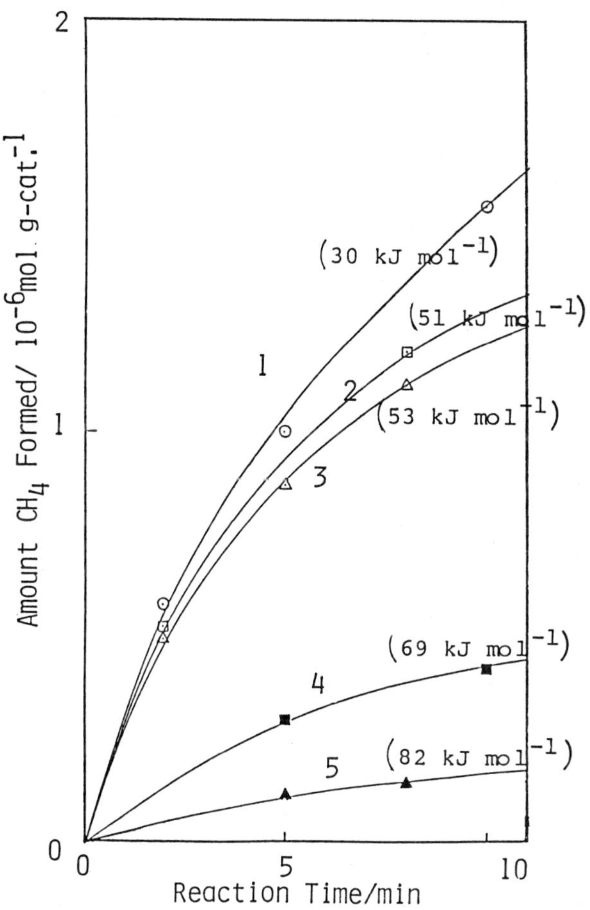

Figure 5. Ethane Hydrogenolysis on Attached and Impregnated Rh Catalysts;
catalyst : 0.33 g(Rh/TiO_2=2.0 wt%),
react. temp. : 373 K,
C_2H_6= 20 Torr, H_2= 40 Torr,
1 : species(2), 2 : species(3)(473 K redn.), 3 : species(3)(773 K redn.), 4 : impreg.cat. (473 K redn.), 5 : impreg.cat.(773 K redn.). The activation energies are shown in ( ).

lost the ability of chemisorptions of H_2 and CO by high temperature reduction(SMSI effect)(Ref.7). On contrary, the TiO_2-attached Rh metal clusters with a positively charged character showed no SMSI behavior(see Fig.4). The Rh particles with positive charge did not exhibit a 1:1 relation between the quantity of Rh atoms exposed and the amount of H atoms adsorbed, the amount of H_2 adsorption being less than expected from the particle size.

The TiO_2-attached Rh-hydride species(2) and Rh-thin clusters(3) showed high catalytic activities for ethene hydrogenation and ethane hydrogenolysis compared with the impregnation Rh catalysts(see Fig.5). The catalytic activity of the impregnation catalyst markedly decreased by the prereduction at 773 K owing to the SMSI effect, but the attache Rh catalyst retained the high activity. Our present work suggests that the non-SMSI behavior of the nobel TiO_2-attached Rh clusters is due to the Rh-O-Ti bonds which are produced in the well defined attachment steps in Scheme 1 and partially maintained after the high-temperature reduction. In general, ethane hydrogenolysis requires multi-metal atoms for reaction. It may be noted that the most active catalyst was the Rh-hydride monomers(2). The activation energy was observed to be the smallest value(30 kJ mol^{-1}) with the catalyst(2). Unsaturated Ti ions located closely to Rh sites may assist the catalytic action of Rh monomers.

3. SiO_2-ATTACHED Mo-DIMER CATALYSTS

3.1. Preparations and Transformations of Mo Structures on SiO_2

The SiO_2-attached Mo-dimer catalysts were synthesized on the basis of the reaction between $Mo_2(\eta^3\text{-}C_3H_5)_4$ and surface OH groups of SiO_2, followed by chemical treatments with H_2 or O_2, as previously reported (Scheme 2)(Refs.8 and 9). The SiO_2 was treated at 473 K for 1.5 h before use and had a specific surface area of 285 m^2g^{-1}. The $Mo_2(C_3H_5)_4$ complex reacted with the silanol in a stoichiometric ratio of 1:4 at 273 K to form the surface complexes(4). The complex(4)

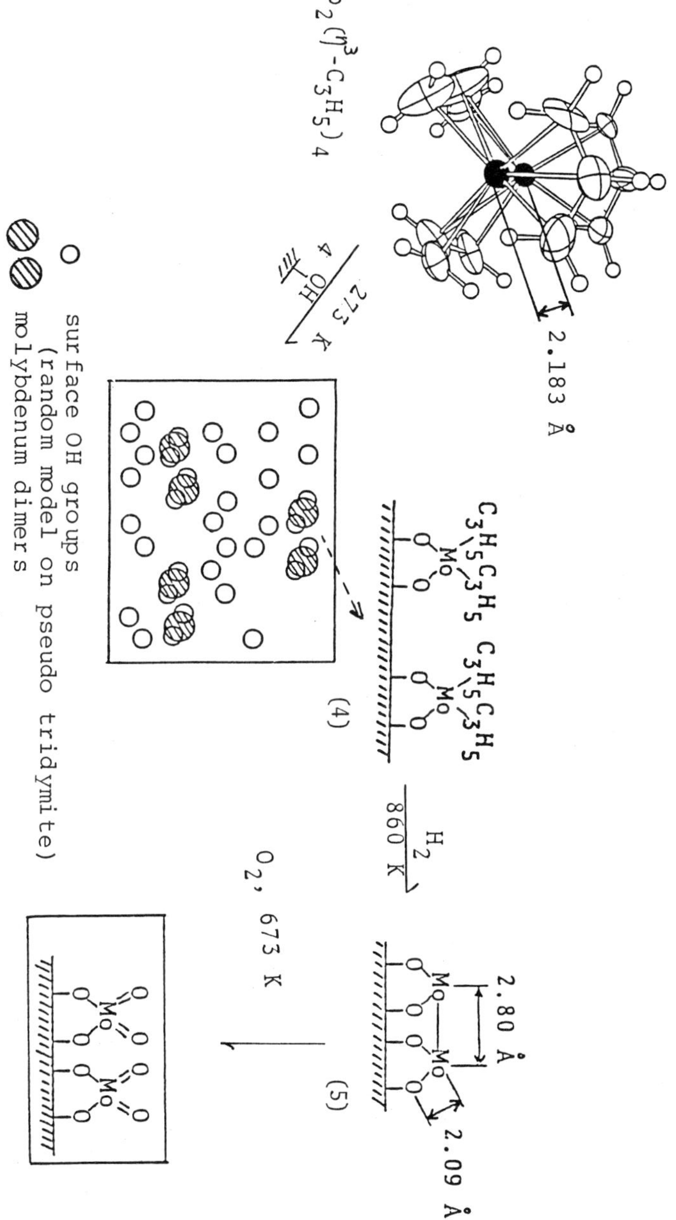

Scheme 3. Attachment and Transformations of Mo-Dimer Structures on SiO_2

was converted to bivalent Mo dimers(5) by removal of the ligands with H_2. The Mo(II) dimers(5) were quantitatively oxidized to dioxomolybdenum(VI) structures(6), uptaking the stoichiometric amount of O_2 at 673 K. Reversely, the species(6) were reduced again to the Mo(II) dimers with H_2 at 853 K, evolving the quatitative amount of water and uptaking the stoichiometric amount of H_2.

3.2. Structures of Mo-Dimer Catalysts

One of the two allyl ligands coordinated to a Mo atom was concluded to be different in bonding feature from the other by the temperature programmed decomposition studies(Ref.8). According to the EXAFS analysis of the structure(5), the two adjacent Mo atoms are directly bonded each other at a distance of 2.80 Å. The Mo-dimers were also bonded to SiO_2 surface through the chemical bonding with surface oxygen atoms at 2.09 Å in a bidentate form(Ref.10). When the Mo(II) dimers were exposed to oxygen, the Mo-Mo bond disappeared but a new peak appeared in the Fourier transform. The new peak was straightforwardly assigned to Mo-O bonds judging from the bond distance. The EXAFS analysis showed the Mo=O double bond length of 1.70 Å and the Mo-O single bond distance of 2.10 Å. When the Mo(VI) structures were reduced with H_2 to the Mo(II) level, the Mo=O bond peak disappeared with the concomitant appearance of the Mo-Mo bond peak again. In these redox treatments the intensity of the Mo-O bond peak remained unchanged. The EXAFS studies clearly confirm the strucutral change of Mo-dimer structures between Mo(II) and Mo(VI) levels as shown in Scheme 3.

3.3. Reaction Mechanism for Ethanol Oxidation on the Basis of Direct Observation of the Mo Structures by EXAFS

The Mo(VI)-dimer catalyst(6) was found to be active for the oxidation of ethanol to acetaldehyde in the presence of oxygen as previously reported(Ref.9). The activity of the catalyst(6) was about one order

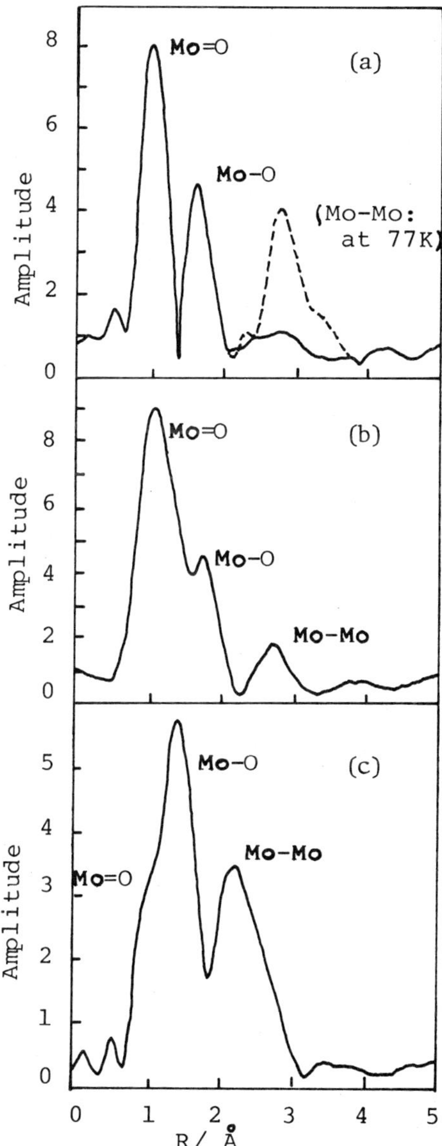

Figure 6. Fourier Transforms of the EXAFS Data of the SiO_2-Attached Mo-Dimer Structures in the Course of Ethanol Oxidation;

(a) Species(6), (b) Species(8), (c) Species(10)

of magnitude higher than that of a conventional Mo catalyst prepared by an impregnation method in the reaction temperature range 398-448 K. It was indicated from the reaction kinetics, IR, UV/Vis diffuse reflectance and the stoichiometries in each elementary reaction steps, that the reaction proceeded by the two-stage redox mechanism. Ethanol oxidation was well explained by cooperative catalysis of the two adjacent Mo atoms. In the present article the direct observation of the Mo structures by means of EXAFS is described in relation to the origin of the cooperative catalysis on a molecular scale(Scheme 4)(Ref.11).

Fig.6 shows the Fourier transforms of the EXAFS data for the Mo dimer catlaysts in the course of ethanol oxidation. The Fourier transform depicts two distinct peaks in the region of 1-2 Å. These two pwaks are straightforwardly assigned to Mo-O judging from the bond distances(Fig.6(a)). This was further confirmed by performing two-wave fitting for the $k^3\chi(k)$ curve obtained by the reverse-Fourier transform of the region of the two peaks. The analysis for the EXAFS data measured at room temperature did not give a confidence for the pairing of two Mo atoms at SiO_2 surface. The dotted curve in Fig.6(a) is the Fourier transform in the region 2-4 Å fo the EXAFS oscillation measured at 77 K. The peak was observed around 3 Å. The detail analysis by three-wave(Mo-O, Mo-O and Mo-Mo) fitting over the region including all the these peaks showed the presence of Mo-Mo bond at a distance of 3.04 Å at 77 K besides Mo=O at 1.70 Å and Mo-O at 2.10 Å. Thus the two adjacent Mo(VI) ions attached on SiO_2 seem to thermally vibrate without fixing at a definite distance in the temperatures above 200 K.

After ethanol vapor was admitted onto the Mo(VI) dimer catalyst(6) at 353 K for 2 min in the absence of oxygen, followed by short evacuation of physisorbed ethanol molecules, the catalyst was quickly cooled down to 200 K to measure the EXAFS spectrum. The Fourier transform of the EXAFS spectrum thus taken exhibits two peaks in the region of 1-2 Å and a small peak around 2.8 Å(Fig.6(b)). The former two a attributable to Mo-O, while the last one is due to Mo-Mo. Since the Mo-Mo bond of the structure(6) was not observed above 200 K, the appearance of Mo-Mo bond in the EXAFS result of the species(8) suggests that Mo-Mo idstance was fixed by the formation of some kind of linkage. The fixation of the Mo-Mo separation at a definite distance of 2.92 Å which is shorter by 0.12 Å than that of (6) may be due to the hydrogen bonding of the methylene-hydrogen of ethoxide group with the double-bond oxygen of the adjacent molybdenum (VI)(Scheme 4).

Acetaldehyde was produced when the species(8) was kept at 433 K under about 2.7 kPa of the partial pressure of ethanol in the absence of oxygen. The reaction stopped at the stage when the amount of acetaldehyde equivalent to the amount of Mo-dimers contained in the catalyst was produced, indicating that one Mo(VI)-dimer acted as one active site. The EXAFS data measured after 60 min-exposure to ethanol vapor at 433 K followed by evacuation of the residual ethanol were much different from thoast of the samples(6) and (8) (Fig.6(c)). The Mo=O bond and Mo-O bond distances were determined to be 1.71 Å and 1.94 Å and the Mo-Mo bond was also intensively observed at a distance of 2.61 Å which is much shorter than 3.04 Å of the Mo(VI) pair. The strong Mo-Mo peak observed in the Fourier transform together wtih the UV/Vis spectrum and the composition of the species(10) indicates the fixation of Mo-Mo distance by oxygen-bridging. The intensity of the Mo=O double bond peak described to about half level of the intensity fro the structure(6) by the hydrogen transfor from ethanol to molybdenum through the intermediate species (8), confirming that the double-bond oxygen acted as a reaction site for the hydrogen abstraction from ethanol molecule(Scheme 4).

The Mo(V)-dimers(10) with a bridging-oxygen atom were found to be oxidized with oxygen at 433 K to form the original dioxomolybdenum (VI)-dimers(6) again. Thus, ethanol oxidation catalytically proceeds on the SiO_2-attached Mo-dimer catalyst in the presence of oxygen.

The EXAFS studies reveal that the structure of Mo sites changes drastically in the steps of the abstraction of two H atoms(dehydrogenation) from ethanol to form acetaldehyde and of the oxidation of the reduced Mo sites with O_2. The Mo-Mo separation is shortened from ca. 3 Å to 2.61 Å and the Mo-O bond distance is also shortened from 2.10 Å to 1.94 Å in the course of ethanol oxidation and subsequently the shrinked Mo-dimer is expanded to the original structure

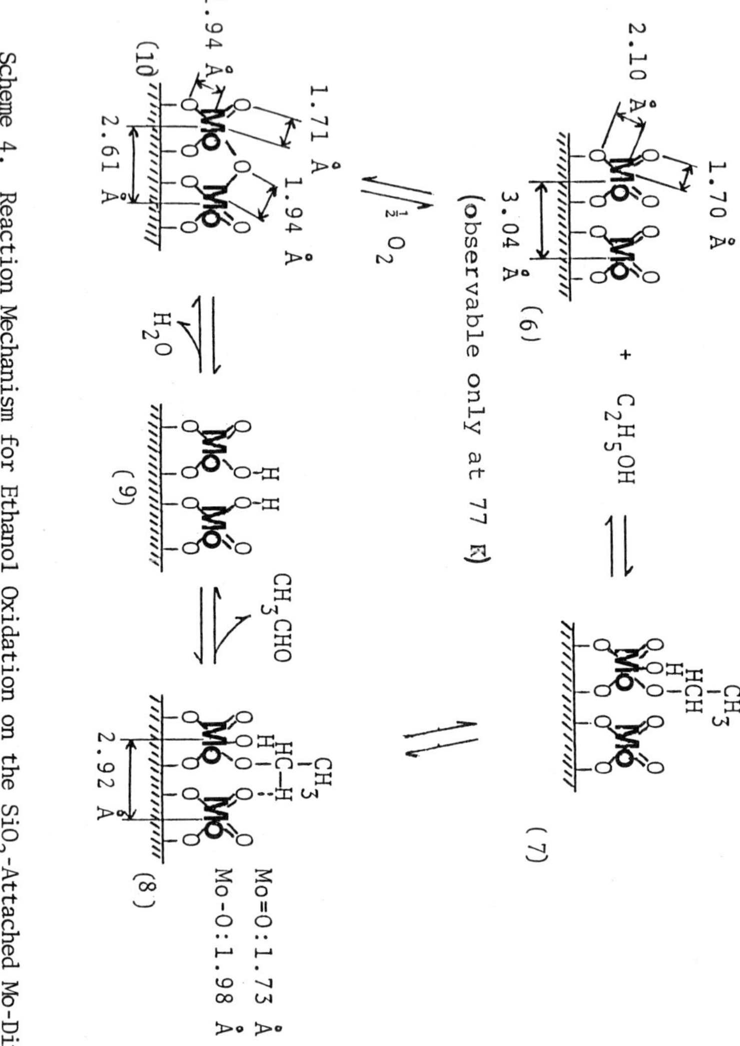

Scheme 4. Reaction Mechanism for Ethanol Oxidation on the SiO_2-Attached Mo-Dimer Catalyst

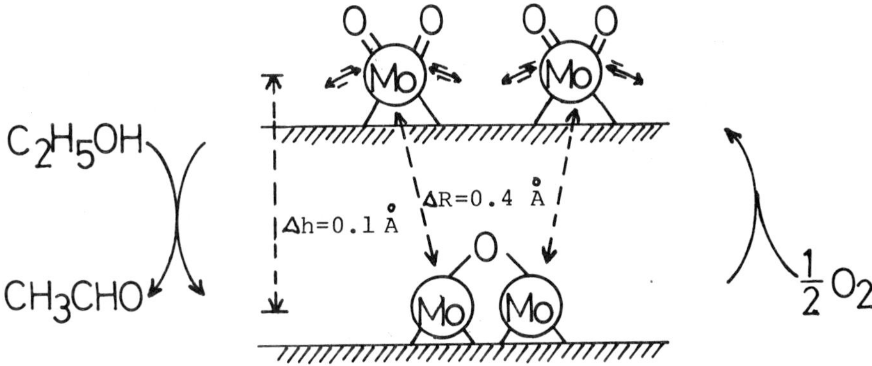

Figure 7. Dynamic Change of the Mo-Dimer Structures during Ethanol Oxidation

by break of the oxygen-bridge(Scheme 4). Thus, ethanol oxidation catalytically proceeds on the SiO_2-attached Mo-dimer catalyst in the presence of oxygen. It is concluded that one acetaldehyde molecule is produced in conjunction with one structural oscillation of the active Mo sites, lateral change(shortening ⇌ expanding) by ca. 0.4 Å and vertical change(up ⇌ down) by ca. 0.1 Å(see Fig.7).

REFERENCES

1. Y.Iwasawa ed., "Tailored Metal Catalysts", D.Reidel Publishing Company, The Netherlands(1986).
2. Yu.I.Yermakov, B.N.Kuznetsov and V.A.Zakharov, "Catalysis by Supported Complexes", Elsevier, Amsterdam(1981).
3. J.Powel and B.L.Shaw, J.Chem.Soc.A, 583(1968).
4. Y.Iwasawa and H.Sato, Chem.Lett., 1985, 507.
5. K.Asakura and Y.Iwasawa, to be published.
6. R.B.Greegor and F.W.Lytle, J.Catal., 63, 476(1980).

7. S.J.Tauster, S.C.Fung and R.L.Garten, J.Am.Chem.Soc., 100, 170 (1978); S.J.Tauster, S.C.Fung, R.T.K.Baker and J.A.Horsley, Science, 211, 1121(1981).
8. Y.Iwasawa and M.Yamagishi, J.Catal., 82, 373(1983).
9. Y.Iwasawa and H.Tanaka, Proc.8th Internat.Congr.Catal.(Berlin), Springer, IV-381(1984).
10. Y.Iwasawa, N.Ito, T.Chiba, H.Ishii and H.Kuroda, Chem.Lett., 1985, 1141.
11. Y.Iwasawa, K.Asakura, H.Ishii and H.Kuroda, Z.Phys.Chem.NF., in press.

MODERN CONCEPTS OF HOMOGENEOUS AND HETEROGENEOUS ACID CATALYSIS: SIMILARITY AND DISTINCTION IN MECHANISMS

V.B. Kazansky
N.D.Zelinsky Institute of Organic Chemistry,
Moscow 117913, USSR

ABSTRACT - Modern ideas of the nature of protonated states in aqueous solutions, nonpolar solvents and on the surfaces of oxide acidic catalysts are considered. A conclusion is made that formation of ionic active intermediates is possible only as a result of the contribution of solvation effects to the energetic parameters of the elementary step of proton transfer. Therefore, mechanisms, similar to those of homogeneous acid catalysis in aqueous solutions, are accomplished in heterogeneous systems for only the catalytic transformations of polar molecules. The catalytic reactions, involving nonpolar molecules, follow the concerted mechanisms via intermediate covalent structures having strongly polarized carbon-oxygen bonds.

1. INTRODUCTION

Acid catalysis is usually regarded as a most investigated example of the catalytic action with a postulated common mechanism for homogeneous and heterogeneous acid catalysis. Solvated protons (specific acid catalysis) or undissociated acid molecules (general acid catalysis) are supposed to be active centers of homogeneous catalysis. When they interact with the substrate, active protonated intermediates are formed, secondary reactions of which result in final products of catalytic conversions. Undoubtedly, this mechanism is valid for homogeneous acid catalysis though in each particular case the determination of the active intermediate complex structure on the atomic-molecular level by taking into account their solvation by the medium is not an easy problem.

A mechanism similar to general homogeneous catalysis by acids is usually postulated for heterogeneous acid catalysis on oxides and zeolites. Here active centers are represented by highly acidic surface hydroxyl groups. At the first stage of their interaction with adsorbed molecules hydrogen bonded complexes are formed. It is supposed that a full proton transfer to interacting substrate molecules subsequently takes place along with the formation of surface protonated complexes similar in their reactivity to the active forms of homogeneous catalysis.

Despite that this mechanism is generally accepted and its similarity to homogeneous acid catalysis is emphasized, there are some objections to it. As we already discussed in [1-4], the main one lies in the principal difference between the nature of protonated intermediates in water solutions and on the surface of oxide hetereogeneous catalysts where solvation effects arc of minor importance. Therefore we consider attempts to describe a great variety of heterogeneous acid catalytic reactions by a common scheme to be a rough approximation. Mechanisms of heterogeneous acid catalysis itself are so diverse that can hardly be restricted to the above described general scheme of homogeneous acid catalysis.

The paper deals with similarities and distinctions between homogeneous and heterogeneous catalytic reactions. In the first case we will concentrate on the discussion of solvation effects and the nature of active intermediate solvated complexes. In the case of heterogeneous acid catalysis we will focuse our attention on the discussion of reaction mechanisms in the presence of non-polar molecules where the absence of solvation accounts for a minor possibility of a full proton transfer, and intermediates, similar to those observed in homogeneous systems, are not formed.

2. HOMOGENEOUS ACID CATALYSIS

2.1. Modern Concepts of Proton Transfer in Water Solutions

It is well established that the electrolytic dissociation in water solutions is possible only as a result of solvation. Its contribution to the total energetics of the electrolytic dissociation process is dominating. For instance, heterolytic cleavage of the water molecule in the gas phase into H^+ and OH^- ions requires energy of about 17 eV (1630 kJ/mol). In the liquid phase it is, however, almost totally compensated by the sum of proton and OH ion hydration heats. Therefore the endothermicity of heterolytic dissociation of liquid water is only 0.6 eV (60 kJ/mol). Let us examine the components of the proton hydration thermal effects in liquid water.

Affinity of the proton for the first water molecule is 7.4 eV (710 kJ/mol). After addition of the second and subsequent water molecules, solvation energy monotonously decreases. This process can be described by the following sequence of chemical reactions:

$$
\begin{aligned}
H_3O^+ + H_2O &\longrightarrow H_5O_2^+ - 134\ \text{kJ/mol} \\
H_5O_2^+ + H_2O &\longrightarrow H_7O_3^+ - 96\ \text{kJ/mol} \\
H_7O_3^+ + H_2O &\longrightarrow H_9O_4^+ - 72\ \text{kJ/mol} \\
H_9O_4^+ + H_2O &\longrightarrow H_{11}O_5^+ - 79\ \text{kJ/mol} \qquad (1)
\end{aligned}
$$

Thus total hydration energy of the H_3O^+ ion is about 380 kJ/mol, which accounts approximately half of proton affinity for the first water molecule. The total heat of hydroxyl ion hydration is somewhat lower than that of the proton, and equals 477 kJ/mol.

Principal information about the real structure and geometry of different hydrated proton states in liquid water was obtained on the basis of quantum chemical calculations, from IR spectroscopic data and other physicochemical measurements. This information can be summarized as follows:

The simplest stable state of the hydrated proton in liquid water is symmetrical, easily polarized $O_2H_5^+$ ion (see 2a), rather than H_3O^+. Whatever the nature of the initial acid, it is characterized by a continuous optical absorption in the 1000-35000 cm^{-1} IR spectral range [5-7].

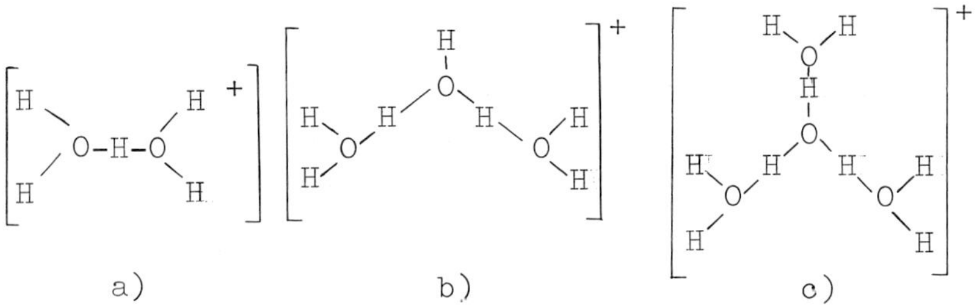

$$\left[H_2O\text{-}H\text{-}OH_2 \right]^+ \quad \left[H_2O \cdots H\text{-}\overset{H}{O} \cdots H \cdots OH_2 \right]^+ \quad \left[\text{structure c} \right]^+ \qquad (2)$$

a) b) c)

In this structure proton is not located close to one of the oxygen atoms, but vibrates between them, that is in good agreement with the calculated H_3O^+ ion average lifetime of $2\cdot10^{-13}$sec in liquid water [5].

Two approaches were proposed for theoretic interpretation of continuous IR absorption: 1) taking into account the interaction of proton transfers with water bond vibrations [6] and 2) the interaction of proton transfers with the phonon spectrum of the medium [7]. Both showed good agreement with experimental results.

If the $O_2H_5^+$ ion adds one more water molecule, an easily polarized structure 2b is formed. A slight distortion of its geometry leads to the localization of the proton in the right or left part of the water molecular chain.

As it follows from equation (1), further hydration of the $H_7O_3^+$ ion also results in the considerable energy gain. It agrees with the fact that the symmetrical $O_4H_9^+$ ion is most stable and, therefore, predominant in water solutions (2c). In this ion, the proton in each particular moment may also be considered as vibrating between two neighbouring oxygen atoms, i.e. as a continuously migrating elementary particle $O_2H_5^+$. This particle is easily formed and decomposed when vibrating in the water molecule hydrogen bonded net which accounts for the high proton mobility in water solutions. In other words, a well-known Grotgus mechanism of proton migration in liquid water is a result of high polarizability of $O_2H_5^+$ solvated fragments.

2.2. Proton Trasfer in Non-Aqueous Media

Some details of the proton transfer mechanism from acid molecules to molecules of a solvent or a substrate are more easy to observe in non-aqueous solutions where broad lines of hydrogen bonded hydroxyl groups do not exist in IR spectra. Therefore, it becomes possible to investigate the effect of the successive addition of substrate molecules on protonation. We conducted such measurements by means of IR spectroscopy, taking HCl and HBr interaction with water, methyl, ethyl, isopropyl alcohols as well as diethyl ether in CCl_4, CH_2Cl_2 nitromethane [8-10] as examples. Diluted solutions used in these works contained 0.02-0.2 gmol/l of substrate molecules.

Fig. 1 illustrates typical IR spectra obtained for HCl complexes with methanol, ethanol and isopropanol in CCl_4. Narrow lines assigned to stretching vibrations of H-Cl, C-H and O-H bonds, are observed on the background of the broad lines from HCl molecules perturbed by hydrogen bonding. Moreover, in the spectra, continuous absorption

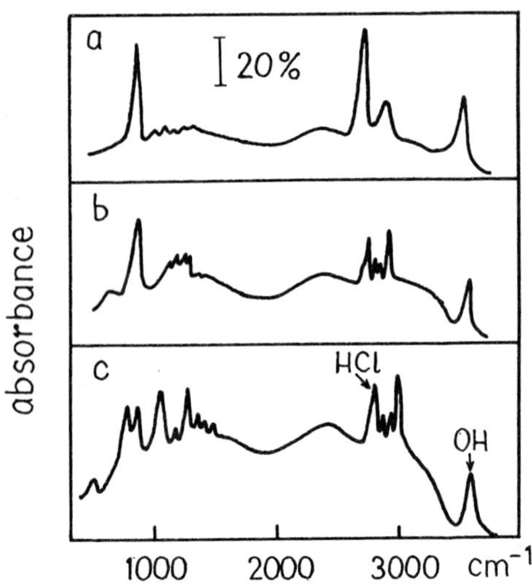

Fig. 1. Absorption spectra of the complexes of methanol (a), ethanol (b) and isopropanol (c) with hydrochloric acid in carbon tetrachloride [8]

is also present, which is similar to that of $O_2H_5^+$ ions in water solutions, i.e. is associated with protons vibrating between two oxygen atoms.

Measurements conducted for various ratios of acid and alcohol demonstrate that the intensity of the broad lines with maxima at 2400 cm^{-1} related to hydrogen bonded molecules of HCl, increase linearly with the growth of the alcohol concentration. At the same time a narrow line assigned to free HCl molecules at 2820 cm^{-1} is linearly decreasing. These effects were attributed to the formation of complexes with the stable hydrogen bond between acid and alcohol molecules of 1:1 composition:

$$\text{Cl-H} + \text{ROH} \longrightarrow \text{Cl-H-O}\begin{matrix} \diagup R \\ \diagdown H \end{matrix} \qquad (3)$$

In contrast to this, the intensity of the continuous absorption measured at 2000 cm^{-1} increased proportionally to the square of the concentration of alcohol molecules. Therefore, it was assigned to the complexes of HCl·2ROH composition. Apparently, even in the case of such strong acid as HCl, a full proton transfer takes place only to two alcohol molecules at once, resulting in formation of complexes similar to $O_2H_5^+$ ions in water solutions, in which the proton vibrates between two neighbouring atoms of oxygen. This leads to the similar continuous absorption in IR spectrum. In contrast to this, affinity of the alcohol molecule for the proton is sufficient only for the formation of hydrogen bonded complexes.

Proton transfer to two water molecules seems to result from the same mechanism, while the interaction with one H_2O molecule causes formation of a complex with a moderately stable hydrogen bond with the maximum of HCl stretching vibration frequency shifted from 2820 to 2500 cm^{-1}.

Proton transfer to two alcohol molecules in non-polar media is probably similar to it's migration in liquid water. First, the formation of one more hydrogen bond in the presence of the second polar molecule takes place:

```
δ- δ+
       R                      R
      /                      /
Cl-H-- O   + ROH ——> Cl-H-O         H
      \                      \     /
       H                      H--O
                                  \
                                   R              (4)
```

Polarization of such an associate due to the interaction with phonons of the medium results then in the full proton transfer to two alcohol or water molecules:

```
          R
         /
Cl-H---O        H              ┌R                H┐+
         \     /               │ \              / │
          H---O    ——>  Cl⁻-   │  O---H---O       │
               \               │ /              \ │
                R              └H                R┘       (5)
```

At the end of this section it should be noted that the conclusion on the proton transfer from strong acids only to two molecules of water at once was also drawn for water solutions of polystyrenesulfonic acid by Zündel [6] and for sulphuric acid by Librovich and Vinnik [11,12].

2.3. Modern Concepts of Homogeneous Acid Catalysis Mechanism

The above results emphasize a predominant role of the interaction with a number of water or substrate molecules in the proton transfer elementary step. Therefore, it is quite obvious that in the discussion of the mechanism of homogeneous acid catalysis in water solutions the consideration of the interaction only with hydroxonic ions or undissociated molecules of acid is not sufficient. Instead it is necessary to consider a possibility of the formation of still more complicated solvated complexes with several water molecules. This conclusion is in good agreement with the results obtained in the investigation of the hydrolysis kinetics of esters catalyzed by strong acids in water solutions by Librovich and Vinnik [11,12] who showed that proton is involved in the activated complex as $O_2H_5^+$ ion rather than as the hydroxonic H_3O^+ion. Let us discuss a possible mechanism of this reaction in more detail.

It seems probable that its first stage is proton transfer from the $O_2H_5^+$ ion to the molecule of the ester, similar to the proton migration along the chains of hydrogen bonded water molecules:

$$RCOOR' + O_2H_5^+ \longrightarrow \left[\begin{array}{c} R' \\ | \\ O \\ RC \quad\quad H \\ \| \quad\quad\quad \backslash \\ O \quad\quad O{-}{-}H{-}{-}O \begin{array}{l} H \\ H \end{array} \\ H \end{array} \right]^+ \longrightarrow \left[\begin{array}{c} R' \\ \backslash \\ RC{-}O{-}H{-}{-}O \begin{array}{l} H \\ H \end{array} \\ \| \\ O \end{array} \right]^+ + H_2O$$

Then the positively charged ion of the protonated ester is solvated by one more water molecule, resulting in the localization of the positive charge on fragment R':

$$H_2O + \left[\begin{matrix} R' \\ (O{=}CR){-}O \end{matrix} \cdots H \cdots OH_2 \right]^+ \rightarrow \left[H_2O \cdots R' \cdots O(H)(C(=O)R) \cdots H{-}O{-}H \right]^+ \rightarrow \left[H_2O \cdots R' \cdots O(H){-}C(=O)R \right]^+ + H_2O \qquad (7)$$

Dissociation of such a complex with the participation of the water molecule leads to the formation of acid and protonated alcohol which loses the proton in the reaction with one more water molecule:

$$H_2O + \left[H_2O \cdots R' \cdots O(H){-}C(=O)R \right]^+ \rightarrow \left[H_2O \cdots H \cdots O(H)(R') \cdots O(H){-}C(=O)R \right]^+ \rightarrow \left[H_2O \cdots H \cdots O(H)R' \right]^+ + RCOOH$$

$$H_2O + \left[H_2O \cdots H \cdots O(H)R' \right]^+ \rightarrow \left[H_2O \cdots H \cdots OH_2 \right]^+ + ROH \qquad (8)$$

Thus a kind of a switch process of initial and newly formed chemical bonds takes place, caused by the proton migration in solvated activated complexes. This mechanism differs from that investigated in the works [11,12], though we believe that it agrees better with modern data on the hydrated proton state in water solutions and its transfer mechanism in liquid water. This is also confirmed by the concepts of "chains of redistributing bonds" developed by Poltorak for enzymatic catalysis [13] and by proton transfer mechanisms in biological systems discussed in the recent works by Zündel [14].

3. HETEROGENEOUS ACID CATALYSIS MECHANISM

3.1. Nature of Protonated States on the Surface of Oxides

As we have mentioned above, electrolytic dissociation in solutions is possible exclusively in the presence of solvation effects, and it is the solvation that leads to the formation of substrate protonated forms in homogeneous catalysis. On the contrary, the tendency to proton transfer in high temperature gas-phase heterogeneous catalytic reactions with the participation of non-polar hydrocarbon molecules which are most interesting in connection with oil-refining processes, is much more vague.

Here the situation is to some extent similar to the described above proton transfer in non-polar solvents where solvation effects are minimal and the substrate protonation requires considerable energy consumption. For most adsorbed molecules proton transfer is essentially endothermic and becomes advantageous only when highly acidic hydroxyl groups interact with strong bases such as ammonia and pyridine. These molecules are usually used as tests on strongly acidic surface OH-groups. However, it should be born in mind that in the case of adsorbed ammonia or pyridine, formation of adsorbed ions from these groups on the surface of acid-type catalysts does not mean that similar ionic states occur at adsorption of olefins or even paraffins whose affinity for the proton is 2-3 eV less than that of ammonia and pyridine. In fact, as a result of hydrocarbon interaction even with highly acidic Brönsted centers either complexes with short hydrogen bonds or polarized covalent structures are formed. Thus, intermediate active forms of heterogeneous acid catalysis differs considerably from protonated ionic states in water solutions and other polar solvents. They are rather similar to complexes formed as a result of strong acid interaction with weak bases or weak acids with strong bases in non-polar solvents.

This conclusion is confirmed by the quantum-chemical calculations. Recently in the work [15] in STO-3G basis we conducted non-empirical calculations of ammonia interaction with the bridging structural hydroxyl groups of high-silica zeolite. Thus a reasonable ammonia adsorption heat of 159 kJ/mol and a sufficiently high positive charge on the ammonia ion equal to +0.711 e were obtained.

On the other hand, according to similar calculations, proton transfer from the same acid center to the adsorbed ethylene molecule leads to the formation of a surface ethoxy fragment, having a comparatively low positive charge of only +0.26 e, instead of the carbon ion formation.

In the absence of solvation effects, ethylene interaction with sulphuric acid also results in the ester formation, in which the ethyl fragment positive charge is even somewhat lower than in the case of the zeolite (+0.21 e). This fact agrees well with chemical data, since sulphuric acid ester is surely a covalent compound. Therefore, a customary relation of the carbon-ionic mechanism to homogeneous catalytic ethylene hydration reaction is rather due to heterolytic dissociation of the ester at the second stage of the reaction (as a result of hydrolysis by water molecules) than to a high polarity of its carbon-oxygen bond.

3.2. Spectroscopic Investigation of the Potential Energy Profile of the Proton Transfer Elementary Step

The values of low-frequency shifts of fundamental stretching vibrations of OH groups, resulting from the formation of a hydrogen bond with adsorbed molecules, are often used to measure the strength of Brönsted acid centers on surfaces of oxides. However, this is not the only possibility for IR spectroscopy in the investigation of the proton transfer mechanism. In the publications [1-4] we proposed to measure shifts of overtones in addition to fundamental shifts of stretching vibrations. From these values it is possible to reconstruct the lower part of the OH bond potential curve in surface complexes with a hydrogen bond and to estimate the energy required for the formation of protonated intermediates as well as the height of the potential barrier of proton transfer to adsorbed molecules.

Indeed, for the reconstruction of the potential curve for diatomic molecules in a Morse function approximation it is sufficient to know the position of any two vibrational levels. As was demonstrated in (2), surface OH groups due to high characteristics of their stretching vibrations can also be considered as diatomic fragments which do not interact with oxide lattice vibrations. Thus, the lower part of OH bond potential curves in Brönsted acid centers on the surface of oxides can be drawn on the basis of fundamental frequencies and overtones of their stretching vibrations. A full profile of the proton transfer potential energy and the potential barrier height of this elementary step can be obtained as follows:

It is well known that the lengths of stable hydrogen bonds in various complexes fall in the 0.24 - 0.26 nm range. On the other hand, the equilibrium O-H distance in the initial hydroxyl groups is about 0.1 nm. The length of C-H (0.109 nm), N-H (0.101 nm) and of other bonds, formed in protonated intermediates are close to these values. Therefore proton transfer to adsorbed molecules corresponds to the elongation of OH bond by about 0.05 nm and the position of the maximum of the proton transfer potential barrier corresponds to a somewhat less elongation. It is also evident, that the full profile of the potential energy for the proton transfer elementary step can be obtained by placing the OH bond potential curve, calculated from spectroscopic data, and the potential curve of a newly formed bond in such a way that the distance between their minima would be about 0.05 nm.

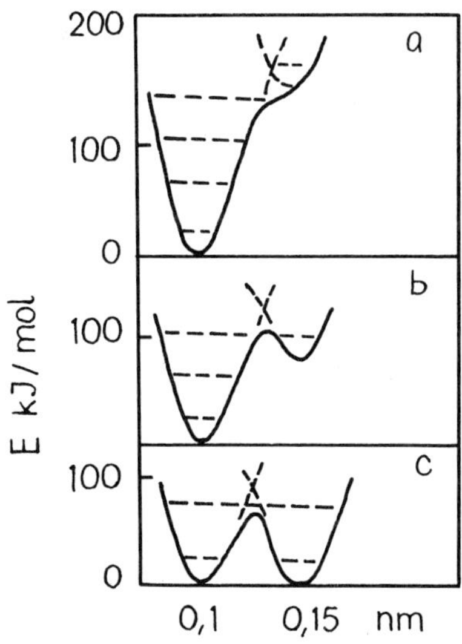

Fig. 2. Potential energy profiles for endothermic elementary steps of proton transfer on the surfaces of oxide catalysts

As it is seen from the potential energy profiles for endothermic proton transfer stages, obtained by this way and presented in Fig. 2, the formation of stable protonated intermediates is possible only in the case of thermoneutral or weakly endothermic reactions. Even at the endothermicity of 100-150 кJ/mol of this elementary step the local minimum on the potential curve which corresponds to the protonation of the substrate is either absent, or its depth is so small, that the protonated state becomes metastable and decomposes after a few vibrations. This is caused by a relatively small elongation of the OH bond corresponding to a full proton transfer and to a comparatively large width of the lower part of its potential curve in surface complexes. In this respect heterogeneous acid catalysis differs greatly from homogeneous acid catalysis in water solutions,where the reaction coordinate does not mean a simple OH bond stretching, but includes rearrangement of many molecules of the solvent that leads to the space separation of cations and anions. Therefore in water solutions, protonated states can exist even in the case when proton transfer is essentially endothermic.

It is possible to formulate several spectral criteria of a real tendency to proton transfer, from Brönsted centers to adsorbed molecules. The most evident one is an increase of anharmonicity of the potential curve for the OH bond even when surface complexes are formed. Figuratively, the anharmonicity value allows one to analyze the response of acid hydroxyl groups to OH bond excitation at higher and still higher vibrational levels and, consequently, to the proton approach to the adsorbed molecule. The anharmonicity increase is an indication of proton attraction to the substrate molecule and its decrease indicates the repulsion.

The isotopic effects, that is, ratios of vibrational frequencies of light and heavy hydroxyl groups excited by hydrogen bonding can also be used as another criterion of the tendency to the formation of protonated states. When anharmonicity increases and the tendency to proton transfer is positive, this ratio becomes lower. On the contrary, it increases if the proton is repulsed by the substrate. Besides, the tendency to proton transfer manifests itself either in the absence or in a great perturbation of the first overtone and in great low-frequency shifts of fundamental stretching vibrations in hydrogen bonded complexes (up to 400-600 cm^{-1} or higher).

In the publications [2-4] the above stated criteria were applied to a model system to study the interaction of silanol groups with a wide spectrum of adsorbed molecules on silica. A good agreement of the above-discussed simple theory with the experimental data was found.

For weak bases with the proton affinity in the 6-8 eV range (cyclohexane, carbon tetrachloride, fluorobenzene, methyl chloride, chlorobenzene, benzene and nitromethane) no appreciable tendency to the formation of protonated intermediate states when interacting with weakly acidic sylanol groups was observed. The anharmonicity values in surface complexes are lower than in free hydroxyl groups and the values of low frequency shifts of their stretching vibrations are rather small (100-200 cm^{-1}). The bands of the first overtone in IR spectra are only slightly perturbed. Apparently, the potential energy profile of proton transfer corresponds to the (a) case in Fig. 2.

A real tendency to proton transfer appears only for molecules with the proton affinity higher than 8 eV (methyl cyanide, acetone), where values of low-frequency shifts of fundamental OH bond stretching vibrations increase up to 400 cm^{-1} and their anharmonicity increases for acetone up to $6 \cdot 10^{-2}$ instead of $2.3 \cdot 10^{-2}$ for a free silanol group.

According to the above concept the band of the first overtone is considerably broadened. Most likely this accounts for the fact that the second vibrational level of OH bond is close to the maximum of the proton transfer potential barrier, its height being, therefore, equal to 80-100 kJ/mol (case (b) in Fig. 2).

Finally, for the adsorption of ammonia with a proton affinity of about 9 eV, the value of a low-frequency shift of a silanol group increases up to 800 cm^{-1}. In this case even the first overtone is not observed, from which it follows that the top of the proton transfer potential barrier is located between the first and the second vibrational levels and the value of its height falls in the 40-80 kJ/mol range (case (c) in Fig. 2).

In spite of the fact that the tendency to proton transfer naturally increases in the case of such strong solid acids as zeolites or amorphous silica alumina catalysts, an application of the above spectral criteria shows that of course not all intermediates formed due to proton transfer at their surfaces are of ionic character, in other words, are contact ionic pairs, as is commonly suggested. For instance, though in benzene complexes with acidic hydroxyl groups of Y zeolite the fundamental stretching frequency decrease is by far higher than that characteristic of silica and is equal to 350 cm^{-1}, it is still not enough for a full proton transfer. The anharmonicity coefficient increases only to about 4, the potential energy profile of proton transfer being close to that in the case of the interaction of silanol groups with methyl cyanide or acetone (Fig. 2b). This is an indication of only a moderate tendency to protonation. When HY zeolite acidic hydroxyl groups interact with ethylene, their perturbation and, consequently, the tendency to proton transfer is even less (the value of the low-frequency shift amounts only to about 300 cm^{-1})

The formation of carbonium ions is likely even when ethylene interacts with highly acidic hydroxyl groups of high-silica ZSM zeolites. This is confirmed by moderate shifts of OH-vibrational frequencies in hydrogen bonded complexes as well as by a linear structure of ethylene oligomers formed at room temperature. This structure cannot be related to the shape selectivity since propylene polimerization on the same contacts results in the formation of branched oligomers in their channels.

Apparently, according to the above-discussed quantum-chemical calculations highly polarized covalent structures rather than ionic ones are formed as a result of ethylene interaction with such highly acidic centers. The problem does not seem so clear for propylene and especially for branched heavier olefins with their higher proton affinity. Thus, the concept of carbonium ions as applied to heterogeneous systems seems too formal and too general. Therefore, the very conception of "adsorbed carbonium ion" may have quite different meaning in each particular case. From our standpoint it should rather imply the presence of active intermediates on the surface which are similar in their reactivity to carbonium ions in the liquid phase than a real existence of contact ionic pairs.

Finally, it should be emphasized that the transition from hydrogen bonded complexes to polarized covalent structures and further to adsorbed contact ionic pairs sharply depends on the energetics of the proton transfer elementary step. Indeed, as it was mentioned above, for silica the range of proton affinity, where the tendency to proton transfer becomes positive, lies within 1.3-1.5 eV. This is also true for highly acidic OH-groups of Y zeolite where protonation does not occur for benzene (the proton affinity 7.5 eV), though contact ionic pairs are undoubtedly formed at the interaction with ammonia (the proton affinity 9 eV). Thus, for the majority of weak bases additional energy required for the production of adsorbed ionic structures is as low as 1-1.5 eV. In water and other polar solvents this lack of energy is compensated by solvation. In high temperature catalytic conversions of hydrocarbons this source of additional energy is not available. It also cannot be substituted for the activation energy, since for endothermic stages of proton transfer there are no minima characteristic of protonated ionic structures on the potential energy surface. From the energy standpoint this is the reason for the difference between mechanisms of homogeneous and heterogeneous acid catalytic processes.

3.3. Heterogeneous Acid Catalysis Mechanism

3.3.1. Low Temperature Catalytic Reactions With the Participation of Polar Molecules. The mechanism of low temperature heterogeneous-catalytic conversions with the participation of polar molecules, for example, hydration of olefins into alcohols, alcohol esterification, hydrolysis of esters, etc., is most similar to homogeneous acid catalysis. Methanol catalytic conversions of high-silica zeolites of ZSM type may also be related to this type of mechanisms.

If the reaction temperature is not sufficiently high and the formation of associates comprising a number of molecules is possible, effects similar to solvation with the participation of polar molecules in liquids come into force, and similar protonated forms of the substrate act as active intermediates. This situation to a great extent resembles homogeneous catalysis in water solutions where solvation effects are a dominating contribution to the production of protonated states and to the proton transfer elementary step. It can be more precisely identified as two-dimensional homogeneous acid catalysis in thin liquid films condensed on the heterogeneous catalyst surface. In particular, a possibility of this mechanism for olefin hydration was demonstrated by Chirkov, Vinnik et al in [16].

Apparently, a similar mechanism is realized in enzymatic catalysis as well. However, here the interaction of water molecules or substrate with polar groups on the biopolymer surface has its specific nature. It results in structurization of the adsorbed film, which affects the solvation energetics as well as the proton transfer mechanism. The main idea of a real existence of protonated intermediates in these systems remains, however, unchanged.

In this connection it appears interesting to draw an analogy between enzymatic catalysis with the participation of polar molecules and their catalytic conversions in zeolite narrow pores, where effects of mutual orientation of adsorbed molecules and structurization of condensed liquid may also be rather important.

4. SYNCHRONOUS OR CONCERTED MECHANISM OF HETEROGENEOUS ACID CATALYSIS

Another limiting case of heterogeneous acid catalysis is represented by conversions of non-polar molecules with the minimal proton affinity, since in this case solvation effects are practically absent. This is best illustrated by isotopic and heteroisotopic exchanges of hydrogen with deuterium (these molecules have the proton affinity as low as 4.5 eV) as well as acid catalysis with the participation of paraffins (their affinity for the proton varies from 5.3 eV for methane to 6.5 eV for higher paraffins). These molecules also only slightly perturb highly acidic hydroxyl groups on the surface of oxide catalysts. For example, frequency of OH group stretching vibrations in Y zeolite decreases as a result of the interaction with molecular hydrogen only by 20-30 cm^{-1} and by 100-150 cm^{-1} for paraffins. Thus, the absence of the tendency to proton transfer to these molecules is also confirmed by spectral data.

Let us consider the mechanism of heterogeneous acid catalytic reactions in these systems. One of the possibilities to avoid the formation of energy-disadvantageous and unstable protonated intermediates is connected with synchronous mechanisms, where a new bond of the molecule or its fragment with the catalyst surface is formed simultaneously with proton transfer. Here the active center acts as an acid-base pair, and the energy released for the production of a new bond partially compensates the disadvantage of a full proton transfer similar to solvation effects in polar liquids.

For a series of heterogeneous-catalytic reactions with the participation of Brönsted acid centers a possibility of such synchronous mechanisms was confirmed by quantum-chemical calculations [17]. Particularly, the mechanism of a heteroisotopic exchange between terminal hydroxyl groups on the silica surface amd molecular deuterium is as follows:

$$\mathrm{H{-}O} + D_2 \rightarrow \mathrm{H \cdots D \cdots D \cdots O} \rightarrow \mathrm{H \cdots D \cdots D \cdots O} \rightarrow \mathrm{O{-}D} + HD. \qquad (9)$$

A symmetrical transition state is formed here as a result of polarization of the deuterium molecule and the increase of its interatomic distance up to ~ 1Å. After positive charges of the proton of the hydroxyl group and of a deuterium atom become equal, this transition state decomposes without any additional energy consumption. If the temperature is high enough to avoid the condensation of adsorbed vapor in pores of a catalyst, similar synchronous mechanisms appear also most probable for the heteroisotopic exchange of surface hydroxyl groups with polar water molecules and alcohols.

The proton-catalyzed cracking of paraffins on zeolites and amorphous silica aluminas also is most likely to proceed according to a similar synchronous mechanism. We believe formation of intermediate non-classic carbonium ions with penta-coordinated carbon which is postulated sometimes, is hardly probable because of weak proton affinity of paraffins. An alternative synchronous mechanism with a simultaneous proton transfer to the CH_3 group, C-disruption in the paraffin molecule and formation of a new carbon-oxygen bond with the surface is more probable:

$$\mathrm{(O)_2Al(O)(OH)} + C_3H_8 \rightarrow \mathrm{(O)_2Al(O)(O\cdots CH_2(CH_3)-CH_3\cdots H)} \rightarrow \mathrm{(O)_2Al(O)(O-CH_2-CH_3)} + CH_4 \quad (10)$$

In this case the transition state resembles a non-classic carbonium ion, however, it is not a really existing intermediate structure.

Synchronous mechanisms are also possible in catalytic conversions of olefins, the proton affinity of which is higher than that of paraffins. As a matter of fact, even the formation of polarized alcoxyl structures when olefins interact with acidic hydroxyl groups in zeolites and silica aluminas is a synchronous reaction, since a new carbon-oxygen bond is formed simultaneously with proton transfer:

$$\mathrm{(O)_2Al(O)(OH)} + C_2H_4 \rightarrow \mathrm{(O)_2Al(O)(OH\cdots CH_2{=}CH_2)} \rightarrow \mathrm{(O)_2Al(O)(O\cdots CH_2{=}CH_2\cdots H)} \rightarrow \mathrm{(O)_2Al(O)(O-CH_2-CH_3)} \quad (11)$$

Synchronous mechanisms for the double bond isomerization in olefins are also often considered. In this case the proton of the acid center displaces another proton from the adsorbed molecule, thus reducing the energy of the formation of a corresponding transition state:

$$1-C_4H_8 + \underset{|}{\overset{H}{O}} - CH_2 \cdots \overset{CH}{} \cdots CH - CH_3 \longrightarrow 2-C_4H_8 + \underset{|}{\overset{H}{O}} \qquad (12)$$

Synchronous or concerted mechanisms can be of great importance for some other acid catalytic conversions of olefins, for example, in their skeletal isomerization, cracking, alkylation by isoparaffins, etc. It is not our purpose, however, to discuss these reactions in detail in the present paper.

5. CONCLUSION

The above results indicate that apart from a number of similarities between homogeneous and heterogeneous catalytic processes, there are great distinctions caused by the absence of solvated effects in these latter. Instead of classic stepwise schemes, various versions of synchronous mechanisms are widely spread in heterogeneous acid catalysis. They permit one to avoid the formation of protonated intermediates which from the energy point of view are disadvantageous in the absence of solvation. For this reason the intermediates and mechanisms characteristic of homogeneous catalysis cannot be transferred in one-to-one manner to heterogeneous systems.

REFERENCES

1. V.B. Kazansky, A.M. Gritscov, V.M. Andreev, G.M. Zhidomirov, J.Molec.Catalysis, 4, 135 (1978).
2. V.B. Kazansky, Kinetics and Catalysis, 21, 1, 159 (1980).
3. V.B. Kazansky, Proc.Ac.Sc.Bulgaria, 13, 1, 19 (1980).
4. V.B. Kazansky, Zh.Fiz.Khim.(USSR), 3, 318 (1982).
5. R.P. Bell, Proton in Chemistry, Chapman and Hall, London 1973.
6. G. Zündel, Hydration and Intermolecular Interaction, Academic Press, New York and London 1969.
7. N.B. Librovich, V.P. Sakun and N.D. Sokolov, Teor. Eksp.Khim.(USSR), 14, 4, 1358 (1978).
8. A.A. Pankov, V.Yu. Borovkov and V.B. Kazansky, Dokl. Akad. Nauk SSSR, 258, 902 (1981).
9. A.A. Pankov, V.Yu. Borovkov and V.B. Kazansky, Zh. Prikl.Spektr.(USSR), 37, 2, 291 (1982).
10. A.A. Pankov, V.Yu. Borovkov and V.B. Kazansky, Zh. Prikl.Spektr.(USSR), 37, 5, 824 (1982).
11. M.I. Vinnik, N.B. Librovich, Izv.Akad.Nauk SSSR, 10, 2211 (1975).
12. N.B. Librovich, and M.I. Vinnik, Zh.Fiz.Khim.(USSR), 12, 2993 (1979).
13. O.M. Poltorak, Zh.Fiz.Khim.(USSR), 46, 1361 (1972).
14. V. Kristof, and G. Zündel, Biopolymers, 21, 25 (1982).
15. I.N. Senchenya, I.D. Mikheikin, G.M. Zhidomirov and V.B. Kazansky, Kinetics and Catalysis, 22, 5, 1174 (1981).
16. N.M. Chirkov, M.I. Vinnik, S.G. Entelis and V.I. Tsvetkova, Kinetics and Catalysis Problems . (Publ. AN SSSR, M., 496 (1955)).
17. I.D. Mikheikin, I.A. Abronin, G.M. Zhidomirov and V.B. Kazansky, J.Molec.Catal., 3, 435 (1977/78).

MULTIPLE ROLES OF PALLADIUM IN LIQUID PHASE OXIDATION

James E. Lyons, George Suld, and Chao-Yang Hsu
Applied Research and Development Department
Sun Refining and Marketing Company, P O. Box 1135,
Marcus Hook, PA 19061 USA

ABSTRACT--Vinylic oxidation products or their stable isomers, including acetaldehyde, acetone, and vinyl acetate, are manufactured by related reactions involving nucleophilic attack on a Pd(II) olefin π-complex. Other productive pathways are available to coordinated olefins, however. Olefins coordinated to an electrophilic Pd(II) center form π-allyl complexes which may undergo nucleophilic attack to give allylic rather than vinylic products. In this way, a family of industrially important α,β-unsaturated alcohols, esters and acids can be produced in catalytic liquid phase oxidations. Oxidative addition of olefinic substrates to supported palladium catalysts may give surface π-allylic species which undergo nucleophilic attack. A catalytic application of this type is the mild palladium catalyzed oxidation of propylene to acrylic acid in water. Oxidation of aromatic substrates by both soluble and supported palladium catalysts gives phenols, aryl acetates or products of oxidative dimerization. Electrophilic attack on the ring by Pd(II) is the likely first step in these reactions.

1. INTRODUCTION

One of the most synthetically useful and industrially important classes of catalytic oxidation reactions involves nucleophilic attack on palladium-olefin π-complexes to give vinylic oxidation products and reduced palladium species which can be re-oxidized *in situ* by a number of oxidants including Cu(II) (1,2). In this manner, vinylic oxidation products or their stable isomers including acetaldehyde, acetone and vinyl acetate are manufactured by a series of related catalytic reactions. These reactions occur either in solutions of palladium complexes or over heterogeneous supported palladium catalysts. It is interesting to note that the palladium catalyzed oxidation of ethylene to acetaldehyde was initially carried out over a supported palladium catalyst, but is now practiced in solution whereas the oxidation of ethylene to vinyl acetate was first practiced in the liquid phase and now industrial production of vinyl acetate employs supported palladium catalysts (1).

Other synthetically important pathways are available to palladium-olefin complexes which produce not vinylic, but allylic, oxidation products. Palladium π-allyl intermediates can form by oxidative addition to the metal center. Nucleophilic attack at the termini of the π-allyl ligand gives rise to allylic oxidation products and reduced palladium species. As with Wacker chemistry, these reactions can be made to occur catalytically in the presence of palladium either in homogeneous solution or over heterogeneous supported catalysts. The development of this chemistry for the production of a large and important family of allylic oxidation products, including α,β-unsaturated alcohols, esters, ketones, and carboxylic acids, is possible; yet until recently, only limited application of these reactions to catalytic syntheses has been made.

Another instance in which palladium catalysts, either supported or in solution, can be used to advantage is in the selective oxidation of the aromatic ring. To date, most of these reactions have required strong oxidants and high oxidation state palladium aryl intermediates have been implicated. The selective air-oxidation of aromatics to phenolic derivatives is possible, however, and could be of considerable utility.

In this paper, the role of supported palladium catalysts in the allylic oxidation of olefins will be discussed and those factors which favor allylic oxidation over vinylic oxidation will be considered. Relationships between homogeneous and heterogeneous catalysts will be developed and analogies between solution and surface species will be drawn. Evidence will be presented which is consistent with the formation of π-allyl intermediates in both homogeneous and heterogeneous catalytic reactions. In addition, the results of oxidations of aromatic rings using molecular oxygen in the presence of palladium catalysts will be presented. The data suggests the involvement of high oxidation state palladium aryl intermediates. Methods for the control of reaction selectivity will be given.

2. RESULTS AND DISCUSSION

Oxidation of Olefins. There are several pathways available for the oxidation of α-olefins catalyzed by Pd(II) in solution, eq. 1.

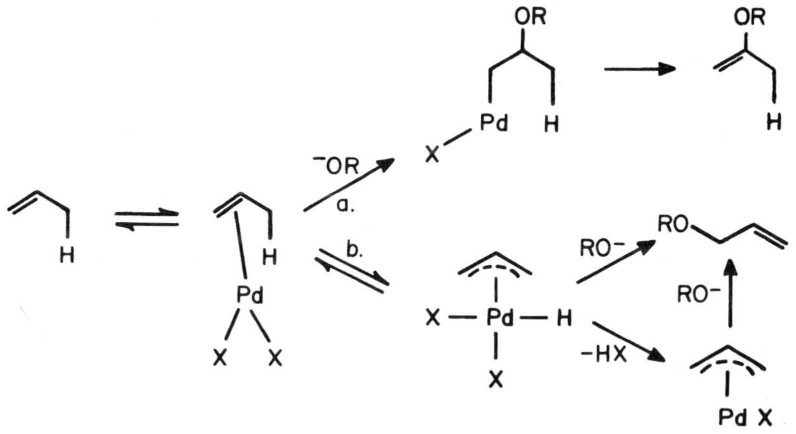

(1)

The pathway which predominates can be controlled by the nature of the Pd(II) complex which is used as the catalyst. For example, palladium acetate or palladium chloride can catalyze predominant vinylic oxidation, eq. 1a, whereas palladium trifluoroacetate or π-allyl palladium trifluoroacetate gives predominant allylic oxidation, eq. 1b. It has been suggested that Pd(II) complexes with strongly electron withdrawing ligands are highly electrophilic and oxidatively add an allylic C-H bond, resulting in Pd(IV) π-allyl intermediates (3).

A catalytic application of the use of palladium trifluoroacetate to the formation of an allylic oxidation product is the production of allyl acetate in high yield from propylene, eq. 2 (4).

$$CH_2{=}CH{-}CH_3 + \tfrac{1}{2}O_2 + HOAc \xrightarrow{CAT.} CH_2{=}CH{-}CH_2{-}OAc + H_2O \quad (2)$$

$$CAT. = (\pi\text{-allyl}){-}PdO_2CCF_3 \text{ or } Pd(O_2CCF_3)_2$$

There appears to be another means of encouraging allylic over vinylic oxidation of α-olefins. This involves the use of relatively weak inorganic bases to assist proton removal. Thus the addition of sodium acetate, carbonate or bicarbonate promoted formation of palladium(II)π-allyl complexes from either $PdCl_2$ and propylene or di-μ-chloro-dichloro<u>bis</u>propenedipalladium (5). It is interesting that the products of the stoichiometric oxidation of propylene by palladium acetate shift from predominantly vinylic to predominantly allylic as sodium acetate is progressively added to the solution, Table 1 (6). Furthermore, it has been found that addition of sodium acetate to solutions of palladium(II) gives fair yields of allyl acetate during the catalytic oxidation of propylene in acetic acid (7,8). Low oxidation state palladium has been implicated in these reactions (8) and heterogeneous catalysts such as Pd/SiO_2 treated with sodium or potassium acetate are very effective for catalyzing the oxidation of propylene to allyl acetate at elevated temperature in the vapor phase (9).

TABLE 1. THE EFFECT OF [NaOAc] ON PROPYLENE OXIDATION PRODUCTS AT 25°C[a,b]

NaOAc M	3-Acetoxy-Propene, %	2-Acetoxy-Propene, %	Cis- + Trans-1-Acetoxypropene, %
Nil	2	96	2
0.098	13	86.5	0.5
0.49	71	26	3
0.90	94	6	<0.5

a) [Pd(II)] = 0.0033M, reaction time was 45-65 min.
b) Winstein, McCaskie, Lee and Henry, 1976

Finally, another route to catalytically active π-allyl intermediates proceeds *via* oxidative addition to a coordinatively unsaturated palladium(0) center. This usually requires a good leaving group in the allylic position. A cationic π-allyl complex can be isolated and shown to be a reaction intermediate, Scheme 1, eq. 3 (10). When the leaving group is H, as it is in an α-olefin, the reaction is thermodynamically disfavored unless the hydrogen which would be produced can be oxidatively converted to water.

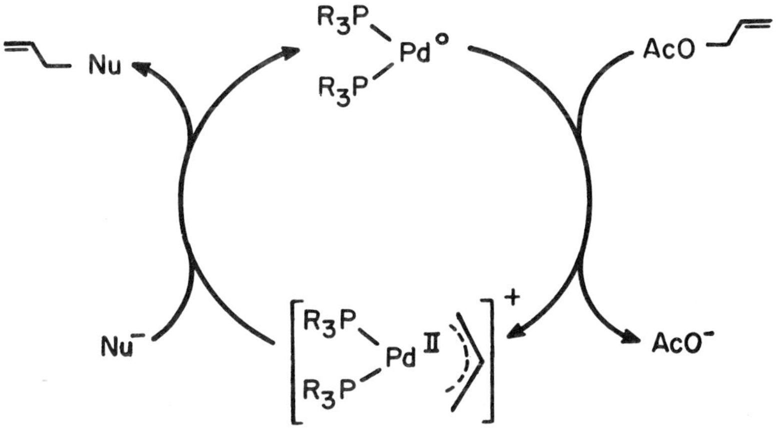

Nu
R_3P Pd^o R_3P
AcO
Nu^-
$[(R_3P)_2Pd^{II}(\pi\text{-allyl})]^+$
AcO^-

Scheme 1. Catalytic Cycle for Palladium-Catalyzed Allylation

$$CH_2{=}CHCH_2OAc + HNu \longrightarrow CH_2{=}CHCH_2Nu + HOAc \qquad (3)$$

Heterogeneous catalysts are known to activate the C-H bond of α-olefins and, hence, it became of interest to determine whether an efficient catalytic olefin oxidation could be accomplished in a manner analogous to that shown in Scheme 1 over a supported palladium(0) center, Scheme 2.

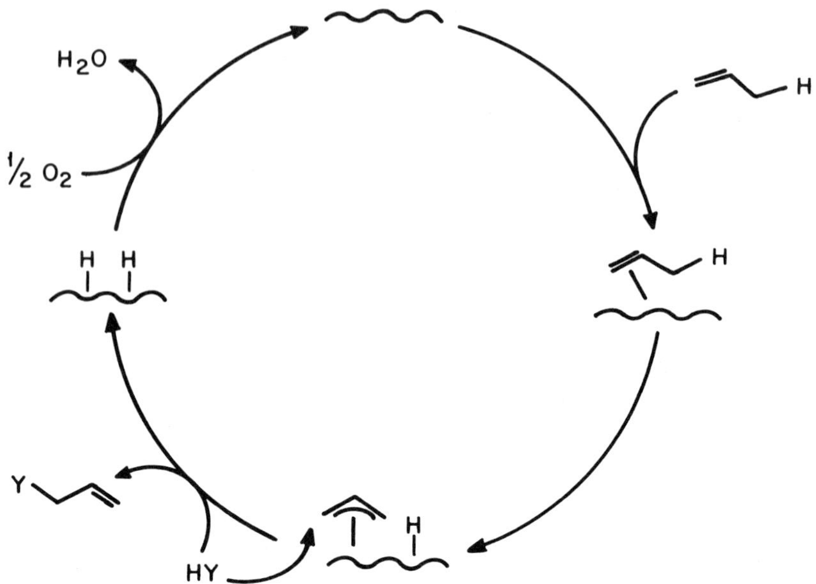

Scheme 2. Oxidative Addition of Propylene to a Pd Surface

$$\frac{1}{2}O_2 + \text{propylene} + HY \xrightarrow{Pd/S} \text{CH}_2\text{=CHCH}_2Y + H_2O \qquad (4)$$

A route of this kind was proposed (HY=HOH) for the conversion of propylene to acrylic acid _via_ allyl alcohol (11-14). Low conversions (turnover numbers <1) were observed under mild conditions (11) and poor selectivities (<68%, T.O.N. ~10) were observed under forcing conditions (13).

Propylene Oxidations over Supported Palladium Catalysts. We have examined Reaction 4 (HY=HOAc, HOH) and find that it occurs over heterogeneous palladium catalysts under mild conditions in the liquid phase to give high yields of allylic oxidation products. An efficient catalyst for these reactions is 10% palladium on carbon, Table 2. When HY is acetic acid, allyl acetate is the major product, Table 3, and when HY is water, acrylic acid is produced in high yields, Table 4. Presumably, in the latter case, allyl

TABLE 3. PALLADIUM CATALYZED OXIDATION OF PROPYLENE TO ALLYL ACETATE[a)]

Catalyst[b)]	NaOAc moles/l.	Activation Time, Min.	Gas Uptake Mmoles	Wt. % in Product Allyl Acetate	Wt. % in Product i-Propenyl Acetate	Selectivity to Allyl Acetate, %
$Pd(OAc)_2/CuCl_2$	0.167	0	47	>0.4	0.75	<8.0
$Pd(OAc)_2/CuCl_2$	0.667	0	65	2.33	3.36	9.1
10% Pd-C	0.0	0	15	1.69	1.70	29.9
10% Pd-C	0.167	0	20	1.89	1.75	39.1
10% Pd-C	0.333	0	38	5.79	1.65	69.7
10% Pd-C	0.0	15	32	4.94	0.83	85.6
10% Pd-C	0.167	30	41	16.04	1.71	90.6
10% Pd-C	0.333	30	50	17.57	0.98	94.7
10% Pd-C	0.500	30	49	17.40	0.38	97.9
10% Pd-C	0.667	30	53	18.57	TR	>99.0

a) Glacial acetic acid, 30 ml, and the catalyst were charged to a 100 ml Fisher-Porter aerosol tube, flushed 3 times with propylene and activated under propylene for the designated time. A 65/35 $C_3^=/O_2$ mixture was then added and the reaction allowed to proceed for 5 hours at 65°C.

b) 10% Pd-C = 1.0 gram; $Pd(OAc)_2$ = 0.224 gram; $CuCl_2$ = 0.072 gram

TABLE 4. EFFECTS OF ADDITIVES AND ACTIVATION TEMPERATURE ON RATE AND SELECTIVITY OF PALLADIUM-CATALYZED OXIDATIONS OF PROPYLENE IN WATER[a)]

Activation T, °C	Reaction T, °C	Additive mmoles	Reaction Products[b)], mmol. CO_2	Acetaldehyde	Acetic Acid	Acetone	Acrolein	Acrylic Acid	Turnover No. mmoles Acrylic/ g.atom Pd/hr
80	65	--	2.3	0.5	0.8	0.8	0.2	12.9	3.2
80	50	--	0.7	0.2	0.5	0.8	0.3	7.7	1.9
80	40	--	0.4	0.1	0.4	0.4	0.1	8.3	1.2
40	40	--	--	--	--	--	--	--	--
50	50	--	--	--	0.3	0.3	--	--	--
65	65	--	2.3	0.4	0.7	1.0	0.3	14.4	3.5
65	65	BHT[c)]	2.4	0.4	1.0	0.6	0.2	17.3	4.3
--[d)]	65	--	2.5	0.4	1.1	1.1	0.3	9.7	2.4
65	65	LiCl[e)]	na	--	--	3.4	--	0.1	--
65	--[f)]	--	0.25	--	0.3	0.7	0.1	0.9	--

a) One gram 10% Pd-C and 30 ml H_2O were added to an aerosol tube, flushed 3 times with propylene, heated under 50 psi of propylene at "activiation T" for 30 minutes, then brought to reaction T. The propylene was replaced with 100 psi of a 60/40 $O_2/C_3^=$ gas mixture and stirred for 4 hours. As gas was taken up, the 60/40 mix was periodically added. b) After cooling the reaction mixture, the gas was captured and analyzed by GC. The liquid was also analyzed by GC using propionic acid as an internal standard. All volatile reaction products were analyzed and the amounts (mmoles) reported in the table above. c) The radical inhibitor, 2,6-di-*tert*-butyl-*p*-cresol, 0.1 g, was added to the reaction mixture. d) No activation period--60/40:$O_2/C_3^=$ was admitted initially. e) LiCl, 1M, was present in aqueous solution. f) Product analysis directly after activation period prior to adding oxygen.

alcohol is the initial product which is rapidly converted to acrylic acid on the surface of the catalyst, eq. 5.

$$CH_2=CH-CH_3 \longrightarrow CH_2=CH-CH_2OH \longrightarrow CH_2=CH-CHO \longrightarrow CH_2=CH-COOH \qquad (5)$$

TABLE 2. PROPYLENE OXIDATIONS OVER SUPPORTED PALLADIUM CATALYSTS IN WATER [a)]

Catalyst	Gas Uptake Mol/g. atom Pd/hr	Sel to Acrylic Acid Wt. % in Liquid
1% Pd-C	12.4	31
5% Pd-C	10.8	72
10% Pd-C	10.1	88
20% Pd-C	7.5	91
5% Pd-Al_2O_3	4.2	78
5% Pd-SiO_2	0.8	--

a) One gram of the catalyst and 30 ml of water were added to a Fisher-Porter aerosol tube, flushed 3 times with propylene and heated for 30 minutes at 65°C under 50 psig of propylene. The propylene was then replaced with 100 psi of a 60/40 $O_2/C_3^=$ gas mixture and stirred for 5 hours. As gas was consumed, it was continually replaced with the 60/40 mix. Liquid products were analyzed by standardized GLPC.

Activation of the Catalyst. These reactions have several interesting characteristics which must be taken into account if high rates and selectivities are to be achieved. First, it is very beneficial to pre-treat the catalyst with propylene in the absence of oxygen at a temperature of at least about 65°C prior to conducting the oxidation. If this is done, higher rates and selectivities are observed and the catalyst is made active for low temperature oxidations. Catalysts which were inactive at 40°C or 50°C become active for allylic oxidations at these temperatures, Table 4.

During the period when the catalyst is treated with propylene, some allylic oxidation product is formed but a considerable amount of vinylic oxidation product is also formed. For example, over 70%

of the by-product acetone (0.67 moles per g. atom Pd present) produced in the oxidation of propylene in water over 10% Pd-C arose during the propylene pre-treatment period, Table 4. Very little acetone was formed during the rest of the run.

A possible explanation for these observations is that during the propylene pre-treatment period, surface oxygen (peroxide or oxide) is removed and oxidized palladium sites are reduced to Pd(0), eq. 6. The low oxidation state metal centers may then oxidatively add the olefin to form surface π-allyl intermediates, eq. 7, which go on to produce allylic oxidation products, Scheme 2. Whether or not this is actually occurring, the propylene pre-treatment serves to form a more active catalyst for selectively producing allylic oxidation products.

$$\text{"PdO"} + CH_2{=}CHCH_3 \longrightarrow Pd + CH_3COCH_3 \quad (6)$$

$$Pd + CH_2{=}CHCH_3 \longrightarrow (\pi\text{-allyl})Pd \quad H{-}Pd \quad (7)$$

Several things are particularly detrimental to both the activity and selectivity in allylic oxidations over supported palladium catalysts. The presence of a strong oxidant such as MnO_4^-, $S_2O_8^=$, AgO, Ag + O_2, NO_3^-, etc. slows the rate of allylic oxidation and causes vinylic oxidation to either compete or predominate. Pre-treatment of the palladium catalyst with oxygen in the absence of propylene at elevated temperatures prior to reaction also has a deleterious effect on allylic oxidation. The presence of halide ion is extremely detrimental to the reaction, promoting instead a Wacker reaction which forms acetone to the virtual exclusion of allylic oxidation, Table 4.

These observations suggest that the active catalyst for these allylic oxidations could be low oxidation state palladium, and that conditions or reagents that cause higher oxidation state palladium to be formed on the catalyst surface, lead to vinylic oxidations.

Radical Inhibitors. Addition of a free radical inhibitor such as 2,6-di-*tert*-butyl-*p*-cresol, BHT, has been found to have a highly beneficial effect. Higher yields of acrylic acid result from reactions conducted in the presence of BHT, Table 4. This implies that the allylic oxidation is not free radical in nature. However, deleterious free radical reactions, such as further oxidation or polymerization of acrylic acid might be inhibited.

By properly activating the catalyst in water and conducting the reaction at 65°C in the presence of BHT, it is possible to smoothly oxidize propylene to acrylic acid in greater than 90% yield, eq. 8. Turnover numbers exceed 4 moles acrylic acid produced per gram atom of palladium per hour under these conditions. No leaching of

$$CH_2{=}CH{-}CH_3 + \tfrac{3}{2}O_2 \xrightarrow[65°C,\ 5\ BARS]{10\%\ Pd/C,\ BHT} \underset{>90\%}{CH_2{=}CH{-}CO_2H} \qquad (8)$$

catalyst into the aqueous phase was detected and a clear water-white solution was obtained on filtration which exhibited no activity toward oxidation under reaction conditions in the absence of added catalyst. Thus, palladium on carbon is an efficient, long-lived catalyst for the selective generation of acrylic acid from propylene in aqueous solution under very mild conditions.

Allyl Acetate Formation. Propylene pre-treatment is useful in preparing an active catalyst for allylic oxidation in acetic acid to form allyl acetate as well, Table 3. Care to prevent over-oxidation of the catalyst is even more crucial for reactions in acetic acid than for those carried out in water. Addition of acetate ion is beneficial for allylic oxidations over supported palladium catalysts in acetic acid just as was found with soluble catalysts (7,8) and with vapor phase oxidations over Pd/SiO_2 (9). Palladium slowly leaches from the catalyst imparting low-level catalytic activity to the solution. As we have seen, however, acetate ion will direct even Pd(II)-catalyzed oxidations toward allylic products (6-8).

Butene Oxidations. Butene-1, *cis*- and *trans*-butene-2 and isobutylene were oxidized over supported palladium catalysts, both in water and in acetic acid. Butenes gave crotonic acid in water but, as had been noted earlier (11), the α,β-unsaturated ketone and aldehyde were also major products, eq. 9. Isobutylene gave considerably more methacrolein than methacrylic acid, eq. 10. Perhaps the 2-methyl group causes desorption of the intermediate from the surface before it oxidizes further to the acid.

$$\text{CH}_3\text{CH=CHCH}_3 \xrightarrow[60°\text{C}, H_2O]{O_2,\ 10\%\ \text{Pd-C}} \text{CH}_2\text{=CHCOCH}_3 + \text{CH}_3\text{CH=CHCHO} + \text{CH}_3\text{CH=CHCO}_2\text{H} \quad (9)$$

$$(\text{CH}_3)_2\text{C=CH}_2 \xrightarrow[60°\text{C}, H_2O]{O_2,\ 10\%\ \text{Pd-C}} \text{CH}_2\text{=C(CH}_3\text{)CHO} + \text{CH}_2\text{=C(CH}_3\text{)CO}_2\text{H} \quad (10)$$

Oxidation of either *cis*- or *trans*-2-butene over palladium in acetic acid gave three times as much of the branched 1-butene-3-acetate as the linear *cis*- and *trans*-2-butene-1-acetates, Table 5. Although *cis*-2-butene-1-acetate was initially produced from *cis*-2-butene, and *trans*-2-butene-1-acetate was formed from *trans*-2-butene, geometrical isomerization occurred with time. It was found that supported palladium catalysts catalyze the isomerization of the products preventing highly regio- and stereoselective reactions. The 3/1 ratio of branched to linear acetates formed during butene oxidations over palladium in acetic acid is similar to the product mixture which is obtained by oxidizing palladium π-crotyl complexes with either $Tl(OAc)_3$ or $Pb(OAc)_4$.

TABLE 5. THE OXIDATION OF CIS- AND TRANS-2-BUTENE OVER PALLADIUM IN ACETIC ACID

2-Butene	HOAc, mls	Catalyst	Grams Acetate Produced	Selectivity, % Monoacetates Linear Allylic	Branched Allylic	Vinylic	Diacetates
Cis	30	10% Pd-C	2.9	15.4	45.1	9.0	30.5
Trans	30	10% Pd-C	4.0	17.9	43.2	7.2	31.3
Cis	30	5% Pd/Al_2O_3	2.4	17.0	57.3	5.3	19.2
Trans	30	5% Pd/Al_2O_3	3.6	15.1	48.5	7.4	26.9

a) Butene, 4 grams, was stirred under 5 atm pp of O_2 at 65°C over 1 gram of the catalyst for ~8 hours in 30 ml of acetic acid.

Oxidation of Aromatics using Strong Oxidants. Palladium acetate catalyzes the oxidation of aromatics by strong oxidants in acetic acid to give aryl acetates and coupling products under mild conditions, eq. 11 (16). Unusual regioselectivity has been reported for this reaction (17). It has been suggested that this reaction proceeds via electrophilic attack by $Pd(OAc)_2$ on the

$$C_6H_6 \xrightarrow[\text{HOAc}]{[Ox],\ Pd(OAc)_2} C_6H_5\text{-OAc} + C_6H_5\text{-}C_6H_5 \qquad (11)$$

$[Ox] = Cr_2O_7^{=}$, MnO_4^{-}, $S_2O_8^{=}$, etc.

aromatic ring to give a palladium(II) aryl intermediate, eq. 12, which can either oxidize to give phenyl acetate or couple to give biphenyl (18). This mechanism is supported by the observations that: a) partial rate factors for oxidation of substituted aromatics are consistent with electrophilic palladation of the ring, b) isotope effects for forming both phenyl acetate and biphenyl are the same suggesting a common intermediate, and c) the phenyl acetate/biphenyl ratio is proportional to the concentration of strong oxidant used.

$$C_6H_6 + Pd(OAc)_2 \xrightarrow{-HOAc} C_6H_5Pd(II)OAc \quad \begin{cases} \xrightarrow{[Ox]} C_6H_5Pd(IV)(OAc)_3 \longrightarrow C_6H_5OAc + Pd(OAc)_2 \\ \xrightarrow[-HOAc]{C_6H_6} (C_6H_5)_2Pd(II) \longrightarrow C_6H_5\text{–}C_6H_5 \end{cases} \qquad (12)$$

A number of investigations have shown that catalytic oxidative dimerization can be made to predominate over acetoxylation in solvents other than pure acetic acid (19). Although a comprehensive study of the factors affecting this reaction has not been done to date, reaction conditions and solvents play a major role.

Aromatic Oxidations using O_2. It would be of practical synthetic interest to carry out these reactions using air or oxygen as the oxidant, rather than requiring stoichiometric consumption of expensive strong oxidants. However, attempts to use molecular oxygen--even in the presence of Cu(II) in hopes of reoxidizing palladium as in the Wacker reaction--were unsuccessful under mild conditions. Oxygen was successfully used as the oxidant for reaction 11 when high temperatures were employed (20). In this case, both $Pd(OAc)_2$ and supported palladium catalysts were effective.

Little work has been done on the acetoxylation of substituted aromatics using molecular oxygen as the oxidant. We have explored the palladium-catalyzed oxidation of phenyl acetate as a route to derivatives of the industrially important dihydroxyaromatics: catechol, resorcinol and hydroquinone. We were interested in finding conditions under which O_2 could be used as the oxidant and in determining those factors which controlled acetoxylation and those which controlled dimerization. Furthermore, we were interested in elucidating and understanding reaction regioselectivity.

Palladium acetate was found to catalyze the oxidation of phenyl acetate at 145°C using 800 psig of an oxygen-containing gas, Table 6, eq. 13. Phenylene diacetates and diacetoxy biphenyls were formed as major reaction products. In all cases, the three possible phenylene diacetates were formed. Only two of the six possible diacetoxybiphenyls were produced in significant quantities: o,m′-diacetoxybiphenyl and o,p′-diacetoxybiphenyl. The yields of major reaction products vary considerably with temperature, oxygen partial pressure, phenyl acetate concentration and added acetate.

OAc

$$+ \ {}^{1}\!/_{2}O_2 \xrightarrow[HOAc]{Pd(OAc)_2} \quad + \qquad (13)$$

OAc OAc OAc OAc

Effect of Temperature and Pressure. Although $Pd(OAc)_2$ catalyzes the oxidative dimerization of aromatics under mild conditions, acetoxylation of the ring becomes important at higher temperatures. Acetoxylation selectivity increases with increasing oxygen partial pressure as well. It is interesting to compare this effect with the increase in ring acetoxylation on increasing oxidant concentration in the $Pd(OAc)_2$ catalyzed oxidation of benzene in acetic acid, Table 7. The fact that higher temperatures and greater oxygen partial pressures favor acetoxylation relative to oxidative dimerization is consistent with the requirement that palladium be in a higher oxidation state for acetoxylation than for oxidative dimerization in cases where molecular oxygen is the oxidant.

Effects of Metal Acetates. The presence of alkali metal acetate has little effect on the rate of acetoxylation at low partial pressure of oxygen, but has a pronounced inhibiting effect

TABLE 6. OXIDATION OF PHENYL ACETATE CATALYZED BY $Pd(OAc)_2$ IN ACETIC ACID [a]

$\%$ O_2 In Gas	Added Substance, mmoles	Product Analyses, mmoles Produced: $C_6H_4(OAc)_2$ [b] o-	m-	p-	Total	$(AcOC_6H_4)_2$ o-,m′-	o-,p′-	Total	AcC_6H_4OAc [b] o-	m-	p-	Total
4	--	1.9	2.3	2.5	6.7	37.4	21.2	58.6	0.8	Tr	Tr	0.8
21	--	6.8	7.2	7.8	21.8	33.3	19.2	52.5	0.5	Tr	Tr	0.5
40	--	7.0	9.1	11.7	27.8	33.4	19.0	52.4	0.8	Tr	Tr	0.8
4	KOAc, 15.6	1.9	2.5	3.4	7.8	20.5	10.5	31.0	0.9	0.5	Tr	1.4
4	CsOAc, 15.6	1.9	2.0	3.1	7.0	8.1	4.1	12.2	1.0	Tr	Tr	1.0
40	CsOAc, 15.8	3.4	5.1	6.5	15.0	9.3	4.8	14.1	Tr	1.1	0.9	2.0
4	Ac_2O, 159	4.8	2.6	3.7	11.1	35.1	18.8	53.9	26.5	1.3	0.7	28.5
4	Ac_2O, 53	2.5	2.3	3.2	8.0	37.9	20.6	58.5	6.7	0.7	0.3	7.7

a) PhOAc, 1284 mmoles; $Pd(OAc)_2$, 3 mmoles; HOAc, 160 ml; 800 psi gas, sparged through solution at 500 ml/min., for for 420 min. at 145°C.

b) Some hydrolysis to o-,m-,p-$C_6H_4(OH)OAc$ and AcC_6H_4OH occurred during reaction and the amount of these products are added to the corresponding $(C_6H_4OAc)_2$ or AcC_6H_4OAc.

TABLE 7. PRODUCT DISTRIBUTIONS FOR THE $[Pd(OAc)_2]$-CATALYZED OXIDATION OF BENZENE BY DICHROMATE[a] AND PHENYL ACETATE BY OXYGEN[b]

Benzene Oxidation			Phenyl Acetate Oxidation		
$Na_2Cr_2O_7$, mmole	% Yield[c] C_6H_5OAc	$(C_6H_5)_2$	O_2 Partial Pressure, psig	% Yield[c] $C_6H_4(OAc)_2$	$(C_6H_4OAc)_2$
0	<1	22			
1	99	23	32	223	1953
3	190	33	168	726	1750
7.5	353	19	320	927	1747

a) Reactions conducted at 90 C for 16 hours using benzene (56 mmoles) and $[Pd(OAc)_2]$ (1 mmole) in in acetic acid (25 ml). Ref. 18.

b) Reactions conducted at 145 C for 3 hours under conditions given in Table 6.

c) Calculated on the basis of $[Pd(OAc)_2]$.

on oxidative dimerization. It is seen, Table 6, in oxidations carried out at high oxygen partial pressure in the presence of a large excess of CsOAc, that although the acetoxylation yield is reduced less than two-fold, the yield of oxidative dimerization product is lowered by a factor of ~4.

Effects of Phenyl Acetate Concentration--Predominant Acetoxylation. As would be expected, decrease in phenyl acetate concentration disfavors oxidative dimerization more than acetoxylation. Thus, those conditions which favor acetoxylation over oxidative dimerization are high temperature, high oxygen partial pressure, high acetate ion concentration, and low phenyl acetate concentration. Under these conditions, (175°C, 320 psi O_2, 0.78M PhOAc, excess CsOAc) selective acetoxylation occurs. The molar ratio of the phenylene diacetates to biphenyl diacetates formed under these conditions was 94.5/1. Unfortunately, the low phenyl acetate concentration, coupled with high CsOAc concentration, depressed the overall rate to a rather low level.

Regioselectivity and Mechanism. The relative amounts of the three isomeric phenylene diacetates produced during acetoxylation of phenyl acetate catalyzed by $Pd(OAc)_2$ using O_2 as oxidant are in the order *o*-<*m*-<*p*-; however, the differences in yields are slight. This pattern is similar to that found for $Pd(OAc)_2$-catalyzed acetoxylation using strong oxidants which are believed to involve unselective electrophilic palladation of the aromatic ring (18).

The oxidative dimerization reaction which occurs during $Pd(OAc)_2$-catalyzed oxidation using O_2 is both unique and selective: *o,m*′- and *o,p*′ -biphenyl diacetates are formed in approximately a 2:1 ratio. Neither *m,m*′- nor *m,p*′- were detected and only traces of *o,o*′- and *p,p*′- were formed.

A mechanism which is consistent with the observed product profile is given in Scheme 3.

Supported Palladium Catalysts for Aromatic Oxidation. Since benzene can be oxidized in acetic acid using both homogeneous and heterogeneous palladium catalysts (20), we compared supported palladium catalysts to $Pd(OAc)_2$ for oxidation of phenyl acetate. Supported catalysts required temperatures of >175°C for significant

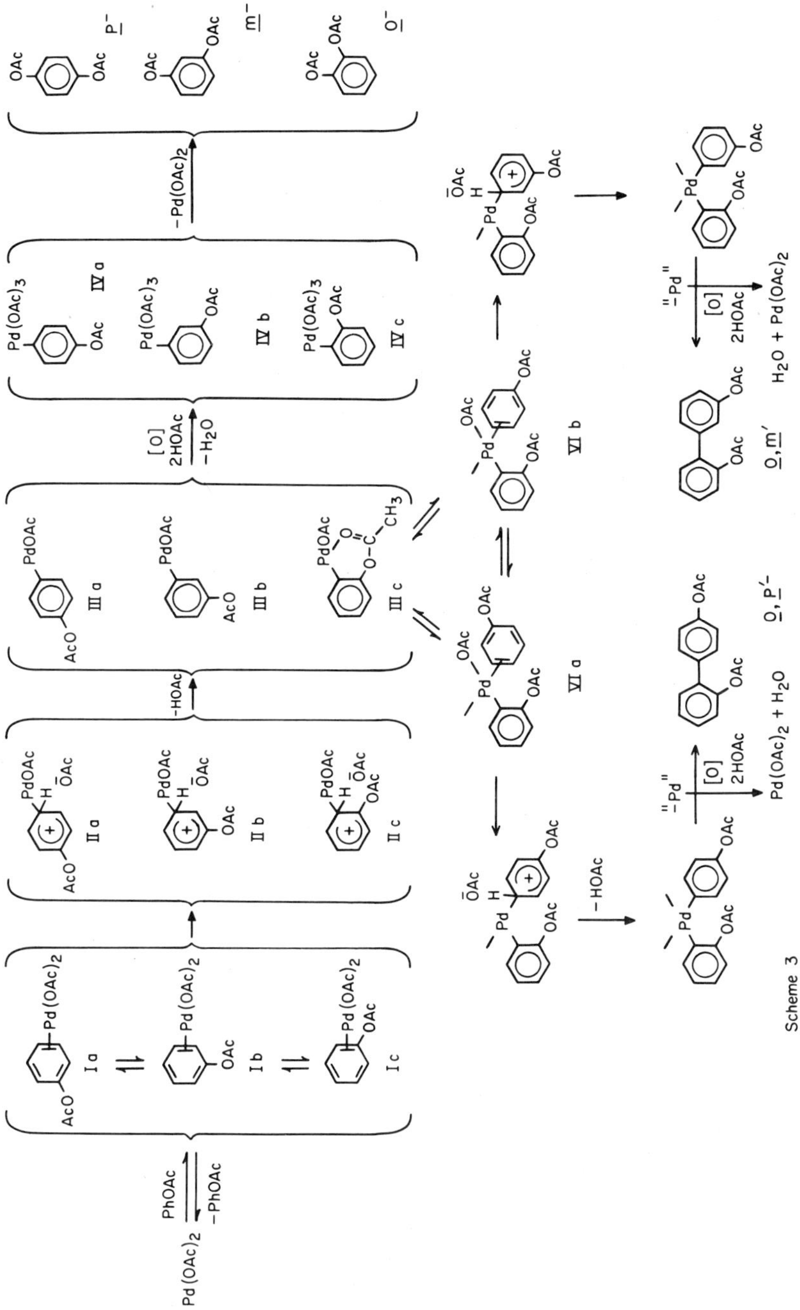
Pd(OAc)2
PhOAc
-PhOAc
Ia
Ib
Ic
IIa
IIb
IIc
-HOAc
IIIa
IIIb
IIIc
[O]
2HOAc
-H2O
IVa
IVb
IVc
-Pd(OAc)2
p-
m-
o-
VIa
VIb
"Pd"
Pd(OAc)2 + H2O
o,p'-
o,m'
H2O + Pd(OAc)2

Scheme 3

activity using O_2 as the oxidant. At these temperatures carbon was not a suitable support. Palladium (5%) on alumina catalyzed oxidation at high temperature (175°C) and high oxygen partial pressure (320 psig) to give predominant ring acetoxylation. Even at high temperatures over three-hour reaction periods, phenyl acetate conversion was low (5%); however, regioselectivity was considerably different from the homogeneous system. The major reaction product over 5% Pd/Al_2O_3 was p-phenylene diacetate, eq. 14, with very little o- and only traces of the m- isomer.

$$AcO\text{-}C_6H_5 + O_2 \xrightarrow[\substack{175°C,\ 3\ hrs \\ 500\ psi \\ 40\%\ O_2\ in\ N_2}]{\substack{Pd/Al_2O_3 \\ HOAc}} AcO\text{-}C_6H_4\text{-}OAc \qquad (14)$$

It has long been known that $Pd(OAc)_2$ catalyzes the oxidation of toluene in acetic acid to give benzylic oxidation products (21). The catalyst appears to be more active when reactions are run in the presence of activated carbon (21). Low oxidation state palladium has been implicated in benzylic oxidations (22). We have found that palladium on carbon is effective for benzylic oxidations under mild conditions, eq. 15, under which little or no ring oxidation occurs. This encourages one to speculate, in close analogy with olefinic systems, that over supported metal catalysts, higher oxidation state palladium species lead to ring oxidation, whereas lower oxidation state species are responsible for benzylic oxidation.

$$C_6H_5\text{-}CH_3 \xrightarrow[\substack{100°C \\ 5\ ATM}]{10\%\ Pd\text{-}C} \begin{cases} \xrightarrow{H_2O} C_6H_5\text{-}CO_2H \\ \xrightarrow[HOAc]{} C_6H_5\text{-}CH_2OAc \end{cases} \qquad (15)$$

SUMMARY

Nucleophilic attack on a Pd(II) π-complex gives an intermediate which can lead to a vinylic oxidation product or its stable isomer.

If carried out in the presence of a suitable oxidant which is regenerable in air, the reaction is catalytic. Allylic oxidation products also arise via palladium catalyzed oxidation of olefins, but these catalytic reactions have not been extensively investigated. Several factors favor the allylic oxidation of olefins over vinylic oxidation. In homogeneous solutions of Pd(II) π-complexes, weak inorganic bases aid formation of π-allyl complexes which give rise to allylic oxidation products. Strongly electron withdrawing anionic ligands promote oxidative addition of C-H bonds of coordinated olefin to give π-allyl intermediates which can undergo nucleophilic attack to produce allylic, rather than vinylic oxidation products. Supported palladium catalysts which can be maintained in a low oxidation state appear to oxidatively add α-olefins to produce π-allyl surface species and give allylic oxidation products, whereas the addition of strong oxidants, chloride ion or use of forcing conditions promotes formation of vinylic oxidation products over these catalysts.

Palladium catalyzed aromatic oxidations proceed in two distinct directions--ring oxidation or oxidative coupling. In solution, oxidative coupling seems to be favored by low oxidation state ($Pd^{II} \rightleftharpoons Pd^{0}$) catalysis, whereas ring oxidation is favored by high oxidation state ($Pd^{IV} \rightleftharpoons Pd^{II}$) systems. Alkyl aromatic compounds may undergo competitive oxidation at the alkyl side chain. Benzylic oxidation is favored by low oxidation state palladium, whereas more electrophilic high oxidation state species attack the ring.

By applying these considerations, it becomes possible to direct olefin oxidations along new and interesting allylic oxidation pathways. We are also learning how to promote aromatic oxidations using palladium catalysts. Ring oxidation, oxidative coupling or side chain oxidation can be accomplished selectively by proper choice of catalyst and conditions.

REFERENCES

1. P. M. Henry, "Palladium Catalyzed Oxidation of Hydrocarbons", D. Reidel Publishing Company, Boston, 1980.

2. P. M. Maitlis, "The Organic Chemistry of Palladium, Vol. II Catalytic Reactions", Academic Press, New York, 1971.

3. B.M. Trost and P.J. Metzner, J. Am. Chem. Soc, 102, 3572 (1980).

4. L. Saussine, J.-P. Laloz and H. Mimoun, French Patent 2,450,802 (1980).

5. A. D. Ketley and J. Braatz, Chem. Commun., 169 (1968).

6. S. Winstein, J. McCaskie, H.-B. Lee and P. M. Henry, J. Am. Chem. Soc., 98, 6913 (1978).

7. British Patents 1,101,055; 1,101,056; 1,027,396.

8. I. P. Stolyarov, M. N. Vargaftik, O. M. Nefedov and I. I. Moiseev, Kinetika i Kataliz, 23, 376 (1982).

9. W. Swodenk and G. Scharfe, U. S. Patent 3,925,452 (1975).

10. P. R. Auburn, P. B. Mackenzie and B. Bosnich, J. Am. Chem. Soc., 107, 2033 (1985).

11. T. Seiyama, N. Yamazoe, J. Hojo and M. Hayakawa, J. Catal., 173 (1972).

12. K. Fujimoto and T. Kunugi, J. Jpn. Petrol. Inst., 17, 739 (1974).

13. R. David and J. Estienne, U. S. Patent 3,624,147, (1971).

14. J. A. Hinnenkamp, U. S. Patent 4,435,598, (1984).

15. W. Kitching, T. Sakakiyama, Z. Rappoport, P. D. Sleezer, S. Winstein, and W. G. Young, J. Am. Chem. Soc., 94, 2329 (1972).

16. Henry, P. M., J. Org. Chem., 36, 1886 (1971).

17. a. Eberson, L. and Gomez-Gonzalez, L., Act. Chem. Scand., 27, 1162 (1973); 27, 1249 (1973); 27, 1253 (1973). b. Eberson, L. and Jonsson, L., Act. Chem. Scand., 28, 597 (1974); 28, 771 (1974); 30, 361 (1976). c. Eberson, L. and Jonsson, L., Act. Chem. Scand., 28, 771 (1974); Liebigs Ann. Chem., 233 (1977).

18. Stock, L. M., Tse, K., Vorvick, L. J., and Walstrum, S. A., J. Org. Chem., 46, 1759 (1981).

19. a. Van Helden, R. and Verberg, G., Rec. Trav. Chim., 84, 1263 (1965). b. Unger, M. O. and Fouty, R. A., J. Org. Chem., 34, 18 (1969). c. Clark, F.R.S., Norman, R.O.C., Thomas, C.B., and Willson, J.S., J.C.S. Perkin I, 1289 (1974). d. Iataaki, H. Yoshimoto, H., J. Org. Chem., 38, 76 (1973). e. Itatani, H. and Yoshimoto, H., Chem. Ind., 674 (1971). f. Rudenkov, A.I., Mennenga, G.U., Rachkovskaya, L.N., Matveev, K.I., and Kozhevnikov, I.V., Kinet. i Katal., 18, 915 (1977). g. Itahara, Y., Chem. Ind., 330 (1982).

20. Arpe, V.H.-J. and Hornig, L., Erdol und Kohle, 23, 79 (1970).

21. Bryant, D.R., McKeon, J.E. and Ream, B.C., Tetrahedon Lett., 3371 (1968); J. Org. Chem., 33, 4123 (1968).

22. Starchevskii, M.K., Vargaftik, M.N. and Moiseev, I.I., Kinetika i Kataliz, 20, 1163 (1979).

ACTIVATION OF C-C BONDS IN SATURATED HYDROCARBONS BY METAL COMPLEXES

M.E. Vol'pin and I.S. Akhrem
Institute of Organoelement Compounds, USSR Academy of Sciences, Moscow 117813, USSR

ABSTRACT - New possibilities of creation of active catalysts for breaking alkane and cycloalkane C-C bonds on the basis of both new acid systems and coordinatively unsaturated transition metal complexes are discussed. Novel highly active homogeneous catalysts for the electrophilic activation of saturated hydrocarbons on the basis of complexes of acyl halides with aluminium halides (1:2) have been found. At room temperature these complexes perform the catalytic cracking of alkanes, their isomerization and the oxidative coupling of cyclohexane or methylcyclopentane with formation of dimethyldecalines. Ziegler-type systems based on rhenium carbonyls or carbonyl-hydrides and organoaluminium compounds are shown to be capable of catalyzing the cleavage and hydrogenolysis of C-C bonds in alkanes and cycloalkanes. The regularities of these reactions have been studied.

1. INTRODUCTION

Reactions of saturated hydrocarbons accompanied by the cleavage of C-C bonds form the basis of cracking and hydrocracking, skeletal isomerization and other important processes of hydrocarbon transformations.

Two essentially different ways of alkane C-C bond activation, which give rise to their splitting, are most important. The first way consists of the initial cleavage of a C-H bond under the action of electrophilic reagents and formation of carbocations or cation-like intermediates followed by their conversion with the cleavage of C-C bonds.

The second way of the activation of C-C bonds in saturated hydrocarbons is their interaction with metal catalysts which may be represented formally as proceeding via a stage of the oxidative addition of an alkane C-C bond to the metal.

Both ways of alkane activation have been investigated in detail for the high temperature gas-phase conversions of hydrocarbons on heterogeneous catalysts. At present an ever increasing interest is shown in the ability of metal complexes to affect similar liquid-phase processes at lower temperatures. Application of metal complexes as catalysts provides new possibilities to carry out selectively the reactions of alkanes and cycloalkanes under mild conditions.

In this work both above-mentioned possibilities of activation of saturated hydrocarbons under the action of metal complexes have been studied.

2. HIGHLY EFFECTIVE CATALYSTS FOR ELECTROPHILIC ACTIVATION AND TRANSFORMATIONS OF ALKANES AND CYCLOALKANES UNDER MILD CONDITIONS*

Since the basicity of saturated hydrocarbons is low, even strong protic and aprotic acids affect their conversions at room temperature very poorly.

One of the methods for increasing activity of aprotic acids is the transfer of the reaction centre from a covalently bound metal atom of a Lewis acid to another atom with the formation of a partial or total positive charge on the latter. This principle is the basis of the

---

* The participation of A.V. Orlinkov in this part of investigation is gratefully acknowledged.

action of promotors (AB) which as weak nucleophiles are converted to charged salt-like systems on interaction with Lewis acids:

$$AB + AlX_3 \longrightarrow A^+AlX_3B^-,$$

A = H, R, RCO, etc.; B = Hal, OH, etc.

However, even promoted Lewis acids are frequently insufficiently active, and their reactions with saturated hydrocarbons require either long time or elevated temperatures and, as a rule, have low selectivity.

Thus the development of new active and selective catalysts, which are readily available, non-toxic and convenient for work, is an important problem. Using the above principle of increasing Lewis acid activity, we found that the systems $RCOX\cdot 2AlX_3$ (X = Cl, Br), obtained by mixing acyl halides with aluminium halides in the molar ratio 1:2, are highly active in respect to saturated hydrocarbons and effective in providing transformations at room temperature and, in some cases, below 0 °C.

Note that at room temperature neither complexes $RCOX\cdot AlX_3$, nor AlX_3 alone cause marked conversions of saturated hydrocarbons. Thus, an addition of one more mole of AlX_3 to the classical Friedel-Crafts systems, $RCOX\cdot AlX_3$, changes their properties dramatically and converts these systems, inert to paraffins, into the most active reagents exceeding in their reactivity all reported systems based on Lewis acids.

Thus, it was found that linear alkanes, C_5-C_{18}, can readily react with the $RCOX\cdot 2AlX_3$ systems. The reactions proceed rapidly and quantitatively at room temperature, with cracking being the main direction of paraffin conversions (Table 1). Products of the alkane cracking are isobutane, isopentane, ketones and oligomers.

It is noteworthy that pure and promoted Lewis acids

have no noticeable effect on paraffin cracking at room temperature. The most active "super-acid" systems also are inferior in acidity and selectivity to the systems found.

TABLE 1. Cracking of n-alkanes under the action of a stoichiometric amount of $MeCOBr \cdot 2AlBr_3$ (30 min, 100% conversion

n-RH	Reaction products (wt.% to RH)				
	$i\text{-}C_4H_{10}$	$i\text{-}C_5H_{12}$	$i\text{-}C_4H_9COMe$	Oligomers	i-RCOMe
C_5H_{12}	36	12	8	21	12
C_6H_{14}	36	5	10	24	13
$C_{12}H_{26}$	44	7	20	26	absent
$C_{18}H_{38}$	29	9	8	46	absent

One may suppose that in this case, the scheme of alkane cracking does not differ from the usual scheme of the cracking under the action of electrophilic reagents and includes stages of the carbenium ion generation, its isomerization and subsequent β-elimination with the formation of a lower hydrocarbon and olefin. Polymerization of the latter gives rise to oligomers.

$$n\text{-}C_8H_{18} \xrightarrow{E^+} sec\text{-}C_8H_{18}^+ \rightleftarrows Me_3C\text{-}CH_2\text{-}\overset{+}{C}Me_2$$

$$\downarrow$$

$$CH_2{=}CMe_2 + Me_3C^+$$

$$CH_2{=}CMe_2 \rightarrow \text{oligomer} \qquad Me_3C^+ \rightarrow Me_3CH$$

An essential feature of our systems is their ability to provide cracking at room temperature, in certain cases catalytically (Table 2).

TABLE 2. Catalytic cracking of n-alkanes at 20 °C under the action of $MeCOBr \cdot 2AlBr_3$

RH	RH/cat	Time, h	RH conversion, %	Reaction products*, (wt.%)		
				$i\text{-}C_4H_{10}$	$i\text{-}C_5H_{12}$	Oligomer
C_6H_{14}	5/1	0.5	96	59	20	18
C_6H_{14}	10/1	0.75	89	58	20	20
C_6H_{14}	20/1	1.0	57	58	20	21
C_6H_{14}	50/1	5.0	29	59	21	20
C_8H_{18}	5/1	0.5	98	44	14	33
C_8H_{18}	10/1	2.0	74	38	14	16
$C_{12}H_{26}$	5/1	2.0	58	27	12	20
$C_{12}H_{26}$	3/1	2.0	44	25	11	26

* Products of C_8-C_{18} alkanes conversion also contain up to 20% of C_6- C_8 alkanes and 6% of C_{11}-C_{13} alkanes

For example, in the presence of catalytic quantities of the systems, octane undergoes 74% cracking during 10 min at room temperature to be converted to lower alkanes and a mixture of oligomers. In the catalytic process the proportions of isobutane and isopentane are markedly increased. Thus, these hydrocarbons constitute 80% of hexane conversion products.

Oligomer products of cracking of various alkanes do not differ significantly. The molecular masses of these products are about 700. According to NMR the oligomers are characterized by a high degree of branching.

In the presence of the systems found cyclododecane behaves similarly, the only direction of its conversion being cracking. At room temperature this hydrocarbon converts rapidly and quantitatively to isoalkanes C_4-C_5, ketones and a mixture of oligomers.

$\longrightarrow$ i-C_4H_{10} + i-C_5H_{12} + i-C_4H_9COMe +
27* 3 25

+ C_8H_{15}COMe + Oligomer
21 24

*Wt.%, 100° conversion, 20 °C, 30 min.

A quite different pattern is observed under the action of RCOX·2 AlX_3 on 5- and 6-membered cycloalkanes. In this case the yields of cracking products do not exceed 5%. Depending on a ratio of reagents and on a hydrocarbon structure, reactions of five- and six-membered cycloalkanes (cyclohexane, methylcyclopentane, methylcyclohexane and cyclopentane) with RCOX·2$AlBr_3$ (R = Me, Pr; X = Cl, Br) proceed mostly (or exclusively) along one of the following main directions.

The reaction without solvent in the presence of an excess of cycloalkane results in the facile coupling of cyclohexane or methylcyclopentane, with dimethyldecalines being formed in high yields.

2 $\xrightarrow{\text{RCOX·2 AlBr}_3}$ $\xleftarrow{\text{RCOX·2AlBr}_3}$ 2

100% to RCOX

An increase in concentration of the acylating agent suppresses this process and favours the acylation of cycloalkanes. In the case of cyclopentane, ketones are formed in high yields, this reaction route being the only one under any conditions.

+ RCOX·2$AlBr_3$ $\longrightarrow$ COR

80%, 20 °C, 2 h

Using an excess acylating system, at lowering the temperature it is possible to perform dehydroacylation of a cycloalkane with the formation of α, β-unsaturated ketones.

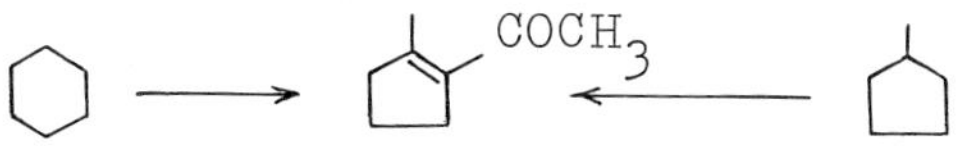

$COCH_3$

40%, -10 °C, 1h

It may be supposed that the key stage of the reactions is the formation of a carbocation from the hydrocarbon. Primary and secondary carbonium ions are known to rearrange readily to thermodynamically stable tertiary ions. If the carbocation formed is sufficiently stable to β-elimination, one may expect isomerization to be the main conversion of a hydrocarbon. Indeed, we have found that the systems catalyze efficiently the isomerization of n-butane to isobutane. The reaction is highly selective, and the equilibrium concentration of isobutane in the mixture is observed already after 3 h. Under similar conditions in the presence of $AlBr_3$ the equilibrium concentration of isobutane is not attained even after 1000 h.

Table 3 shows that the systems found are more active than the most active catalysts of n-butane isomerization, including "super-acid" systems with the highest Hammett acidity function which have recently been reported.

Another example of the isomerization by the systems found is the conversion of trimethylenenorbornane to adamantane, which also proceeds readily at room temperature:

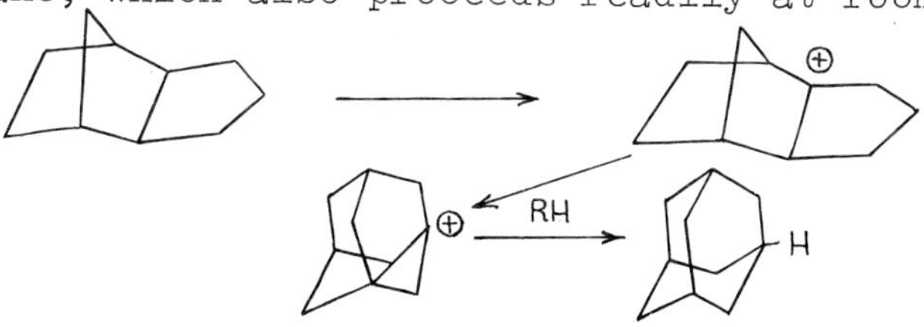

RH

H

TABLE 3. Isomerization of n-butane at room temperature

Catalyst	$n\text{-}C_4H_{10}$/ catalyst	Time, h	$[i\text{-}C_4H_8]$ [a]	References
$MeCOBr \cdot 2AlBr_3$	11/1	0.5	23	
	11/1	3	76	
	34/1	6.5	80	
$MeCOCl \cdot 2AlCl_3$	7/1	3	22	
$MeCOBr \cdot AlBr_3$	21/1	4	4	
$AlBr_3$	20/1	1000	73	b)
$AlCl_3 \cdot HCl$	9/1	64	21	c)
$ZrO_2 \cdot H_2SO_4$ ($H_o \leqslant -16$)		24	30	d)
$CF_3SO_2OH_2^+ B(OSO_2CF_3)_4^-$ ($H_o \leqslant -20$)	1/1.2	24	40	e)

a) $[i\text{-}C_4H_8]$ - concentration of isobutane in the mixture of butanes, (%);
b) US Patent 2288477 (1942);
c) US Patent 2249366 (1941);
d) A. Usubillaga, G. de Caltellano, V. Zabel and W.H. Watson, J. Chem. Soc.,Chem. Commun.,851 (1980).
e) G.A. Olah, Kh. Laali and O. Farooq, J. Org. Chem., 49 4591 (1984).

It is obvious that the considered reactions of alkanes and cycloalkanes have one common stage, viz., the generation of a carbocation from the hydrocarbon.

Formation of the reaction product observed is explained by further conversions of the carbenium ion and by the olefin reactions which achieve an equilibrium (isomerization, β-elimination, alkylation, acylation).

$$RH + E^+ \rightleftharpoons R^+ + EH$$

$$\gt C{=}C\lt \rightleftharpoons R^{+} \begin{cases} \xrightarrow{\text{isomerization}} \text{tert-}R^{+} \longrightarrow \text{tert-}RH \\ \xrightarrow{\beta\text{-elimination}} R_1^{+} + \gt C{=}C\lt \\ \xrightarrow[\gt C=C\lt]{\text{alkylation}} R{-}\overset{|}{\underset{|}{C}}{-}\overset{|}{\underset{+}{C}} \longrightarrow \text{oligomer and others} \end{cases}$$

$$\gt C{=}C\lt \begin{cases} \xrightarrow{\text{oligomerization}} (\overset{|}{\underset{|}{C}}{-}\overset{|}{\underset{|}{C}})_n \\ \xrightarrow{\text{acylation}} {-}\overset{\oplus}{C}{-}\underset{COR}{C}{-} \begin{cases} \xrightarrow{RH} \gt CH{-}\overset{\vee}{C}{-}COR \\ \longrightarrow {-}\overset{|}{C}{=}\overset{|}{C}{-}COR \end{cases} \end{cases}$$

In our opinion, a high activity of the systems found in respect to saturated hydrocarbons is accounted for by a remarkable ability of these systems to initiate the stage of carbenium ion formation.

What is the reason of their activity in the process? To answer this question, we have studied structures of two systems: an active system, $MeCOBr{\cdot}2AlBr_3$, and an inactive model system, $MeCOBr{\cdot}AlBr_3$. These systems were investigated by ^{1}H, ^{13}C and ^{13}O NMR in solutions (in collaboration with V.I. Bakhmutov and P.V. Petrovsky) and in a solid state (in collaboration with E.T. Lipmaa and T.N. Pekhk) as well as by NQR (in collaboration with Yu.I. Amiantov and T.L. Khotzjanova) and vibrational spectroscopy (in collaboration with L.A. Leites and S.S. Bukalov). It was shown that the first system in the solid state had the structure of an acetylium salt, RCO^{+}, with a dimeric anion, $Al_2Br_7^{-}$, whereas in solution it existed as a mixture of two equilibrium forms, viz., acetylium salt and donor-acceptor complex.

In the solid state the system of 1:1 composition has the structure of acetylium salt with a monomeric anion, $AlBr_4^{-}$; but it should be emphasized that in solution this inactive system does not produce acetylium ions

$$\mathrm{MeCOBr} + 2\mathrm{AlBr_3} \longrightarrow \underset{\substack{|\\ \mathrm{Br}}}{\mathrm{MeC{=}O}} \rightarrow \mathrm{Al_2Br_6} \rightleftharpoons \underset{\text{in solution}}{\mathrm{MeCO^+\ Al_2Br_7^-}}$$

$$\mathrm{MeCOBr} + \mathrm{AlBr_3} \longrightarrow \underset{\substack{|\\ \mathrm{Br}\\ \text{in solution}}}{\mathrm{MeC{=}O}} \rightarrow \mathrm{AlBr_3} \not\rightarrow \mathrm{MeCO^+\ AlBr_4^-}$$

Thus the decisive difference between these two systems is that only the system with an excess AlX_3 gives acetylium cations in solution.

The data presented allow us to suppose that an unusually high activity of the systems found is due to the presence of the acetylium ions stabilized by the complex anion Al_2X_7-.

The mechanism of the C-H bond cleavage under the action of these systems is not clear yet. For example, one cannot exclude that in an equilibrium with the acetylium salt there is a very low concentration of the complex, wherein a positive charge of the carbon atom is increased due to coordination with aprotic acid and namely this complex takes part in the reactions with hydrocarbons:

$$\mathrm{MeCO^+Al_2Br_7^-} \rightleftharpoons \mathrm{Me\overset{+}{C}{=}O} \rightarrow \mathrm{Al_2Br_7^-} \quad .$$

It is, however, obvious that in either case a high activity of the system is due to the presence of acetylium cations with the dimeric anion.

To confirm this conclusion, we have compared the activities of the systems studied in catalytic cracking of octane with the concentration of acetilium cations in these systems measured by NMR. According to Fig. 1, both curves change in a similar way, and on passing from 1:1 to 1:2 complex, a jumpwise change of both curves is observed.

Thus, we have found novel highly active systems, which promote conversion of saturated hydrocarbons under

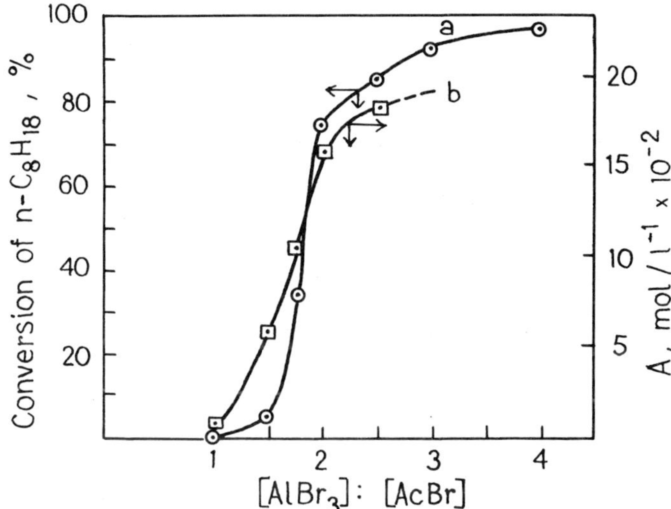

Figure 1. Concentration of acetylium salts in systems AcBr n $AlBr_3$ compared with activity of these systems a) n-octane conversion vs. the $AlBr_3$/AcBr ratio (AcBr= 0.29 mol/l, C_8H_8 = 2.7 mol/l, 20 °C, 10 min, CH_2Br_2); b) acetylium salt concentration, A, vs. the Al_2Br_6/AcBr ratio (NMR data) (AcBr = 1.10 mol/l, -30 °C, CH_2Br_2).

extremely mild conditions. This opens up new possibilities in synthesis based on alkanes and cycloalkanes and in preparation of efficient catalysts for the conversion of saturated hydrocarbons.

3. ACTIVATION OF C-C BONDS IN ALKANES UNDER THE ACTION OF TRANSITION METAL COMPLEXES*

All known catalytic reactions of alkanes and cycloalkanes under the action of transition metal complexes occur with the participation of C-H bonds while ordinary C-C bonds remain unchanged.

On the contrary, from data on bond energies it follows

* The participation by S.V. Reznichenko and V.V. Grushin in this part of our investigation is gratefully acknowledged.

that of two possible reaction routes, (a) and (b), the route (b) (oxidative addition of the hydrocarbon C-C bond to metal) should be even more favourable than the metal insertion into a C-H bond (route (a))

$$RCH_2\text{-}CH_3 + M \begin{cases} \xrightarrow{a} R\text{-}CH_2\text{-}CH_2\text{-}M\text{-}H \\ \xrightarrow{b} RCH_2\text{-}M\text{-}CH_3 \end{cases}$$

The main reason of inertness of C-C bonds in hydrocarbons seems to be steric screening of carbon atoms of these bonds. Steric hindrances considerably increase the activation energy of the attack of a reagent on C-C bonds, thus facilitating the attack on less screened C-H bonds.

The minimizing of steric overcrowding in the moiety attacking the hydrocarbon results frequently in equalizing the reactivities of C-H and C-C bonds. Moreover, the cleavage of C-C bonds in hydrocarbons is predominant in certain cases.

It is known that some atomic metals, such as Ni, Zr and Ti and also metal ions in a gas phase, such as Co^+, Fe^+, Ni^+ and Sc^+ cause the cleavage of both C-H and C-C bonds in alkanes. However, in all cases the ligand addition to a metal atom or ion suppresses the C-C bond cleavage.

On the basis of these data, we have formulated the following main criteria for the search of transition metal complexes capable of activating and splitting C-C bonds in hydrocarbons:

(a) low oxidation state of a metal, i.e. its ability to a facile oxidative addition;

(b) low coordination number, which ensures the least steric hindrances in the attack on C-C bonds;

(c) presence of small ligands, such as H and CO that cause the least steric hindrances, at the metal atom.

Besides, the ligands should prevent the formation of a free metal and its transformation to a compact metal;

(d) participation of a reagent reacting irreversibly with the product of metal insertion into the C-C bond and thus shifting the equilibrium to the C-C bond cleavage. Hydrogen can act as such reagent.

$$\begin{array}{ccccc} RCH_2CH_2\text{-}M\text{-}H & \longleftarrow & RCH_2CH_3 + M & \longrightarrow & RCH_2\text{-}M\text{-}CH_3 \\ \downarrow H_2 & & & & \downarrow H_2 \\ RCH_2CH_3 + MH_2 & & MH_2 \rightarrow M + H_2 & & RCH_3 + CH_4 + MH_2 \end{array}$$

In the search for catalytic systems that would meet these requirements, we concentrated our attention on Ziegler-type systems and studied their reactions with alkanes under hydrogen pressure.

After much efforts, we have found that rhenium carbonyl and carbonyl-hydride complexes, such as $Re_2(CO)_{10}$, $H_3Re_3(CO)_{12}$ and $H_4Re_4(CO)_{12}$ combined with organoaluminium compounds (Et_3Al, iBu_2AlH, etc.), in fact, can activate C-C bonds in alkanes and cycloalkanes, such as octane, cyclopentane, methylcyclopentane, cyclohexane and cyclododecane, as well as catalyze their hydrogenolysis. The reaction proceeds in a medium of the corresponding hydrocarbon at 150-180 °C and the initial hydrogen pressure of 50 atm. The hydrocarbon conversion after 15 hours is 30 to 100%, with one mole of the rhenium compound catalyzing the conversion of hundreds moles of hydrocarbon (examples see in Table 4).

The activity of the catalytic system largely depends on a components' ratio. Thus, the activity of the catalyst formed by interaction of $Re_2(CO)_{10}$ with iBu_2AlH is maximum at Al/Re $\sim$ 3; with increasing the ratio to 6 or decreasing it to 2 the catalyst loses its activity. Note that neither organoaluminum compounds alone nor rhenium complexes, such as $Re_2(CO)_{10}$, $H_3Re_3(CO)_{12}$, $H_4Re_4(CO)_{12}$,

TABLE 4. Catalytic hydrocracking of alkanes and cycloalkanes (180 °C, 15 h, 50 atm H_2)

RH	Catalyst[a)]	Conversion of RH %	Conversion of RH $\frac{\text{mole RH}}{\text{mole cat}}$	Products[d)], moles per mole of reacted RH	
n-C_8H_{18}	$Re_2(CO)_{10}$+A	48	91	C_7 0.27 C_5 0.16 C_3 0.10	C_6 0.18 C_4 0.15 C_1 2.08
cyclopentane	$Re_2(CO)_{10}$+B	92	94	C_5 0.46 C_3 0.08 C_1 0.61	C_4 0.14 C_2 0.12
cyclopentane[b)]	$Re_2(CO)_{10}$+B	29	59	C_5 0.63 C_3 0.02 C_1 0.06	C_4 0.06 C_2 0.02
cyclopentane[c)]	$H_4Re_4(CO)_{12}$+B	84	165	C_5 0.58 C_3 0.13 C_1 0.98	C_4 0.07 C_2 0.16
methylcyclopentane (CH_3)	$Re_2(CO)_{10}$+B	83	57	C_6 0.53 C_4 0.07 C_2 0.05	C_5 0.19 C_3 0.02 C_1 0.49
cyclodecane ($(CH_2)_5$)	$Re_2(CO)_{10}$+A	78	21	C_{10} 0.14 C_8 0.06 C_6 0.04 C_4 0.06 C_1 0.93	C_9 0.14 C_7 0.04 C_5 0.06 C_3 0.06

a) A - Et_3Al, B - i-Bu_2AlH; b) 150 °C; c) 7 h; d) n-alkanes, in all cases excluding methylcyclopentane

$CpRe(CO)_3$ and Na_2ReH_9 in the absence of the organoaluminum compounds, show activity in the alkane cleavage.

Special experiments demonstrated rhenium black to exert no activity in this reaction up to 300 °C.

The main products of the conversion of n-alkanes

catalyzed by the systems found are methane and linear hydrocarbons with one carbon atom less than the starting paraffin. In addition, lower alkanes, which are the products of subsequent hydrogenolysis, are also formed. Thus, the cleavage of the terminal CH_2-CH_3 bonds is the main direction of the reactions.

The main products of the conversion of unsubstituted cycloalkanes are linear alkanes with the same number of carbon atoms, as well as the products of their further stepwise hydrogenolysis. Under these conditions the hydrogenolysis of methylcyclopentane produces three isomeric hexanes, viz., 2-methylpentane, 3-methylpentane and n-hexane in a ratio 2:1:0.04 and also n-pentane and isopentane.

Hence, in methylcyclopentane the least screened C-C bonds are cleaved: (c) > (d) ≫ (b)

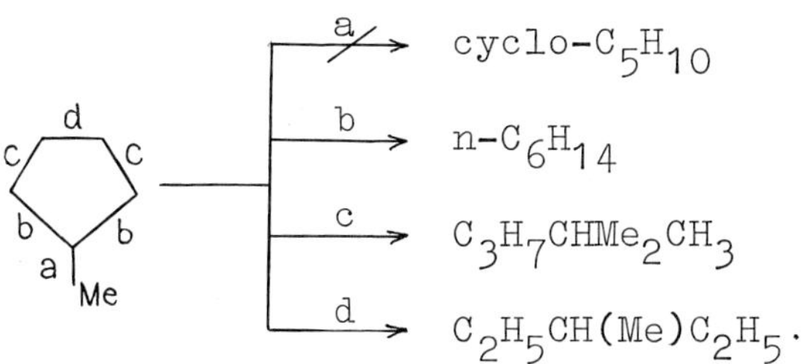

Thus, the present investigation provides the first example of the catalytic activation of ordinary C-C bonds in saturated hydrocarbons by the Ziegler-type systems.

One may hope that further investigations in this direction will lead to discovery of a lacking link between classical heterogeneous metal catalysts, which split C-C bonds at high temperatures, on the one hand, and highly active atomic metals and metal ions affecting these processes at low temperatures in the gas phase, on the other.

The results of the present investigation indicate wide possibilities in the field of development of new active catalysts for the cleavage of very inert C-C bonds in saturated hydrocarbons on the basis of both traditional acid systems and complexes of transition metals in low oxidation states.

MECHANISM OF FORMATION OF ACTIVE METAL-CARBENE INTERMEDIATES IN PHOTOREDUCED SILICA-MOLYBDENA CATALYSTS FOR PROPENE METATHESIS

B.N. Shelimov, I.V. Elev and V.B. Kazansky
N.D. Zelinsky Institute of Organic Chemistry, Academy of Sciences of the USSR, Moscow 117913, USSR

ABSTRACT - It is shown that preliminary photoassisted reduction of molybdenum (VI) ions supported on silica (1 wt.% Mo) under u.v.-irradiation in CO atmosphere is an effective way to produce highly active catalysts for propene metathesis. Specific activities of photoreduced catalysts referred to one Mo ion exceed those of thermally reduced samples by a factor of 6-10 and are still higher (by a factor of 2-3) than those of the catalysts with silica-"anchored" complexes tetraallylmolybdenum. High activity of photoreduced Mo/SiO_2 is associated with the presence of coordinatively unsaturated molybdenum(IV) ions characterized by some specific features. Molybdenum(IV) ions are able to form with ethylene and propylene stable π-complexes, a certain fraction of which already at 300 K isomerizes to carbene structures - active intermediates in the metathesis reaction. The isomerization is supposed to occur via an intramolecular 1,2-transfer of H atom.

1. INTRODUCTION

Preliminary reduction of Mo^{6+} ions in H_2, CO or hydrocarbons in impregnated catalysts Mo^{6+}/SiO_2 is known to increase considerably their catalytic activity for the propene metathesis reaction and to lower the reaction temperature (Refs. 1 & 2). However, for this type of catalysts it is rather difficult to find out which oxidation state of molybdenum ions is responsible for the metathesis reaction, since the process of their thermal reduction occurs non-selectively resulting in formation of various valence and coordination states.

In our previous works (Refs. 3,4) it was shown that Mo^{6+}/SiO_2 low temperature photoreduction is a very effective and selective way to produce reduced molybdenum ions in a certain oxidation state (Mo^{4+} or Mo^{5+}). Therefore, it was thought that photoreduced samples could be suitable objects for the determination of a valence state responsible for the activity in the metathesis reaction.

The mechanism of the formation of active centers for this reaction involving reduced molybdenum ions is supposed to be another interesting problem. Currently, it is generally accepted that a propene metathesis reaction proceeds by the chain carbene mechanism via a successive interchange of $Mo{=}CH_2$ and $Mo{=}CHCH_3$ carbene species on the same center. However, the problem of their formation mechanism is still under discussion. Generally speaking, the analysis of the initial products of the olefin interaction with reduced molybdenum ions (in particular, Mo^{4+}), which are supposed to be precursors of the active centers, could allow one to make a choice in favour of any of alternative reaction schemes. However, the number of such centers is necessary to be as high as possible and they must be homogeneous. Mo^{4+}/SiO_2 catalysts photoreduced in CO and containing high concentrations of coordinatively unsaturated Mo^{4+} ions meet at most all these demands.

2. EXPERIMENTAL

The samples of Mo^{6+}/SiO_2 (1 weight % of Mo) were obtained by impregnation of silica-gel "KSK" with S = 360 m^2g^{-1} with an aqueous solution of ammonium paramolybdate. The pretreatment procedure consisted of heating the sample in a circulating flow of dry oxygen at 1073 K for 1-1.5 h, then in vacuum at the same temperature for 1.5-2.0 h. After repeated admission of oxygen the sample was cooled down to 423 K and evacuated to 10^{-4} Torr. Photoreduction of Mo^{6+}/SiO_2 (0.2 g) in CO or H_2 at pressures of 10 to 30 Torr was performed in a quartz reactor at room temperature under u.v.-irradiation of a high pressure mercury lamp DRSh-1000 (1 kW). Irradiation time was varied from 5 to 270 min. The number of photoreduced Mo^{4+} ions was determined by measuring the amount of CO_2 formed during photoreduction in CO atmosphere.

Catalytic activities of prereduced Mo/SiO_2 samples for the propene metathesis were measured at room temperature at circulation of a gaseous mixture through the catalyst. The total reaction volume (including the circulating system) was 570 cm^3; the initial propene pressures were in the range of 30 to 90 Torr. Quantitative analyses of gaseous mixtures were made by UTI-100C quadrupole mass spectrometer. Mixtures of olefins of known compositions were used for calibration. U.v. and visible diffuse reflectance spectra were recorded using a Hitachi Model M-340 spectrometer supplied with an integrating sphere attachment.

3. RESULTS AND DISCUSSION

3.1. Comparison of Activities of Silica-Molybdena Catalysts Activated in Different Ways

Unreduced Mo^{6+}/SiO_2 samples exhibited no appreciable catalytic activity for the propene metathesis at room temperature (Figure 1, line a). However, preliminary reduction of these samples in different ways resulted in appearance of catalytic activity. Figure 1 shows typical dependences of propene concentration in reaction mixtures as a function of contact time for the Mo/SiO_2 preactivated by thermal reduction at 773 K in H_2 (curve b),

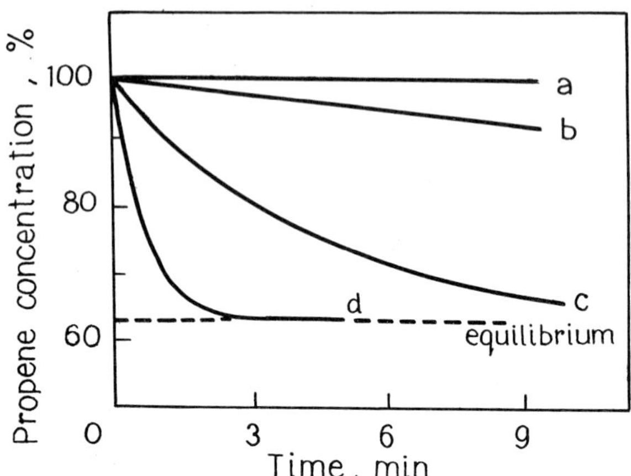

Figure 1. Propene concentration in reaction mixture vs time: a - unreduced sample; b - sample reduced by H_2 at 773 K; c - sample reduced by CO at 773 K; d - sample photoreduced at 300 K and CO atmosphere.

in CO (curve c) or by photoreduction in CO at 300 K. (curve d). It is apparent from Figure 1 that the highest catalytic activity was achieved in the latter case.

As a measure of specific catalytic activity of the reduced Mo/SiO_2 catalyst we used "turnover numbers" (T.O.N.) expressed in numbers of C_3H_6 molecules converted over one molybdenum ion per one second(Table 1). To determine T.O.N., linear transformations of the kinetic curves of Figure 1 should be used, which could be obtained by applying the Langmuir-Hinshelwood model to the kinetic data.

Now some comments on Table 1 should be made. It is seen that the activity of a photoreduced sample considerably exceeds that of Mo/SiO_2 catalysts activated by thermal reduction. Note that in this study we did not try to attain the highest possible T.O.N. By photoreduction it was possible to obtain even a more active catalyst than those containing "anchored" organo-molyb-

TABLE 1. Comparison of activities of silica-molybdena catalysts for the propene metathesis reaction

Activation pretreatment	Reaction temperature, K	Initial C_3H_6 pressure, Torr	T.O.N., s^{-1}	Refs.
Photoreduction of Mo^{6+}/SiO_2 in CO at 300 K	300	85	0.42	this work
Thermal reduction of Mo^{6+}/SiO_2 in CO at 773 K	300	64	0.070	this work
Thermal reduction of Mo^{6+}/SiO_2 in H_2 at 773 K	300	48	0.003	this work
The same as above	363	120	0.007	(Ref. 1)
$Mo(\eta_3\text{-}C_3H_5)_2/SiO_2$, no activation	363	120	0.17	(Ref. 1)
Oxidation of $Mo(\eta_3\text{-}C_3H_5)_2/SiO_2$ in O_2 at 673 K, reduction in H_2 at 723 K, and oxidation in O_2 at 300 K	363	120	0.13	(Ref. 1)
The same as above	273	21	0.075	(Ref. 5)

denum complexes. The data of Table 1 demonstrates that even the most active $Mo(\eta_3\text{-}C_3H_5)_2/SiO_2$ is considerably inferior in this respect to the photoreduced Mo/SiO_2. As was shown in (Refs. 1 & 5), Mo^{4+} ions bonded to the silica surface could be obtained by an oxidizing-reducing treatment (see Table 1). However, for these catalysts T.O.Ns. were also found to be smaller than those of photoreduced silica-molybdenum.

The data given in Table 1 and in Figure 1 were obtained at the first propene admission onto preliminary reduced Mo/SiO_2 samples. After pumping off the reaction mixture at room temperature and the repeated propene admission to the catalyst, propene conversion to ethylene and butene again took place. The catalysts kept their activities also at subsequent admissions of new propene portions without any additional activation. For the most active photoreduced catalysts a decrease in activity was observed which is likely to be associated with partial poisoning the active sites. After the heat treatment of Mo/SiO_2 in O_2 and in vacuum at 773 K and subsequent photo- or thermal reduction activities of the catalysts were fully regenerated.

3.2. Nature of Active Sites for the Propene Metathesis Reaction over Photoreduced Catalysts

Figure 2 shows T.O.N. related to one Mo^{4+} ion (T.O.N.$_{Mo^{4+}}$) as a function of the number of Mo^{4+} ions ($N_{Mo^{4+}}$) in the photoreduced catalyst. $N_{Mo^{4+}}$ could be determined by measuring the amounts of CO_2 evolved during the photoreduction process in CO. It is seen that T.O.N.$_{Mo^{4+}}$ remained practically constant at almost tenfold variation of $N_{Mo^{4+}}$. This constancy of T.O.N.$_{Mo^{4+}}$ can be regarded as a convincing evidence for the participation of Mo^{4+} ions in the propene metathesis. These ions are formed in the photoassisted two electron reduction process according to the scheme proposed in our earlier works (Refs. 3,4):

$$\mathrm{(O{=})_2Mo^{6+}(O\!-)_2} \xrightarrow{h\nu} \left[\mathrm{O{=}Mo^{5+}(\bar{O})(O\!-)_2}\right]^* \xrightarrow{+CO} \mathrm{O{=}Mo^{4+}(\square)(O\!-)_2} + CO_2.$$

This process occurs with the selectivity close to 100%. An analysis of optical spectra of the photoreduced

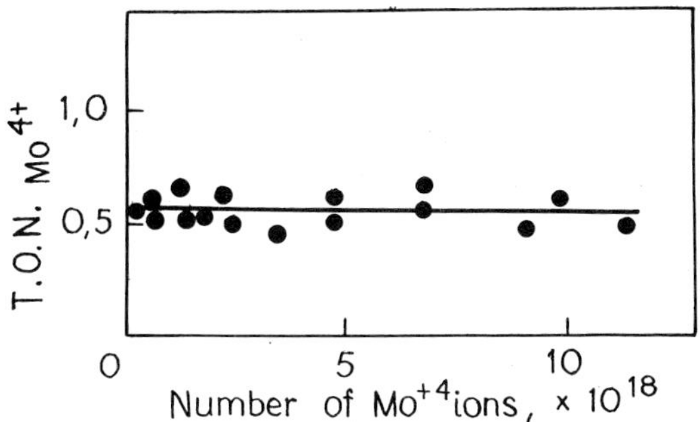

Figure 2. Turn-over numbers related to one Mo^{4+} ion vs the number of Mo^{4+} ions in the photoreduced catalyst

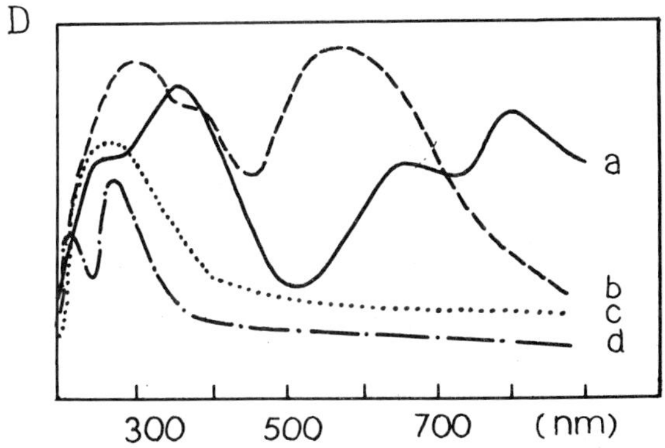

Figure 3. Optical spectra of molybdenum catalysts: a - sample photoreduced at 300 K in CO; b - the same (a) sample after adsorption of propene; c - sample thermally reduced at 773 K by CO; d - the same (c) sample after adsorption of propene.

samples in the region of d-d transition showed that Mo^{4+} ions were in a distorted tetrahedral or trigonal coordination (Figure 3).

The samples containing Mo^{5+} ions could be prepared upon exposing Mo^{6+}/SiO_2 to radiation with a wavelength longer than 360 nm in H_2 atmosphere at room temperature. Such samples showed no activity for the propene metathesis. However, upon further irradiation with the unfiltered light in addition to Mo^{5+} ions, Mo^{4+} ions, which are formed from Mo^{5+} ions by a consequent photoreduction, were accumulated in the samples. These samples exhibited high catalytic activity.

The conclusion about the correlation between the catalytic activity and the presence of Mo^{4+} ions is further confirmed by selective poisoning of the activity by oxidizing gases. At room temperature N_2O oxidizes Mo^{4+} ions quantitatively and selectively to Mo^{6+} state, and NO forms paramagnetic complexes $Mo^{6+} \ldots NO^{2-}$. At $N_{poison} > N_{Mo^{4+}}$ the catalyst was completely poisoned, while at $N_{poison} < N_{Mo^{4+}}$ T.O.N.$_{Mo^{4+}}$ remained unchanged although the total activity drastically decreased.

3.3. Mechanism of Formation of Molybden-Carbene Intermediates

Ethylene and propene are strongly chemisorbed on the photoreduced Mo^{4+}/SiO_2 samples at room temperature, apparently, forming π-complexes with Mo^{4+} ions. At the same time drastic changes take place in optical spectra (Figure 3, curve b). These complexes are rather stable since olefin desorption occurs only at 573-623 K, Mo^{4+} ions in the original coordinative state being regenerated. In the case of the thermally reduced samples no profound changes are observed in the spectra following the olefin adsorption (Figure 3, curves c and d). Obviously, that could be due to the lower concentration of coordinatively unsaturated Mo^{4+} ions in such catalysts or to their coordinative inhomogeneity.

A certain proportion of chemisorbed olefin molecules undergoes further conversion in Mo^{4+} coordination sphere even at room temperature. This is evidenced by the change of a gas phase composition observed during the initial stages of Mo^{4+} interaction with olefins.

When C_2H_4 comes in contact with photoreduced Mo^{4+}/SiO_2 samples at $N_{C_2H_4} : N_{Mo^{4+}} \approx 1$ the C_3H_6 concentration in the

TABLE 2. The interaction of photoreduced Mo^{4+}/SiO_2 samples with propene

Exp. number	Conditions of gas phase analysis	$N_{C_3H_6}/N_{Mo^{4+}}$	Gas phase composition, %			
			C_2H_4	C_3H_6	C_4H_8	C_5H_{10}
1	Propene used in the study		0.3	99.4	0.3	$<5\cdot10^{-3}$
2	Equilibrium mixture of C_3H_6 metathesis products at 293 K		18	63	18	$<5\cdot10^{-3}$
3	C_3H_6 contacted with photoreduced Mo^{4+}/SiO_2 at 293 K	0.6	13	52	33	2.0
4	C_3H_6 contacted with photoreduced Mo^{4+}/SiO_2 at 293 K	0.73	11	67	21	1.2
	The same as above	0.94	15	67	18	0.4
	The same as above	1.27	18	67	15	0.2
	The same as above	3.3	16	67	17	$5\cdot10^{-3}$

gas phase noticeably increases as compared to the blank experiment with unreduced Mo^{6+}/SiO_2. The C_3H_6 percentage in the gas phase increases even more when chemisorbed C_2H_4 is desorbed in the H_2 flow or in vacuum at 673 K. Thus C_2H_4 interaction with photoreduced samples results in the production of propene.

The gas phase composition changes considerably when propene comes in contact with the photoreduced Mo^{4+}/SiO_2 (Table 2). It is quite natural, since such samples are highly active catalysts for the C_3H_6 metathesis. However, it should be noted that for $N_{C_3H_6}: N_{Mo^{4+}} < 1$ ratios less than unity (Experiment 3) a considerable deviation from the equilibrium gas phase composition is observed: under these conditions the degree of propene conversion and the butene fraction exceed the corresponding equilibrium values. Moreover, in contrast to the equilibrium mixture, the C_4H_8 C_2H_4 ratio is much higher than unity. Thus at the initial stages of C_3H_6 interaction with Mo^{4+} ions butenes are predominantly produced. In addition, in this case pentenes evolved in the gas phase are also observed, the amount of which reaches a maximum at $N_{C_3H_6} : N_{Mo^{4+}} < 1$. And then with the growth of this ratio pentene concentration tends to decrease (Experiment 4).

The data obtained on the product composition of the early stages of the olefin interaction with Mo^{4+} ions can be regarded as a good evidence of the formation of surface carbenes. Most likely these species are produced by an intramolecular 1,2-H atom transfer in π-complexes, which is known from homogeneous metal-complex catalysis:

$$Mo^{4+} + H_2C{=}CHR \rightleftarrows \underset{Mo}{\overset{H_2C{=}CHR}{\downarrow}} \rightleftarrows \left[\begin{array}{c} H{-}C{-}\cdot{-}C(H)(R) \\ |\quad \dot{H} \\ Mo \end{array}\right] \rightleftarrows Mo{=}CH{-}CH_2R, \qquad (1)$$

where R=H or CH_3.

This scheme of the formation of active intermediates is confirmed by the kinetic isotopic effect in the propene metathesis on photoreduced catalysts: when C_3H_6 is replaced by C_3D_6 the reaction rate lowered by a factor of two. This effect is indicative of the occurrence of reaction steps involving cleavage and/or formation of C-H bonds.

From scheme (1) it follows that propene should result from Mo^{4+} interaction with C_2H_4:

$$Mo{=}CH{-}CH_3 + C_2H_4 \rightarrow \left[\begin{array}{ccc} H_2C & - & CH_2 \\ | & & | \\ Mo & - & CH{-}CH_3 \end{array}\right] \rightarrow Mo{=}CH_2 + C_3H_6 . \qquad (2)$$

Carbene Mo=CH-CH_2-CH_3 formed by the C_3H_6 interaction with Mo^{4+} reacts with the second propene molecule in two ways, leading to the production of butene and pentene:

$$Mo{=}CH{-}CH_2{-}CH_3 + C_3H_6 \nearrow \left[\begin{array}{l} H_2C{-}CH{-}CH_3 \\ \quad | \qquad | \\ \quad Mo{-}CH{-}CH_2{-}CH_3 \end{array}\right] \rightarrow \begin{array}{c} CH_2 \\ \| \\ Mo \end{array} + C_5H_{10}$$

$$Mo{=}CH{-}CH_2{-}CH_3 + C_3H_6 \searrow \left[\begin{array}{l} CH_3{-}CH{-}CH_2 \\ \qquad\quad | \qquad | \\ \qquad\quad Mo{-}CH{-}CH_2{-}CH_3 \end{array}\right] \rightarrow \begin{array}{c} CH{-}CH_3 \\ \| \\ Mo \end{array} + C_4H_8.$$

The formation of ethylene is possible only at the approach of the third C_3H_6 molecule to the active site:

$$Mo{=}CH_2 + C_3H_6 \rightleftharpoons Mo{=}CH{-}CH_3 + C_2H_4$$

though butene can also be formed:

$$Mo{=}CH{-}CH_3 + C_3H_6 \rightleftharpoons Mo{=}CH_2 + C_4H_8.$$

Thus the experimental data are in good agreement with scheme (1) for the carbene formation.

An interesting feature of photoreduced Mo^{4+}/SiO_2 samples that can be explained in terms of a carbene mechanism is their abiliyy to convert ethylene to propene at higher temperatures: $3\ C_2H_4 \rightarrow 2\ C_3H_6$.

At circulation of ethylene (30-50 Torr) over the photoreduced Mo^{4+}/SiO_2 sample at temperatures above 520 K propene concentration in the gas phase is continuously growing (Figure 4 a). The rate of propene formation is increased with raising the temperature. The kinetics of C_3H_6 production from C_2H_4 at 623 K is shown in Figure 4 b. Under similar conditions no propene is produced over the unreduced Mo^{6+}/SiO_2 sample (Figure 4 c). It is seen from Figure 4, that the number of C_3H_6 molecules formed considerably exceeds that of Mo^{4+} ions in the

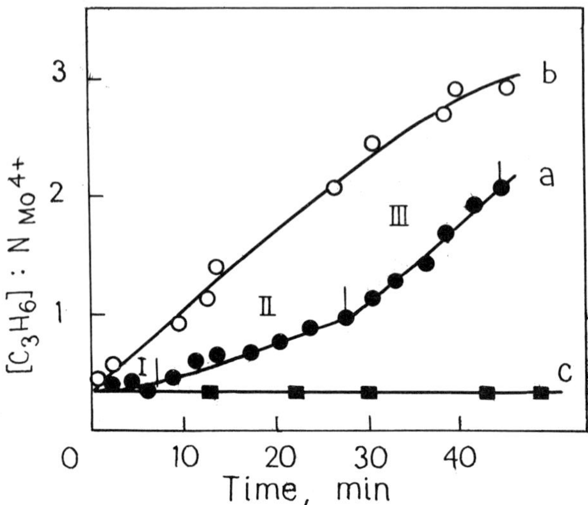

Figure 4. The change of amount of propene related to the number of Mo^{+4} ions vs time at ethylene circulation over molybdenum catalysts: a - photoreduced sample; reaction temperature: I - 470 K, II - 520 K, III - 570 K; b - photoreduced sample, reaction temperature - 623 K; c - non-reduced sample, reaction temperature - 623 K.

sample ($N_{Mo}4+$), in other words, the process is catalytic. It should be noted that the formation of propene from ethylene is not accompanied by the change in the valence state of Mo^{4+} ions since when the reaction mixture had been pumped off at 623 K the initial optical spectrum of Mo^{4+} ions was restored and the sample retained high catalytic activity in the propene metathesis.

The formation of C_3H_6 in amounts exceeding the number of Mo^{4+} ions in the catalyst seems to be associated with the following reaction, which occurs along with reactions (1) and (2) at high temperatures:

$$\begin{matrix} CH_2 \\ \| \\ Mo \end{matrix} + C_2H_4 \longrightarrow \left[\begin{matrix} CH_2 - CH_2 \\ | \qquad | \\ Mo - CH_2 \end{matrix}\right] Mo^{4+} + \begin{matrix} CH_2 - CH_2 \\ \diagdown \; \diagup \\ CH_2 \end{matrix} \longrightarrow Mo^{4+} + C_3H_6 .$$

In blank experiments it was found that at temperatures above 570 K cyclopropane quickly isomerizes into propene on Mo^{4+}/SiO_2. Evidently, for this reason a mass-spectroscopic analysis does not indicate the presence of cyclopropane in the gas phase when C_2H_4 comes in contact with Mo^{4+}/SiO_2.

REFERENCES

1. Yu.I. Yermakov, B.N. Kuznetsov and V.A. Zakharov, in Catalysis by Supported Complexes, Elsevier, Amsterdam (1981).
2. K. Tanaka and K. Miyahara, Intern. Congr. Catal., 7-th (1980), Prepr. B 47, Tokyo.
3. B.N. Shelimov, A.N. Pershin and V.B. Kazansky, J. Cata , 64, 426 (1980).
4. A.N. Pershin, B.N. Shelimov and V.B. Kazansky, Kinet. Katal., 20, 1298 (1979); 21, 494 (1980).
5. Y. Iwasawa, H. Ichinose and S. Ogasawara, J. Chem. Soc., Faraday Trans., 1, 77, 1763 (1981).

THE ROLE OF ORGANOMETALLIC INTERMEDIATES IN THE CATALYSIS OF HYDROCARBON TRANSFORMATIONS

B.A. Dolgoplosk and I.A. Oreshkin
A.V. Topchiev Institute of Petrochemical Sunthesis of the USSR Academy of Sciences, Moscow, 117912, USSR

ABSTRACT - The catalytic reactions that occur via the steps of formation of organometallic compounds (polymerization, oligomerization, metathesis and carbonylation) are considered. As found, the low temperature of oxidative addition of various hydrocarbons to transition metal atoms and ions in the gas phase occurs through the formation of organometallic intermediates. The routes of the decomposition of various transition metal organometallic compounds have been studied. The mechanism of elementary acts of heterogeneous catalytic conversions of hydrocarbons (hydrogenation, dehydrogenation, hydrogen-deuterium exchange, hydrogenolysis and skeletal isomerization) is discussed.

1. INTRODUCTION

Differences in the initial physical states of the catalysts (complexes or organometallic compounds in homogeneous catalysis, metals or their oxides and salts in heterogeneous catalysis) account for different approaches to the study of the chemical nature of the elementary acts in catalytic transformations of hydrocarbons. The ideas used in heterogeneous catalysis usually imply the dissociative adsorption of the reactants on the surface of a catalyst which produce radical-like species or atoms attached by some way to the transition metal atoms. These species are capable of migrating from one active center to another. As a rule the multicentered nature of the active sites is suggested. Sometimes the formal routes where the active centers of an idefinite chemical nature Z_1, Z_2, etc. and intermediates CH_x, CH_y are used.

In our report, the chemical nature of hydrocarbon transformations in heterogeneous catalysis will be discussed in terms of organometallic intermediates that arise on a single metal atom. This fact determines the identity of approaches to the mechanisms of both heterogeneous and homogeneous catalysis.

Recent advances in our knowledge of the mechanisms of catalytic transformations of hydrocarbons make it clear that catalytic reactions of the same type occur via the same intermediates (Refs. 1,2).

Let us discuss first the mechanisms of well-studied reactions that proceed through intermediates that have been established to be organo-transition metal compounds.

1.1. <u>The Stereospecific Polymerization of Dienes and Olefins</u> by Ziegler systems containing transition metal salts and organoaluminium compounds and polymerization induced by transition metal oxides (higher oxides of chromium, etc.) were earlier thought to take place on the surface of insoluble crystals ($TiCl_3$, Cr_nO_m, etc.) with their direct participation. Further developments in this area have shown that homogeneous systems using organo-transition metal compounds (Ti^{3+}, Ti^{4+}, Ni^{2+}, Cr^{2+}, Cr^{3+}, W^{4+}, Mo^{4+}, Nd^{3+}, etc.) are considerably more efficient in polymerization than heterogeneous systems based on the corresponding metal, the stereospecificity being the same. These studies reveal the fact that high stereospecificity is not due to the heterogeneity of Ziegler systems. The decisive factor is the structure of the terminal unit of the growing chain, which can be different (1,4-cis, 1,4-trans, 1,2-, 3,4-) depending on the nature of the metal and ligands and on the life time of the terminal unit (Ref. 3). The chain growth proceeds through the C-M bond. The mechanism of stereoregulation was discussed in detail in a number of articles (Refs.4, 5) and in the monograph (Ref. 3).

1.2. <u>The Metathesis Reaction</u> first discovered for olefins in heterogeneous systems (supported oxides of molybdenum, tungsten and rhenium) (Ref. 6) has been extensively studied for homogeneous systems. Olefin metathesis and ring-opening polymerization of cycloolefins were found to occur by a chain mechanism involving carbene complexes of transition metals. The role of the carbene species and the chain nature of the process was first established in (Refs. 7,8). This reaction may be represented by the following scheme:

$$RCH{:}MX_n + \begin{matrix} R'\diagdown \\ CH \\ \| \\ CH \\ R'\diagup \end{matrix} \longrightarrow RCH{=}CHR' + {'RCH{:}MX_n} \text{ etc.} \qquad (1)$$

$$RCH{:}MX_n + \bigcirc \longrightarrow \left[\begin{matrix} RCH{:}MX_n \\ \vdots \\ \bigcirc \end{matrix}\right] \longrightarrow RCH{=}CH(CH_2)_3CH{:}MX_n \text{ etc.} \qquad (2)$$

The initiation and chain propagation involve peculiar organometallic intermediates, complexes of W, Mo, Re and Ta being the most effective.

1.3. <u>Dimerization (Oligomerization)</u> of ethylene, styrene, and other olefins occurs via the same organometallic intermediates in homogeneous and heterogeneous systems

$$HNiX + CH_2{=}CH_2 \rightleftharpoons CH_3CH_2NiX \xrightarrow{C_2H_4} CH_3(CH_2)_2CH_2NiX \longrightarrow$$

$$\longrightarrow CH_3CH_2CH{=}CH_2 + HNiX \qquad (3)$$

For example, π-allylnickel halides induce dimerization (oligomerization) of styrene and ethylene in solution (Refs. 9,10). Similar reactions are caused by nickel subhalides incorporated into $NiCl_2$ crystals or those which can be formed on the support surface (Refs. 11,12).

1.4. Chain Transfer Reactions Involving Hydrogen.

It is well known that molecular hydrogen transfers a chain in the stereospecific polymerization of dienes and olefins with Ziegler systems at low temperatures that proceeds via organometallic compounds of Ti,Cr,Ni,Co, and other metals. Hence, it follows that the insertion of H_2 into the C-M bond is preceded by the coordination of H_2 to the metal, as it is assumed for the polymerization of olefins and dienes:

$$\geqslant C{-}MX_n + H_2 \longrightarrow \left[\begin{array}{c} H_2 \\ \vdots \\ \geqslant C{-}MX_n \end{array}\right] \longrightarrow \geqslant CH + HMX_n \xrightarrow[\text{olefin}]{\text{diene}} PMX_n \longrightarrow \text{polymerization} \quad (4)$$

In fact, the insertion of diene (olefin) into the C-H bond and hydrogen into the C-M bond is the essence of hydrogenation of olefins by transition metals in homogeneous systems. In this case, the coordinated molecule of hydrogen plays the role of a reactant.

Homogeneous hydrogenation through the successive formation of hydride and alkyl derivatives of Pt, Rh, and Mo has been established in many cases (Ref. 13). The reaction involves the addition of a metal hydride to the double bond of coordinated alkene (Refs. 14-17). The reversibility of this reaction completely agrees with the β-elimination of alkyl derivatives of transition metals in hydrocarbon media. Similar reactions of dienes with stoichiometric hydrides of Ni, Cr and Co produce π-crotyl complexes (Ref. 18)

1.5. Oxo and Carbonylation Reactions.

Oxo reactions with hydrocarbonyls of Co and Rh are well known to proceed through the addition of an olefin to the C-M bond, the carbonylation of the C-M bond and the hydrogenation of the $-\overset{\underset{\|}{O}}{C}-M$ bond to regenerate the original active center (Refs. 19,20).

$$HCo(CO)_3 \xrightarrow{C_2H_4} CH_3CH_2Co(CO)_3 \xrightarrow{CO} CH_3CH_2\overset{O}{\overset{\|}{C}}Co(CO)_3 \xrightarrow{H_2}$$

$$\rightarrow CH_3CH_2\overset{O}{\overset{\|}{C}}H + HCo(CO)_3. \qquad (5)$$

Another example to illustrate the role of organometallic intermediates in catalysis is the carbonylation of methanol to acetic acid catalyzed by compounds of Rh and other metals. The main stage of this reaction is the oxidative addition of CH_3I to $RhI_2(CO)_2$ to form a CH_3-Rh bond, followed by the insertion of the CO group into the C-M bond (Refs. 21,22).

2. OXIDATIVE ADDITION OF HYDROCARBONS TO THE ATOMS AND IONS OF TRANSITION METALS IN THE GAS PHASE

Recently the role of organometallic intermediates in the catalysis of hydrocarbon transformations has been clearly shown by the reactions of a variety of hydrocarbons with atoms and univalent ions of transition metals at room temperature.

Skell et al. (Ref. 23) first established that the interaction of zirconium atoms evaporated in a high vacuum with vapored isobutane and neopentane started in the matrix at $-196^{o}C$. Matrix warm-up and subsequent hydrolysis of the reaction products produce significant amounts of hydrogen and a variety of hydrocarbons resulted from the decomposition of the corresponding organometallic intermediates. In the case of isobutane, isobutene and methane are the main reaction products. The products mentioned are formed by the oxidative addition of Zr^0 via the C-H or C-C bonds of hydrocarbons followed by the spontaneous decomposition or hydrolysis of the organometallic compound

$$CH_3CH(CH_3)CH_3 + Zr^\circ \begin{cases} \rightarrow CH_3CH(CH_3)CH_2ZrH \rightarrow CH_3C(CH_3){=}CH_2 + H_2 \\ \rightarrow CH_3CH(CH_3)ZrCH_3 \xrightarrow{H_2O} C_3H_8 + CH_4 . \end{cases} \qquad (6)$$

For neopentane Skell suggests the following reaction scheme:

$$CH_3{-}C(CH_3)_2{-}CH_3 + Zr^\circ \begin{cases} \rightarrow HZrCH_2C(CH_3)_3 \\ \rightarrow CH_3ZrC(CH_3)_3 \end{cases} \qquad (7)$$

$$CH_3ZrC(CH_3)_3 \downarrow \quad H(CH_3)Zr{\cdots}CH_2{-}C(CH_3)_2 \rightarrow CH_4 + CH_2{=}C(CH_3)_2 .$$

The oxidative addition of hydrocarbons in the conditions mentioned was also established by Skell for the reactions of isobutane with atoms of V, Ti, Cr, Mn, Fe, Ni, Co, Mo and W (Ref. 23). Organometallic compounds are formed during cocondensation of the metal on the hydrocarbon surface at -196°C as well as subsequent warm-up of the matrix. In the case of nickel, organometallic compound is produced only during the matrix warm-up (Ref. 24). It is natural that the fraction of the metal addition across the C-H and C-C bonds depends on the nature of hydrocarbon and, in particular, on the nature of the metal.

In recent years, the gas-phase reactions of hydrocarbons with univalent ions of transition metals in a high vacuum at room temperature were extensively studied. The ions were obtained by laser desorption or by other methods. Freizer et al. (Ref. 25) first examined the product distribution for the primary interactions of Rh^+ with hydrocarbons using mass spectrometry (Table 1).

As shown in Table 1, C_2-C_8 paraffins as well as cyclopentane and cyclohexane produce mainly complexes of olefins, conjugated dienes and trienes with Rh^+ which are the result of the dehydrogenation of hydrocarbons. In the case of neopentane only, the reaction proceeds in two directions resulting in skeletal isomerization and dehydrogenation (11^a and 11^b) or elimination of methane (11^c and 11^d). In the latter case the insertion of Rh^+ into the C-C bond is admitted.

It is noteworthy that cyclopropane gives rise to carbene complexes $Rh{=}CH_2+$(11%) and ethylene along with Rh RhC_3H_4+(76%).

The authors suggest the mechanism that involves the primary oxidative addition of alkanes to Rh^+ to give organometallic intermediates that decompose to yield the products listed in Table 1. For example, the following intermediate stages are proposed for propane and neopentane:

$$CH_3CH_2CH_3 + Rh^+ \rightarrow \left[\begin{array}{c} CH_3\underset{|}{C}HCH_3 \\ Rh^+\text{-}H \end{array}\right] \rightarrow \left[\begin{array}{c} CH_3CH\text{-}CH_2 \\ \diagdown \ \diagup \\ H\diagup \ Rh^+ \diagdown H \end{array}\right] \rightarrow$$

$$\rightarrow \underset{\vdots \atop Rh^+}{CH_3CH{=}CH_2} + H_2 \qquad (8)$$

TABLE 1. Products resulting from the primary reactions of Rh^+ with hydrocarbons in the gas phase at room temperature

N	Alkane	Reaction product RhA^+	Yield of RhA^+, %	Neutral(s)
1	C_2H_6	$RhC_2H_4^+$	100	H_2
2	C_3H_8	$RhC_3H_6^+$	94	H_2
3	n-C_4H_{10}	$RhC_4H_6^+$	100	$2H_2$
4	n-C_5H_{12}	$RhC_5H_8^+$	88	$2H_2$
5	i-C_5H_{12}	$RhC_5H_8^+$	97	$2H_2$
6	n-C_6H_{14}	a) $RhC_6H_{10}^+$	49	$2H_2$
		b) $RhC_6H_8^+$	33	$3H_2$
7	n-C_7H_{16}	$RhC_7H_{10}^+$	94	$3H_2$
8	n-C_8H_{18}	$RhC_8H_{12}^+$	91	$3H_2$
9	c-C_5H_{10}	$RhC_5H_6^+$	100	$2H_2$
10	c-C_6H_{12}	a) $RhC_6H_8^+$	65	$2H_2$
		b) $RhC_6H_6^+$	35	$3H_2$
11	neo-C_5H_{12}	a) $RhC_5H_{10}^+$	15	H_2
		b) $RhC_5H_8^+$	29	$2H_2$
		c) $RhC_4H_8^+$	13	CH_4
		d) $RhC_4H_6^+$	34	H_2+CH_4
12	$(CH_3)_2CHCH(CH_3)_2$	$RhC_6H_{10}^+$	100	$2H_2$

$$\mathrm{CH_3{-}\overset{CH_3}{\underset{CH_3}{C}}{-}CH_3 + Rh^+ \rightarrow \left[H{-}Rh^+{-}CH_2{-}\overset{CH_3}{\underset{CH_3}{C}}{-}CH_3\right] \rightarrow \left[\begin{array}{c} CH_3\ \ CH_3 \\ H_2C \overset{\cdots}{=\!=} C{<} \\ \vdots \qquad CH_3 \\ Rh^+{-}H \end{array}\right] \rightarrow}$$

$$\mathrm{\rightarrow \left[\begin{array}{c} H_2C{=}C(CH_3)_2 \\ \vdots \\ Rh^+ \\ CH_3 \quad\ H \end{array}\right] \rightarrow \begin{array}{c} H_2C{=}C{<}^{CH_3}_{CH_3} \\ \vdots \\ Rh^+ \end{array} + CH_4} \qquad (9)$$

Though the suggested intermediates shown in brackets seem to be a matter of speculation, no doubt that organometallic intermediates are involved in these reactions.

Lately a number of studies appeared examining the gas-phase reactions of alkanes and alkenes at room temperature with the metal ions formed by electron impact on volatile metal compounds, transition metal carbonyls being used more frequently. Table 2 lists the relative rate constants of reactions of Cr^+, Fe^+, Co^+ and Ni^+ with n- and is-perdeuterobutanes reported by Freas and Ridge (Ref. 26).

TABLE 2. Relative rate constants of reactions of M^+ with butanes. (M^+ was formed by electron impact on metal carbonyls)

Reactant ion	n-C_4D_{10} product ions				i-C_4D_{10} product ions	
M^+	$MC_2D_4^+$	$MC_3D_6^+$	$MC_4D_6^+$	$MC_4D_8^+$	$MC_3D_6^+$	$MC_4D_8^+$
Cr^+	0.22	0	0.16	0.16	0	0.09
Fe^+	0.53	0.07	0.04	0.04	0.46	0.09
Co^+	0.53	0.09	0	0.13	0.82	0.15
Ni^+	0.14	0.02	0	0.07	0.29	0.04

The authors postulate the following reaction mechanism:

$$Fe^{+} + i\text{-}C_4H_{10} \rightarrow \left[CH_3\text{-}Fe^{+}\text{-}CH(CH_3)_2\right] \rightarrow \left[\begin{matrix} H \\ | \\ Fe^{+} \\ | \\ CH_3 \end{matrix} \cdots \begin{matrix} CH_2 \\ \| \\ CH \\ \diagdown CH_3 \end{matrix}\right] \rightarrow$$

$$\rightarrow \left[CH_4 \cdots Fe^{+} \cdots \begin{matrix} CH_2 \\ \| \\ CH \\ \diagdown CH_3 \end{matrix}\right] \rightarrow Fe^{+} \cdots \begin{matrix} CH_2 \\ \| \\ CH \\ \diagdown CH_3 \end{matrix} + CH_4. \quad (10)$$

In the case of n-butane, M inserts mainly into the middle C-C bond

$$M^{+} + n\text{-}C_4H_{10} \rightarrow \left[C_2H_5M^{+}C_2H_5\right] \rightarrow \left[C_2H_6 \cdots M^{+} \cdots \begin{matrix} CH_2 \\ \| \\ CH_2 \end{matrix}\right] \rightarrow$$

$$\rightarrow C_2H_6 + M^{+} \cdots \begin{matrix} CH_2 \\ \| \\ CH_2 \end{matrix}. \quad (11)$$

Ti^{+} inserts only into a C-H bond (Ref. 27). Naturally, the reaction pathway strongly depends on the kinetic energy of ions. In all cases, however, organometallic compounds are formed which decompose to yield the final reaction products.

Another way to produce organometallic compounds is the oxidative addition to aldehydes and ketones to reduced metals. It gives rise to carbene complexes which subsequently yield recombination products. Thus, the interaction of bis(tolyl)titanium - arene complex of titanium with acetophenone and acetone at 75-80°C forms quantitatively tetraphenylethylene and tetramethylethylene (Ref. 28), and the reaction of benzaldehyde with the reduced tungsten yields up to 76% of stilbene.

The oxidative addition of hydrocarbons to the reduced metals was illustrated by the reactions of p-xylene and toluene with Cp_2W (Ref. 29)

$$Cp_2W + CH_3C_6H_5 \longrightarrow \begin{cases} Cp_2W(H)(CH_2C_6H_5) \\ Cp_2W(H)(C_6H_4\text{-}CH_3) \end{cases} \qquad (12)$$

and by the reaction of addition of triphenylphosphine to Pd° (Ir°, Rh°, Pt°) at room temperature

$$Ph_3P + Pd^{\circ} \longrightarrow Ph_2P - Pd - Ph. \qquad (13)$$

The above data show that all the reactions typical for transformations of the corresponding alkanes at high temperatures that are interpreted in heterogeneous catalysis by multicentered mechanism take place on a single metal atom in the gas phase.

There is no question that the "unsaturated" metal atom at the interface which retains its maximum ability for coordination, making it analogous to a zero-valent metal, is the most active in heterogeneous systems.

To understand the chemical nature of elementary acts in catalytic transformations of hydrocarbons, it is necessary to discuss the pathways of decomposition of organo-transition metal compounds prepared *in situ*.

3. PATHWAYS OF DECOMPOSITION OF ORGANO-TRANSITION METAL COMPOUNDS PREPARED *IN SITU* AT ROOM TEMPERATURE

Kharash and many others who studied the products of decomposition of organo-transition metal compounds prepared *i n s i t u*, come to a conclusion about the free radical stages of the reactions (Refs. 30-35). Metal alkyls form an equimolecular amounts of R_{-H} and R_{+H}, and in the case of benzyl and other aryl derivatives, recombination products ($ArCH_2CH_2Ar$), for example:

$$\begin{aligned} &C_2H_5Cl + Co^{\circ} \longrightarrow [C_2H_5CoCl] \longrightarrow {}^{\cdot}CoCl + C_2H_5^{\cdot} \\ &2\ C_2H_5^{\cdot} \longrightarrow C_2H_4 + C_2H_6 \\ &(C_6H_5)_2M \longrightarrow C_6H_5 - C_6H_5 + M. \end{aligned} \tag{14}$$

Gilman et al. (Ref. 36) found that the reaction of CH_3Li with a variety of transition metal halides results in mainly methane or ethane or their mixture, depending on the nature of the metal:

$$CH_3Li + MX_n \begin{cases} \xrightarrow{Ti,\,Zr,\,Mn,\,Cr,\,La} \text{mainly } CH_4 \\ \xrightarrow{Ag,\,Cu} \text{mainly } CH_3CH_3 \\ \xrightarrow{Au,\,Fe,\,Ni} \text{mixture of } CH_4 \text{ and } CH_3CH_3 \end{cases} \tag{15}$$

According to Gilman, the products mentioned are formed by the reactions of a free methyl radical resulting from the homolytic decomposition of methyl derivatives of transition metals. Many authors throw doubt upon the radical nature of the intermediates. In particular, the formation of equimolar amounts of C_2H_4 and C_2H_6 cannot be explained in terms of radicals taking into account the high activity of ethyl radicals in the H-atom abstraction from the solvent. The same is true for the methyl radical which produces only methane but not ethane in solution. Lappert noted (Ref. 37) that the energy of a carbon-transition metal bond is usually in the range from 25 to 60 kcal/mol, thus, a homolytic decomposition at room temperature is not possible. The ability of the compounds mentioned needs a special explanation.

The absence of radical stages in the reaction of ethyllithium with $CoCl_2$ and $TiCl_4$ was shown when the reaction was carried out in the presence of α-methylstyrene or other acceptors of free radicals (Refs. 38,39). Ethane and ethylene were found in equimolar yields. Similar data were obtained for the addition of C_2H_5Br to Co° and Ni° included in the ethylmagnesium bromide matrix in the presence of α-methylstyrene (Ref. 40).

According to the current theory the decomposition of alkyl derivatives of transition metals proceeds mainly via β-elimination:

$$\begin{aligned} &a)\ RCH_2CH_2MX_n \longrightarrow RCH{=}CH_2 + HMX_n \\ &b)\ (RCH_2CH_2)_2MX_m \longrightarrow RCH{=}CH_2 + RCH_2CH_3 + MX_m. \end{aligned} \qquad (16)$$

The reaction 16[b] is believed to occur through the hydride intermediate (β-elimination):

$$(RCH_2CH_2)_2MX_m \rightarrow RCH{=}CH_2 + RCH_2CH_2\overset{\displaystyle H}{\overset{|}{M}}X_m \rightarrow RCH_2CH_3 + MX_m \qquad (17)$$

The hydrogen atom at the β-carbon always takes part in these reactions.

β-Elimination was observed even in the decomposition of phenyl derivatives of some metals that gives rather unstable dehydrobenzene (Ref. 41)

$$C_6H_5{-}MX_n \longrightarrow C_6H_4 + HMX_n. \qquad (18)$$

When there is a quaternary carbon atom, atoms of silicon or germanium in a β-position to the carbon, or in the case of methyl or benzyl derivatives incapable of β-elimination, α-elimination occurs giving carbene species and the products of recombination of the original groups bound to the metal.

In our laboratory we studied the decomposition products for the reactions of CD_3Li with various transition metal chlorides in dibutyl ether. Insoluble salts were used in 5-fold excess with respect to lithium compound. In all cases the reaction was completed in 2 hours. Table 3 shows gaseous products of these reactions.

It can be seen that a significant amount of CD_3H is produced only in some cases. As examplified by the reactions of $CrCl_3$, $ReCl_5$, and $NiCl_2$ the yield of CD_3H remains almost unchanged in the presence of α-methylstyrene. It is probably the result of metallation of dibutyl ether by an organo-transition metal compound. The data presented show that the formation of CH_4 and CH_3CH_3 found by Gilman is not due to free radical stages. Since CD_4 may be produced in the interaction of two CD_3groups, the total balance of products summarizes the amount of CD_3-CD_3, CD_3H and doubled CD_4

$$[(CD_3)_2MX_n \longrightarrow CD_4 + CD_2{:}MX_n \text{ or } 2CD_3MX_m \longrightarrow CD_4 + CD_2{:}MX_m + MX_m].$$

In the case of molybdenum, rhenium, tungsten and chromium compounds, the total amount of the products significantly exceeds 100%. The reaction of CH_3Li with $MoCl_5$ was found to generate some amount of carbine species and carbides (Ref. 42)

$$(CD_3)_3MX_m \longrightarrow 2CD_4 + CD{:}MX_m$$

$$(CD_3)_4MX_n \longrightarrow 3CD_4 + CMX_n \quad (19)$$

TABLE 3. Products of the reaction of CD_3Li with metal chlorides in dibutyl ether at 20°C for 2-5 hours

MCl_n	CD_3Li / MCl_n (mol %)	Reaction products, % based on R groups				Symbolic balance of CD_3 groups
		CD_4	CD_3H	C_2D_6	C_2D_4	
$CoCl_2$	0.2	16.4	2.0	66.0	2.3	100.8
$NiCl_2$	0.2	15.1	2.9	60.0	2.1	93.1
$NiCl_2$*	0.2	10.9	2.0	68.0	1.2	91.8
$CrCl_3$	0.2	41.7	22.8	6.8	0.72	113.0
$CrCl_3$*	0.2	30.1	28.9	9.0	traces	98.1
$MoCl_5$	1.0	60.0	0	4.2	3.8	124.2
WCl_6	1.0	11.0	0	66.0	0.8	88.0
WCl_6	2.0	20.0	0	46.0	1.5	86.0
WCl_6	4.0	40.0	0	35.0	0.8	115.0
$ReCl_5$	0.33	56.1	9.9	0.7	1.0	122.8
$ReCl_5$*	0.33	45.6	14.1	1.4	2.8	106.7
$RuCl_3$	0.2	24.9	2.8	40.8	1.4	93.4
$RuCl_3$	0.2	13.4	3.5	66,3	1.5	96.6
$PdCl_2$	0.2	6.44	5.36	84.0	traces	102.2
$OsCl_4$	1.0	34.8	11.6	15.9	4.0	97.1
$PtCl_4$	0.2	36.7	9.8	10.0	traces	93.2

* in the presence of 50 mol α-methylstyrene per 1 mol of CD_3Li

We have studied in detail the decomposition pathways of transition metal compounds of the general formula R_mMCl_n prepared *in situ* at room temperature, where $R=CD_3$, $C_6H_5CH_2$, $(CH_3)_3SiCH_2$, $(CH_3)_3GeCH_2$, $C_6H_5-C(CH_3)_2-CH_2$, and M=Mo, W, Re, Ta, Nb, Cr, Pt, Rh, Co, Ru,

Pd, Os. These metals are widely used in various fields of heterogeneous catalysis. As a rule, various catalytic transformations of hydrocarbons are carried out at high temperatures when the homolytic decomposition of organometallic intermediates with the formation of free radicals is possible. But the high selectivity of the catalysts means that the role of these stages in the chemical transformations of hydrocarbons is not important. Experimental data on the products formed at room temperature may be regarded as a basis to consider the chemical nature of the reactions at higher temperatures though the fraction of the particular decomposition pathway may change greatly.

Experimental data on the products of the decomposition of various organo-transition metal compounds are summarized in Table 4. Minimum and maximum yields of products at different RLi(RMgX):transition metal halides ratios are listed.

Only three possible pathways of the reaction are taken into account because all R groups used are devoid of β-hydrogens:

$$(RCH_2)_mMX_n \begin{cases} \xrightarrow{a} RCH{:}MX_n + RCH_3 \\ \xrightarrow{b} RC{\vdots}MX_n + 2RCH_3 \\ \xrightarrow{c} R\text{-}R \end{cases} \qquad (20)$$

For most metal chlorides (except for halides of palladium, nickel and cobalt), α-elimination, which gives carbene and carbine species, is the main mode of decomposition. In the case of molybdenum and yttrium, some organometallics stable in hydrocarbon media are obtained in almost quantitative yields. These are the compounds of the carbene (carbine) type. Hydrolysis with deuterated water of the products of the $(C_6H_5C{=}Nd\text{-})_n$type produces

TABLE 4. Products of the reactions of organolithium and organomagnesium compounds with transition metal halides at 20°C

System $RLi(RMgX)+MCl_n$	Solvent	Total yield of products, % based on R groups introduced: eq.20[a],20[b]	eq.20[c]	Reference
CD_3Li + $MoCl_5$	Bu_2O	99	1	[42]
WCl_6		22-38	35-66	[43]
$CrCl_3$, $ReCl_5$		83-91	1-9	[44]
$OsCl_4$, $PtCl_4$		70-73	10-16	[44]
$RuCl_3$		50	41	[44]
$NiCl_2$, $CoCl_2$, $RhCl_3$		27-32	60-68	[44]
$PdCl_2$		13	84	[44]
CD_3MgI + $NiCl_2$, $CoCl_2$, $RuCl_3$, $RhCl_3$, WCl_6	Bu_2O	80-91	1-14	
$(CH_3)_3SiCH_2Li$ + $MoCl_5$, WCl_6, $TaCl_5$, $RhCl_3$, $NbCl_5$	C_6H_6	70-100	0-25	[45,46]
$ReCl_5$, $PtCl_4$, $CoCl_2$, $PdCl_2$		39-45	55-61	[45,46]
$C_6H_5CH_2MgCl$ + $MoCl_5$, WCl_6, $ReCl_5$, $RhCl_3$, $PtCl_4$, $CrCl_3$	Et_2O, Bu_2O	68-86	14-32	[47]
$C_6H_5CH_2Li$ + $NdCl_3$, YCl_3	Et_2O	96	4	[48]
$C_6H_5-C(CH_3)_2-CH_2Li$ + YCl_3	Et_2O	100	0	[48]

quantitative amounts of $C_6H_5CD_3$. In the reaction of benzylmagnesium chloride with $PtCl_4$ stilbene (9-10%) - the product of recombination - is formed. When the reaction is carried out in deuterated solvents (trideuterotoluene, perdeuterobenzene, perdeuterodiethyl ether), no deuterium atoms are found in the reaction products. This is reasonably good evidence in support of the nonradical pathway of decomposition. α-Elimination is also available, to a limited extent, for alkyl derivatives of transition metals bearing β-hydrogens.

4. ON THE CHEMICAL NATURE OF ELEMENTARY ACTS IN CATALYTIC TRANSFORMATIONS OF HYDROCARBONS

The above data on the reactions of the oxidative addition of hydrocarbons to active metals in the reduced state and on the decomposition pathways of organometallic compounds make it possible to examine the chemical nature of elementary acts in some catalytic transformations of hydrocarbons.

4.1. Reactions of Hydrogenation and Dehydrogenation of Hydrocarbons

It has been mentioned that unsaturated compounds such as ethylene and styrene quantitatively react at room temperature with the H-Ni bond to produce organometallic intermediates. Similar reactions have been found for lanthanide hydrides as well. Hydride forms of the metal rather than dissociatively adsorbed hydrogen atoms take part in these reactions. In the chemical reaction of molecular hydrogen with an active metal at the surface, the most probable result is a transition from Ni°, Co°, Pt°, Pd° or some other metal to the appropriate dihydrides with the metal in a more stable divalent form. An olefin is added to the hydrides that have been formed

$$RCH{=}CH_2 + H\text{-}M\text{-} \longrightarrow RCH_2CH_2M\text{-}. \qquad (21)$$

Hydrogenation as well as polymerization is preceded by the preliminary formation of $M...H_2$ and $H\text{-}M...\overset{CH_2}{\underset{CHR}{\|}}$ complexes. The next step is the insertion of H_2, coordinated to the organometallic compound, into the C-M bond:

$$RCH_2CH_2\overset{H_2}{\overset{\vdots}{M}}\text{-}H \longrightarrow RCH_2CH_3 + H\text{-}M\text{-}H \qquad (22)$$

The reversibility of this reaction is due to the possible β-elimination of an organometallic compound

$$RCH_2CH_2MH \rightleftarrows RCH{=}CH_2 + H\text{-}M\text{-}H \qquad (23)$$

If two molecules of an olefin react with the starting H-M-H, the reaction may proceed via an R_2M intermediate.

According to the existing theory, the hydrogenation of hydrocarbons proceeds through the dissociative adsorption of separate fragments of a molecule (H-atoms and hydrocarbon "radical") on various metal centers. Above we have offered some examples to illustrate hydrocarbon hydrogenation in the gas phase on a <u>single</u> transition metal. These reactions were found to occur via the organometallic intermediates produced by the oxidative addition of hydrocarbons to a reduced metal.

The hydrogenation of hydrocarbons may be represented by the following schemes:

a) $$CH_3CH_2CH_2CH_3 + M^\circ \longrightarrow CH_3CH_2CH_2CH_2M\text{-}H \xrightleftharpoons{\beta\text{-decay}} CH_3CH_2CH{=}{=}CH_2 + \underset{\displaystyle \downarrow \atop \displaystyle M + H_2}{MH_2}$$

b) $$CH_3CH_2CH{=}CH_2 + M^\circ \longrightarrow CH_3\underset{M\text{-}H}{\underset{|}{C}}HCH{=}CH_2 \xrightleftharpoons{\beta\text{-decay}} CH_2{=}CHCH{=}CH_2 + \underset{\displaystyle \downarrow \atop \displaystyle M + H_2}{MH_2}$$

c)

$$C_6H_5CH_2CH_3 + M^\circ \longrightarrow C_6H_5\underset{\displaystyle M{-}H}{\underset{|}{C}}HCH_3 \overset{\beta\text{-decay}}{\rightleftharpoons} C_6H_5CH{=}CH_2 + MH_2 \quad (MH_2 \downarrow M + H_2) \qquad (24)$$

When the proportion of methylene groups in alkanes is small, the addition of a metal should occur through the C-H bond of the methyl and β-elimination should lead to the primary formation of α-olefin, which may undergo isomerization to β-olefin. For olefins or alkanes bearing electronegative substituents (C_6H_5-, -CN, and others), the addition of a metal to the C-H bond in the group adjacent to either the substituent mentioned or the double bond is most advantegeous energetically because of the conjugation of the C-H bond with the double bond or the phenyl ring. This is the reason for the increase in the rate of oxidative addition of M° in the series: cyclohexane-cyclohexene-cyclohexadiene.

(cyclohexane) → (cyclohexene) → (cyclohexadiene) → (benzene) (25)

The latter stage is most favorable because the C-M bond in the intermediate obtained is conjugated with two C=C bonds, and β-elimination leads to conjugation with two aromatic nucleus. Hence, the addition of M° to cyclohexane is the rate-determining step.

4.2. Deuterium-Exchange Reactions of Hydrocarbons.

The mechanism proposed for the exchange reactions of alkanes and alkenes and their derivatives usually involve the idea of polyadsorbed species on the assembly of metal atoms. This is especially true for multiple exchange reactions. The experimental information and theory of exchange reactions of hydrocarbons have been summarized by Ozaki in "Isotopic studies of heterogeneous catalysis" (Ref. 49). The effects at the initial stages of the reactions reported by this author are of special interest for discussing elementary acts in catalysis.

We shall consider the mechanism of deuterium exchange reactions of hydrocarbons in terms of organometallic intermediates. Alkanes with two or more carbon atoms in the chain quickly exchange at the initial stages of the reaction. Thus, the initial products of a propane exchange on Pd^o at 140°C consist of C_3D_8(80%), C_3D_7H (12%) and $C_3D_6H_2$ (4%) (Ref. 50). If there is a quaternary carbon or oxygen atom in the chain, the isotopic distribution will have maxima corresponding to the number of hydrogens in groups with more than one carbon atom. Thus, 2,2,3-trimethylpentane produces $(CH_3)_3C\text{-}CD(CD_3)_2$, and ethylbutyl ether on Rh gives maxima of d_5 and d_9, respectively, corresponding to full hydrogen exchange in the groups bound to oxygen (Ref. 49). Partial exchange is observed only with methane or methyl groups in pentane. These facts can be explained in terms of the decomposition of organometallic intermediates via β-elimination.

$$RCH_2CH_2CH_3 + M^o \rightarrow RCH_2CH_2CH_2M\text{-}H \rightarrow RCH_2CH{=}CH_2 + MH_2$$

$$RCH_2CH{=}CH_2 \xrightarrow{M^o} RCH{\cdots}CH{\cdots}CH_2 \text{ (allylic, bound to } M\text{–}H) \qquad MH_2 \rightarrow M + H_2 \qquad (26)$$

The formation of the double bond and its migration in the chain of allylic conjugation are the main factors responsible for multiple hydrogen exchange. This accounts for the termination of the chain by quaternary carbon and oxygen. Methyl and neopentyl derivatives can decompose only via α-elimination, thus allowing only a partial exchange of hydrogen for deuterium.

$$CH_4 + M \longrightarrow CH_3MH \longrightarrow \begin{cases} \xrightarrow{D_2} CH_3D + DMH \\ \longrightarrow CH_2:MH_2 \xrightarrow{D_2} CH_2D_2. \end{cases} \quad (27)$$

Hirota et al. (Refs. 51,52) have found that propylene selectively exchanges on copper to give $CH_3CD{=}CH_2$ and on Pd, Rh and Pt to give a mixture of $CH_3CD{=}CH_2$ and $CH_3CH{=}CHD$. The preferential exchange of the vinyl hydrogen, especially at C_1, as mentioned by this author, directly results from adding propylene to the M-D bond, followed by the decomposition of the organometallic compound,

$$CH_3CH{=}CH_2 + D\text{-}M\text{-} \longrightarrow CH_3CHDCH_2M \longrightarrow CH_3CD{=}CH_2 + H\text{-}M\text{-}. \quad (28)$$

The same idea may be used to explain the mechanism of ethylene exchange in deuterium hydrogenation:

$$CH_2{=}CH_2 + D\text{-}M\text{-} \longrightarrow CH_2DCH_2M\text{-} \longrightarrow CHD{=}CH_2 + HM\text{-}. \quad (29)$$

If a phenyl or a carbonyl group is next to the double bond, the reaction products are enriched with deuterium at the β-position (Ref. 53). This is due to the energetic advantage of the addition of the metal to the α-carbon because of the conjugation of the C-M bond with the phenyl ring or the C=O group:

$$C_6H_5CH{=}CH_2 + D\text{-}M\text{-}D \rightarrow C_6H_5\underset{\displaystyle MD}{\underset{|}{CH}}\text{-}CH_2D \rightarrow C_6H_5CH{=}CHD + HMD. \quad (30)$$

Special attention should be paid to the autohydrogenation reactions first studied by Turkevich (Refs. 54,55). The reaction of ethylene with nickel wire gives ethane and, even in the presence of some D_2, $C_2H_6(d_o)$ is the main primary reaction product. Similar transformations of cyclopentene into cyclopentane on Fe (Ref. 56) and butadiene into butane on Cu (Ref. 49) have been observed. The mechanism of these reactions may be considered in terms of the oxidative addition of an olefin to a metal followed by the decomposition of an organometallic intermediate:

a)

$$CH_2{=}CH_2 + M^o \rightarrow CH_2{=}CHMH \xrightarrow{C_2H_4} CH_2MCH_2CH_3 \rightarrow CH{\equiv}CH + M^o + C_2H_6$$

b)

$$\text{(cyclopentene)}CH_2 + M^o \rightarrow \text{(cyclopentene)}CHMH \xrightarrow{\text{(cyclopentene)}} \text{(cyclopentene)}CH\text{-}M\text{-}CH\text{-}CH_2\text{(ring)} \rightarrow \text{(cyclopentadiene)} + \text{(cyclopentane)} + M^o$$

c)

$$CH_2{=}CH\text{-}CH{=}CH_2 + M^o \rightarrow CH_2{=}CH\text{-}CH{=}CHMH \xrightarrow{C_4H_6} CH_2{=}CHCH{=}CH\text{-}M\text{-}$$

$$\rightarrow CH_2{=}CHCH{=}CH\text{-}M\text{-}CH_2{=}CH\text{-}CHCH_3$$

$$\downarrow$$

$$CH_2{=}CHC{\equiv}CH + C_4H_8 \qquad (31)$$

The organometallics formed decompose *via* β-elimination to produce acetylenic derivatives or cyclopentadiene with more reactive C-H bonds that are responsible for the formation of carbides (in the case of ethylene or resins (for cyclopentene or butadiene).

4.3. Hydrogenolysis Reactions

The mechanism of hydrogenolysis usually involves the successive abstraction of H-atoms, with the participation of many active centers (metal atoms). The formation of organometallic intermediates in the oxidative addition of hydrocarbons to atoms or reduced ions of transition metals is likely to be the most probable explanation of the mechanism of hydrogenation because the addition of hydrogen to the C-M bond yields the final reaction products:

$$RCH_2CH_2R' + M^o \longrightarrow RCH_2\text{-}M\text{-}CH_2R' \xrightarrow{H_2} RCH_3 + R'CH_3. \qquad (32)$$

Depending on the particular metal and hydrocarbon, the site of insertion may be approximately equal to all C-C bonds, thus leading to a statistic distribution of various products of hydrogenolysis, or the reaction may proceed more selectively, for example, across the terminal bond in aliphatic hydrocarbons or the C-C bond in isoalkanes.

The above data on the mechanism of metathesis and the pathways of formation of carbene species in the decomposition of organo-transition metal compounds show the possible role of these reactions in hydrogenolysis. Ethane hydrogenolysis catalyzed by Pt or Pd is ten times slower than with the same metals promoted by reduced compounds of W or Mo (Ref. 57). When there is no W or Mo compounds the increasing hydrogen concentration causes a decrease in the reaction rate. If these compounds are added, the inhibiting action of hydrogen decreases considerably or even disappears. This dependence may be explained in terms of the role of metathesis in the hydrogenolysis of hydrocarbons.

For compounds of Pt and Pd the dehydrogenation reaction is most typical

$$CH_3CH_3 + Pt^o \longrightarrow [CH_3CH_2PtH] \rightleftharpoons CH_2{=}CH_2 + H_2 + Pt^o. \qquad (33)$$

If organometallic intermediates decompose to give carbene complexes, ethylene metathesis proceeds through the formation of methylene complexes which add the hydrogen molecule

$$CH_2{=}CH_2 \xrightarrow{MX_n} CH_2{=}MX_n \xrightarrow{H_2} CH_4. \qquad (34)$$

The increasing rate of hydrogenolysis under the influence of reduced compounds of W and Mo is in accord with their high efficiency as metathesis catalysts.

4.4. Skeletal Isomerization.

Neopentane on Pt/SiO_2 in the presence of H_2 undergoes two reactions (Ref. 58):

$$(CH_3)_4C \begin{cases} \xrightarrow{a} i\text{-}C_5H_{12} \\ \xrightarrow[b]{+H_2} CH_4 + i\text{-}C_4H_{10} \end{cases} \qquad (35)$$

As shown by Skell (Ref. 23) for the reaction of $(CH_3)_4C$ with Zr^0 and by Freizer (Ref. 25) for the reaction of Rh^+ in the gas phase, direction 35^b proceeds via insertion of a metal into a C-C bond with the subsequent adding H_2 to a C-M bond. The addition of a metal to a C-H bond in neopentane should lead to carbene species through α-decomposition. The specific pathway of decomposition of organometallic compounds devoid of β-hydrogens allows one to suggest the following scheme of the skeletal isomerization of neopentane:

$$(CH_3)_4C + Pt \rightleftarrows CH_3\text{-}\overset{CH_3}{\underset{CH_3}{C}}\text{-}CH_2PtH \longrightarrow \left[CH_3\text{-}C(CH_3) \overset{1}{\searrow} \overset{CH_2}{\underset{2}{\uparrow}} CH_2 \right] +$$

$$\downarrow$$

$$CH_3\text{-}\overset{CH_3}{\underset{CH_3}{C}}\text{-}CH{:}PtH_2 \longrightarrow \uparrow$$

$$+ PtH_2 \longrightarrow CH_3\underset{CH_3}{C}HCH_3CH_3 + Pt \quad (36)$$

The cleavage of cyclopropane intermediate via addition of hydrogen in "1" and "2" directions leads to the skeletal isomerization. The similar mechanism was proposed in (Ref. 59) for the cationic polymerization of n-butane to isobutane:

$$\underset{④}{CH_3}\underset{③}{\overset{+}{C}H}\underset{②}{CH_2^{13}}\underset{①}{CH_3} \longrightarrow \left[\underset{④}{CH_3}\text{-}\underset{③}{CH} \overset{^{13}CH_2\,①}{\underset{H^+}{\uparrow}} \underset{②}{CH_2} \right] \quad (37)$$

The addition of H^+ to C_1 with the cleavage of C_1-C_3 bond to the starting cation; the addition to C_2 with cleavage of C_2-C_3 bond produces the normal hydrocarbon skeleton with ^{13}C migrating to the neighbouring position; the addition of H^+ to C_1 with C_1-C_2 bond cleavage leads to skeletal isomerization. These ideas are used to explain the mechanism of skeletal isomerization of n-pentane to isopentane, the protonated cyclopropane derivative being an intermediate

REFERENCES

1. B.A. Dolgoplosk, Kinet.Katal.,22, N 4,807-825 (1981).
2. B.A. Dolgoplosk, I.A. Oreshkin and E.I. Tinyakova, Kinet.Katal.,25, N 2, 306-315 (1984).
3. B.A. Dolgoplosk and E.I. Tinyakova, Organometallic catalysis in polymerization processes, Nauka, Moscow (1982).
4. B.A. Dolgoplosk, Vysokomol.Soed., A-13, N 2,325-347 (1971).
5. B.A. Dolgoplosk and E.I. Tinyakova, Usp. Khim., 53, N 1, 40-64 (1984).
6. R.L. Banks and G.C. Bailey, Ind.Eng.Chem., Prod.Res. Dev.,3, N 3, 170-173 (1964).
7. B.A. Dolgoplosk, T.G. Golenko, K.L. Makovetsky, E.I. Tinyakova and I.A. Oreshkin, Dokl.Akad.Nauk SSSR, 216, N 4, 807-808 (1974).
8. B.A. Dolgoplosk, K.L. Makovetsky and E.I. Tinyakova, Dokl.Akad.Nauk SSSR, 202, N 4, 871-874 (1972).
9. G. Wilke, B. Bogdanovich, P. Hardt et al., Z.Org. Chem., 78, N 3, 157-172 (1966).
10. L.I. Redkina, K.L. Makovetsky, E.I. Tinyakova and B.A. Dolgoplosk, Dokl.Akad.Nauk SSSR, 186, N 2, 397-398 (1969).
11. V.Sh. Feldblum, T.I. Baranova, R.B. Svitich et al., Z.Org.Khim., 8, N 2, 650-651 (1972).
12. I.Ya. Ostrovskaya, K.L. Makovetsky, G.P. Karpacheva et al., Dokl.Akad.Nauk SSSR, 197, N 6,1344-1347 (1971).
13. F.J. McQuillin, Homogeneous hydrogenation in organic chemistry, D.Reidel Publishing Company, Dordrecht, Holland, 1976.
14. J. Chatt and B.L. Shaw, J.Chem.Soc., N 12, 5075-5084, (1962).
15. H.C. Clark and C.S. Wong, J.Am.Chem.Soc., 96, N 13, 7213-7221 (1974).
16. H.C. Clark and C.R. Jablonski, Inorg.Chem,, 13, N 9, 2213-2218 (1974).
17. J. Chatt, R.S. Coffey, A. Gough and D.T. Thompson, J.Chem.Soc.A., N 1, 190-194 (1968).
18. O.K. Sharaev, A.V. Alferov, E.I. Tinyakova and B.A. Dolgoplosk, Dokl.Akad.Nauk SSSR, 164, N 1, 119-121, (1965).
19. M. Orchin and W. Rupilius, Catal.Rev., 6, N 1, 85-131 (1972).
20. D. Evans, J.A. Osborn and G. Wikinson, J.Chem.Soc.A., N12, 3133-3142 (1968).
21. D. Forster, J.Am.Chem.Soc., 98, N 3, 846-848 (1976).
22. R.P. Lowry and A. Aguilo, Hydrocarbon Process, 53, N 11, 103-113 (1974).
23. R.J. Remick, T.A. Asunta and P.S. Skell, J.Am.Chem. Soc., 101, N 5, 1320-1322 (1979).
24. S.C. Davis, S.J. Severson and K.J. Klabunde, J.Am. Chem.Soc., 103, N 11, 3024-3029 (1981).

25. G.D. Byrd and B.S. Freizer, J.Am.Chem.Soc., 104,N 22, 5944-5950 (1982).
26. R.B. Freas and D.P. Ridge, J.Am.Chem.Soc., 102, N 23, 7129-7131 (1980).
27. G.D. Byrd, R.C. Burnier and B.S. Freizer, J.Am.Chem. Soc., 104, N 13, 3563-3569 (1982).
28. H. Ledon, I. Tkachenko and D. Young, Tetrahedron Letters, N 2, 173-176 (1979).
29. L.H. Green, Pure Appl.Chem., 50, N 1, 27-35 (1978).
30. M.S. Kharash and W. Urry, J.Org.Chem., 13, N 1, 101-109 (1948).
31. M.S. Kharash, D.W. Lewis and W.B. Reynolds, J.Am. Chem.Soc., 65, N 4, 498-500 (1943).
32. W. Urry, Mand M.S. Kharash, J.Am.Chem.Soc., 66, N 9, 1438-1440 (1944).
33. M.S. Kharash, D.W. Lewis and W.B. Reynolds, J.Am.Chem. Soc., 65, N 4, 493-495 (1943).
34. M.S. Kharash and M. Kleiman, J.Am.Chem.Soc., 65, N 4, 491-493 (1943).
35. O.A. Reutov, I.P. Beletskaya, G.A. Artamkina and A.N. Kashin, Reactions of organometallic compounds as redox processes, Nauka, Moscow, 196-245 (1981).
36. H. Gilman, R.G. Jones and L.A. Woods, J.Am.Chem.Soc., 76, N 14, 3615-3617 (1954).
37. P.J. Davidson, M.P. Lappert and R. Pearce, Chem.Rev., 76, N 2, 219-242 (1976).
38. B.A. Kropachev, B.A. Dolgoplosk, N.M. Geller et al., Izv.Akad.Nauk SSSR, OKhN, N 6, 1044-1048 (1960).
39. Van-Fo Sun, B.L. Erusalimsky and B.A. Dolgoplosk, Izv.Akad.Nauk SSSR, OKhN, N 3, 459-473 (1960).
40. I.L. Kershenbaum, K.L. Makovetsky and B.A. Dolgoplosk, Izv.Akad.Nauk SSSR, Ser.Khim., N 5, 1170-1172 (1973).
41. J. Dvorak, R.J. O'Brien and W. Santo, J.Chem.Soc., Chem.Commun., N 7, 411-412 (1970).
42. S.A. Smirnov, I.A. Oreshkin, B.G. Zaikin and B.A. Dolgoplosk, Dokl.Akad.Nauk SSSR, 245, N 6, 1400-1403 (1979).
43. S.A. Smirnov, I.A. Oreshkin and B.A. Dolgoplosk, Dokl.Akad.Nauk SSSR, 239, N 6, 1375-1376 (1978).
44. N.N. Kostitsyna, I.A. Oreshkin, B.A. Dolgoplosk et al., Dokl.Akad.Nauk SSSR, 277, 1348-1350 (1984).
45. I.L. Kershenbaum, L.N. Grebenyak, I.A. Oreshkin et al., Dokl.Akad.Nauk SSSR, 238, N 2, 359-360 (1978).
46. I.L. Kershenbaum, L.N. Grebenyak, I.A. Oreshkin et al.,Dokl.Akad.Nauk SSSR, 259, N 1, 119-120 (1981); 256, N 6, 1400-1401.
47. L.S. Cherezova, V.A. Yakovlev and B.A. Dolgoplosk, Dokl.Akad.Nauk SSSR, 254, N 6, 1410-1411 (1980);258, N 6, 1380-1382 (1982); 262, N 6, 1403-1405.
48. I.Sh. Guzman, N.N. Chigir, O.K. Sharaev et al., Dokl. Akad.Nauk SSSR, 249, N 4, 860-862 (1979).
49. A. Ozaki, Isotopic studies of heterogeneous catalysis, Kodansha Ltd.Tokyo, Acad.Press New York, San Francisco

London, A Subsidiary of Harcourt Jovanovich, Publishers, 1977.

50. F.G. Gault and C. Kemball, Trans.Farad.Soc., 57, N 466, 1781-1794 (1961).
51. T. Ueda, J. Hara, K. Hirota et al., Z.Phys.Chem.N.F., 64, 64-70 (1969).
52. Y. Hironaka and K. Hirota, Bull.Chem.Soc.Jpn., 38, N 9, 1558-1559 (1965).
53. G.V. Smith and J.F. Deany, J.Catal., 6, N 1, 14-19, (1966).
54. J. Turkevich, D. Schissler and P. Irsa, J.Phys.Colloid.Chem., 55, N 6, 1078-1084 (1951).
55. J. Turkevich, F. Bonner, D. Schissler et al., Disc. Faraday Soc., N 8, 352-356 (1950).
56. J. Erkelens, A.K. Gaiwey and C. Kemball, Proc.Roy. Soc., Ser.A, 260, N 1300, 273-292 (1961).
57. Yu.I. Yermakov, B.N. Kuznetsov and Yu.A. Ryndin, Kinet. Katal., 19, No 1, 169-176 (1978).
58. B.N.Kuznetsov, Yu.I. Yermakov, M. Boudart and J. Collman, Kinet. Katal., 18, No 3, 559-560 (1977).
59. D. Brouwer and H. Hogeveen, Prog. Phys. Org. Chem., 9, 179-240 (1972).

ORGANIZED MOLECULAR ASSEMBLIES IN CATALYSIS

A.E. Shilov

Institute of Chemical Physics, USSR Academy of Sciences, Chernogolovka 142432, USSR

ABSTRACT - Organized molecular assemblies are multicomponent catalytic systems characterized by an optimal spatial arrangement as well as by the energy and orbital correspondence between the system and substrate. Ideal systems of these type are biological catalysts (enzymes). By using the principles of enzyme action it is possible to solve complicated catalytic problems such as involving chemically inert molecules in the catalytic cycle, accomplishing chemical conjugation of redox reactions, and conducting photocatalytic reactions with accumulation of energy of the light.
A succesful application of organized molecular assemblies is examplified by the development of catalytic processes of nitrogen reduction, saturated hydrocarbon oxidation as well as of the photosynthesis modeling.

1. INTRODUCTION

Modern catalysis science faces ever more complicated and interesting problems. These are connected with the urgent necessity to create new selective industrial processes, which are echologically pure, involve chemically inert molecules, consume minimum energy, etc.

At least some of these problems can be solved successfully by creating organized molecular assemblies (OMAs). This term may be applied to multicomponent catalytic systems with optimum space arrangement, energy and orbital correspondence of the system components and the substrates.

The creation of OMAs is based on modern knowledge of

the mechanism of reactions constituting a catalytic cycle. Though the reactions involved are usually homogeneous, the OMAs, in most developed cases combine homogeneous, heterogeneous and heterophase reactions.

Generally speaking, any catalytic system presumes a certain organization of its components. For example, definite space organization is clearly seen in the case of bifunctional catalysis. However, the term "OMA" is more appropriate to use for much more complicated systems. An example of a perfect organization is provided by enzymes in living nature. We could try to follow this example and construct a purely chemical systems similar to what we know about the enzymes. Evidently, to form a catalytic system repeating all complications of an enzyme is at present impossible, but experimental evidence shows that we can successfully use the <u>principles</u> of enzymatic mechanisms to create quite simple systems, which have certain elements of organization and are sufficiently effective from the viewpoint of a chosen catalytic process.

Let us consider some examples of catalytic processes which may be substantially improved using OMA principles.

1. A number of redox processes proceed according to the schemes:

$$nD + S \xrightarrow{Cat.} nD \overset{e}{\overset{\frown}{\cdots}} Cat. \overset{ne}{\overset{\frown}{\cdots}} S \longrightarrow nD_{ox} + S_{red}$$

or

$$nA + S \xrightarrow{Cat.} nA \overset{ne}{\overset{\frown}{\cdots}} Cat. \overset{e}{\overset{\frown}{\cdots}} S \longrightarrow nA,$$

where the role of a catalyst (Cat.) consists of transforming a one-electron process provided by an electron donor (D) or acceptor (A) into a two- or n-elctron reduction (oxidation) of a substrate (S) by the catalyst.

If in the initial state the catalyst is a united

system (cluster) involving several transition metal ions, n-electron transfer becomes possible or strongly facilitated in a single elementary process.

The rate of the catalytic process may be determined by the electron transfer from the donor to the catalyst (or to the acceptor from the catalyst). Then the process will be greatly accelerated, if the catalytic complex is attached near the donor (or the acceptor). This process can be further facilitated in the presence of certain electron transfer agents, which at best should have redox potential intermediate between those of the external donor (acceptor) and the catalyst and spatially situated in-between.
Thus even in this fairly simple case, an effective catalytic system must be an organized molecular assembly.

2. The oxidation of the substrate may be coupled with the oxidation of the external donor, the energy of the latter process providing the possibility of creating particularly reactive intermediate forms of a catalyst capable to oxidize even comparatively inert substrate molecules. The mechanism of external donor action involves the reduction of the catalyst. The subsequent action of the oxidant such as molecular oxygen, produces the catalyst active state capable of oxidizing the substrate. There could also be coupling with an acid-base process which also increases the activity of an intermediate catalytic state. What is the way of avoiding "useless" reactions such as the direct oxidation of the donor by the oxidant or acid-base process uncoupled with the redox process we are interested in? To solve this problem, we can also use OMA with a definite space organization, e.g. place the catalyst between the donor and the acceptor.

3. Organized molecular assemblies are definitely necessary for the photocatalytic processes with the increase in the free energy when the initial process con-

sists of the electron transfer from the weak donor to the weak acceptor to form strong acceptor and donor, which subsequently take part in catalytic processes producing stable substances with the energy higher than that of the initial reagents. In the absence of a special space and energy organization, fast back reactions will follow the initial electron transfer, first, the return of an electron to the primary donor; thus, more or less effective process of this kind is virtually impossible.

Let us consider examples of catalytic transformations studied in our laboratory where the principle of OMA has demonstrated its effectiveness.

2. REDUCTION OF MOLECULAR NITROGEN (Refs. 1,2)

Enzymatic reduction of dinitrogen with participation of nitrogenase proceeds with the initial formation of ammonia according to the stoichiometry

$$H_2 + 8\ e + 8\ H^+ \xrightarrow[N_2ase]{} 2\ NH_3 + H_2\ .$$

It is particularly important for this case to accumulate electrons in the catalytic center, since one- and two-electron processes require very strong reducing agents. 2 Mo, 34 Fe and 26-28 S atoms participate in the nitrogenase protein complex. The electron transfer from an external donor is accompanied by ATP hydrolysis, this coupled process providing the energy to shift the system redox potential to a more negative value. Activation and reduction of an N_2 molecule is believed to proceed at the molybdenum site of molybdenum-iron protein. We have witnessed the development of the mechanism of the nitrogenase action from the very beginning. The mechanism is in full agreement with the conception, which was put forward by Likhtenstein and the author as long as in 1969, of the necessity of multi-electron reduction of dinitrogen for a comparatively weak reducing agent. How could we explain without this suggestion such a large number of metal ions which are known to be situated in close neighbourhood capable to accept and transfer electrons to the reaction centre? The most general conclusion, which is valid also for non-enzymatic chemical systems activating and reducing N_2, is that the weaker is the reducing agent present in the system, the more developed must be its organization and the larger must be its electron capacity. It follows from the nature of metal complexes activating dinitrogen that binuclear complexes are more suitable for N_2 activation than mononuclear ones. An optimum electron configuration of metal ions forming the complex seems to be d^2 or d^3. If the ions are bound by the common ligand this leads to better entropy factors in the process of the complex formation with N_2

Thus, a definite organization is important for N_2 reduction even at the level of an intermediate complex. Let us follow the development of a catalytic system discovered several years ago in our laboratory.

The system includes sodium amalgam or mercury cathode as a reducing agent and molybdenum(III) complex as a catalyst and reduces N_2 at room temperature. Without any special organization the system demonstrates very low rates ($v \approx 1\text{x}10^{-4}$ mol N_2 per one mol of the catalyst per 1 s) and yields (0.5 mol N_2H_4 per 1 mol of Mo or $4\text{x}10^{-3}$% with respect to the reductant) at N_2 pressure in the range of 50-100 atm. It is only a very sensitive analysis for hydrazine that allows us to definitely conclude that the reaction does take place at all. The system becomes more stable at the introduction of Mg^{2+} ions forming a complex Mo(III). In this case $v = 3\times10^{-4}$, the yield 2.5 mol of N_2H_4 per mol of Mo (0.02% with respect to the reductant). Further significant progress is reached with the introduction of one of phospholipids, that is, phosphatidylcholine (lecithin). At 70 atm N_2 the yield reaches 25 cycles per 1 Mo and 0.25% with respect to the reductant. This gives us the opportunity to investigate the reaction at 1 atm N_2. The rate is increased up to 5×10^{-2} mol of N_2H_4 per mol of Mo per 1 s. The yield rises not so dramatically, but even at 1 atm N_2 reaches 3.5 mol N_2H_4 per mol of Mo, which corresponds to 3×10^{-2}% per the reductant.

New increase has been attained at the introduction of organic phosphines, which act only in the presence of the phospholipid. Now the rate is up to 0.4 mol N_2H_4 per mol of Mo per 1 s, i.e. even at 1 atm N_2 we have virtually equal specific rates of model and enzymatic processes. The yield per the reductant reaches several per cents at atmospheric pressure and $\sim$30% at 70 atm N_2. The yield per molybdenum corresponds to 10 000 cycles at a high pressure and close to 1 000 cycles at atmospheric N_2 pressure. What is the cause of this striking catalytic effect? The mechanism of the catalytic reaction as we see it now is presented in Fig. 1. The catalytic complex containing at least three molybdenum atoms is incorporated in a phospholipid film covering the surface of the amalgam. The electrons move from the donor to the catalyst situated closely nearby. The molecules of a proton-containing solvent (methanol or water) diffuse through the hydrophobic film to the catalytic complex attached to the surface. N_2 is activated by two molybdenum ions of the complex. Phosphine stabilizing low-valency state of the molybdenum atom (not bound directly to an N_2 molecule), facilitates the

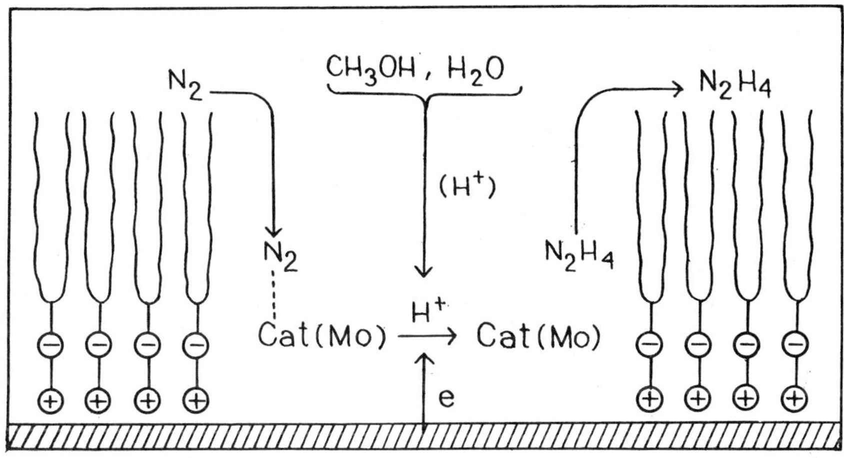

Figure 1. The scheme of nitrogen reduction catalyzed by Mo complex, immobilized on phospholipid membrane

electron transfer to the catalytic complex. As a result, we have a complex molecular assembly, which ensures the reaction rate higher at least 5 orders of magnitude as compared with an unorganized catalyst. It should be noted that the main factor of success was due to the use of the phosphatidylcholine, which allowed the further self-organization of the system. We were unable so far to achieve any comparable effect not only with other surface-active materials but even with other phospholipids. It should be suggested that phosphatidylcholine, which forms natural membranes, has some special properties suitable for formation of OMAs.

3. COUPLED OXIDATION OF C-H BONDS IN HYDROCARBONS (Ref. 3)

Oxidation of a non-activated C-H bond, for example, in alkanes, proceeds in nature via enzymatic processes catalyzed by monooxygenases according to the following scheme:

$$2\ e + 2\ H^+ + RH + O_2 \longrightarrow ROH + H_2O$$

Here electrons (2 e) are provided by an electron donor, i.e., NADH or NADPH, which are oxidized in a coupled process with RH oxidation. This process was studied in *more detail* for a group of monooxygenases known as cytochrome P 450. The central role in the mechanism of the process is played by the iron-porphyrin complex, which participates in the catalytic cycle (see Scheme 1).

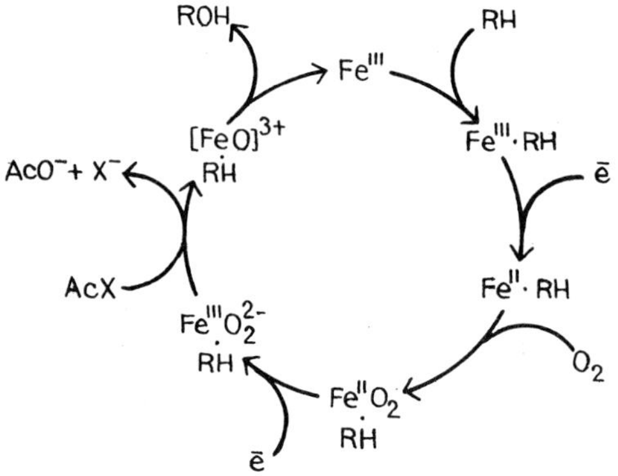

Scheme 1

The significance of the electron donor participation may be explained in the following way. Transferring two electrons to the porphyrin complex of Fe(III) we reduce it to Fe(I), which is then oxidized by four-electron oxidant O_2 to $[Fe^VO]^{3+}$. The latter is sufficiently active to react with a strong C-H bond transforming it to C-OH by way of insertion of the O atom. In fact, the reduction of Fe(III) does not result in Fe(I), since even in the state of Fe(II) it adds the O_2 molecule and the subsequent reduction brings about the compound, which may be considered as a peroxocomplex, $Fe(III)O_2^{2-}$, rather than $Fe(I)O_2$. In this state the complex reacts with the O^{2-} acceptor (e.g. acylation reagent) and the resulting complex is $[P^{+\cdot}Fe^{IV}O]^{3+}$ rather than $Fe^{V}O^{3+}$, since one electron is transferred from the porphyrin ring to the iron centre.

However, the simplified picture remains essentially valid. The energy provided by the donor oxidation is used to create the particle possessing a high redox potential. An estimate of the potential of the porphyrin complex oxidized form shows that, in principle, it may be much higher than the molecular oxygen potential and reach, for example, +2.5 V.

Shteinman, Khenkin et al. in our laboratory have obtained high-valency iron complex by the action of $K^+O_2^-$ at the iron porphyrin complex PFe^{3+} with the formation of $PFeO_2^+$. Acylation of the latter by acetic anhydride produces an active particle capable to hydroxylate alkanes, e.g. cyclohexane. A special investigation shows that hydroxylation proceeds by nonfree-radical mechanism. Thus, hydroxylation of the tertiary C-H bond in cis-1,2-dimethylcyclohexane results in the respective alcohol with considerable retention of configuration. A free-radical mechanism would have led to the loss of space specificity (i.e. racemization).

These results allowed us to try to construct the full model of cytochrome P 450. Porphyrin iron complex was deposited on a carbon cathode and acetic anhydride as an acylating agent and an acceptor of O^{2-} was added to the solution (acetonitrile was used as a solvent). Oxidation of cyclohexane by molecular oxygen proceeds catalytically when the electric current goes through the solution with the current efficiency reaching about 10% (the scheme is shown in Figure 2).

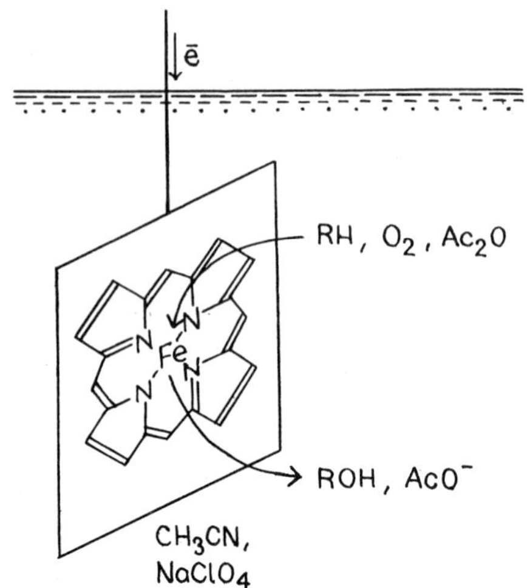

Figure 2. Coupled oxidation of hydrocarbons with participation of supported porphyrin iron complex (a model of cytochrome P 450).

In another system, also realized in the work of Khenkin et al., zinc amalgam was chosen as an electron donor with the porphyrin complex present in the solution. An electron transfer agent is needed for the functioning of the system. Methyl viologen was successfully used for this purpose (Scheme 2).

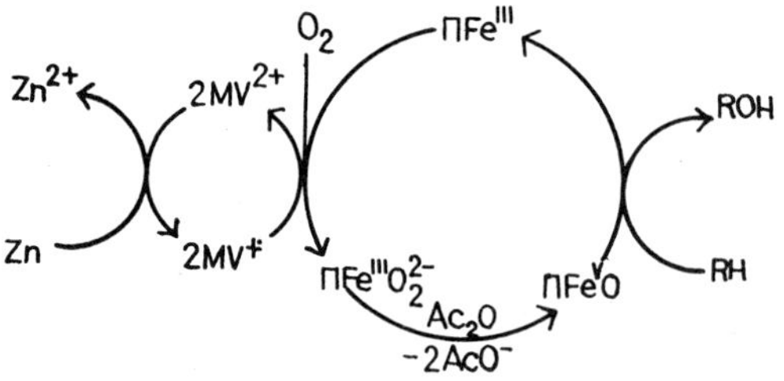

Scheme 2

Thus, the full functional model of cytochrome P 450 was realized in this work for the first time.

4. CHEMICAL MODELS OF PHOTOSYNTHESIS (Refs. 4,5)

Photosynthesis in plants is one of the most important processes in living nature. It is a complex OMA fulfilling the following functions:

1. Photoseparation of charges. Under the action of light the one-sided electron transfer takes place from the weak donor (chlorophyll) to the weak acceptor (oxidized ferredoxin).

2. Catalytic reduction of CO_2. Carbon dioxide is reduced in Calvin cycle to carbohydrates in a dark enzymatic process.

3. Catalytic oxidation of water. The acceptor formed in the primary charge separation (P^+_{680}) oxidizes water catalytically also in a dark process to form molecular oxygen. The process as a whole is believed now to proceed according to the so-called Z scheme (Fig. 3).

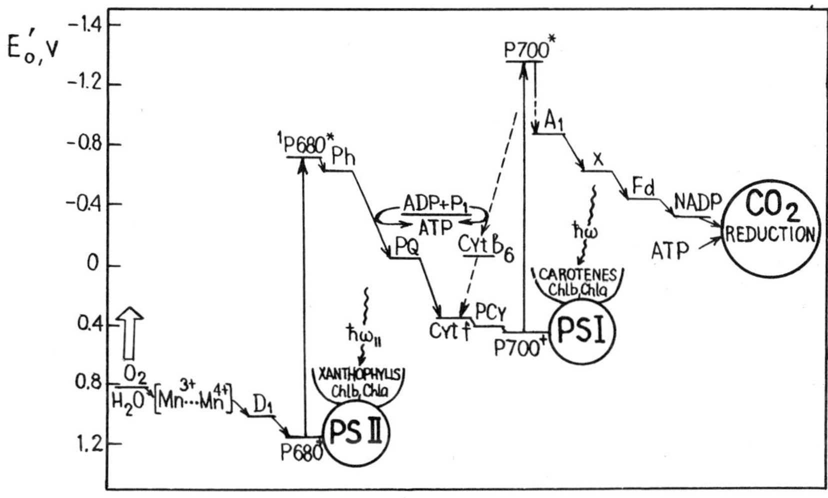

Figure 3. Z scheme of photosynthesis.

At present to solve the problem of photosynthesis means to create its acting chemical model, since only in this case we can say that we correctly understand its mechanism. The development of chemical models may in future lead to the construction of new effective ways of transformation of solar energy to that of chemical substances. The principal solution of the problem may involve (instead of photocatalytic synthesis of carbohydrates from CO_2 and water) photocatalytic decomposition of water to molecular hydrogen and oxygen. In the latter process nearly the same amount of energy is stored, and the mechanism of the process can be similar. Indeed, the reduction of CO_2 to carbohydrates and that of water to hydrogen require almost the same energy (per one electron). Some plants are known to evolve dihydrogen, and this process can be made predominant by using isolated chloroplasts with the addition of enzyme hydrogenase, which catalyzes dihydrogen evolution. Before creating the full model it was natural to solve three problems mentioned above separately, in other words, to perform

1) photochemical reduction of water to produce dihydrogen;

2) photochemical oxidation of water to form dioxygen;

3) photoseparation of charges

under visible light in chemical models.

Strangely, not a single problem of these three had been solved before 1975, when we started this work, which was suggested to us by N.N. Semenov. We began with the first problem, which was the easiest one, and found the solution soon. Everything was in fact prepared at the time we started the work.

All stages involved in the OMA were already known, we had only to unite them in a unified system. The following scheme shows the first purely chemical system, which turned out to be able to evolve dihydrogen from water under the action of visible light. It was created in the work of Dzhabiev and Koryakin in our laboratory in 1977 (Fig. 4).

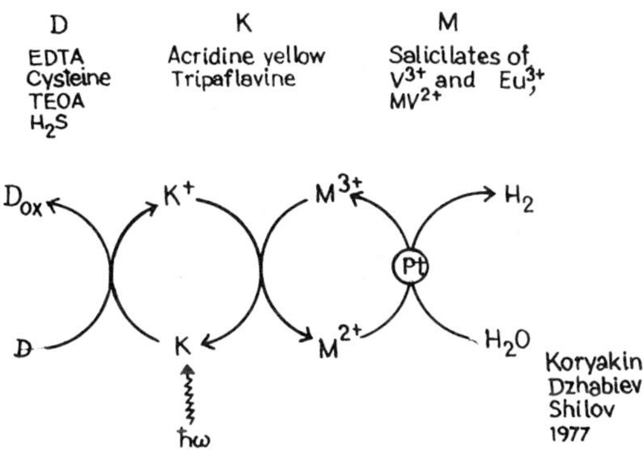

Figure 4. Photocatalytic formation of hydrogen

The system is almost entirely homogeneous with the exception of catalytic formation of dihydrogen at metallic platinum. Electron donors, e.g., EDTA, triethanolamine or hydrogen sulfide, are irreversibly oxidized in this and similar systems. H_2S is known to be the possible donor in photosynthesis performed by certain bacteria, thus this simple system may be regarded as simple functional model of bacterial photosynthesis (or the model of photosystem I of plant photosynthesis).

Since the creation of this model, many similar systems have been developed in several laboratories in different countries. They all include so-called sacrificial donors, which are irreversibly oxidized in the process.

The problem of harvesting the whole light of visible range was also solved by creating systems analogous to "antenna" in the plant and bacterial photosynthesis.

Methyl and other viologens are often applied as electron transfer agents. Catalysts for dihydrogen production were thoroughly investigated in Novosibirsk Institute of Catalysis, but colloidal platinum remains a satisfactory catalysts and almost always can be included in the acting systems.

It was somewhat more difficult to find a similar system for production of dioxygen. The point is that dioxygen formation is a four-electron process, and a catalyst must be capable of transforming its consecutive one-electron oxidation to the four-electron oxidation of the substrate (water) by the catalyst in order to be able to use an acceptor of minimum strength with the redox potential only slightly higher than that necessary for four-electron oxidation of water (E_{O_2/H_2O} = 1.23 V).

First catalytic system which could produce dioxygen under the action of the visible light was found in 1979. A sacrificial acceptor was used again in this and similar systems, e.g., Co(III), with a well-known trisbipyridyl Ru(II) complex as a sensitizer. In this case the catalyst may include the four-valent ruthenium complex RuO_2 and, as Shafirovich et al. have shown, also hydroxocomplexes of cobalt, iron and copper. In photosynthesis the role of the oxygen-forming catalyst is played by a manganese complex, which we still don't know much about. Therefore, it was interesting to learn whether we can use a manganese compound as a catalyst for oxygen formation from water. This was carried out in our work with Shafirovich and Khannanov in 1981. Manganese dioxide was found to be able to serve as a catalyst in the system according to the Scheme (see Figure 5).

This system may be considered as a chemical model of photosystem II of plant photosynthesis. Naturally, the catalyst for O_2 formation in plants is not MnO_2, however finely divided. Besides, if so many compounds can be used as catalysts for oxygen formation, why was it manganese which was chosen by nature to be the

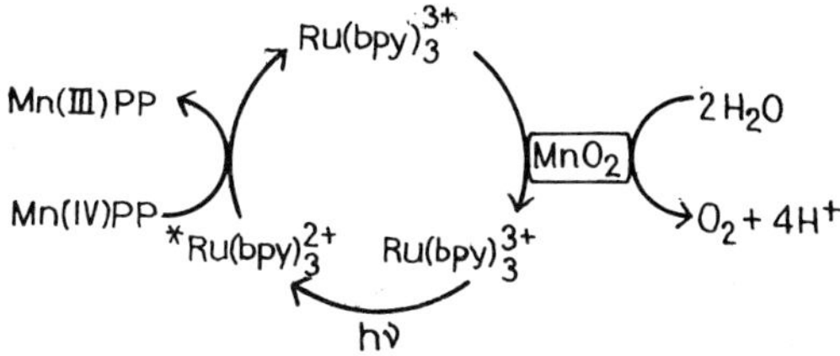

Figure 5. Photocatalytic formation of oxygen

catalyst in plants? MnO_2 is not the best catalyst among those mentioned as regard to catalysis of O_2 formation.

In this connection it is very interesting, from my point of view, that the incorporation of manganese ions in vesicles, built of phosphatidylcholine, facilitates catalytic oxygen evolution in the system found in the work of Shafirovich and Knerelman (Fig. 6).

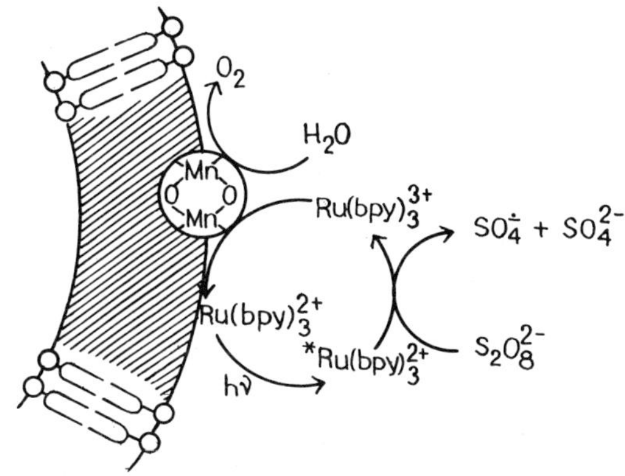

Figure 6. Formation of oxygen with the use of catalyst incorporated into vesicle.

Now we can imagine how oxygen-evolving system was created in living nature. The salt of bivalent manganese must have been in the neighbourhood with a membrane of lecithin type and an acceptor of chlorophyll cation-radical type. Again as in the case of nitrogen fixation, phosphatidylcholine membrane turned out to be suitable for the organization of a molecular assembly.

Finally, consider the problem of photoseparation of charges. As it was mentioned above, it is difficult (or impossible) to realize the electron transfer against thermodynamic potential in homogeneous solution with the formation of long-living molecules because of fast reactions of reduced and oxidized products with each other.

In the systems involving membranes such an electron transfer is possible; indeed, it is in the membrane that primary electron transfer takes place in photosynthesis.

OMA creates the possibility of one-way electron transfer. The principle of a cascade is realized in the natural system with the energy level of acceptors somewhat decreasing with each acceptor in succession. The reverse process is, therefore, very slow, since it requires crossing the barrier, if it takes the same steps as in the direct process. The under-barrier transfer is also slow, since the distance between the primary donor and a comparatively long-living acceptor is already too long. The fullfilment of this principle leads to the quantum yield equal to unity in the natural process.

Starting from the pioneering work by Mangel in 1976 vesicles are widely used in chemical models of photosynthesis. The traditional system involves a donor (or an acceptor) of an electron placed in the inner part of the vesicle, a photosensitizer incorporated in the wall of the vesicle and an acceptor (or donor) present in the outside solution (Fig. 7).

On irradiating the system the products of the photochemical reaction, D^+ and A^-, become divided by the

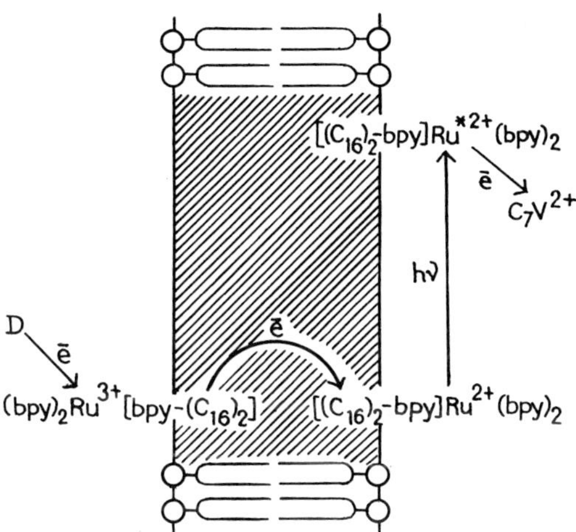

Figure 7. Photochemical transfer of electron through bilayer membrane with incorporated photosensitizer

membrane, which prevents their recombination. The systems of this kind were thoroughly investigated in the Institute of Catalysis and in the Institute of Biochemistry of the USSR Academy of Sciences. Without going into mechanistic details of the electron transfer (they can be found in the published reviews, e.g. (Ref. 4)), note that in such systems the back process becomes possible with the accumulation of A^- and D^+; in other words, when A^- and D^+ become the donor and the acceptor of electrons, respectively.

A more satisfactory approach to the problem of charge separation must include a stabilization of the oxidant and reductant formed in the first stage. A possible stabilization mechanism can be considered for the example of $*Ru(bpy)_3^{2+}$ reaction with methyl viologen, which is often used in model photosystems. $Ru(bpy)_3^{3+}$ and $MV^{+\cdot}$ formed in the primary photoreaction can be further used in reactions of catalytic oxidation and reduction of water, respectively.

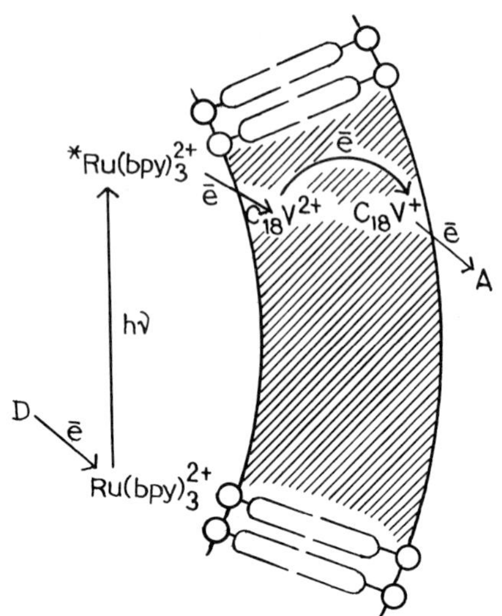

Figure 8. Photochemical transfer of electron through a bilayer membrane with photosensitizer inside of vesicle

In order to stabilize the separation of charges the third component must be introduced in the system such as an oxidant A_2 rapidly reacting with $MV^{+\cdot}$.

The method suggested for stabilization of charges can be realized in OMA containing the photosensitizer in the inner part of the vesicle, the first electron acceptor (which is viologen with two long alkyl groups) incorporated in the vesicle wall and the second acceptor present in the outside solution (Fig. 8).

The electron transfer occurs through the vesicle wall with a comparatively high quantum yield ($\varphi = 0.1$) in the presence of a sacrificial electron donor rapidly reacting with $Ru(bpy)_3^{3+}$ in conditions of stationary photolysis (Fig. 8). This system was realized almost simultaneously by Shafirovich et al. in the Institute of Chemical Physics and by Zamaraev and Parmon in the Institute of Catalysis. The application of pulse methods including a laser excitation source made it possible to determine and thoroughly investigate individual stages of the process. Based on the kinetic data obtained the effectiveness of charge separation (η) in the system without an electron donor has been calculated. The η-value can be determined as the ratio of the quantity of reduced oxidant in the outside solution (A^-) to the quantity of viologen cation-radical formed in the primary photochemical reaction. A precise analysis of the recombination kinetics for $Ru(bpy)_3^{3+}$ with viologen radical and trans-membrane electron transfer (oxidation of viologen radical by the outside oxidant occurs as a non-limiting step) has shown that depending on the temperature and phase condition of the bilayer, η-value is in the range from 0.06 to 0.19. From these data the limiting quantum yield of the charge separation can be determined as $\Phi = \Phi_c \cdot \eta$, where Φ_c is the probability of the escape of $Ru(bpy)_3^{3+}$ and viologen radical from the ion-radical pair. In the system under consideration Φ reaches quite noticeable value 0.008-0.027, allowing the hope to use the reductant and oxidant formed in the subsequent processes. Thus, even this comparatively simple OMA will permit us (hopefully in not too distant future) to solve the problem of photodecomposition of water under irradiation by visible light. The future development of more perfect systems probably will involve more sophisticated OMAs similar to triads, carotine-phosphirine-quinone, united in a single system and incorporated in the membrane.

We have chosen probably the most difficult processes for realization, viz. dinitrogen fixation, oxidation of C-H bonds in alkanes, artificial photosynthesis. In all cases the most successful approach to the problem in the creation of a least energy-consuming catalytic system involved the construction of a certain OMA. It turns out sometimes unexpectedly, that even self-organization leads to sufficiently effective catalysis.

It may be hoped that a wide use of this approach will lead to the solution of other catalytic problems and will be used in the future also for practical application.

REFERENCES

1. A Treatise on Dinitrogen Fixation, part I, G. Henrici-Olive et al. (Ed.), Part II, P. Hardy et al.(Ed),
2. A.E. Shilov, in Energy Resources through Photochemistry and Catalysis, M. Grätzel (Ed.), p. 535, Acad. Press, 1983.
3. A.E. Shilov, in Activation of Saturated Hydrocarbons by Transition Metal Complexes, D. Reidel Publ. Co., 1984.
4. K.I. Zamaraev and V.N. Parmon, in Energy Resources through Photochemistry and Catalysis, M. Grätzel (Ed.), Acad. Press, 1983.
5. V.Ya. Shafirovich and A.E. Shilov, Zh. Vses. Khim. Obshch. im. D.I. Mendeleeva, 31, No 4 (1986).

CATALYTIC OXIDATION OF CARBON MONOXIDE IN GAS PHASE HOMOGENEOUS AND GAS-SOLID HETEROGENEOUS REACTION SYSTEMS IN PRESENCE OF Ni, Fe, and Cr CATALYSTS

Shimpei Matsuda, Takahiro Tachi, Akira Kato, Hisao Yamashita
Hitachi Research Laboratory of Hitachi Ltd, Hitachi, Ibaraki, Japan 319-12

ABSTRACT - The catalytic oxidation of CO was investigated

(1) in the gas phase homogeneous system in the presence of Ni,Fe, and Cr vapor using a shock tube technique,

and

(2) in the gas-solid heterogeneous system in the presence of NiO, Fe_2O_3, and Cr_2O_3 using an ordinary flow reactor.

The kinetic data from the two systems showed a notable similarity and the reaction mechanism was discussed.

1. INTRODUCTION

It has long been known that the oxidation of carbon monoxide is proceeded by a solid catalyst such as platinum metals and transition metal oxides (Cr_2O_3, Fe_2O_3, and NiO etc). in a temperature range of 500 and 800 K. (Ref. 1) But little has been reported on the gas phase homogeneous catalysis by transition metal vapors mainly because of experimental difficulties. A shock tube technique made it possible to investigate the oxidation of CO at high temperatures (1200 - 2400 K) in presence of Cr, Fe, or Ni vapor (Refs.2,3,4). It is a main objective of the present paper to compare the catalytic effect of Cr, Fe, and Ni in the gas phase homogeneous and in the gas-solid heterogeneous system.

2. SHOCK TUBE REACTION APPARATUS

A shock tube technique has been widely used to study chemical kinetics in a gas phase homogeneous system at high temperatures, 1000 - 3000 K. (Ref. 5) A shock tube is an experimental apparatus in which a gas is uniformly heated to a high temperature by an adiabatic compression of shock waves.

A 10 cm inner diameter shock tube coupled with two

infrared detectors (IRST apparatus) is schematically shown in Figure 1. A shock tube is divided by a diaphragm into two sections, i.e., a driver section (high pressure) and a test section (low pressure). Upon the rupture of diaphragm, a shock wave is initiated and propagated in the test section where the gas is heated instantly (in a few micro seconds). The infrared emissions at 4.20 and 5.03 μm are monitored to measure the concentrations of CO_2 and CO in the shock heated gas, respectively.

A 2.5 cm inner diameter shock tube coupled to a time-of-flight mass spectrometer (TOFST apparatus) was also used to detect gas phase species of transition metal compounds. Mass spectra of the gas streaming through a pin hole in the end wall were recorded photographically at 20 μsec intervals over the mass range m/e=40 - 110 during a time of 150 μsec. The ionization energy was 22.5 $\pm$ 1.0 eV in all shock experiments.

3.GAS PHASE HOMOGENEOUS OXIDATION OF CARBON MONOXIDE

The oxidation of carbon monoxide in shock heated gas was studied in the absence and presence of $Cr(CO)_6$. Figure 2 shows an oscillogram of infrared signals from CO and CO_2 when a gas mixture containing CO and O_2 (without $Cr(CO)_6$) was heated to 2180 K and 0.15 atm. It is seen that the CO_2 signal grows rapidly only after a long induction period, about 900 μsec. Figure 3 shows an oscillogram of the infrared signals when a gas mixture containing CO, O_2, and $Cr(CO)_6$ was heated to 1863 K and o.145 atm. The decomposition of $Cr(CO)_6$ is fast under all conditions, giving rise to Cr atom vapor in the gas upon shock heating. It is clearly seen that the CO_2 signal grows linearly with time without any induction time. Intensities of the CO_2 signals in Figure 2 and 3 are converted into an absolute CO_2 concentration, and plotted in Figure 4. It is obvious that the oxidation of CO is

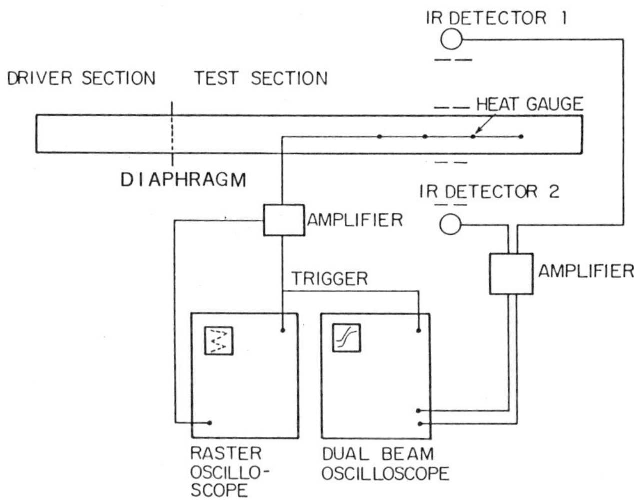

Fig. 1. SHOCK TUBE—IR DETECTOR APPARATUS

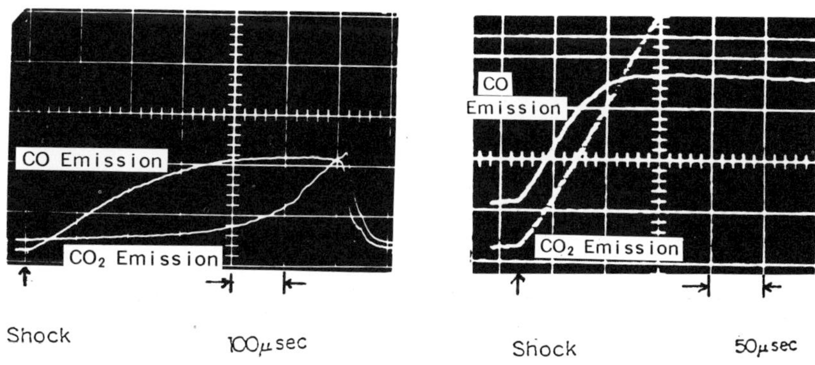

Fig. 2. Thermal reaction of CO with O_2 in shock waves.
T=2180°K, P=0.15atm,
Gas Mixture; CO 4.3%, O_2 2.1%, Ar 93.6%

Fig. 3. Oxidation of carbon monoxide in presence of chromium carbonyl in shock waves
T=1863°K, P=0.145atm,
Gas Mixture; $Cr(CO)_6$ 212ppm, CO 4.15%, O_2 1.70%, Ar balance

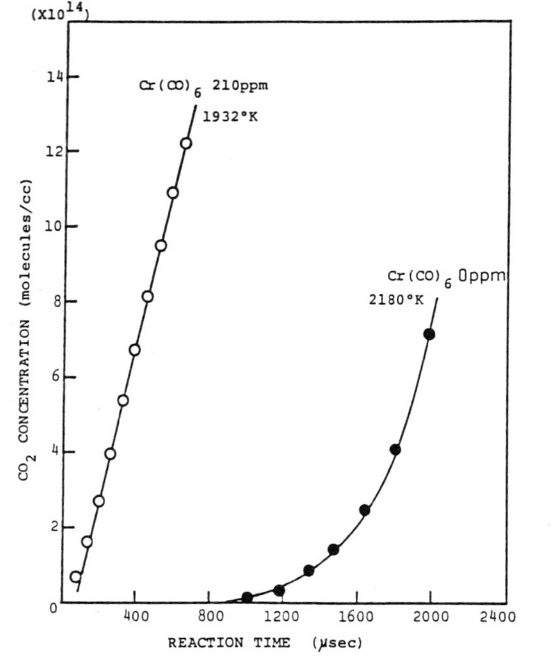

Fig.4. Formation of CO_2 in $CO-O_2$ reaction in presence or absence of $Cr(CO)_6$ in shock waves.
Gas Mixture : CO 4.3%, O_2 2.1%, Ar 93.6%
Pressure : 0.15 atm

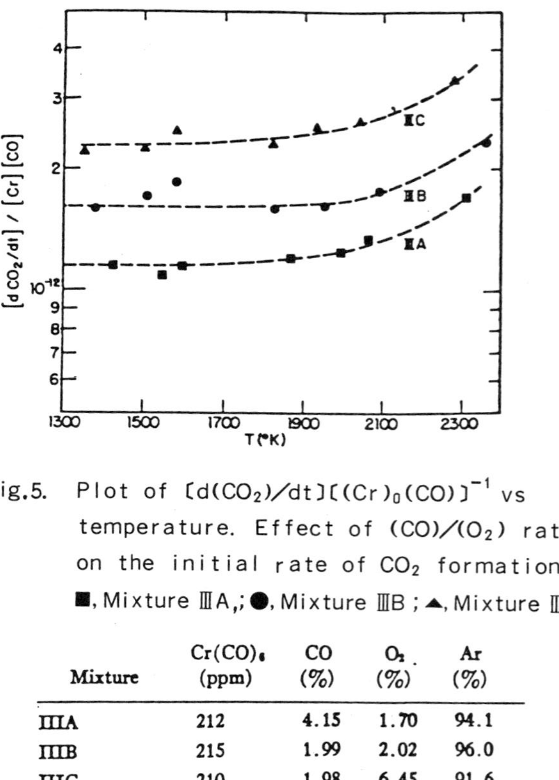

Fig.5. Plot of $[d(CO_2)/dt][(Cr)_0(CO)]^{-1}$ vs temperature. Effect of $(CO)/(O_2)$ ratio on the initial rate of CO_2 formation. ■, Mixture IIIA; ●, Mixture IIIB; ▲, Mixture IIIC

Mixture	$Cr(CO)_6$ (ppm)	CO (%)	O_2 (%)	Ar (%)
IIIA	212	4.15	1.70	94.1
IIIB	215	1.99	2.02	96.0
IIIC	210	1.98	6.45	91.6

catalyzed by the presence of Cr vapor.

Several series of experiments were performed under variety of temperatures, total pressures, and partial pressure of $Cr(CO)_6$. It was found that the rate of CO_2 formation is proportional to the initial concentration of $Cr(CO)_6$. The effect of $[CO]/[O_2]$ ratio on the rate of CO_2 formation was investigated and plotted in Figure 5. It is seen that there is little dependence of the rate of CO_2 formation on temperature between 1400 and 2000 K. The rate equation is expressed in an empirical formula,

$$\frac{d\,[CO_2]}{dt} \propto [Cr]_o [CO]^{0.66} [O_2]^{0.34} \qquad (1)$$

A similar series of experiments were performed using a gas mixture containing CO, O_2, and $Fe(CO)_5$, or $Ni(CO)_4$. In both cases the linear formation of CO_2 was observed and the rate was proportional to the initial $Fe(CO)_5$ or $Ni(CO)_4$ concentration. The rate equations are expressed as follows;

$$\text{Fe : } d[CO_2]/dt \propto [Fe]_o [CO]^{0.52} [O_2]^{0.48} \quad (1500\text{-}2400\ K) \qquad (2)$$

$$\text{Ni : } d[CO_2]/dt \propto [Ni]_o [CO]^{0.80} [O_2]^{0.20} \quad (1500\text{-}2400\ K) \qquad (3)$$

Using the shock tube coupled with a TOF mass spectrometer, transition metal species in a shock heated $CO\text{-}O_2$ gas mixture containing the carbonyls were investigated. Figure 6 (a),(b), and (c) show consecutive mass spectra obtained in three experiments using gas mixtures containing 0.4% $Cr(CO)_6$, 0.5% $Fe(CO)_5$, and 0.5% $Ni(CO)_4$, respectively. The gas mixtures contained 2.0 - 2.5 % CO from the decomposition of the metal carbonyls.

In the $Cr\text{-}CO\text{-}O_2$ reaction system the mass spectra show the formation of Cr, CrO, CrO_2, and CrO_3. The peak height of CrO_3^+ became very small when the $[CO]/[O_2]$ ratio in the gas mixure was high. It seems that CrO_n (n=0,1,2,3) attained a steady state concentration instead of a thermal equilibrium value. A mechanism involving the oxidation and reduction of CrO_n is proposed (Ref. 4),

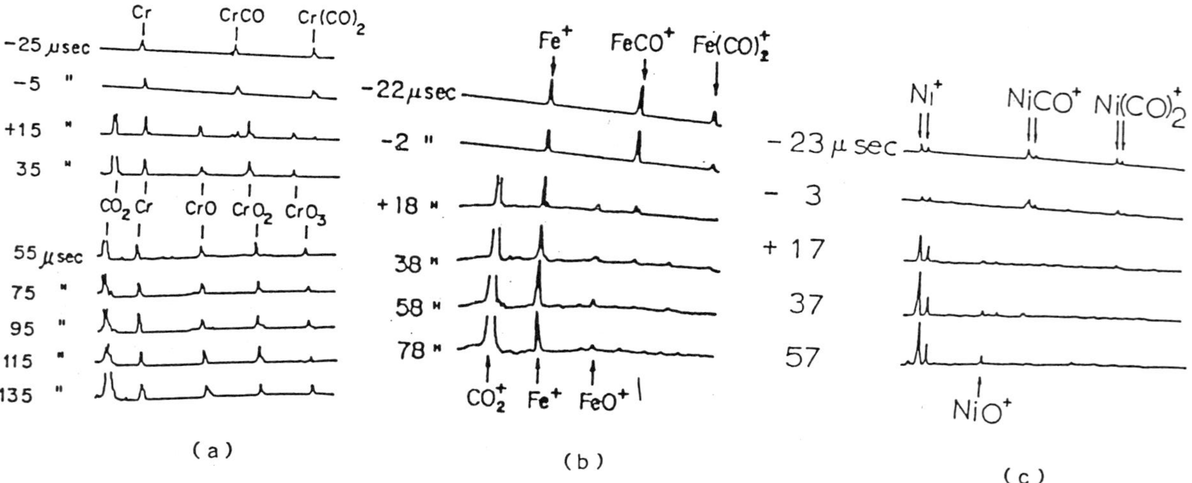

Fig.6. Tracing of consecutive mass spectra from a typical shock tube experiment coupled with a time-of-flight mass spectro meter. Time with respect to the shock reflection is given at the left hand side of each mass spectrum

(a) T=2044°K, P=0.36atm, Gas Mixture; $Cr(CO)_6$ 0.4%, O_2 8%, Ne balance.

(b) T=2290°K, P=0.35atm, Gas Mixture; $Fe(CO)_5$ 0.5%, O_2 8%, Ne balance.

(c) T=2060°K, P=0.31atm, Gas Mixture; $Ni(CO)_4$ 0.5%, O_2 8%, Ne balance.

$$CrO_n + O_2 \rightarrow CrO_{n+2} \qquad (n = 0, 1) \qquad (4)$$

$$CrO_m + CO \rightarrow CrO_{m-1} + CO_2 \qquad (m = 1, 2, 3) \qquad (5)$$

In the Fe-CO-O_2 reaction system the only ions observed were Fe^+ and FeO^+, the peak height of FeO^+ being less than 10% of Fe^+. The intensity of Fe^+ in mass spectra was much smaller than in Cr or Ni reaction system. Thus it is highly probable that partial pressures of Fe and FeO in this reaction system are rather small and species containing Fe must be mainly high molecular weight iron oxides, such as Fe_3O_4 and $(FeO)_3$ (Ref. 6). A similar reaction mechanism involving the oxidation and reduction of iron oxides is proposed (Ref. 3).

In the Ni-CO-O_2 reaction system mass peaks due to Ni^+ (m/e=58 and 60) and NiO^+ were observed, the peak height of NiO^+ being only 10% of Ni^+. Although very small $NiCO^+$ peaks were observed frequently in mass spectra, the existence of NiCO in appreciable amounts could not be conclusively shown. The following reaction mechanism is proposed (Ref. 2),

$$Ni + CO \rightleftarrows NiCO \qquad (6)$$

$$NiCO + O_2 \longrightarrow NiO + CO_2 \qquad (7)$$

$$NiO + CO \longrightarrow Ni + CO_2 \qquad (8)$$

4.GAS-SOLID HETEROGENEOUS OXIDATION OF CARBON MONOXIDE

The oxidation of carbon monoxide in the presence of a solid catalyst was studied in a temperature range 473 - 773 K and at normal pressure. The solid catalysts containing Cr_2O_3, Fe_2O_3, and NiO were prepared through impregnation method using Al_2O_3 as a carrier. The content of active component was 10 wt%. The catalytic activities were measured using a conventional flow-type reaction apparatus. The space velocity was 30,000 vol/vol/h, the contact time being 120 msec. The CO conversion was calculated by monitoring CO and CO_2 concentration in the exit gas by a gas chromatograph.

The CO conversion vs temperature for the three catalysts are shown in Figure 7. The Fe_2O_3 and Cr_2O_3 catalysts had roughly the same activity, and the NiO catalyst a lower activity. In a series of experiments the rate of CO_2 formation was measured using gas mixtures containing CO and O_2 at various ratio. The rate dependence on [CO] and $[O_2]$ are shown in Figure 8(a) and (b), respectively. The rate is stronger dependent on [CO] than on $[O_2]$, and strongest in the case of NiO catalyst.

Using an empirical form the rate of CO oxidation is given by,

$$Cr_2O_3 : \frac{d[CO_2]}{dt} \propto [CO]^{0.66}[O_2]^{0.12} \tag{9}$$

$$Fe_2O_3 : \frac{d[CO_2]}{dt} \propto [CO]^{0.50}[O_2]^{0.29} \tag{10}$$

$$NiO : \frac{d[CO_2]}{dt} \propto [CO]^{1.01}[O_2]^{0.32} \tag{11}$$

Comparing the two sets of rate expressions (Equations (1),(2), and(3), and Equations (9),(10), and (11)), one finds rather remarkable resemblance of the dependency between homogeneous and heterogeneous catalysis. In both reaction systems the rate dependence on [CO] decreases in the order

$$Ni > Cr > Fe$$

and the magnitudes are fairly in good agreement. It might be said in the present reaction system that the catalytic effects of individual gaseous metal atom is retained in the respective solid metal oxide.

5. CONCLUSION

The oxidation of CO was studied in the gas phase homogeneous system in the presence of Cr,Fe, and Ni vapor, and in the gas-solid heterogeneous system in the presence of Cr_2O_3, Fe_2O_3, and NiO catalysts . It is found that

(1) The oxidation of CO is greatly accelerated in the presence of trace amounts of Cr, Fe, and Ni vapor in the

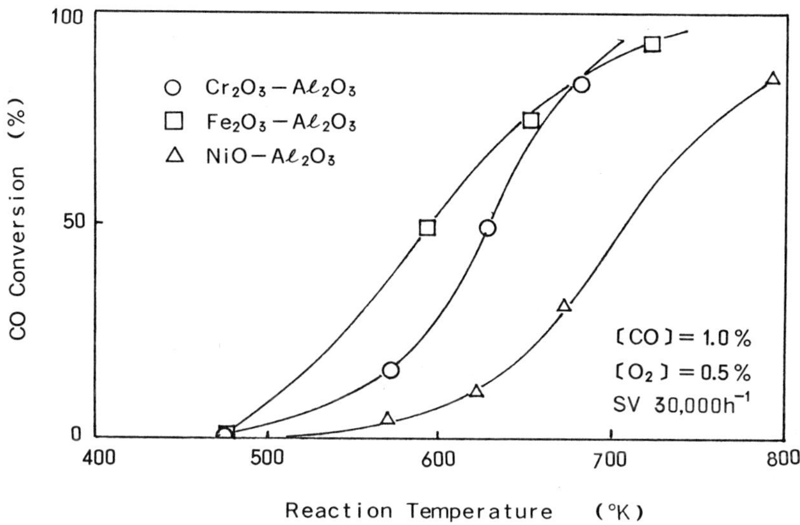

Fig.7. Oxidation of CO in presence of Cr_2O_3, Fe_2O_3 and $NiO-Al_2O_3$ catalyst.

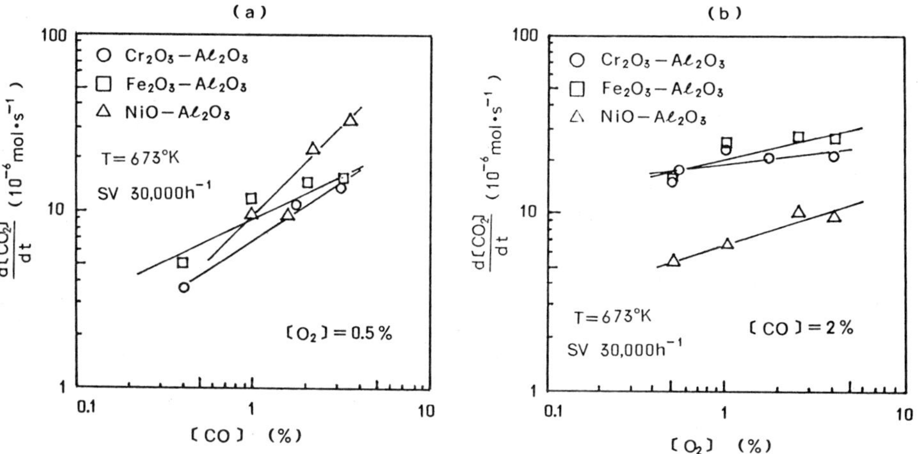

Fig.8. Dependence of d[CO]/dt on [CO] at 0.5% O_2 (a), and on [O_2] at 2% CO (b).

homogeneous system. The rate of CO_2 formation depends linearly on the metal vapor pressure.

(2) A mechanism involving the oxidation-reduction cycle of metal oxide molecules is proposed to account for the gas phase homogeneous catalysis.

(3) The rate dependence on [CO] and $[O_2]$ is similar in both homogeneous and heterogeneous reaction systems.

REFERENCES

(1) M.Katz, Advan. Catal. 8, 177 (1953)
(2) S.Matsuda, J.Phys.Chem. 76, 2833 (1972)
(3) S.Matsuda, J.Chem.Phys. 57, 807 (1972)
(4) T.Izod,G.Kistiakowsky, S.Matsuda, J.Chem.Phys. 56, 1083 (1972)
(5) J.N.Bradley, "Shock Waves in Chemistry and Physics " John Wiley and Sons, Inc. New York (1962)
(6) O.Glemser and H.H.Weizenkorn, Naturwiss. 48, 715 (1961)

COMPUTER-AIDED PROGNOSIS OF THE CATALYTIC PROPERTIES OF METAL ATOMS IN THE GAS-PHASE HYDROGENATION OF CARBON MONOXIDE

V.A. Likholobov, L.I. Mayer, N.N. Bulgakov,
A.V. Fedotov and L.G. Steingauer
Institute of Catalysis, Novosibirsk 630090, USSR

ABSTRACT - Evaluation of the catalytic properties of metal atoms with regard to the gas-phase hydrogenation of carbon monoxide has been done using the "COMSYCAT" program developed here. Structural and energy characteristics of intermediates and products that can form during the catalytic hydrogenation of CO to C_1- and C_2-organic compounds are described. Metals, which catalyze the reaction predominantly in the direction of production of methane, methanol, formaldehyde, ethane, ethanol, acetic acid and ethyleneglycol have been found.

1. INTRODUCTION

In chemical research, computers have been most commonly used for modeling catalytic processes, calculating chemical reactors and modes of their optimum control, as well as for conducting the various computations related to data and documentation processing, storage and extraction of information. In recent years, however, the further development of computer techniques and theory of programming has opened up a possibility to use computers for solving such problems as the retrieval of patterns of organic syntheses, the elucidation of the "structure - reactivity" correlation, and the establishment of molecular structures on the basis of spectroscopic data. Of special interest seem to be the problem of the retrieval of patterns of organic syntheses and based on it a more general problem of the generation of chemical reaction mechanisms. The pioneering works in this area are those by Corey et al. (Ref. 1) and Wipke et al. (Ref.2). The results reported by several teams of scientists (Refs. 3-8) concern the construction of computer programs to be used for determining an optimum route of the synthesis of an organic compound, the so-called "computer-synthesis".

The programs proposed are based on (i) the use of experimental data and information about organic reactions, which are comprehensively studied and are of synthesis interest (information-oriented empirical programs (Refs. 1-5)); or (ii) the retrieval of chemical reactions, founded on logic, that is, on combinatorial selection of the possible types of the redistribution of the bonds between atoms of a given structure (logical structurally oriented programs (Refs. 6-8)). Most existing programs, successfully applied to synthesis, belong to the first type and are restricted to definite problems and definite known reactions such as Dils-Alder's reaction and synthesis of peptides and proteins.

An empirical approach in the "computer synthesis" has two serious limitations: (i) experimental chemical information, typically in letters or graphs, must be transformed to suit computer operation, and (ii) results of the "computer synthesis" often do not present a novelty for an expertised, well-informed chemist having great powers of intuition. While the former limitation refers to the efficiency of programs and becomes less important with the progress in technique and theory of programming, the latter limitation is a typical feature of information-oriented systems. Along with the retrieval of an optimal chain of known reactions for the synthesis of compounds of a given structure, we outline the spectrum of other important problems usually facing the chemist-

-experimentator;
(1) Determination of possible chemical reactions in the system of a given structure (this problem is significant by itself as a preliminary "computer experiment" prior to an actual study of the reaction of interest);
(2) Determination of possible mechanisms of given chemical transformations;
(3) Search for reactions new in principle, which meet the requirements for a given class of compounds;
(4) Determination of possible catalysts for given chemical reactions.
These problems can be solved most efficiently with the use of programs based on logic.

To realize the logically oriented programs, one can resort to an algorithm proposed in Refs. 6,7 for the planning of the organic synthesis and for the prognosis of the composition of products. The basis of this algorithm is a mathematical model of structural chemistry, in terms of which (Ref. 7), ensembles of molecules or molecules themselves are represented as the so-called bond-electron matrices (BE matrices after (Ref. 7)). Rows and columns of this matrix indicate the types of atoms in molecules; diagonal elements give the number of free electrons in a given atom, while nondiagonal ones - the bond order (number of electron pairs) between the respective atoms. The BE matrix is symmetrical, and the sum of elements in a row or column indicates the overall number of electrons in the external electron layer (orbital) of an atom.

In terms of representation (Ref. 8), a chemical reaction corresponds to the transformation of the BE matrix of parent substances X to that of products Y. This transformation is carried out by the reaction matrix $\|R\|$. In matrix representation the chemical reaction $\|X\| \rightarrow \|Y\|$ is written as:

$$\|X\| + \|R\| = \|Y\|.$$

Note some features of the $\|R\|$ -matrix. Nondiagonal elements $\pm r_{ij}$ indicate the number of bonds between atoms A_i and A_j that are formed $(r_{ii} > 0)$ and dissociated $(r_{ij} < 0)$. Diagonal elements r_{ij} indicate how many free electrons are obtained $(r_{ii} > 0)$ or lost $(r_{ii} < 0)$ by the atom A_i. The sum of all elements of the matrix $\|R\|$ is zero, which accounts for the fact that the overall number of valence electrons remains unchanged during the reaction. Since the chemical reaction is complicated and involves several steps, the $\|R\|$ -matrix can show both a single rearrangement of the bonds and their totality, i.e. $\|R\| = \sum_i R_i$. The reaction matrix can be transformed by several methods: the transformation of the $\|R\|$ -matrix by the basis vectors (modeling of heterolytic reactions or by combinatorial selection.

The transformation of the reaction matrix provides the possibility of estimating the totality of elementary rearrangements of bonds and resulting intermediates leading to a desired product. However, as is readily seen, the number of combinations of the reactive matrix transformation can be very high. How can one then choose a most optimal combination?

The criteria of selection of elementary rearrangements and intermediates were considered in (Refs. 8-10). For example, in (Ref. 8) restrictions imposed on the number of atoms and electrons, simultaneously participating in the reaction (complexity of the reaction), as well as the rules of selection related to the matrix transformation, were used as criteria. To develop the criteria, such physico-chemical characteristics as atomic mass, valence and nonvalence interactions (Ref. 9), steric parameters (Refs. 9,10), atomic charge and electronegativity (Ref. 10) were also used. These physico-chemical parameters were introduced in (Ref. 9) as a notion "information content" of the molecular system. In terms of this notion, of a wide variety of generated compounds, only those are selected which satisfy the requirements of a maximum structural identity to the starting substances or reaction products; in other words, an intermediate should "contain maximum information" about the compound, from which it has been formed or about the compound into which it has been transformed. The necessity of a thermodynamic selection of generated final reaction products was mentioned in (Ref. 10). The selection was made via assessment of the reaction heat using thermochemical data and Allen's additive schemes (Ref. 11).

To solve the problem posed by us on the development of the algorithm that would allow the realization of the "computer synthesis of catalysts", we used an energy criterion as a criterion of the selection of intermediates. With this criterion, each elementary bond rearrangement resulting in a given reaction product was estimated taking into account not only the number of simultaneously forming and breaking bonds , but also the energy consumed by the system for such rearrangement. The principles mentioned above were used for constructing the structurally oriented program "COMSYCAT" ("computer synthesis of catalysts") (Refs. 12-14). The results of the realization of this program for the case of the gas-phase reaction, $nCO + mH_2$ (n = 1,2, m = 1,2), which takes place in the absence of catalysts, were described by us in (Ref. 14).

In this work, this reaction is regarded as catalytical, the catalyst being an isolated metal atom having one free valency (designated by $M^{\bullet}$ in the following discussion) and modeling, for example, a mononuclear active center belonging to the metal crystal face. The objective of this consideration in terms of the program "COMSYCAT" was to obtain data on the effect of the nature of the metal atom as catalyst upon the mechanism of the formation and the composition of the products of CO hydrogenation.

2. DESCRIPTION OF THE ALGORITHM OF THE PROGRAM "COMSYCAT"

The block scheme of the program developed by us is represented in Fig. 1. Logical and computational procedures of the program are as follows.

1. Chemical reaction, catalyzed by M, is described in terms of the mathematical model of structural chemistry (MMSC) as a sequence

$$\{X + M\} \longrightarrow \{ZM\} \longrightarrow \{Y + M\},$$

where $\{X + M\}$, $\{ZM\}$ and $\{Y + M\}$ are sets of the starting substances, intermediates and reaction products, respectively.

2. Molecular system $\{X + M\}$ is given and is described by a structural bond electron matrix (BEM), which, for example for the $CO + H_2 + M^{\cdot}$ system, will have the form:

$$\begin{array}{c|cc|cc|c} & C & O & H & H & M \\ \hline C & 2 & 2 & 0 & 0 & 0 \\ O & 2 & 4 & 0 & 0 & 0 \\ \hline H & 0 & 0 & 0 & 1 & 0 \\ H & 0 & 0 & 1 & 0 & 0 \\ \hline M & 0 & 0 & 0 & 0 & 1 \end{array} \equiv \begin{array}{c|c|c|c} & CO & H\ H & M \\ \hline C,\ O & G_1 & 0 & 0 \\ \hline H,\ H & 0 & G_2 & 0 \\ \hline M & 0 & 0 & G_3 \end{array} \qquad G_1 = \begin{vmatrix} 2 & 2 \\ 2 & 4 \end{vmatrix} \quad G_2 = \begin{vmatrix} 0 & 1 \\ 1 & 0 \end{vmatrix} \quad G_3 = |1|$$

where G_1, G_2 and G_3 are the group BE matrices for CO, H_2 and M, respectively.

3. Transformation of starting compounds to products ${}^{r}Y$ is described by the reactive matrix ${}^{r}R$:

$$\| {}^{r}R \| = \sum_{s} \| {}^{r}R_{s} \| ,$$

where r is the number of the route leading to product ^{r}Y; s is the number of the step of route r.

Matrices $^{r}R_{s}$ are constructed by a logical analysis of the combinations of the group block matrices (ρ) that describe formally possible rearrangements of the bonds upon a monomolecular decomposition or bimolecular collision. The main types of the group reactive matrices can be found in (Ref. 14).

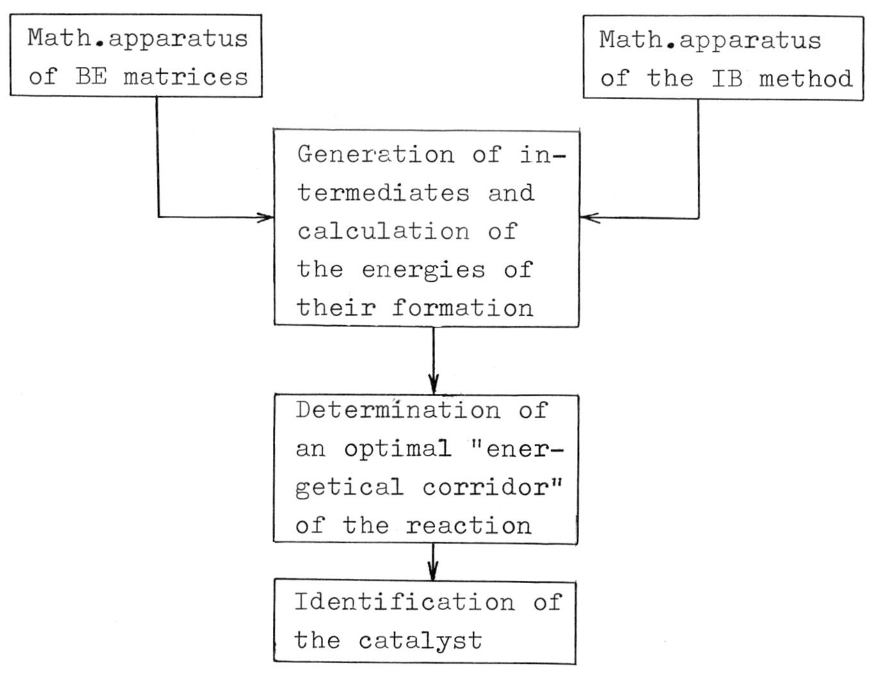

Figure 1. Block scheme of the program COMSYCAT

4. The structure of intermediate $^{r}Z_{s}$, formed at the step s of route r, was determined from its BEM calculated by the equation

$$\|^{r}Z_{s}\| = \|^{r}Z_{s-1}\| + \|^{r}R_{s}\|$$

5. Selection of most probable generated intermediates was made on the basis of the estimates of formation energies, $E({}^{r}Z_{s})$. In this case to each molecular system, described by BEM, corresponds the energy matrix of the bonds, $H({}^{r}Z_{s})$. Elements of this matrix, H_{ii} and H_{ik}, are calculated by the interacting bonds method (IBM) (Ref.19)

$$H_{at} = \sum_{i} \nu_i (2-\nu_i) E_i - \sum_{i}\sum_{k} \nu_i \nu_k \Delta ik$$

taking into account nondiagonal elements of the structural matrix Z_{ik} by the following equations:

$$H_{ii} = 2E_i(\nu_i - 1)$$
$$H_{ik} = Z_{ik}\,\Delta_{ik}\nu_k,$$

where $i = 1,\ldots,L$, $k = 1,\ldots,L$, matrix $H({}^{r}Z_{s})$ has dimension L equal to the number of various types of bonds in molecular system ${}^{r}Z_{s}$; ν_i and ν_k are vectors of bond coefficients whose values are calculated via a system of linear equations, as described in (Ref. 14). E_i and Δ_{ik} are semiempirical parameters that characterize, respectively, the energy of the formation of the i-th bond and the energy of the interaction between the i-th and k-th bonds.

The formation of an intermediate with a maximum atomization energy seemed to be most probable. The sequence of intermediates between X and Y that are characterized by a maximum atomization energy composed an "energy corridor" of the reaction.

6. Logical operation of the variation of the nature of the metal catalyst was performed by varying given values of parameters E_{M-H} (where A = C, H and O atoms) within the region from minimum to maximum value with a step e_{M-A}. Thus, by dividing the interval of E_{M-C} parameters into n parts, that of E_{M-H} into m parts

and of E_{H-O} into l parts, for each reaction route resulting in product ^{r}Y and involving intermediates with M-C, M-H, and M-O bonds, we obtain n x m x l energy paths that differ in energies of the formation of intermediates. Regarding the transformation rate, an optimal combination was that of parameters E_{H-C}, E_{M-H} and E_{M-O}, with which the reaction energy $X \longrightarrow {}^{r}Y$ was redistributed most uniformly over the steps of the reaction. For example, for an N-step route, most optimal was the combination (ξ) of parameters E_{M-A}, for which values λ_s (s = 1 ... i, ...k...N)

$$\lambda_s = |E(^{r}Z_s) - E(^{r}Z_{s+1})| - \left|\frac{E_x - E_y^r}{N}\right|$$

did not differ significantly from each other, that is,

$$\lim(\xi)(\lambda_i - \lambda_k) = 0.$$

7. Logical operation of the identification of catalysts was performed by using the table of parameters E_{M-A}, calculated for particular metals from experimental data on the formation heats of their oxides, hydrides and carbonyls. The metal, whose E_{M-A} values were the closest to the optimal E_{M-A} values, found from item 6, was considered to be an optimal catalyst of X transformation to product ^{r}Y.

The program is written in FORTRAN, calculations were made on an EC-1050 computer.

In the calculations, the parameters E_{M-A} were varied within the ranges: E_{M-C}= 50-150 kcal/mol, E_{M-H}= 20-120 kcal/mol; E_{M-O}= 50-200 kcal/mol with e_{M-A}= 10 kcal/mol. The bond parameters , E_{M-A}, of particular metals, determined from experimental heats of the formation of carbonyls, hydrides and oxides of these metals, are listed in Table 1.

TABLE 1. E_{M-A} parameters for some metals determined from the formation heats of their complexes

Me-	E_{M-C}	E_{M-H}	E_{M-O}	Starting compound for calculation	Refs.
Ti	112.8	80.0	140.5	$TiC(c)$, $TiH_2(c)$, $TiO_2(c)$	15
Zr	120.8	87.0	145.3	$ZrC(c)$, $ZrH_2(c)$, $ZrO_2(c)$	15
Hf	122.3	-	149.2	$HfC(c)$, $HfO_2(c)$	15
V	98.2 (108.5)*	-	131.1	$V(CO)_6(g)$, $VO(c)$	15
Nb	125.0	-	151.5	$NbC(c)$, $NbO_2(c)$	15
Ta	130.0	-	-	$TaC(c)$	15
Cr	95.6	66.5	114.7	$Cr(CO)_6(c)$, $CrH(c)$, $CrO_2(c)$	15
Mo	112.6 (113.4)*	-	135.8	$Mo(CO)_6(g)$, $MoO_2(c)$	15
W	127.1 (129.7)*	-	151.8	$W(CO)_6(g)$, $WO_2(c)$	15
Mn	110.5	-	105.7	$Mn(CO)_5(g)$, $MnO(c)$, $MnO_2(c)$	15
Re	102.6	-	143.7	$Re_2(CO)_{10}(g)$, $ReO_3(c)$	15
Fe	91.9	-	111.3	$Fe(CO)_5(g)$, $FeO(c)$	15
Ru	-	-	120.5	$RuO_2(c)$ 15	15
Os	-	-	162.0	$OsO_4(c)$	15
Co	101.8	56.80	110.3	$Co(CO)_4(g)$, $Co_2(CO)_8(g)$ $HCo(CO)_4(g)$, $CoO(c)$	15
Rh	-	-	108.5	$Rh_2O_3(c)$	15
Ir	-	-	118.1	$IrO_2(c)$	15
Ni	98	68	110.0	$Ni(CO)_4(l)$, $NiO(c)$	15
				$Ni_2H(c)$	16
Pd	102.8	67.5	86.0	$PdO(c)$	15
				$Pd(CO)_4(g)$	17
				$Pd_2H(c)$	16
Pt	107.5	80.2	94.1	$Pt(CO)_4(g)$	17
				$Pt_2H(c)$, $PtO(c)$, $Pt_3O_4(c)$	16

TABLE 1 (continued)

Cu	-	66.4	90.0	CuH(c), CuO(c), Cu_2O(c)	15
Ag	-	53.0	69.0	AgH(c)	16
				Ag_2O(c), Ag_2O_2	15
Au	-	-	71.5	Au_2O_3(c)	18
Zn	74.8	-	92.5	$Zn(CH_3)_2$(g), ZnO(c)	15
Cd	87.5	-	83.1	$Cd(CH_3)_2$(g), CdO(c)	15
Hg	77.7	-	-	$Hg(CH_3)_2$(g)	15
Al	100.5	99.4	127.8	$Al(CH_3)_3$(c),	15
				$Al(C_2H_5)_2H$(l)	15
				Al_2O_3(c)	15
Ga	90.0	-	103.0	$Ga(CH_3)_3$(g), Ga_2O(c)	15
In	101	-	99.6	$In(CH_3)_3$(g), In_2O_3(c)	15
Tl	-	-	79.0	Tl_2O_3(c)	15
Ge	94.7	85.0	112.8	$Ge(C_2H_5)_4$(l),	15
				$HGe(C_2H_5)_3$(l), GeO_2(c)	15
Sn	84.3	70	76.7	$Sn(CH_3)_4$(g),	15
				$HSn(CH_3)_3$(g), SnO(c)	15
Pb	82.7	-	81.7	$Pb(CH_3)_4$(g), PbO_2(c)	15

* In brackets E_{M-C} parameters calculated from metal carbides, M_2C, are given.

3. RESULTS OF COMPUTATION AND THEIR DISCUSSION

3.1. Hydrogenation of CO Giving C_1 Organic Compounds

Fig. 2 shows the scheme of the mechanism of the formation of C_1 organic compounds from CO and H_2 catalyzed by the metal atom M, which was calculated using the algorithm described above. This mechanism consists of the two reaction routes: (a) via a primary formation of carbonyl complexes of M, and (b) via a primary formation of hydride complexes of M. The contributions of these routes depend upon the parameters E_{M-C} and E_{M-H}. In general, the scheme indicates the possibilty of occurrence of successive transformations:

$$CO+H_2 \longrightarrow H_2CO \xrightarrow{H_2} CH_3OH \xrightarrow{H_2} CH_4.$$

The calculations have shown that the formation of a thermodynamically unfavourable formaldehyde requires that the values of parameter E_{M-H} lie within the range from 20 to 60 kcal/mol and those of E_{M-C} within the range from 90 to 120 kcal/mol. The formation of methanol may occur via two different routes: (1) via metal formyl and then hydroxymethyl complexes (E_{M-C}= 80-100 kcal/mol and E_{M-H}= 60-70 kcal/mol) and (2) via the formation of metoxy intermediates (E_{M-O}= 90-115 kcal/mol). The formation via the route (1) occurs with E_{M-C}= 90-100 kcal/mol, whereas route (2) is more probable for metals having the maximum values of parameter E_{M-O}, that is, equal to or higher than 130-150 kcal/mol.

The optimal values of E_{M-A} for C_1-products are given in Table 2. Metals, for which these values are more typical, are also listed in the table.

TABLE 2. Ranges of optimal parameters E_{M-A} for the formation of C_1-products and metals corresponding to them

Products	E_{M-C}	E_{M-H}	E_{M-O}	Metals
CH_4	90-130	50-80	80-150>	Ni, Pt, Ir, Co, Pd, Fe, Ru, Cr, Ta, Re, Os, Ti, W, Nb
CH_3OH	80-100	60-70	90-120	Cu, Cr, Pd, Zn
H_2CO	90-110>	40-60	100-110	Rh, Fe, Co

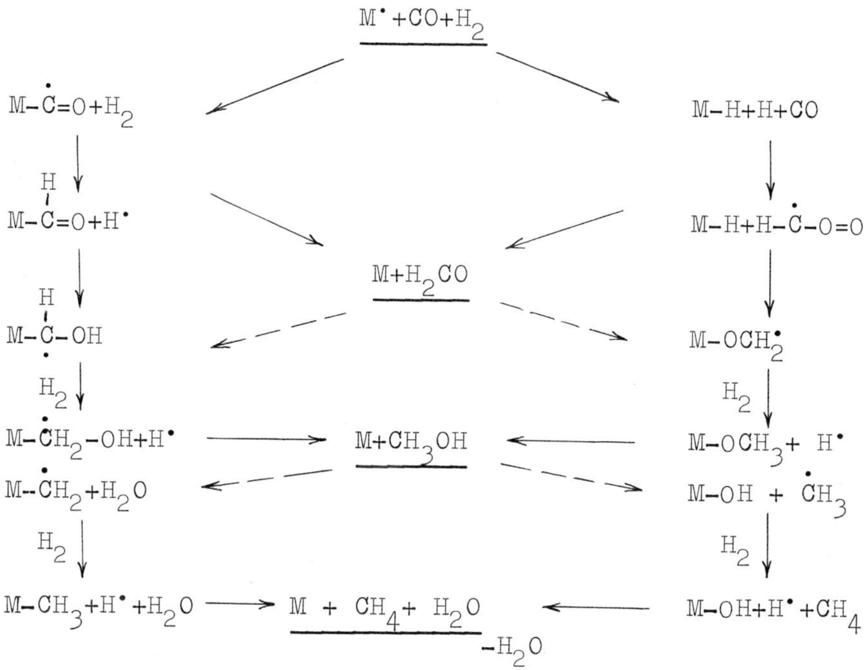

Figure 2. Scheme of the formation of C_1-products in the $CO+nH_2$ reaction.

3.2. CO Hydrogenation Giving C_2-Organic Compounds

The mechanism of the formation of C_2-organic products obtained by computation is represented in a schematic fashion in Fig. 3. The calculations evidence that the formation of ethylene glycol is most probable in the case of low values of parameter $E_{M-H} < 40-60$ kcal/mol and values of $E_{M-C} > 100$ kcal/mol, that is, within the range of E_{M-A} values optimal for the formation of formaldehyde. Thus ethylene glycol may be the direct product of formaldehyde transformation, the formation of the latter demanding the greatest energy consumptions.

The formation of other C_2-products occurs through the steps of intramolecular rearrangements of carboxyformyl complex M-CH(OH)-$\dot{C}$=O; moreover, depending upon the value of parameter E_{M-O}, the routes resulting either in acetic acid ($E_{M-O} < 110$ kcal/mol) or in ethanol and ethane ($E_{M-O} > 110$ kcal/mol) may be realized.

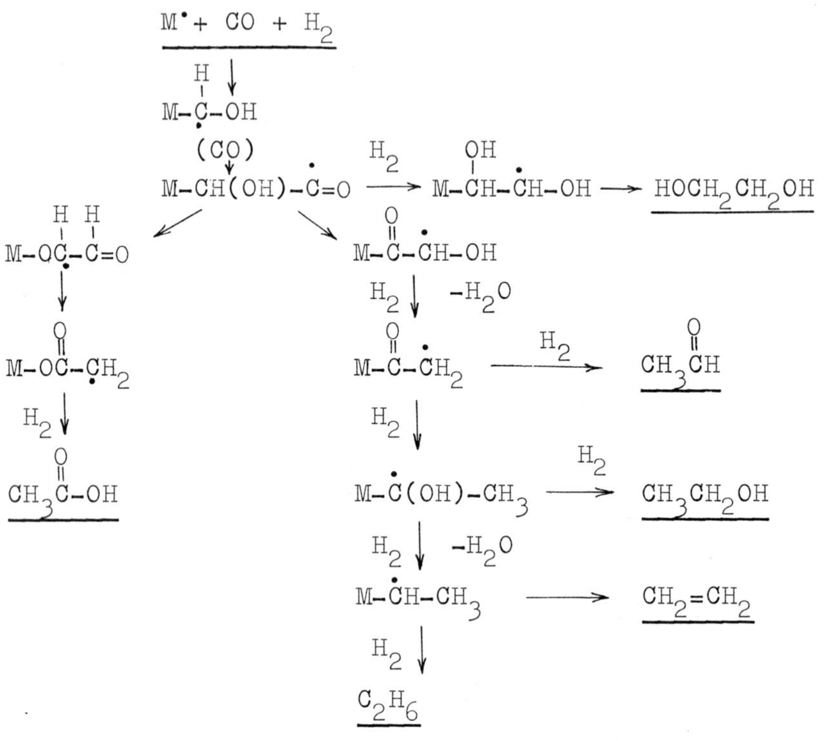

Figure 3. Mechanism of the formation of C_2-products in the 2CO + nH_2 reaction

The values of parameter E_{M-A} that are optimal for a given C_2-product are listed in Table 3 together with the metals for which these values are most typical.

CONCLUSION

The results of the computer-aided modeling of the hydrogenation of carbon monoxide on an isolated metal atom prove this reaction to be possible on a monoatomic active center. The factor determining the direction of the reaction is the relation of the values of the energies of M-C, M-H and M-O bonds. Interestingly, during the process of computer generation of intermediates, at very high values of parameter E_{M-O} (characteristic metals are Hf, Zr, Hb and Ti), stable (with respect to the molecular system "M + nCO + mH_2") complexes of M with the products of a "soft" CO hydrogenation, namely, formaldehyde or glyoxale, are produced. An organic radical in these complexes is bound to a metal through an oxygen atom.

TABLE 3. Ranges of optimal values of parameter E_{M-A} for the formation of C_2-products and metals corresponding to them

Products	E_{M-C}	E_{M-H}	E_{M-O}	Metals
C_2H_6	90-100	50-70	110-120	Ru, Ni, Co, Fe, Ir, Ge, In
C_2H_5OH	~100	60-70	90-120	Cr, Mn, Cu
CH_3COOH	90-100	40-60	<110	Co, Fe, Rh
$C_2H_6O_2$	>100	40-60	110-130	Rh, Co, Ir, Ru, V

Due to a low value of parameter E_{M-H} for the above-mentioned metals, no further hydrogenolysis of a M-O bond takes place. Thus, these complexes may be regarded as products of a stoichiometric interaction of M with CO and H_2 that are not involved in the subsequent catalytic process.

Most optimal, from the standpoint of the catalytic process occurrence, are metals with E_{M-C}= 90-110 kcal/mol, E_{M-H}= 60-80 kcal/mol and E_{M-O}= 90-120 kcal/mol. Metals showing such parameters are Mn, Fe, Ru, Co, Rh,

Ir, Ni, Pd and Pt. These metals are most typical catalysts of carbon monoxide hydrogenation, as ascertained by experiments (Ref. 20).

REFERENCES

1. E.J. Corey and W.T. Wipke, Science,166,178 (1969).
2. W.T. Wipke and P. Gund, J.Am.Chem.Soc.,96,99 (1974).
3. H. Gelernter, H.S. Sridharan and A.J. Hart, Topics Curr.Chem.,41,113 (1973).
4. J.B. Hendrickson, J.Am.Chem.Soc.,23,6847 (1971).
5. M. Bersohn and A. Esack, Chem.Rev.,76,279 (1976).
6. I. Ugi, J. Bauer, J. Brandt, J. Friedrich, J. Gasteiger, C. Jochum and W. Schubert, Angew.Chem.Int.Ed. Engl.,18,2,111-122 (1979).
7. J. Dugundji and I. Ugi, Topics Curr.Chem.,39,3 (1973).
8. Y. Yoneda, Bull.Chem.Soc.Jpn.,52,1,8-14 (1979).
9. S.V. Dozmorov, React.Kinet.Catal.Lett.,19,3-4,289-295 (1982).
10. J. Gasteiger, O. Jacob and V. Strauss, Tetrahedron, 35,139 (1979); 1419 (1979).
11. T.L. Allen, J.Chem.Phys.,31,1039 (1959).
12. L.I. Mayer, L.G. Shteingauer, A.V. Fedotov, G.S. Yablonskii and V.A. Likholobov, Proc. VIIth All-Union Conf."Use of computers in spectroscopy of molecules and in chemical research" (in Russian), Novosibirsk, 6-8 September, 1983, pp. 106-107.
13. L.I. Mayer, L.G. Shteingauer, A.V. Fedotov and V.A. Likholobov, Proc.All-Union Meeting "Chemical Synthesis Based on C_1 Molecules" (in Russian), Moscow, 3-7 December, 1984, p. 8.
14. L.I. Shtokalo, L.G. Shteingauer, V.A. Likholobov and A.V. Fedotov, React.Kinet.Catal.Lett.,26,3-4,227-233 (1984).
15. Thermal Constants of Substances. Ed.V.R. Glushko (1972-1974).
16. V.A. Sobyanin, N.N. Bulgakov and V.V. Gorodetskii, React.Kinet.Catal.Lett.,6,2,125-132 (1977).
17. M. Moskovits and G.A. Ozin, J.Mol.Struct.,32,11,71 (1976).
18. V.A. Sobyanin, V.V. Gorodetskii and N.N. Bulgakov, React.Kinet.Catal.Lett.,7,3,285-290 (1977).
19. N.N. Bulgakov, Yu.A. Borisov and V.V. Popovskii, Kinet.Katal.,14,468 (1973).
20. V. Bonec, Catalysis,5,48-79 (1982).

DESIGN OF HIGHLY SELECTIVE OLEFIN METATHESIS CATALYSTS AND REACTION INTERMEDIATES ON THE DESIGNED CATALYSTS

Ken-ichi Tanaka[1)] and Katsumi Tanaka[2)]

1) The Institute for Solid State Physics, The University of Tokyo, Roppongi, Minato-ku, Tokyo 106, Japan
2) Research Institute for Catalysis, Hokkaido University, Sapporo 060, Japan

ABSTRACT: Novel olefin metathesis catalysts were obtained by treating supported molybdenum oxides with $Sn(CH_3)_4$. Some of the catalysts are very active for olefin metathesis reaction but promote no hydrogen scrambling reaction of olefins. A mixture of cis-2-butene-d_o and cis-2-butene-d_8 underwent the metathesis reaction and yielded cis-2-butene-d_4 and trans-2-butene-d_4 in the ratio of 8 to 2, and a mixture of trans-2-butene-d_o and trans-2-butene-d_8 gave trans-2-butene-d_4 selectively. A cross metathesis between Z-1-d_1-propene ($^{D}_{H}{>}C{=}C{<}^{CH_3}_{H}$) and d_6-propene underwent a selective metathesis reaction which yielded E-1-h_1-propene ($^{D}_{H}{>}C{=}C{<}^{CD_3}_{H}$) and Z-1-h_1-propene ($^{H}_{D}{>}C{=}C{<}^{CD_3}_{D}$) in the ratio of 8 to 2. Conformation retaining metathesis in the cis or trans olefin metathesis reaction as well as the transfer of Z-1-d_1 conformation of methylene in the cross metathesis reaction are explained by a preferable conformation originated by the 1,3-disubstituents of metallacyclobutanes. From these results, it was concluded that the productive metathesis of propene, 2 C=C-C $\longrightarrow$ C=C + C-C=C-C, is brought about by the alternative reaction of propene with methylidene and ethylidene, but the degenerate metathesis of propene, C=C-C + $C^*{=}C^*{-}C^*$ $\longrightarrow$ $C^*{=}C{-}C$ + $C{=}C^*{-}C^*$, is caused by the predominant reaction of propene with ethylidene.

1. INTRODUCTION

Synthesis of a non-Fischer type carbene complex of $(CO)_5W{=}C(ph)_2$ by Casey et al.(1) in 1973 made a begining of a new metal alkylidene chemistry, and a significant progress was the synthesis of a terminal methylidene complex of $(\eta\text{-}Cp)_2Ta(CH_3)CH_2$ by Schrock et al.(2) and a bridged methylidene complex of $(\eta\text{-}Cp)_2(CO)_4Mn_2(\mu^2\text{-}CH_2)$ by Herrmann et al.(3) in 1975. Since then, numerous number of transition metal alkylidene complexes have been synthesized in this decade, and characteristic reaction of alkylidene ligands has been studied in relation to the olefin metathesis , olefin homologation, and the Fischer-Tropsch hydrocarbon synthesis reactions.

If olefin metathesis reaction over the heterogeneous catalysts occurs by alkylidene and metallacyclobutane mechanism which has been accepted in the homogeneous catalysis, the catalyst surfaceshould be active for the formation of alkylidene species from olefins at least in the initial stage of the reaction. In other word, if the surface is defective in the formation of alkylidene species from olefins, no metathesis reaction will occur on this surface. However, if such a defective function of the surface could be recovered by furnishing alkylidene species, the inactive surface will change to be active for the olefin metathesis reaction. Based on this thought, the authors tried to design a catalyst which is only active for olefin metathesis reaction but is inactive for the hydrogen scrambling reaction of olefins.

2. EXPERIMENTAL

A series of supported MoO_3 catalysts were prepared by immersing support oxides in a solution of ammonium paramolybdate, and were followed by calcination and oxidation with O_2 at 500 C. A partly reduced catalysts, MoO_{3-x}/support, were prepared in a closed circulation system byperforming following redox reactions; reduced with H_2 at 500 C for 1 hour and reoxized with a mixture of N_2O and H_2 at 200 C. By this procedure, we can adjust the molybdenum oxide surface at a certain level of reduction (4). MoO_3/support or MoO_{3-x}/support of 10 to 15 mg prepared in a closed circulation system of about 280 ml. was exposed to a 60 Torr of He containing about 1 Torr of $Sn(CH_3)_4$ at room temperature for 30 minutes. After then, the catalyst was evacuated for 30 minutes at room temperature and was subjected to the olefin metathesis reaction. Gas phase composition was analyzed by using on-line gas chromatography, and the deuterium labelled olefins were analyzed by the mass spectrometer with low ionization voltage (15 eV). The analysis of Z-1-d_1, E-1-d_1, Z-1-h_1, and E-1-h_1-propene was carried out by the microwave spectroscopic analysis

3. RESULTS AND DISCUSSION

3.1. Preparation of Functionally Designed Catalysts

Currently proposed favoured mechanism for the olefin metathesis reaction suggests that the reaction involves metal alkylidene and metallacyclobutane intermediacies. If the olefin metathesis reaction on heterogeneous catalyst is caused by the same mechanism, the catalyst surface should fulfill the following two functions:

(i) Initiation process which involves the formation of alkylidene species from olefins.

(ii) Propagation process in which in which olefin reacts with alkylidene species to form metallacyclobutanes.

Accordingly, if the surface is defective for the formation of alkylidene species from olefins, no metathesis reaction is brought about even if the surface is active for the propagation process such as shown in Scheme 1.

However, if the defective function of the surface for the formation of alkylidene species could be supplied by furnishing alkylidene species, the surface will change into metathesis active catalyst. Based on this concept, we tried to prepare a novel catalyst which is active for the olefin metathesis reaction but is inactive for the isomerization reaction as well as for the hydrogen scrambling reaction of olefins. The MoO_3/supports listed in Table 1 were inactive for olefin metathesis reaction as well as for the isomerization reaction. We found that some of these inactive oxides changed into very active olefin metathesis catalysts by treating with $Sn(CH_3)_4$ at room temperature as listed in Table 1.

It should be emphasized here that the treatment with $Sn(CH_3)_4$ does not enhance the activity for the isomerization as well as for the hydrogen scrambling reactions of olefins. In this sense, we prepared " Functionally Designed Surface " by furnishing methylidene species by the adsorption of $Sn(CH_3)_4$. More pronounced activation was attained by treating partly reduced catalysts, MoO_{3-x}/supports, with $Sn(CH_3)_4$ as shown in Table 1. In this case, the activity was enhanced almost

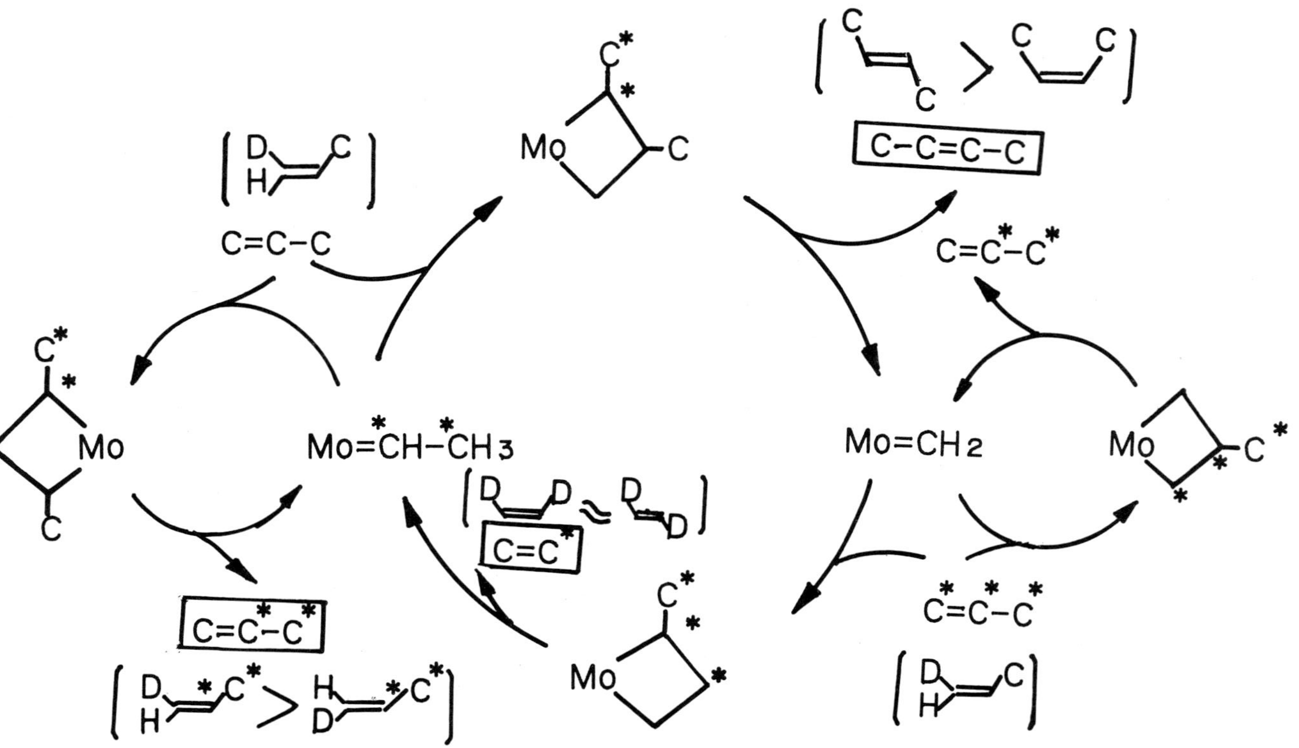

Scheme 1. Catalytic reaction cycles for productive metathesis and degenerate metathesis of propene.

10^3 times by the treatment with $Sn(CH_3)_4$.

TABLE 1. Relativistic activity of MoO_3/support and MoO_{3-x} catalysts treated and untreated with $Sn(CH_3)_4$ for propene metathesis reaction.

	Supports						
	$-Al_2O_3$	SiO_2	TiO_2	ZrO_2	SnO_2	MgO	ZnO
MoO_3	-	-	-	-	-	-	-
MoO_3 + $SnMe_4$	++++	+++	+++	+	+++	-	-
MoO_{3-x}	++	++	+*	+	-	-	-
MoO_{3-x} + $SnMe_4$	+++++	++++	++++	++++	-	-	-

* Number of + sign shows the order of relative turnover frequency to MoO_{3-x}/TiO_2; 2.0×10^{-7} molecules/Mo·sec.

3.2. Preferable Conformation of Metallacyclobutane Intermediates

As was pointed out in the preceeding paper(5), " Stereo Selectivity " and " Conformation Transfer Selectivity " in olefin metathesis reaction are distinguishable properties as described by the following equations;

(i) Stereo selectivity in olefin metathesis;

$$2\ R\text{-}CH{=}CH_2 \longrightarrow CH_2{=}CH_2 + R\text{-}CH{=}CH\text{-}R\ (trans)\ \text{or}\ R\text{-}CH{=}CH\text{-}R\ (cis)$$

(ii) Conformation transfer selectivity in olefin metathesis;

$$2\ R\text{-}CH{=}CH\text{-}R'\ (cis) \longrightarrow R\text{-}CH{=}CH\text{-}R\ (cis) + R'\text{-}CH{=}CH\text{-}R'\ (cis)$$

$$2\ R\text{-}CH{=}CH\text{-}R'\ (trans) \longrightarrow R\text{-}CH{=}CH\text{-}R\ (trans) + R'\text{-}CH{=}CH\text{-}R'\ (trans)$$

In order to obtain an intrinsic conformation transfer selectivity, the metathesis reaction between 2-butene-d_o and 2-butene-d_8 was performed over the three catalysts, MoO_3/TiO_2 + $Sn(CH_3)_4$, MoO_{3-x}/TiO_2, and $MoO_{3-x}/TiO_2 + Sn(CH_3)_4$. If a 1 : 1 mixture of cis-2-butene-d_0 and cis-2-butene-d_8 or of trans-2-butene-d_0 and trans-2-butene-d_8 was added onto the catalysts, 2-butene-d_4 was mainly formed by metathesis reaction. That is, the metathesis reaction proceeds with little hydrogen scrambling. Fraction of cis-2-butene-d_4 in the 2-butene-d_4 produced by the metathesis reaction is shown in Figure 1.

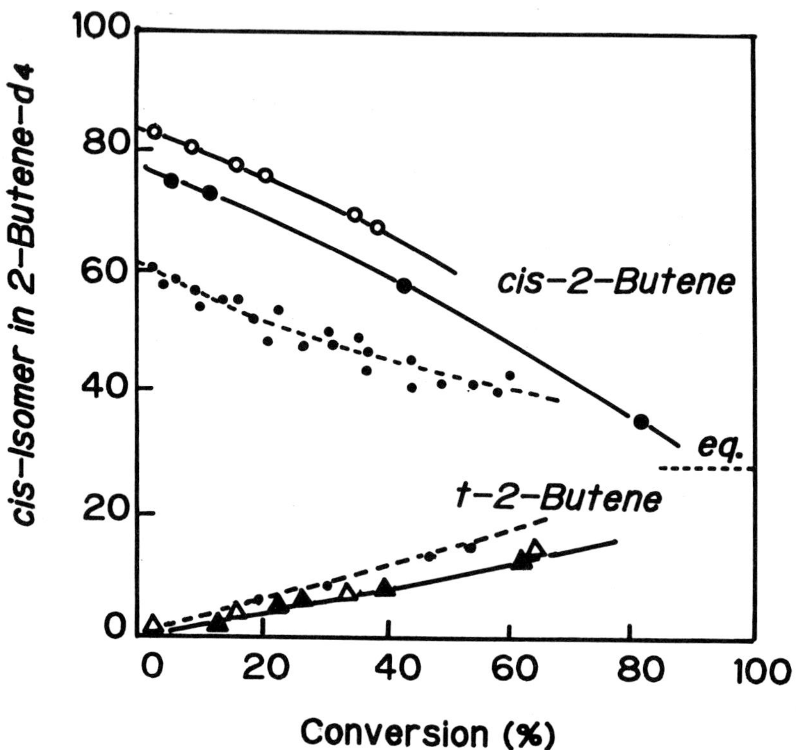

Figure 1. Fraction of cis-isomer in 2-butene-d_4 formed by metathesis reaction between cis-2-butene-d_0 and cis-2-butene-d_8 (1), and that for trans-2-butene-d_0 and trans-2-butene-d_8 (2) over the three catalysts.
$MoO_3/\beta\text{-}TiO_2 + Sn(CH_3)_4$; —○—, —△—
$MoO_{3-x}/\beta\text{-}TiO_2 + Sn(CH_3)_4$; —●—, —▲—
$MoO_{3-x}/\beta\text{-}TiO_2$; ---●---

It is known that the cis-olefin prefers to give cis-products and the trans-olefin prefers to give trans-products. Furthermore, such the

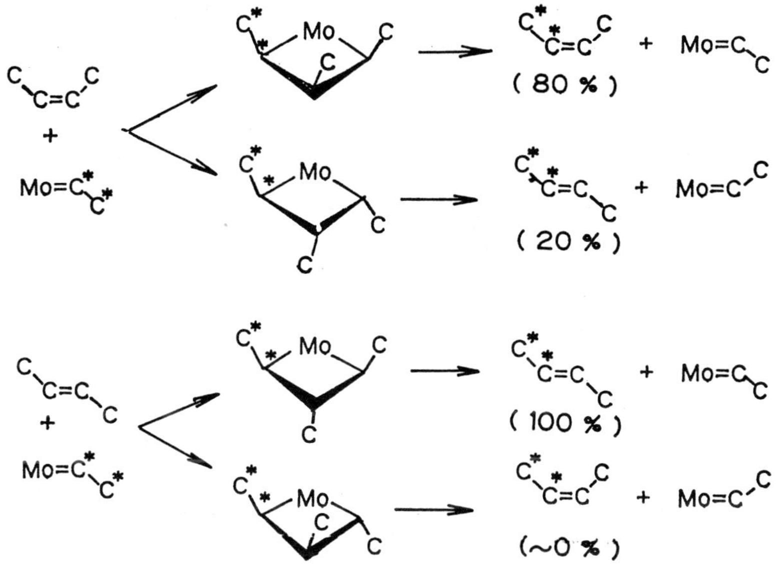

(80 %)

(20 %)

(100 %)

(~0 %)

Scheme 2. Mechanism for selective conformation transfer in the metathesis of cis-2-butene and of trans-2-butene.

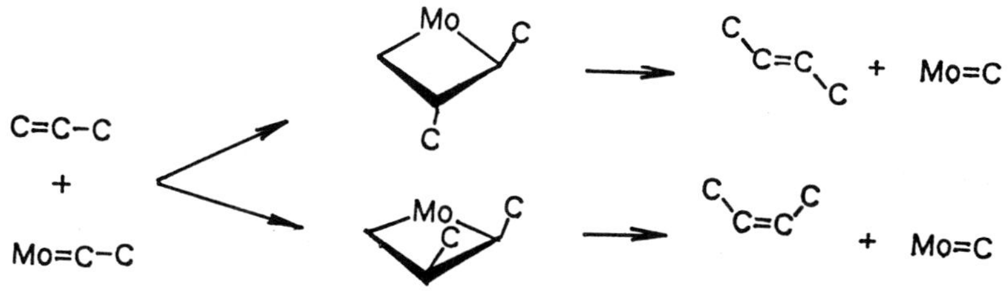

Scheme 3. Mechanism for the stereo-selective metathesis of propene.

selectivity as retaining the cis or trans conformation of starting olefins is remarkably improved by the treatment with $Sn(CH_3)_4$. Accordingly, we could conclude that the treatment with $Sn(CH_3)_4$ makes change molybdenum oxide surfaces to highly active and highly selectiv olefin metathesis catalyst (6). The conformation transfer selectivit can be explained by a preferential conformation of 1,3-dimethyl metal cyclobutanes as illustrated in Scheme 2. It should be noticed that the stereo selectivity for α-olefin metathesis reaction is caused by the relative contribution of the two conformations of the 1,2-dimethy metallacyclobutanes as shown in Scheme 3.

It is worthy of note that the stereo selectivity for the metathesis reaction of propene on MoO_3/TiO_2 + $Sn(CH_3)_4$ was quite different from that on the other two catalysts, MoO_{3-x}/TiO_2 and MoO_{3-x}/TiO_2 + $Sn(CH_3$ that is, the former gave the value of cis/trans = 1/2 while the latte was cis/trans = 1/5. This fact seems to indicate that the extent of reduction affects the relative stability of the 1,2-dimethyl substitu metallacyclobutanes but the treatment with $Sn(CH_3)_4$ influences little the stability of the 1,2-dimethyl metallacyclobutanes. Contrary to this, the cis or trans conformation transfer selectivity is markedly improved by the treatment with $Sn(CH_3)_4$ but is influenced little by the extent of reduction as attained the same selectivity on MoO_3/TiO_2 + $Sn(CH_3)_4$ and MoO_{3-x}/TiO_2 + $Sn(CH_3)_4$ catalysts. Based on these fact it may be deduced that the interaction of 1,3-dimethyls in metallacyclobutane might be more effective to relative stability of metallacyclobutane derivatives, and that of 1,2-dimethyls might be the secor factor for it (5).

3.3. Reactivity of $Mo=CH_2$ and $Mo=CH-CH_3$ Intermediates with Propene

As illustrated in Scheme 1, the productive metathesis reaction of propene, 2 $CH_2=CH-CH_3 \longrightarrow CH_2=CH_2 + CH_3-CH=CH-CH_3$, proceeds b alternative reaction of propene with $Mo=CH_2$ and $Mo=CH-CH_3$ intermedia Then, both ethylidene and methylidene species should coexist on the catalyst surface during the productive metathesis reaction of propen On the other hand, the degenerate metathesis of propene is brought about by reacting propene with either Mo=CH_2 or Mo=CH-CH_3. Taking account of these facts, it is an interesting question that either the Mo=CH_2 or Mo=CH-CH_3 is actual intermediate of the degenerate metathesis of propene. Similar question was proposed by Casey et al. (7) for the methylene exchange reaction between Z-1-d_1-dec-1-ene and 1,1-d_2-oct-1-ene in a homogeneous catalyst system.

In this paper, a cross metathesis between Z-1-d_1-propene and d_6-propene was performed on the two catalysts, MoO_3/TiO_2 + $Sn(CH_3)_4$ and MoO_{3-x}/TiO_2 + $Sn(CH_3)_4$. In this case, the two types of cross metathesis will proceed simultaneously.

(i) Cross metathesis between Z-1-d_1-propene and d_6-propene.

$$(D)(H)C{=}C(CH_3)(H)\ (\text{Z-1-}d_1) + (D)(D)C{=}C(CD_3)(D) \rightarrow \begin{cases} (D)(H)C{=}C(CD_3)(D)\ (\text{E-1-}h_1) + CD_2{=}CH{-}CH_3 \\ (H)(D)C{=}C(CD_3)(D)\ (\text{Z-1-}h_1) + CD_2{=}CH{-}CH_3 \end{cases}$$

(ii) Self-cross metathesis of Z-1-d_1-propene.

$$(D)(H)C{=}C(CH_3)(H) + (D)(H)C{=}C(CH_3)(H) \longrightarrow (D)(H)C{=}C(CH_3)(H)\ \text{or}\ (H)(D)C{=}C(CH_3)(H)$$

The results are shown in Figure 2. It is known that the fraction of Z-1-d_1-propene in the d_1-propene decreases linearly with the conversion being proportional to the self-cross metathesis reaction. Contrary to this, the fraction of E-1-h_1-propene in the d_5-propene does not change in the initial stage of the reaction. Taking account of the purity of Z-1-d_1-propene used as a reactant (about 90 %), it was evaluated that the Z-1-d_1 conformation of methylene was retained about 80 % in the cross metathesis reaction, that is, the cross metathesis between Z-1-d_1-propene and d_6-propene proceeded selectively to yield E-1-h_1-propene. This fact indicates that the intermediates for the degenerate metathesis should be explainable this Z-1-d_1 conformation retained methylene transfer in the metathesis reaction.

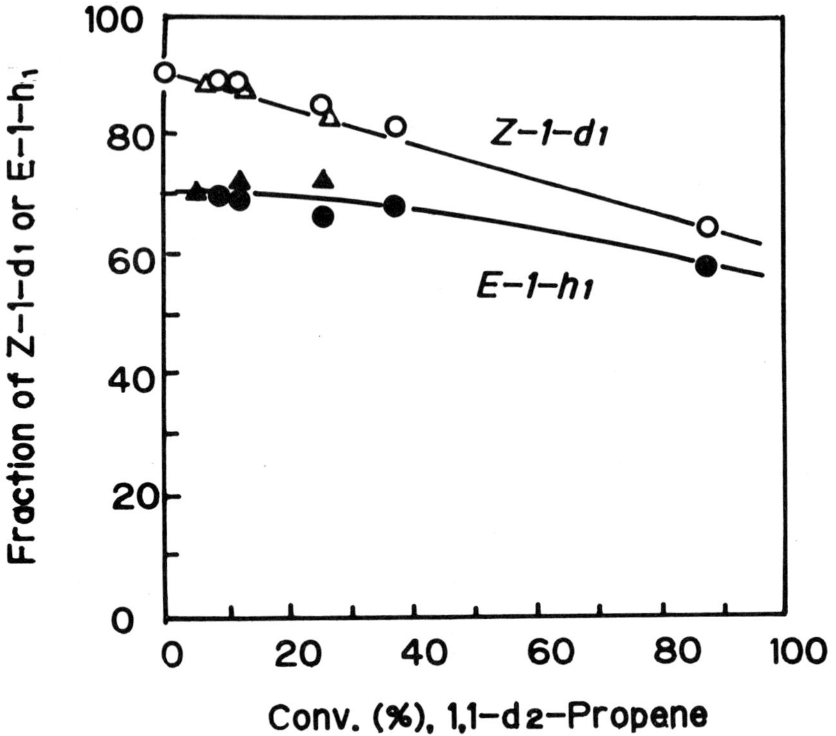

Figure 2. Cross metathesis of Z-1-d_1-propene and d_6-propene on the two catalysts. MoO_3/TiO_2 + $Sn(CH_3)_4$ (△, ▲); MoO_{3-x}/TiO_2 + $Sn(CH_3)_4$ (○, ●)
Z-1-d_1 corresponds to $\frac{D}{H}{>}C{=}C{<}\frac{CH_3}{H}$ and E-1-h_1 corresponds to $\frac{D}{H}{>}C{=}C{<}\frac{CD_3}{D}$.

If the degenerate metathesis of propene is caused by the reaction of propene with ethylidene intermediates, the reaction proceeds via 1,3-dimethyl substituted metallacyclobutanes as described in Scheme 4. By considering the fact that the 1,3-substitution influences the stability of the metallacyclobutanes, the cross metathesis of propene via ethylidene intermediates will retain the conformation of Z-1-d_1 methylene by the mechanism of predominant contribution of the prefer- conformation of metallacyclobutanes as described schematically in Scheme 4. Taking account of the fact that the metathesis of trans-2-butene retains its conformation in nearly 100 % as shown in Figure

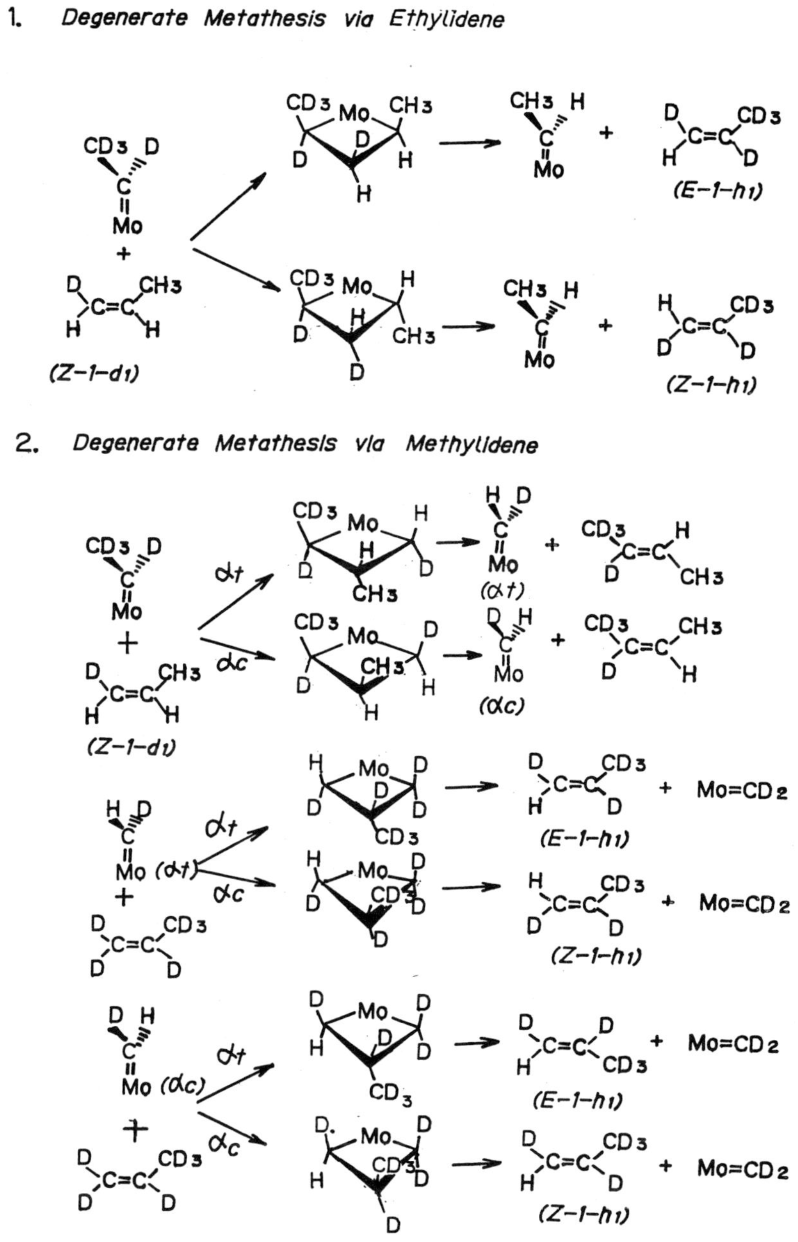

1. *Degenerate Metathesis via Ethylidene*

(Z-1-d1) (E-1-h1) (Z-1-h1)

2. *Degenerate Metathesis via Methylidene*

(Z-1-d1) (αt) (αc) (E-1-h1) (Z-1-h1) (E-1-h1) (Z-1-h1)

Scheme 4. Mechanism for Z-1-d_1 methylene transfer in the degenerate metathesis of Z-1-d_1-propene and d_6-propene.
(1) Degenerate Metatehesis via Ethylidene Intermediates.
(2) Degenerate Metathesis via Methylidene Intermediates

and Scheme 2, the maximum value for expectable conformation retaining of the Z-1-d_1 methylene in its transfer process would be close to 100

On the other hand, if the degenerate metathesis of propene would proceed through methylidene intermediates, the retain of Z-1-d_1 conformation in the methylene transfer process might occur only when the active sites are asymmetric, on which the two Mo=CHD species denoted as αt and αc in Scheme 4 are distinguishable in the reaction with propene. The αt and αc types of Mo=CHD are formed by the reaction of Z-1-d_1-propene with Mo=CD-CD_3, then the relative population of αt type Mo=CHD and αc type Mo=CHD may correspond to the cis/trans ratio in the 2-butene formed by the productive metathesis. If this is the case, the maximum value for expectable conformation retaining in the Z-1-d_1 methylene transfer could be given by $\alpha t^2 + \alpha c^2$. In the case of the present experiments, the highest stereo selectivit was cis/trans = 1/5. By using this value, we can evaluate the maximu expectable value for the Z-1-d_1 conformation transfer as to be 72 %. Obtained experimental value was 80 %, which is much higher than the expectable maximum value of 72 %.

Accordingly, we can conclude that the degenerate metathesis of propene proceeds mainly through Mo=CH-CH_3 route instead Mo=CH_2 route. This fact suggests that methylidene formed on a series of molybdenum oxide catalysts might be more reactive than ethylidene species, and as a result, the population of methylidene becomes very low comparing to ethylidene duriing the metathesis reaction of propene. By this reason, the main route for the degenerate metathesis of propene is the reaction of propene with ethylidene intermediates, and the productive and the degenerate metathesis of propene are competitive in the reaction of propene with ethylidene intermediates. The fact that the degenerate metathesis is almost one order faster than the productive metathesis on our catalysts seems to indicate nucleophilic reaction of the carbon at the carbene coordinated to the active sites

ACKNOWLEDGEMENTS. The authors are indebted to Nippon Sheet Glass Foundation (Tokyo, Japan) and to Iwatani Naoji Foundation for the financial support of the part of this work.

REFERENCES

1. C.P. Casey and T.J. Burkhart, J.Am.Chem.Soc., 95, 5833 (1973)
2. R.R. Schrock, J.Am.Chem.Soc., 97, 6577 (1975)
3. W.A. Herrmann, B. Reiter and H. Biersack, J. Organometal. Chem., 97, 245 (1975)
4. K. Tanaka, K. Tanaka and K. Miyahara, J.C.S., Chem.Comm., 314 (1979)
5. K. Tanaka, K. Miyahara and K. Tanaka, J. Molecular Catal., 15, 133 (1982)
6. K. Tanaka and K. Tanaka, J.C.S., Chem. Comm., 748 (1984)
7. C.P. Casey and H.E. Tuinstra, J.Am.Chem.Soc., 100, 2270 (1978)
8. K. Tanaka and K. Tanaka, Unpublished result.

APPENDIX

Selective transfer of cis or trans conformation of olefins in the metathesis reaction is responsible to the generation of preferable orientation of the 1,3-disubstituents in metallacyclobutane intermediates, and stereo-selectivity in α-olefin metathesis reaction is responsible to preferable orientationof the 1,2-disubstituents.
In this paper, it was assumed that mono-substituted metallacyclobutane has no preferable orientation of the substituent. This hypothesis was recently proved by performing the following productive metathesis of Z-1-d_1-propene (8).

$$2\ \text{(D)(H)C=C(CH}_3\text{)(H)} \longrightarrow \text{(D)(H)C=C(D)(H)}\ \text{or}\ \text{(D)(H)C=C(H)(H)}\ +\ \text{C–C=C–C (cis)}\ \text{or}\ \text{C–C=C–C (trans)}$$

That is, the cis/trans for the 1,2-d_2-ethene was nearly unity while the cis/trans ratio for the 2-butene was 1/5. These results are displayed by parenthesizing in Scheme 1.

CATALYTIC OXIDATION OF CO TO CO_2 BY DIOXYGEN IN THE PRESENCE OF PALLADIUM AND HETEROPOLYCOMPLEXES $H_xPM_{12-n}V_nO_{40}^{-(3+n-x)}$

L.I. Kuznetsova, S.N. Pavlova, E.G. Zhizhina,
K.I. Matveev, V.V. Popovskii, V.A. Sazonov,
A.V. Pashis and V.B. Fenelonov
Institute of Catalysis, Novosibirsk, 630090, USSR

ABSTRACT - The catalytic oxidation of CO to CO_2 has been studied in aqueous solutions of $PdSO_4$ and heteropolycomplexes with the composition of $Na_xH_{3+n-x}PM_{12-n}V_nO_{40}$ (M=Mo, W, n=0÷4). By supporting these complexes and Pd(II) on SiO_2 it is possible to obtain catalysts of CO oxidation by dioxygen, which are active at 15-40°C in the presence of water vapour. Reaction mechanisms of CO oxidation on supported and homogeneous catalysts are similar. The support pore structure has a pronounced effect on the catalytic activity. Intermediate active complexes of Pd(I) are formed on the surface both from Pd(II) and metallic Pd.

1. INTRODUCTION

Oxidation of CO to CO_2 by dioxygen plays an important role in pollution control efforts. As an essential part of these efforts is the development of novel low temperature catalysts resistable to poisoning by water vapour. Conventional catalysts of CO oxidation containing metallic palladium on oxide supports are active at temperatures above 100-150°C (Ref. 1), while oxidation of CO by palladium complexes in solution occurs under mild conditions (below 100°C) (Ref. 2). An analysis of the known mechanisms of CO oxidation by dioxygen assisted by homogeneous or supported palladium catalysts has suggested a way to create efficient supported catalysts based on palladium complexes and oxidants (Ref. 3). For example, the CO oxidation occurs rapidly at 20°C in the presence of $PdCl_2$ and $CuCl_2$ in acidic aqueous solutions (Refs. 4,5) or on the catalysts obtained by supporting the same components (Refs. 6-9). However volatility of HCl makes application of such catalysts quite difficult.

In solutions containing $PdSO_4$ and heteropolycomplexes (HPCs) CO oxidizes rapidly to CO_2 (Ref. 10). Since the reduced forms of some HPCs are oxidized by O_2, the catalytic oxidation of CO by dioxygen can occur on this system:

$$CO + H_2O + OX \xrightarrow{Pd^{2+}_{aq}} CO_2 + H_2OX, \quad (1)$$

$$H_2OX + 1/2O_2 \longrightarrow H_2O + OX, \quad (2)$$

$$CO + 1/2O_2 \longrightarrow CO_2 \quad (3)$$

(here OX and H_2OX are the oxidized and reduced forms of HPCs.

In an attempt to develop low temperature catalysts, we made a comparison of the mechanisms of CO oxidation by dioxygen on silica-supported Pd+HPCs catalysts and on similar $PdSO_4$+HPCs catalysts in aqueous solutions.

2. EXPERIMENTAL

The rate of the homogeneous oxidation of CO was determined from the amount of the gas (CO or CO+O_2) absorbed by $PdSO_4$+HPC solutions at atmospheric pressure as described in (Ref. 10).

The catalytic activity of supported samples was measured in a flow-circulation system consisting of a reactor, a gas-purification and dosage system. The reaction was carried out in a mixture of moist air with 1% CO; the catalyst volume was 1 cm^3, the gas and circulation rates were 5 and 500 dm^3/h, respectively. These conditions provided the reaction occurrence in the kinetic region.

$H_3PMo_{12}O_{40}$(I) solutions were obtained from solid reagent purified by ether extraction. HPCs solutions with the composition of $Na_xH_{3+n-x}PMo_{12-n}V_nO_{40}$(n=1÷4) (II-V) were prepared as described in (Ref. 11). HPCs with the composition of $Na_xH_{3+n-x}PW_{12-n}V_nO_{40}$ (n=1÷4) (III-XI) and $Na_{3-x}H_xPW_3Mo_3O_{40}$ (VI) and $Na_{3-x}H_xPW_6Mo_6O_{40}$ (VII) were prepared from the stoichiometric amounts of reagents.

To measure the redox potentials, solutions of HPCs (0.025 mol/dm^3) containing 0.5 M Na_2SO_4 were electrochemically reduced on a Pt electrode at a potential 0.0 V (relative to a silver chloride electrode) in argon. The reduction states of HPCs were checked by titration with $KMnO_4$ and were characterized by the mean number of electrons (m) transferred into the molecule of HPC. Supported catalysts were prepared in vacuum by impregnation of SiO_2 (BET surface 300 m^2/g, ΣV_{pore}= 1.05 cm^3/g, r_{pore}= 60--75 Å) with HPC solution followed by drying and impregnation with a solution of a palladium salt.

3. RESULTS AND DISCUSSION

3.1. Oxidation of CO to CO_2 in Aqueous Solutions of $PdSO_4$ and HPC

To study the mechanism of the overall catalytic reaction, the reactions (1) and (2) were tested separately. The mechanism of reaction (1) was shown to depend on the strength and the nature of the oxidant (Refs. 10, 12). The oxidation of CO by strong oxidants such as $Cr_2O_7^{2-}$ and Ce (IV) occurs stepwise. The rate-determining step is one of the steps of complex reaction (4) which is followed by a rapid oxidation of Pd(O) by the oxidant (reaction (8)). In the case of moderate oxidants such as HPCs, having potentials within the range 0.7 to 1 V, the reaction occurs via steps (4)-(7) with the participation of Pd(I) complexes and is one order of magnitude faster than the reaction in presence of strong oxidants (Ref.10). The rate-determining step seems to be step (7)

$$Pd^{2+}_{aq} + CO + H_2O \longrightarrow Pd(0) + CO_2 + 2H^+, \qquad (4)$$

$$Pd(0) + Pd^{2+}_{aq} + CO \rightleftharpoons \underset{A}{[Pd\text{—}CO\text{—}Pd]}, \qquad (5)$$

$$A + OX \rightleftharpoons \underset{B}{[Pd\text{—}CO\text{—}Pd]\cdot OX}, \qquad (6)$$

$$B + H_2O + CO \longrightarrow [Pd\text{—}CO\text{—}Pd] + CO_2 + H_2OX, \qquad (7)$$

$$Pd(0) + OX + 2H^+ \longrightarrow Pd^{2+}_{aq} + H_2OX \qquad (8).$$

To prove the above ideas concerning the reaction mechanism, we have studied the reaction rate as a function of the redox potential of the HPA (E). The changes in E were achieved by varying the composition of the ligand sphere of the heteropolyanion. As shown by ^{31}P and ^{51}V NMR (Ref. 13) under experimental conditions, all the HPCs studied were either individual heteropolyanions of the series 1:12 ($PMo_{12}O_{40}^{-3}$, $PW_{12-n}V_nO_{40}^{-(3+n)}$ ($n=1\div4$)) or the mixture of isostructural ions of the series 1:12 with different n ($PMo_{12-n}V_nO_{40}^{-(3+n)}$ and $PMo_{12-n}W_nO_{40}^{-3}$). At pH<3 the $PMo_{12-n}V_nO_{40}^{-(3+n)}$ (n>2) ions partly dissociate to produce the equilibrium mixture of VO_2^+, phosphate ions and HPCs of the series 1:12 with a low vanadium content.

The reaction of CO oxidation occurs with an induction period, corresponding to accumulation of complex B at the steps (4)-(6), after which a constant high rate is established and eventually the reaction stops as a result of Pd precipitation (see Fig. 1). The reaction rate and the maximum reduction degree of HPC depend on the compo-

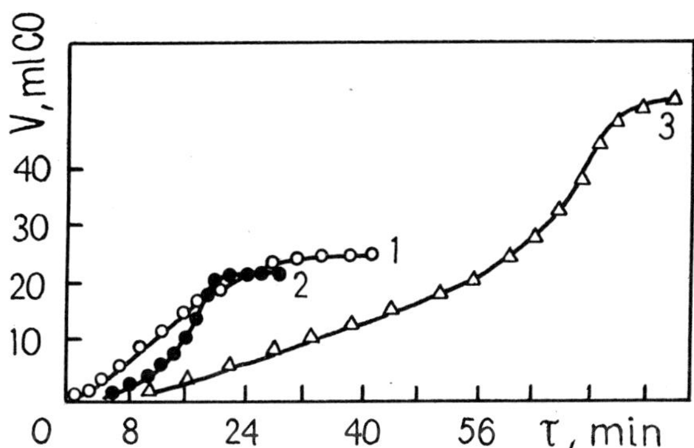

Fig. 1. The quantity of CO absorbed by solutions of Pd^{2+}_{aq} and HPC vs time; $[PdSO_4]=5\cdot10^{-5}$(1) and $1\cdot10^{-4}$ mol/dm^3(2,3); $[PW_{10}V_2]$=0.05 mol/dm^3(1); $[PMo_8V_4]$=0.025 mol/dm^3(2,3); $[NaVO_3]$=0.1 mol/dm^3(3); pH 1.0, 25°C, V=20 cm^3.

sition of the HPC (see Table 1).

The rate of CO oxidation by various HPCs depends on their potentials measured under the same conditions. The dependence is the curve having a maximum (see Fig. 2). For many HPCs the logarithm of the rate depends linearly on the potential of the HPC. At pH 1 after a maximum value of W has been achieved, the rate drastically falls with increasing the HPC potential for PMo_9V_3 and PMo_8V_4. The redox potentials of the HPCs and the rates of CO oxidation are similar to the corresponding values for VO_2^+ ions (see Fig. 2B).

In the region of the linear correlation between lgW and E for a series of heteropolyanions of the same type, the rate-determining step is that of electron transfer from CO into the HPC in the carbonyl complex of Pd(I) (step (7)). A further increase in the potential of the oxidant leads to acceleration of step (8). Thus, the

TABLE 1. Rates of CO oxidation by heteropolycomplexes in the presence of Pd^{2+}_{aq} ($PdSO_4$)= = 2 10^{-4} mol/dm^3, (HPC)=0.05 mol/dm^3, pH_o=1.5, V=25 cm^3, 25°C

	Mean composition of heteropolyanions	CO + Ar (2:1)		CO + O_2 (2:1)				O_2
		W, cm^3/min (m=0.5 $\bar{e}$)	m, $\bar{e}$ (maximum)	W, cm^3/min	m, $\bar{e}$ (maximum)	W_{stat}, cm^3/min	m_{stat}, $\bar{e}$	W, cm^3/min (m=1.0 $\bar{e}$)
I	PMo_{12}	2.2	1.9	2.2	1.9	-	-	0
III	$PMo_{10}V_2$	1.6	1.5	-	-	-	-	0.07
IV	PMo_9V_3	1.9	2.8	-	-	4.6	1.2	0.12
V	PMo_8V_4	1.9	3.2	-	-	3.9	1.0	0.08
VI	PMo_9W_3	3.7	1.35	3.7	1.4	-	-	0
VII	PMo_6W_6	3.7	0.95	3.8	1.0	-	-	0
VIII	$PW_{11}V_1$	2.8	0.95	2.8	0.9	-	-	0
IX	$PW_{10}V_2$	1.85	21.2	1.8	1.25	-	-	0
X	PW_9V_3	1.6	1.4	1.5	1.4	-	-	0
XI	PW_8V_4	1.4	3.0	-	-	1.1	0.7	0.3

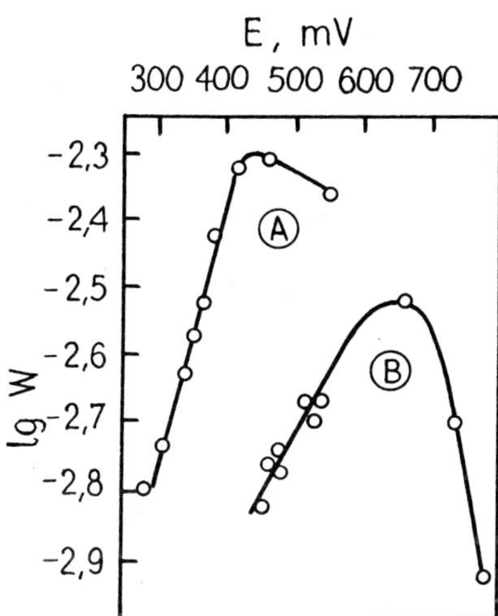

Fig. 2. Logarithm of the CO oxidation rate vs HPC potential (relative to hydrogen electrode): A - pH 3.0 (40% CH_3COOH), B - pH 1.0 (H_2O); in E measurements (HPC) = = 0.025 mol/dm^3, $[Na_2SO_4]$ = 0.3 mol/dm^3, in W measurements (HPC) = 0.05 mol/dm^3, $[PdSO_4]$ = $5 \cdot 10^{-5}$ mol/dm^3, $[Na_2SO_4]$ = 0; m = 0.5 ē, 25°C.

stationary concentration of active complexes B decreases, and the reaction slows down. In reaction (9) the VO_2^+ ions are more active than HPCs (Ref. 12). In solutions of P-Mo-V HPCs (pH < 3) having VO_2^+ ions, the reaction is determined primarily by the action of the VO_2^+ ions as more active oxidants of Pd(0) than the HPCs (steps (4) and (8)), and the reaction occurs slowly. In fact, addition of the VO_2^+ ions to the solution of HPCs inhibits the oxidation of CO (see Fig. 1, curve 3).

Only four of the HPCs studied, namely, PW_8V_4, $PMo_{10}V_2$, PMo_9V_3 and PMo_8V_4 coupled with Pd^{2+}_{aq} can catalyze the oxidation of CO. Stationary values of the reaction rate and the reduction state of these HPCs are listed in Table 1.

The reduced forms of only tabulated HPCs can be oxidized by dioxygen (see Table 1). Consequently, the process of CO oxidation by dioxygen in aqueous solutions of $PdSO_4$+ HPCs is described by two individual complex reactions (1) and (2). Both reactions are accelerated as the pH grows from 1 to 3 (see Fig. 2 (Ref. 11)).

Thus, P-Mo-V HPCs ($pH \geq 1$) are active electron carriers, because they provide the reaction occurrence with the participation of active carbonyl complexes of Pd(I) and the reduced forms of these HPCs are rapidly oxidized by dioxygen.

In contrast to aqueous solutions, reduction of $PdCl_2$ on solid supports does not lead to catalyst deactivation, since the supported metallic Pd can be oxidized by dioxygen (Refs. 15, 6). To compare the mechanisms of the action of supported catalysts and those in solution, as well as to elucidate the properties of the supported catalysts, we have examined the state of Pd in them and the kinetic behaviour of the reaction.

3.2. Supported Catalyst Based on Pd and HPCs for CO Oxidation by Dioxygen

The catalysts, which are active in CO oxidation by O_2 at 15-40°C and a high relative moisture content of the gas (more than 70%), were obtained by supporting Pd(II) and HPCs over SiO_2, TiO_2 or Al_2O_3. Prereduction of the samples in H_2 atmosphere does not lead to catalyst deactivation, although Pd_{met} is formed during the process (see Fig. 3, curves 3-6). Similar samples having no HPCs are far less active (see Table 2). Only water soluble salts of the HPCs are active towards CO oxidation. Samples obtained by precipitating water insoluble salt, $Cs_5H_2PMo_8V_4O_{40} \cdot nH_2O$, are inactive.

As follows from ESCA spectra of the samples reduced with H_2 or CO and of the samples after operation, the binding energy of Pd $3d_{5/2}$ is 335-336 eV, which indicates

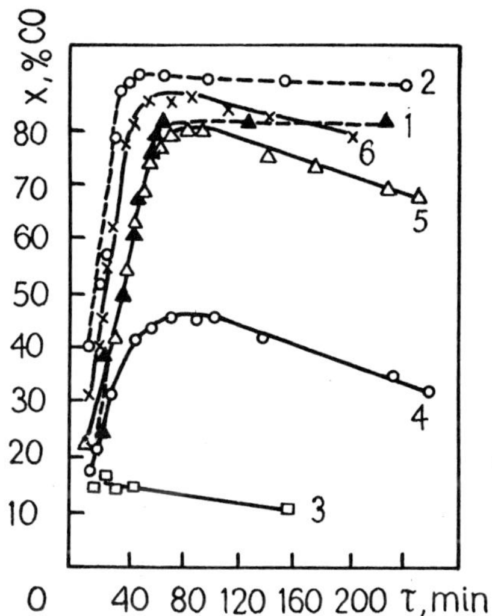

Fig. 3. Activity of the supported catalyst at various relative moisture contents: 1 - 75%; 2 - 90% (unreduced samples); 3 - 35%; 4 - 50%; 5 - 75%; 6 - 90% (reduced samples; 1.45% Pd + 7.8% HPC/SiO_2; Vgas= 5 dm^3/h, inlet CO concentration is 1% by volume, 30°C, HPC = $Na_5H_2PMo_8V_4O_{40} \cdot 30H_2O$.

that Pd_{met} is the main palladium state in the catalyst (Ref. 16). The line from the support Si 2p with a binding energy of 103.5 eV served as an internal reference. In the study of the catalyst granules (0.25 - 0.5 mm), the Pd:Si atomic ratio calculated from the integral line intensity ratio was 0.032 (at 3% wt Pd) and fell by a factor of 2-4 depending on the support particle size. ESCA spectra reveal the presence of the atoms near the grain surface. A decrease in the Pd line intensity of the powdered samples is evidence that Pd_{met} is localized on the surface of granules. The metal particle size is 30 to 40 Å, as ascertained by electron microscopy data.

As shown by the study of the effect of the HPC composition on the rate of CO oxidation by dioxygen, only those HPCs, whose reduced forms are oxidized by O_2, were

TABLE 2. Effect of the composition of the SiO_2-supported catalyst on the activity of unreduced samples: V_{gas}= 5 dm^3/h, V_{cat}= 1 cm^3, 30°C, moisture 95%

Pd, wt. %	HPC, wt. %	Pd:HPC, molar ratio	pH	τ , min [a)]	X_{CO}, % (maximum)
1.55	0	-	-	-	0
1.45	7.8	3.8:1	2.0	10	82
1.32	14.5	1.9:1	2.0	25	80
1.15	25	1:1	2.0	40	65
0.72	7.8	2:1	2.0	40	70
0.36	8.0	1:1	2.0	60	42
1.45	7.5	3.8:1	1.3		66
1.45	7.9	3.8:1	3.0		48

[a)] Time of achieving the maximum rate

found to be active. The activity of the HPC changed in the order: $PW_8V_4 < PMo_{10}V_2 < PMo_9V_3 \approx PMo_8V_4$, which corresponds to the order of variations in their catalytic activity and in the rate of oxidation of the HPCs reduced with dioxygen in solution (see Table 1).

Data on the effect of the content of Pd and HPC on the catalytic activity is given in Table 2. The most active samples are those containing 7.5-14.5% wt of HPC and 1.5% wt of Pd, which corresponds to the molar ratio Pd/HPC= 2÷4. In other words, on supporting Pd and HPC on the carrier their ratio, corresponding to the maximum activity, changes in comparison with a homogeneous system in which an excess HPC (HPC:Pd ≈1000) is needed to avoid deposition of Pd_{met}. The rate of CO oxidation on the supported catalyst, like in aqueous solutions, depends on the pH of the solution of the HPC to be supported (see Table 2).

Initially the reaction rate tends to increase gradually both for the samples containing Pd(II) in the initial state and for those reduced with H_2 (see Fig. 3). To elucidate the nature of an induction period, we have made a series of experiments in which the catalyst reduced with H_2 was subjected to one-hour pretreatment at 30° in the flow of moist gases: N_2, O_2 and CO (moisture content 90 and 75%). Then the catalyst was tested under the same reaction conditions. The pretreatment resulted in shortening of the induction period at moisture content 90%, but the shape of the kinetic curves remained unchanged (see Fig. 4). Obviously, similarly to the homogeneous system, Pd(I) complexes, which form under the action of all components of the catalytic system, are active on the support surface.

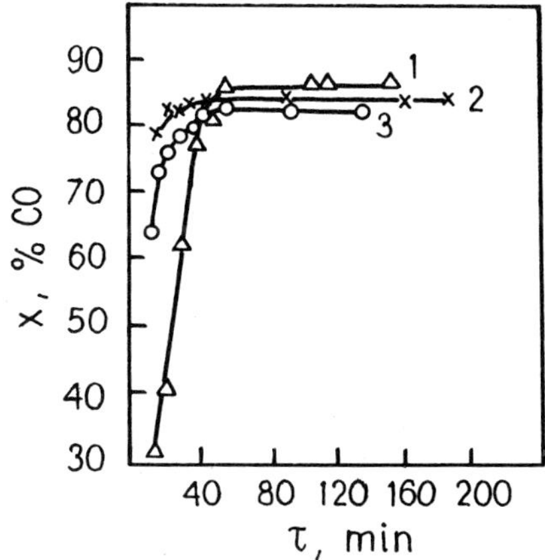

Fig. 4. Effect of the catalyst pretreatment on CO conversion: 1.45% Pd + 7.8% HPC/SiO_2; V_{gas} = 5 dm^3/h, inlet CO concentration 1% vol., 30°C; HPC=$Na_5H_2PMo_8V_4O_{40}\cdot 30H_2O$; 1 - without treatment, 2 - treatment with the mixture 1% CO + N_2+ H_2O, 3 - treatment with the mixture air+H_2O; moisture content 90%

Water plays an important role in the oxidation of CO by dioxygen on the supported catalyst. In the case of the homogeneous reaction, water is directly involved in

CO oxidation in the coordination sphere of palladium (Ref. 2):

$$\mathrm{L{-}\underset{L}{\overset{L}{Pd}}{-}OH_2} \overset{-H^+}{\rightleftharpoons} \mathrm{L{-}\underset{L}{\overset{L}{Pd}}{-}OH} \underset{-L}{\overset{+CO}{\rightleftharpoons}} \mathrm{L{-}\underset{L}{\overset{CO}{Pd}}{-}OH} \xrightarrow{+L} \mathrm{L{-}\underset{L}{\overset{L}{Pd}}{-}COOH} \longrightarrow \mathrm{Pd(0)} + \mathrm{CO_2} + \mathrm{H^+} + 3\mathrm{L^-}. \qquad (9)$$

This mechanism is different from that of a high temperature oxidation of CO by oxygen in the presence of metallic Pd on the support (Ref. 3). The mechanism of the reaction assisted by the Pd+HPC supported catalyst (Eqs. (1) and (2)) is essentially similar to the mechanism of CO oxidation in the presence of Pd complexes in solution. Therefore, H_2O participates in the reaction on the supported catalyst too. However, in the supported catalyst water is necessary also for the formation of the active phase, as ascertained by a comparison of the reaction rate with the amount of H_2O absorbed in the catalyst pores at various H_2O concentrations. At moisture content 35%, when the catalyst surface is covered with only adsorbed film of H_2O, the reaction rate is slow (see Fig.3, curve 3). At water content more than 70%, a capillary condensation of H_2O occurs in the catalyst pores, and the reaction rate markedly increases (see Fig. 3, curve 5). Taking into account that only water-soluble HPCs are active in the reaction, and the HPCs do not interact with the support surface, one can conclude that during catalytic reaction the HPCs are in the aqueous solution formed in the pores. This conclusion is confirmed by that reaction (2), that is, oxidation of the reduced HPCs by dioxygen, is a multi-electron process preceded by the equilibria of electron exchange leading to the formation of species with a high content of V(IV) atoms. These equilibria are rapidly established in solution (Ref. 17). The dissolution of HPC in water makes the interaction of the HPC with the surface complexes of Pd via route (1) more facile. The other catalyst component, namely palladium (in the form of Pd_{met} and palladium carbonyl intermediates) is on the support surface. This leads to the change in the rate of Pd_{met} oxidation by the HPC in comparison with solution.

Thus, the oxidation of CO by dioxygen in the presence of supported Pd and HPCs is an example of the homogeneous-heterogeneous catalytic reaction discussed in (Ref. 6). Our study has shown that in general the mechanism of CO oxidation by dioxygen in the presence of Pd and HPCs is similar to that in solution. An important difference in the mechanisms is the possibility of the formation of surface active carbonyl intermediate complexes of Pd(I) both from Pd(II) and from Pd_{met}. The pore structure of the support and the distribution of the active component in the catalyst are important parameters determining the operation of the supported catalyst.

REFERENCES

1. G.I. Golodets, Heterogeneous Catalytic Reactions Involving Molecular Oxygen, p. 291, Naukova Dumka, Kiev (1977).
2. V.I. Spitsyn, I.V. Fedoseev, Carbonyl Complexes of Platinum Metals, Nauka, Moscow (1980).
3. L.I. Kuznetsova, K.I. Matveev, E.G. Zhizhina, Kinet. Katal., 26, 1029 (1985).
4. W.G. Lloyd, D.R. Rowe, US Patent n 3849336, 1974.
5. V.A. Golodov, E.L. Kuksenko, D.V. Sokolskii, Dokl. Akad.Nauk SSSR, 272, 3, 628 (1983).
6. V.A. Golodov, D.V. Sokolskii, N.F. Noskova, J.Molec. Catal., 3, 51 (1977/78).
7. T.L. Ratitskaya, V.Ya. Paina, A.A. Ennan, XIII All-Union Chugaev Conf.on Chemistry of Complex Compounds, (1978), p. 332, Moscow.
8. K.S. Rabindra, Great Britain Patent n 1498571, 1978.
9. M.N. Desai, J.B. Butt, J.S. Dranoff, J.Catal.,79, 95 (1983).
10. E.G. Zhizhina, K.I. Matveev, L.I. Kuznetsova, Kinet. Katal., 26, 2, 461 (1985).
11. L.I. Kuznetsova, E.N. Yurchenko, R.I. Maksimovskaya, N.P. Kirik, K.I. Matveev, Koord.Khim., 3, 51 (1977).

12. E.G. Zhizhina, L.I. Kuznetsova, E.N. Yurchenko, K.I. Matveev, Koord. Khim., 6, 1846 (1980).
13. R.I. Maksimovskaya, Thesis, Novosibirsk (1980).
14. W. Manchot, J. König, Ber., 59, 883 (1926).
15. K. Fujimoto, K. Juchi, T. Kunigi, Int.Chem.Eng., 12, 4, 741 (1972).
16. V.I. Nefedov, I.A. Zakharova, I.I. Moiseev et al., Zh.Neorg.Khim., 18, 12, 3264 (1973).
17. V.M. Berdnikov, L.I. Kuznetsova, K.I. Matveev, N.P. Kirik, E.N. Yurchenko, Koord.Khim., 5, 78 (1979).

Heteropoly Acid as Solid-Acid Catalysts

Yoshio Ono[1] and Toshihide Baba[2]

[1] Department of Chemical Engineering, Tokyo Institute of Technology, Ookayama, Meguro-ku, Tokyo 152, Japan

[2] Department of Industrial Chemistry, Kochi College of Technology, Monobe, Nankoku-shi, Kochi 783, Japan

ABSTRACT - Heteropoly acids are strong acids. In homogeneous phases, they are much more active catalysts than conventional acids such as sulfuric acid, because of the interaction of heteropoly anions with protonated intermediates. Heteropoly acids are also very active catalysts as solid acids for many reactions, i.e., the vapor-phase nitration of benzene and the conversion of methanol into hydrocarbons. For some reactions, the silver or copper salts are more active than the parent acids. The mechanism of the acid sites formation in these salts is discussed. Interaction of gaseous hydrogen with group VIII metals induces strong acidity in solid heteropoly acids as evidenced by the high activity of the systems for methanol conversion and alkane isomerization.

1. INTRODUCTION

Heteropoly acids with Keggin-type structure are strong acids and are active catalysts for a number of "acid-catalyzed" reactions which proceed in aqueous phase or in some organic solvents. At the same time, heteropoly acids can be used as catalysts for the oxidation reactions due to their redox properties. In some cases, the combination of acidic and redox properties of the acids is indispensable for their catalytic functions. While heteropoly acids are active catalysts in homogeneous phases, they can also be used as solid catalyst in a neat form or in a supported form over suitable carriers. This paper will review the nature of heteropoly acids as solid-acid catalysts with special emphasis on the systems with heteropoly acids in conjunction with transition metals or transition metal cations.

2. CHARACTERISTICS OF HETEROPOLY ACIDS AS HOMONEGEOUS CATALYST

Before going directly to the problems specific to the properties of heteropoly acids as solid-acid catalysts, it seems appropriate to briefly summarize some of the characteristics of the acids which are common to both homogeneous and heterogeneous catalysis and quite different from conventional acids. Heteropoly acids like dodecatungstophosphoric acid($H_3PW_{12}O_{40}$) are completely dissociated in aqueous solutions. In many cases, however, the reaction proceeds much faster with heteropoly acids than with conventional acids (Ref. 1 & 2). A typical example is seen in the dehydration-cyclization of 1,4-butanediol into tetrahydrofuran(THF) in aqueous solutions (Ref. 3).

$$HOCH_2CH_2CH_2CH_2OH \longrightarrow \text{THF} + H_2O \qquad (1)$$

As is shown in Table 1, the activities of heteropoly acids for the dehydration-cyclization are much higher than the conventional acids at the same concentration of protons. Thus, the rate with dodecatungstosilicic acid ($H_4SiW_{12}O_{40}$) is more than three times higher than that with sulfuric acid. The detailed kinetic study has revealed that the rate can be expressed by a following equation.

$$\frac{d[THF]}{dt} = (k_1 + k_2[SiW_{12}O_{40}{}^{4-}])[H^+][1,4\text{-Butanediol}] \qquad (2)$$

where k_1 and k_2 are $(2.4 + 0.1) \times 10^{-5}$ dm^3 mol^{-1} s^{-1} and $(3.1 + 0.2) \times 10^{-3}$ dm^6 mol^{-2} s^{-1}, respectively, at 430 K. Thus, the rate can be expressed as the sum of two terms. The first term is due to the catalysis by the conventional mechanism with specific acid, while the second term is unique to the catalysis by heteropoly acids, indicating the positive participation of the heteropoly

TABLE 1. The catalytic activities of heteropoly acids and conventional acids for the dehydration-cyclization of 1,4-butanediol.

Catalyst	d[THF]/dt / mol dm^{-3} s^{-1}
$H_4SiW_{12}O_{40}$	3.2×10^{-6}
$H_3PW_{12}O_{40}$	2.3×10^{-6}
$H_3PMo_{12}O_{40}$	1.9×10^{-6}
HNO_3	1.4×10^{-6}
H_2SO_4	0.9×10^{-6}
H_3PO_4	0.4×10^{-6}

Reaction conditions; raction temperature: 430 K, $[H^+] = 1.49 \times 10^{-2}$ mol dm^{-3}, [1,4-Butanediol] = 5.66 mol dm^{-3}, ionic strength = 7.46 $\times 10^{-2}$ mol dm^{-3}, solvent: water.

anions. Based upon the kinetics, we have proposed a reaction mechanism involving two parallel reaction paths for dehydration-cyclization of 1,4-butanediol.

Path I

$$\underset{\mathbf{1}}{HOCH_2CH_2CH_2CH_2OH} + H^+ \overset{K_1}{\rightleftharpoons} \underset{\mathbf{2}}{HOCH_2CH_2CH_2CH_2\overset{+}{O}H_2} \xrightarrow[-H^+]{k'} \underset{\mathbf{3}}{\text{THF}} + H_2O$$

$$\mathbf{2} \overset{K_2,\ SiW_{12}O_{40}^{4-}}{\rightleftharpoons} \underset{\mathbf{4}}{[HOCH_2CH_2CH_2CH_2\overset{+}{O}H_2]\,SiW_{12}O_{40}^{4-}} \xrightarrow[-H^+,\ -SiW_{12}O_{40}^{4-}]{k''} \mathbf{3}$$

Path II

With Path I, the reaction proceeds with a conventional mechanism through protonated alcohol **2**. Path II is a characteristic of dehydration with heteropoly acids, and involves the interaction of protonated alcohol with heteropoly anion to form the reaction intermediate 4. If the rate determining step is the step giving THF in both reaction paths, the overall rate of THF formation will be expressed by eq. (2), by putting $k_1 = k'K_1$ and $k''K_1K_2$. The activation energies for k_1 and k_2 were determined to be $E_1 = (69 + 5)$ kJ mol^{-1} and $E_2 = (54 + 4)$ kJ mol^{-1}, respectively. The lower activation energy for Path II compared with Path I indicates that the protonated 1,4-butanediol 2 is stabilized by forming complex 4 with heteropoly anion.

Izumi et al. (Ref. 4) have found similar kinetics for hydration of 2-methylpropene (eq. (2)) and proposed the mechanism involving a protonated olefin-heteropoly anion complex. Recently Aoshima et al. (Ref. 5) found that the rate of hydration of 2-methylpropene is the second order with respect to the concentration of the catalyst, when the highly concentrated solution of heteropoly acid was used. In this case, the second term of eq. (2) is predominant due to the very high concentration of the heteropoly acid (anion).

The examples given above clearly show that the heteropoly anions have a profound effect on the rate of the acid catalyzed reactions. Though this effect by no means can be demonstrated explicitly in the heterogeneous systems, there is no doubt that the interaction of the protonated intermediates with heteropoly anion have a decisive role in determining the rate even when the heteropoly compounds are used as solid-acid catalysts.

The second feature of heteropoly acids in contrast to conventional acids is the high affinity for the polar organic solvents. They are usually soluble in acetone, alcohols, and other organic solvents. A number of catalytic reactions proceed with heteropoly acids in organic solvents. For the dehydration-cyclization of 1,4-butanediol, the reaction proceeds much faster in 1,4-dioxane than in water. In 1,4-dioxane, heteropoly acids do not dissociate, and the reaction rate is simply proportional to the concentration of $H_4SiW_{12}O_{40}$. This indicates that the protonated diol-heteropoly anions complexes are directly formed from the diol and the non-dissociated acid. Water seems to work as an inhibitor.

The rate drops sharply as the content of water in dioxane increases.

These facts show that water in the system has complex roles in the catalysis. (1) Water is a good solvent for heteropoly acid. (2) The acid is dissociated in water. (3) Water can be an inhibitor for the acid-catalyzed reactions. These features may be important in considering the catalytic properties of heteropoly compounds as heterogeneous catalysts.

3. VAPOR PHASE NITRATION OF BENZENE OVER HETEROPOLY ACIDS SUPPORTED ON SILICA

Nitration of benzene has been carried out in a liquid phase with a mixed acid which is a mixture of nitric acid and concentrated sulfuric acid. This process has still some problems left unsolved which are peculiar to the liquid-phase process such as treatments of waste sulfuric acid and of waste water. Vapor-phase nitration of benzene using nitrogen dioxide as a nitrating agent was studied over heteropoly acids supported on silica. The reaction can be expressed by the following equation.

$$C_6H_6 + 3/2\ NO_2 \longrightarrow C_6H_5NO_2 + 1/2\ NO + 1/2\ H_2O \quad (3)$$

The reaction was carried out with a flow-type reactor operating at atmospheric pressure. Fig. 1 shows the effect of the reaction temperature on the yield of nitrobenzene over $H_3PW_{12}O_{40}$ supported on silica. At the reaction temperature below 523 K, the rate decreased gradually with time on stream while steady activity was obtained from the beginning. Nitrobenzene was the only product and the formation of dinitrobenzene was not observed. The rate increased with the supported amount up to 30 wt%, and was virtually independent of the supported amount over 30 wt%.

The influence of the partial pressure of nitrogen dioxide on the nitrobenzene yield was very extensive, and the presence of water in the feed enhanced the rate of the reaction rate. The yield increased with the partial pressure of water up to 20 kPa and remained almost constant above 20 kPa. The effects of the partial pressures of nitrogen dioxide and water on the rate are illustrated in Fig. 2. Thus, the yield of nitrobenzene was 50 % when the ratios in the feed was $NO_2/C_6H_6/N_2$ = 4/1/17 , while the yield was 14 % with the ratio in the feed of $NO_2C_6H_6/N_2$ = 2/1/19. The yield was further increased by adding water in the feed. Thus, the nitrobenzene yield of 56 % with a steady activity was obtained with the feeding ratio of $NO_2/ C_6H_6/H_2O/N_2$ = 1/4/5/12.

This example shows that heteropoly acids are potential catalysts for reactions which are usually operated as homogeneous reactions.

4. CONVERSION OF METHANOL OVER HETEROPOLY ACIDS

Zeolites, especially ZSM-5, are very effective catalysts for converting methanol into hydrocarbons. Ono and Mori (Ref. 6) have studied the mechanism of the conversion and have concluded that Bronsted acid sites are the sole responsible sites for the conversion. This led to the idea that methanol conversion should proceed also over solid substances other than zeolites, if they are highly acidic. In fact, it has been demonstrated that $H_3PW_{12}O_{40}$ and $H_4SiW_{12}O_{40}$ are very active for methanol conversion (Ref. 6-8).

The high catalytic activity of heteropoly acids seems to be rather unusual, considering the small surface area of the acids (4 m^2 g^{-1}). To determine the effective number of active sites of $H_3PW_{12}O_{40}$ under the reaction conditions, the effect of pyridine sorption was examined. The reaction was started by feeding methanol and nitrogen at 573 K. After 2 h, pyridine(or 2,6-di-t-butylpyridine) was added to the feed, which was finally changed back to the starting mixture. Pyridine was almost entirely sorbed by the catalyst. As is shown in Fig. 3, the activity after pyridine sorption decreased linearly with the amount of sorbed pyridine, the complete loss of the activity being observed when three molecules of pyridine per Keggin unit was sorbed. Thus, the effective number of acid sites thus estimated indicates that pyridine molecules and also the methanol molecules are accessible to all the protons in the solid of $H_3PW_{12}O_{40}$.

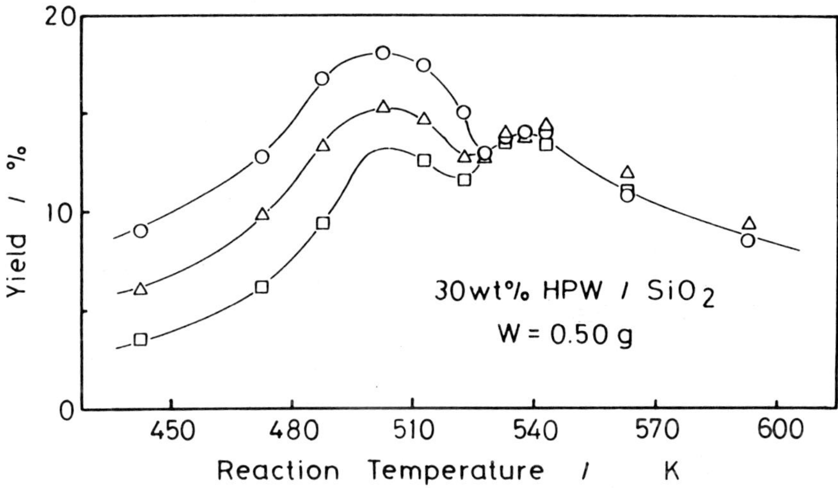

Figure 1. Dependence of nitrobenzene yield on the reaction temperature in vapor phase nitration of benzene. time on stream,. (O) 1 h, (Δ) 2 h, (□) 3 h; W/F = 8.14 g h mol^{-1}, $NO_2/C_6H_6/N_2$ = 2/1/19.

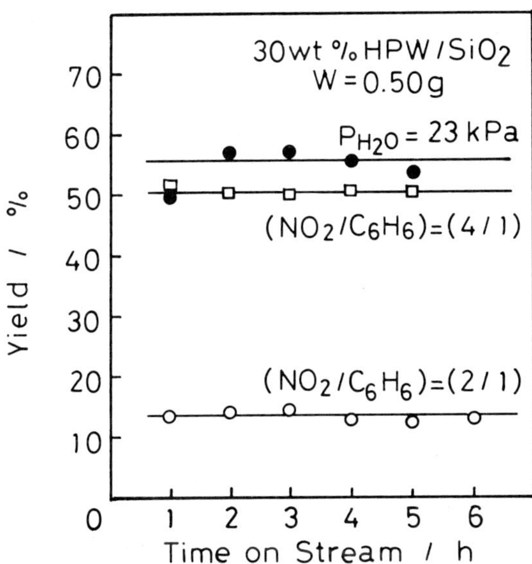

Figure 2. Effect of the partial pressure of nitrogen dioxide and water on the nitrobenzene yield in vapor phase nitration of benzene at 543 K. W/F = 6.15 g h mol^{-1}, C_6H_6 = 4.6 kPa.

To confirm that the phenomenon is not specific to the conversion of methanol, the number of acid sites effective for the dehydration of 1-butanol over $H_3PW_{12}O_{40}$ at 433 K was determined in a similar manner. The number of acid sites responsible for the dehydration is again three times as large as the number of the acid molecules.

As mentioned earlier, heteropoly compounds have a high affinity for polar organic molecules. Misono et al. (Ref. 9 & 10) found that about six molecules of pyridine, ammonia or ethanol are irreversibly absorbed at room temperature, while the amount of sorption of nonpolar molecules such as benzene and ethylene is very slight. They propose a "pseudo-liquid phase" concept for the catalytic reaction of polar molecules. The solid with sorbed molecules behaves like a solution with a very high concentration of heteropoly anions. Thus, the polar molecules can penetrate into the bulk of the solid heteropoly acid and undergoes the catalytic reaction in the bulk of the acids. The highly effective catalysis of heteropoly acids found in the conversion of methanol and the dehydration of 1-butanol could be fully accounted by the "pseudo-liquid" concept.

5. ACIDIC PROPERTIES OF METAL SALTS OF HETEROPOLY ACIDS

Besides heteropoly acids themselves, some metal salts of heteropoly acids are also active for various reactions for which Bronsted acid sites are supposed to be active centers. Niiyama et al. (Ref. 11) found that the catalytic activities for dehydration of 2-propanol of metal salts of $H_3PMo_{12}O_{40}$ have a good correlation with the electronegativities of the metal salts of $H_3PMo_{12}O_{40}$ except for alkali-metal salts, and they proposed that water coordinated to metal cation (M^{n+}) yielded protons by the following reaction.

$$M^{n+} + m\ H_2O \longrightarrow [M(H_2O)_m]^{n+} \longrightarrow [M(H_2O)_{m-1}(OH)]^{(n-1)+} + H^+ \qquad (4)$$

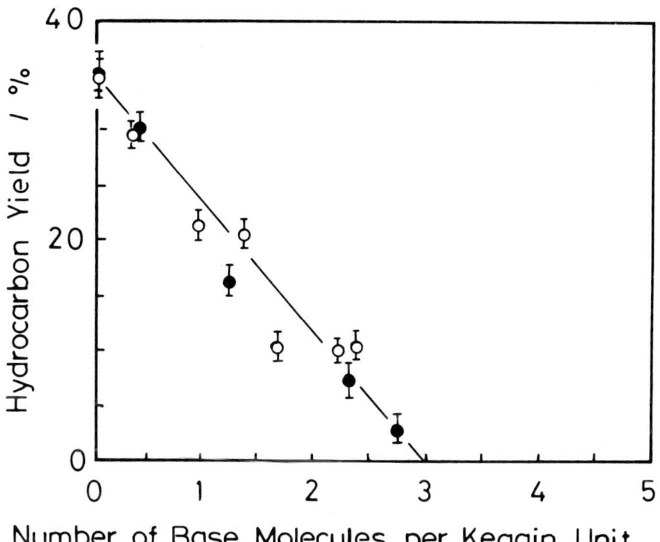

Figure 3. Change in the hydrocarbon yield with the amount of the sorbed pyridine (○) and 2,6-di-t-butylpyridine (●). Reaction conditions: 573 K, methanol pressure 50.7 kPa, W/F = 55.0 g h mol^{-1}.

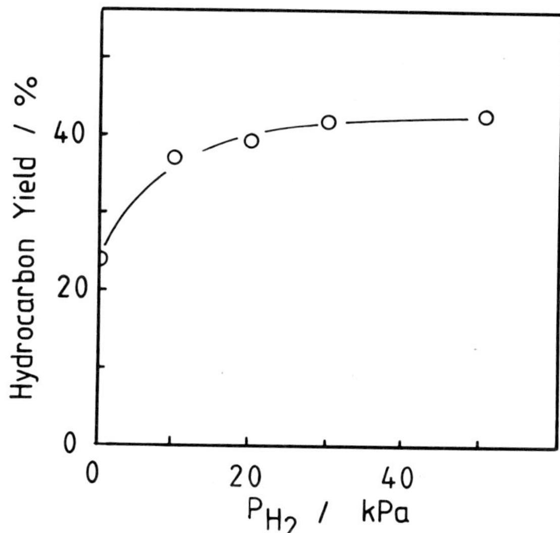

Figure 4. Effect of hydrogen pressure on the hydrocarbon yield in methanol conversion over $Ag_3PW_{12}O_{40}$ supported on active carbon. 573 K, CH_3OH = 51 kPa, W/F = 50 g h mol^{-1}.

The catalytic activities of various metal salts of $H_3PW_{12}O_{40}$ and $H_4SiW_{12}O_{40}$ for the conversion of methanol were examined at 573 K (Ref. 12 & 13). The activities of the series of metal salts of $H_3PW_{12}O_{40}$ and $H_4SiW_{12}O_{40}$ are listed as follows.

For salts of $H_3PW_{12}O_{40}$

Ag(98) > Cu(60), H(60) > Fe(48) > Al(36)

For salts of $H_4SiW_{12}O_{40}$

Ag(79) > Cu(61) > H(39) > Fe(24) > Al(15)

The numbers in parentheses indicate the hydrocarbon yield at 2-6 h of running time. The distribution of hydrocarbons over various metal salts are very similar to that over heteropoly acids, indicating that the reaction mechanism is common to parent heteropoly acids and their metal salts. Thus, the active sites for methanol conversion should be common, and they are presumably Bronsted acid sites. It should be noted that the silver and copper salts are most active among metal salts, and even more active than the parent acids. The high catalytic activity of silver and copper salts cannot be explained by the mechanism of the proton formation proposed by Niiyama et al. (eq. (4)), since the electronegativity of silver and copper cations is not so large. Therefore, another mechanism of the acid sites formation should be sought at least for these two metal salts. This does not mean that the mechanism (4) is not operative for other metal salts of heteropoly acids. Actually, infrared studies confirmed that the number of Bronsted sites increased when the aluminum salt of $H_3PW_{12}O_{40}$ was treated in water at 573 K (Ref. 14).

6. GENERATION OF ACID SITES IN SILVER AND COPPER SALTS OF HETEROPOLY ACIDS

In the methanol conversion over copper (II) salt of $H_3PW_{12}O_{40}$ at 523 K, a long induction time was observed (Ref. 15). Since the induction time is often related to the formation of active centers, the examination of the factors influencing the induction time may give a clue for the mechanism of the formation of active centers. The effect of hydrogen was examined as a possible source of protons, since small amounts of hydrogen were always found in the reaction products. The catalyst was kept in a hydrogen stream at 523 K for 0.5 or 3 h prior to the reaction, and then the reaction was started. The pretreatment for 0.5 h with hydrogen shortened the induction time to 4 h. The pretreatment for 3 h led to an almost complete disappearance of the induction time. The induction was not influenced at all by the pretreatment of water vapor at 573 K, suggesting that the reaction (4) is not operating here.

It is clear that hydrogen plays an essential role in the formation of Bronsted acid sites. Protons may be generated by the reaction of Cu(II) cations with hydrogen molecules.

$$\begin{aligned} Cu(II) + 1/2\ H_2 &\rightleftharpoons Cu(I) + H^+ \\ Cu(I) + 1/2\ H_2 &\rightleftharpoons Cu(0) + H^+ \end{aligned} \qquad (5)$$

During the methanol conversion, hydrogen molecules (or atoms) may be provided by the decomposition of methanol. Induction time is supposed to be the period which is required for the establishment of the equilibrium of reaction (5). The induction time was also observed in methanol conversion over Ag(I) salt of $H_3PW_{12}O_{40}$ at 513 K and it disappeared by the pretreatment of the Ag(I) salt of hydrogen at 523 K (Ref. 16). Thus, the mechanism similar to reaction (5) is operative also in the Ag(I) salt.

$$Ag(I) + 1/2\ H_2 \rightleftharpoons Ag(0) + H^+ \qquad (6)$$

Reaction (5) and (6) explains why Ag(I) and Cu(II) salts are the most active among the metal salts of heteropoly acids, since these salts are known to be the ones which are most easily reduced by hydrogen.

The generation of Bronsted acid sites by the interaction of hydrogen and Ag(I) or Cu(II) cations of the salts is confirmed by examining the catalytic activity for the isomerization of o-xylene, which is the reaction catalyzed by Bronsted acids sites (Ref. 14). The reaction was carried out at 573 K by using the Ag

(I) or Cu(II) salt of $H_3PW_{12}O_{40}$ (30 wt%) supported on activated carbon as a catalyst. Both the Ag(I) and the Cu(II) salts were totally inactive for the isomerization, but the activity was developed when the catalysts were pretreated in a hydrogen or methanol stream for 2 h at 573 K. These facts show that the Ag(I) and the Cu(II) salts as prepared, have no intrinsic acidity, and the acid sites are generated only after the interaction of the salts with hydrogen or methanol.

The formation of the Bronsted acid sites by the treatment of the Ag(I) and Cu(II) salts by hydrogen or methanol at 573 K was confirmed also by infrared spectroscopy (Ref. 14).

7. ENHANCEMENT OF THE CATLYTIC ACTIVITY OF Ag(I) SALT BY GASEOUS HYDROGEN

While hydrogen pretreatment eliminates the induction period in the methanol conversion, the presence of hydrogen in the gas phase was found to enhance the rate of the methanol conversion (Ref. 16). The methanol conversion was carried out at atmospheric pressure with the $Ag_3PW_{12}O_{40}$ (30 wt%) on active carbon as a catalyst with the initial pressure of methanol of 51 kPa and with a varying ratio of hydrogen and nitrogen for the remainder. As shown in Fig. 4, the hydrocarbon yield increased with the rise in the partial pressure of hydrogen. Thus, without hydrogen, the hydrocarbon yield was 24 %, while it was 43 % under the hydrogen pressure of 51 kPa. The effect of hydrogen was reversible. Thus, after carrying out the run under a hydrogen pressure of 51 kPa for 2 h, hydrogen was replaced by nitrogen. The hydrocarbon yield was reduced to the value which would be expected when the reaction had been started without gaseous hydrogen. Then, nitrogen was again replaced by hydrogen, the hydrocarbon yield returning back to the original value. The distribution of hydrocarbons was not affected by the presence of hydrogen, only the activity being enhanced. The effect of hydrogen on the catalysis is not restricted to methanol conversion. The catalytic activities of the Ag(I) salt for the vapor-phase synthesis of methyl t-butyl ether from methanol and 2-

methylpropene was also greatly enhanced by the pretreatment of hydrogen and also by the presence of gaseous hydrogen in the system (Ref. 17).

The effect of gaseous hydrogen on the acid site formation could be again understood by assuming the reversibility of eq. (6). Thus, the presence of gaseous hydrogen shifts eq.(6) to the right-hand side to increase the number of acid centers (H^+).

8. THE EFFECT OF HYDROGEN ON TRANSITION METAL-HETEROPOLY ACID CATALYSTS

When methanol conversion was carried out over palladium salt of $H_3PW_{12}O_{40}$ supported on silica, a greater enhancement by gaseous hydrogen was observed. As shown in Table 2, the conversion of methanol was 75.6 and 11.7 % in the presence, and in the absence, of hydrogen, respectively. The distribution of products also differed greatly. In the presence of hydrogen, no alkenes were observed, in the products, as expected from the high hydrogenation activity of palladium metal, which should be formed by the reduction of the Pd(II) cations under the reaction conditions. This is in sharp contrast to the hydrogen effect on silver salts, where the product distribution was not altered. The activity of the Pd(II) salt under hydrogen was much higher than those of the Ag(I) salt under hydrogen and of the parent acid, whose activity was not influenced by the presence of gaseous hydrogen. In the case of Pd salt, carbon monoxide was also formed.

The effect of hydrogen on the reaction rate was completely reversible. When the gas fed with methanol was changed from nitrogen to hydrogen or from hydrogen to nitrogen after running 2 h, the hydrocarbon yield gradually changes to the one which was supposed to be obtained if the reaction was started by cofeeding the second gas from the beginning.

The effect of hydrogen pressure was examined. The yield of hydrocarbons increased almost linearly with the partial pressure of hydrogen while the extent of methanol decomposition did not depend on the pressure. This indicates that, as expected, the

TABLE 2. The activity and the product distribution of the conversion of methanol into hydrocarbons over salts of dodecatungstophosphate.

Cation	Pd	Pd[a)]	Ag	H
Product Distribution / %				
CH_3OH	0.0	0.0	5.7	0.0
CH_3OCH_3	15.7	82.5	51.7	87.7
CO	8.7	5.8	0.0	0.0
Hydrocarbons	75.6	11.7	42.6	12.3
Hydrocarbons Distribution / %				
CH_4	13.6	17.9	7.5	8.9
C_2-C_3	38.4	20.5	5.9	4.1
C_2'-C_3'	0.0	25.6	26.3	45.5
C_4^+	48.0	36.0	60.3	41.5

Reaction conditions: 573 K, $CH_3OH = H_2$ = 51 kPa, W/F = 50 g h mol^{-1}, a) $CH_3OH = N_2$ = 51 kPa.

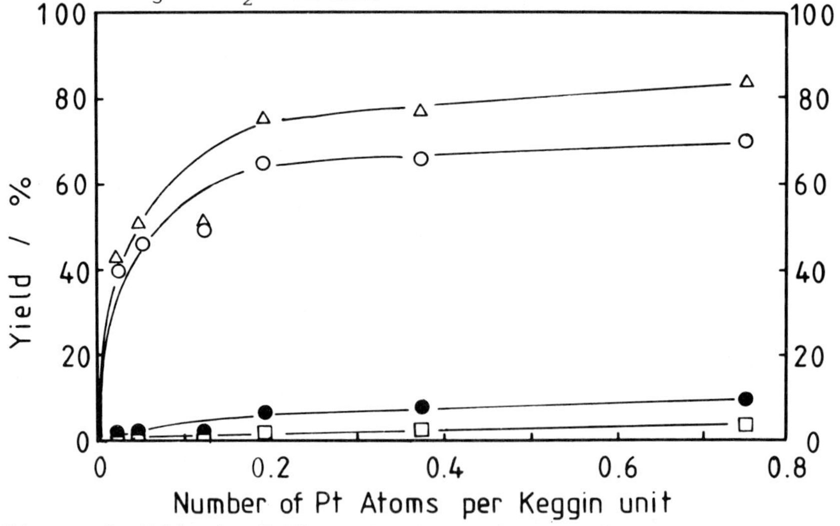

Figure 5. Effect of Pt content on the catalytic activity for methanol conversion over Pt-$H_3PW_{12}O_{40}/SiO_2$ catalysts. total conversion (Δ), yields of hydrocarbons (○), methane (●), and carbon monoxide (□); 573 K, CH_3OH = hydrogen = 51 kPa, W/F = 50 g h mol^{-1}.

active centers for hydrocarbon formation is different from those for methanol decomposition. The former must be Bronsted acid sites, the latter being palladium metal. Gaseous hydrogen strongly affects the acidic properties of the palladium salt; and metallic palladium should have a decisive role in generating acidic centers; as hydrogen did not give any enhancement on the parent acid.

The catalytic activity of $Pd_xH_{3-2x}PW_{12}O_{40}$ supported on silica was examined. The yield of hydrocarbons greatly increased with a very small content of palladium (x = 1/16). Only a slight increase in the hydrocarbon yield was attained by a further increase of x. The yield of carbon monoxide increased with the rising content of palladium, confirming that the decomposition of methanol proceeds on the metallic palladium. These results show that the enhancement by hydrogen is not a stoichiometric reaction of hydrogen with palladium species, though its presence in the system is essential.

A plausible mechanism for the enhancement of the activity may be as the following. Palladium cations are completely reduced to the metal by the pretreatment with hydrogen. Hydrogen molecules from the gas phase may dissociate into hydrogen atoms over the metal, and hydrogen atoms thus formed spill-over and interact with surrounding heteropoly anions to be converted into protons. The process is reversible.

$$H_2 \rightleftarrows 2\ H \quad \text{(over Pd metal)}$$
$$H + [PW_{12}O_{40}]^{3-} \rightleftarrows H^+ + [PW_{12}O_{40}]^{4-} \qquad (7)$$

In this mechanism, metallic palladium is a catalyst for the interconversion of hydrogen molecules and protons.

According to the above mechanism, the catalysts are not necessarily prepared from the metal salts of heteropoly acid. The point is the coexistence of the heteropoly anions and the metal which can dissociate hydrogen molecules. The methanol conversion was carried out with the catalyst by supporting $H_3PW_{12}O_{40}$ on Pd (2.5 wt%)/SiO_2. Thus, the palladium suppported on silica was prepared by palladium nitrate on silica and reducing it under flowing hydrogen at 593 K. To Pd-SiO_2 thus formed, $H_3PW_{12}O_{40}$ (28 wt%) was supported with an impregnation method. The catalyst, Pd-$H_3PW_{12}O_{40}$ /SiO_2 has a comparable catalytic activity as the Pd(II) salt/SiO_2. Pd metal supported on silica has negligible activity for hydrocarbon formation, a small amount of CO and CH_4 being formed. The $H_3PW_{12}O_{40}$/ SiO_2 catalyst showed much smaller activity than Pd-$H_3PW_{12}O_{40}$ as described earlier. Another catalyst examined was a 1:1 physical mixture of Pd/SiO_2 and $H_3PW_{12}O_{40}$/SiO_2. Again, the catalyst showed a much higher activity than the $H_3PW_{12}O_{40}$/SiO_2 catalyst. These results clearly demonstrate the synergetic effect of heteropoly acid and transition-metals.

The Pt-$H_3PW_{12}O_{40}$/SiO_2 catalysts were also prepared by an impregnation method. The impregnation of the SiO_2 support was made with an aqueous solution containing both $H_3PW_{12}O_{40}$ and H_2PtCl_6. The catalyst was reduced by hydrogen prior to the reaction.

The amount of platium was varied in $H_3PW_{12}O_{40}$-Pt/SiO_2 catalyst, where the supported amount of $H_3PW_{12}O_{40}$ was kept constant (30 wt%), and its effect on the catalytic activity for methanol conversion was examined. The results are given in Fig. 6. Activity increases very sharply with a rather small amount of platinum (0.2 Pt/Keggin anion). The further increase in Pt content resulted in only a small increase in activity. These results again conform well with the results of the $Pd_xH_{3-2x}PW_{12}O_{40}$ catalysts. When Pt content was below 0.5 %, the formation of carbon monoxide (catalysis by the metal) was almost completely depressed, and yet the reasonable enhancement of methanol conversion to hydrocarbons (catalysis by acids) was attained. These results show that by selecting the proper ratio of heteropoly anion to a metal component, a highly acidic catalyst without intervention by the catalysis by metals could be obtained.

9. ISOMERIZATION OF ALKANES OVER TRANSITION METAL-HETEROPOLY ACID CATALYSTS

Isomerization of alkanes is an industrial process, in which the catalyst is platinum in combination with acidic carriers such as fluorinated alumina and zeolites. As for the mechanism, dual functionality is generally accepted.

Since Pd salts of heteropoly acids or Pd-heteropoly acid systems were found to be highly active for the acid-catalyzed reactions, (especially in the presence of gaseous hydrogen) as seen in the previous section, the activities of these catalyst systems in relation to the isomerization of alkanes was examined.

The total conversion of hexane and the distribution of the products in hexane isomerization over Pd salt of $H_3PW_{12}O_{40}$ at three different temperatures are given in Table 3 (Ref. 18). Hexane (30 kPa) was fed with hydrogen of 71 kPa. The Pd salt gave the total conversion of 75.5 % under the same conditions, while the parent acid gave the hexane conversion of 7.6 %. Thus, the activity of the Pd salt is much higher than the parent acid. The presence of hydrogen is essential for isomerization. Fig. 6 show the effect of the partial pressure of hydogen on the total conversion of hexane. The isomerization did not practically proceed without gaseous hydrogen in the system.

The rate depends greatly on hydrogen pressure. The role of hydrogen in the catalysis may be multiple. Primarily, hydrogen enhances the acidic property of the catalytic system as has been observed in the methanol conversion. Thus, hyrogen may dissociate on the surface of the metallic palladium and spill-over to interact with heropoly anions to give protons. Secondly, the spilled-over hydrogen atoms may hydrogenate the coke precursors, and thus ensure higher steady rate and longer catalyst life.

Isomerization of pentane and heptane was also carried out under the partial pressures of the alkane of 30 kPa and hydrogen of 71 kPa. In the case of pentane isomerization, the conversion of 40 % with selectivity to isopentane of 97 % was obtained at 453 K, those being 92 % and 58 % at 473 K. In the isomerization of heptane, the selectivity to isomers was 70 % at the conversion of 20 % at 453 K.

TABLE 3. Product distribution in hexane isomerization over $Pd_{3/2}PW_{12}O_{40}$ supported on silica(A) and $H_3PW_{12}O_{40}$ supported on Pd/carbon(B).

Catalyst	A	A	A	B
Reaction temperature /K	443	483	523	523
Conversion /%	27.1	53.0	53.8	77.9
Selectivity to isomers /%	94.2	89.8	78.5	96.4
Product distribution /%				
Methane	0.0	0.0	0.1	0.1
Ethane	0.0	0.1	0.3	t
Propane	0.3	1.7	6.1	0.7
Butanes	1.1	4.2	9.6	0.9
Pentanes	0.6	3.4	5.1	0.5
2,2-Dimethylbutane	2.7	4.3	3.1	5.2
2,3-Dimethylbutane 2-Methylpentane	63.7	59.1	51.5	49.0
3-Methylpentane	27.8	26.4	23.9	20.9
Methylcyclopentane	1.6	0.5	0.2	0.3
Cyclohexane	2.2	0.2	0.1	0.1
Heptanes	0.0	0.1	0.1	0.1

Hexane 30 kPa, Hydrogen 71 kPa, W/F = 100 g h mol^{-1}.

In order to show that the combination of metallic palladium and heteropoly anions are essential for the catalysis, $H_3PW_{12}O_{40}$ was supported on Pd(5 %)-active carbon obtained from a commercial source. The $H_3PW_{12}O_{40}$-Pd/C catalyst revealed better activity and selectivity for the isomerization. Thus, the selectivity to isomers was 97 % at hexane conversion of 78 % at 523 K, while they were 78.5 and 53.8 % over Pd salt of $H_3PW_{12}O_{40}$ at the same temperature (Table 3).

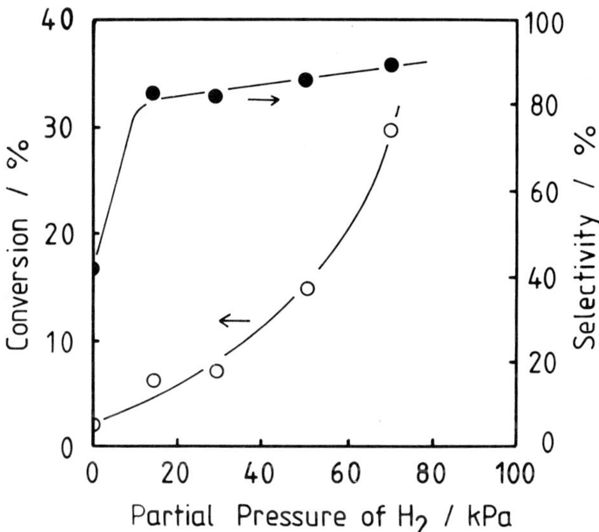

Figure 6. Effect of hydrogen pressure on hexane isomerization over Pd salt of $H_3PW_{12}O_{40}$. 483 K, hexane = 30 kPa, W/F = 40 g h mol^{-1}.

REFERENCES

1. Y. Izumi, K. Matsuo, and K. Urabe, J. Mol. Catal., 18, 299 (1983).
2. I. V. Kozhevnikov, and K. I. Matveev, Applied Catal., 5, 135 (1985).
3. T. Baba and Y. Ono, J. Mol. Catal., Submitted.
4. Y. Izumi, K. Matsuo, and K. Urabe, "The Chemistry and Uses of Molybdenum, Proceeding, 4th International Conference on Catalysis", Section V-B, Golden (1982) p.289.
5. A. Aoshima, S. Yamamatsu, T. Yamaguchi, Shokubai, 27, 386 (1985).
6. Y. Ono and T. Mori, J. Chem. Soc., Faraday Trans. I, 77, 2209 (1981).
7. Y. Ono, T. Mori, and T. Keii, Proc. 7th Intern. Congr. Catal., 1980, Tokyo, Kodansha-Elsevier (1981) p.1414.
8. T. Baba, J. Sakai, H. Watanabe, and Y. Ono, Bull. Chem. Soc. Jpn., 55, 2555 (1982).
9. M. Misono, K. Sakata, Y. Yoneda, and W. Y. Lee, Proc. 7th Intern. Congr. Catal., 1980, Tokyo, Kodansha-Elsevier (1981) p.1047.
10. T. Okuhara, A. Kasai, N. Hayakawa, M. Misono, and Y. Yoneda, J. Catal., 83, 121 (1983).
11. H. Niiyama, Y. Saito, S. Yoshida, and E. Echigoya, Nippon Kagaku Kaishi, 1982, 569.

12. Y. Ono, T. Baba, J. Sakai, and T. Keii, J. Chem. Soc., Chem. Comm., 1982, 400.
13. T. Baba, J. Sakai, and Y. Ono, Bull. Chem. Soc. Jpn., 55, 2637 (1982).
14. T. Baba and Y. Ono, J. Phys. Chem., 87, 2406 (1983).
15. T. Baba and Y. Ono, Applied Catal., 8, 315 (1983).
16. Y. Ono, M. Kogai, and T. Baba, Proc. Pab-Pacific Synfuel Conf. Tokyo 1982, Tokyo, Japan Petroleum Inst. (1982) Vol.1, p.115.
17. Y. Ono and T. Baba, Proc. 8th Intern. Congr. Catal., 1984, Verlag Chemie Vol. 5, p.405 (1984).
18. S. Suzuki, M. Kogai, and Y. Ono, Chem. Lett., 1984, 699.

EPOXIDATION OF CYCLOHEXENE BY MOLECULAR OXYGEN OVER A COMBINED SUPPORTED PLATINUM AND DISSOLVED MANGANESE TETRAPHENYLPORPHYRIN CATALYST

J. Van Beijnum, A.J. Van Dillen and J.W. Geus
Department of Inorganic Chemistry, University of Utrecht, Croesestraat 77a;
W. Drenth
Department of Organic Chemistry, University of Utrecht, Transitorium 3, Padualaan 8, 3584 CH, Utrecht, The Netherlands

ABSTRACT - Cyclohexene is epoxidized with molecular oxygen, applying hydrogen as co-oxidized species. The catalyst consists of platinum on silica and homogeneously dissolved manganese(III) tetraphenylporphyrin. The predominant reaction is the formation of water. The epoxidation selectivity, however, is increased by alloying the platinum with silver.

1. INTRODUCTION

Many research groups are investigating the epoxidation of alkenes catalysed by manganese porphyrins (Ref.1). The epoxide oxygen atom is supplied by a Mn=O porphyrin intermediate. This intermediate is formed by oxidizing Mn(III) porphyrin with sodium hypochlorite or other single oxygen donors. Since molecular oxygen is cheap and does not give rise to undesired by-products, it is the logical oxidant of choice. Moreover, in natural systems, molecular oxygen is bound to iron porphyrins. A co-reducing species removes one of the two atoms of the coordinated molecular oxygen. The remaining oxygen atom subsequently reacts with an alkene to the epoxide.

In model systems, the use of molecular oxygen as an oxidant in porphyrin catalysed epoxidations, has until recently been unsuccessful. Mansuy et al.(Ref. 1 e) could use molecular oxygen by combining manganese porphyrin with ascorbic acid as a co-reductant. Tabushi et al. utilized a metal porphyrin as catalyst and as a co-reductant they used either sodium borohydride (Ref.2) or, more successfully, H_2/Pt (Refs. 3-5). In the latter case, platinum was precipitated onto an organic polymer, and suspended in the solution.

Previously, we reported on the epoxidation of alkenes by sodium hypochlorite in the presence of manganese porphyrins as catalysts (Meunier system (Ref. 1 c)). We especially studied the kinetics and the effect of site isolation which enabled us to extend the mechanistic picture of the catalytic reaction (Ref. 1 h). Here we report on preliminary experiments done with molecular oxygen as the oxidant. The catalytic system consists of manganese(III) tetraphenylporphyrin and silica-supported platinum.

2. RESULTS AND DISCUSSION

In a first series of experiments, the manganese(III) tetraphenylporphyrin (Mn(III).TPP) was homogeneously dissolved in methanol. The solution also contained 1-methylimidazole, the substrate cyclohexene, and suspended platinum on silica. The suspension was shaken vigorously and kept in contact with a 1:1 mixture of hydrogen and oxygen. The activity of platinum in the oxidation of hydrogen to water is so high, that it predominates and only a minor fraction of the hydrogen adsorbed by platinum is able to act as co-reductant in the epoxidation. As a result, the epoxidation proceeds very slowly. It took three weeks at room temperature to get appreciable amounts of epoxide. With respect to the substrate, the yield of 1,2-epoxycyclohexane was approximately 70%. In view of the presence of a ring opened product, viz. 2-methoxycyclohexanol, in the reaction mixture, the initial yield of the epoxide is presumably higher. Minor by-products were cyclohexanone, 1,1-dimethoxycyclohexane, a cyclohexenol, and 2-cyclohexenone. Owing to the predominating reaction to water, the yield of epoxide with respect to reacted oxygen was 1-2% only.

To suppress the reaction to water we substituted silver for platinum. Whereas silver dissociatively adsorbs oxygen at room temperature, the dissocition of hydrogen does not occur. However, no reaction of cyclohexene was observed. Apparently, dissociated hydrogen is required in our reaction.

To decrease the activity of platinum in the reaction to water, we alloyed the platinum surface with silver. To cover the 2 nm platinum particles uniformly with silver, we applied a special procedure in which reduction of dissolved silver is catalysed on the surface of the platinum particles (Ref. 6). We compared the activity and selectivity in reactions with three different co-catalysts:

A. 6% Pt on silica
B. 5.9% Pt and 2.0% Ag on silica
C. 5.7% Pt and 5.5% Ag on silica.

Figure 1 reveals that with increasing amounts of silver the activity decreases. However, the selectivity increases in the same direction: with co-catalyst C the selectivity of epoxide formation with respect to converted oxygen is approximately 15%.

At room temperature, silver takes up oxygen slowly and to a limited extent. The adsorbed oxygen atoms react with hydrogen to adsorbed hydroxyl groups (Ref. 7). The hydroxyl groups do not readily desorb at room temperature and thus block the silver surfaces. The active platinum sites will, therefore, be separated by sil-

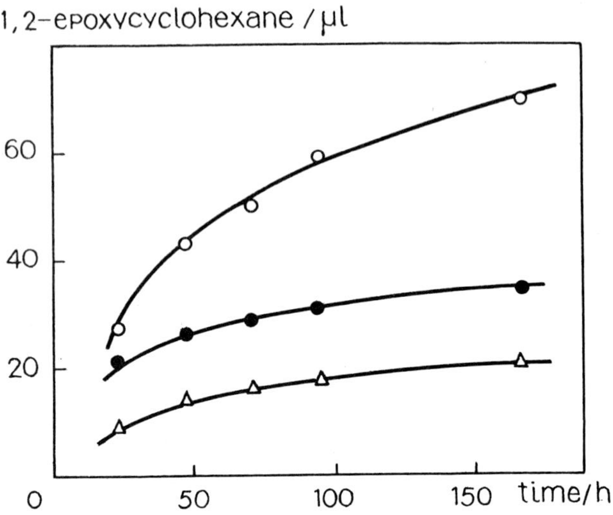

Figure 1. Formation of epoxide as a function of time; O A, no Ag; ● B, 2.0% Ag; Δ C, 5.5% Ag.

ver patches. In particular, the mobility of hydrogen adatoms over the metal surfaces will be significantly restricted by alloying with silver. As a result the reaction between adsorbed hydrogen and oxygen will be less rapid, whereas the reaction of adsorbed hydrogen atoms with oxygen molecules bonded to manganese porphyrin remains relatively unaffected. To restrict the mobility and hence the reactivity of adsorbed hydrogen atoms with adsorbed oxygen atoms, a fair amount of silver must be present in the alloy surface. On the other hand, silver lowers the number of sites capable of dissociatively adsorbing hydrogen, which causes the activity to drop with an increasing fraction of silver in the surface.

According to current opinion (Ref. 8) the first step in the catalytic process will probably be the reduction of manganese(III) to manganese(II). This view is supported by our observation that an absorption peak at 428 nm appears when hydrogen gas is bubbled through a methanolic solution of manganese(III) porphyrin in which the co-catalyst, Pt/SiO_2, has been suspended. In the next step, molecular oxygen will coordinate to manganese(II) and subsequently, the O-O bond in Mn-O-O will be broken by the action of an additional electron and two protons, resulting in Mn(V)=O.TPP and H_2O. Formation of the high valent metal ion is facilitated by the electron donating ligand 1-methylimidazole. Under our present conditions the Mn(V)=O species could not be detected.

3. EXPERIMENTAL

Mn(III).TPP was prepared as described earlier (Ref.9). As platinum catalyst the so-called Standard Euro platinum catalyst, produced by Johnson Matthey Chemicals, was used. It contains 6% of Pt, precipitated on SiO_2. The epoxidation reaction was performed in a round bottomed flask using a magnetic stirrer. The reaction mixture consisted of 13 mg Mn-TPP, 80 µl of 1-methylimidazole, 5 cm^3 of cyclohexene, 20 mg co-catalyst and 45 cm^3 of CH_3OH. A 1:1 mixture of H_2 and O_2 was in contact with the solution. The consumption of the gas mixture in the reaction was recorded. Product analyses were performed by gas chromatography, applying a Carbowax 20-M column and flame ionisation detection.

The silver co-catalyst (20% Ag on SiO_2) was prepared as described by De Jong (Ref. 6). In the same way, the Ag/Pt co-catalysts were obtained by precipitation of Ag on the Euro catalyst.

ACKNOWLEDGEMENT. The authors thank Dr. R.J.M. Nolte for helpful discussions.

REFERENCES

1. (a) J.T. Groves and T.E. Nemo, J. Am. Chem. Soc., 105, 5786 (1983);
(b) M.F. Powell, E.F. Pai and T.C. Bruice, J. Am. Chem. Soc., 106, 3277 (1984);
(c) B. Meunier, E. Guilmet, M.E. De Carvalho and R. Poilblanc, J. Am. Chem. Soc., 106, 6688 (1984);
(d) J.A. Smegal and C.L. Hill, J. Am. Chem. Soc., 105, 3515 (1983);
J.A. Smegal, B.C. Schardt and C.L. Hill, J. Am. Chem. Soc., 105, 3510 (1983);
(e) D. Mansuy, M. Fontecave and J.F. Bartoli, J. Chem.Soc., Chem. Commun., 253 (1983);
(f) J.P. Collman, J.I. Brauman, B. Meunier, T. Hayashi, T. Kodadek and S.A. Raybuck, J. Am. Chem. Soc., 107, 2000 (1985);
(h) J.A.S.J. Razenberg, R.J.M. Nolte and W. Drenth, Tetrahedron. Lett., 25, 789 (1984);
J.A.S.J. Razenberg, A.W. van der Made, J.W.H. Smeets and R.J.M. Nolte, J. Mol. Catal.,(1985) in press;
(i) J.R. Lindsay Smith and P.R. Sleath, J. Chem. Soc., Perkin Trans. II, 1009 (1982).
2. I. Tabushi and N. Koga, J. Am. Chem. Soc., 101, 6456 (1979).
3. I. Tabushi and A. Yazaki, J. Am. Chem. Soc., 103, 7371 (1981).
4. I. Tabushi and K. Morimitsu, J. Am. Chem. Soc., 106, 6871 (1984).
5. I. Tabushi, M. Kodera and M. Yokoyama, J. Am. Chem. Soc., 107, 4466 (1985).
6. K.P. de Jong and J.W. Geus, in Preparation of Catalysts III, G. Poncelet, P. Grange and P.A. Jacobs (Eds.), p. 111, Elsevier, Amsterdam (1983).
7. V.Sh. Gruver, A.V. Khasin and G.K. Boreskov, Kinet. Katal., 12, 156 (1971).
8. See e.g. a detailed discussion for Fe.TPP in Ref. 5.
9. See e.g. Ref. 1 (h).

Part 2

Catalysis by Immobilized Metal Complexes

SYNTHETIC POLYMERS IN DESIGN OF METAL-COMPLEX CATALYTIC SYSTEMS

V.A. Kabanov and V.I. Smetanyuk
Institute of Petrochemical Synthesis, USSR Academy of Sciences, Moscow, USSR

ABSTRACT - Main results of the study of novel catalysts for liquid-phase processes are reported. The catalysts contain metal-complex active centers fixed on a special class of polymeric carriers capable of limited swelling in the reaction medium. These gel-immobilized catalytic systems (GICSs) can combine some properties of homogeneous and heterogeneous catalysts. General principles of design, morphology and methods of synthesis of the GICSs are considered. Basic properties and peculiarities of the behaviour of the GICSs related to their specific mosaic structure are discussed using catalytic reactions of olefin dimerization and ethylene polymerization as examples.

1. INTRODUCTION

Interest in theory and application of catalysis by metal complexes has greatly increased in last thirty years. Metal-complex catalysts (MCCs) per se can be dissolved (homogeneous MCCs) or finely dispersed (microheterogeneous MCCs) in liquid reaction media. A considerable increase in MCCs stability and productivity is achieved by means of their deposition on solid inorganic or polymeric carriers (second-generation MCCs). About ten years ago we suggested that a special class of polymeric carriers capable of limited swelling in the reaction media be used for the fixation of metal complexes. In the opertaing state such catalytic systems represent gel-like particles of the given sharp and dimensions. Active centers are distributed and immobilized in the bulk of these particles, the latter being permeable to the solvent, reagents and reaction products. These catalytic systems, which can combine some properties of homogeneous and heterogeneous MCCs, were called gel-immobilized catalytic systems (GICSs) (Refs. 1-4).

2. PRINCIPLES OF DESIGN, SYNTHESIS AND STRUCTURE OF GEL-IMMOBILIZED CATALYTIC SYSTEMS

These principles are dictated by the need to have the following preprogrammed catalytic properties:

1) The catalyst must consist of specifically formed particles of a separate phase, insoluble but sufficiently swellable in the reaction medium, i.e., permeable to molecules of the substrate and the solvent;

2) The reaction products must easily diffuse into the catalyst phase and pass from this into solution, making room for new portions of substrate;

3) The catalyst particles in the operating state must be mechanically strong, elastic and morphologically stable, in order not to be crushed and not to adhere during the process;

4) The catalytic centers must be sufficiently uniformly distributed throughout the whole volume of the catalyst system and must be attached sufficiently strongly to the chains of the polymeric carrier, so as not to be washed out by the streams of reagents.

The combination of the above-mentioned requirements places specific limitations on the composition and structure of the polymeric carrier.

It is evident that ordinary homopolymers, which coordinatively bind with transition metal ions, are not suitable for this purpose because the corresponding polymer complexes, as a rule, are insoluble and do not swell. As might first appear, the simplest solution could consist of the synthesis of an ordinary random copolymer, in whose macromolecules the units capable of forming complexes with transition metal compounds are diluted by inert units showing affinity for only the solvent and substrate. However, it can easily be proved that even such copolymers can hardly serve as carriers for the design of catalyst systems with the above-mentioned properties. Even when the content of randomly distributed units saturated with transition metal ions reaches 10 mole %, the system as a whole also loses the capability of swelling because of hardening which results from additional cross-linking of macromolecules by complex bridges. Thus, homopolymers and random copolymers, as well as inorganic sorbents, are actually suitable only for the production of the second-generation catalyst systems mentioned above.

A different situation should be expected when block or graft copolymers are used as carriers. It is known that already at the stage of synthesis of block or graft macromolecules, the incompatibility or limited compatibility of molecular chains, which differ in chemical composition or structure, leads to their microsegregation (microseparation). No macroscopic segregation takes place; this is prevented by chemical bonds between different chains. As a result, the material built up of block and graft copolymers represent a microheterogeneous mosaic system, in which one of the components is dispersed in the other and forms domains of certain dimensions and shape. It is significant that these materials can easily be made morphologically stable by producing a three-dimensional vulcanized network within them. In this case their swelling in the solvent proceeds only as a result of a reversible affine deformation of the chemically stabilized mosaic structure. To synthesize an optimal carrier, one of the macromolecular components forming the domains must be selected so that to provide its sufficiently easy bonding with transition metal atoms. The role of a polymeric ligand to this component will be assigned. The other (a basic polymer), which forms a high-molecular dispersion medium and is inert toward a transition metal, must be capable of swelling in the reaction solution. After treatment of a combined carrier of this type with a soluble transition metal compound, this latter is spontaneously fixed in the domains of the polymeric ligand. The chemically unaffected swollen inert layers of another macromolecular component can serve to transport the reagents and reaction products in the body of the catalyst system. Our task was to develop GICSs for olefin dimerization and polymerization on nickel and titanium complexes, respectively (Refs. 5-8). In accordance with the above requirements, the polymer basis for the GICss must be inert toward immobilized transition metal compounds and must show an affinity for olefins and solvents (hydrocarbons, chlorohydrocarbons), which are usually used for polymerization and oligomerization. As the polymer basis for the synthesis of the GICSs, it is advisable that nonpolar elastomers be used. Of these, ethylene-propylene rubber (EPTR) - a ternary copolymer of ethylene, propylene and unconjugated diene capable of dissolving hydrocarbons and inert to transition metal compounds, is preferable. 1,2-polybitadiene (PB), 4-vinylpyridine (PVP), polyallyl alcohol (PAAL) or polymetacrylic acid (PMAA) (Refs. 5,9) were grafted on EPTR as polymeric ligands by a radical mechanism. Then, the materials obtained were shaped and vulcanized to form a continuous three-dimensional network from the basic macromolecules. Finally, the shaped particle represents a chemically cross-linked system capable of only limited swelling in the solvent.

As carriers for nickel-containing GICSs, grafted copolymers EPTR and PVP(EPTR-PVP) and EPTR, PB and PVP (EPTR-PB-PVP) with a PVP content of 5-40 wt% and a PB content of 10-20 wt% were used.

The carrier so obtained was placed in a hydrocarbon solvent and, in a swollen state, was treated with soluble divalent nickel compounds, and the material was kept at room temperature for 8-24 h. Then the complex formed was carefully washed to remove an excess nickel compound and was activated by treating with alkylaluminium halides (Ref. 6). The gel-form complex based on EPTR-PB-PVP was treated with $HAlR_2$ and then with $RAlCl_2$. As a result, the GICS was obtained in which both catalyst components were immobilized (Ref. 8).

As carriers for titanium GICS grafted copolymers EPTR-PAA with a PAA content of 5-10 wt% and EPTR-PMAA with a PMAA content of 5-20 wt% were used. Hydroxyl or carboxyl groups of grafted macroligands of the carrier were then modified by means of their treatment with an excess Grignard's reagent or by other organomagnesium compounds; titanium tetrachloride was chemisorbed on them,and the complex obtained was activated with organoaluminium compounds (Refs. 9-10).

The morphology of the typical GICS carriers before and after their treatment with transition metal compounds has been studied in every detail by electron-microscopic methods (Refs. 11-12).

The electron microphotographs of thin layers of the carrier (EPTR-PVP) before and after their treatment with nickel acetylacetonate are presented in Fig. 1. One can see that the distribution and shape of globular inclusions which, in one case, contain a polymeric ligand and in the other, corresponding nickel complexes, are essentially identical. Microphotographs presented in Fig. 2 were obtained at a larger magnification after the surface layers had been etched by O_2-plasma. They show a thin structure of globular inclusions, which undergoes no cardinal changes after metal binding. Thus, the morphological structure of the GICS can be described by the scheme shown in Fig. 3.

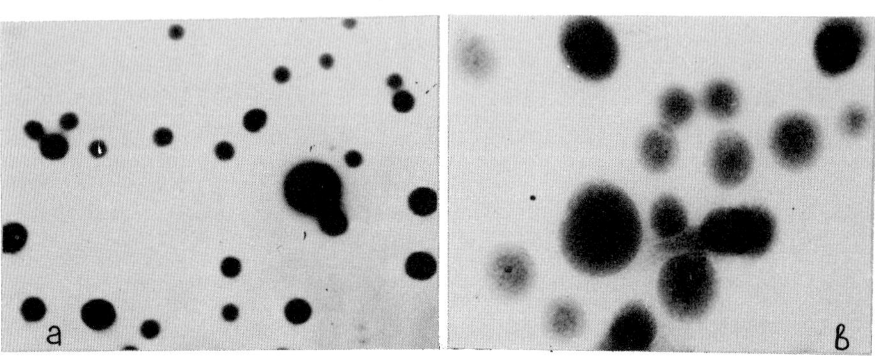

Figure 1. Microphotographs of a thin layer of the polymer carrier EPTR-PVP-40: a) untreated with nickel salts, 30000 times magnified; b) treated with nickel acetylacetonate, 40000 times magnified.

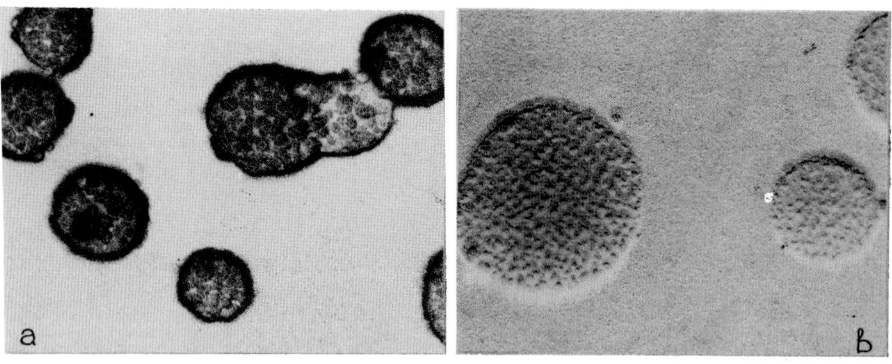

Figure 2. Microphotographs of thin layers of polymer carrier EPTR-PVP-40 etched by O_2-plasma: a) 100000 times magnified; b) treated with nickel acetylacetonate, 100000 times magnified.

The region defined by the outline is part of a random cross section of a GICS granule. The shaded region represents a pure elastomer, capable of swelling in the reaction medium. Here the transport of reagents and reaction products takes place. The particles with the size of about 10^3 Å, (dark spots in the figure) are inclusions of a grafted copolymer, dispersed in the pure elastomer.

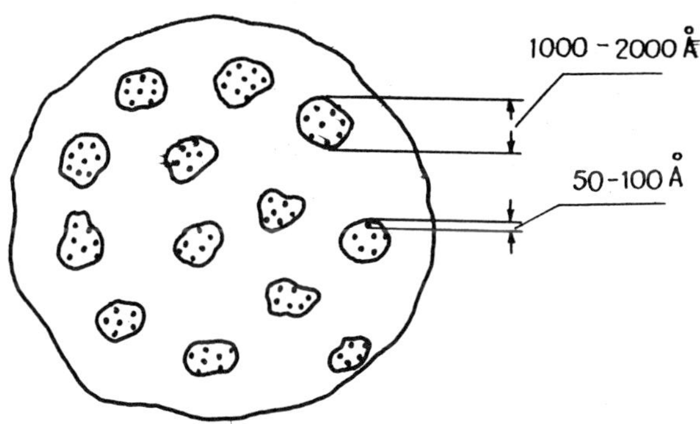

Figure 3. Schematic representation of GICS morphology.

These inclusions, in turn, contain domains of the polymeric ligand, formed as a result of microsegregation of the grafted branches incompatible with the base polymer. The size of each domain does not exceed values of the order 10-10^2 Å. During the treatment of the carrier with the soluble starting components of the catalyst, chemisorption of the transition metal compound and formation of immobilized, catalytically active centers take place in these domains.

3. BASIC PROPERTIES AND PECULIARITIES OF GICS BEHAVIOUR IN CATALYTIC REACTIONS

Should GICSs be referred to homogeneous or heterogeneous systems? This question is not trivial. On the one hand, as it was already mentioned, the GICSs can be formed as macroscopic bodies. In the operating state, they preserve their shape and are characterized by a certain elasticity. These properties, by definition, are inherent in all solids, including solid heterogeneous catalysts. From this point of view, the same basic principles can be used in the technological development of processes with the use of GICS and the same basic advantages can be obtained that are generally related to heterogeneous contact catalytic systems. On the other hand, the carriers for the GICSs were prepared by grafting the chains of a polymeric ligand to an ethylene-propylene rubber which, in all cases, was present in the reaction system in a sufficiently large excess. Therefore, the final grafting product is actually a colloidal dispersion of the graft copolymer in an excess pure elastomer. The latter serves as a polymeric dispersion medium.

Thus, the bodies of macroscopic dimensions permeable to the solvent, in the form of which the GICS is formed, are filled with the catalyst, brought close to the upper limit of its degree of dispersity. In this sense, GICSs are similar to homogeneous or microheterogeneous catalytic systems. The characteristic double behaviour of the aggregate state of the GICSs is a key factor that determines their unique properties.

3.1. Stability of the Catalytic Activity

In the absence of catalytic poisons GICSs, like some other heterogenized MCCs, preserve constant catalytic activity for thousands of hours (Refs. 9,10,13). To illustrate this, the dependences of ethylene (1,2) and propylene (3,4) dimerization on nickel-containing GICSs in a flow reactor are shown in Fig. 4. When the olefins, containing about 10 ppm of O_2 and H_2O as impurities, are used, the catalytic activity begins to decrease in a few tens of hours. But if some portion of isobutyl aluminium sesquichloride (IBASC) is added, the activity is restored up to the initial level (curves 2,4). Regeneration cycles of the GICSs can be repeated.

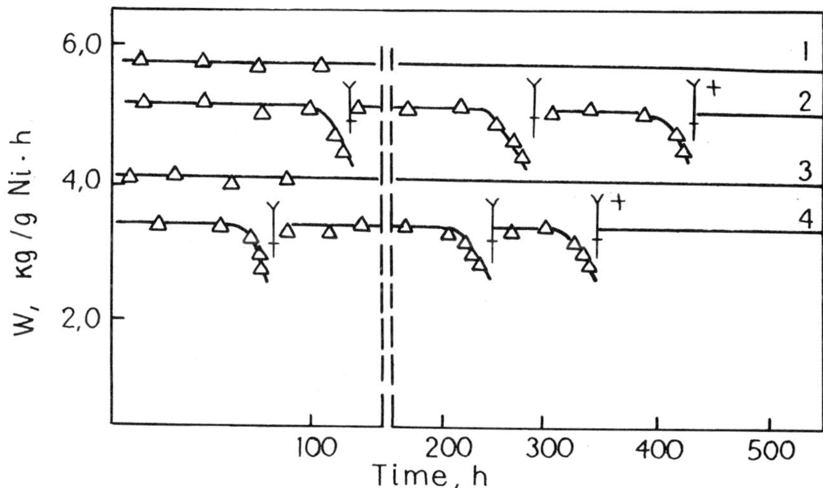

Figure 4. Olefin dimerization rate (W) on nickel-containing GICS vs. time. Solvent - olefin dimers, cocatalyst - IBASC, T = 60°C, ↓ - addition of IBASC portion, ↓$^+$ - the beginning of continuous feeding of IBASC 1 - ethylene (after additional purification), P = 8 atm., [IBASC] = 0.01%; 2 - ethylene (without additional purification), P = 6 atm., [IBASC] = 0.1%; 3 - propylene (after additional purification), P = 7 atm., [IBASC] = 0.01%; 4 - propylene (without additional purification), P = 6 atm., [IBASC] = 0.1%.

The dependences of the rate of ethylene polymerization on titanium-containing GICS are shown in Fig. 5. One can see that after a certain period, the rate of polyethylene formation is constant. It remains constant during the experiment, even when the temperature reaches 150-200°C. At these temperatures a high-molecular polyethylene formed inside the GICS swollen granules, is continuously fed into the surrounding solvent. We do not know any other titanium-containing catalyst which could polymerize polyethylene for a long time at such high temperatures.

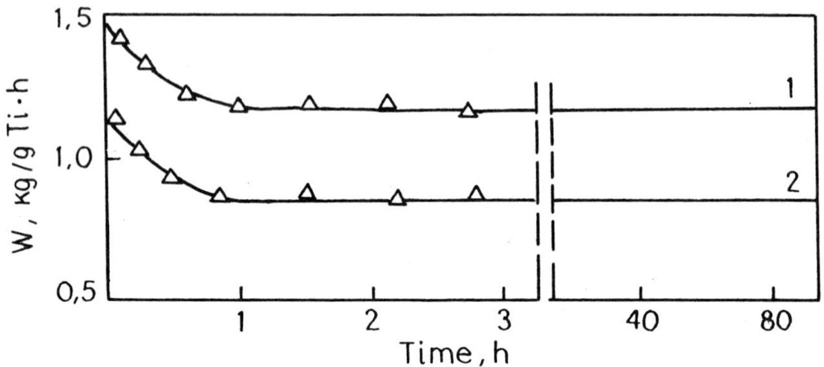

Figure 5. The rate of ethylene polymerization (W) on titanium-containing GICS vs. time.
1 - EPTR-PMAA-3RMgX-$TiCl_4$-$HAlR_2$, T=160°C, P=20 atm.;
2 - EPTR-PAAl-RMgX-$TiCl_4$-AlR_2Cl, T=145°C, P=10 atm.

One of the reasons for the increased thermal stability of the active centers of GICS, as compared with those of analogous homogeneous or microheterogeneous metal-complex catalysts, is undoubtedly related to the limited progressive diffusion of the active centers, due to their attachment to the carrier. It is known, for example, that the main reaction of thermal deactivation of titanium-containing catalytic centers for the polymerization of ethylene and α-olefins consists of disproportionation of hydrocarbon radicals, leading to the disappearance of the Ti-C bonds responsible for introduction of monomer molecules. The biomolecular disproportionation reaction

$$2\ RTiX_n \longrightarrow 2\ TiX_n + R_{-H} + R_{+H} \qquad (1)$$

naturally requires the approach of two initial RTiX particles, and consequently, cannot take place in a system where these particles are spatially separated from each other.

Spatial fixation of active centers in the GICS, apparently, also causes their above-mentioned capacity for an easy and almost complete regeneration after deactivation by ordinary catalyst poisons such as H_2O and O_2. In fact, the poisoning of an active center of a metal-complex catalyst, for example, by moisture traces is due to hydrolysis of Me-C or Me-H bonds:

$$\text{Me-R} + H_2O \longrightarrow \text{Me-OH} + \text{RH} \quad (2)$$

$$\text{Me-H} + H_2O \longrightarrow \text{Me-OH} + H_2. \quad (3)$$

In a homogeneous or microheterogeneous system, the initial acts of hydrolysis are accompanied by secondary conversions of the hydroxides obatined, which necessarily include the stages of their aggregation with each other, in the final analysis, to practically irreversible deactivation processes. In contrast with this, the immobilized centers, even after the hydrolysis described by reactions (2) and (3), remain structurally separated from each other and, consequently, do not undergo any more extensive physicochemical changes. They can therefore be easily regenerated by suitable alkylating or hydrogenating agents. If macromolecular ligands are used for the immobilization of transition metal compounds, as in the case of GICSs, a very important factor of cooperative stabilization of the complexes becomes valid. The structure of the polymer-metal complex can be represented by the following general scheme:

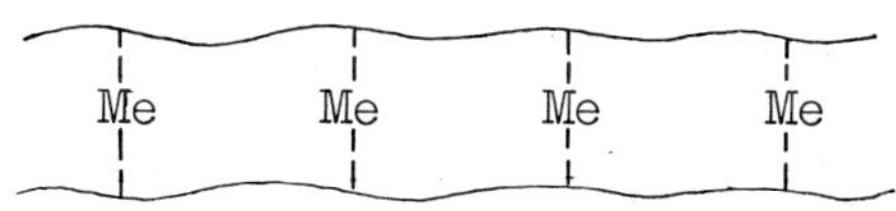

The rupture of one of the bonds between the metal and the unit of a polymeric ligand, shown by the broken line in the above scheme, requires approximately the same amount of energy as the rupture of a bond in an analogous complex with a low-molecular-mass ligand. However, in the former case, the expenditure of this energy is not compensated by new progressive degrees of freedom in the system, i.e. is not accompanied by an increase in its configurational entropy. The probability of destruction of a polymer-metal complex, under otherwise identical conditions, is therefore sharply reduced. The cooperative stabilization of macromolecular structures, including active centers, seems to play an important role in increasing thermal and chemical stabilities of metal-complex catalysts, particularly of the GICSs, deposited on polymeric ligands. It should be noted that in the kinetic aspect, the state of the swollen gel is extremely favourable for the mutual rearrangement of macromolecules required for the formation of the polymer-metallic structure shown above. The ease of structural-conformational adaptation in the complexing of polymeric ligands with transition metal compounds is one of the unique properties of the GICSs. Therefore, the mosaic polymeric carriers of the GICSs are optimal systems for the production of stable polymer-metal complexes.

It is significant that the values of the reversible deformation of GICs granules, which are elastomers swelling in the reaction medium, are higher by several orders of magnitude than the destructive deformations of ordinary solid catalyst carriers, and the moduli of elasticity are several orders of magnitude lower. The GICSs, used even in a contact catalytic process designed in the usual way, can therefore be expected to provide additional advantages resulting from their increased ability to adapt to the different types of mechanical and thermal actions that they undergo in the reaction medium.

Among these actions we should mention, primarily, collisions and friction of the catalyst particles with each other. Because of their high elasticity, the GICSs rather easily withstand deformations that are catastrophic to conventional heterogeneous catalyst carriers. Mechanical stresses in the GICSs are rapidly relaxed, as in any rubber bodies. The possibility of heat transfer by diffusion and convection of the liquid component in the body of the polymer gel also helps to prevent local overheating.

3.2. Selectivity

In several possible routes of a given catalytic reaction, an increase of the relative rate of one of the routes is an increase of the catalytic reaction selectivity. Since each of the possible routes of catalytic conversion has its own specific type of active center, the creation of highly selective catalysts requires formation of a single type of centers at the stage of synthesis and activation of the catalytic system and, later on, prevention of their chemical rearrangement of active centers during a catalytic reaction. As applied to metal-complex catalysts, the optimum method for solving these problems is the use of polymeric ligands as carriers. For the bonding of transition metals to polymeric ligands in solutions, because of structural characteristics and conformational limitations of the corresponding macromolecules, only one or few of the several types of complexes possible in principle are frequently produced. Effects of this type have been observed, for example, in the complexing of ions of divalent copper with PVP (Ref. 14), with linear polyethylenimine (Ref. 15), and also in a number of other cases. It was noted above that regardless of the Ni/PVP ratio during adsorption of nickel compounds on an EPTR-PVP carrier each transition metal atom in all probability, is bonded to two pyridine rings:

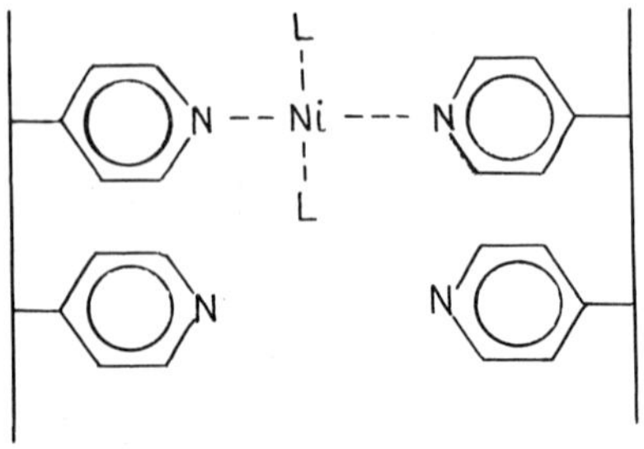

It should be emphasized again that, in the medium of a swollen polymer gel, kinetically favourable conditions are established for conformational selection during the formation of complexes including macromolecules. The indicated circumstances, in conjunction with the consideration presented above with respect to the cooperative stabilization of polymer-metal complexes, permit us to hope that the gel immobilization of transition metal compounds with an adequate selection of polymeric ligands, will open up new ways for producing highly selective catalytic systems.

The data presented in Table 1, permitting a comparison of the output of isomeric products of ethylene dimerization on soluble nickel and nickel-containing GICS complexes, can serve as an example. The same is illustrated in Fig. 6.

As follows from these data, the primary product of dimerization in all cases is 1-butene, which then is, probably, isomerized on catalytic centers of different compositions. This secondary process is almost eliminated on the GICS which along with a transition metal, contains also a polymeric organoaluminium compound (curve 3)

TABLE 1. Isomeric composition of butenes obtained from the dimerization of ethylene on different catalytic systems

Product No	Ligand	Cocatalyst	Composition of butenes, %		
			1-butene	trans-2-butene	cis-2-butene
1	4-ethylpyridine (homogeneous)	AlR_2Cl	15	61	24
2	EPTR-PVP	AlR_2Cl	37	37	26
3	EPTR-PB-PVP	None	91	5	4

Reaction conditions: T=40°C, P=2 atm., Al/Ni=10, n-heptane.

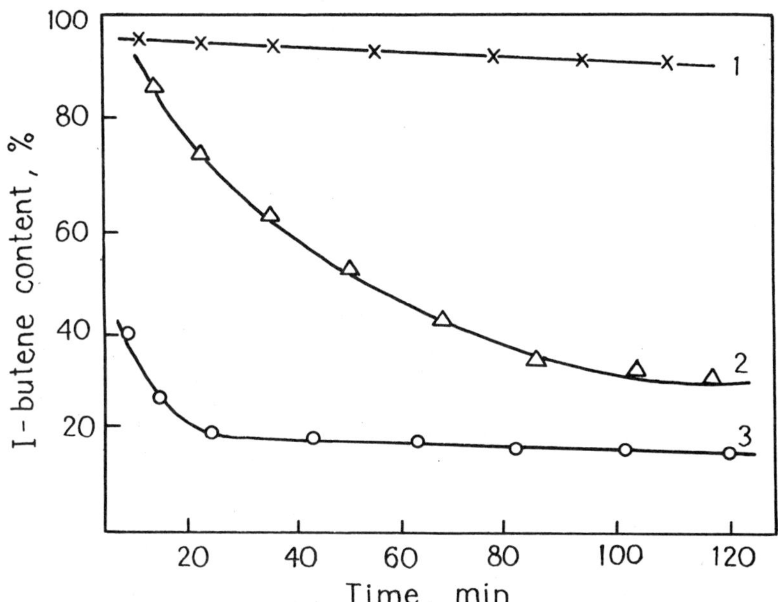

Figure 6. Dependence of 1-butene content in a butene mixture on time during ethylene dimerization on different nickel-containing catalytic systems. Dimerization conditions and catalyst numeration are presented in Table 1.

3.3. Properties Determined by a Specific Nature of the Carrier

Thermal dependences of the rates of dimerization and polymerization of ethylene on the GICSs are anomalous: they are characterized by the maximum (Figs. 7 and 8). Dependences are completely reversible with temperature, i.e. the rate decrease in the region of high temperatures is not determined by thermal deactivation of catalytic centers. Using the method of paramagnetic probes for swollen cross-linked EPTR as well as for EPTR-PVP GICS-carrier, we showed that microsynergesis of polycarbon gel takes place in the carbon solvent environment within the studied temperature region, and is, naturally, reversible. At the same time, accessibility of active centers to olefin molecules is most likely to reversibly reduce (local concentration of the monomer reduces in the vicinity of active centers). This factor can additionally contribute to the increase in thermal stability of the GICS active centers.

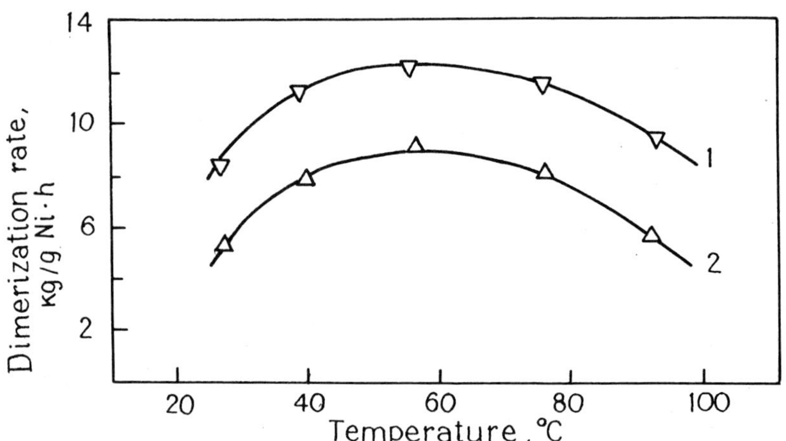

Figure 7. Effect of temperature on the stationary dimerization rate; olefin concentration 1.6 M, Al/Ni=10. 1. ethylene, EPTR-PVP-NiX_2-AlR_2Cl; 2. propylene, EPTR-PVP-NiX_2-AlR_2Cl.

In fact, in the case of local overheating in some micro-area of a catalytic granule, a temporary retardation of exothermic reaction immediately takes place there. This reversibility ensures GICS "self-protection" against irreversible thermal deactivation and prevents self-heating of the reaction system as a whole. It is clear that the capacity for self-regulation is determined by the specific behaviour of the base polymer in the reaction medium and by the peculiarities of a GICS granule construction.

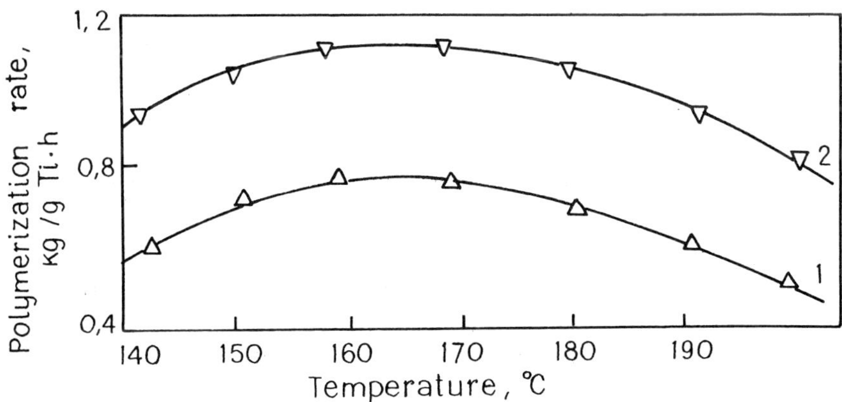

Figure 8. Effect of temperature on stationary rate of ethylene polymerization; ethylene concentration 0.25 M, Al/Ti=10. 1. EPTR-PMAA-3RMgX-$TiCl_4$-AlR_3, 2. EPTR-PAAl-RMgX-$TiCl_4$-AlR_3.

Ethylene polymerization reveals another unique possibility to regulate GICS function. As it was already mentioned, the kinetic curves of polymerization are characterized by initial non-stationary periods. Kinetic behaviour at these periods depends on the physical state of the catalytic system at the start of the experiment (Fig. 9). For catalyst systems in the form of dry particles brought into contact with the reaction medium, the reaction rate during a nonstationary period increases

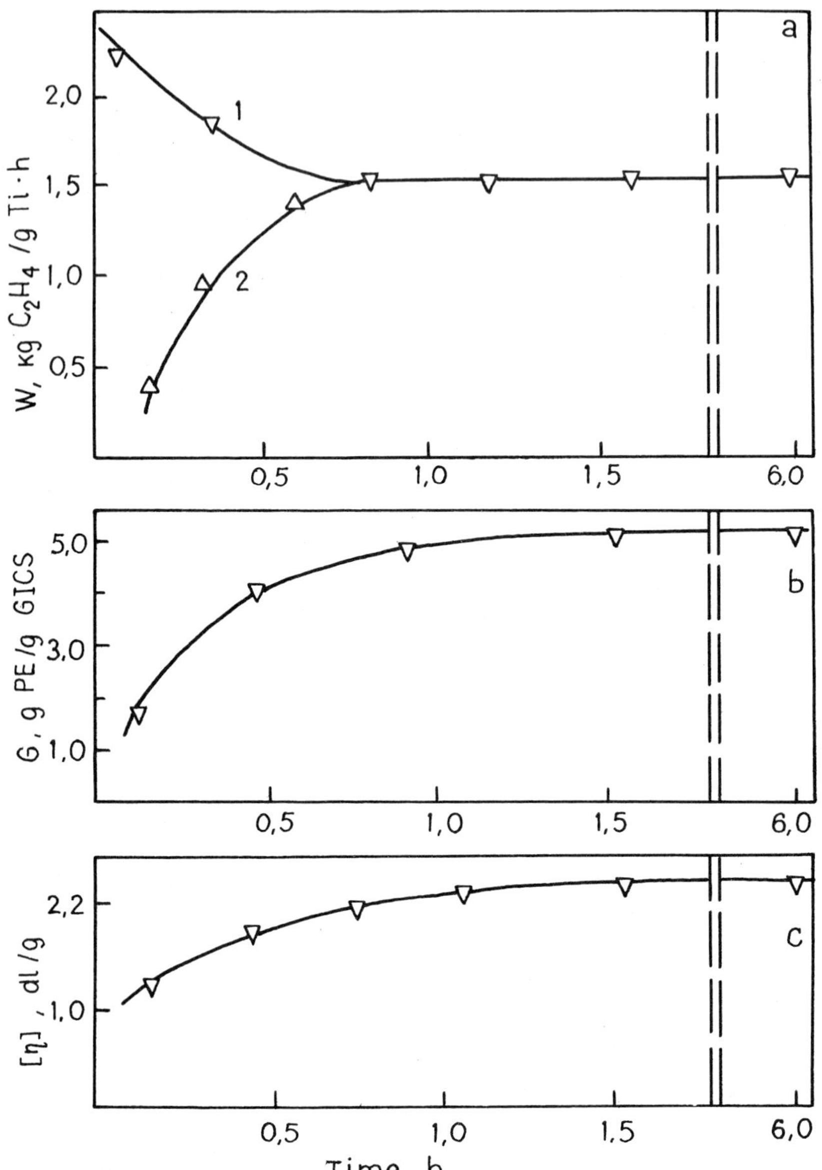

Figure 9. Ethylene polymerization: a - ethylene absorption rate (W); b - PE amount in GICS volume (G); c - PE intrinsic viscosity (η); T=160°C, P=20 atm., solvent - n-heptane. GICS: EPTR-PMAA-3RMgX-$TiCl_4$-$HAlR_2$.

1 - the GICS granules are added in the swollen state,
2 - the GICS granules are added as dry particles.

with their swelling in the monomer solution. For the same particles, but brought to an equilibrium-degree of swelling in the solvent before contact with the monomer, the polymerization rate, measured by ethylene consumption, decreases during a nonstationary period. The stationary rates are identical in both cases, in other words, are independent of the manner in which the initial particles achieve a stable operating state (compare curves 1 and 2 in Fig. 9 a). The GICS particles, which have undergone equilibrium swelling in the solvent, undergo some additional swelling when brought into contact with the monomer during the nonstationary period, and, in parallel with this, the PE formed within each particle accumulates there, with its amount approaching a limit simultaneously with the establishment of a stationary polymerization rate and a stationary degree of swelling (Fig. 9b). In other words, the "superequilibrium" swelling of cross-linked GICS particles is due to an increase in concentration of macromolecules within the previously established three-dimensional network of the polymeric carrier.[*)] If V_{st} is the volume during a stationary operating state, $V = V_{st} - V_p$, where V_p is the volume of the GICS undergoing equilibrium swelling in the pure solvent, and V is an additional increase in volume due to "superequilibrium" swelling of the three-dimensional structure of the starting gel. The value of V_{st} may be rather high. Under ordinary polymerization conditions, it is several tens of times greater than V_p. The work on expanding the GICS by a value of V,

---

[*)] Additional swelling cannot be explained by the improvement of the solvent quality by means of the monomer introduction into the system since the swelling degree of the non-activated GICS almost does not change in this case.

$$A = \int_{V_p}^{V_{st}} P(V)dV$$

is performed by means of the free energy of ethylene polymerization and causes the appearance of additional stresses in the elastic GICS network.**)

The molecular mass (MM) of the PE formed in the volume of the GICS continuously increases during the nonstationary period (Fig. 9 c). After establishing a stationary polymerization rate of ethylene and a stationary degree of swelling of the GICS particles, the MM of the PE diffusing into the surrounding solution becomes independent of the polymerization time. In other words, the GICS operating under stationary conditions continuously produces macromolecules with a certain average length, that is, the average time of formation of a single macromolecule is much shorter than the time of observation. Dependence of the polymerization rate of ethylene on the degree of GICS equilibrium swelling in heptane, serving as a measure of the density of the network of the swollen gel under otherwise identical conditions is shown in Fig. 10. It can be seen that the specific catalytic activity increases noticeably with an increase in this value.

It is remarkable, however, that the stationary length of macromolecules formed during the high temperature polymerization of ethylene appears to be related to the degree of swelling of the GICS particles. As is seen in Fig. 10 b, the intrinsic viscosities of the PE increase

---

**) Since the value of additional stresses is the function of the PE concentration in cross-linked GICS particles, it becomes clear that in the case of too high polymerization rates, they can easily be destroyed. This factor determines the upper limit of the GICS specific productivity.

with increasing the degree of swelling of the GICS. From a comparison of the intrinsic viscosities of the PE obtained under identical conditions on unswollen heterogeneous catalytic systems and on analogous GICSs, it follows that the immobilization of the active centers in the gel is accompanied by a decrease in the average length of the macromolecules formed (Table 2).

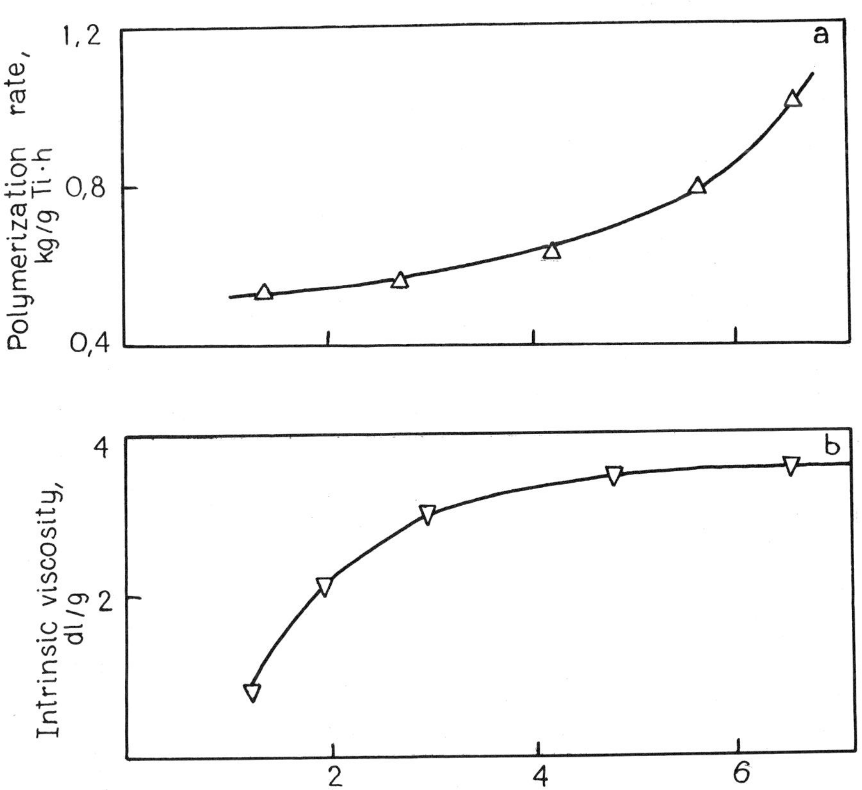

Figure 10. Effect of the degree of GICS swelling on the stationary rate of ethylene polymerization (a) and on the intrinsic viscosity of PE (b).
GICS: EPTR-PMAA-3RMgX-$TiCl_4$-AlR_2Cl; T=140°C, P=10 atm., Al/Ti=10, solvent - n-heptane.

TABLE 2. Comparison of intrinsic viscosities of polyethylene obtained in the presence of GICSs and heterogeneous MCCs

No	Catalyst	Degree of swelling in heptane	Intrinsic viscosity in dl/g
1	PAAl-RMgX-$TiCl_4$	none	5.6
2	EPTR-PAAl-RMgX-$TiCl_4$	4.8	2.2
3	IBA*-3RMgX-$TiCl_4$	none	4.1
4	PMAA-3RMgX-$TiCl_4$	none	4.5
5	EPTR-PMAA-3RMgX-$TiCl_4$	4.8	1.5

Polymerization conditions: T=145°C, P=10 atm., cocatalyst AlR_2Cl, solvent - n-heptane.

* Isobutyric acid

It can be assumed that, in the case of the GICS, some additional mechanism of limitation of the material chains comes into play; this is closely related to the physical nature of the cross-linked polymer gel, which is in the state of "superequilibrium" swelling. The swelling pressure acting on each macromolecule, which increases within the overstressed polymer gel network, attempts to eject this into the region of dilute solution, in other words, to "force it out" from a catalyst granule into the surrounding solvent. It is reasonable to assume that, during this process, an additional stress is also produced at the "point of growth" of a macromolecule, to be more precise, on a Me-C bond between the metal atom of the active center immobilized in the gel and the end of the growing chain. This additional stress may be the source of an increase in the probability of chain termination reactions, i.e., of mechanical activation of elementary steps leading to a separation of the growing macromolecule from the active center. The rate constant (k) of elementary step leading to the separation of the chain end from the immobilized active center, increases according to Eyring's formula:

$$k \sim \exp\left(-\frac{E-\sigma\nu}{RT}\right),$$

where E is the energy of activation in the absence of mechanical stress, σ - the stress affecting the Me-C bond and ν is the element of the volume on the order of the monomer chain volume.

If the above assumption is correct, the molecular mass and molecular-mass distribution can be controlled by varying the degree of cross-linking of the initial polymeric support of the GICS.

4. CONCLUSIONS

From the discussed above it follows that, as compared to homogeneous and microheterogeneous metal-complex catalysts, GICSs offer a number of significant advantages. Some of these can be achieved by the adsorption of catalytic transition metal complexes on ordinary solid carriers. However, some very important advantages and potentialities are unique and are clearly related to the specific mosaic structure and to the properties of the elastic polymer gel.

Basing on the available literature data on the regulated morphology of block copolymers having a regular structure with the given number, length and block distribution in macromolecules, one can outline possible ways of the GICS improvement. In fact, it is shown that depending on the length ratio of homosequences of block copolymers in a macromolecule, block microsegregation leads to the formation of supermolecular structures in which the inclusions of one of the components are regularly packed spheres, cylinders or lamellas. Cross-sectional dimensions of these elements are of the order of 10^2 Å. The morphology of such systems resembles various constructions of conventional contact catalytic reactors used in industry. As is known, such constructions are employed for optimization of conditions of reagent contact with catalyst surfaces, heat removal, mass transfer, etc. It is clear that the use of the specifically designed regular block copolymers opens up the possibilities of such optimization, but only on a microlevel.

It can be added that adsorption of different transition metal complexes on an appropriate carrier, generally speaking, opens up prospective possibilities of realization of parallel, consecutive and consecutive-parallel chemical conversions in a single reactor. In our opinion, the GICSs are most promising systems at this level as well, because of the wide possibilities of optimization of their topochemical characteristics, for example, by an adequate combination of several specific polymeric ligands, grafted to a basic polymer, or by use of tailor-made multi-component graft or block copolymers.

REFERNECES

1. V.A. Kabanov, V.I. Smetanyuk and V.G. Popov, Dokl. Akad.Nauk SSSR, 225, 1377 (1975).
2. V.A. Kabanov, V.I. Smetanyuk and A.I. Prudnikov, Khim.Prom., No 11, 16-21 (1979).
3. V.A. Kabanov and V.I. Smetanyuk, Sov.Sci.Rev., Sect. B, 2, 83-138 (1980).
4. V.A. Kabanov and V.I. Smetanyuk, Makromol.Chem., Suppl 5, 121-154 (1981).
5. Yu.N. Bocharov, V.A. Kabanov, M.A. Martynova, V.G. Popov and V.I. Smetanyuk, USSR Auth. Cert., No492298 Bull. Izobr. No 43 (1975); Belg. Pat. No 818000.
6. Belg. Pat. No 852471; USA Pat. No 4118432.
7. V.A. Kabanov, M.A. Martynova, S.K. Plyzhnov and V.I. Smetanyuk, USSR Auth.Cert. No 692822, Bull. Izobr. No 39 (1979); Belg. Pat. No 852472; USA Pat. No 4118432.

8. A.A. Antonov, Yu.N. Bocharov, V.A. Kabanov, M.A. Martynova, S.K. Pluzhnov and V.I. Smetanyuk, USSR Author. Cert. No 681036, Bull. Izobr. No 31 (1979); Belg. Pat. No 852473; USA Pat. No 4115469.

9. V.A. Kabanov, V.I. Smetanyuk, V.G. Popov, M.A. Martynova and M.V. Ulyanova, Vysokomol. Soedin., Ser. A22, No 2, 335-343 (1980).

10. V.A. Kabanov, S.S. Ivanchev, V.I. Smetanyuk, A.A. Baulin, M.A. Martynova and V.M. Kopylov, Vysokomol. Soedin., Ser. A22, No 2, 345-351 (1980).

11. V.A. Kabanov, I.A. Litvinov, T.V. Budantseva and V.I. Smetanyuk, Dokl. Akad. Nauk SSSR, 262, 1169 (1982).

12. T.V. Budantseva, I.A. Litvinov, V.I. Smetanyuk and V.A. Kabanov, Vysokomol. Soedin., Ser A22, 585 (1984).

13. V.A. Kabanov, M.A. Martynova, S.K. Pluzhnov, V.I. Smetanyuk and R.V. Chednya, Kinet. Katal., 20, 1012-1018 (1979).

14. Yi.E. Kirsh, V.Ya. Kovner, A.I. Kokorin, K.I. Zamaraev, V.Ya. Chernyak and V.A. Kabanov, Eur. Polym. J., 10, 671 (1974).

15. A.B. Zezin, N.M. Kabanov, A.I. Kokorin and V.B. Rogacheva, Vysokomol. Soedin., Ser. A19, 118 (1977).

UNUSUAL POLYMERIZATION POSSIBILITIES WITH HIGHLY ACTIVE METALLOCENE/ALUMINOCANE CATALYSTS

W. Kaminsky and M.H. Miri
Institute of Technical and Macromolecular Chemistry
University of Hamburg, F.R. of Germany

ABSTRACT - Zirconocene or titanocene compounds and methylaluminocane as soluble Ziegler catalysts not only show a very high activity but also produces polyethylene with special properties. Copolymers of ethylene and hexene are synthesized with different hexene concentrations. The regiospecifity of the catalyst leads exclusively to the formation of head - to tail enchainments. The polymerization of propene with bis(cyclopentadienyl)zirconiumdichloride gives only atactic polypropylene. Changing the zirconium compound into the chiral ethylene (bisindenyl)zirconiumdichloride lead to a catalyst which produces only isotactic polypropylene with less than 0,2% of atactic part. Only every 160th propene unit has the opposite configuration from the foregoing. The molecular weights are strongly dependent on reaction temperature.

1. INTRODUCTION

By mixing up bis(cyclopentadienyl)zirconiumdimethyl - or dichloride with aluminoxanes of structure

$$\begin{array}{c} R \\ | \\ (Al - O -)_n \end{array} \qquad \begin{array}{l} R = CH_3;\ C_2H_5 \\ n = 10\text{-}20 \end{array}$$

there are created extremely active, soluble Ziegler catalysts.[1)2)3)] Analogous titanium or hafnium compounds can work as well as transition metals, but they show a smaller polymerization activity however.

In Table 1 there is a comparison among several Ziegler-catalysts.[4)] From this it becomes apparent that homogeneous catalysts are more active than heterogeneous catalysts, mainly due to their high molar ratios of active transition metal to overall employed transition metal (C*).

TABLE 1. Comparison of effective polymerization rates (kp·C*) among several Ziegler catalysts for polymerization of ethylene (C*: molar ratio of active to overall employed transition metal; values for heterogeneous catalysts from literature[4]).

System	$\frac{k_p \cdot C^*}{l \cdot mol^{-1} \cdot s^{-1}}$	C*	$\frac{T}{^oC}$
$Cp_2Zr(CH_3)_2$/Methylaluminoxane	105 000	(1)	90
Cp_2TiCl_2/Methylaluminoxane	800	(0,7)	20
$TiCl_3/AlEt_2Cl$	2,7	0,005	50
$TiCl_4/Mg(OEt)_2/Al(i\text{-}Bu)_3$	280	0,023	60
$TiCl_4/RMgCl/AlEt_3$	4 680	0,36	70

This high polymerization activity enables the synthesis of copolymers from ethylene and several α-olefines and diolefines (e.g. propylene, butene, hexene, ethylidennorbornene, octadiene); like pure polyethylene they stand out for special characteristics.

Natural substances like starch, cellulose and lignin can, due to their free water, form aluminoxane analogous structures on addition of trimethylaluminium. After addition of the metal compound starch grains or cellulose fibers treated in this manner have a considerable polymerization activity when suspended in toluene. By this procedure physical and chemical properties of natural and synthetic polymers can be combined.[5) 6)]

In the same manner it is possible to cover cellulose, lignin or inorganic materials like $CaCO_3$, $CaSO_4 \cdot 0{,}5\ H_2O$ or Al_2O_3 with polyethylene or other polyolefins.

2. ACTIVITIES OF DIFFERENT SOLUBLE CATALYSTS

By application of bis(cyclopentadienyl) zirconiumdichloride as transition metal compound in polymerizations of ethylene there can be achieved activity-values of $5 \cdot 10^{-5}$s, provided that every zirconium atom forms an active complex.

At a polymerization temperature of -20°C a macromolecule is formed during app. 10 min. without occurrence of any transfer - or breaking off reactions. From the concentration of the transition metal, the amount of produced polymer and the mean molecular weight (number average) it can be calculated that app. 70% of all zirconium atoms are active in the polymerization reaction. With increasing polymerization temperatures this portion increases as well. For the highest productivity value at 70°C and 5.10^{-10} mol/lit so there could be registered $4.25 \cdot 10^{8}$ g polyethylene per g zirconium over 120 h. Here every zirconium atom has inserted more than $1.3 \cdot 10^{9}$ ethylene molecules and by doing this it has generated app. 35,000 macromolecules with a mean molecular weight above 10^{6}.

A comparison among different homogeneous systems is depicted in Table 2. It becomes obvious that zirconium systems are more active

TABLE 2. Polymerization of ethylene with different catalysts in toluene and $1,5 \cdot 10^{-2}$ mol al/l aluminoxane

Transition metal compound	Cokatalyst	Activity (kg/PE mol M·h·bar)	Temperature (°C)
$Cp_2Ti(CH_3)_2$	methyl alumoxane	24	20
Cp_2TiCl_2	"	4 300	20
Cp_2ZrCl_2	"	820	20
		242 000	80
$(Cp_2ZrCl)_2O$	"	203 000	80
$Cp_2Zr(CH_3)Cl$	"	198 000	80
$(Cp_2Zr(CH_3))_2O$	"	192 000	80
Cp_4Zr	"	160 000	80
$Cp_2Zr(CH_3)_2$	"	152 000	80
$Cp_2ZrH \cdot AlEt_3$	"	142 000	80
$Cp_2ZrCl(OBu)$	"	73 000	80
$(CpMe_5)_2ZrCl_2$	"	31 000	80
$CP_2Hf(CH_3)_2$	"	10 700	70
Cp_2ZrCl_2	ethyl-aluminoxane	2 100	60
$Cp_2Zr(CH_3)_2$	tetraisobutyldi-aluminoxane	15 900	70

than analogous titanium- and hafnium systems. On the cocatalyst-side methylaluminoxane is more effective than ethyl- or (only dimeric) isobutylaluminoxane. Chloride containing transition metals are more active than systems entirely free from chloride, especially at low polymerization temperatures.

2.1 Properties of the Polyethylene

With decreasing zirconium concentration the mean molecular weight of polyethylene increases. From this it can be concluded that in the breaking off reaction there is engaged more than one zirconium complex.

Polyolefines produced via this catalyst distinguish themselves by several characteristic features (Table 3).

TABLE 3. Characteristic features of several polymers produced by the catalyst system bis(cyclopentadienyl)zirconiumdimethyl/ methylaluminoxene

	polyethylene 1	polyethylene 2	copolymer from ethylene a. hexene 1	2
portion of hexene monomer units in mol%			4,6	7,7
$\bar{M}_\eta$	107 000	147 000	88 000	41 000
$\bar{M}_w/\bar{M}_n$	1,93	2,19	2,64	3,15
specific gravity in $\frac{g}{cm^3}$	0,97	0,98	0,92	0,90
melting point in ^{o}C	140	139	115	105
transition temperature in ^{o}C			-45	-45

Molecular weight distributions of $\bar{M}_w/\bar{M}_\eta=2$ are rather narrow, compared to those for polymers which are produced with heterogeneous Ziegler-catalysts. With respect to this copolymers from ethylene and hexene display a greater irregularity.

Freshly produced, non molten polyethylene crystals have a high specific gravity up to 0.98 g/cm^3 and a high melting point of 140^oC.

When this material is transformed to plates by extrusion then specific gravity diminishes to values around 0.95 g/cm^3 without any alteration of melting point. Increasing contents of comonomers lead to lower values for specific gravity and melting points.

3. EHTYLENE - HEXENE - COPOLYMERS

By incorporation of hexene-1 into the polyethylene matrix LLDPE is produced. In here there is a remarkable dependence on the activity upon the hexene/ethylene proportion (Figure 1). If the ratio

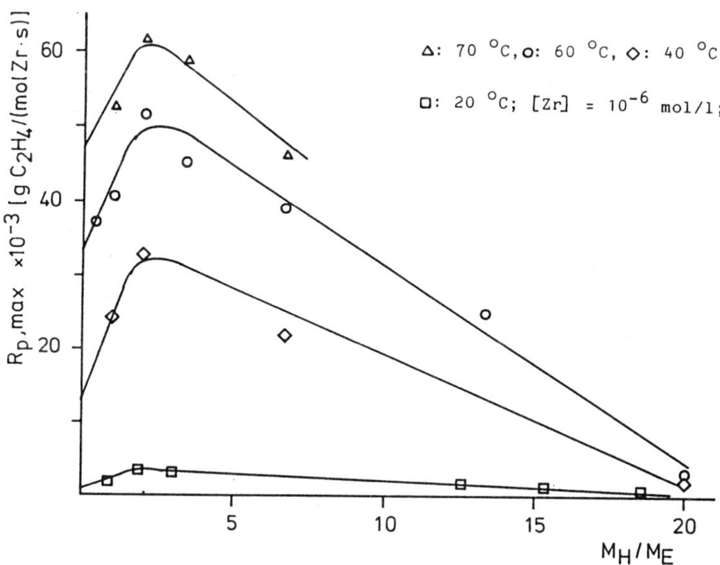

Figure 1. Polymerization rate R_p,max in dependence of the hexene/ ethylene ratio in the solution and by different temperatures; ethylene pressure 4 bars.

is nearly 2 there is a maximum in the polymerization rates, at all temperatures which are higher than that of the ethylene homopolymerization. Further increase of the hexene proportion leads to a decrease in the polymerization rate.

The activation energy for the copolymerization depends on the mol ratio of hexene and ethylene in the starting solution and reaches for M_H: M_E = 0,5 - 2, E_A = 59 EJ/mol and

for M_H: ME = 2 - 15 , E_A = 48 EJ/mol.

The co-catalyst, the methylaluminoxane, likewise influences the activity. The increase of the aluminoxane concentration from 0.006 to 0.022 mol/l leads to an increase of the reaction rate from 2.5 to nearly 10g cop/h. As shown in Figure 2, the double logarithmic apply of both magnitudes leads to a straight line whose slope is

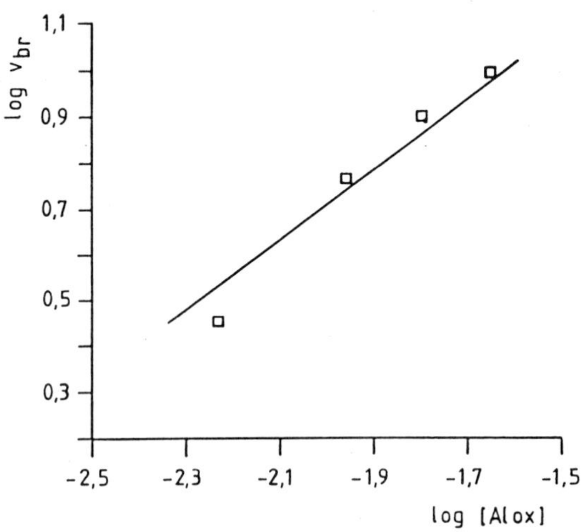

Figure 2. Double logarithmic apply of the polymerization rate vs. the methylaluminoxane concentration; hexene concentration 100g/l.

corresponding to the order of reaction and to the methylaluminoxane respectively. For copolymerizations with 100g/l hexene, it turned out an order of reaction of 0.8 which is slightly increased in comparison to the corresponding order of reaction of 0.6 in case of ethylene polymerization.

3.1 Spectroscopic Measurements

The spectroscopic estimation of the hexene contents in the copolymeres was carried out by IR and NMR spectroscopic techniques. Figure 3 shows the characteristic IR peaks at 1378cm^{-1} for different copolymeres. Figure 4 gives an expression for the proton

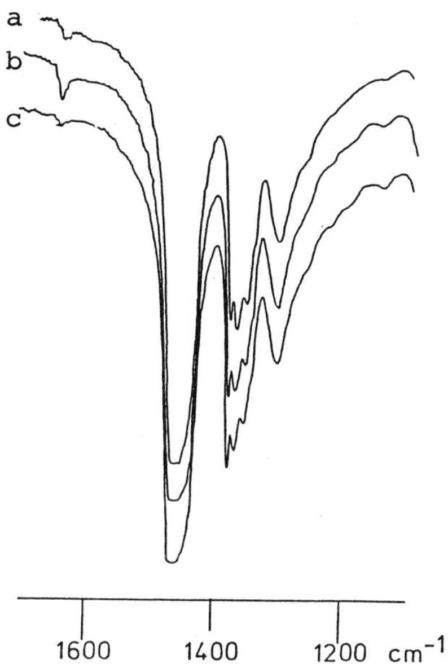

Figure 3. IR-spectra of different ethylene/hexene copolymers with different hexene content. a: 3,5 mol%, b: 6 mol%, c: 13 mol%

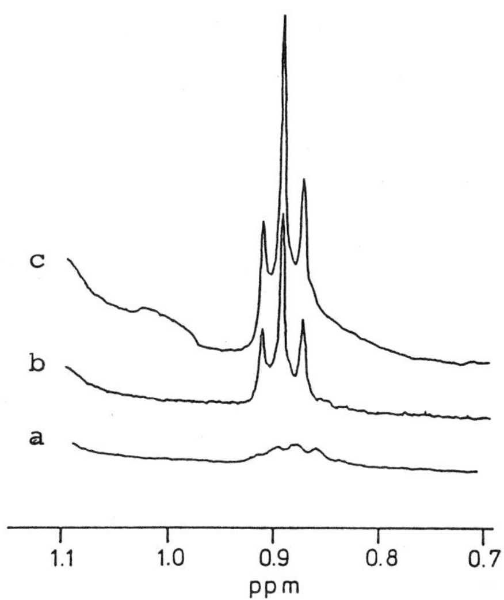

Figure 4. H-NMR-spectra (360 MH_2, 120°C in 1,2-dideuterotetrachlorethane) of ethylene-hexene (E/H) copolymers (methyl signals)
a: polyethylene >0,2 methyl groups/1000 C
b: E/H-copolymer 10 methyl groups/1000 C
c: E/H-copolymer 13,5 methyl groups/1000 C

spectra of the copolymers. Only the methyl protons are shown for the different hexene concentrations.

Another striking feature of hexene as comonomer is its tendency for interrupting chains. 50 mol-% hexene lead at 60°C to a molecular weight of 100 000 instead of 200 000 for ethylene homopolymerization. This is achieved without any loss of activity. Right along with this decrease in molecular weight there is an increase in the proportion of vinyliden groups in the copolymer (Table 4).

TABLE 4. Concentration of unsaturated endgroups in the ethylene-hexene-copolymers (IR-measurements)

Hexene in Copolymer mol-%	Transvinyl/ 1 000 C	Vinyl/ 1 000 C	Vinylidene/ 1 000 C
0	0,03	0,12	0,02
1	0,02	0,15	0,06
1,9	0,03	0,20	0,15
4,6	0,04	0,14	1,00
7,7	0,04	0,11	1,20

It is supposed that the unsaturated endgroups are formed in predominantly via chain transfer towards the monomer units and that hydrogen transfer to the transition metal is only of minor importance. Molecular weight distributions of copolymers are relatively narrow; a typical figure of $\frac{M_w}{M_n}$ is 2, which corresponds well with the predominance of one kind of active centers.

In addition to IR-, ^{1}H-, ^{13}C-nmr spectrochemical investigations there have been performed DSC experiments which lead to correlation diagrams between melting points and comonomer amounts. Insertion of 4,6 weight-% hexene causes a drop of the melting point by 21 degrees to 115°C (see Table 3).

Wide angle X-ray scattering is a profound means for investigating the crystallinity of copolymers. Although the samples have been drastically frozen from 160°C to 0°C during the production of films, the effect of the butyl side chains was still manifest. Copolymers with amounts of hexene higher than 20 mol-% are nearly completely amorphous.

3.2 Copolymerization Parameter

Reaction rates are strongly dependent on the molar ratios 1-hexene/ethylene in the reaction mixture. Based on the Mayo-Lewis-concept for binary copolymerizations

$$\text{Cat} -- M_1 + M_1^x \xrightarrow{k_{11}} \text{Cat} -- M_1^x - M_1-$$

$$\text{Cat} -- M_1 + M_2^x \xrightarrow{k_{12}} \text{Cat} -- M_2^x - M_1-$$

$$\text{Cat} -- M_2 + M_2^x \xrightarrow{k_{22}} \text{Cat} -- M_2^x - M_1-$$

$$\text{Cat} -- M_2 + M_1^x \xrightarrow{k_{21}} \text{Cat} -- M_1^x - M_2-$$

with

$r_1 = k_{11}/k_{12}$

$r_2 = k_{22}/k_{21}$

M_1 = Ethylene

M_2 = 1-Hexene

the reaction parameter could be determined. Table 5 gives the dependence of the polymer compositions on the hexene concentrations.

According to the method of Fineman-Ross[7)] the copolymerization parameters r_1 and r_2 were calculated from the insertion rates.

The copolymerization parameters and rate constant for all four single steps is recorded in Table 6.

At 20°C they were found to be $r_1 = 55$ for ethylene and $r_2 = 0{,}005$ for hexene. This means that ethylene is 55 times quicker incorporated than hexene.

Figure 5 shows the copolymerization diagram for ethylene/hexene copolymers.

In this copolymerization diagram the statistical incorporation of propene units into the polymer chain becomes evident by the

TABLE 5. Hexene concentration in dependence of the hexene concentration in toluene as solvent; methylaluminoxane 10^{-2} mol Al/l, ethylene pressure 4 bars.

Hexene in solvent (mol %)	(Zr) (mol/l)	T (^{o}C)	Hexene in polymer (mol %)
25	10^{-7}	60	0,4
40	10^{-7}	60	1,0
50	10^{-7}	60	1,3
67	10^{-7}	60	2,2
77	10^{-7}	60	4,0
87	10^{-7}	60	6,5
93	10^{-7}	60	12,5
95	10^{-7}	60	19,0
95	10^{-6}	40	23,0

TABLE 6. Resulting kinetic data of the copolymerization of ethylene/hexene.

T	r_1	r_2	k_{11} (l/mol·s)	k_{22} (l/mol·s)
20°C	55	0,004	150	0,2
40°C	54	0,005	1 450	0,6
60°C	52	0,005	3 460	1,5
70°C	79	0,005	5 750	-

shape of the curves.

The hexene units are distributed at random into the polyethylene matrix (Figure 6). The content of hexene diads is relatively low. At 70 - 80°C there is even less insertion of the comonomer (r_1 = 80). A drop in zirkonocene concentration by a factor of 10 will lead to less incorporation of the α-olefin.

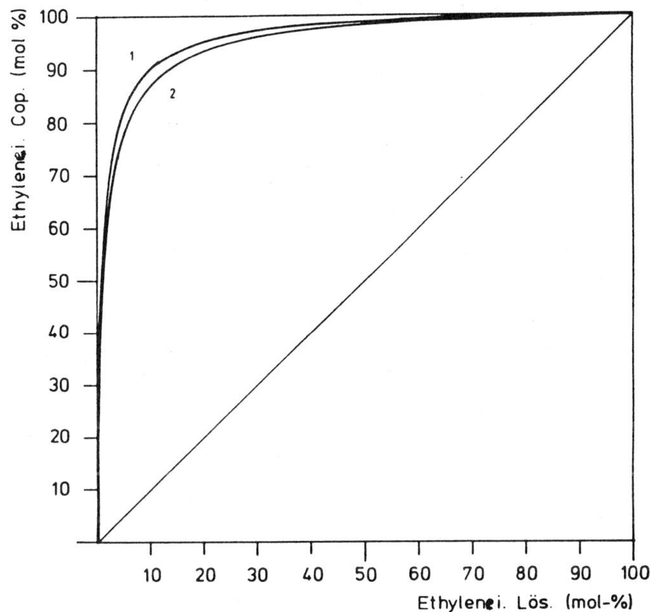

Figure 5. Copolymerization diagram for ethylene propene copolymers. Curve 1 is valid for temperatures between 20 and 60°C, curve 2 for 70°C.

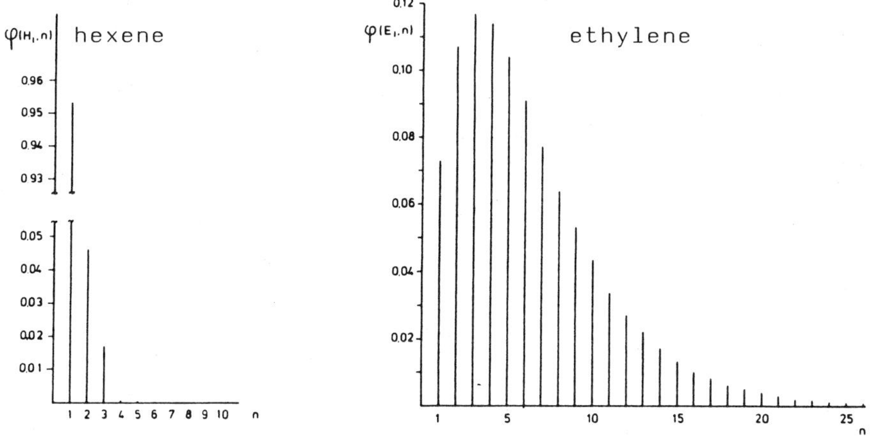

Figure 6. Monomer distribution of hexene and ethylene in the copolymer.

The Figure 7 shows the dependence of the melting point vs. the hexene content in the copolymer.

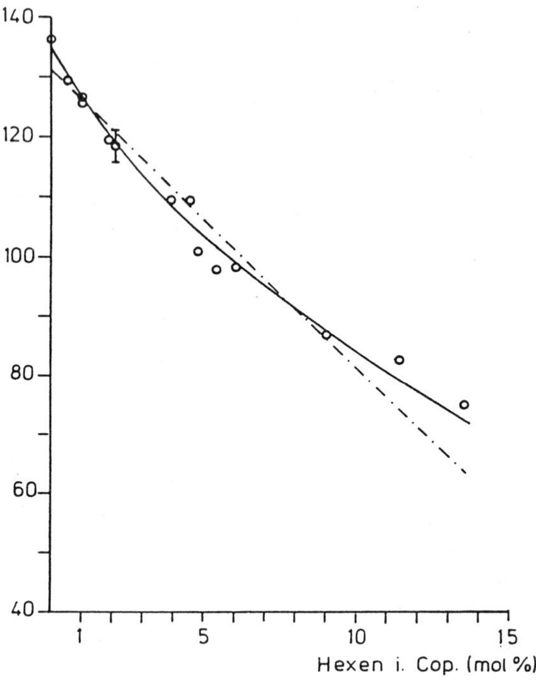

Figure 7. Dependence of the melting point of the hexene content in the ethylene/hexene copolymer, line calculated by Shirayana.[8)]

4. EPDM ELASTOMERES

In the same way as ethylene/hexene copolymers EPDM terpolymers with ethylidene norbornene as diene monomer could be prepared[9) 10)] The overall activities lie between 100 and 1.000 kg EPDM/mol Zr·h·bar. This polymerization could be carried out at zirconium concentrations as low as $5 \cdot 10^{-7}$ mol/l. Propylene sequences produced by the bis(cyclopentadienyl)zirconium compounds are totally atactic as was observed in the case of polypropylene. In the polymers which were produced $r_1 \cdot r_2 = 0{,}3$ was measured for the EPDM samples, $r_1 \cdot r_2 = 0{,}25$ for the ethylene/propylene copolymers, corresponding to a random distribution of ethylene and propylene.

Molecular weight distributions (M_w/M_n) given by GPC-curves have an average value of only 1,7. This narrow distribution implies that only one active center is predominantly present.

5. ATACTIC POLYOLEFINS

Olefins can be formed into atactic or isotactic polymers by using catalysts, whose ligands are varied at the transition metals. Bis(cyclopentadienyl) compounds of zirconium and titanium also substituted compounds as

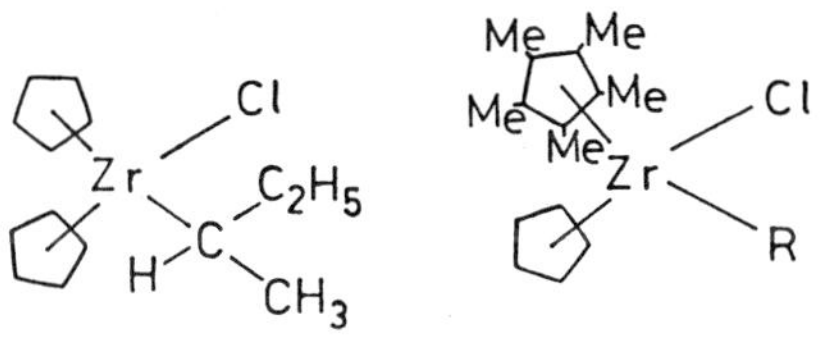

yield pure atactic polypropylene. The concentration of transition metals are in the range of 10^{-7} and 10^{-6} mol/l as by the ethylene polymerization.[11] It can be polymerized in solution as well as in liquid propene. Under similar conditions, the zirconium compound becomes more active than the titanium compound. Methyl substitution at the Cp-ligand also decreases the activity. Atactic polymers with molecular weights up to 590 000 are available. Polymerization of 1-butene however remains only low-molecular polybutene, that is absolutely atactic. All atactic polymers show a very small molecular weight distribution between 1.6 and 2.5.

6. ISOTACTIC POLYOLEFINS

Since a few years even chiral, soluble zirconium compounds are reported. If those compounds or analogous titanium structures together with methylaluminoxane are used, e.g. rac-ethylene-(bisindenyl)zirconiumdichloride[12)13], propene is polymerized to highly isotactic polypropylene (Table 7)[14].

The (indenyl)zirconcene is more active than the (tetrahydroidenyl) zirconocene or the titanocene. The activity increases with increasing polymerization temperature and reaches a value of 31 200 kg isotactic polypropylene per mol Zr and h.

TABLE 7. Polymerization of 70 ml propylene with different transition metal compounds and 10^{-2} mol Al/l methylaluminoxane in toluene.

Catalyst	Temperature °C	Activity kg PP/mol M·h	M_η	Isotactic part (insoluble in toluene)
$en(ind)_2ZrCl_2$	35	31 200	38 000	100
$en(thind)_2ZrCl_2$	20	2 140	47 000	99,6
$(S)en(thind)_2ZrCl_2$	37	326	27 000	100
$en(ind)_2TiCl_2$	30	3	167 000	18,2
$en(thind)_2TiCl_2$	14	6	50 000	69

1-Butene can also be polymerized highly effectively with the zirconocene/methylalumoxane catalyst. Thus, polymerization activity is still 2640 kg of polybutene per hour per mole of Zr at 20°C. Although the polymers have molecular weights of 150 000 or 50 000, they are soluble in toluene. The consistency is wax-like to crystalline, but not clear and viscous like atactic polybutene, which was produced with bis(cyclopentadienyl)zirconium compounds and measured ^{13}C-NMR spectroscopically. The toluene soluble part consists of less than 0,1 weight percent and therefore lies below that of isotactic polypropylene produced with supported catalysts.

The IR-spectroscopically determined isotacticity index of 99% is also very high. The crystallinity of the highly isotactic polypropylene is largely determined by the stereochemical uniformity of the polymerization. High-resolution ^{13}C-NMR spectroscopy affords an insight into the pattern of the sequences in the polymer chain.

A profound analysis of the number and intensity of the methyl peaks leads to a specification about the distribution between isotactic percentage I and isodiotactic sequential lengths n_{iso} and n_{syn}, respectively. Typical figures are found as following: 99% I; 330 n_{iso}. With these figures in mind, it can be stated that polypropylene produced via the ansa-zirconocene compound consists of an extraordinarily high degree of isotacticity. Only every 160th propene unit has the opposite configuration from the foregoing. The molecular weight distribution M_w/M_n is small 2 to 2,5. This is a characteristic value for homogeneous catalysts with only one type of active centers[15]. The molecular weights are strongly dependent on reaction temperature.

By variation of reaction temperature in the range from $-20^{\circ}C$ to $60^{\circ}C$ molecular weights from 300,000 down to 1,500 are yielded.

It is also possible to separate the racemic mixture of the zirconocene compound into the optically active pure enantiomeric forms and use these as catalysts for the isotactic polymerization (see Table 7)[16].

REFERENCES

1. H. Sinn, W. Kaminsky, H.-J. Vollmer, R. Woldt; Angew. Chem. Int.Ed.Engl. 19, 390 (1980).
2. H. Sinn, W. Kaminsky; Adv.Organomet.Chem, 18, 99 (1980).
3. J. Herwig, W. Kaminsky; Polymer Bull. 9, 464 (1983).
4. K.H. Reichert, Angew.Makromol.Chem. 94, 1 (1981).
5. J. Dutschke, W. Kaminsky, H. Lüker, in Polymer Reaction Engineering, ed. K.H. Reichert a. W. Geiseler, Hanser Publ., Munich (1983), p. 207.
6. W. Kaminsky; Nachr.Chem.Techn.Lab. 29, 373 (1981).
7. M. Fineman a. S.D. Ross, J.Polym.Sci., 5, 259 (1950).
8. K. Shirayama, S.-I. Kita a. H. Watabe, Makromol.Chem. 151, 97 (1972).
9. W. Kaminsky, M. Miri, H. Sinn, R. Woldt; Makromol.Chem.Rapid Commun., 4, 464 (1983).
10. W. Kaminsky, M. Miri; J.Polym.Sci.Polym.Chem.Ed., 23, 2151 (1985).
11. W. Kaminsky in MMI Press, Symposium Series Vol. 4, Transition Metal Catalyzed Polymerization, ed. R.P. Quirk, Harwood Academic Press Publ., New York (1983), p. 225.
12. F.R.W.P. Wild, L. Zsolnai, G. Huttner and H.H. Brintzinger; J.Organomet.Chem. 232, 233 (1982).
13. F.R.W.P. Wild, M. Wasiacionek, G. Guttner and H.H. Brintzinger; J.Organomet.Chem., 288, 63 (1985).
14. W. Kaminsky, K. Külper, H.H. Brintzinger, F.R.W.P. Wild; Angew.Chem. 97, 507 (1985).
15. K. Soga, K. Izumi, S. Ikeda and T. Keii, Makromol.Chem. 178, 337 (1977).
16. W. Kaminsky, in History of Polyolefins, ed. R.B. Seymour a. T. Cheng, Reidel Publ. Co. (1986), p. 257.

CATALYSIS BY OXIDE-SUPPORTED PORPHYRINE AND PHTHALOCYANINE COMPLEXES

B.V. Romanovsky
Chemistry Department, Moscow University, Moscow 117234, USSR

ABSTRACT - Methods of preparation of porphyrine and phthalocyanine metal complexes attached to oxide and zeolite supports and physico-chemical properties of these complexes as well as their states on the support surfaces are considered. On the one hand, heterogeneized porphines preserve their specific properties typical of molecular solutions; on the other hand, their interaction with the support has marked effect on their efficiency as centers of redox catalytic reactions.

1. INTRODUCTION

Porphyrine and phthalocyanine complexes of transition metals are of great interest not only as mild catalysts for selective oxidation but also as functional models of enzymes. In particular, they can be useful for the modelling of reversible binding of molecular oxygen. Some papers have been published on synthesis of sterically hindered porphyrines which imitate active centers of respiratory enzymes as well as environmental situation of working haem (Refs. 1-4). However, the study of porphin complexes as catalysts is hampered by two factors. First, they are almost insoluble in most of the solvents; second, they tend to form clusters and molecular adducts when used in solutions. Therefore, the heterogeneizing of these metal complexes onto polymer or oxide supports seems to be very promising. The heterogenization would permit one to solve many experimental problems, in particular, that of reaction media for catalytic performance.

Due to a lack of superficial mobility, the supported complexes cannot cluster into great particles. Hence, high dispersion may be achieved, and, as a consequence, high efficiency in adsorption and catalysis processes are reasonable to expect.

Though having rather poor ability to undergo functionalisation, the oxide supports offer some advantages over polymeric ones, that is, they possess a rigid and easily controllable pore structure and do not swell in organic solvents; they have good thermal and chemical stability. The latter fact is of particular importance in porphin supporting, since these metal complexes show surprisingly high thermostability and resistance to acids and bases.

Among oxide supports, zeolites constitute a particular group. These crystalline aluminosilicates, which possess a very regular system of channels and cages, provide an almost unique possibility to produce highly ordered inclusion compounds, i.e. to obtain molecular dispersion of an active component. The most characteristic of these guest-host systems is that the guest molecule has neither covalent nor ionic bonding with host lattice but due to spatial restrictions is held topologically. In other words, it turns out to be captured within zeolite cavity. It is quite obvious that in this case any loss of active moiety could be avoided. Combination of molecular sieve properties of zeolite supports with the specificity of porphin complexes as catalytic species would seem to explain enormous interest in these new catalysts and, primarily, in their practical applications, as clearly seen from the recent publication (Ref. 5).

The topological inclusion could be carried out by some nontrivial techniques, such as intracrystalline "in situ" synthesis, which are to be discussed in what follows. Of course, in spite of being topological, this mode of supporting does by no means exclude strong interaction of complex with support. Up to date this interaction was interpreted only in terms of phenomenological models (Ref. 6).

The purpose of the present paper is to review recent works on the heterogenizing of porphyrine (Pn), phthalocyanine (Pc) complexes of transition metals (TM) onto some oxide matrices as well as on their physico-chemical and catalytic properties.

2. SYNTHESIS OF OXIDE-SUPPORTED PORPHIN COMPLEXES

Four different methods of preparing the supported Pn and Pc metal complexes have been developed:

(1) covalent bonding as a result of chemical reaction between functional groups on the support surface and of supporting species;

(2) anchoring by coordination of surface atom into the 5th axial position of a complex, the support playing a role of an extraligand;

(3) adsorption of complex compound from its solution or its vapour onto support;

(4) surface assembling of complex molecule from either TM-containing compound (salt, complex, etc.) or TM itself and appropriate chelating agent.

Regarding oxide supports, the latter two methods of heterogenizing are most often used when Pn or Pc complexes are to be supported. The adsorption method of supporting suffers from two serious limitations, i.e., from poor solubilities of porphin which do not permit one to obtain reasonable surface concentrations of the complex, and from spatial restrictions when entry of bulky molecules of complex into the support internal cavities is excluded. Obviously it is the case when porphin complexes are to be heterogenized onto molecular sieve matrices such as zeolites. Indeed, the channel openings of these latter are in the range from 0.4 to 0.9 nm, whereas the critical diameter of the complex molecule is not less than 1.2 nm.

Surface assembling method as just mentioned could be considered, in a sense, as "in situ" synthesis of porphin molecule upon the surface or within the cavity of supports which contain TM species previously introduced into the matrix in some way.

If zeolite supports with a rigid geometry of intracrystalline cavities are involved, the steric factor becomes crucial. There are two determining ratios of molecular sizes: (1) of zeolite channel opening to that of entering potential ligand, and (2) of zeolite cavity to that of forming chelate species. Of course, these conditions are evident to be necessary but not sufficient, and success of "in situ" synthesis depends, after all, on stability constants of TM complex which are extremely high in the case of TM porphin complexes (Ref. 7).

If the formation of the complex molecule is allowed thermodynamically and sterically, the question arises on whether the complex species once formed will be held topologically or can leave freely the mother cage. All these considerations are illustrated in Fig. 1 where a scheme of "in situ" synthesis of zeolite-included PcM is presented.

The key point of these syntheses is that the molecules of dicyanobenzene (DCB) as chelating agents may enter freely a large cage of zeolite where the TM atoms reside. Inside the cage, the tetramerization of DCB occurs, and thus the tetradentate macroligand formed chelates the TM atom into metal phthalocyanine molecule. This latter, because of a greater size than the cage window, cannot leave the cage and, hence, is to stay there forever unless its thermal decomposition at elevated temperatures happens.

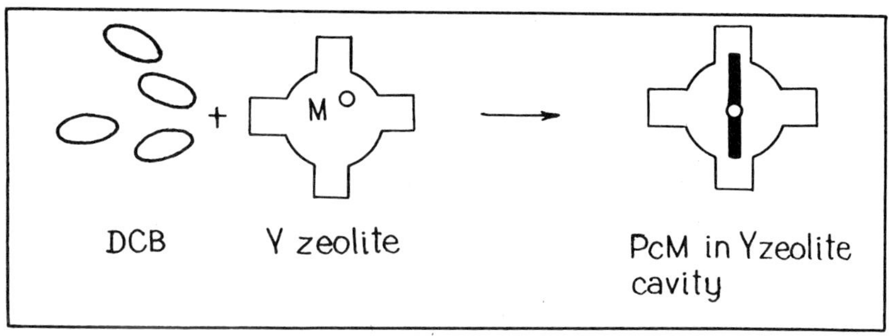

Figure 1. "In situ" synthesis of PcM in Y zeolite supercage.

Interestingly, this scheme of "in situ" synthesis, in general, reproduces the so-called "phthalogenic" technique of dyeing well-known in textile industry when phthalocyanine pigment forms directly on a cloth previously impregnated with one of the reagents (Ref. 8).

The "in situ" synthesis of zeolite-included PcMs just described was carried out by us as early as in 1977 (Ref. 9) and afterwards reproduced by Schulz-Ekloff et al. (Ref. 10) and by Herron et al. (Ref. 11). In all these works the synthesis of PcMs in X or Y zeolite matrices have been performed using TM cation-exchanged forms as a starting material by heating this latter in either DCB vapour or molten DCB at 250-300°C.

It is assumed in Ref.10 that formation of the PcM in zeolite cavity is accompanied by a two-electron oxidation of residual water molecules:

$$M^{2+} + 4\ DCB + H_2O = PcM + 0.5O_2 + 2H^+.$$

Quite analogously, the "in situ" synthesis can be carried out with any other oxide support containing TM atoms ions or some compounds of TM which can be re-complexed. As an example, the preparation of silica-supported PcCu complexes using Cu ion exchanged silica could be cited (Ref. 12).

However, the surface assembling of Pc complexes via metal ions in the zeolite does not provide a complete transformation of all M^{2+} ions in the matrix to the corresponding complex. For example, in the case of X and Y zeolites, the degree of the TM inclusion into Pc complexes may vary over a wide range (see Table 1).

TABLE 1. Contents of transition metal atoms in the initial zeolite (M) and of the complexes (PcM) formed by these atoms in elementary cell and degree of complexing (%)

Sample	M	PcM	%	Refs.
PcCoX	8	6	75	10
PcCoY	4	1	25	13
PcCoY	11	3	27	13
PcCoY	16	7	44	10
PcNiX	8	2	25	13
PcNiY	4	1	25	13
PcNiY	8	4	50	13
PcNiY	16	9	56	13
PcCuX	8	1	13	10

Another point to be mentioned is that depending on conditions of treating the zeolite with the DCB reagent PcM compounds could form on the external surface of matrix crystallites. Using scanning electron microscopy technique, the authors (Ref. 10) were able to observe well-shaped PcCu needles residing on the X zeolite crystal faces. This shortcoming of the "in situ" synthesis via the cationic forms of zeolites made us look for an alternative method of the metal introduction into the zeolite matrix. The introduction of the TM component in the form of volatile carbonyls or other organometallic compounds which, upon further treatment with DCB, give Pc complexes was found to be most advantageous (Ref.14). This ligand substitution reaction is strongly shifted towards the formation of Pc complexes due to very high stability constants of these latter.

In the case of X and Y zeolites a fairly wide set of mononuclear metal carbonyls can successfully be used. An advantage of zeolites as microporous supports is that metal carbonyls adsorb almost irreversibly and, upon heating in an inert medium or in vacuum, produce highly dispersed metal particles (Ref. 15). We have synthsized Cr, Fe and Ni phthalocyanines attached to Y zeolites:

$$Cr(CO)_6 + 4\ DCB = PcCr + 6\ CO$$
$$Fe(CO)_5 + 4\ DCB = PcFe + 5\ CO$$
$$Ni(CO)_4 + 4\ DCB = PcNi + 4\ CO.$$

It is essential that the use of the ligand substitution for the "in situ" synthesis of heterogenized Pc complexes makes it possible to low the reaction temperature by almost 200° and, thus, to practically exclude the metal migration onto the zeolite outer surface. In addition, a more uniform distribution of complexes in the zeolite matrix is ensured. Finally, there is no more necessity to perform a further ion exchange, in other words, to remove transition metal cations which are not incorporated in the complexes. Also, the possibility of the formation of acid centers, which in many cases decrease the selectivity of redox reactions, is precluded.

All of the above refers only to phthalocyanine complexes; we have failed to obtain heterogenized porphyrine complexes by the "in situ" synthesis: the complexes formed only on the outer surface.

3. PHYSICO-CHEMICAL PROPERTIES OF HETEROGENIZED PORPHIN COMPLEXES

The state of included complex molecules depends, in general, on two factors: (1) intermolecular, i.e., interaction between metal complexes and (2) interaction between supported molecule and support surface.

The first type of effects is more remarkable when adsorption technique of supporting is applied. This problem has been studied for PcCo complexes supported on SiO_2 (Refs. 16,17). As electron spectra revealed, single PcCo species exist only at low coverages. At moderate surface concentrations though far from monolayer, the PcCo molecules tend to cluster. The same problem seems to exist with heterogenized Pn complexes though there is no direct evidence and the question is still open (Refs. 6,18,19).

More numerous are the manifestations of the interaction of the second type, that is, between the supported TM complex and support surface atoms or atomic groups. In this respect, spectral data are obviously most informative. As seen from Fig. 2, the remarkable changes in ESR spectra are observed on going from a benzene solution of Co tetraphenylporphyrine (Co-TPP) to titania-supported Co-TPP sample (Ref. 19). It was noted that the intensity of an ESR signal for Co-TPP/TiO_2 sample was enhanced on raising vacuum treatment temperatures, meanwhile the NO adsorption led to depression of that. Analogous changes were found for oxide-supported Pn complexes (Ref. 19), see Fig. 3, where electron spectra are shown. In the case of SiO support, changes in intensities of the Soret band (at 400 nm) and the band at 590 nm are observed. In contrast to silica, the supporting of Co-TPP on titania results in essential changes of the spectra pattern.

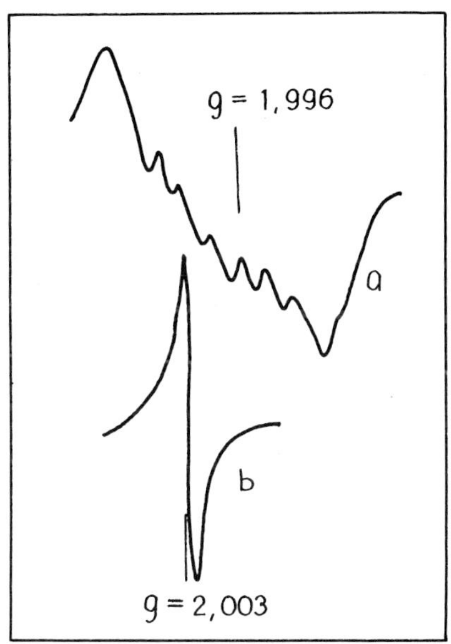

Figure 2. ESR spectrum at 77 K of Co-TPP in benzene solution (A) and supported on TiO_2 (B).

Supporting of Co-TPP on TiO_2 decreases the $Co2p_{3/2}$ binding energy from 781.0 to 780.2 eV. On the contrary, the N1s binding energy increases from 397.3 to 398.6 eV, as follows from XPS data (Ref. 20). The very fact of anion-radical formation with spin density located on a macrocycle (Ref. 20) taken in conjunction with the changes in effective charges on chelate atoms provides strong evidence that there is a severe perturbation provoked in a TPP molecule by electron transfer. The perturbating influence of SiO_2 on supported complex molecules, as seen from Fig. 3, is much smaller.

On the other hand, we found that on supporting Co-TPP from its benzene solution onto SiO_2 the colour of the resulting sample depends on the temperature of vacuum heating of the support. The sample is reddish at 80°C treatment and green at 200°C. Note, that the Co-TPP exhibits the corresponding colours in aprotic and protic media. Electron spectra for these samples are given in Fig. 4.

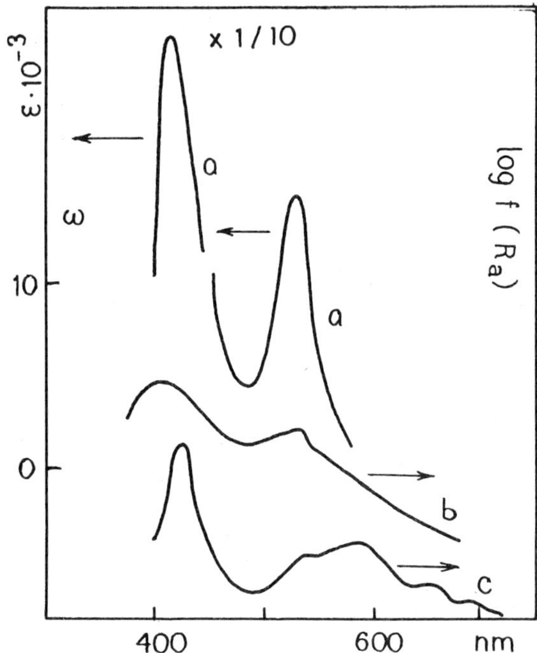

Figure 3. Electron spectra of Co-TPP in benzene (a), on silica (b) and on titania (c).

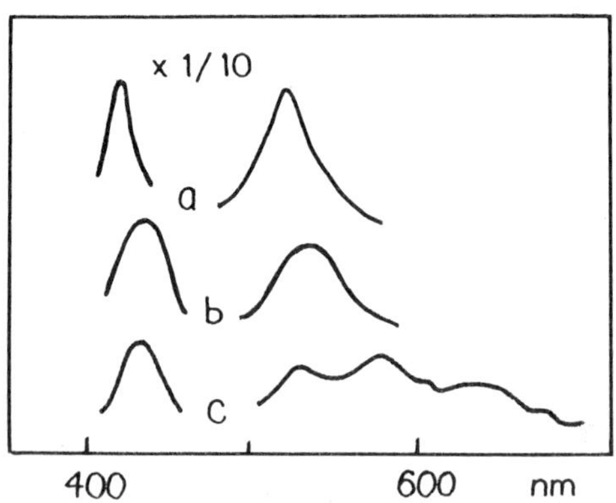

Figure 4. Electron spectra of Co-TPP in benzene (a) and on SiO_2 evacuated at 80°C (b) and at 200°C (c).

Interestingly, the spectra pattern for Co-TPP/SiO_2-200 appears to be closer to that for H_2-TPP/SiO_2 than for Co-TPP itself. Since there seem to be some acid centers on silica support, as it was evidenced from rather high (about 100 kJ/mole) values of the initial adsorption heats of NH_3, one cannot exclude a possibility of reversible surface protonation of the complexes:

$$\text{Co-TPP} + 2H^+ = H_2\text{-TPP} + Co^{2+}.$$

The reversibility of this reaction is hardly surprising since, in contrast to PcM complexes, metal atoms in PnMs are very labile. Hence, it is quite understandable that benzene extraction of the Co-TPP/SiO_2-200 sample results in a solution whose spectra are similar to the spectrum a in Fig.4. Of course, this acid-base model of the changes observed in spectra of TPPs on supporting them onto oxides does not negate the redox mechanism proposed by Mochida et al. (Refs. 19,20), but it adds some new insight into intimate guest-host relations.

In the case of PcMs, the situation differs from that just described for PnMs. This is because of higher stability of metal atoms in phthalocyanine macroligand coordination. Therefore, the protons might attack on nucleophilic nitrogen mesoatoms. As a result, up to 7 bands could arise in 600-800 nm region from different protonated forms $PcMH^+$-$(PcMH_4)^{4+}$ (Ref. 21). Similarly, there are 6 separate bands in the diffuse reflectance spectra of silica-supported PcCu (Ref. 12), which are supposed to be due to the presence of several protonated forms on the support surface.

The same concept of protonated phthalocyanines on zeolite supports has been proposed by Diegruber and Plath (Ref. 22). They studied electron spectra of PcCo complexes in X zeolite matrix. Not only the positions of spectral bands but also the distances between them differ essentially, as follows from a comparison of PcCo in the gas phase, a PcCo film on diamond and PcCo on the outer surface of a single crystal of X zeolite, see Fig. 5.

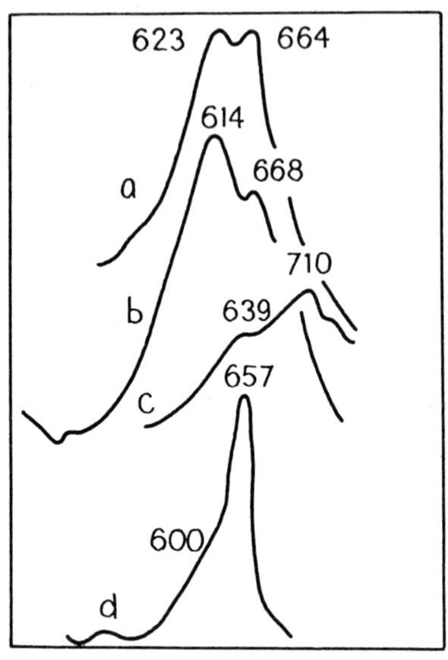

Figure 5. Electron spectra of PcCo within supercage (a), on outer surface (b) of X zeolite, on diamond surface (c) and in gas phase (d) (Refs. 22, 23).

A mean sfift of spectral bands corresponds to ca.5 kJ/mole. These results can be understood in terms of a strong interaction between supported complexes and the matrix. In fact, the presence of more than one PcM molecule in a supercage of faujasite-type zeolite is impossible. Consequently, the changes in the spectra cannot be attributed to the intermolecular influence. The very fact of the band splitting in low frequency region of spectra indicates the symmetry loss. Since a distortion of a planar configuration of the PcM molecule, for example D_{4h} to C_{4v} transition, cannot result in band splitting, there remains a sole possibility of protonation of PcCo species on two nitrogen mesoatoms (D_{4h} to

D_{2h} transform), i.e. the formation of phthalocyanine dication on account of acid centers of zeolite:

$$PcCo + 2\ H^+ = (PcCoH_2)^{2+}.$$

Indeed, there are two distinct bands at 664 and 700 nm of about the same intensity in hydrogen PcH_2 complexes (Ref. 24). Further, the dissolving of PcCo in concentrated H_2SO_4 led also to the splitting of the low frequency band, and, finally, for PcCo in X zeolite matrix the splitting of the band disappears at 250°C (Ref. 22). Consistent with the concept of protonation of PcMs in zeolite matrices are the observations by ESCA method (Ref. 13). These showed that bond energies (E_b) as well as line HWFM for zeolite-included PcMs and for individual complexes are very close. By contrast, the narrow singlet of N1s for individual PcM transforms into poorly resolved doublet with a distance between its components of about 1.5 eV for zeolite-included PcM complexes. An analogous shift, which indicates unequivalency of nitrogen atoms, has been observed for H_2-TPP (Ref. 25) and for protonated PcMs (Ref. 26) as well.

Then the question arises as to the nature of a proton source on oxide surfaces. While the acidic nature of TM cation-exchanged zeolites is beyond any doubt, the presence of acid sites on SiO_2, TiO_2 and other oxides seems to be not so evident. However, as mentioned above, these do protonate the supported PcMs and PnMs. Then it appears that in these cases chemisorbed H_2O molecules might deliver protons to attack guest species. This is supported by findings of Mochida et al. (Ref. 20) who observed essential changes in catalytic behaviour of titania-supported Co-TPP on evacuating the sample at 100°C and at 200°C. The loss of water in this temperature interval is of 2% whose readsorption restores the catalytic activity.

In the net, the support surface affects substantially the state of immobilized Pn and Pc complexes. This influence is obvious to account for the observed dependence of attached complex properties on the support nature, which is clearly manifested in adsorption and catalysis processes.

The fact that the adsorption of molecular dioxygen on supported Pn and Pc complexes was most studied is quite understandable. Molecular adducts of O_2 with chelates of TMs are well known for homogeneous systems. The adducts of 1:1 as well of 1:2 stoichiometries have been found for dioxygen and porphyrines in non-aqueous media.

The lacking of mobility of supported complexes as a result of their anchoring remarkably changes the situation. For example, in the case of zeolite-included PcMs in which only one complex molecule, if any, can reside within the supercage the stoichiometry of molecular oxygen adducts formed at temperature as low as $-78^{o}C$ was found to be 2:1 (Ref. 27). It is noteworthy that the ability of zeolite-included PcMs to fix molecular O_2 could be enhanced by preadsorbed pyridine base, the analogous effect being well known for solutions. The molecule of base ligated at the 5th axial position increases an electron density on the metal atom and thereby promotes its ability to back-donation toward the substrate molecule situated trans, i.e. at the 6th axial site.

The detailed investigation on the state of dioxygen species in molecular adducts with alumina-supported PcCo has been made by Morazzoni et al. (Ref. 29). They reported that Al_2O_3 support activates the PcCo molecule (other supports were found to be inert (Ref. 30)); the complex -surface binding is accomplished via electron donation from the d_{z2}-AO of the metal to the vacant AO of the surface Al atom playing the role of the 5th extra-ligand. The chemisorption of dioxygen on the anchored PcCo is activated and occurs at $70-150^{o}C$. Another important fact was also reported in (Ref. 29), that is, a pronounced dependence of the strength of dioxygen adducts on acid properties of the surface: at a low acidity the superoxide anion O_2^- formed can even stabilize on a Lewis center rather than on PcCo; the high acidity of the surface promotes dioxygen binding with the chelating agent (the same has also been observed in (Refs. 30, 31)). This fact has been established previously for the charged ethylenediaminate complexes of Co in Y zeolites in different decationation states (Ref. 32).

On the contrary, $CoTPP/TiO_2$ is inactive toward adsorption of molecular O_2, but adsorbs well up to 2 molecules of NO per metal atom of the complex at 50°C. Some decreas in $Co2p_{3/2}$ bond energy occurs, as has been mentioned above (Ref. 20). The adsorption of molecular H_2 is observed only on TiO_2-supported cobalt and metal-free TPP, the amount of adsorbed hydrogen being increasing with increasing evacuation temperature. At the same time, close values of H_2 adsorption on Co- and H_2-forms of TPP (19.6 and 13.5 Mmole/g, respectively) indicate the participation of the macrocycle in molecular hydrogen fixation (Ref. 20).

Finally, it is to be mentioned that the stoichiometry of NH_3 adsorption on silica-supported H_2-TPP is quite unexpected, since it amounts to ca.4 molecules of NH_3 per complex. This is, however, in agreement with the concept of protonation of TPP complex which results, as described above, in the formation of $(H_4\text{-TPP})^{2+}$ species. Support for such an interpretation comes from our observation that the green colour of H_2-TPP/SiO_2 sample changes for reddish after ammonia adsorption.

The examples just cited on the adsorption of various substances on supported Pn and Pc complexes clearly show how largely different and complex would be the interactions in triple, substrate-complex-support systems. On the other hand, the experimental data on the subject are rather poor so that any detailed pictures of the mechanism seem to be ambiguous.

4. CATALYSIS BY SUPPORTED PORPHIN COMPLEXES

The possibilities of the use of such systems in catalysis practice depends primarily on their resistance to oxidation, hydrogenolysis and thermolysis. Zeolite-included PcMs were found in (Ref. 33) to be stable up to 350°C in hydrogen, but they begin to burn above 100°C in air. In inert atmosphere, the supported PcMs are stable up to 550°C. Under temperature-programmed conditions (5 °/min), the zeolite-supported PcMs oxidize at a maximum rate at 340°C, as was shown by DTA and DTG methods (Ref. 34).

The stability limits established determine the area of potential applications of supported porphin complexes as catalysts. Catalytic processes where these systems could be most advantageous are the following: (i) liquid -phase oxidation and hydrogenation, (ii) vapour-phase dehydrogenation and elimination.

Up to date, more than 20 reactions catalyzed by oxide -supported Pn and Pc complexes have been studied. They are listed in Table 2.

Most reactions given there are also known for unsupported analogs, except for gas-phase oxidation of carbon monoxide and ammonia. These latter have been first demonstrated by us for PcM complexes.

The data obtained on catalysis by supported Pn and Pc complexes permit one to elucidate three main peculiarities of these systems.

Due to an outstanding stability of metal porphin molecules, their heterogenizing does not result in any substantial change in catalytic behaviour, in other words, they retain as a rule their characteristic properties in solution with a molecular dispersion of the complex compound.

TABLE 2. Reactions catalyzed by oxide-supported metal porphins

Reaction	Catalyst	Refs.
Oxidation of		
alkanes	PcFeX,Y	11
olefins	PcCoX, PcMX,PcM/SiO_2	34-36
cycloalkanes	PcFeX,Y	11
alkylbenzenes	PcNiY, PcFeY	14,35
cystein	PcCoY	32
butylmercaptane	PcMY	37
isopropanol	PcCo/SiO_2	38
carbon monoxide	PcMY, PcM/SiO_2	12,39
ammonia	PcM/SiO_2	12
Dehydrogenation of		
cyclohexane	PcMY	39
ethylbenzene	PcNiY	33
isopropanol	PcMY	39
Dehydration of		
cyclohexane epoxide	PcMY	39
Dehydrothiolation of sulfur-containing compounds		32
butylmercaptane	PcMY	32
thiophene	PcNiY	32
Hydrogenolysis of		
butylmercaptane	PcMY	32
thiophene	PcMY	32
Reduction of NO with		
hydrogen	PcNiY, PnCo/TiO_2	20,32
carbon monoxide	PcRuY, PnCo/TiO_2	13,20
Decomposition of	Co-TPP oxides	
hydrogen peroxide	PnCo, PnCu/SiO_2	6,18
nitrous oxide	PnM oxides, Co-TPP/TiO_2	20
Decarboxylation of oxalic acid	PcCo/SiO_2	40

Undoubtedly, the most important advantage of homogeneous catalysts is an ideal uniformity of the active centers. This valuable property is preserved upon heterogenizing metal porphins on oxides. For example, during the dehydrogenation of cyclohexane on PcNiY catalysts different in amount of the active phase, the atomic catalytic activity remained almost unchanged even upon a ten times change in the PcM content in the sample (Ref. 39).

For PcM in zeolites, in which the metal atom is accessible from both sides of the molecular plane, a well-pronounced kinetic trans effect, that is, the change in catalytic reaction rate under the action of the 5th axial ligand, is observed. For example, it has been found that the activity of PcMY in CO oxidation, isopropanol and cyclohexane dehydrogenation tends to increase due to an extra-coordination of pyridine (Ref. 39). The same effect has been detected for $PcCo/SiO_2$ in oxidative decarboxylation of oxalic acid (Ref. 40). An additional example of this effect is provided by a typical volcano-type shape of the dependence of the activity of Co phthalocyanine fixed in Y zeolite on the pH of solution during the oxidation of cysteine (Ref. 32).

Due to the high stability of Ni and Co atoms in the chelating unit of Pc complexes, these latter can be used as stable catalysts of hydrogenolysis and decomposition of sulfur-containing compounds that do not require presulphuration (Ref. 27).

Moreover, anchoring of the complexes on the supports, along with quite obvious advantages of heterogeneous systems, allows one to make use of the "effect of micromedia", that is, the specific effect of the environment of the active center - the immobilized molecule of metal porphin. Clearly, such effect is most pronounced when using zeolites as matrices, and these are molecular sieve properties of the zeolite support that cause an almost two-fold decrease in oxidation state when passing from n-pentane to n-octane, as well as from cyclohexane to cyclodecane (provided that instead of a homogeneous PcFe whose oxidation rate is independent of the molecule size, Fe phthalocyanine attached to Y zeolite is used (Ref.11)).

Incorporation of PcM into zeolite cavities, which excludes washing the active component by the solvent, has allowed us to obtain a stable catalyst of the liquid-phase oxidation of cumene (note that on any other supports the catalytic activity was inevitably lost as a result of the leaching of the complex into the solution).

Another advantage of zeolite matrices is a high adsorption capacity in the cavities, which remains high after fixation of heterogenized complexes in the cavities. As a result, substrates are strongly concentrated in the cavities, and eventually, the atomic catalytic activity of the heterogenized molecule of the complex increases, as is well illustrated in Table 3.

TABLE 3. Atomic catalytic activity (ACA, conventional units) of homogeneous and zeolite-supported Pc complexes

Reaction	ACA of homogeneous complex	ACA of supported complex
Oxidation of cumene	0.23	5.1
Oxidation of cysteine	7.7	20.7
Oxidation of CO	2.8	8.4

The presence of not more than one molecule of the hetereogenized PcM in the zeolite cavity at a distance between individual cavities of about 1.5 nm excludes those routes of the catalytic reaction that are associated with the participation of two neighbouring active centers in one or several steps. For example, when the reduction of NO with molecular hydrogen occurs on PcNiY, the only reaction product is ammonia, while on the reduced Ni-zeolite this is dinitrogen. Assuming that formation of molecular N_2 is assisted by two mononitrogen intermediates chemisorbed on neighbouring sites, which do not exist on PcNiY but, most probably, exist on metallic nickel, the "reversal" of the selectivity on the two systems becomes quite clear.

Finally, note that in some cases the specific interaction of the fixed molecule of metal porphin with the support surface gives rise to the properties that are not typical for homogeneous systems. We have already mentioned the data obtained by Morazzoni et al. (Refs, 29-31) on activation of PcCo via interaction of the chelating unit with Al atoms of the oxide surface. A similar activation of Co-TPP upon its supporting on oxides was also observed during the decomposition of H_2O_2 (Ref. 6). It is usually assumed that tetracoordinated Co(II)is inactive in this reaction. In fact, we have found that the rate of peroxide decomposition in the presence of individual porphyrines almost does not differ from the rate of the homogeneous noncatalytic reaction. However, supporting on oxides leads to a sharp increase of the reaction rate, which is clearly dependent on the support nature (Ref. 6). Additional evidence for the specificity of Co-TPP activation by the support is a noticeable difference in the effect of the matrix on the decomposition of H_2O_2 and the reduction of NO (Refs. 6,19). While in the first reaction the highest efficiency of the Co-TPP molecule is observed on nickel oxide, in the second reaction, on titania.

It seems likely that on supporting metal porphins on oxides (except for zeolites) their molecules are anchored on a definite group of surface atoms of the support that show affinity either to the central atom of the chelating unit or to a microcyclic ligand and that are capable of donor-acceptor and/or back-donation interactions (Refs.

19, 29-31). For example, the surface of silica has inhibitory effect on the decomposition of H_2O_2; however, after the supporting of inactive FeC1TPP on this surface, the reaction rate becomes as high as the rate of the homogeneous noncatalytic reaction. Once supported on the surface, the TPP seems to selectively suppress the sites of reaction inhibition.

CONCLUSIONS

This review is an attempt to give some insight into the mechanism of catalysis by supported porphin complexes. Although no commercial use of these have been claimed as yet, the utility of the oxide-supported porphyrines and phthalocyanines of TMs as promising catalysts for a wide variety of reactions appears to be quite evident. These systems, on the one hand, retain their ability to extracoordinate into the 5th and the 6th axial positions for a number of potential substrates and thereby to activate them. On the other hand, being influenced by a support surface as environmental factor, the chelates can display the catalytic behaviour different from that of their homogeneous analogs.

It is obvious that zeolites which offer a possibility for topological encapsulation of TM complexes are of particular interest. In fact, in view that porphin molecules play a key role in biochemical processes, there is every reason to believe that the inclusion of these into zeolite cavities of well-defined geometry and chemistry would represent a valuable starting point in the design of artificial enzymes.

Although there is, in general, only limited information on the included porphin-support interaction, this latter is thought to be one of the efficient means to modify activites and selectivities of heterogenized complexes, in other words, to "tune-up" smoothly an active species under desirable reaction route.

In conclusion, the review of the studies on oxide-supported porphins done in the last five years indicates that the growing interest in these systems is encouraging for the future contribution of the field to cross-fertilization of homogeneous and heterogeneous catalysis.

REFERENCES

1. M. Momenteau, and B. Lock, J.Mol.Catal.,7,315 (1880).
2. J.P. Collman, Acc.Chem.Res., 10, 265 (1977).
3. M. Momenteau, J. Mispelter, B. Lock and E. Bisagni, J.Chem.Soc.Perkin Trans.,1, 189 (1983).
4. M. Momenteau and D. Lavalette, J.Chem.Soc.Chem.Commun., 311 (1982).
5. Chem.Eng.News, 30 (1984) Sept.17.
6. I. Mochida, A. Yasutake, H. Fujitsu and K. Takeshita, J.Phys.Chem., 86, 3468 (1982).
7. B.D. Berezin, Coordinating compounds of porphins and phthalocyanines, p. 280, Nauka, Moscow (1978) (in Russian).
8. O.M. Lifentsov and B.N. Melnikov, in Dyeing and Printing of Cloth Pigment Synthesis on a Fibre, Legkaya Prom-st, Moscow (1973) (in Russian).
9. V.Yu. Zakharov and B.V. Romanovsky, Vestnik Moscow University, Ser.Khim., 18, 2, 143 (1977).
10. G. Meyer, D. Wörle, M. Mohl and G. Schulz-Ekloff, Zeolites, 4, 1, 30 (1984).
11. N. Herron, C.A. Tolman and G.D. Stucky, Abstr.Int. Conf.Coord.Chem.,23rd, p. 111, Boulder, July 29-Aug. 3, (1984).
12. T.G. Borisova, Thesis, Moscow University (1982).
13. E.S. Shpiro, G.V. Antoshin, O.P. Tkachenko, S.V. Gudkov, B.V. Romanovsky and Ch.M. Minachev, Stud.Surf. Sci.Catal., 18, 31 (1984).
14. T.V. Korolkova, A.N. Zakharov and B.V. Romanovsky, Vestnik Moscow University, Ser.Khim.,25, 4, 362 (1984).
15. Th. Bein, P.A. Jacobs and F. Schmidt, Stud.Surf.Sci. Catal.,12, 111 (1982).
16. A.N. Khefni, L.I. Nekrasov and S.A. Borisenkova, Zh. Fiz.Khim., 53, 7, 2071 (1979).
17. A.N. Khefni, N.A. Mamleeva, L.I. Nekrasov and S.A. Borisenkova, Zh.Fiz.Khim., 59, 8, 2326 (1979).
18. I. Okura, N. Kim-Thuan and T. Keii, J.Mol.Catal.,5, 293 (1979).
19. I. Mochida, K. Tsuji, K. Suetsugu and K. Takeshita, J.Phys.Chem., 184, 3159 (1980).

20. I. Mochida, K. Suetsugu, H. Fujitsu and K. Takeshita, J.Phys.Chem., 87, 1524 (1983).
21. U. Ahrens and H. Kuhl, Z.phys.Chem.(N.F.),37,1 (1963).
22. H. Diegruber and P.J. Plath, Z.phys.Chem., Leipzig, 266, 4 (1985).
23. L. Edwarts and M. Gouterman, J.Molec.Spectr., 33, 292 (1970).
24. L.E. Lyons, J.R. Walsh and J.W. White, J.Chem.Soc., 167 (1960).
25. J.P. Macquet, M.M. Millard and F. Theophamidis, J.Am. Chem.Soc.,100, 4741 ( ).
26. T. Kamai, M. Soma and Y. Tatsumoto, Chem.Phys.Lett., 37, 37 (1976).
27. B.V. Romanovsky, Proc. 8th Int.Congr.Catal., Verlag Chemie, W.Berlin, 4,,657 (1984).
28. F.A. Walker, J.Am.Chem.Soc., 95, 1796 (1973).
29. M. Barzaghi, T. Beringhelli and F. Morazzoni, J.Mol. Catal., 357 (1982).
30. F. Campadelli, F. Cariati, P. Carniti, F. Morazzoni and V. Ragaini, J.Catal., 44, 167 (1976).
31. G. Mercati, F. Morazzoni, M. Barzaghi, P. Carniti, V. Ragaini and F. Campadelli, J.Chem.Soc.Faraday Trans., 75 1857 (1979).
32. O. M. Zakharova, Thesis, Moscow University (1982).
33. S.V. Gudkov, Cand.diss., Moscow University (1981).
34. H. Diegruber, P.J. Plath, G. Schulz-Ekloff and M. Mohl, J.Mol.Catal., 24, 115 (1984).
35. I.I. Ioffe and T.A. Zapevalov, Catalysts containing supported complexes, p. 203, Institute of Catalysis, Novosibirsk (1977) (in Russian).
36. V. Ragaini, P. Carniti, U. Rainaldi and F. Morazzoni, Chi.e ind., 58, 473 (1976).
37. B.V. Romanovsky, Proc.Int.Symp.Zeolite Catal., Siofok, 215 (1985)
38. N.A. Khefni, Thesis, Moscow University (1980).
39. V.Yu. Zakharov, Thesis, Moscow University (1977).
40. S.A. Borisenkova, N.E. Davidenko and A.P. Rudenko, Catalysts containing supported complexes, Institute of Catalysis, Novosibirsk, p. 129 (1977).

RECENT DEVELOPMENTS IN THE USE OF RADIATION POLYMERIZATION TO IMMOBILIZE ENZYMES AND HETEROGENIZE HOMOGENEOUS METAL COMPLEXES

J.L. Garnett, S.V. Jankiewicz, R. Levot and M.A. Long
University of New South Wales, Kensington, N.S.W. Australia 2033

D.F. Sangster
C.S.I.R.O., Division of Chemical Physics, Sutherland, N.S.W., Australia 2232

ABSTRACT - The technique of radiation grafting is shown to be useful for reagent insolubilization processes including enzyme immobilization and heterogenization of homogeneous metal complexes. In a typical process, a monomer such as p-nitrostyrene is grafted to a backbone polymer, the resulting copolymer than chemically converted to the appropriate functionality (isothiocyanato for enzymes) to which reagent can be attached. Initiation of polymerization by both ionising radiation and UV are shown to be useful in this work. In both systems, the lowest possible radiation doses for grafting should be utilized to minimize degradation of the backbone polymer. Novel additives for lowering the total additive reactions are proposed. Effect of temperature on the additive process is examined. The preparation of typical polymer support systems for enzyme immobilization and heterogenization reactions using the additive techniques is discussed. Future trends of the technique are considered.

1. INTRODUCTION

Radiation polymerization can be used in two general ways to insolubilize chemical reagents (Refs. 1-3). Firstly monomer containing appropriate functional groups can be radiation grafted to backbone polymers. The reagent to be insolubilize can then be chemically attached to the graft copolymer via the incorporated functional group (Refs. 4,5). This method has been successfully used to immobilize enzymes (Ref. 3), heterogenize catalytically active homogeneous metal complexes (Ref. 3) and anchor analytic reagents to polymer surfaces to form novel ion exchange resins (Ref. 4).

In the second procedure, the reagent, in this instance an enzyme, has been immobilized by direct radiation polymerization of a monomer containing the enzyme either in suspension or in solution (Ref. 2). In this alternative method, a range of biologically active materials have been insolubilised including antibodies with a wide potential application in medical research (Refs. 2,5).

In the former method for insolubilization using radiation grafting, the technique can be extended to a larger range of backbone polymers if the radiation doses used in the grafting process are relatively low. Under these conditions naturally occurring backbone polymers readily available in the environment, such as cellulose and derivatives, which are radiation sensitive can be utilized in these reactions.

In the present manuscript the use of novel additives for enhancing radiation grafting will be discussed. Preliminary communications of the effects of these additives have been accepted for publication (Refs. 6,7). The inclusion of these materials in the grafting solution significantly lowers the total radiation dose required to achieve a particular percentage graft. This additive technique is thus of value in radiation grafting to insolubilize chemical reagents such as enzymes and catalytically active homogeneous metal complexes.

In addition to the use of ionizing radiation for initiating grafting reactions of value in reagent insolubilization, UV can also be utilized for the same purpose (Refs. 3,4). In this instance, inclusion of sensitizers in the grafting solution has proved to be of value for speeding up rates of copolymerization, thus giving reasonable grafting yields in several hours of UV exposure. Additives have been used to accelerate these UV grafting processes and typical results from such UV additive work will also be discussed. The effect of temperature on these grafting processes will also be examined. Finally in the paper, the relative merits of utilizing ionizing radiation and UV for grafting in reagent insolubilization processes will be compared and future trends of the technique with recently available radiation sources will be predicted. Cellulose, polyethylene, polypropylene and PVC will be used as representative backbone polymers with styrene and acrylate derivatives as typical monomers.

2. PRINCIPLE OF GRAFTING FOR REAGENT INSOLUBILIZATION

The procedure involves modifying a suitable backbone polymer by radiation grafting a monomer containing an appropriate functional group to which is attached the reagent to be insolubilized. For biological applications, such as enzyme immobilization, p-nitrostyrene has been used as typical monomer with polypropylene or polyethylene as backbone polymer. The nitro group in the graft copolymer is then converted chemically into its isothiocyanato derivative to which is bonded the enzyme e.g. trypsin (Eq. 1). As an alternative to directly

$$\sim CH_2-CH_2-C_6H_4-NO_2 \xrightarrow[HCl]{Sn} \sim CH_2-CH_2-C_6H_4-NH_2 \xrightarrow{SCCl_2} \sim CH_2-CH_2-C_6H_4-NCS \xrightarrow{NH_2\text{-protein}} \sim CH_2-CH_2-C_6H_4-NH-CS-NH\text{-protein} \qquad (1)$$

grafting a substituted monomer, such as p-nitrostyrene, a monomer containing no substituent, such as styrene itself, can be grafted to the trunk polymer, then chemical substitution reactions such as nitration performed to give a nitrated copolymer. Technically any functional group can be incorporated into the graft copolymer, the choice of functionality being determined by the chemical structure of the reagent to be insolubilized. Thus, for heterogenization reactions involving the anchoring of catalytically active homogeneous metal complexes to polymer surfaces, the phosphine group is a useful chemical entity for attachment since many catalytic complexes are phosphine derivatives. Heterogenization is then achieved by either direct complexation on the surface (Eq. 2) or by ligand exchange (Eq. 3). Most of the insolubilization work reported previously

$$\left[\text{~~}C_6H_4\text{–}PPh_2 \right]_2 \xrightarrow{PdCl_2} (\text{~~}C_6H_4\text{–}PPh_2)_2PdCl_2 \quad (2)$$

$$\text{~~}C_6H_4\text{–}PPh_2 \xrightarrow{RhCl(PPh_3)_3} \text{~~}C_6H_4\text{–}PPh_2RhCl(PPh_3)_2 + PPh_3 \quad (3)$$

(Refs. 3,5,8) has been performed with graft copolymers prepared by ionizing radiation initiation, however, more recent developments (Refs. 4,7) include sensitized UV systems. For laboratory scale work, such UV grafting reactions possess advantages in insolubilization reactions, thus experiments with both cobalt-60 sources and high pressure UV lamps for grafting will be discussed in this paper.

3. EXPERIMENTAL

3.1. Grafting Procedures

Two types of grafting methods have been utilized in these studies, namely the simultaneous and pre-irradiation techniques. Films of the relevant backbone polymers have been used to demonstrate the quantitative aspects of the additive effects in grafting whilst powders of these same polymers were utilized for the actual insolubilization experiments. The grafting methods adopted were modifications of previous procedures (Ref. 3). For the simultaneous work using ionizing radiation initiation strips (40 x 20 mm) of PVC (0.20 mm), low density polyethylene (0.12 mm) and Whatman ashless 41 filter paper were added to freshly prepared grafting solutions contained in stoppered vessels, then irradiated in a cobalt-60 facility. For those experiments performed in the presence of acid, especially the cellulose runs, the backbone polymer samples after irradiation were washed in cold methanol prior to soxhlet extraction in an appropriate solvent. For the powder grafting work, polypropylene (200-400 mesh) and PVC (100μ) were reacted in stoppered pyrex bottles thus enabling the monomer solutions to be stirred continuously during irradiation. Graphted powders were collected in cellulose thimbles immediately after irradiation, the copolymer washed with appropriate cold solvent, extracted in a modified soxhlet for 72 hours, dried in air to remove most of the solvent then in a ventilated oven at 45°C.

For the preirradiation grafting technique (Ref. 9), cellulose samples were dried in an oven at 80°C for 24 hours before sealing under high vacuum in a breakseal apparatus for irradiation. This process is necessary to give reproducible data since cellulose contains a finite equilibrium moisture content which must be kept constant during irradiation. Following irradiation, the cellulose was reacted in evacuated vessels with the monomer grafting solution in the dark at 20°C. After reaction, copolymerized celluloses were treated as for the simultaneous method.

In the UV work, a Philips 90 W high pressure mercury vapour lamp fitted with a quartz envelope was utilized. Sample preparation was similar to the ionizing radiation method except that tubes with larger volumes (25 ml) of grafting solution containing sensitizer were used. The tubes were positioned on a motor driven ventilated circulating drum at distances of up to 30 cm from the source. The polymer films were so positioned that, during irradiation, the surfaces of the films were perpendicular to the incident radiation. After irradiation, films and powders were treated as for the gamma radiation system. Actinometry was performed with the uranyl nitrate-oxalic acid system.

Molecular weight determinations were performed on the polystyrene component of the grafted celluloses by the following acid digestion procedure. The styrenated celluloses were allowed to swell for 24 hours in a mixture of acetone, dimethylformamide and water (15 : 4 : : 1 by volume). An equal volume of aqueous H_2SO_4 (75% v/v) was added and the mixture allowed to stand for 2 hours. This mixture was then diluted to a 6% H_2SO_4 concentration and refluxed for 24 hours. On cooling, the digest mixture was extracted with toluene and the polymer precipitated by dropwise addition of methanol to the extract. The polymer was collected on a sintered glass filter, dried to constant weight at 40°C and analyzed by g.p.c. on a Waters Associates Model ALG/GPS

201 instrument.

3.2. Enzyme Immobilization Methods

For the attachment of enzyme, nitrostyrene copolymer was converted to the isothiocyanato derivative. For the styrene copolymers, the graft was firstly nitrated with a mixture of concentrated nitric and sulfuric acids at 35-60°C to give a nitrostyrene copolymer. The copolymer produced from the nitration reaction or by the direct grafting of p-nitrostyrene was reduced with stannous chloride and hydrochloric acid in dimethylformamide, or with the PVC based copolymers in absolute ethanol. The resulting polyaminostyrene graft copolymer, suspended in dimethylformamide, was treated at 0°C successively with carbon disulfide, triethylamine and ethylchloroformate to give the isothiocyanato derivative. The number of isothiocyanate groups in the copolymer was determined by a modification of the method of Karten and Ma (Ref. 10). The covalent attachment of the trypsin to the copolymer-isothiocyanate derivative was achieved by suspending the support in a solution of trypsin (0.2 %) in a bicarbonate buffer (0.5 M, pH 9.6) and left overnight at 5°C (Ref. 11). The amount of protein covalently bound to the support was determined by acid hydrolysis (Ref. 12) followed by spectrophotometric analysis of the amino acids in the hydrolysate (Refs. 13,14). The esterase activity of the bound trypsin was determined by a spectrophotometric method using N-α-benzoyl-L-arginine ethyl ester, a synthetic low molecular weight substrate (Ref. 15).

3.3. Heterogenization of Metal Complexes

For the insolubilization of the catalytically active

homogeneous metal complexes, procedures analogous to those previously reported (Ref. 3) were used. This p-nitrostyrene was radiation grafted to polypropylene powder and the nitro group converted to the appropriate functionality for attachment of complex. In like manner, the styryl diphenyl phosphine copolymers with polypropylene and PVC powders were prepared by direct radiation grafting of the monomer to the relevant trunk polymers. The copolymers were then directly reacted with the homogeneous complexes dissolved in appropriate solvents. The heterogenized complexes were washed, dried and hydrogenations performed under the conditions described in the tables.

4. RESULTS AND DISCUSSION

There are three classes of additives which enhance radiation and UV grafting reactions and thus are of importance in the following insolubilization processes. The additives discussed in this section are inorganic acids, salts such as lithium perchlorate and polyfunctional monomers. Four backbone polymers have been studied in these processes, namely cellulose, polyethylene, polypropylene and PVC. Of these, cellulose is the least radiation stable (Refs. 8,16,17) and thus any additive which will reduce the total radiation dose to achieve a particular percentage graft will be valuable for copolymerization with this backbone polymer.

4.1. Acids and Salts as Additives in Radiation Grafting

Inclusion of mineral acid, such as sulfuric, in the grafting solution is shown to enhance the copolymerization yield for reaction of styrene in methanol to cellulose using the simultaneous radiation technique. The increase in yields is observed at the lower monomer concentrations (Table 1) especially at 25% styrene where the peak in accelerated grafting is observed

TABLE 1. Comparison of acid with salt as additives in the Simultaneous Radiation Grafting of Styrene in Methanol to Cellulose[a]

Styrene (% v/v)	Graft (%) Additive Free		H_2SO_4 (0.1M)		$LiClO_4$ (0.2M)	
15	32	(0.35)	34	(0.48)	54	(0.37)
20	66	(0.5)	120	(0.98)	155	(0.8)
25	106	(0.77)	153	(1.0)	159	(1.0)
30	112	(1.2)	95	(0.86)	96	(0.9)
35	110	(0.92)	60	(0.76)	80	(0.7)
40	82		52	(0.95)	72	(1.2)

[a]Irradiated at 3.3×10^4 rad/hr to total dose of 5.0×10^5 rad at 24°C. Data in brackets, number average molecular weight ($\bar{M}_n \times 10^{-6}$) of copolymer.

(Trommsdorff effect). This gel effect is important since the data for Mn values confirm that the length of grafted chain corresponding to this peak is highest whereas the shortest grafted chains are observed at the lowest styrene concentration studied (15%). The length of the grafted chain can influence the resulting insolubilization process, since the chain should be long enough to avoid polymer surface problems, but not too long otherwise entropy effects may minimize the efficiency of the actual reagent attachment.

Similar grafting enhancement to acid is observed when inorganic salts such as lithium perchlorate are included in the above grafting solutions, the presence of the salt generally leading to higher yields than with acid (Table 1). Again, the presence of the salt affects the length of grafted chain, the shortest chains being observed at the lowest monomer concentrations. In particular the lengths of grafted chains in the 15% monomer solution for both salt and additive free solutions are comparable whereas the grafting yield in the salt system is significantly higher.

When the same two additives are used for grafting of styrene in methanol to a synthetic trunk polymer, such as polyethylene, similar increases in copolymerization to the cellulose system are observed. With polyethylene film, under the radiation conditions used, the presence of acid leads to grafting enhancement at all monomer concentrations studied, whereas the copolymerization yields are only improved at the lower monomer concentrations with lithium perchlorate as additive (Table 2).

TABLE 2. Comparison of acid with salt as additives in radiation grafting of styrene in methanol to polyethylene film[a]

Styrene (% v/v)	Graft (%) Additive Free	H_2SO_4 (0.1M)	$LiClO_4$ (0.2M)
15	31	32	44
20	64	70	81
25	103	148	192
30	187	240	196
35	193	212	140
40	150	157	114

[a]Irradiated at 3.3×10^4 rad/hr to total dose of 2.0×10^5 rad.

Acid is also more efficient than the salt under the experimental conditions used in Table 2.

4.2. Mechanism of Acid and Salt Enhancement in Grafting

In previous work (Refs. 3,4,7) on the mechanism of the acid effect on radiation grafting of styrene (MH) in methanol (SH) to cellulose (Cell-OH) the sequence of events (Eqs. 4-12) was proposed to explain the phenomenon. Thus, cellulose can undergo bond needs to be qualified.

TABLE 3. Comparison of acid with salt as additives in preirradiation grafting of styrene onto cellulose in methanol[a]

Styrene (% v/v)	Graft (%) Additive Free	H_2SO_4 (0.1M)	$LiClO_4$(0.2M)
10	3 (4)	34 (19)	1
15	5 (6)	42 (58)	3
20	6 (8)	54 (91)	58
25	6 (11)	42 (82)	180
30	6 (18)	36 (78)	242
35	10 (22)	40 (56)	229
40	51 (84)	36 (45)	147
45	46 (76)	27 (36)	98

[a]Irradiated at 3.3×10^4 rad/hr to total dose of 5.0×10^5 rad followed by reaction for 72 hr in dark at 20°C. Data in brackets for 1.0×10^6 rad total dose.

Based on the above results and more extensive evidence to be published elsewhere (Ref. 7), a new model is proposed to explain the effect of acids and metal salts such as $LiClO_4$ in enhancing radiation grafting. In any grafting system at any one time, there is an equilibrium concentration of monomer absorbed within the grafting region of the backbone polymer. This grafting region may be continually changing as grafting proceeds. Thus in grafting styrene to cellulose, during the initial part of the reaction, the grafting region will be essentially cellulosic in nature, however, as reaction proceeds, the grafting region will become more styrenated. The degree to which monomer will be absorbed by this grafting region will therefore depend on the chemical structure of the region at the specific time of grafting. Experimental data now available indicate that increased partitioning of monomer occurs in the graft region when ionic solutes are dissolved in the bulk grafting solution. Thus higher concentrations of monomer are available for grafting at a particular backbone polymer site in the presence of these additives. The extent of this improved monomer partitioning depends on the polarities of monomer, substrate and solvent and also on the concentration of ionic solute. Metal salts such as $LiClO_4$ are more efficient than acids in enhancing grafting due to the overall monomer partitioning effect. It is thus the effect of these ionic solutes on partitioning which is essentially responsible for the observed increase in radiation grafting yields in the presence of these additives. Because of all the parameters involved, the enhancement process is mechanistically complicated. It is obvious that radiolytic effects will also have made some contribution to the overall reaction, however, on the basis of the present work, such processes will not be involved in the predominant pathway.

4.3. Synergistic Effect of Acid and PFMs in Grafting

In addition to acid and inorganic salts, a third series of additives have proved useful for both ionizing radiation and photografting processes. These materials are polyfunctional monomers (PFMs), typified by trimethylol propane triacetylate (TMPTA), when used in amounts up to 1% v/v in the grafting solution. In the gamma ray nitrated grafting of styrene in methanol to polyethylene film (Table 4), TMPTA is as effective as acid in increasing the grafting yield. The pattern of enhancement, however, appears to be different suggesting that each additive operates via a separate mechanistic pathway. Thus with PFM's like TMPTA/in additive amounts, branching of the grafted polystyrene can occur

TABLE 4. Synergistic effect of acid and TMPTA as additives in UV and ionising radiation grafting of styrene to polyethylene film[a]

Styrene (% v/v)	Graft (%) Ionising Radiation[b] No Additive	H^+	TMPTA	H^++TMPTA	UV[c] No Additive	H^+	TMPTA	H^++TMPTA
20	14	19	-	-	28	14	28	41
30	37	51	39	54	101	126	52	78
40	76	81	73	106	189	193	321	266
50	109	134	137	181	124	107	412	525
60	89	73	105	101	37	31	133	188
70	68	62	59	95	32	39	-	-

[a]TMPTA = trimethylol propane triacrylate; low density polyethylene film (0.12 mm)

[b]Dose rate of 4.1 x 10^4 rad/hr to 2.4 x 10^5 rad with sulfuric acid (0.2M) and TMPTA (1% v/v)

[c]Irradiated at 24 cm from 90W high pressure UV lamp for 24 hr. Other conditions as in footnote b. Benzoin ethyl ether (1% w/v).

when one end of the TMPTA, immobilized during grafting, is bonded to the growing chain. The other end is unsaturated and free to initiate new chain growth by scavenging reactions. The new branched polystyrene may eventually terminate, cross-linked by reacting with an adjacent polystyrene chain or immobilized TMPTA radical. Grafting is thus enhanced mainly through the branching of the grafted polystyrene chain.

Because the mechanisms for grafting enhancement differ for acid and PFMs, it is not surprising to find synergistic effects when both additives are included in the same solution (Table 4). Such synergism is observed in both gamma ray and UV initiated processes (Table 4), however, in the photografting of styrene to polyethylene film sensitised by benzoin ethyl ether (BEE), TMPTA is more efficient as an additive than acid and the combined effect of acid and TMPTA is greater in the UV system than with the gamma ray process.

4.4. Effect of Acid and PFMs in Photografting to PVC

When the backbone polymer is changed to PVC film, similar acid (Table 5) and PFM (Table 6) enhancement ef-

TABLE 5. Effect of acid on photografting styrene in methanol to PVC film with biacetyl as photosensitiser[a]

Styrene (% v/v)	Graft (%) Additive Free	H_2SO_4 (0.1M)
20	4	12
40	9	24
60	9	13
80	5	7

[a]Samples irradiated at 30 cm from 90W high pressure source for 24 hr with biacetyl (1% w/v).

TABLE 6. Effect of polyfunctional monomers on photografting of styrene in methanol to PVC film[a]

Styrene (% v/v)	Graft (%) Additive Free	DVB (1% v/v)	TMPTA (1% v/v)
20	14	15	-
30	15	19	16
40	34	47	30
50	51	71	58
60	27	30	44
80	25	24	28

[a]Samples irradiated at 30cm from 90W high pressure source for 24 hr with benzoin ethyl ether (1% w/v) as sensitiser.

fects in sensitized photografting of styrene are observed. With biacetyl (BAC) as sensitizer inclusion of sulfuric acid in the monomer solution leads to increased UV grafting yields to PVC at all styrene concentrations examined (Table 5). Similar results are obtained with BEE as sensitized (Table 6) and the two PFMs namely TMPTA and divinyl benzene (DVB), thus these additive effects appear to be of general applicability in overall radiation induced and UV grafting processes.

4.5. Effect of Acid and PFMs in Photografting MMA to Cellulose

The acrylates are useful monomers in these grafting reactions. Thus methyl methacrylate (MMA) can impart physical strength to a backbone polymer whilst hydroxyethyl acrylate can change the polarity of a substrate. In Table 7, MMA is used to represent this class of monomer in photografting to cellulose. Using uranyl nitrate as sensitizer, it is observed that acid and TMPTA rupture by direct irradiation effects yielding sites for grafting (Eq. 4). Hydrogen abstraction reactions involving radiolytic fragments from the solvent can also activate the backbone polymer (Eqs. 5-7). Hydrogen ions from

TABLE 7. Synergistic effect of acid and polyfunctional monomers as additives in photografting MMA to cellulose[a]

MMA (% v/v)	Graft (%)			
	UO_2^{2+}	UO_2^{2+} + H^+	UO_2^{2+} + TMPTA	UO_2^{2+} + H^+ + TMPTA
30	51	186	3	433
40	95	228	160	580
60	67	373	160	1043
80	10	227	32	1417

[a]Irradiations at 20°C using uranyl nitrate (UO_2^{2+}, 1% w/v) for 16 hr at 24 cm from 90W source; MMA = methyl methacrylate; H_2SO_4(0.2M); TMPTA = trimethylol propane triacrylate (1% v/v); solvent methanol.

$$\text{Cell-OH} \rightarrow \text{Cell-O}^{\cdot} + \text{H} \quad (4)$$

$$\text{SH} \rightarrow \text{S}^{\cdot} + \text{H} \quad (5)$$

$$\text{Cell-OH} + \text{H} \rightarrow \text{Cell-O}^{\cdot} + \text{H}_2 \quad (6)$$

$$\text{Cell-OH} + \text{S}^{\cdot} \rightarrow \text{Cell-O}^{\cdot} + \text{HS} \quad (7)$$

$$\text{Cell-OH} + \text{H}^{+} \rightarrow \text{Cell-OH}_2^{+} \quad (8)$$

$$\text{Cell-OH}_2^{+} + e \rightarrow (\text{Cell-OH})^{*} + \text{H} \quad (9)$$

$$(\text{Cell-OH})^{*} \rightarrow \text{Cell-O}^{\cdot} + \text{H} \quad (10)$$

$$\text{SH} + \text{H} \rightarrow \text{SH}_2^{+} \quad (11)$$

$$\text{SH}_2^{+} + e \rightarrow \text{SH} + \text{H} \quad (12)$$

the added acid can equilibrate with the cellulose to form ionic species which can capture thermalized electrons from the primary radiation act to give excited cellulose molecules and H atoms (Eqs. 8-9). Decomposition of the excited cellulose molecules yields radical sites for grafting and additional H atoms (Eq. 10). In the presence of methanol solvent, additional hydrogen atoms can be formed by reactions depicted in Eq. 11 & 12. Overall the acid effect was therefore attributed to increased G(H) yields in acidified solvent upon irradiation, leading to enhanced H atom scavenging and the creation of more grafting sites in the trunk polymer by abstraction reactions. Radical scavenger styrene can then react with the above species leading to the observed acid enhancement in both grafting and homopolymerization processes.

More recent data, typified by the results in Table 3, indicate that the above mechanism of the acid effect needs to be modified. Thus the fact that inorganic salts such as lithium perchlorate which do not contain hydrogen ions yet give similar grafting enhancement to acid by the simultaneous method suggests that hydrogen ions are not the predominant mechanistic pathway for the observed increase in yields. In Table 3, very large acid and salt enhancement in grafting of styrene in methanol to cellulose by the preirradiation technique is observed. In this pre-irradiation system hydrogen ions are only present in the grafting solution after irradiation, thus the proposal that the increase in copolymerization yield is due to an increase in G(H) due to reaction between H^+ species and secondary electrons enhance the UV grafting of MMA to cellulose. As with the previous styrene data, a synergistic effect is evident when the two additives are included in the same MMA grafting solution, thus confirming that these enhancement effects are also relevant to monomers other than styrene.

4.6. Effect of Temperature on Acid Enhancement in Grafting

Increasing the temperature at which the grafting of styrene in methanol to cellulose is performed also leads to direct increase in grafting yield both in the two additive free monomer solutions studied and also when acid is included (Table 8). A similar situation is observed when radiation grafting styrene in methanol to polyethylene film (Table 9). In a preparative context these results are important since any method for increasing grafting yields for the same radiation conditions will be useful for insolubilization processes of the type described in this paper.

4.7. Immobilization of Enzymes on Grafted Supports

The data in Table 10 show that radiation graft copolymers prepared from polypropylene powder, PVC powder and cross-linked polystyrene beads are useful in enzyme immobilization work. The systems with the highest retained activities have both used p-nitrostyrene as graf-

TABLE 8. Effect of temperature on acid enhancement in radiation grafting of styrene in methanol to cellulose[a]

Temperature (^{o}C)	Graft (%)			
	20% Styrene(v/v)		30% Styrene(v/v)	
	Additive Free	H_2SO_4 (0.1M)	Additive Free	H_2SO_4 (0.1M)
0	17	25	33	56
15	40	62	85	94
20	51	70	101	112
25	60	121	-	-

[a]Irradiated at 3.3×10^4 rad/hr to total dose of 5.0×10^5 rad

TABLE 9. Effect of temperature on acid enhancement in radiation grafting of styrene in methanol to polyethylene film[a]

Temperature (^oC)	Graft (%) 20% Styrene (v/v)		30% Styrene (v/v)	
	Additive Free	H_2SO_4(0.1M)	Additive Free	H_2SO_4(0.1M)
14	42	44	135	172
20	64	70	187	240
29	84	118	244	267

[a]Irradiated at 3.3×10^4 rad/hr to total dose of 2.0×10^5 rad.

TABLE 10. Activity of immobilised trypsin on radiation graft copolymers[a]

Run No	Description	mEq CNS/ g support	mg Enzyme g Support	Activity µmoles/min/ g support	Activity µmoles/min/ mg/enzyme	% Retained Activity
1	Free trypsin	-	-	-	18.0	100
2	1% NS-PP	0.33	2.1	2.4	0.9	5
3	19% NS-PP	-	4.9	2.2	0.4	2
4	33% NS-PP	1.46	10.4	8.5	0.8	4
5	13% S-PP	0.92	1.8	2.2	1.0	5
6	32% S-PP	1.64	12.8	1.1	0.1	1
7	6% NS-PS20	-	0.1	2.4	5.9	33
8	17% NS-PVC	-	0.4	2.8	3.1	17

[a]NS-PP indicates direct nitrostyrene graft to polypropylene powder; S-PP indicates direct styrene graft to polypropylene powder followed by chemical nitration; PS20 is polystyrene beads 20% cross-linked with DVB (ex-Strem Chemicals); all nitrostyrene copolymers converted to isothiocyanato derivatives for enzyme attachment.

ted monomer as distict from utilizing styrene followed by nitration reactions on the resulting copolymer. It has been suggested that if the latter process is used (i) depolymerization and oxidation of the backbone polymer may occur during nitration with fuming nitric acid (ii) polynitration may also occur during this process leading to a distribution of nitro groups in the copolymer whereas with the direct grafting of p-nitrostyrene only one nitro group is present and (iii) substitution of nitro groups into sterically hindered ortho and meta positions may occur whereas with p-nitrostyrene grafts all nitro groups are in the same chemical environment and readily accessible to enzyme. The data in runs 2-6 of Table 10 indicate that the relative reactivities of p-nitrostyrene and styrene grafts on polypropylene are comparable for the immobilization process, however the levels of retained activity are considerably lower in these examples when compared with runs 7 and 8. Other enzyme systems in addition to the trypsin work reported here have been successfully immobilized with these radiation graft copolymers so the process appears to be generally applicable to a wide range of enzymes (Ref. 18).

4.8. Heterogenization of Complexes on Graft Copolymers

In addition to being used for enzyme immobilization work, radiation graft copolymers can also be utilized for the heterogenization of catalytically active homogeneous metal complexes. Typical of the activities to be expected from such heterogenization processes are the data shown in Tables 11 and 12 where a variety of supports, including PVC, polypropylene and cross-linked polystyrene, have been investigated. The results show that either styryl diphenyl phosphine can be directly

TABLE 11. Activity of heterogenised RhCl $(PPh_3)_3$ on radiation graft copolymers

Run No	Description[a]	Phosphine Content[b]	Metal Content[c]	Reaction Products[d]: Hexane	trans-2-Hexene	cis-2-Hexene	Total Products
1	7% SDP-PS20	0.18	-	23	19	15	57
2	31% SDP-PS20	0.58	0.056	34	22	14	70
3	85% SDP-PS20	0.73	0.048	70	24	3	97
4	28% SDP-PS2	0.62	0.061	75	21	3	99
5	36% SDP-PS2	0.74	0.041	75	22	3	100

[a]SDP=styryl diphenyl phosphine; PS20=20% DVB cross-linked polystyrene beads (ex-Strem); PS2=2% DVB cross-linked polystyrene beads.

[b]mEq/g support.

[c]Rhodium metal present as RhCl $(PPh_3)_3$ mEq/g support

[d]Hydrogenation conditions:hexene-1 (1M) in benzene (2ml), catalyst (0.2g) sealed under H_2 pressure (700 Torr) and shaken 16 hrs at 40^oC. Products expressed as percentage of non-aromatic hydrocarbon in reaction mixture on completion of reaction.

TABLE 12. Hydrogenation of olefins with chlorotris (triphenylphosphine) rhodium (I) heterogenised on various radiation graft copolymers[a]

Run	Support	Graft (%)	Olefin Conversion (%)[d]
1	SDP/PVC	23	3
2	S/PP[b]	50	88
3	SDP/PS2[c]	16	25
4	PS2[c]	0	0

[a]Benzene used to heterogenise the catalysts. SDP=styryl diphenyl phosphine
[b]Styrene grafted to polypropylene powder (PP), then copolymer phosphinated
[c]PS2=polystyrene beads 2% cross-linked with DVB (ex-Strem)
[d]Hydrogenation with cyclohexene for 65 hr at 80°C for run 1; hexene-1 used in run 2 (16 hr at 45°C) and 3,4 (16 hr at 17°C).

grafted to the support or alternatively styrene can be copolymerized and the resulting graft chemically phosphinated. Both types of support have been successfully used in these studies. Preliminary work on the leaching of the metal complex from the surface of the polymer indicates that under ideal radiation conditions loss of complex during reaction will be low.

5. FUTURE OF RADIATION GRAFTING IN INSOLUBILIZATION REACTIONS

The potential of the radiation polymerization process for reagent insolubilization is enormous since there is a wide variety of monomers and backbone polymers available for use (Refs. 3 & 19). In one step, it is not only possible to achieve insolubilization of reagent, but the polarity of the system can also be simultaneously altered by including an appropriate monomer. For example, if inert polyethylene is used as backbone polymer, inclusion of hydroxyethyl acrylate in the grafting solution will lead to a copolymer more polar then the original material. In the same process, a suitable monomer for insolubilization can be simultaneously grafted with the hydroxyethyl acrylate, thus two properties of the backbone polymer can be altered in the one operation. Finally, one of the most attractive areas of research in this field is the speeding up of the polymerization process. Recent work (Ref. 20) indicates that this aspect of the field will be very valuable in future applications of radiation polymerization for insolubilization of reagents.

ACKNOWLEDGEMENTS. The authors wish to thank the Australian Institute of Nuclear Science and Engineering, the Australian Research Grants Scheme and the International Atomic Energy Agency for the support of this research.

REFERENCES

1. J.L. Garnett, R.S. Kenyon and M.J. Liddy, J.Chem. Soc.Chem.Comm., 735-736 (1974).
2. I.Kaetsu, M. Kumakura, M. Yoshida, M. Asano, M. Himei, M. Tamura and K. Hayashi, Rad.Phys.Chem., 14, 595-602 (1979).
3. J.L. Garnett, Rad.Phys.Chem., 14, 79-99 (1979).
4. C.H. Ang, J.L. Garnett, R. Levot, M.A. Long and N.T. Yen, in Biomedical Polymers, Polymeric Materials and Pharmaceuticals for Biomedical Use, E.P. Golberg and A. Nakajima (Eds.), pp. 299-334, Acad. Press, N.Y., (1980).
5. A.S. Hoffman, Rad.Phys.Chem., 18, 323-342 (1981).
6. J.L. Garnett, S.V. Jankiewicz, R. Levot and D.F. Sangster, Rad.Phys.Chem., in press.
7. J.L. Garnett, S.V. Jankiewicz, M.A. Long and D.F. Sangster, J.Polym.Sci.Polym.Lett.Ed., in press.
8. A. Hebeish and J.T. Guthrie, The Chemistry and Technology of Cellulosic Copolymers, Springer-Verlag, Berlin and Heidelberg (1981).
9. S. Dilli and J.L. Garnett, Aust.J.Chem., 24, 981-987 (1971).
10. B.S. Karten and T.S. Ma, Microchem.J., 3, 507-514 (1959).
11. L. Goldstein, Biochem.Biophys.Acta, 315, 1-9 (1973).
12. S. Blackburn, Amino Acid Determination, E. Arnold, London (1968).
13. J.P. Greenstein and M. Winitz, Chemistry of the Amino Acids, Wiley, N.Y., (1961).
14. E.W. Yemm and E.C. Cocking, Analyst, 80, 209-215 (1955).
15. H. Bergmeyer, Methods of Enzymatic Analysis, Acad. Press, N.Y., (1963).
16. R.B. Phillips, J. Quere, G. Guiroy and V.T. Stannett, Tappi, 55, 858-867 (1972).
17. R.J. Demint, J.C. Arthur, Jr., A.R. Markezich and W.F. McSherry, Text. Res.J. 32, 918-927 (1962).
18. J.L. Garnett and R. Levot, unpublished work.
19. F.R. Hartley and P.N. Vezey, Adv.Organomet.Chem., 15, 189 (1977).
20. J.L. Garnett and I. Kaetsu, unpublished work.

PREDICTION OF STABILITY OF M-L-SiO_2 TYPE HETEROGENEOUS METAL COMPLEX CATALYSTS ON THE BASIS OF THE DATA ON COMPLEXATION IN SOLUTIONS

G.V. Lisichkin and G.V. Kudryavtsev
Chemical Department, State University, Moscow
119899, USSR

ABSTRACT - Different aspects of complex formation of transition metal ions with ligands fixed on silica have been studied: methods for calculation of composition and stability constants of forming complexes in terms of a "polydentate bonding" model; routes to predict the stability of complexes formed on modified silica surfaces and mechanism of formation of heterogeneous metal complex catalysts on silica supports with grafted carboxy, amino and iminodiacetic groups. The reasons of possible instability of heterogeneous metal complex catalysts and low reproducibility of catalytic activity have been analysed.

1. INTRODUCTION

Most heterogeneous metal complex catalysts (HMCs) studied hitherto are the systems in which metal complex fixation on surfaces is carried out by means of coordination bonds (Refs. 1,2). In particular, interest in these catalysts arose out of the assumption that in this case activity and selectivity of homogeneous metal complexes as catalysts do not undergo substational changes.

One of the restrictions preventing wide application of the catalysts based on coordinatively fixed metal complexes is their insufficient stability at reaction conditions. Generally two factors determine instability: catalyst poisoning by reaction products and leaching of some part of the supported complex into solution. To a certain extent leaching of complex is always observed in systems with coordination bonds due to the equilibrium of complex formation. For the purpose of quantitative description of this process and prediction of catalyst stability under conditions of catalysis it is necessary to know composition and stability constants of the complexes anchored on the surface of modified carriers.

The first aspect of the problem of HMC stability in reaction has been studied quite completely while the data on stability of the surface complexes are contradictory and not numerous.

This paper is the review of our investigations of the regularities of complexation of transition metal ions with ligands chemically bonded to silica surface. One of the main objectives of the work was an attempt to correlate complexation in solution and on modified carrier in order to use theoretical and experimental material on complexation in homogeneous systems to predict the surface complex stability, in particular.

2. PECULIARITIES OF COMPLEXATION WITH MACROLIGANDS AND SPECIFIC PROPERTIES OF SILICA CARRIERS

To elucidate regularities of metal complex formation in a modified carrier phase and a correlation between homogeneous and heterogeneous processes, it is necessary to develop methods of calculation of thermodynamic stability constants of complexes forming on the carrier. We consider these methods to be necessarily required to provide an inverse task solution, that is, using data on complexation in solution for calculation of bonding strength of ions (complexes) on the surface.

In spite of numerous works the task of calculation of stability constants for the complexes on modified carrier has not been as yet solved.

For organic polymers, which are most common types of HMC carriers, there are a lot of problems in calculating. The first problem is the necessity to take into account the polymer chain mobility. In the extreme case of a linear straight polymer, a macromolecular form changes with ion bonding, for example, a globular conformation transforms into a linear one. Interaction of metal ions with a few grafted ligands leads to a polymer chain strain decreasing the complex stability. Due to the difference in the chain strain energy in various chain segments and its dependence upon the degree of the carrier coverage with metal ions, in the general case, complexes with different stability constants form in the polymer matrix. The second problem is that the concentration of ions in the polymer phase and in solution differs according to the Donnan ion exclusion. This effect is most pronounced in the case of highly cross-linked polymers.

The situation is more optimistic for modified silicas. First, carriers based on silica do not swell and strain, nor they change volume during sorption because of a silica rigid framework, thus making it possible to disregard matrix configuration entropy changes. Second, most of chemically modified silicas are produced on the base of wideporous matrices with an average pore diameter of over 10 nm or nonporous samples, for example, aerosiles. Effect of the Donnan ion exclusion is small in the first case and absent in the second one. These considerations determined selection of silica carriers for elucidation of a relation between complexation in homogeneous and heterogeneous systems.

The presence of surface species, namely, functional groups and complexes, results in dependence of apparent equilibrium constants, in particular, apparent stability constants, on the degree of surface coverage with metal complexes. For calculation of the thermodynamic stability constants it is necessary either to carry out experiment under conditions excluding this dependence or to introduce corrections considering it into the calculated values. An analysis (Ref. 3) shows that the influence of grafted groups in modified silicas on equilibrium parameters is determined mainly by a single factor, that is, by a change in free energy electrostatic component due to ionization of the surface.

3. THEORETICAL CONSIDERATION

Most methods described in the literature for calculation of stability constants for complexes of metal ions with ligands fixed on the carrier use the mass action law in the form identical with complexation in homogeneous media. Thus, for basic anchored ligands, have

$$M + n\bar{R} \overset{\beta_n}{\rightleftharpoons} \overline{MR}_n; \qquad \beta_n = [\overline{MR}_n] / [M][\bar{R}]^n, \tag{1}$$

(vinculum over a particle symbol denotes that the particle is bonded with the carrier; charges on a metal ion and a complex are omitted).

This approach meets with a number of substantional objections. First, such a model requires a complete analogy between complexation in homogeneous system and in heterogeneous sorbent-solution system; anchored ligands are considered to be statistically independent and uniformly distributed in the total volume of the system, but these assumptions are not valid. Second, it is not clear, what volume we must select for calculation of the concentrations of fixed ligands and complexes in relation (1).

An analysis of relation (1) shows that uncertainty in selection of this volume disappears when $n = 1$. Therefore, there is a need to select such a form for the mass action law that a ligand concentration would always be in the first power. Moreover, the model selected must take into account the main specific feature of the system - the fixation of the ligand on the solid carrier surface.

We consider a "polydentate bonding" model to be the best to meet these requirements. According to the model the reaction of metal ion with a few (n) anchored ligands is considered to be the reaction with a single n-dentate ligand.

In terms of the polydentate bonding model equilibrium of formation of surface complexes can be expressed in the following way:

$$M + [\bar{R}_n] \xrightleftharpoons{\beta_n} \overline{MR}_n; \qquad \beta_n = [\overline{MR}_n]/[M][\bar{R}_n]. \tag{2}$$

For β_n calculation by Eq. (2) it is necessary to know the value of $[\bar{R}_n]$; we suggest that it should be found using two alternative approaches. The first one is based on the model of "fixed polydentate centres" (FPCs). According to this model the surface, containing C_R^S ligands, is divided into C_R^S/n regions containing n ligands each. The second approach is based on the model of "statistical polydentate centres" (SPCs). In this case the total quantity of groups of n ligands capable to form complexes with metal ions is calculated for the predetermined ligand distribution on the surface.

3.1. Fixed Polydentate Centres Model

Let us consider the interaction of M^{z+} metal ion with a carrier containing C_R^S mmole/g grafted basic groups $\bar{R}$. An apparent constant of $\bar{R}$ groups protonation is assumed to be equal to K and independent of the degree of their protonation and sorbent coverage with metal ions. Let us separate C_R^S/n surface regions containing n ligands each and consider an n ligand set as n-basic compound with equal protonation constants. For the sake of simplicity the equations of the equilibrium constants will be written in a concentrational form; particle equilibrium concentrations in the system will be written in square brackets.

According to the equilibria of protonation and complexation with participation of anchored ligands and mass balance equations we obtain:

$$[\bar{R}_n] = C_R/n \left\{1 + \sum_{i=1}^{n} K^i [H^+]^i + \beta_n [M]\right\}^{-1}. \qquad (3)$$

The equation for the sorption value Γ (mmol/g) is as follows:

$$\Gamma = \beta_n \frac{C_R}{n} [M] \left\{1 + \sum_{n=1}^{n} K^i [H^+]^i + \beta_n [M]\right\}^{-1}. \qquad (4)$$

Knowing the sorption dependence on M or H^+ it may be possible to determine the composition of complexes formed (n) and stability constant (β_n) from Eq. (4).

When analyzing the process of stepwise surface complexing we assume that an equilibrium constant of the interaction of a metal ion with (n-m) groups in the n-dentate ligand is independent of the type of ions bonded to remainder m groups. In this case we obtain the following equation for sorption isotherm:

$$\Gamma = \frac{C_R}{n} \; \frac{\sum_{i=1}^{n} i \, \tilde{\beta}_{n-i+1} [M]^i}{1 + \sum_{i=1}^{n} K^i [H^+]^i + \sum_{i=1}^{n} \tilde{\beta}_{n-i+1} [M]^i} \qquad (5)$$

Expressions for $\tilde{\beta}_{n-i+1}$ take into account equilibria of formation of all surface complexes, including i metal ions and n anchored ligands. For example, for threedentate centres (n=3) we have:

$$\tilde{\beta}_3 = \beta_3 + \beta_2 K[H^+] + \beta_1 K^2 [H^+]^2;$$
$$\tilde{\beta}_2 = \beta_2 \beta_1 + \beta_1^2 K [H^+] ; \qquad (6)$$
$$\tilde{\beta}_1 = \beta_1^3 .$$

When the values of Γ, $[M]$, K, $[H^+]$, n and C_R^S are known, for β_i calculation we obtain the system of linear equations which can easily be solved with the help of computers. Then it would be possible to calculate the values of $\tilde{\beta}_i$ using Eqs. (6).

3.2. Statistical Polydentate Centres Model

In this section we focus our attention on the formation of the $\overline{MR}_2$ complexes. Let it be a certain flat surface containing C_R^S mmol/g uniformly located anchored ligands. Around each ligand there are $\varkappa_2$ ligands spaced at a distance providing formation of complexes of the composition MR_2. In this case ligand pairs quantity $[\bar{R}_2]$ is equal to $C_R^S \varkappa_2/2$. In the case of basic anchored ligands, for example, one must consider an acid-base equilibrium and calculate the quantity of unprotonated ligand pairs spaced at a distance providing the formation of $\overline{MR}_2$ complexes. Using the known statistical methods (Ref. 4), we find that

$$[\bar{R}_2] = [\bar{R}]^2 \varkappa_2 / 2C_R^S, \qquad (7)$$

where $[\bar{R}_2]$ and $[\bar{R}]$ particle equilibrium concentrations as well as C_R^S total ligand concentration refer to sorbent mass.

In the general case of $\overline{M}R_n$ surface complex formation

$$[R_n] = [R]^n \ \varkappa_n / n(C_R^S)^{n-1}; \qquad (8)$$

where $\varkappa_n$ is the quantity of n-ligand groups, including the fixed ligand, capable of forming complexes of $\overline{M}R_n$ composition. As a first approximation we can, for example, let $\varkappa_3$ be as follows: $\varkappa_3 = \varkappa_2(\varkappa_2 - 1)$.

Using the mass balance equations for ligand and relation (8) it is possible to obtain the following equation for the Γ sorption value for the general case of formation on the surface of a mixture of the $\overline{M}R$, ..., $\overline{M}R_n$ complexes:

$$\Gamma = \sum_{i=1}^{n} [\overline{M}R_i] = \sum_{i=n}^{n} \beta_i \frac{[\bar{R}]^i [M] \varkappa_i}{i(C_R^S)^{i-1}} . \qquad (9)$$

The following method can be recommended to calculate the values of $[\bar{R}]$. From Eq. (9) it follows that

$$\sum_{i=1}^{n} \beta_i \frac{[\bar{R}]^i [M] \varkappa_i}{(C_R^S)^{i-1}} = [M][\bar{R}] \frac{d(\Gamma / [M])}{d[\bar{R}]} . \qquad (10)$$

Substituting this expression in the mass balance equation for the ligand, after separation of arguments and integration we get:

$$\int_0^{[M]} [M] d(\Gamma / [M]) = C_R^S lg[\bar{R}] - C_R^S lg \frac{C_R^S}{1 + K[H^+]} - (1 + K[H^+]) \cdot$$

$$\cdot [\bar{R}] + C_R^S \qquad (11)$$

The integral value in the left-hand side of the equation is to be found by graphical integration from the experimental isotherm; then we can solve transcendental equation (11) for the known C_R^S, K and $[H^+]$ by computational methods. Substituting the values of $[\bar{R}]$ thus obtained into Eq. (9) we get for a set of experimental Γ and [M] a set of linear equations for β_i calculation.

Derived equations hold true for monobasic anchored ligands. In case of a polybasic ligand having stepwise constants of protonation, K_1, ..., K_n, it is necessary to introduce the term

$$\sum_{i=1}^{n} \left(\prod_{j=1}^{i} K_j \right) [H^+]^i$$

instead of the term $K[H^+]$ in all equations. This type of equations remains the same as applied to acidic ligands, and K is the inverse constant of anchored ligands activity.

When calculating stability constants by the derived equations, in all cases the product of β_i by $\varkappa_i$ is to be found; for determination of β_i it is necessary to know the value of $\varkappa_i$.

Now consider the possibility of using data on complexation in solution for the calculation of the strength of metal ion bonding by modified silicas. Within the framework of the polydentate bonding model such a transition can take place under the assumption that when fixed on the carrier surface, a ligand does not change its complex-forming ability. In this case the values of β_i on surface and in solution are numerically equal. This conclusion can easily be drawn when using a standard procedure for comparing complex-forming properties of mono- and polydentate ligands in solution at a zero chelate effect. Note that stability constants for surface complexes, β_n, have $l \cdot mole^{-1}$ dimension for each n; therefore dimension of stability constants β_n for complexes on surface and in solution differ for $n > 1$.

3.3. Acid-Base Equilibria Calculation

The calculation of equilibria for modified carrier surface is complicated since apparent equilibrium constants depend upon the process proceeding degree; for example, apparent protonation constants and stability of complexes depend upon the ionization degree of functional groups. This results in the necessity, for example, for polybasic groups, to know both the values of apparent stepwise protonation constants and the law of changes of all these apparent constants due to ionization degree increase. In most cases o it is impossible to obtain this information.

We have developed (Refs. 3,5) a new method for calculating acid-base characteristics of ligands fixed on a carrier. The analysis of the sorption isotherm equations shows that for the calculation there is no need to know the values of each constant K_i; it is sufficient to know the value of function F_n of all these constants and activity (concentration) of hydrogen ions. For the basic grafted groups this function is

as follows: $F_n = 1 + \sum_{i=1}^{n} (\prod_{j=1}^{i} K_j) [H^+]^i$.

The calculation method is based on the following theorem: for two systems with stepwise apparent protonation constants, $K_1', \ldots, K_n'$ and $K_1'', \ldots, K_n''$, related as

$$K_1'/K_1'' = K_2'/K_2'' = \ldots = K_n'/K_n'' \quad (12)$$

the values of function F_n are equal at the equal titration degrees, α .

This theorem can be applied to sorbents, that is to systems with changing apparent stepwise protonation constants, if the relation between apparent stepwise constants remains constant and unaffected by their changing.

The calculation procedure for function F_n for sorbents at a given pH value is as follows. Some "comparison system" with unchanging protonation constants is taken and we find for it the dependence of function F_n on α (for example, on the basis of the known values of K_i). Then the value of α corresponding to the given pH is to be found from the sorbent titration curve. The value of F_n for the sorbent is equal to the value of function F_n for the "comparison system" at the equal values of α.

The procedure described is most convenient to use when specific interactions of anchored groups with the carrier surface are absent. For such sorbents a homogeneous analog solution can serve as comparison system.

3.4. Effect of the Interaction of the Surface Species

For modified wideporous silicas we can assume that interference of fixed functional groups and complexes is primarily associated with a change in free energy electrostatic component during surface ionization. The change in apparent equilibrium constants can be explained by that ionization of surface groups occurs in the electrostatic field of the groups which have already reacted. At an arbitrary value of the ionization degree (α) the value of ΔG^o_α is described as

$$\Delta G^o_\alpha = \Delta G^o_{\alpha=0} + \Delta G_\alpha ,$$

where: ΔG_α is Gibbs' energy change caused by the electrostatic interaction.

The change in ΔG_α leads to a change in apparent equilibrium constants related to ΔG^o_α by equation $\ln K = -\Delta G^o_\alpha / RT$.

Now we use this approach to analyze acid-base equilibria with participation of a grafted dibasic compound (Ref. 6). At an arbitrary titration degree, α , the following equations can be written down:

$$\ln K_1 = -(\Delta G_1^o)_{\alpha=0}/RT - \Delta G_\alpha/RT$$

$$\ln K_2 = -(\Delta G_2^o)_{\alpha=0}/RT - \Delta G_\alpha/RT$$

$$\ln(K_1/K_2) = -(\Delta G_1^o)_{\alpha=0}/RT + (\Delta G_2^o)_{\alpha=0}/RT.$$

Thus, regardless of the change in apparent stepwise protonation constants, the relation between them must remain constant, which is the necessary condition of applicability of the calculation method described in 3.3.

At present it is impossible to obtain in explicit form the theoretical dependence of ΔG_α (and thus of ΔG_α^o) on α , I and other system parameters, so approximate methods should be used. We consider the virial expansion (Ref. 4) to be most perspective:

$$\Delta G_\alpha^o = \Delta G_{\alpha=0}^o + \sum_{n=1}^{\infty} A_n \Gamma^n, \qquad (13)$$

where A_n - virial coefficients. Instead of Γ in Eq.(13) we can use α or θ (surface coverage).

4. ACID BASE-PROPERTIES

Investigation of acid-base equilibria on a modified silica surface is the first and the necessary step when finding stability constants of surface complexes. Primarily it is caused by the fact that protonation (acidity) constants for surface functional groups are introduced in the equations for β (5,6,9,11). Besides, a comparison of the derived values with the data for similar homogeneous systems permits one to draw a conclusion about the carrier effect on properties of surface groups; calculation of protonation constants for different ionization degrees gives information about groups' interaction on the surface.

Investigation of acid-base properties of modified silicas in aqueous media has been carried out by pH-metric titration at a constant ionic strength of the solution. In order to determine the carrier effect on the properties of surface groups, titration of a homogeneous analog and a model system and silica slurry in a homogeneous analog solution were performed under identical conditions. Apparent protonation constants were calculated by the equation $\lg K = pH + \lg\alpha/1-\alpha$.

It has been shown (Refs. 5-9) that all modified silicas, similarly to functionalized organic polymers, are characterized by the dependence of apparent protonation (acidity) constants on the titration (ionization) degree of surface groups. When the solution ionic strength is nearly zero, apparent protonation constants decrease by 2-4 orders with α increasing from 0 to 1. As the solution ionic strength increases, the effect of α on the value of K decreases, and when I = 1 an apparent protonation constant remains constant up to an ion surface concentration of 0.6-0.8 ion/nm^2.

In accordance with the general equation (13) for evaluation of interaction of ionized surface groups we have found the values of virial coefficients for a number of modified silicas. Thus, for aminosilicas, having the surface concentration of functional groups of 1.6-3.4 group/nm^2 and different specific surface areas, K changes mainly as a function of the term with α^2. The dependence is plotted in Fig. 1. Deviation is observed for aminosilica with a specific surface area of 600 m^2/g and a pore diameter of 6 nm.

Data on protonation (acidity) constants for a number of anchored to silica functional groups as well as for their homogeneous analogs and model systems when I = 1

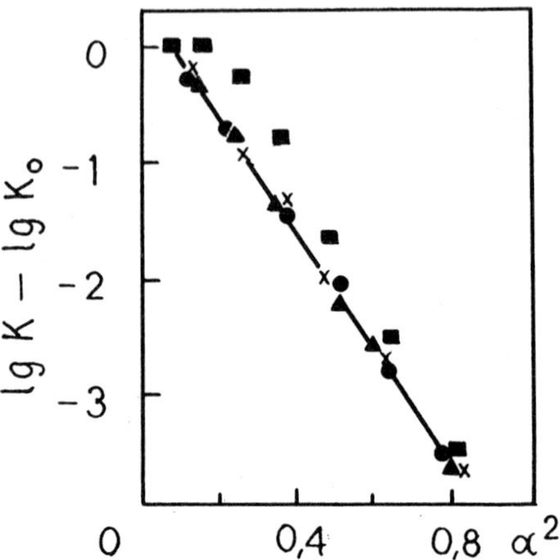

Figure 1. Plot of apparent protonation constants for aminosilicas vs. α^2. I = 0.01, specific surface area of original silicas (S): 120 (× , ●), 250 (▲), 600 m^2/g (■); group surface concentration: 2.28 (×), 3.40 (●), 2.91 (▲), 1.64 (■) group/nm^2.

are given in Table 1. It can be seen that fixation of acidic groups on silica does not affect acid-base characteristics. For the basic surface groups protonation constants decrease; this effect lowers as the basicity of the compound being anchored drops; for fixed aniline protonation constants are equal in solution and on surface (lgK = 4.52).

Calculation of stepwise apparent protonation constants for diamine modified silicas shows (Ref. 6) that despite that K_1 and K_2 decrease as α increases, the K_1/K_2 ratio remains constant, thus confirming the conclusion made in Section 3.4.

The decrease in basicity of surface functional groups as compared with that of homogeneous analog results from their interaction with residual surface silanol groups. This is illustrated on Scheme 1 for anchored mono- and diamines. It is shown that additional (blocking of residual silanol groups by trimethylsilyl groups.

TABLE 1. Effect of anchoring on the properties of acidic and basic groups (I = 1) on silica surface

Functional group	Modified silica	pK_a	Homogeneous analog	pK_a a)	pK_a b)
$-COOH$	$(SiO_2)-(CH_2)_2COOH$	$pK_a=4.78$	C_2H_5COOH	$pK_a=4.87$	$pK_a=4.82$
	$(SiO_2)-(CH_2)_{10}COOH$	$pK_a=5.05$	$C_{10}H_{21}COOH$	$pK_a=4.91$	-
$-N(CH_2COOH)_2$	$(SiO_2)-(CH_2)_3N(CH_2COOH)_2$	$pK_{a1}=2.12$	$CH_3N(CH_2COOH)_2$	$pK_{a1}=2.15$	-
			$HN(CH_2COOH)_2$	$pK_{a1}=2.54$	$pK_{a1}=2.51$
$-P(O)(OH)_2$	$(SiO_2)-(CH_2)_2-C_6H_4-P(O)(OH)_2$	$pK_{a1}^{c)}=1.9$	$CH_3-C_6H_4-P(O)(OH)_2$	$pK_{a1}=1.83$	-
		lgK		lgK	
$-NH_2$	$(SiO_2)-(CH_2)_3NH_2$	lgK=7.60	$C_4H_9NH_2$	lgK=10.61	lgK=8.35
$-NHCH_2CH_2NH_2$	$(SiO_2)-(CH_2)_3NHCH_2CH_2NH_2$	$lgK_1=7.7$	$NH_2CH_2CH_2NH_2$	$lgK_1=10.1$	$lgK_1=8.5$
		$lgK_2=6.2$		$lgK_2=6.6$	$lgK_2=6.6$
$-NH(CH_2)_6NH_2$	$(SiO_2)-(CH_2)_3NH(CH_2)_6NH_2$	$lgK_1=7.7$	$NH_2(CH_2)_6NH_2$	$lgK_1=10.6$	$lgK_1=8.3$
		$lgK_2=6.3$		$lgK_2=9.9$	$lgK_2=6.6$
2-pyridyl	$(SiO_2)-(CH_2)_4$-2-pyridyl	lgK =6.8	CH_3-2-pyridyl	lg K=5.97	lgK =5.3

a) For anchored ligands

b) For soluble analogs + SiO_2

c) Extrapolated to $\alpha = 0$.

lowers the effect of basicity change after fixation.

Carboxy groups do not tend to produce hydrogen bonds with silanols in aqueous systems. Nevertheless drying carboxy-containing silicas in vacuum at 140 °C leads to a decrease in pK_a of grafted groups by 0.4 as compared with that of the air-dried sample; for regeneration of acid-base characteristics modified silica must be kept in an aqueous solution with the pH > 6 for 1-6 h. The results obtained are caused by equilibrium shown on Scheme 1.

Scheme 1

```
                          NH2
                          <
      __                __NH         __        H2O, pH>6       COOH
     /  \ NH2          /   :        /  \     <=========>        /
    /    :            /    :       /    \     vacuum,           \
  -Si    H         -Si     H.   -Si     C=O   140 °C          -Si     H
   /\               /\           / \      \                    /\
  O    O          O    O        O    O                        O    O
   \   |           \   |         \   |                         \   |
////Si////Si//// ////Si////Si//// ////Si//////Si////          ////Si////Si////
```

5. COMPLEXATION WITH TRANSITION METAL IONS

This section concerns the regularities of complexation on a modified silica surface, methods for determination of composition of complexes formed and the results of application of the calculation methods described earlier in Section 2.

Silica, chemically modified by iminodiacetic acid $(SiO_2)-(CH_2)_3N(CH_2COOH)_2$, will be considered first. Catalysts on its basis are active in oxidation and epoxidation of unsaturated organic compounds (Ref. 10).

Transition metal ions and grafted molecules of iminodiacetic acid interact to give complexes of composition 1:1, so the results resemble the data for similar organic polymers. Unlike the latter ones, direct calculation of stability constants is impossible for silica carriers because we cannot find the value of pK_{a2} for grafted groups in the presence of low-acidic silanol groups of the carrier. However, it is possible to calculate the value of function F_2 of grafted iminodiacetic acid for various solution acidities using the conclusion about invariable acid-base properties of acidic groups fixed on the silica surface and the approach described in Section 3.3. Then with the values obtained we can fiid stability constants for complexes formed. The calculated data in comparison with stability constants for imino-diacetic complexes in solution are given in Table 2.

TABLE 2. Stability constants of Cu(II), La(III) and Eu(III) complexes with iminodiacetic acid on the silica surface and in aqueous solution (I = 1) (Refs. 8,11)

lg β \ ion	Cu(II)	La(III)	Eu(III)
Surface	10.5±0.1	5.9±0.1	7.3±0.1
Solution	10.55	5.9	7.26

According to the electrostatic model of surface groups complexes' interaction, as the solution ionic strength decreases the stability of iminodiacetic surface complexes increases and, at $I \rightarrow 0$ lg β, rises up to 11.9 for Cu(II), 10.7 for Zn(II) and Eu(III), 10.2 for La(III).

Now we consider the regularities of surface complexing for carboxy-containing silica $SiO_2)-(CH_2)_2COOH$; Cu(II), Zn(II), Cr(III), La(III) are used as complex-forming ions.

When surface coverage with metal ions is low, it is possible to neglect their concentration as compared with the total ligand concentration and assume the formation of complexes having a single $\overline{MR}_n$ composition with a maximum ligand : metal ratio. Then for FPC and SPC models we get the following equations in logarithmic forms, respectively:

$$\lg D = \lg \beta_n + \lg(C_R^S/n) - \lg F_n$$

$$\lg D = \lg(\beta_n \varkappa_n) + \lg(C_R^S/n) - n\lg F_1.$$

From the dependence of ion bonding degree upon medium acidity we have determined the complex composition (n) and the values of stability constants. The results obtained are summarized in Table 3. In the table we give also the literature data on acetate complexes in aqueous solutions. It can be seen from Table 3 that when calculating according to the models above-described good correlation of the calculated stability constants with the literature data for homogeneous analogs is observed. Correlation proved to be the best for the data calculated according to the SPC model on assumption that $\varkappa_2 = \varkappa_3 = 1$. This confirms our conclusion about similarity in properties of free acidic groups and those anchored on silica.

Now we consider complexation over a wide range of surface coverage with metal ions. The dependence of the quantity of copper(II) ions bonded to the surface on their equilibrium concentration in solution is plotted in Fig. 2. The isotherms calculated in the framework of SPC and FPC models using the values of β_1 and β_2 for copper(II) acetate complexes are shown in this figure as well. A quality difference between the experimental and calculated data are clearly seen. In particular, a break on the experimental curve at a coverage of 10-15% is always absent on calculated isotherms whatever values of β_1 and β_2 were given for calculation. As copper(II) concentration further increases, the experimental sorption value increases less than that of the calculated one. These facts permit us to assume that conditions of metal complex formation on various surface regions differs. As a first approximation we can assume that on some part of the surface (θ) there are surface ligands

TABLE 3. Composition and stability of complexes with carboxylated ligands anchored on silica at low coverage

Metal ion	n	I	FPC model		SPC model		lg β_n (acetate complexes in solution
			lg β_n	r	lg β_n	r	
		1	2.54±0.08	0.886	2.63±0.08	0.909	2.71
Cu(II)	2	0.1	2.77±0.06	0.924	3.03±0.10	0.848	3.09
		0.01	3.35±0.10	0.860	3.63±0.10	0.867	3.63
Zn(II)	2	1	1.74±0.05	0.929	1.86±0.04	0.946	1.90
La(III)	3	1	3.76±0.12	0.850	4.04±0.10	0.861	4.12
Cr(III)	3	1	10.4±0.2	-	10.7 ±0.2	-	9.60

sterically available for formation of complexes of $\overline{MR}_2$ composition, whereas on the other surface part (1- θ) ligands are "isolated" and can form only complexes of MR_1 composition. Polydentate islets can be formed from those surface groups which are located in the region of micropores, structural defects always being on amorphous silica surface. When concentration of the anchored compounds, C_R^S, is constant, a simultaneous coordination of a few functional groups by metal ion will be higher in the region of microdefects.

We have succeeded in achieving the most adequate description of the experiment both for Cu(II) and for Zn(II) and Cr(III) using the SPC model for "islets" description with $\theta = 0.25$ and $\varkappa_2$ for this surface part equal to 4. For calculation of $\varkappa_3$ in the case of Cr(III) the approximation $\varkappa_3 = \varkappa_2(\varkappa_2 - 1)$ proved to be good. By way of example the calculated curve for Cu(II) is plotted in Fig. 2.

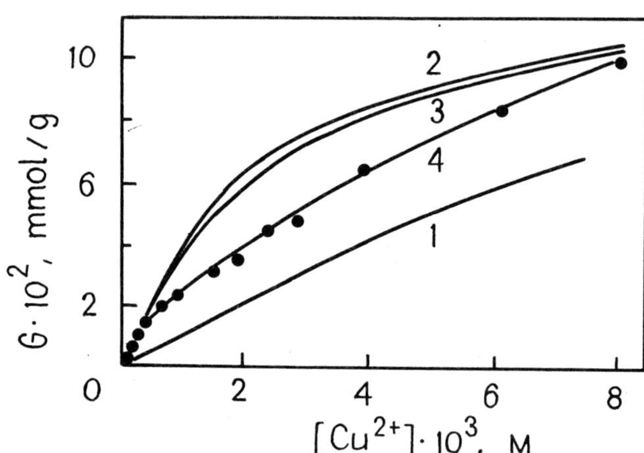

Figure 2. Isotherms of Cu(II) ion sorption of carboxy-containing silica. 1 - formation of complexes of 1:1 composition; 2 - SPC model; 3 - FPC model; 4 - SPC model considering islet distribution; • - experimental points, I - 1, pH = 6.0.

TABLE 4. Composition of complexes and the ratio of complexes with different composition (in mmol/g) for Cr(III) supported on carboxy-containing silica

Complex quantity				Ratio
ΣMR_i	MR_1	MR_2	MR_3	$MR_1 : MR_2 : MR_3$
7.7	0.7	0.7	6.3	1 : 1 : 9
20.0	6.5	1.9	11.6	3.5 : 1 : 6
45.0	28.5	3.5	13.0	8 : 1 : 4
77	59	5	13	12 : 1 : 2.6
111	92	6	13	15 : 1 : 2
147	129	7	11	19 : 1 : 1.6

Not considering all other peculiarities of complex formation on carboxy containing silica surface, we note only two moments that are of interest to catalysis. Given in Table 4 are the absolute quantities of surface complexes of various compositions at various surface coverages and their relation for Cr(III) complexes. The change in relation between surface complexes of various compositions must result in the catalyst activity and selectivity variations as functions of metal coverage.

Cu(II) ion sorption isotherms at various values of solution ionic strength are given in Fig. 3. Ion fixation on the surface is usually carried out in an excess metal salt and, consequently, at a high ionic strength. Catalytic experiment is performed in the absence or at low electrolyte concentrations. Therefore, the catalysis process will result in substantial (up to 50%) leaching of supported ions in solution due to difference in catalyst production and application conditions. Neglect of concentration control of electrolyte, which is not involved in the process, will lead to low reproducibility of the results.

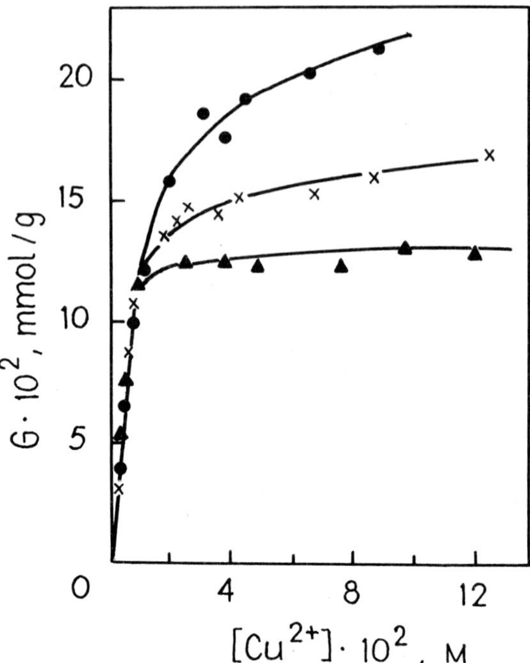

Figure 3. Isotherms of Cu(II) sorption on carboxy-containing silica for I = 0.05. (▲), 0.3 (×), 1.0 (●).

Other regularities of surface complexation are observed when using silicas with basic surface groups. As an example we shall consider the interaction of metal ions and aminosilica $(SiO_2)-(CH_2)_3NH_2$.

Bonding of metal ions by aminosilica is characterized by the absence of the stepwise complexation processes. Sorption measurements under conditions of excess metal ions give average stoichiometry of bonding, $M:NH_2=1:2$. A similar composition of forming surface complexes is obtained from linearization of the dependences of bonding degree on solution acidity at low surface coverages with metal ions (Ref. 8) according to the equations derived in Section 3.

The constancy of the complex composition within a

TABLE 5. ESR-spectra parameters for complexes of copper ions with aminosilica

$\bar{r}$, nm	$g_{\perp}$ ± 0.005	$g_{\parallel}$ ± 0.005	g_{av} ± 0.01	A, Oe ± 4	Linewidth ΔH, Oe, ± 2
1.4	2.052	2.317	2.14	150	89
1.7	2.052	2.317	2.14	150	82
2.0	2.056	2.324	2.14	150	74
2.8	2.054	2.321	2.14	152	64
3.5	2.055	2.321	2.14	151	56
4.0	2.054	-	-	-	54
5.5	2.054	-	-	-	56
7.5	2.056	-	-	-	53

wide range of coverages is confirmed by the data obtained by physical methods such as ESR-, electron and photoacustic spectroscopies. Thus, ESR spectra of copper(II) complexes on aminosilica surface have a satisfactorily resolved hyperfine structure from copper nuclei. The calculated spectra parameters are given in Table 5. The concentration of copper ions was expressed by $\bar{r}$, an average distance between copper ions assumed to be uniformly distributed on the surface. As can be seen from Table 5, in the range of $\bar{r}$ = 1.4-3.5 nm, where it is possible to define $g_{\perp}$, $g_{\parallel}$, g_{av} and A with a sufficient accuracy, their values within the error limits remain constant; the value of $g_{\perp}$ remains constant up to maximum $\bar{r}$ = 7.5 nm. Calculated g_{av} = 2.14 $\pm$ 0.01 resembles the g = 2.14 $\pm$ 0.02 for $Cu(NH_3)_2(H_2O)_4^{2+}$ in aqueous solutions. In turn, the values of $g_{\perp}$ and $g_{\parallel}$ are within the range for crystalline complexes, $Cu(NH_3)_2X_2$.

ESR method also permits evaluation of the character of the surface distribution of the substance on the basis of the dipole-dipole interaction value. Fig. 4 displays the dependences of (ESR-spectra) line broadening on $\bar{r}$ for copper(II) complexes with monoamine and polyamine $NH_2(CH_2-CH_2NH)_4H$ anchored on silica as well as for ammonia complexes of copper(II) in solution under identical conditions. It was assumed for $\bar{r}$ = 7-8 nm that the dipole-dipole interaction has almost no effect on the linewidth. In this case, the broadening in oersteads was equal to: $\Delta\Delta H = \Delta H_{\bar{r}} - \Delta H_{\bar{r}=7.5\ nm}$. As can be seen from Fig. 4, when $\bar{r} < 3$ nm the dipole-dipole interaction

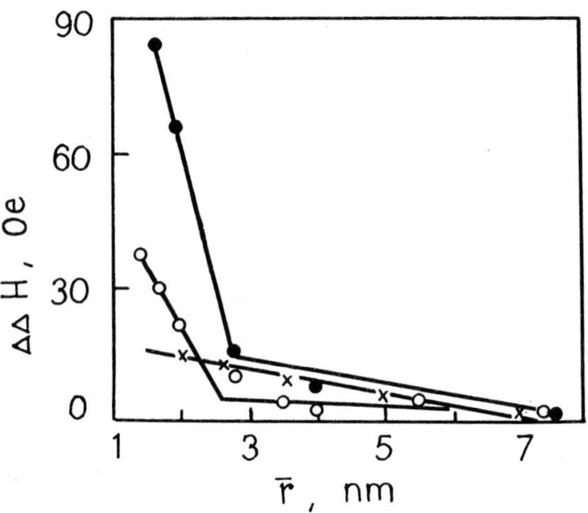

Figure 4. ESR spectrum linewidth dependence on the average distance between copper ions on silica surface with mono- (•) and polyamine (o) groups and in ammonia solution (x).

intensity sharply increases for both modified silicas. The results obtained make it possible to assume that when $\bar{r} < 3$ nm the character of ion distribution on the surface changes.

CONCLUSIONS

The represented material allows us to draw following

general conclusions.

1. For characterization of heterogeneous metal complex catalysts it is insufficient to present data on total surface concentrations of surface groups and quantities of supported metal ions. The data obtained show that the amount and relation between complexes with various compositions on the surface and, consequently, catalyst activity and selectivity are determined also by a sample biography. Maintenance of constant surface coverage by metal ions during catalyst synthesis does not guarantee the reproducibility of the catalytic activity. Neglect of inessential, as might first appear, differences in conditions of catalyst synthesis and application, in particular, in concentrations of an electrolyte which is not involved in the process, will lead to substantial leaching of a supported complex into the solution and, consequently, to a low catalyst stability.

2. In spite of a number of significant differences between complex formation in homogeneous and heterogeneous systems, it is possible to develop such mathematical description of complex formation of solid carrier surface that provides a quantitative comparison of stabilities of analogous complexes in solution and on the carrier surface. Experimentally obtained regularities of the complex formation on a modified silica surface show that the presence of a sufficiently long (2-3 methylene groups) hydrocarbon chain between the carrier and anchored functional groups prevents the electronic influence of the carrier. The data obtained leads us to conclude that it is possible to predict the stability of complexes formed on the modified silica surface on the basis of data valid for formation of complexes in solutions.

REFERENCES

1. G.V. Lisichkin and A.Ya. Yuffa, in Heterogeneous Metal Complex Catalysts (in Russian), Khimia, Moskva (1981).
2. Yu.I. Yermakov, V.A. Zakharov and B.N. Kuznetsov, in Fixed Complexes on Oxide Supports in Catalysis (in Russian), Nauka, Novosibirsk (1980).
3. G.V. Kudryavtsev, Zh. Fiz. Khim., 56, 1996-1999 (1982).
4. A.A. Lopatkin, in Theoretical Principles of Physical Adsorption (in Russian), MGU, Moscow (1983).
5. G.V. Kudryavtsev, Zh. Fiz. Khim., 57, 503-505 (1983).
6. G.V. Kudryavtsev, G.V. Lisichkin and V.M. Ivanov, Zh. Vses. Khim. Obschestva (USSR), 26, 345-347 (1981).
7. G.V. Kudryavtsev and G.V. Lisichkin, Zh. Fiz. Khim., 55, 1352-1353 (1981).
8. G.V. Kudryavtsev, G.V. Lisichkin and V.M. Ivanov, Zh. Anal. Khim., 38, 22-32 (1983).
9. G.V. Kudryavtsev and G.V. Lisichkin, Adsorb. i Adsorbenty, No 12, 33-39 (1984).
10. V.V. Berentsveig, Chan Bik Nga, T.V. Barinova and G.V. Lisichkin, Kinet. Katal., 25, 1477-1480 (1984).
11. R.D. Voronina, G.V. Kudryavtsev, G.V. Lisichkin and V.K. Runov, Zh. Fiz. Khim., 59, 505-507 (1985).

DIMERIZATION OF ETHYLENE AND PROPYLENE WITH TAILOR MADE NICKEL COMPLEXES SUPPORTED ON SILICA

F.X. Cai, C. Lepetit, M; Kermarec, D. Olivier
Laboratoire de Réactivité de Surface et Structure, UA 1106, CNRS,
4, Place Jussieu, Tour 54-55, 2ème étage, 75252 Paris Cédex 05

ABSTRACT : Supported $Ni^{+}Ln$ complexes (L = PEt_3, PMe_3) are prepared and characterized : $Ni^{+}L_4(O_s^{2-})$, $Ni^{+}L_2(O_s^{2-})_2$, $Ni^{+}L(O_s^{2-})_2$. The pentacoordinated species are inactive in C_2H_4 and C_3H_6 dimerization. The square planar $Ni^{+}L_2(O_s^{2-})_2$ and the unsaturated $Ni^{+}L(O_s^{2-})_2$ species are both active and selective for C_2H_4 dimerization. $Ni^{+}L(O_s^{2-})_2$ are acactive and selective for C_3H_6 dimerization. The presence of phosphine ligands increases the electron density on supported Ni^{+} ions and thus the lability of the coordination sphere. This electronic effect allows an easier desorption of the reaction products ; the consequences are : i) a strong increase of the selectivity for the dimerization and of the stability of the catalytic sites poisonned by heavy oligomeres in the phosphine free catalysts. ii) a kinetic control of the selectivity for the dimerization of C_2H_4 into 1-butene which reaches 100 % with $Ni^{+}(PEt_3)_2(O_s^{2-})_2$ catalysts.

1. INTRODUCTION

Dimerization, oligomerization and similar reactions of olefins have been reported to be catalyzed by systems containing nickel complexes (1,2). Most of the studies have been concerned with homogeneous catalysts, but recently articles have appeared describing heterogeneous systems (3-6).

Our approach differs from that generally used in supported homogeneous catalysis : from an isolated precursor Ni^{2+} cation bonded only to the surface oxygen atoms acting as ligands, we fashion the active entity and its coordination sphere. This catalyst tailoring implies

1 - a good dispersion of the precursor cations
2 - their insaturation obtained by an adapted thermal pretreatment
3 - their reduction to the expected oxidation state using mild

conditions to eliminate migration phenomena

4 - The introduction in the coordination sphere of a given number of Lewis bases such as phosphine which have influence on both the activity and selectivity of the catalytic system.

The resolution of the two first points has been reported in an earlier work (7-9) ; this report is concerned with the study of the two last ones. The preparation and the characterization of SiO_2 supported nickel complexes Ni^+L_n(L = alkyl phosphine ; n = 1 to 4) are described. In order to investigate the effect of the ligand phosphine on the selectivity of the heterogeneous catalytic process we have decided to examine the reaction of supported Ni^+ ions with $P(CH_3)_3$ and $P(C_2H_5)_3$. The reactivities of the catalysts are compared in the dimerization of C_2H_4 and C_3H_6.

2. EXPERIMENTAL

The studies samples are nickel exchanged silica prepared using the competitive cation exchanged method (7-9) the percentage of nickel by weight lies in the range 1.7 to 5.1 %. The reduction of supported Ni^{2+} ion into Ni^+ is carried out using alkyl phosphine (L) as reductor (10) at 300 K, either in gaseous phase ($P_L \sim$ 50-100 Torr) or in solution with the sample in suspension in toluene ($|L|/|Ni| \sim 6$).

The reduction degree of Ni^{2+} into Ni^+ is calculated from :

i) the Ni^+ spin concentration using Cu^{2+} as standard

ii) the decrease of the intensity of the electronic transitions of the Ni^{2+} entities.

The reduction degree is found to be nearly equal to 100 %.

3. RESULTS AND DISCUSSION

3.1. Preparation and spectroscopic characterization of supported Ni^+Ln complexes (L = PEt_3, PMe_3)

The reflectance spectra and the EPR spectra of the reduced samples are followed first in function of phosphine pressure at room temperature, then after outgassing at increasing temperature.

The symmetry of the supported Ni^+ complexes is determined from the comparison of the reflectance spectra with those observed for Ni^+ complexes in coordination chemistry (11-14). The discussion of the EPR spectra is based on a self consistent analysis of :

i) the g tensor components wich give the symmetry of the environment of the paramagnetic $3d^9$ Ni^+ ions

ii) the relative order of the g tensor components which lead to the probable ground state of the Ni^+ ions

iii) the superhyperfine (shf) structure due to $^{31}P(I = 1/2)$ from which the number of phosphine ligands can be deduced.

TABLE 1. Spectroscopic data for the supported Ni^+-Ln complexes $L = PMe_3$, PEt_3

Symmetry	L	Experimental Conditions	Diffuse	Reflectance Spectra λ(nm)	
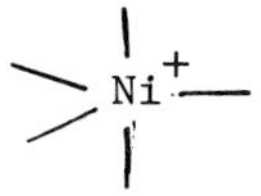 Ni^+ (trigonal bipyramid sketch)	PMe_3	p > 5 Torr	1175	370	275
	PEt_3	p > 10 Torr	1400	370	282
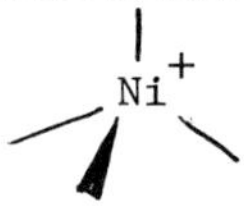 Ni^+ (tetrahedral sketch)	PMe_3	vacuum RT	850		
	PEt_3	vacuum RT	870		
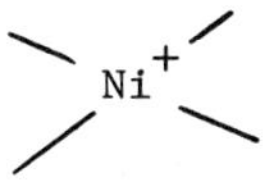 Ni^+ (square planar sketch)	PMe_3	vacuum 200 °C		450	327
	PEt_3	vacuum 100 °C		500	350
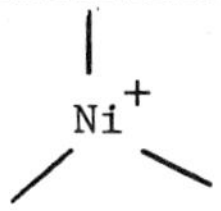 Ni^+ (trigonal sketch)	PMe_3	vacuum 350 °C		400	
	PEt_3	vacuum 250 °C		400	

Table I and II summarize the results obtained from the analysis of respectively the reflectance and EPR spectra.

TABLE II. ESR parameters for the supported Ni^+Ln species
($L = PMe_3$, PEt_3)

Complexes	L	Ground state	g tensor $g_{//}$	$g_{\perp}$	A tensor (gauss) $A_{//}$	$A_{\perp}$
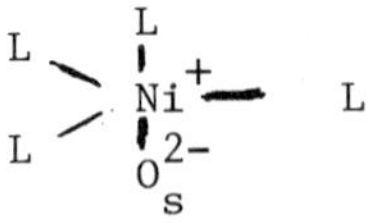 L, L, L, L, Ni^+, O_s^{2-}	PMe_3	$d_{x^2-y^2}$	2.21	2.052	40	50
	PEt_3		2.27	2.065	40	50
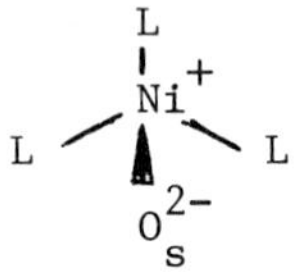 L, L, L, Ni^+, O_s^{2-}	PMe_3	Species non obtained pure				
	PEt_3					
					$A_{\perp}1$	$A_{\perp}2$
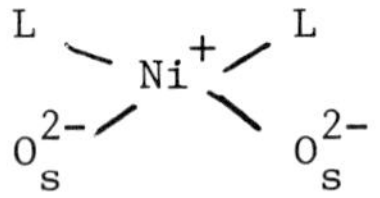 L, L, Ni^+, O_s^{2-}, O_s^{2-}	PMe_3	$d_{x^2-y^2}$	2.26	2.077	60	75
	PEt_3		2.32	2.083	62	85
						$A_{\perp}$
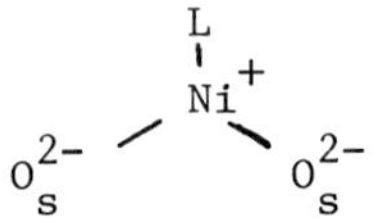 L, Ni^+, O_s^{2-}, O_s^{2-}	PMe_3	d_{z^2}	2.015	2.120		55
	PEt_3		2.019	2.124		50

From this study it is shown that it is possible to introduce selectively 4, 2 or 1 phosphine molecules in the coordination sphere of supported Ni^+ ions. The saturated pentacoordinated species $Ni^+(O_s^{2-})L_4$ (O_s^{2-} is a surface oxygene atom) are stable under a phosphine pressure higher than 10 Torr.

The square planar complexes $Ni^+(O_s^{2-})_2L_2$ are prepared by desorption at 100 °C for $L = PEt_3$ or 200 °C for $L = PMe_3$, whereas the unsaturated tricoordinated species $Ni^+(O_s^{2-})_2L$ are obtained after prolongated outgassing at 200 °C for $L = PEt_3$ and 300 °C for $L = PMe_3$. A tetrahedral species, probably $Ni^+(O_s^{2-})L_3$ appears stable under vacuum at room temperature.

3.2. Electron donor properties of phosphorus ligands

The electron donor properties of phosphorus ligands have been investigated using CO as a probe molecule : 5 Torr CO pressure were successively adsorbed onto a mixture of :

i) $Ni^+(PEt_3)_3(O_s^{2-})$ and $Ni^+(PEt_3)_2(O_s^{2-})_2$ entities

ii) $Ni^+(PEt_3)_2(O_s^{2-})_2$ species

iii) $Ni^+(PEt_3)\ (O_s^{2-})_2$ species

Only one ν_{CO} absorption is observed. On figure 1, ν_{CO} frequency is reported versus the number of phosphine ligands : successive additions of phosphine ligand cause ν_{CO} to change by a nearly constant

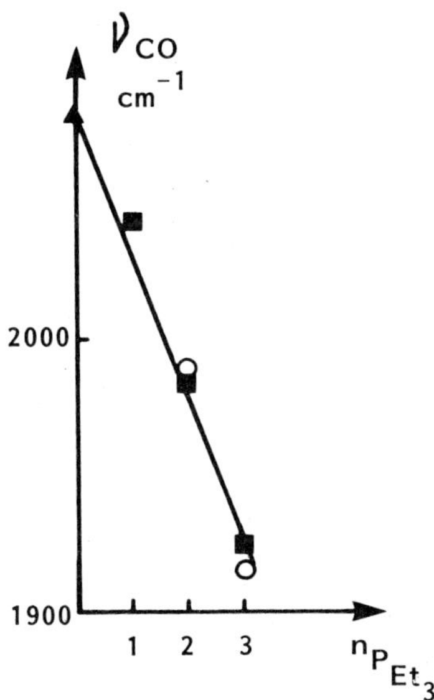

Figure 1. Infrared ν_{CO} frequencies of supported $Ni^+(CO)(PEt_3)_n$ complexes.

Increment. It is possible to assign to each number of phosphine ligand a contribution to the ν_{CO} streching frequency, but this correlation should be only evaluable for phosphines of the same basicity with the same substituents on the phosphorus atom.

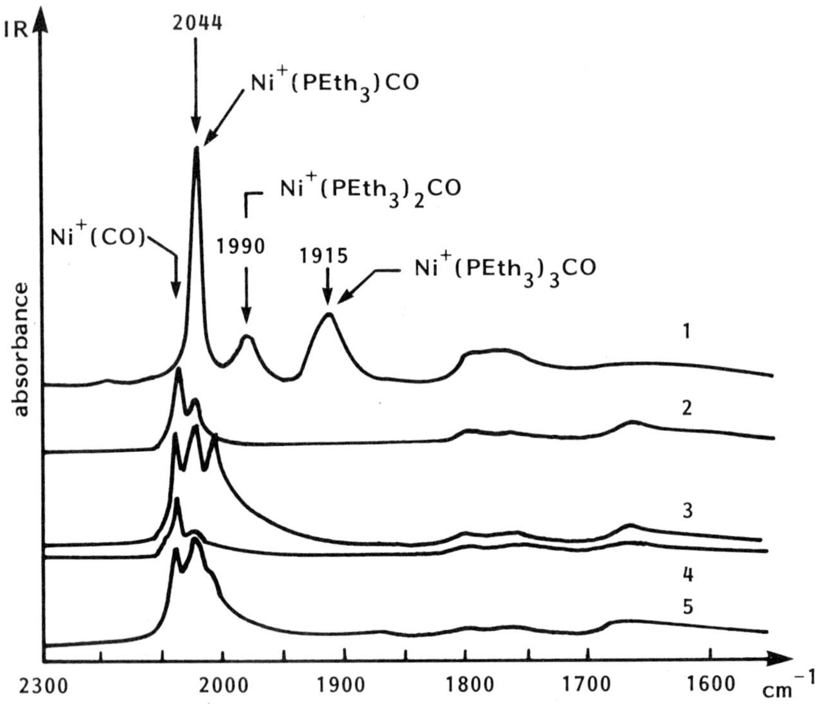

Figure 2. Infrared spectra of Ni/SiO_2 sample after photoassisted reduction under CO and adsorption of 4 Torr PEt_3 pressure
1 : outgassed at 25 °C
2 : outgassed at 150 °C 1 hour
3 : 2 + adsorption of 5 Torr CO pressure
4 : 3 + outgassed at 200 °C 1 hour
5 : 4 + adsorption of 5 Torr CO pressure

The photoassisted reduction under CO of supported Ni^{2+} ions has been reported in another work (15). The adsorption of 4 Torr of PEt_3 onto the photoreduced sample in the presence of a residual CO pressure of about 20 Torr, followed by outgassing at 25 °C, leads to a mixture of Ni^+ species containing one (2044 cm^{-1}), two (1990 cm^{-1}) and three (1915 cm^{-1}) phosphine ligands (figure 2). In these conditions the formation of metal is not observed as shown by the absence of a ferromagnetic signal. Bands in the range 1850-1750 cm^{-1} have been assigned (15) to superficial carbonate species. The bands ν_{CO} characteristic of $Ni^+(PEt_3)_2CO$ and $Ni^+(PEt_3)_3$ CO are removed by evacuation at 150 °C whereas the intensity of the band assigned to $Ni^+(PEt_3)CO$ only decreases. The ν_{CO} band at 2080 cm^{-1} is characteristic of $Ni^+(CO)(O_s^{2-})_3$ species (16). A subsequent adsorption of 5 Torr CO shows that it remains only one or two phosphine ligands on Ni^+ surface species. Outgassing at 200 °C removes all the ν_{CO} bands except that of $Ni^+(O_s^{2-})_3$. After a new adsorption of 5 Torr CO it appears that the principal surface species are $Ni^+(PEt_3)$ CO and $Ni^+(CO)(O_s^{2-})_3$. From these results we conclude that i) the increase of the number of phosphine ligands in the coordination sphere of Ni^+ increases the electronic charge of the metal thus the back donation of electron from Ni^+ to CO. ii) the increases of the number of phosphine ligands allows a easier desorption of the CO molecules.

3.3. Dimerization of C_2H_4

The reaction was first followed in a batch reactor under 20 bars C_2H_4 pressure at 300 K, and the reactivity of the supported Ni^{2+} precursors was compared to that of the reduced samples Ni^+ and $Ni^+(PEt_3)_2$. An induction time was observed in case of the unreduced catalyst corresponding to the reduction of the Ni^{2+} species into Ni^+. The reduction of Ni^{2+} into Ni^+ was evidenced by EPR after reaction : this result confirms that Ni^+ ions are the active entities (8,17).

Figure 3 summarizes the experimental results for C_2H_4 dimerization followed in a flow reactor at 313 K under 20 bars C_2H_4 pressure at a conversion degree of about 5 %.

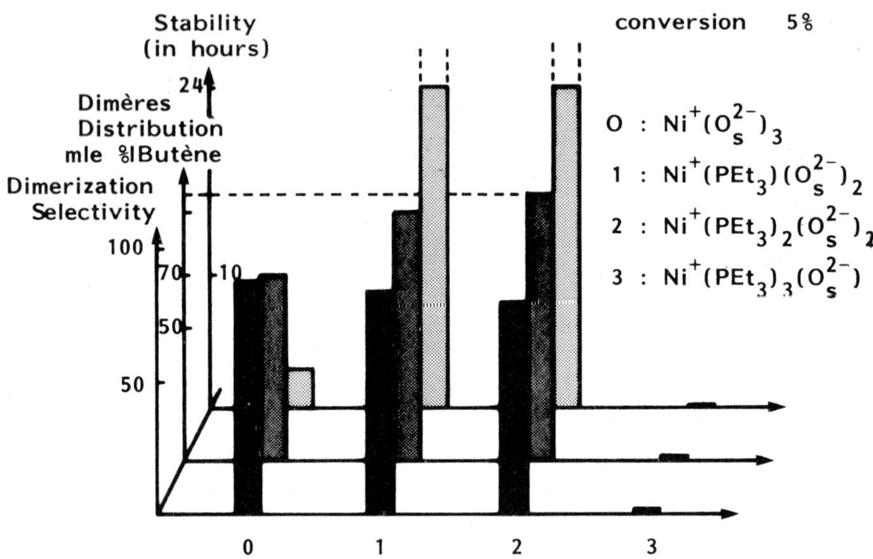

Figure3. Dimerization of C_2H_4 in a flow reactor, under 20 bars C_2H_4 pressure at 40 °C, at a conversion degree of $\sim$ 5 %.

The activity of the phosphine modified Ni^+ species remains stable after more than 24 hours while in the absence of phosphine a loss of about 86 % of the activity is observed after about 3 hours.

In an earlier work (8) we have shown that :

i) the higher coordination number of Ni^+ is five

ii) the case of two olefin molecules bonded to the same Ni^+ is favorable for the dimerization to take place

iii) in connection with this it is observed that the saturated $Ni^+(PEt_3)_4$ (O_s^{2-}) (and also $Ni^+(PEt_3)_3(O_s^{2-})$ species) are inactive.

The selectivity for dimerization initially good for all the catalysts decreases rapidly in the absence of phosphine. The percentage of hexenes and higher oligomeres increases with the decrease of the reaction temperature.

The composition of the dimere fraction is strongly dependent on the presence of phosphine : using supported $Ni^+(PEt_3)_2(O_s^{2-})_2$ species a <u>selectivity of 100 % for 1-butene</u> is achieved whereas in the case of the phosphine free catalyst the percentage of 1-butene in the dimere fraction never exceeds 76 %.

The reactivity of the phosphine modified catalyst was followed by EPR : after adsorption of C_2H_4 at 298 K under 600 Torr the reaction was allowed to proceed at 298 K for 24 hours and the EPR spectra recorded before and after outgassing (the EPR spectra recorded during the reaction will be discussed elsewhere). The conversion of C_2H_4 was found of about 100 %. The EPR signal of $Ni^+(PEt_3)_2(O_s^{2-})_2$ was the one and only signal observed after and before outgassing and the intensity is equal to that observed before reaction with C_2H_4. This result indicate that the reaction products are rapidly desorbed whereas with the phosphine free catalysts adsorbed butenes and hexenes have been evidenced both by EPR (8) and IR spectroscopies.

The IR results have shown that one of the function of the phosphine Lewis base is to increase the charge on the nickel thereby decreasing the affinity of the Ni^+ cation for soft donors such as CO and olefin molecules : it thus appears that the effect of the phosphine ligands is to increase the lability of the coordination sphere. The nature and the distribution of the products of olefin dimerization depends on the relative rate constants of the insertion step ki versus the displacement step kd (2). If $k_d > k_i$ the olefine dimerization will predominate. It has been assumed that the displacement reaction takes place via a pentacoordinated intermediate having two olefin molecules bonded to the metal atom. It would be expected on the basis of this mechanism that the rapid desorption of the butenes from the nickel atoms would be in favor of the formation of such coordinated species, whereby the insertion step will be suppressed because this step is more associated with an increase of the steric hindrance.

A high rate of isomerization of olefin dimeres has generally been observed using homogeneous catalysts modified by PMe_3 and/or PEt_3. Figure 3 shows that a selectivity of 100 % for 1-butene is achieved in the presence of $Ni^+(PEt_3)_2(O_s^{2-})_2$: the selectivity of the dimerization reaction thus appears kinetically controlled, the rate of desorption of 1-butene being higher than the rate of isomerization.

3.4. Dimerization of C_3H_6

The reaction was followed in a batch reactor under about 600 Torr of C_3H_6 at 300 K. Figure 4 summarizes the experimental results.

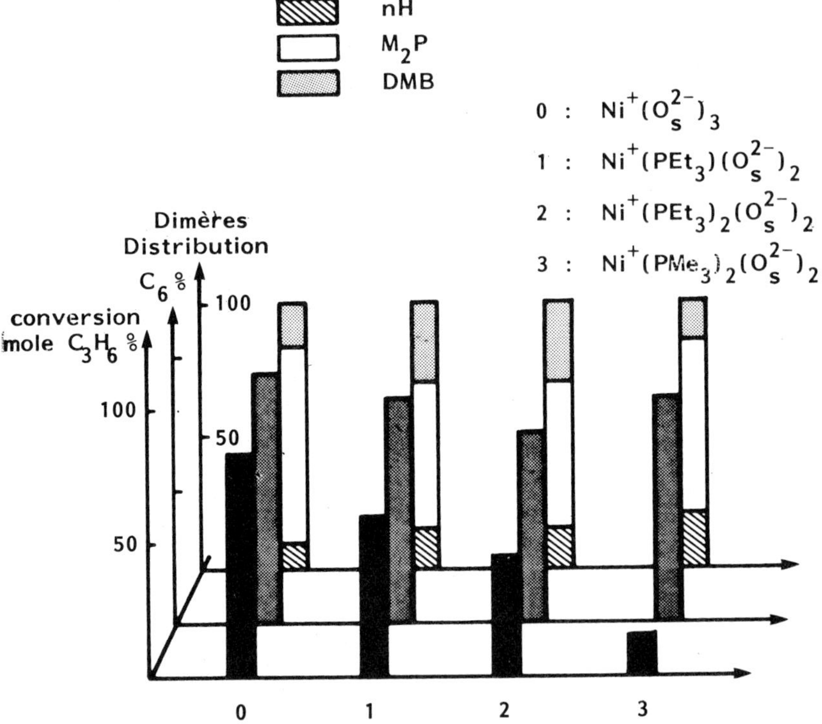

Figure 4. Dimerization of C_3H_6 in a batch reactor, under 600 Torr C_3H_6 pressure at 25 °C.

The conversion degree decreases when the number of phosphine ligands increases : the conversion of propylene appears more sensible to the steric hindrance than the conversion of ethylene. Furthermore when the steric hindrance increases ($Ni^+(PEt_3)_2(O_s^{2-})_2$) the selectivity for dimerization decreases from 95 to 70 % ; about 30 % of a mixture of butenes and pentenes are observed, and no higher oligomere is observed in the presence of phosphine. The presence of light oligomers may be explained by the formation of a metallocyclobutane derivative from the addition of one propene molecule to the Ni^+ complexes following the scheme proposed by Green and al (18). This mechanism may be relevant as an initiation mechanism for olefin methathesis reactions.

As in case of C_2H_4 dimerization, the presence of phosphines strongly increases the catalyst stability.

REFERENCES

1. P.W. Jolly and G. Wilke, The organic chemistry of nickel, VII, Acad. Press., NY (1975), p. 1.
2. B. Bogdanovic, Advances in organometallic chemistry, 17, Acad. Press., NY, (1979), p. 105.
3. U.A. Kabanov, M.A. Martynova, S.K. Pluzkov, V.I. Smetanyuk, R.V. Chedya, Kinet. Kat., 20, 1012, (1979).
4. S.B. Echmaev, I.N. Ivleva, N.M. Braraya, A.D. Pomogaïlo, Yu. G. Borodko, Kinet. Kat., 21, 1530, (1980).
5. W. Shupinski, S. Malinovski
   a) J. Org. Metal. Chem., 99, 465, (1975).
   b) J. Org. Metal. Chem., 117, 183, (1976).
6. Yu L. Yermakov, B.N. Kuznetzov, Yu P. Grabowski, A.M. Lazutkin, V.A. Zakharov, A.I. Lazutkina, J. Mol. Catal., 1, 93, (1975).
7. L. Bonneviot, D. Olivier, M. Che, J. Chem. Soc. Chem. Com., 952, (1982).
8. L. Bonneviot, D. Olivier, M. Che, J. Mol. Cat., 21, 415, (1983).
9. D. Olivier, L. Bonneviot, F.X. Caï, M. Che, P. Gihr, M. Kermarec, C. Lepetit, B. Morin, Bull. Soc. Chim., 3, 370, (1985).
10. D. Olivier, F.X. Caï, M. Kermarec, C. Lepetit, French Patents, D 11288/230422 (1985).
11. L. Sacconi, A. Orlandini, S. Midollini, Inorg. Chem., 13, 2850, (1974).
12. D. Dapporto, G. Fallani, L. Sacconi, Inorg. Chem., 13, 2847, (1974).
13. A. Gleizes, A. Kerkeni, M. Dartiguenave, Y. Dartiguenave, H.F. Klein, Inorg. Chem., 20, 2372, (1981).

14. M. Arista, C.F. Nobile, A. Sacoo, Inorg. Chem. Acta, 12, 167, (1975).
15. L. Bonneviot, F.X. Caï, M. Che, M. Kermarec, O. Legendre, C. Lepetit and D. Olivier, J. Phys. Chem., (submitted).
16. M. Kermarec, D. Delafosse, M. Che, J. Chem. Soc. Commun., 412, (1983).
17. S. Sibille, J. Coulombeix, J. Perichon, J.M. Fuchs, A. Mortreux and F. Petit, J. Mol. Catal., 32, 239, (1985).
18. M. Ephretikine, M.L.H. Green, R.E. Mackenzie, J.C.S. Chem. Comm., 619, (1976).

VAPOR SYNTHESIS OF HIGH-ACTIVITY ZIEGLER-NATTA CATALYSTS - STEREOREGULARITY

S. D. Ittel[∓], R. Mulhaupt, A. P. Shreve, and U. Klabunde
Central Research & Development Department*
E. I. du Pont de Nemours & Company
Experimental Station
Wilmington, DE 19898 U.S.A.

ABSTRACT - High activity, $MgCl_2$-supported olefin polymerization catalysts can be prepared by a versatile new technique which employs monomeric, gas-phase, metal halide vapors. As an example of this new technique, condensation of $MgCl_2$ into a heptane/$TiCl_4$ matrix gives a high-surface-area material which is an active ethylene or propylene polymerization catalyst when activated with aluminum alkyls. Catalyst activity is controlled by rates of deposition, the reactivity of the condensation matrix, and the presence of other metal halides. These catalysts are generally non-stereospecific catalysts for 1-olefin polymerization. High activity, high stereospecificity catalysts are prepared by addition of sterically hindered Lewis bases to both the solid catalyst component and to the aluminum alkyl activator. Equilibria involving the aluminum alkyl activator and the Lewis base modifiers have been studied in detail and a phenol adduct has been structurally characterized.

1. INTRODUCTION

In an attempt to combine the best features of homogeneous and heterogeneous catalysts, homogeneous systems are often "heterogenized" by attaching them to solid supports.[1] When potentially active transition metal sites are inactive because they are buried inside a catalyst particle, the activity is often increased by dispersing or supporting the active catalyst on a support. In both applications, the supports, be they inorganic or organic, are rarely, if ever, inert.

∓ Author to whom correspondence should be addressed

* Contribution No.: 3974

These observations are especially relevant to catalysts for the polymerization of olefins. Homogeneous ethylene polymerization catalysts are often less active than heterogeneous Ziegler-Natta systems[2] though they do serve as excellent models more amenable to physical studies.[3] Until recently,[4] homogeneous systems displayed no stereoregularity[5] in the polymerization of propylene and other 1-olefins. Current generation high-activity catalysts for ethylene polymerization are based upon supported systems. The Phillips catalysts are based on chromium compounds on silica.[6] Du Pont has developed a catalyst system involving tetra(neophyl)zirconium on alumina.[7] New generation Ziegler-Natta systems for slurry polymerization are based upon $TiCl_4$ supported on $MgCl_2$.[8] In each case, the choice of catalyst support is critical and the support has become an integral part of the catalyst.

Previous meetings of ISRHHC have included lectures on the use of vapor synthesis to prepare catalysts on inorganic supports.[9] These lectures did not include vapor synthesis of the supports. We have discussed the versatility of the metal halide vapor route to activate supports for $MgCl_2$-supported $TiCl_4$ ethylene polymerization catalysts.[10,11] We will now discuss the application of this technique to propylene polymerization. Our objective is to understand the fundamental structure/property relationships of the catalyst and to identify the origins of high catalyst activity and steric control. While stereospecificity is not a concern in ethylene polymerization, its study provides an additional tool for characterization of catalyst sites. The nature and uniformity of catalyst sites significantly influences molecular weight distributions and the incorporation of higher 1-olefins in ethylene copolymerization.

2. EXPERIMENTAL

Our equipment and a general description of our vapor synthesis technique have appeared elsewhere.[12] Application to olefin polymerization catalysis has also been discussed.[11]

Solutions of the various catalyst modifiers in hydrocarbon diluents were introduced by means of a peristaltic pump connected to a cannula inserted into the shower head of the evaporator. The preparation conditions and results of propylene polymerization reactions are listed in Table 1. In the initial polymerizations, the solid catalyst components were activated with triethyl aluminum. Later runs were activated with aluminum alkyl phenoxides prepared from sterically hindered phenols. Polymerizations were generally carried out at 30 psig propylene pressure in heptane solvent at 60-65°C for 2 hr. After deactivation with n-butanol, the polymer was precipitated by addition to methanol with vigorous stirring.

The aluminum alkyl phenoxides were prepared by mixing the appropriate compounds in pentane. NMR scale reactions of phenoxides with aluminum alkyls were carried out in benzene-d_6.

3. RESULTS

Evaporation of $MgCl_2$ leads primarily to monomeric $MgCl_2$ in the gas-phase with small quantities of higher oligomers.[13] Condensation of these species into a dilute, non-reactive, cryogenic matrix such as heptane leads to isolated $MgCl_2$ and higher oligomers. If the matrix is allowed to anneal or melt, the metal halide molecules oligomerize and eventually condense to bulk $MgCl_2$. The surface area of this material can be very high and x-ray diffraction indicates a high-defect structure and very small crystallite size. These are highly desirable properties for materials to be used as supports for high-activity Ziegler-Natta catalysts.[14] The ultimate nature of the $MgCl_2$ is determined by a number of factors including the concentration of the matrix, the rate of evaporation during matrix build-up, and the rate of matrix warm-up. Table 1 presents preparative details and catalytic activities for the vapor-synthesized catalysts.

TABLE 1

Cocondensation Matrix[b]	Catalyst Post-Treatment[b]	Activator[b]	Productivity gPP/gTi·H·Bar	% Heptane Insoluble
Heptane[a]	$TiCl_4$	$AlEt_3$	3300	41
Heptane/$TiCl_4$[a]	None	$AlEt_3$	2200	37
$TiCl_4$[a]	None	$AlEt_3$	3200	41
Toluene/$TiCl_4$[a]	Toluene	$AlEt_3$	5800	35
Heptane	EB/$TiCl_4$	$AlEt_3$	2000	51
Heptane	26DMP/$TiCl_4$	$AlEt_3$	2700	42
Heptane	IAE/$TiCl_4$	$AlEt_3$	3900	44
Heptane	TMEDA/$TiCl_4$	$AlEt_3$	580	49
Heptane/Et	$TiCl_4$	$AlEt_3$	6000	42
Heptane/26DMP	$TiCl_4$	$AlEt_3$	4000	51
Heptane/IAE	$TiCl_4$	$AlEt_3$	3000	59
Heptane/TMEDA	$TiCl_4$	$AlEt_3$	230	43
Heptane/BHT	$TiCl_4$	$AlEt_3$	3700	49
Heptane/BHT	$TiCl_4$	$AlEt_3$/0.5 EB	640	91
Heptane/BHT	$TiCl_4$	$AlEt_3$/0.5 ET	750	90
Heptane/BHT	$TiCl_4$	$AlEt_2$BHT	7800	49
Heptane/BHT	$TiCl_4$	$AlEt_2$BHT/$AlEtBHT_2$	10000	50
Heptane/BHT	$TiCl_4$	$AlEt_2$BHT/0.01 ET	16500	60
Heptane/BHT	$TiCl_4$	$AlEt_2$BHT/0.05 ET	12400	78
Heptane/BHT	$TiCl_4$	$AlEt_2$BHT/0.24 ET	5700	91
Heptane/BHT	$TiCl_4$	$AlEt_2$BHT/0.51 ET	1200	92

[a] From reference 10

[b] Abbreviations: ET = ethyl toluate; EB = ethyl benzoate; 26DMP = 2,6-dimethylphenol; TMEDA = tetramethylethylenediamine; IAE = di(isoamyl)ether; BHT = 2,6-di-*t*-butyl-4-methylphenolate or phenol as appropriate

More important than these physical factors is the chemical nature of the matrix. The use of heptane as a matrix gives catalysts of only moderate activity. Activity suffers if $TiCl_4$ is included in the diluent matrix. The $MgCl_2$ may be evaporated directly into a neat $TiCl_4$ matrix, but there is little advantage to this approach. When aromatic solvents and $TiCl_4$ are combined in the matrix, the highest activity for non-oxygen-donor systems results. Unfortunately, in all of these examples, the stereospecificity of the catalyst is low, giving polypropylenes containing around 40% of the desirable heptane-insoluble, isotactic polymer (Table 1).

The incorporation of Lewis base, electron-donor molecules into catalyst preparation is a well-established method to achieve the desired high level of isotacticity.[15] Our first approach was to prepare the activated $MgCl_2$ by evaporation into heptane, treat with electron donor, and then treat with $TiCl_4$. The resulting solid component was then activated with triethyl aluminum. If no Lewis base is added to the aluminum alkyl activator, ethyl benzoate gives only marginal improvement in tacticity while 2,6-dimethylphenol and diisoamylether give no measurable improvement. Tetramethylethylenediamine (TMEDA) gives some improvement in tacticity but at greatly reduced activity. Addition of ethyl benzoate to the aluminum alkyl activator using the ethyl benzoate modified solid catalyst component gives significant improvement in tacticity but only at tremendous expense to catalyst activity as depicted in Figure 1.

Our second approach was to evaporate the $MgCl_2$ into heptane containing the electron donor molecule. The resulting activated $MgCl_2$ was then treated with $TiCl_4$ to give the solid catalyst components. Again there were slight improvements in tacticity, but they were far short of the desired levels.

Evaporation of $MgCl_2$ into a heptane/BHT matrix (where BHT is "butylated hydryoxytoluene" or 2,6-di(*t*-butyl)-4-methylphenol) gives the most interesting catalyst system.

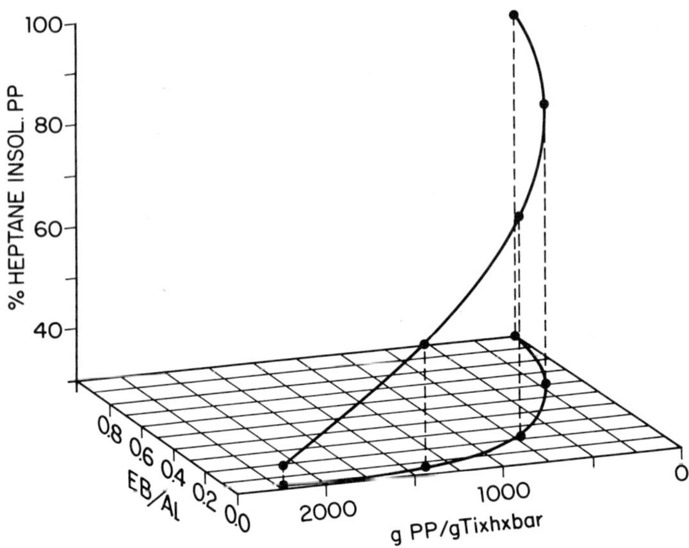

Figure 1. The effect of added ethyl benzoate (EB) on catalyst productivity and polypropylene isotacticity for $MgCl_2$ supported $TiCl_4$/EB catalyst activated with triethylaluminum (Al).

Activation with the standard activator, triethyl aluminum gives only a moderate yield of low isotacticity polypropylene. Addition of half an equivalent of ethyl toluate to the triethyl aluminum activator improves stereospecificity significantly, but again at considerable expense of activity.

Activation with phenoxide-modified activators gives activity much higher than that obtained with triethyl aluminum. These activators are prepared by reaction of triethyl aluminum with one or two equivalents of the same sterically hindered phenol, BHT. An NMR study of the reaction indicates rapid sequential reaction of the aluminum alkyls with the phenol to give first $Et_2Al(BHT)$ and then $EtAl(BHT)_2$. This behavior is shown in Figure 2. There is no evidence for the existence of $Al(BHT)_3$ even in the presence of a large excess of BHT.

When $Et_2Al(BHT)$ is used to activate the phenoxide-modified $MgCl_2/BHT/TiCl_4$ catalyst, the activity is twice that obtained

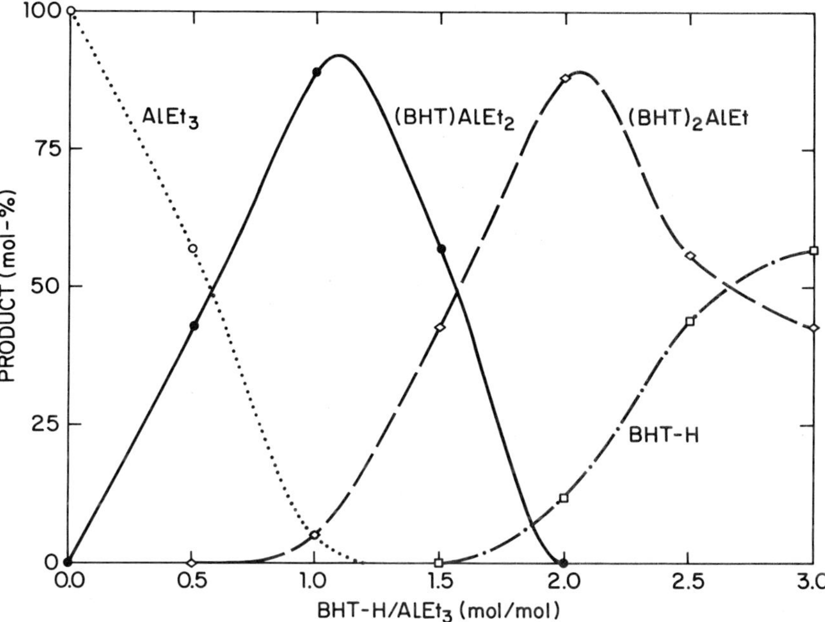

Figure 2. The product distribution upon sequential addition of 2,6-di-*t*-butyl-4-methylphenol (BHT) to triethylaluminum.

with $AlEt_3$, with no change in tacticity. When a 1:1 mixture of $Et_2Al(BHT)$ and $EtAl(BHT)_2$ is used for activation, activity is almost three times that observed with $AlEt_3$, again with no change in tacticity.

Addition of a trace quantity of ethyl toluate to the phenoxide activator system had the remarkable effect of further enhancing the rate of polymerization while simultaneously improving tacticity! Further additions up to about 0.25 equivalents continued to improve tacticity to the point that 91% heptane insolubles were obtained. This was at the expense of the highest activities, but still at an activity higher than that obtained with normal $AlEt_3$ activation which gave low isotactic yields. Additions of ethyl toluate beyond 0.25 equivalents further decreased activity with no further improvement in isotacticity.

The reaction of ethyl benzoate with modified aluminum alkyl activators has been investigated in some detail. In contrast to the reaction with $AlEt_3$ which involves reaction to form an aluminum alkoxide,[16] the reaction with the phenoxide activators is simple Lewis acid-Lewis base adduct formation with coordination of the aluminum to the ester carbonyl group. The x-ray crystal structure of one such adduct is shown in Figure 3. The coordination geometry around aluminum is essentially tetrahedral and obviously very congested.

4. DISCUSSION

Conventional routes to $MgCl_2$-supported $TiCl_4$ Ziegler-Natta catalysts fall into two categories. The first starts with bulk $MgCl_2$ and subjects it to mechanical comminution to achieve a high surface-area, high defect structure. While the layered

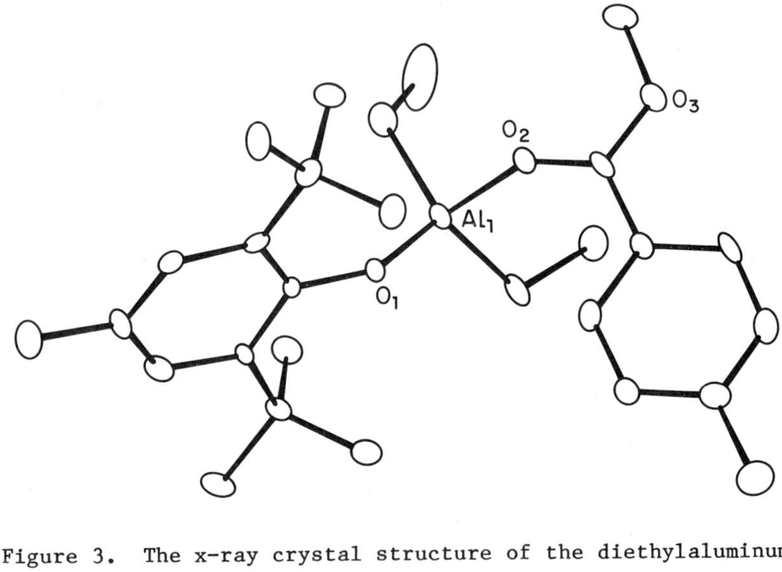

Figure 3. The x-ray crystal structure of the diethylaluminum-(2,6-di-t-butyl-4-methylphenoxide) Lewis acid-base adduct with methyl toluate.

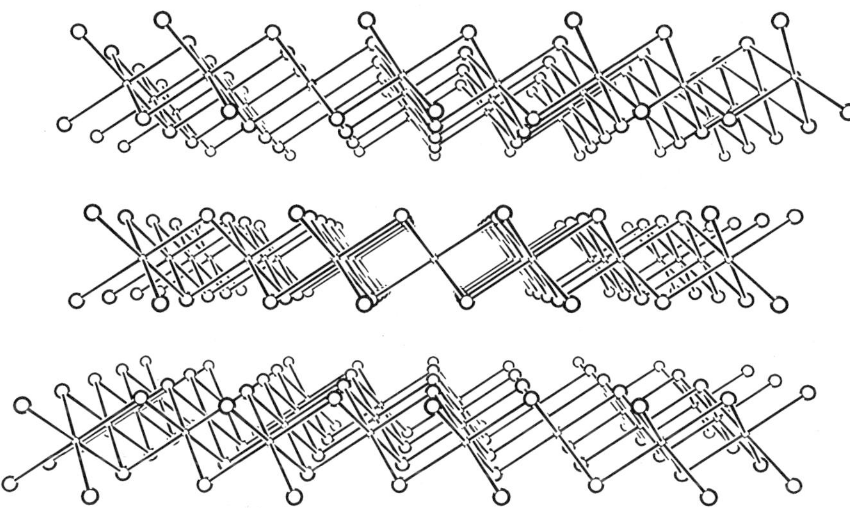

Figure 4. Perspective view of the $MgCl_2$ crystal lattice. The $TiCl_3$ lattice is similar, but one-third of the metal sites are unoccupied.

lattice, shown in Figure 4, is subject to facile cleavage between the layers, prolonged ball-milling is limited in the reduction of size particle it can achieve. Studies have shown that once a certain particle size and activity is reached, continued ball-milling leads to no further improvement.[17] While grinding is a low-energy process in the bulk, extremely high local temperatures are achieved at the point of impact and there can be considerable annealing of defects before the energy is redistributed throughout the lattice. Thus, the continued milling does not lead to further reduction of the primary crystallites.

The second approach to these catalysts is through chemical reaction. For example, dialkylmagnesium reagents are reacted with a halogen donor in the presence of a titanium source.[18] Though the reactions may be highly exothermic, they are thermally activated processes, and the $MgCl_2$ formed has an opportunity to grow to appreciable crystallite size.

Vaporization of magnesium atoms into a halide source has been used to prepare highly active catalysts,[19] but we have taken the approach of vaporizing the $MgCl_2$ directly into a diluent. In a dilute, solid matrix, the predominant species is monomeric $MgCl_2$,[13] but in a concentrated matrix or in a matrix which has been allowed to anneal, there is considerable formation of oligomers. Upon matrix melting, these oligomers continue to condense, giving a solid component, but there is little opportunity to form a well ordered, cystalline lattice. A typical x-ray diffraction pattern, shown in Figure 5, likewise indicates primary crystallite dimensions of 20 Å and less. Typical surface areas range from 20 to as high as 350 $M^2\ g^{-1}$, but we have observed little correlation between surace area and catalyst activity a long as surface area is above 10 $M^2\ g^{-1}$. The chemical nature of the matrix is far more important.

Aliphatic hydrocarbons such as heptane give relatively low productivities. For some unknown reason, aromatic hydrocarbons

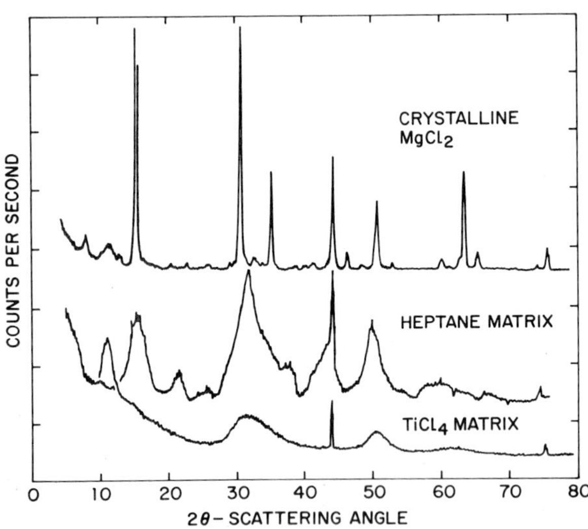

Figure 5. X-ray diffraction patterns of crystalline $MgCl_2$ and $MgCl_2$ prepared by the vapor deposition process.

give a considerable increase in activity. Colors of the matrices are highly influenced by physical properties of the matrix so it is difficult to make a definitive statement, but the aromatic matrices were unique in that they were orange or red, possibly indicating a charge transfer complex between the $MgCl_2$ and the aromatic π-cloud. Whatever the cause, upon warm-up, the color is discharged as the $MgCl_2$ condenses to a bulk phase, and catalysts based upon this material are relatively active.

All of the above catalysts give low yields of the desired heptane-insoluble isotactic polypropylene, and it was clear that Lewis base electron donors were going to be required to achieve the high stereospecificity desired. The series of catalysts prepared by cocondensing $MgCl_2$ in a heptane matrix and treating the resultant solids with $TiCl_4$ and a series of Lewis bases showed only a slight increase in stereospecificity. Incorporation of the Lewis base into the cocondensation matrix gave similar results. These catalysts respond to the addition of Lewis base to the activator alkyl in much the same manner as catalysts prepared by conventional techniques. A typical response is shown in Figure 1. Atactic sites are apparently more Lewis acidic than isotactic sites, so added Lewis base will show some selectivity in poisoning the atactic sites. The net result is improved tacticity at the expense of total and isotactic activity. This phenomenon has been demonstrated on many occasions in propylene polymerization and represented no improvement over the state-of-the-art. It can be generalized to all of catalysis as the traditional trade-off between productivity and selectivity.

The $MgCl_2$ prepared by cocondensation into a heptane/BHT matrix showed the same behavior as the previous catalysts when ethyl benzoate or ethyl toluate were added to the triethyl aluminum activator. Polypropylene which was $\sim$90% isotactic could be obtained at catalyst activities which were approximately one-fifth the original activity. Activation of the catalyst with a modified activator, $Et_2Al(BHT)$, gave catalyst activity which was more than twice that obtained with triethyl aluminum. Activation with a mixture of $Et_2Al(BHT)$ and $EtAl(BHT)_2$ gave activity which was almost three times that obtained with $AlEt_3$. Unfortunately, in each of these cases, there was virtually no change in the tacticity of the polymer. These phenoxide-modified aluminum alkyl activators are milder reducing agents than the trialkyl aluminum compounds. Presumably, the stronger reducing agents were over-reducing the catalysts, thereby decreasing the number of active sites, but as we were to find, they play a more important but less well-understood role.

Addition of just a trace of ethyl toluate to the BHT-modified aluminum alkyl activator caused another dramatic <u>increase</u> in activity while simulanteously <u>improving</u> isotacticity. Continued addition of ethyl toluate improved isotacticity further but at the expense of activity so that by one-quarter equivalent of ethyl toluate 90% isotacticity was achieved at one-third the maximum activity, but at an activity level which was comparable to that obtained with no ethyl benzoate and considerably higher than that obtained with the trialkyl aluminum activators. Continued addition did not improve isotacticity but did continue to decrease activity.

The details of this unusual catalyst behavior remain unclear. It is apparent that the BHT remains associated to the initially formed $MgCl_2$ crystallites from the cocondensation step; these dramatic improvements in activity occur only with the BHT-modified $MgCl_2/TiCl_4$ systems. We do not observe similar behavior when BHT is absent from the initial preparation of activated $MgCl_2$. At present, it is unclear what form the BHT is in as it is initially associated with the $MgCl_2$. Reaction of BHT with transition metal chlorides such as $TiCl_4$ evolves HCl yielding $TiCl_3BHT$ and related species.[20] This reaction is not known to occur for bulk $MgCl_2$, but it may well occur to some slight degree in the cocondensation matrix during warm-up.

Alternatively, the phenoxide may be acting as a Lewis base, coordinated to surface Mg^{2+} ions on the $MgCl_2$ crystallites.

Beyond the milder reduction of supported $TiCl_4$ to an active form, the role of the Et_2AlBHT is even less well defined, though it does seem that the steric bulk of the modified activator must play an important role because other, less congested but equally mild alkyls did not show the same behavior. This observation would seem to be in contrast to the earlier report that the activator alkyls do not influence the active sites.[21] We have not yet carried out the necessary ^{13}C NMR investigation of the stereoisometric composition of the resulting polypropylenes, but we have already noted the substantial difference in behavior of trialkyl aluminums and phenoxide-modified activators which may be more like the alkyl aluminum iodides used to demonstrate the role of ligands on the active titanium centers.[22]

Finally, though we have clearly demonstrated adduct formation between $Et_2Al(BHT)$ and an aromatic ester, it is not clear that the ethyl toluate remains coordinated to aluminum during the activation and polymerization. It could well be dissociated from the aluminum and coordinated to a more Lewis acidic site on the catalyst particle during the actual polymerization.

While this behavior was originally discovered in the vapor-synthesized catalysts, we have been able to extend it to $MgCl_2$-supported catalysts prepared by more conventional methods. Thus is seems to be a general behavior for catalysts of this general type. These catalysts are not only active for propylene polymerization, but also for higher 1-olefins. The responses of polymerization rates for ethylene, propylene, and other olefins are not the same, and we have found it possible to selectively modify reactivity ratios. Thus it is also possible to prepare copolymers with a great deal of control over monomer incorporation. This work will be reported elsewhere.[23]

5. CONCLUSIONS

The metal halide vapor route to $MgCl_2$-supported Ziegler-Natta catalysts allows the controlled, stepwise construction of heterogeneous catalyst particles entirely from molecular species. The observed activity and selectivity is a function of the composite structure; none of the molecular components show the desired activity by themselves. Enhanced selectivity and activity of the composite catalyst particles is a function of the order of addition and the nature of the molecular components. On that basis, it becomes clear that major portions of the molecular identity must be retained in the active sites of the composite. Thus if we are to further enhance catalyst activities and selectivites, we must pay more attention to the interface between homogeneous and heterogeneous systems. This represents a formidable challenge to our ability to characterize and understand heterogeneous particles, especially at their catalytically active surfaces. Nonetheless, our developing ability to custom design heterogeneous catalysts rather than continuing to rely on trial and error represents one of the most significant opportunities for chemistry in the coming decade.

REFERENCES

1. Yu. I. Yermakov, ISRHHC IV, J. Mol. Catal., 21, 35 (1983).
2. a) J. Boor, Jr., "Ziegler-Natta Catalysts and Polymerization", Acad. Press (1979).
   b) B. Rebenstorf, and R. Larsson, ISRHHC III, J. Mol. Catal. II, 247 (1981).
3. P. L. Watson and G. W. Parshall, Accnts. Chem. Res., 18, 51 (1985).
4. a) J. A. Ewen, J. Am. Chem. Soc., 106, 6355 (1984).
   b) W. Kaminsky, K. Kulper, H. H. Brintzinger, and F. R. W. P. Wild, Angew. Chem. Int. Ed. Engl., 24, 507 (1985).
5. H. Sinn, w. Kaminsky, Adv. Organomet. Chem., 18, 99 (1980).
   b) W. Kaminsky, M. Miri, H. Sinn, and R. Woldt, Makromol. Chem. Rapid Commun., 4, 417 (1983).
6. A. Clark, J. P. Hogan, R. L. Banks, W. C. Lanning, Ind. Eng. Chem., 48, 1152 (1956).
7. R. A. Setterquist, F. N. Tebbe, and W. G. Peet, p. 167 in "Coordination Polymerization", C. C. Price and J. Vandenberg, Eds. Plenum (1983).
8. B. L. Goodall in "Transition Metal Catalyzed Polymerizations - Alkenes and Dienes", p. 355 (1981).
9. a) K. J. Klabunde and Y. Tanaka, IV ISRHHC Proceedings, J. Mol. Catal., 21, 57 (1983).
   b) L. F. Nazar, G. A. Ozin, F. Hugues, J. Godber, and D. Rancourt, IV ISRHHC Proceedings, J. Mol. Catal., 21, 313 (1983).
10. R. Mulhaupt, U. Klabunde, and S. D. Ittel, J. Chem. Soc. Chem. Commun., 1745 (1985).
11. S. D. Ittel, R. Mulhaupt, and U. Klabunde, J. Poly. Sci., Chem. Ed., in press (1986).
12. S. D. Ittel, F. A. van Catledge, and C. A. Tolman, Inorg. Chem., 24, 62 (1985).
13. a) S. P. Randall, F. T. Greene, and J. L. Margrave, J. Phys. Chem., 63, 758 (1959).
    b) D. L. Cocke, C.-A. Chang, K. A. Gingerich, Appl. Spectrosc., 27, 260 (1973).
14. X. Youchang, G. Linlin, L. Wanqi, B. Naiyu, and T. Youqi, Sci. Sinica, 22, 1045 (1979).
15. a) G. A. Razuvaev, K. S. Minsker, G. T. Fedoseeva, V. K. Bykhovskii, Polym. Sci. USSR, 2, 299 (1961).
    b) P. Pino, B. Rotzinger, and E. von Achenbach, Makromol. Chem. Suppl., 13, 105 (1985).
    c) P. Pino, R. Mulhaupt, Angew. Chem. Int. Ed. Engl., 19, 857 (1980).
16. J. C. W. Chien, J.-C. Wu, J. Poly. Sci., Chem. Ed., 20, 2445 (1982).
17. a) J. C. W. Chien, J.-C. Wu, and C.-I. Kuo, J. Polym. Sci., Polym. Chem. Ed., 21, 737 (1983).
    b) B. Rotzinger, Thesis, E. T. H. (Zurich).
    c) R. Mulhaupt, Thesis E. T. H. (Zurich).

18. D. E. Gessell, Ger. Offen 2,517,567 (1975).
19. a) M. A. Bruzzone, 190th A.C.S. National Metting, Chicago, Macromol. Sec. Abst. No 12 (1985).
b) M. Corbellini and A. Greco, U.S. patent 4,194,992 to Snamprogetti, S.p.A. (1980).
20. a) S. L. Latesky, J. Keddington, A. K. McMullen, I. P. Rothwell and J. C. Huffman, Inorg. Chem., 24, 995 (1985).
b) A. P. Shreve, S. D. Ittel, and R. Mulhaupt, Organometallics (to be submitted).
21. G. D. Bukatov, V. A. Zakharov, Yu.I. Yermakov and A. Zambelli, Makromol. Chem., 179, 2093 (1978).
22. M. C. Sacchi, P. Locatelli, and I. Tritto, Makromol. Chem., Rapid Commun., 6, 597 (1985).
23. S. D. Ittel and R. Mulhaupt, presented at Int. Symp. Trans. Met. Catal. Polymn., Akron, Ohio (1986).

STRUCTURE AND REACTIVITY OF HETEROGENIZED METAL COMPLEX CATALYTIC SYSTEMS

F.S. Dyachkovskii and A.D. Pomogailo
Institute of Chemical Physics of the USSR
Academy of Sciences, Moscow Region, Chernogolovka, 142432, USSR

ABSTRACT - Principal approaches to the creation of immobilized catalysts, i.e. systems with a transition metal compound anchored on the polymer support surface, are analyzed.

Using different physicochemical methods the distribution of ions of the anchored metal on support is studied, the cooperative character in such systems is established and peculiarities of immobilization of heterocomplexes are elucidated. These effects are considered in ethylene polymerization.

The main distinctive features of polymerization processes under the influence of anchored metal complexes are summarized. Kinetic behaviour of CH_3TiCl_3 anchored on PE-inoculated-poly(4-vinylpyridine) is considered as the active center model on surface. It is shown that stability and the high activity of immobilized catalytic systems are mainly due to a sharp inhibition of consistent processes in the coordination sphere of a transition metal.

1. INTRODUCTION

Heterogenization of metal complexes is performed using both physical and chemical methods. At present some new ways of immobilizing transition metal complexes (MX_n) on nonorganic, organic supports and in the volume of a polymer matrix (in swelling gels) have been developed (Refs. 1-3). The chemical ways of immobilization are based on the interaction of MX_n with reactive groups of carbo- or heterochain polymers. It is very important for a number of catalytic processes that the reaction centres be brought onto support surface. If a macromolecular matrix has no functional groups, its surface may be functionalized by various polymer-analogous reactions or by modifying the surface of three dimensional cross-linked polyligands. To this end, special techniques have been developed, for example, grafting of small amounts of the monomer with appropriate functional groups on polyolefin surfaces (Refs. 4,5).

An approach based on the formation of reaction species (free radicals and ions) via chemical or radiation-chemical treatment of powdered polymers, such as polyethylene, polypropylene, polyvinylchloride, polystyrene, has proved to be convenient from the experimental point of view. MX_n immobilization is carried out using the slurry of such support, porosity of the grafted layer provides penetration of reagents into the depth about 300 Å. The main types of such supports and immobilized complexes are given in Table 1 (Ref. 6).

2. THE STATE OF HETEROGENIZED METAL COMPLEXES ON POLYMER MATRICES

One of the main problems in the synthesis of immobilized metal complexes is the determination of the character of active centres distribution over the surface or in the volume of a polymer support. Is it possible to achieve a uniform distribution of active centres or does a formation of associates and clusters take place and what is their role in the catalytic processes? To be able to answer these questions it is necessary, first, to investigate the topochemistry of the functionalized layer of polymer supports. The functionalization processes are sometimes followed by destruction and cross-linking of the polymer matrix, as well as by the change of the support porosity.

Using spin-label technique, the topochemistry of the

TABLE 1. Character of immobilized MX_n

Polymer support	Index of polymer matrix	Content of grafted fragments, 10^4 mol/g	MX_n	Content of immobilized metal 10^4 g-at/g
1. poly(4-vinylpyridine)	PE-gr-P4VP	6.0	$TiCl_4$	2.04
- " -		6.0	VCl_4	0.70
2. poly(acrylonitrile)	PE-gr-PAN	6.4	$TiCl_4$	0.40
- " -		11.9	VCl_4	0.70
3. poly(methylmetacrylate)	PE-gr-PMMA	12.0	VCl_4	1.14
- " -		12.0	$TiCl_4$	0.77
4. poly(methylvinylketone)	PE-gr-PMVK	6.0	VCl_4	1.50
- " -		6.0	$TiCl_4$	2.70
5. poly(vinylacetate)	PE-ge-PVAc	12.8	$TiCl_4$	1.70
6. poly(diallylsulfide)	PE-gr-PDASu	1.1	$TiCl_4$	0.18
- " -		1.1	VCl_4	0.25
7. poly(dibutylether of phospho-vinyl acid)	PE-gr-PDBEPVA	1.1	VCl_4	0.50
8. poly(trifluorostyrene)	PE-gr-PTFS	2.3	VCl_4	0.45
9. poly(vinylpyrrolidone)	PE-gr-PVPd	8.6	VCl_4	2.60
10. poly(diallylamine)	PE-gr-PDAA	2.6	VCl_4	2.50
11. poly(acrylic acid)	PE-gr-PAA	11.0	VCl_4	1.90
12. poly(allyl alcohol)	PE-gr-PAAL	6.2	$TiCl_4$	1.20
- " -		6.2	VCl_4	3.00
- " -		6.2	$(C_5H_5)_2TiCl_2$	0.10
- " -		3.4	$Ti(OC_4H_9)_4$	0.11
- " -		3.4	$VO(OC_2H_5)_3$	0.84

functionalized layer has been studied, for example, on the PE-grafted-poly(allylamine) systems (Refs. 7,8). The temperature dependence of the rotation times for radical correlation (τ_c) in the functionalized layer consists of two linear regions, the bending point corresponding to the polymer support melting. Above it the activation energy (E_a) is equal to 13.5 Kcal/mol, below it is 5 Kcal/mol. This means that the spin rotation needs not to overcome any considerable energy barriers at low mobility (at low temperature) of polymer segments. On the contrary, when the polymer segment mobility is sufficiently high (at elevated temperatures) the spin rotation is hindered, which leads to an increase in E_a rotation. These facts are important for our understanding of the high catalytic activity and stability of heterogenized systems.

In turn, the functional group distribution in a polymer and on its surface determines the topochemistry of MX_n immobilization, which proceeds through a gradual penetration of the reagents into the surface layer depth. The nature of this process, naturally, depends on tempretaure. At small ratios of MX_n:L (L is the functional group) the content of the fixed metal is proportional to the content of the grafted fragments. The functional groups of the surface layer react first, followed by MX_n penetration inside the surface layer. The near-surface layers are experimentally shown to have more of the active component than the remote ones.

A double character of the MX_n distribution over the surface and in the near-surface layers of functionalized polymers has been proved for the $TiCl_4$ - PE-grafted (polydiallylamine) system. A considerable portion of aggregates involving 11-25% mol $TiCl_4$ are formed on polymer surface, whereas in the near-surface layer (100-300 A) a larger portion (as compared with the surface) of isolated centres is formed, an average effective distance between them being of 20 to 30 A. It may be noted that some other examples are known where the surface has less metal than the volume.

Investigation of the intromolecular mobility of reactive macroligands and immobilized complexes is of particular interest. The immobilization process leads to a considerable reorganization of the support and to the decrease of the correlation of spin-labeled metal containing polymers. Thus the ESR data indicate that the motion (at 293 K) of the polymer ligand (its segments) in the poly(I-vinylimidazol)-Cu(II) solution is largely retarded as compared with the low molecular analogue (Ref. 10).

The NMR relaxation technique shows (Ref. 11) that the acrylic acid grafting on polyethylene results in disintegration of its γ-region (surface layer). Immobilization of the metal complex reduces the surface layer mobility: it becomes more rigid, an increase of the bonded metal concentration enhancing this effect.

It is very often that the immobilized complexes are regarded as ensembles of isolated centres. Actually, as has been mentioned above, the interaction of the MX_n-polymer support systems are of cooperative character observed both in the mutual influence of the functional groups and MX_n. Really, the characteristic feature of magnetic behaviour of bonded paramagnetic complexes of Mo(V), Co(II) and Ni(II) is the presence of the "break" at T = 110-120 K on the temperature dependent magnetic susceptibility plot ($\varkappa^{-1} = f(T)$). The Curie-Weiss constant change from Θ_1, to Θ_2 (Θ_1 for complexes Mo and Co is within the range of 60-112 K) suggests (Ref. 12) the presence of spin-spin antiferromagnetic interaction between unpaired electrons of the fixed metal ions. This also indicates the appearance of cluster species even at the stage of metal complex immobilization. Quite important is here the characteristic of the polymer support surface and metal complex surface concentration. Thus, the components in the ESR-spectrum become less resolvable with increasing the VCl_4 surface concentration on PE-gr-P-4VP (Ref. 13); along with a well resolved 8-component in the ESR spectrum of the isolated V(IV) complexes, there is observed an appearance and increase of a broad signal with the same value of g-factor, which is the spectrum of the same compound extended by a dipole-dipole interaction. At the concentration of $0.15 \cdot 10^{-4}$ g.at. V(IV)/g the complexes represent a cooperative system involving cluster species. The upper limit of the local concentration of the isolated V(IV) complexes (Ref. 13) is estimated as $C_{isol} = 2\ 10^{-4}$ g.at/cm^3 and in aggregates as $C_{aggr} = 2 \div 6 \cdot 10^{-3}$ g.at//cm^3. The mean distances between the isolated complexes $\bar{r}_{isol} > 22$ Å, and in clusters $\bar{r}_{aggr} \approx 6.8$-$9$ Å. A portion of the isolated particles decreases sharply (Fig.1) with increasing concentration of the immobilized V(IV) and vice versa, the V(IV) portion incorporated in the cluster increases.

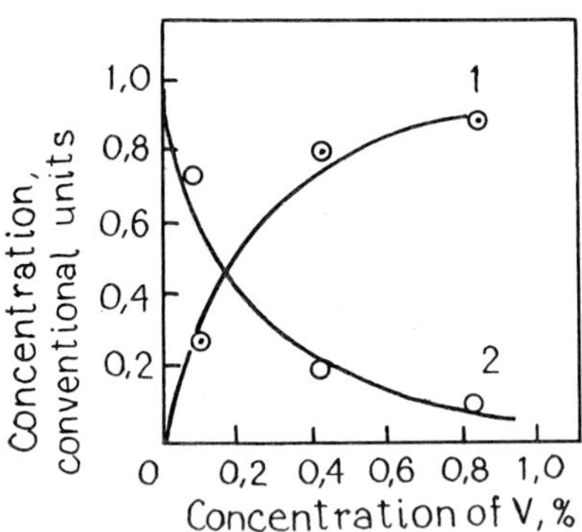

Figure 1. Relative concentration of cluster species (curve 1) and isolated ions (curve 2) versus surface concentration of V(IV) on PE-gr-P4VP.

To some extent, a similar situation is observed for the PE-gr-PAAL-Cu(II) as well (Ref. 14). With concentrations of 0.15 mg.at/g practically all copper is bonded in the form of clusters, their local concentration is (6.0-40.6)$\cdot 10^{20}$ at.Cu(II)/cm^3. The available experimental data allow us to represent the exchange interaction mechanism in clusters as follows. It may be assumed that of significance for aggregation are not only the dipole-dipole interactions, but also the favourable distance distribution of transition metal ions, in view of the fact, that the dipole-dipole interaction energy is much less then kT at the distances of 6-9 Å, when the cooperative effects begin to show (Ref. 15). For this reason, the first step of the immobilization is the MX_n binding with the support in a noncooperative way. An ion bonded in such a manner forms a second bond with a neighbour or non-neighbour centres of a certain polymer chain. These first intra- and intermolecular net is the basis for the further cross- linking and cooperative binding. Naturally, a significant influence is exerted by the polymer ligand nature on the character of transition metal distribution, especially the segment pliability of its macromolecular chain: a hard, poorly swellable polymer determines the space distribution of a metal, whereas in the "soft" swelling matrix the metal space position is determined mainly by the nature of MX_n and its ability to form clusters.

Using the electron microscopy technique (Ref. 16) for the sample of $MoCl_5$ immobilized on PE-gr-PAN, it is possible to observe directly molybdenum distribution on the support surface and to identify three types of aggregates: with a diameter of less than 10 Å, of the order of 10 Å, and those approaching 1000 Å and composed of separate blocks. An increase in "clusterization" has been noted to occur with a rise of the fixed molybdenum concentration, while an increase in the total electron density is indicative of the existence of the homogeneously distributed molybdenum ions as individual complexes. An estimation of cluster content in such immobilized molybdenum complexes gives a value of 30%.

Taking into consideration the exchange interactions and processes of cluster formation in the polymer-MX_n systems it may be assumed that the size of the polymer support segments binding MX_n in complexes of various types is rather large. The general picture of MX_n distribution may be represented in the following way: isolated ions, associations (aggregates), in which the metal ions are not bonded chemically, and clusters with strong exchange interactions due to the chemical bonds of metal ions.

The situation is much more complicated when the compounds of two transition metals (MX_n and $M'X'_m$) are immobilized on polymer supports. There are two possible ways of immobilization the bimetallic complexes (Ref.17): controlled layer-by-layer deposition (immobilization of MX_n and $M'X'_m$ by covalent or donor-acceptor interaction with similar or different functional groups) and immobilization through the heterocomplex additional buildup on the support.

The possibilities of the first method are demonstrated by the immobilization of Ti(IV), V(V), V(IV) and Ni(II), Co(II) on PE-gr-PAC (Table 2).

The methods of the layer-by-layer deposition exclude an effective control of the distribution of M and M' on a polymer, nevertheless, the required approximate ratio between them is achieved by the choice of conditions of immobilization. Moreover, even with statistic distribution of transition metals of different types the components in heterometallic system are not separated with respect to electron interaction. Most probably, the associates formed are polycentre complexes produced as a result of bridging interactions and charge transfer.

TABLE 2. Immobilization of the MX_n - $M'X'_m$ systems on PE-gr-PAA (1.1 mmol/g)[a)]

Conditions of immobilization of $M'X'_m$		Conditions of immobilization of MX_n		Content of metals	
$M'X'_m$	M'/ COOH	MX_n	M/ COOH	M'	M
$Ni(OCOCH_3)_2 \cdot 4H_2O$	0.25	$TiCl_4$	0.8	0.24	0.07
$Ni(OCOCH_3)_2 \cdot 4H_2O$	0.30	$VOCl_3$	0.7	0.28	0.16
$Ni(OCOCH_3)_2 \cdot 4H_2O$	0.15	VCl_4	0.1	0.10	0.12
$Ni(OCOCH_3)_2 \cdot 4H_2O$	0.15	VCl_4	0.15	0.10	0.18
$Ni(OCOCH_3)_2 \cdot 4H_2O$	0.20	VCl_4	0.5	0.18	0.23
$NiCl_2 \cdot 6H_2O$	0.60	$TiCl_4$	0.4	0.55	0.10
$NiCl_2 \cdot 6H_2O$	0.55	VCl_4	0.4	0.50	0.08
$Ni(OCOC_{17}H_{35})_2$	0.10	$TiCl_4$	0.4	0.04	0.06
$Ni(naphth)_2$	0.10	VCl_4	0.25	0.02	0.04
$Ni(AcAc)_2$	0.40	VCl_4	0.30	0.20	0.17
$Co(OCOCH_3)_2$	0.30	VCl_4	0.1	0.25	0.08
$Co(AcAc)_2$	0.5	VCl_4	0.5	0.15	0.14

a) in various cases temperatures were chosen between 323 and 343 K and interaction time between 2 and 7 hrs.

The mechanism of their formation is not clear, the $M'X'_m$ "attachment" to the already fixed MX_n should obviously produce conformational changes in the macrochain leading to the destruction of previous structures. Some interaction of cooperative character takes place between paramagnetic metals. These processes are considerably enhanced in the case of the MX_n and $M'X'_m$ immobilization with a controlled distribution of paramagnetic metals with respect to each other. This is achieved in different ways (Ref. 18): by the "attachment" of chelates Co(II), Ni(II), Cu(II) etc. to immobilized Ti(IV), V(IV) etc., using polymer supports with chelateforming groups as diketone, diacylamine, enaminoketone, acylsalicylamide and other types.

3. CATALYTIC PROPERTIES OF HETEROGENIZED METAL COMPLEXES IN ETHYLENE POLYMERIZATION

Thus, the analysis shows that the structure of heterogenized metal-complex systems depends on a variety of factors, some of which may be controlled at the stages of immobilization, other should be merely considered when developing such systems. Important are those factors that can affect to some extent the catalytic properties of marcomolecular complexes formed in various reactions (valence state and transition metal distribution, its coordination, unsaturation etc.). The dynamics (molecular and intermolecular mobility) of immobilized complexes should also markedly affect the reactivity (as is known from the examples of reactivities of polymers) (Ref. 19).

The processes of olefin polymerization catalyzed by immobilized metal complexes provide a good example demonstrating their pecularities, the influence of support nature and active centre immobilization character on polymerization kinetics, revealing common and specific features of polymerization reactions under the action of immobilized catalytic systems and their homogeneous analogues, raising the problems of stabilization of catalytic activity and increasing thermostability of catalytic systems.

For polymerization processes the catalysts immobilized on the polymer support are of particular interest. In these cases the polymer formed does not contain any inorganic admixture, while in some special cases it might result in forming new polymer-polymer composition (Ref. 20). Besides, such systems have technological advantages: they allow to exclude a stage of washing the polymer, to turn from slurry processes to the gas-phase ones in the regime of fluidized polymer bed, while special techniques lead to the formation of granule-like polymers in the synthesizing process.

As is known, the homogeneous vanadium containing ($VO(OC_2H_5)_3$, VCl_4, $VOCl_3$, Cp_2VCl_2 and so on) catalytic systems are very unstable in ethylene polymerization. The process is characterized by the high initial rate and a fast loss of activity. An unsteady-state character of ethylene polymerization by these systems is the consequence of a decrease of the active centre concentration with time. The results of the active centre number (n_p) determination by kinetic and radiochemical methods have shown (Ref. 21) that the unsteady state of these systems is associated with the irreversible deactivation of active centres in bimolecular process in the initial stages of polymerization and monomolecular process in the subsequent stages. The following scheme corresponds in the main to the kinetic data on active centre deactivation:

$$\text{I)}\quad >\overset{|}{\underset{|}{M}} + AlR_2Cl \longrightarrow >\underset{|}{M} - R + AlRCl_2$$

$$\text{II)}\quad \geq M - R \longrightarrow R_{(-H)} + > MH$$

$$\text{III)}\quad \geqslant M - R + \geqslant M - R \longrightarrow 2\geqslant M + R_{(-H)} + R_{(+H)}$$

$$\text{IV)}\quad \geqslant M - R + AlR_2Cl \longrightarrow \;>M<^{R}_{R} + AlRCl_2$$

$$\text{V)}\quad >M<^{R}_{r} \longrightarrow \;>M + R_{(+H)} + R_{(-H)}$$

Reactions (I-V) in homogeneous vanadium containing catalytic systems are extremely fast, especially at elevated temperatures. Hence, the immobilization of vanadium complexes on polymer supports should lead, first, to an increase of their reactivity. As was shown (Ref. 22), the stabilizing effect of an oligomeric ligand on the MX_n catalytic activity in ethylene polymerization is weak. The increase of the molecular mass of the ligand chain in oligomer-immobilized systems of VCl_4 macromolecular complexes (Ref. 23) results in an increase of their catalytic activity. Similar results have been obtained when studying the catalytic activity of transition metal macromolecular carbonic salts in ethylene polymerization (Ref. 24). It might be expected that the metal complexes heterogenized on polymer surfaces with a functionalized layer would appear most appropriate for these purposes.

Actually, as is seen from Figs. 2 and 3, the metal complexes immobilized on such supports according to the donor-acceptor or covalent types form active and stable catalytic systems (Ref. 6).

Attention ought to be paid to the high stability of vanadium containing catalysts even under comparatively hard (T=343 K, Al:V=50) polymerization conditions: the activity decrease time is twice as much as 180 min ($\tau_{1/2}$), while for a mobile system $\tau_{1/2}$=5 min (even at T=313 K and Al:V=10).

It is difficult to compare the kinetic parameters of ethylene polymerization by mobile and immobilized systems, since there is no temperature overlapping in

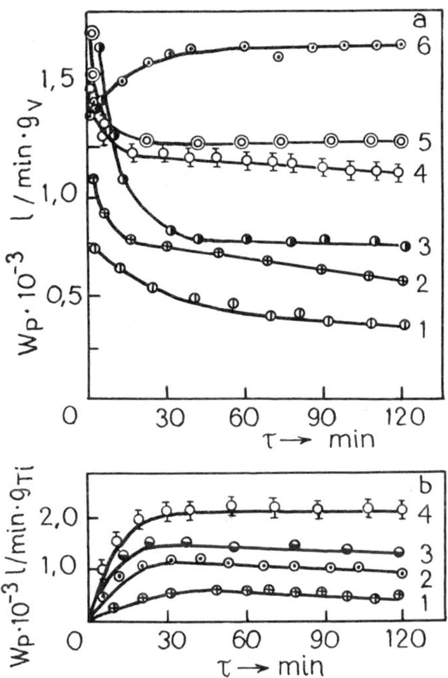

Figure 2. Ethylene polymerization kinetics by the catalytic systems based on VCl_4(a) and $TiCl_4$(b) immobilized on polymer supports by the donor-acceptor type
Polymer supports: a: 1- PE-gr-PDBEVPA, 2- PE-gr-PDASu, 3- PE-gr-PAN, 4- PE-gr-PMMA, 5- PE-gr-PMVK, 6- PE-gr-P4VP. b: 1- PE-gr-PDASu, 2- PE-gr-P4VP, 3- PE-gr-PVAc, 4- PE-gr-PMMA.
Polymerization conditions: solvent - heptane, 343 K, $Al(C_2H_5)_2Cl$, Al:V=5, ethylene pressure: 0.4 MPa.

their catalytic action. The limiting temperature of the ethylene polymerization for the VCl_4 - Al(iso-$C_4H_9)_2Cl$ system is 283-313 K, the expression for the polymerization rate constant is $K_p = 2.2 \cdot 10^8$ exp(-1100/RT)L/mol·s; K_p extrapolation by this equation to T=343 K leads to the value of $K_p^{343} = 41.10^3$ l/mol·s. A comparison of some kinetic parameters for homogeneous and heterogenized systems can be made on extrapolation. Table 3 gives the n_p values found using the radioactive method (^{14}CO) and the K_p values calculated on this basis. It is clear that, within the error of the n_p determination technique,

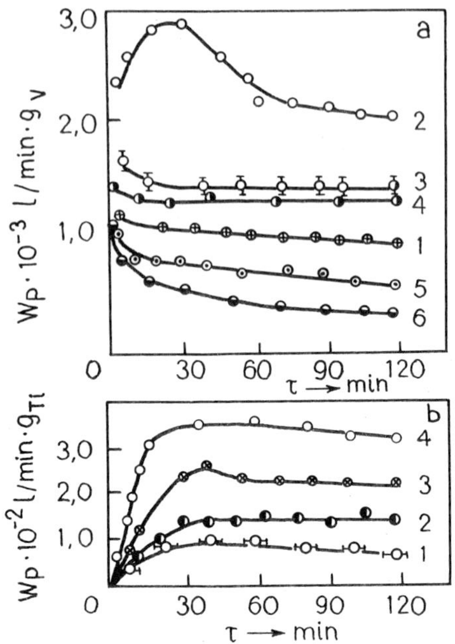

Figure 3. Ethylene polymerization kinetics by the catalytic systems based on VCl_4(a) and $TiCl_4$(b) immobilized by the covalent type.
Polymer supports: a: 1- PE-gr-PAS, 2- PE-gr-PAAl, 3- PE-gr-PAA, 4- PE-gr-PDAA, 5- PE-gr-PAA, 6- PE-gr-PTFS; b: 1- PE-gr-PBAl, 2- PE-gr-PAA, 3- PE-gr-PAAl, 4- PE-gr-PDAA.
Polymerization conditions: solvent - heptane, 343 K, $Al(C_2H_5)_2Cl$, Al:V=5, ethylene pressure: 0.4 MPa.

the K_p values in ethylene polymerization by mobile and immobilized systems are close, or at least of the same order of magnitude. At the same time, the active centre concentration in the immobilized systems is 80-250 times greater than $[n_p]$ in the mobile systems under comparable conditions (Figure 4).

Quite evident is the immobilization effect on heterogenizing the $Ti(OR)_4$ catalytic system: activity is 100-150 times greater here. The kinetic characteristics of this system are also of interest. On immobilization the order with respect to titanium decreases from 1.4

TABLE 3. Kinetic parameters of ethylene polymerization by immobilized VCl_4 systems (polymerization at 343 K)

Polymer system	n_p, mol/mol	$K_p \cdot 10^{-3}$, l/mol·s	E_a^{eff}, kJ/mol
Mobile system	4.58-0.1*	41	14 ± 2
$VCl_4 \cdot$MMA	0.23*	22 ± 5	40 ± 4
PMMA·VCl_4	3.8*	28 ± 6	23 ± 4
PE-gr-PAC	12.5*	34 ± 7	34 ± 4
PE-gr-PDAA	5.0	42 ± 8	27 ± 4
PE-gr-PDBEPVA	3.8	29 ± 6	32 ± 4
PE-gr-PAA	2.1	51 ± 10	38 ± 4
PE-gr-PMVK	4.5	40 ± 8	31 ± 4
PE-gr-PMMA	4.5	34 ± 7	38 ± 4
PE-gr-PAN	4.0	28 ± 6	38 ± 4
PE-gr-PVPd	4.2	31 ± 6	29 ± 4
PE-gr-PAS	3.5	40 ± 8	36 ± 4
PE-gr-PTFS	5.0	48 ± 10	31 ± 4
PE-gr-PDAS	4.2	31 ± 6	29 ± 4

* Polymerization at 313 K. Maximum values of n_p are given .

(for a homogeneous system) to 0.6 - for an immobilized system with the same order with respect to a monomer (1.0), a considerable decrease of the over-all activation energy (down to 11 kJ/mol) is observed as compared with E_a=58 kJ/mol for a homogeneous system. During the $Ti(OC_4H_9)_4$ fixation the most favourable active centre structures are to be formed on the surface and energy barriers for monomer insertion reaction are low. However, K_p is only slightly changed on the active centre immobilization. Hence, the frequency factor of the porpagation rate constant decreases sharply. The compensation effect observed indicates that many near-

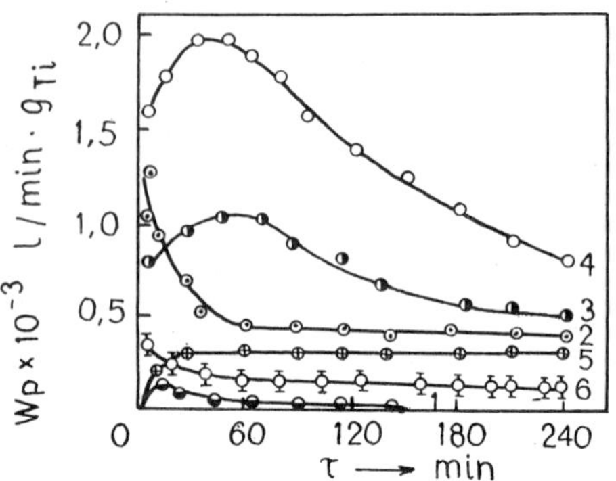

Figure 4. Ethylene polymerization by the heterogenized systems based on $Ti(OC_4H_9)_4$ - $Al(C_2H_5)_2Cl$.
Polymer supports: 1- a homogeneous system, 2- PE-gr-PDAA, 3- PE-gr-PAAm, 4- PE-gr-PAAl, 5- PE-gr-PAA, 6- a hydrolyzed copolymer of styrene with vinylacetate.

neighbour fragments of polymer ligands are involved in the active centre reactions. All this leads to entropy decrease in these reactions.

It follows from an analysis of different types of MX_n immobilization (covalent, donor-acceptor, chelation) that, as a rule, the covalent immobilization results in the formation of more active and stable catalytic systems.

The main regularities of ethylene polymerization kinetics in the presence of immobilized catalytic systems may be represented as follows.

1. Active centre formation in the immobilized systems proceeds more slowly than in the mobile ones. Along with that, the processes of active centre deactivation are considerably retarded. As has been already mentioned, the K_p value in homogeneous and heterogenized systems does not change.

2. Limitations to the diffusion of the active centres is one of the factors of increasing the immobilized system stability. With the temperature rise the polyme-

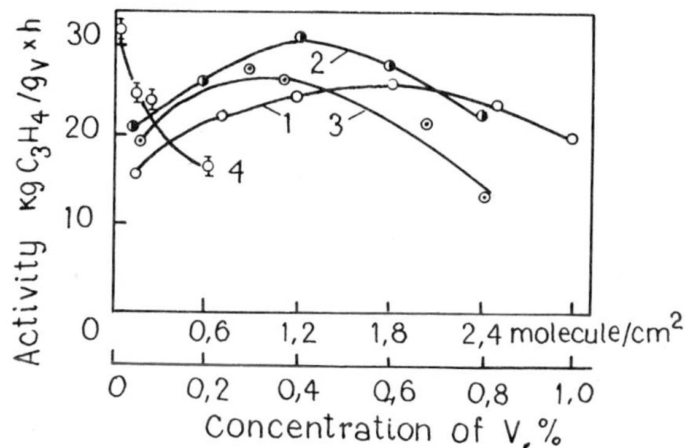

Figure 5. Dependence of the immobilized system catalytic activity in ethylene polymerization on the immobilized metal surface density.
1- VCl_4 PE-gr-P4VP, 2- $VO(OC_2H_5)_3$-PE-gr-PAAl, 3- VO(C $VO(C_2H_5)_3$-PE-gr-PAA, 4- $Ti(OC_4H_9)_4$-PE-gr-PAAl.

rization rate reaches the temperatures of polymer support softening by the Arrhenius law. In the region of those temperatures where the support chain segments possess a certain mobility and the relaxation becomes determining and relative to the reorientation of the whole of the support macromolecule, the polymerization temperature coefficient becomes negative (Ref. 25).

The shape of the dependence curve of the deactivation specific rate (W_t) on the reverse temperature $MW_t = A \exp(-\frac{B}{T-T_o})$ corresponds to the character of the dependence usually observed for the segment movement relaxation times. These result may be used when creating high temperature immobilized catalysts for polymerization processes.

3. One of the peculiarities of the catalytic systems fixed on the polymer support surface is their activity dependence on the MX_n surface concentration (the immobilized MX_n surface density) (see Fig.5). Thus, the maximum activity in the ethylene polymerization by the VCl_4 PE-gr-P4VP-$Al(C_2H_5)_2Cl$ is observed (Ref. 13) with a vanadium content of 0.6 mass%. The relations of this type take place also for many other processes which are catalyzed by immobilized complexes. Such a dependence should obviously be a consequence of the specificity of V(IV) distribution on the support, particularly, of the changes in the correlation between isolated complexes and clusters. The extreme dependence can be connected directly neither with concentrations of isolated ions nor with clusters. At the same time, it is evident, that it is the V(IV) associates concentration that should pass through the maximum, which involves a small number of vanadium complexes (dimers, trimers), hence the immobilized system active centres are assumed to be formed of the isolated vanadium ions stabilized by clusters.

4. One more characteristic feature of the immobilized catalysts under consideration is their behaviour in "ageing" in the absence of a monomer: polymerization rate does not depend on the catalyst "ageing" time (Fig. 6). In mobile systems the "ageing" leads to a substantial decrease of catalytic activity which is connected with a gradual reduction of the transition metal to inactive species. In the MX_n immobilization the reduction of the transition metal by organoaluminum compounds is sharply retarded. For this reason the stabilization of the systems occurs. Besides, the polymerization rate stability shows also that the monomer does not exert any marked effect on the active centre deactivation in these systems.

Therefore, the interaction of the immobilized system components makes possible the following reaction path-

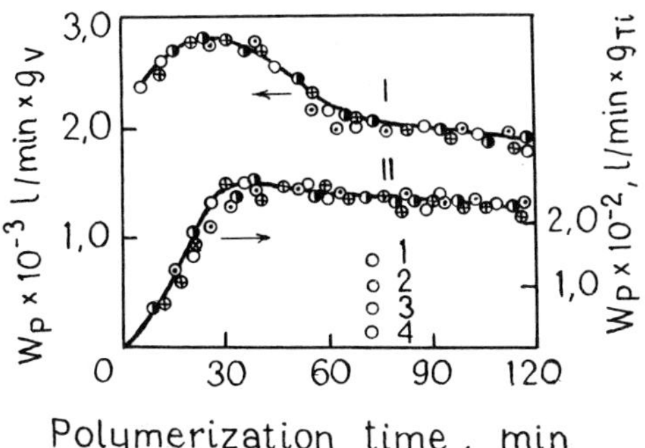

Figure 6. Polymerization curves for the systems "aged" in the absence of a monomer. "Ageing" time, min: 1 - 0; 2 - 10; 3 - 15; 4 - 20. Catalysts: curve I: VCl_4 on PE-gr-PAAl; curve II: $TiClCl_4$ on PE-gr-PAAl.

ways:

$$\text{⫽}\!-L-MX_n + AlR_xCl_{3-x} \longrightarrow \begin{cases} \text{⫽}\!-L-MX_{n-1} - R & \text{a} \\ \text{⫽}\!-L-MX_{n-m} & \text{b} \\ \text{⫽}\!-L-AlR_xCl_{3-x}\cdot MX_n & \text{c} \\ \text{⫽}\!-L-AlR_xCl_{3-x}+MX_{n-1}-R & \text{d} \end{cases}$$

Of these the "a" path is the formation of alkyl derivatives of the transition metals bound with the polymer support-determining active centre n_p formation. The contribution of (a) - (d) reactions depends on the immobilization mode, the reacting components nature and the conditions of their interaction.

4. POSSIBLE REASONS OF HIGHER STABILITY OF CATALYTIC SYSTEMS WITH HETEROGENIZED METAL COMPLEXES

To understand properly the reasons giving rise to the immobilized system stability, it is necessary to compare some results of the transition metal organometallic derivatives decomposition in homogeneous and in heterogenized systems.

The energies of the metal-carbon bond are not very large, but still sufficient to provide for their catalytic systems stability. The dissociation energy of M-C σ-bond energy was shown (Ref. 26) to determine neither the high reactivity of this bond nor the low stability of the transition metal organometallic compounds in solution. The main direction of their decay is the processes within the metal coordination sphere (intrasphere processes) including the β-elimination and disproportionation reactions. The activation energy of these processes is lower than the M-C bond energy.

The general scheme of the reaction of M-C bond with participation of an electron-donor (D) may be represented as follows:

$$RML_n + D \rightleftharpoons RML_n \cdot D$$

$$RML_n \cdot D \begin{cases} \nearrow ML_n\, D + R^{\bullet} \\ \searrow RH + L_nM \cdot D. \end{cases}$$

Despite the differences in the mechanism, it is possible to make an analogy between the olefin interaction with the polymerization active centres and these reactions, since olefins may be regarded as weak organic bases, while the complex catalyst active centres are surely to include the M-R bond.

CH_3TiCl_3 immobilized on PE-gr-P4VP has been chosen as a model of an active centre on the surface:

$$\text{PE}-(CH_2-\underset{\text{Py}}{CH})_n-(CH_2-\underset{\text{Py}}{CH}-CH_2-\underset{\text{Py}}{CH})_m-(CH_2-\underset{\text{Py}}{CH})_p$$

(the two pyridine N atoms of the middle unit both coordinate to CH_3TiCl_3; the N of the last unit coordinates to CH $TiCl_3$)

The CH_3TiCl_3 complex with the pyridine 4-substituted decomposes in the homogeneous system with a Ti-CH_3 bond cleavage to form methane by a concerted reaction in a cycle complex of the titanium coordination sphere, and Ti(IV) in these reactions is not reduced.

A high stability of the complex is observed for the heterogenized system, where, with a temperature increase (more than 80°C), a portion of Ti-C bond homolytic decomposition giving CH_3 radicals increases. This is followed by the Ti(IV) to Ti(III) reduction. The general scheme of the CH_3TiCl_3 decomposition on the support surface can be represented as follows:

$$(CH_3TiCl_3)\cdot \text{PE-gr-P4VP} \begin{array}{l} \xrightarrow{K_1} CH_4 + \text{Ti(IV)} \\ \xrightarrow[K_2]{} CH_3^{\bullet} + \text{Ti(III)}. \end{array}$$

The value of the Ti-CH_3 bond concerted intraspheric decomposition rate constant (K_1) is two orders of magnitute less than that of the corresponding rate constant of the CH_3TiCl_3 complex decomposition with a monomer ligand. The decrease of the reaction rate constant has appeared to be connected with an extremely low value of the frequency factor, since the adjustment (deformations) of the polymer ligand required for the concerted reactions to take place demands considerable entropy changes in the transition state as compared with the low molecular (monomer) analogue. Thus, an entropy part of the immobilized CH_3TiCl_3 decomposition rate constant decreases down to 10^{-2} sec^{-1} (for a homogeneous system 10^3-10^5 sec^{-1}).

It is important that the rate of the Ti-CH_3 group homolysis to give free methyl radicals (rate constant K_2) on the surface and in solution practically does not change. The formation of small quantities of CD_4 in the D_3CTiCl_3 PE-gr-P4VP decomposition serves as a model of bimolecular active centre deactivation. The contribution of this process to reactions of the Ti-C scission is small.

Therefore, the thermodynamic stability of a Ti-C bond, as well as that of the active centres, on the surface and in solution is approximately the same (close values of the Ti-C bond energy). That is why the stability and high activity of immobilized catalytic systems is mainly associated with a sharp breaking of the concerted processes in the coordination sphere of the transition metals.

It is important here to mention one more significant phenomenon. The reactions leading to active centre deactivation (disproportionation, β-elimination, transition metal reduction) obviously require much more rearrangement of the transition metal ligand environment than the reaction of monomer insertion into the M-C bond. It is confirmed by the fact that the elementary chain propagation constant for mobile and immobilized systems are practically the same.

Summarizing the reasons of the high stability of immobilized metal complex it should be noted that the decrease of the rates of the active centre mono- and bimolecular deactivation is confirmed by the experimental facts of retardation of the immobilized transition metal reduction processes as compared with the mobile one. An important reason of the active centre stability increase is represented by cooperative effects, i.e. a collective character of the interaction of the polymer support segments with the immobilized transition metal compounds.

The developed notions are in agreement with the other processes catalyzed by immobilized complexes, for example, α-olefin dimerization, stereoregulated polymerization, etc.

REFERENCES

1. Ju.I. Yermakov, V.A. Zakharov and B.N. Kuznetsov, Complexes fixed on oxide supports in catalysis (in Russian), Novosibirsk, Nauka, p. 248 (1980).
2. V.A. Kabanov and V.J. Smetanuk, Advances in science and technics. Kinetics and catalysis (in Russian), M. VINI.TI, 13, 213 (1984).
3. F.S. Dyachkovskii and A.D. Pomogailo, Catalysts, containing fixed complexes, Yu.I. Yermakov (Ed.), Novosibirsk, Institute of Catalysis, p. 25 (1978).
4. D.A. Kritskaya, A.D. Pomogailo, A.N. Ponomarev and F.S. Dyachkovskii, Vysokomol. Soed. A, 21, 1107 (1979).
5. D.A. Kritskaya, A.D. Pomogailo, A.N. Ponomarev and F.S. Dyachkovskii, Polym. Symp., 68, 21 (1980).
6. A.D. Pomogailo and F.S. Dyachkovskii, Acta Polymerica, 35, 41 (1984).
7. N.M. Bravaya, A.D. Pomogailo and F.S. Dyachkovskii, Vysokomol. Soed. A, 2, 1781 (1979).
8. F.S. Dyachkovskii, A.D. Pomogailo and N.M. Bravaya, J. Polymer Sci., Polym. Chem. Ed., 18, 2612 (1980).
9. N.M. Bravaya, A.D. Pomogailo and F.S. Dyachkovskii, Kinetika i Kataliz, 24, 403 (1983).
10. M. Sato, K. Kondo and K. Takemoto, J. Biol. Chem., 179, 601 (1978).
11. I.I. Nazarova, N.D. Golubeva, N.M. Bravaya, A.D. Pomogailo and S.M. Baturin, in: Modern NMR and EPR methods in chemistry of solid phase (in Russian), Chernogolovka, Institute of Chemical Physics, 102, (1985).
12. Yu.G. Borodjko, I.N. Ivleva, S.B. Echmaev, L.N. Karkil, A.D. Pomogailo and A.B. Raevskii, in: Catalysts containing fixed complexes, Yu.I. Yermakov (Ed.), Novosibirsk, Institute of Catalysis, part I, p. 123 (1980).
13. A.D. Pomogailo, A.T. Nikitaev and F.S. Dyachkovskii, Kinetika i Kataliz, 25, 166 (1984).
14. A.D. Pomogailo and N.D. Golubeva, Kinetika i Kataliz, 26, 947 (1985).
15. K.I. Zamaraev, Yu.N. Molin, K.M. Salikhov, Spin exchange (in Russian), Novosibirsk, Nauka, 320 (1977).
16. S.B. Echmaev, I.N. Ivleva, A.V. Raevskii, A.D. Pomogailo and Yu.G. Borodjko, Kinetika i Kataliz, 24, 1428 (1983).
17. A.D. Pomogailo, N.D. Golubeva, I.N. Ivleva, C.B. Echmaev and F.S. Dyachkovskii, Kinetika i Kataliz, 25, 1145 (1984).
18. I.E. Ufland and A.D. Pomogailo, in: Proceeding of the XV-th Union Chugaev Conference in coordination compound chemistry (in Russian), Kiev, Kiev State University, part II, p. 538 (1985).
19. N.M. Emanuel, V.A. Roginskii and A.L. Buchachenko, Uspechi khimii, 51, 361 (1982).

20. A.D. Pomogailo, A.A. Torosjan, L.N. Raspopov, F.S. Dyachkovskii and N.S. Enikolopyan, Dokl. Akad. Nauk SSSR, 252, 1148 (1980).
21. A.D. Pomogailo, E. Baishiganov, U.A. Mambetov and G.M. Chvostik, Vysokomol. Soed. A, 22, 248 (1980).
22. U.A. Mambetov and A.D. Pomogailo, in: Complex organometallic catalysts for olefin polymerization (in Russian), Chernogolovka, Institute of Chemical Physics, part I, p. 57 (1980).
23. E. Baishiganov, A.D. Pomogailo, D.B. Sokolskii, I.D. Leonov and U.A. Mambetov, in: Complex oil refining (in Russian), Alma-Ata, Nauka, 4, 139 (1972)
24. A.D. Pomogailo, A.P. Lisitskay, A.I. Kusaev and F.S. Dyachkovskii, in: Complex organometallic catalysts for olefin polymerization, Kinetics of polymerization processes (in Russian), Chernogolovka, Institute of Chemical Physics, part I, p. 66 (1980).
25. A.D. Pomogailo, V.I. Irjak, V.I. Burikov, F.S. Dyachkovskii and N.S.Enikolopyan, Dokl. Akad. Nauk SSSR, 266, 1160 (1982).
26. V.E. Lvovskii, E.A. Fushman and F.S. Dyachkovskii, Zh. Fiz. Khim., 56, 1864 (1982).
27. I.V. Serebrynaya, N.E. Khrushch, A.D. Pomogailo and F.S. Dyachkovskii, Kinetika i Kataliz, 27 (1986).

EFFECTS OF CHIRAL AND POLYMERIC SUPPORTS ON STEREOCHEMISTRY OF α-OLEFINS POLYMERIZATION

Carlo Carlini and Francesco Ciardelli
Dipartimento di Chimica e Chimica Industriale, Università di Pisa, Centro del C.N.R. per le Macromolecole Stereordinate ed Otticamente Attive, Via Risorgimento 35, 56100 Pisa(Italy)

SUMMARY

The use of low and high molecular weight chiral components is reported for both conventional and "high yield" heterogeneous Ziegler-Natta catalysts in order to obtain detailed information about the stereochemical control of the α-olefins polymerization process as well as the nature of the active sites.
On the basis of new results concerning the use of monomeric and polymeric optically active Lewis bases in the stereospecific polymerization of chiral α-olefins, the intrinsic dissimmetry of the active sites is confirmed also for $MgCl_2$ supported Ziegler-Natta systems.
The comparison of these data with previous results concerning the stereospecific polymerization of racemic olefins suggests that the origin of stereospecificity and stereoselectivity is governed by similar factors in both conventional and "high yield" α-olefins polymerization catalysts.

INTRODUCTION

Ziegler-Natta catalysts represent an unique case of hybrid catalytic systems; indeed even the conventional system, obtained by reacting $TiCl_4$ with AlR_3 , can be considered as an heterogenized homogeneous catalyst, where a Ti-alkyl derivative complex is supported on $TiCl_3$. Successively, really supported systems have been prepared by attaching titanium active species to inorganic(SiO_2 or Al_2O_3)[1] and organic(polymers)[2] supports. Finally new generation high activity catalysts are normally considered as supported on $MgCl_2$. These systems have been subject of a large number of investigations and many speculations are reported in the literature on the role of the above supports on catalytic activity and stereochemistry.

In this paper the use of chiral components is discussed for the understanding of the stereochemistry of the polymerization process.

Chiral discrimination in conventional Ziegler-Natta catalysts

Isotactic polymers of racemic α-olefins having an asymmetric carbon atom in α or β position to the double bond, obtained in the presence of conventional Ziegler-Natta catalysts prepared by reacting $TiCl_4$ or $TiCl_3$ with aluminum alkyls, can be separated into fractions having opposite optical activity by elution chromatography on an optically active crystalline support, such as poly[(S)-3-methyl-1-pentene][3], thus proving the stereoselectivity(Scheme 1) of the polymerization process.

Moreover co-isotactic copolymers of racemic 3,7-dimethyl-1-octene with ethylene show the same separability as the polymeric product prepared from racemic 3,7-dimethyl-1-octene, thus excluding that the chiral discrimination of the two antipodes of the racemic olefin is determined by the growing chain[4]. These last results clearly demonstrate the intrinsic chirality of the active centers, the catalyst consisting of a "racemic" mixture of sites having opposite absolute

Scheme 1

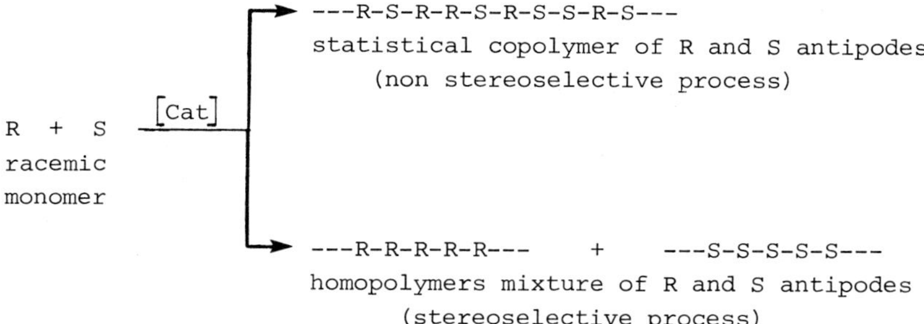

configuration.

More recent confirmation[5,6] comes from the different reactivity of diasterotopic faces of 3-methyl-1-pentene and 3,7-dimethyl-1-octene, as evaluated on the basis of ^{13}C-NMR analysis.

The preparation of conventional Ziegler-Natta catalysts containing active sites of one single chirality has been attempted using optically active metal alkyls. These last systems are indeed able to induce the preferential polymerization of one antipode of the racemic olefin mixture(stereoelective polymerization, Scheme 2), but the stereoelectivity is rather low, the enantiomeric prevalence in the polymeric product being lower than 15-20%[7].

Scheme 2

R + S racemic monomer	optically active ——→ catalyst	---R-R-R-R-R--- homopolymer of R antipode	+	S nonpolymerized S monomer antipode

These results did not allow to demonstrate if the low stereoelectivity observed was due to the prevalence of sites of one single absolute configuration or to the asymmetric perturbation connected with the presence of an optically active alkyl group in the active sites and/or in their environment.

Use of optically active low and high molecular weight Lewis bases as modifiers or supports for conventional Ziegler-Natta catalysts

The addition of a proper amount of an optically active Lewis base to a conventional Ziegler-Natta catalyst offers an additional possibility for obtaining stereoelective polymerization of racemic α-olefins. When the chiral base is added to the preformed catalyst, the active sites of one single absolute configuration can be selectively poisoned; on the other side when the chiral Lewis base is added during catalyst preparation it can in principle induce the formation "in situ" of active sites of one single chirality. Several experiments have been performed changing the order of addition of the catalyst components, the titanium derivative/Lewis base molar ratio and the chemical structure of the chiral base.

Using optically active low molecular weight amines, such as (+)(R)-α-phenylethylamine(PEA), (+)(R)-N,N-dimethyl-α-phenylethylamine(PMA), (+)(S)-N,N-diisobutyl-2-methylbutylamine(MBA) and (-)-sparteine(Sp) in combination with $TiCl_4$ or $TiCl_3$ and $Al(\textit{i}.Bu)_3$, the polymerization of racemic 3,7-dimethyl-1-octene(Table 1) or 4-methyl-1-hexene(Table 2) shows a very low stereoelective character(< 4%), independently of experimental conditions[8,9] (amine/titanium molar ratio, order of addition of catalyst components). However in the presence of (-)-sparteine, even if the stereoelectivity is substantially negligible(Table 2), the stereospecificity of the catalytic system in the polymerization of racemic 4-methyl-1-hexene is markedly increased[9]. In fact the diethylether insoluble polymer fraction shows very high percent weight and isotacticity index I_s, never observed with the same catalyst in absence of the chiral amine(Table 2).

Table 1. Stereoelective polymerization of racemic 3,7-dimethyl-1-octene by conventional Ziegler-Natta catalysts chirally modified by (+)(R)-α-phenylethylamine(PEA)[8]

Catalyst	PEA/Ti	Recovered monomer		Diethylether insol. polymer fraction	
	(mol/mol)	o.p.[a)] (%)	prev. chir.[b)]	o.p.[a)] (%)	prev. chir.[b)]
$TiCl_3$"ARA"/Al(*i*.Bu)$_3$	0.1	∿0	-	∿0	-
$TiCl_3$"ARA"/Al(*i*.Bu)$_3$	3.3	0.2	R	1.4-3	S
$TiCl_4$/Al(*i*.Bu)$_3$	0.1	n.d.[c)]	-	4	R

[a)] Optical purity; [b)] Prevailing chirality; [c)] Not determined

Table 2. Stereoelective polymerization of racemic 4-methyl-1-hexene by conventional Ziegler-Natta catalysts chirally modified by low molecular weight optically active amines(A^{*})[9]

Catalyst	A^{*}	A^{*}/Ti (mol/mol)	Diethylether insol. polymer[a)] %	I_s[b)]	o.p. (%)	prev. chir.
$TiCl_3$"ARA"/Al(*i*.Bu)$_3$	PEA	3.3	4.3	0.92	∿0	-
$TiCl_4$/Al(*i*.Bu)$_3$	PEA	1.0	24.0	0.92	∿0	-
$TiCl_3$"ARA"/Al(*i*.Bu)$_3$	PMA	3.3	64.2	0.87	1.2	R
$TiCl_3$"ARA"/Al(*i*.Bu)$_3$	MBA	3.3	38.4	0.86	0.2	R
$TiCl_3$"ARA"/Al(*i*.Bu)$_3$	Sp	1.5	66.7	0.92	0.3	R
$TiCl_3$"ARA"/Al(*i*.Bu)$_3$	none	-	52.0	0.85	0	-

[a)] Highest stereoregular polymer fraction; [b)] Isotacticity index, evaluated by IR analysis as D_{995}/D_{964}

This result suggests that (-)-sparteine, due to its bulkiness, is able to coordinate preferentially the more exposed active sites which are responsible of the formation of less stereoregular polymer, the isotacticity degree of the overall polymeric product resulting increased.

Analogous results[8,9], as far as stereoelectivity is concerned, have been obtained in the polymerization of racemic 4-methyl-1-hexene and 3,7-dimethyl-1-octene, using low molecular weight optically active ethers and esters, such as (-)-menthylethylether(MtE), (S)-2-methylbutyl-ethylether(MBE) and (-)-menthyl acetate(MtA), respectively, in combination with conventional Ziegler-Natta catalysts(Table 3).

Table 3. Stereoelective polymerization of racemic 4-methyl-1-hexene (4MH) and 3,7-dimethyl-1-octene(DMO) in the presence of conventional Ziegler-Natta catalysts chirally modified by optically active low molecular weight esters and ethers(E^{*})[8,9]

Olefin	Catalyst	E^{*}	E^{*}/Ti (mol/mol)	Diethylether insol. %	I_s	o.p. (%)	polymer prev. chir.
DMO	$TiCl_4/Al(i.Bu)_3$	MtA	0.01	n.d.	n.d.	0.9-2	S
DMO	$TiCl_4/Al(i.Bu)_3$	MtA	0.1	n.d.	n.d.	5	S
4MH	$TiCl_4/Al(i.Bu)_3$	MBE	0.1	62.0	0.84	0.3	R
4MH	$TiCl_4/Al(i.Bu)_3$	MBE	1.0	22.4	0.87	∿ 0	-
4MH	$TiCl_4/Al(i.Bu)_3$	MtE	0.1	37.0	0.85	∿ 0	-
4MH	$TiCl_4/Al(i.Bu)_3$	MtE	1.0	57.5	0.81	∿ 0	-
4MH	$TiCl_3$"ARA"/$Al(i.Bu)_3$	MtE	0.1	42.5	0.86	∿ 0	-
4MH	$TiCl_3$"ARA"/$Al(i.Bu)_3$	MtE	1.0	50.0	0.88	∿ 0	-
4MH	$TiCl_3$"ARA"/$Al(i.Bu)_3$	MtE	3.0	33.7	0.94	∿ 0	-

It is of interest to point out that in the polymerization experiments of racemic 4-methyl-1-hexene, performed in the presence of (-)-menthyl ethylether in combination with the $TiCl_3$"ARA"/$Al(i.Bu)_3$ catalyst(Tab-

le 3), even if the stereoelectivity is absent, the isotacticity degree of the most stereoregular polymer fraction increases with increasing of the MtE/Ti molar ratio, thus suggesting a behaviour similar to that found in the case of (-)-sparteine. The higher stereoelectivity, observed when (-)-menthyl acetate was used as chiral Lewis base(Table 3), must be probably related to the structure of the polymerized olefin, 3,7-dimethyl-1-octene having the asymmetric carbon atom closer to the double bond than 4-methyl-1-hexene.
This hypothesis is confirmed by the comparison of polymerization experiments of 3,7-dimethyl-1-octene and 4-methyl-1-hexene, carried out by $TiCl_3$"ARA"/Al(*i*.Bu)$_3$ catalyst under the same conditions in the presence of (+)(R)-α-phenylethylamine(Tables 1 and 2).
Optically active polymeric Lewis bases, such as poly[(-)-cholesteryl vinylether], crosslinked and linear poly[(-)-menthylvinylether] and poly[(-)-menthyl acrylate], have been used as supports in combination with $TiCl_4$ or $TiCl_3$ and Al(*i*.Bu)$_3$ for the polymerization of racemic 3,7-dimethyl-1-octene and 4-methyl-1-hexene(Table 4).

Table 4. Stereoelective polymerization of racemic 4-methyl-1-hexene (4MH) and 3,7-dimethyl-1-octene(DMO) by conventional Ziegler-Natta catalysts prepared by reacting $TiCl_3$"ARA" or $TiCl_4$ with Al(*i*.Bu)$_3$ in the presence of linear(*l*-PMtVE) or crosslinked(*c*-PMtVE) poly[(-)-menthylvinylether] as chiral Lewis base.

Olefin	$TiCl_n$	Lewis base (B^*)	B^*/Ti (mol/mol)	Recovered monomer o.p. (%)	Recovered monomer prev. chir.	Polymer o.p. (%)	Polymer prev. chir.
4MH	$TiCl_4$	*c*-PMtVE[a)]	1	0	-	0	-
DMO	$TiCl_4$	*l*-PMtVE	1	1.7	S	11.2	R
DMO	$TiCl_4$	*l*-PMtVE	6	0.5	S	37.1	R
DMO	$TiCl_3$"ARA"	*l*-PMtVE	1	0.04	S	0.3	R

a) Crosslinked with 10 mol% of 1,4-divinyloxybutane.

When $TiCl_4$ was added to the linear isotactic poly[(-)-menthylvinyl-ether] in the presence of $Al(\textit{i}.Bu)_3$, a remarkable stereoelectivity (∿37%) was obtained in the polymerization of racemic 3,7-dimethyl-1--octene[9], whereas the use in the same conditions of poly[(-)-menthyl-vinylether] crosslinked with 1,4-divinyloxybutane does not give rise to any stereoelectivity in the polymerization of 4-methyl-1-hexene (Table 4). However, using the $TiCl_4/Al(\textit{i}.Bu)_3$ catalyst chirally modified by crosslinked poly[(-)-menthylvinylether], a remarkable increase of stereospecificity is observed in the polymerization of racemic 4-methyl-1-hexene as in the case of (-)-sparteine. In fact the crude polymer shows an isotacticity degree of the same order of magnitude (I_s = 0.86) as that found for the most stereoregular fraction of the corresponding polymer obtained by conventional Ziegler-Natta catalysts in absence of any chiral agent. The above results, as far as stereoelectivity is concerned, can be explained by assuming that the interaction between $TiCl_4$ and the swelled chiral crosslinked poly[(-)-menthylvinylether] is much lower than that with the corresponding soluble linear polymer, the resulting modified catalyst in the former case being not able to discriminate the antipodes of the racemic olefin. Analogous hypothesis can be invoked for explaining the very low stereoelectivity observed in the polymerization of racemic 3,7-dimethyl-1-octene when $TiCl_4$ is replaced by $TiCl_3$"ARA", still maintaining the optically active linear poly[(-)-menthylvinyl-ether] as third component (Table 4). In fact, whereas in the former case the reduction of Ti^{IV} to Ti^{III} species by $Al(\textit{i}.Bu)_3$ occurs "in situ" after the coordination of the chiral base to $TiCl_4$, in the latter case the crystalline structure of $TiCl_3$"ARA" largely prevents the above coordination, the catalyst obtained by successive addition of $Al(\textit{i}.Bu)_3$ being not chirally modified to an appreciable extent. Also the polymerization of racemic 3,7-dimethyl-1-octene and 4-methyl-1-hexene, in the presence of conventional $TiCl_4/Al(\textit{i}.Bu)_3$ Ziegler-

Natta catalysts modified by poly[(-)-cholesterylvinylether] shows (Table 5) a rather low degree of stereoelectivity(<4%)[10].

Table 5. Stereoelective polymerization of racemic 4-methyl-1-hexene (4MH) and 3,7-dimethyl-1-octene(DMO) by $TiCl_4/Al(i.Bu)_3$ catalytic system in the presence of poly[(-)-cholesterylvinylether] (PChVE) as chiral Lewis base[a)]

Run	Olefin	PChVE/Ti (mol/mol)	Conversion (%)	Recovered monomer o.p. (%)	Recovered monomer prev. chir.	Polymer o.p. (%)	Polymer prev. chir.
1	4MH	1	9.2	0	-	0	-
2	DMO	1	40.1	0.18	S	0.2	R
3	DMO	3	4.0	0.16	S	3.8	R

[a)] Polymerization temperature: 60°C; solvent: *n*.heptane; olefin/Ti and Al/Ti molar ratios are 30 and 3, respectively.

These results can be explained by assuming that, despite of the solubility of poly[(-)-cholesterylvinylether], the large bulkiness of the cholesteryl groups prevents the coordination of side chain ethereal functions to $TiCl_4$, its successive reduction by $Al(i.Bu)_3$ being not able to produce active sites with a prevailing chirality. However, the fractionation with boiling solvents of the overall polymeric product deriving from the racemic olefin and after separation from poly[(-)-cholesterylvinylether] by elution chromatography, has allowed to separate low and high stereoregular polymer fractions having opposite optical activity(Table 6)[10].

These results indicate that in the catalyst both stereospecific and non stereospecific active sites are present, only the former giving rise, due to the chiral perturbation, to the preferential polymerization of one single antipode of the racemic olefin. On the other side the non stereospecific centers polymerize, without any chiral discrimination, the residual olefin which is progressively enriched in the other antipode. As a consequence, the non stereoregular polymer displays optical activity of opposite sign with respect to the isotactic one.

Table 6. Fractionation by boiling solvents of polymers prepared from racemic 3,7-dimethyl-1-octene by catalytic systems obtained from $TiCl_4/Al(i.Bu)_3$ and different amount of poly[(-)-cholesterylvinylether] (PChVE)

Polymer fraction extracted successively with:	Run 2[a]		Run 3[a]	
	%	$\|\alpha\|_D^{25}$ [b]	%	$\|\alpha\|_D^{25}$ [b]
acetone	7.0	+ 1.4	11.9	+ 2.8
diethylether	32.3	+ 0.2	19.7	+ 0.5
cyclohexane	60.6	- 1.4	68.4	- 2.4

[a] PChVE/Ti molar ratio is 1 and 3 for the runs 2 and 3, respectively;
[b] In cyclohexane solution

Use of chiral agents in supported Ziegler-Natta catalysts

Recently Ziegler-Natta catalysts, supported on $MgCl_2$ have been reported to show exceptional catalytic activity combined with a very high stereospecificity[11].

In general the preparation of the above catalysts provides the milling of anhydrous $MgCl_2$ in presence of a proper amount of an aromatic ester or ether, the absorption of $AlEt_3$ and then the addition of $TiCl_4$. Such a system is then activated by a co-catalyst consisting of a mixture of $AlEt_3$ and an aromatic ester. Chien[12] reported the preparation of a "high yield" Ziegler-Natta catalyst which can be schematized as $MgCl_2/EB/PC/AlEt_3/TiCl_4-AlEt_3/MT$, where EB, PC and MT are ethyl benzoate, *p*.cresol and methyl *p*.toluate, respectively.

In such a system the concentration of active sites, evaluated by radiotagging experiments, has been found to be in the range of 5-30% with respect to the added titanium derivative, much higher than in conventional Ziegler-Natta catalysts.

E.P.R. studies have also allowed to evaluate the presence of paramagnetic Ti^{III} species which result about 20-25% of the total amount of Ti^{III} species; in addition possible structures of both paramagnetic and diamagnetic Ti^{III} species are reported[13].

Pino et al.[14] have proved that the polymerization of racemic 4-methyl-1-hexene, in the presence of "high yield" Ziegler-Natta catalysts such as $MgCl_2/TiCl_4$/ethylanisate-$AlEt_3$/ethylanisate, is stereoselective(Table 7).

Table 7. Comparison between the separability(D) of some fractions of the polymerization product of racemic 4-methyl-1-hexene(4MH), obtained by both $MgCl_2$ supported and conventional Ziegler-Natta catalysts, and the separability(D_{st}) of a statistical copolymer of (R)- and (S)-4MH[14]

Catalyst	Polymer fraction	D (%)	D_{st} (%)
$TiCl_3$"ARA"/$Al(i.Bu)_3$	equimolar mixture of poly(R)- and poly(S)-4MH; diethylether insol. benzene sol. fraction	100	-
$MgCl_2/TiCl_4$/EA-$AlEt_3$/EA[a)]	acetone insol. diethylether sol. fraction	13	2.5
$TiCl_3$"ARA"/$Al(i.Bu)_3$	acetone insol. ethyl acetate sol. fraction	13.6	6.0
$MgCl_2/TiCl_4$/EA-$AlEt_3$/EA[a)]	diethylether insol. benzene sol. fraction	30.8	1.0
$TiCl_3$"ARA"/$Al(i.Bu)_3$	diisopropylether insol. isooctane sol. fraction	18.5	2.75

[a)] EA = ethylanisate

In fact the intrinsic separability(D) of some fractions of the polymerization product obtained from racemic 4-methyl-1-hexene, in the presence of the "high yield" $MgCl_2$ supported catalyst, is higher than that(D_{st}) expected for the statistical copolymer of (R)- and (S)-4-methyl-1-hexene having the same molecular weight.
It is of interest to point out that the stereoselectivity of the polymerization process promoted by the $MgCl_2$ supported catalyst seems to be higher than that found for conventional Ziegler-Natta catalysts(Table 7).
Recently[15] the stereoselective character of the polymerization of α-olefins has been confirmed for the "high yield" supported $MgCl_2$/EB/PC/$TiCl_4$-Al(*i*.Bu)$_3$/MT catalyst by copolymerizing racemic 3,7-dimethyl-1-octene with (S)-3-methyl-1-pentene. In fact, according to the Scheme 3, the fractionation of the crude polymer by boiling solvents has allowed to separate the copolymer of (S)-3,7-dimethyl-1-octene with (S)-3-methyl-1-pentene, having positive optical activity, from the homopolymer of (R)-3,7-dimethyl-1-octene, whose optical rotatory power is of negative sign.

Scheme 3

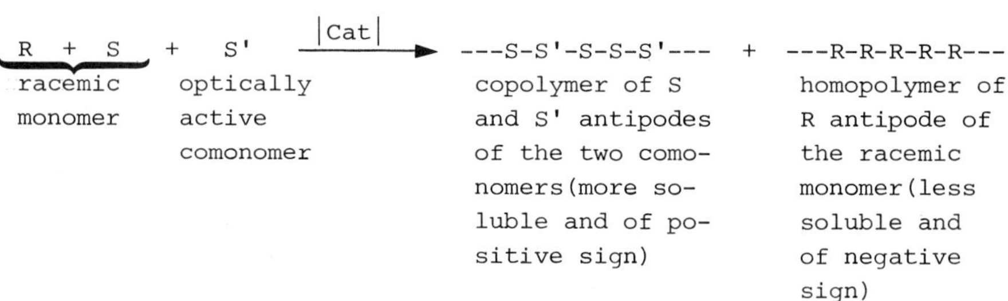

All these results suggest that the active sites of high activity $MgCl_2$ supported catalysts are intrinsically chiral, as those present in conventional Ziegler-Natta catalysts.

In this context the use of chiral Lewis bases, having ether or ester functions, in combination with $MgCl_2$ supported Ziegler-Natta catalysts, appeared very promising for studying the stereochemistry of racemic α-olefins polymerization process. In fact , the high concentration of active sites in such systems allowed to change in a wide range the chiral Lewis base/titanium molar ratio without large decrease of catalytic activity. Moreover, due to the fact that the addition of the activating system(the aluminum alkyl/Lewis base complex) causes the further reduction of residual Ti^{IV} to Ti^{III} species, giving rise to new active sites[13], the formation "in situ" of catalytic centers of one single chirality could also be achieved when using an optically active Lewis base.

Accordingly Pino et al.[16], using the "high yield" $MgCl_2/TiCl_4 \cdot B^*$-$AlEt_3 \cdot B^*$ catalyst, in which the chiral base B^* [(-)-menthylanisate] is added both in the first step of catalyst preparation(internal base) and in the activating system(external base), observed that the polymerization of racemic 4-methyl-1-hexene is stereoelective(Table 8). The optical activity of the recovered monomer was found to increase, in the experiments with similar conversion, with increasing the concentration of B^* in the catalyst(Table 8). In addition, the stereoelectivity of the polymerization process resulted to be higher than that found for the same olefin when chirally modified conventional Ziegler-Natta catalysts were used[8,9].

The fractionation of the crude polymers by boiling solvents allowed to establish that the prevailing chirality of the monomeric units included in the most and in the lowest stereoregular polymer fractions is opposite(Table 8). These results have been explained[16] on the basis of the presence in the "high yield" Ziegler-Natta catalysts of stereospecific active sites having chiral discrimination capability as well as of non stereospecific centers whose enantiomeric discrimination capability is substantially lacking.

Table 8. Stereoelective polymerization of racemic 4-methyl-1-hexene by $MgCl_2/TiCl_4 \cdot B^{*}-AlEt_3 \cdot B^{*}$ catalyst using different $[B^{*}]/([Al]+[Ti])$ molar ratios(r)[a]

r	Conversion (%)	Recovered monomer		Polymerized monomer			
				A[b]		E[c]	
		o.p. (%)	prev. chir.	o.p. (%)	prev. chir.	o.p. (%)	prev. chir.
0.131	88	n.d.	n.d.	0.67	S	0.68	R
0.408	15	0.3	S	0.25	S	3.33	R
0.501	40	1.1	S	0.55	S	1.48	R
0.538	66	2.2	S	1.39	S	1.62	R
0.796	43	2.7	S	0.63	S	1.67	R

[a] B^{*} = (-)-menthylanisate; [b] Acetone insoluble, ethyl acetate soluble fraction having a low degree of stereoregularity; [c] Diisopropylether insoluble, methylcyclohexane soluble fraction having the highest stereoregularity degree.

As mentioned in the previuos section, similar results were obtained with conventional Ziegler-Natta catalysts modified by optically active Lewis bases, thus suggesting that the origin of stereospecificity is similar in both systems.

In this context polymerization experiments of racemic 3,7-dimethyl-1-octene have been performed in the presence of the $MgCl_2$/EB/PC/$AlEt_3/TiCl_4$ catalyst activated by a mixture of $Al(i.Bu)_3$ and an optically active external base such as (-)-menthyl benzoate, poly[(-)-menthylvinylether] and poly[(-)-menthyl acrylate][17].

As expected, no appreciable activity decrease was found using the overmentioned chirally modified catalytic systems. Even if in all cases the chiral discrimination capability in the polymerization process is very low(< 2%), the stereoelectivity in the presence of poly[(-)-menthyl acrylate] is higher than that found in the case of (-)-menthyl benzoate(1.4 and 0.3%, respectively). It is of interest to point out that, using the above polymeric chiral agent, the enantiomeric discrimination capability is opposite to that observed for (-)-menthyl benzoate, even if menthyl groups of the same chirality are present in both systems. These data suggest that the dissymmetric environment of the macromolecule in the polymeric support plays an important role in affecting the chiral discrimination of the active sites. On the other hand the low stereoelectivity observed can be ascribed to the high structural stability of the active sites, the relative amount of centers of opposite absolute configuration being not largely affected by the interaction with the external chiral Lewis base.

The polymerization of racemic 3,7-dimethyl-1-octene was also carried out in the presence of $MgCl_2$/(-)-menthyl benzoate/PC/$AlEt_3$/$TiCl_4$-$Al(i.Bu)_3$/(-)-menthyl benzoate as well as of $MgCl_2$/(-)-menthyl benzoate/PC/$AlEt_3$/$TiCl_4$-$Al(i.Bu)_3$/MT catalysts. In the former system the chiral ester has been used as both internal and external base, whereas in the latter catalyst only as internal base. In both experiments the polymerization process has been found to be stereoelective, the chiral discrimination capability being one order of magnitude higher than that observed in the previous experiment where $MgCl_2$/EB/PC/$AlEt_3$/$TiCl_4$-$Al(i.Bu)_3$/(-)-menthyl benzoate catalyst was used(> 4 and 0.3%, respectively). In addition,in the last two polymerization experiments the chiral discrimination is found to be opposite with respect to that observed in the case of the $MgCl_2$ supported catalyst where (-)-menthyl benzoate was used as external base only, the antipode of racemic 3,7-dimethyl-1-octene prevailingly polymerized being R and S, respectively.

Accordingly, the stereoelectivity found with the $MgCl_2$/(-)-menthyl benzoate/PC/$AlEt_3$/$TiCl_4$-Al(*i*.Bu)$_3$/MT catalyst seems to be higher than that observed with the $MgCl_2$/(-)-menthyl benzoate/PC/$AlEt_3$/ $TiCl_4$-Al(*i*.Bu)$_3$/(-)-menthyl benzoate catalyst(4.9 and 4.0%, respectively). All these data suggest that in the "high yield" supported Ziegler-Natta catalysts two different kinds of active sites exist. The catalytic centers formed in the first step of catalyst preparation, by the addition of the internal optically active base, have much higher chiral discrimination capability and probably are also connected with a very high degree of stereospecificity. The additional active sites, formed in the further reduction of Ti^{IV} to Ti^{III} species by the addition of the aluminum alkyl/chiral external base complex as activator, display a very low degree of stereoelectivity with opposite chirality. These last sites are probably more exposed in terms of steric requirements and should also be connected with a low degree of stereospecificity.

FINAL REMARKS

The previously reported and new data, concerning the polymerization of racemic α-olefins by conventional and new generation supported Ziegler-Natta catalysts, indicate that in both systems active sites capable of chiral discrimination are present, thus suggesting a similar origin of stereospecificity. However distinct differences between the two classes of catalysts are evidenced, suggesting that the heterogeneous support, carrying on its surface the catalytic complexes, not only affects their activity but also the stereochemical features of the polymerization process. In spite of the large number of questions still unanswered, the aim of the present work is that to stress the enormous potentiality of the use of both chiral monomers and catalytic components for obtaining structural and stereochemical information concerning the active sites of the "high yield" Ziegler-Natta polymerization catalysts.

ACKNOWLEDGEMENT

The financial support from Ministry of Public Education(Ricerca Scientifica) and the Research National Council(Rome) are gratefully acknowledged.

REFERENCES

1. Yu. I. Yermakov, V.A. Zakharov, in "Coordination Polymerization. A Memorial to Karl Ziegler", J.C.W. Chien Ed., Academic Press Inc., New York, 1975, p. 91.
2. O.S. Roshchupkina, A.P. Lisitskaya, N.D. Golubeva, Kinetika i Kataliz., 23, 1208(1982).
3. P. Pino, A. Oschwald, F. Ciardelli, C. Carlini, E. Chiellini, in "Coordination Polymerization. A Memorial to Karl Ziegler", J.C. W. Chien Ed., Academic Press Inc., New York, 1975, p. 25.
4. F. Ciardelli, P. Locatelli, M. Marchetti, A. Zambelli, Makromol. Chem., 175, 923(1974).
5. A. Zambelli, P. Ammendola, M.C. Sacchi, P. Locatelli, G. Zannoni, Macromolecules, 16, 341(1983).
6. M.C. Sacchi, I. Tritto, P. Locatelli, D.R. Ferro, Makromol. Chem., Rapid Commun., 5, 731(1984).
7. F. Ciardelli, C. Carlini, G. Montagnoli, L. Lardicci, P. Pino, Chim. Ind.(Milan), 50, 860(1968).
8. C. Carlini, R. Nocci, F. Ciardelli, J. Polym. Sci., Polym. Lett. Ed., 15, 767(1977).
9. C. Carlini, F. Ciardelli, Chim. Ind.(Milan), 63, 486(1981).
10. F. Menconi, Thesis, University of Pisa, 1985.
11. Montecatini-Edison, Ger. Offen., 2,347,577; Chem. Abstr., 83, 59870i(1975).
12. J.C.W. Chien, J.-C. Wu, C.-I. Kuo, J. Polym. Sci., Polym. Chem. Ed., 20, 2019(1982).
13. J.C.W. Chien, J.-C. Wu, J. Polym. Sci., Polym. Chem. Ed., 20, 2461(1982).
14. P. Pino, G. Fochi, A. Oschwald, G. Piccolo, R. Mülhaupt, U. Giannini, in "Coordination Polymerization", C.C. Price and E. J. Vandenberg, Plenum Publishing Co., 1983, p. 207.

15. F. Ciardelli, C. Carlini, A. Altomare, F. Menconi, J.C.W. Chien, Paper presented at Internat. Symp. on "Transition Metal Catalyzed Polymerization", Akron, June 16-20, 1986.
16. P. Pino, G. Fochi, O. Piccolo, U. Giannini, J. Am. Chem. Soc., 104, 7381(1982).
17. C. Carlini, F. Menconi, A. Altomare, F. Ciardelli, Chim. Ind. (Milan), in press.

INTERRELATION BETWEEN HOMOGENEOUS AND HETEROGENEOUS ACID-BASE (ACCEPTOR-DONOR) CATALYSIS. TITANIUM(III) SUPPORTED AND HOMOGENEOUS SYSTEMS

Wincenty Skupiński
Institute of Organic Technology, Faculty of Chemistry, Technical University (Politechnika), 00-662, Warszawa, Poland

ABSTRACT - Two kinds of titanium(III) complexes were prepared:
- supported titanium(III) complexes obtained by BuLi reduction of $CpTiCl_2$-O-M- system, where M= Al or Si of alumina-silica gels with various alumina:silica ratios, and
- homogeneous titanium(III) complexes obtained by BuLi reduction of $CpTiCl_2$-O-Ph-X compounds, where X = CH_3, Cl, NO_2 in para- and meta-positions.

The physico-chemical and catalytic (olefin hydrogenation) properties of both catalytic systems were investigated and compared taking into account the donor-acceptor effect of the ligands: alumina-silica gels and substituted phenoxy rings.

1. INTRODUCTION

Many homogeneous catalytic systems are known in which activity and selectivity are controlled by changing donor-acceptor properties of the ligands attached to the metal core (Ref. 1-3). The same case was described by us (Refs. 4-6) for supported nickel and titanium complexes where the role of "classical" ligands was played by alumina-silica gels having various Al_2O_3/SiO_2 ratios and thanks to that changing donor-acceptor properties.

To compare the influence of inorganic gels on supported complexes with the influence of classical ligands, we synthesized titanium complexes with the following formula: $CpTiCl_2$-O-Ph-X (where X= CH_3, Cl, NO_2 in para- and meta positions), and we assumed that -O-Ph-X group could play the role of -O-M- parts of the supported cyclopentadienylo titanium complexes: $CpTiCl_2$-O-M-(where M= Si, Al of alumina-silica gel surface).

The both systems were reduced with BuLi and the physico-chemical and catalytic properties in olefin hydrogenation were determined.

2. EXPERIMENTAL

The preparation and physico-chemical properties of the heterogenized cyclopentadienylotitanium(IV) complexes having formula: $CpTiCl_2$-O-M- and the homogeneous complexes having formula: $CpTiCl_2$-OPhX were described in our previous papers (Ref. 6-8).

Acid-base (acceptor-donor) properties of alumina-silica, containing 100, 80, 64, 46, 18 and 0% of Al_2O_3, used as supports of the investigated titanium complexes were determined by Benesi and Tanabe methods (Ref. 9 & 10) by titration with butylamine (acid centers) and bensoic acid (basic centers) in the presence of the Hammett indicators.

The donor-acceptor properties of the substituted phenoxy rings are determined by the substituent Hammett factors.

The both kinds of cyclopentadienyltitanium(IV) complexes were reduced with a 3 M hexane solution of BuLi with a constant BuLi/Ti ratio of 10:1 in benzene at room temperature for 2 hours.

EPR measurements were carried out on a 200 Wrocław spectrometer using Mn^{2+} in MgO as an internal standard at room temperature.

The catalytic reactions were carried out in static reactors at room temperature with the ratios: olefin/Ti= 4 : 1, H_2/olefin = 8 : 1.

Reaction progress was monitored by measuring pressure drop and GLC. All operations were carried out using dry and deoxidated argon and solvents.

3. RESULTS AND DISCUSSION

3.1. Ligands Characterization

Donor-acceptor properties of the alumina-silica gels were measured using Benesi-Tanabe method. The strength and the amount of acid (acceptor) and basic (donor) centers are presented in Figure 1 and Figure 2.

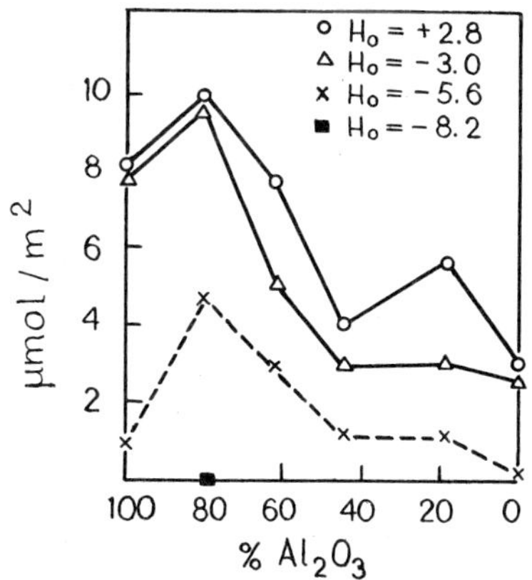

Figure 1. The acidity (Hammett scale) of silica-alumina

The amount of the basic centers increases with increasing Al_2O_3 content in the gels. The amount of the acid (acceptor) centers shows two maxima: the first (greater) is for 80% and the second for 18% of Al_2O_3 gels. All these dependences are typical for alumina-silica gels (Refs. 11-12).

Donor-acceptor properties of the substituted phenoxy ligands are determined by the Hammett factors which are presented in Table 1.

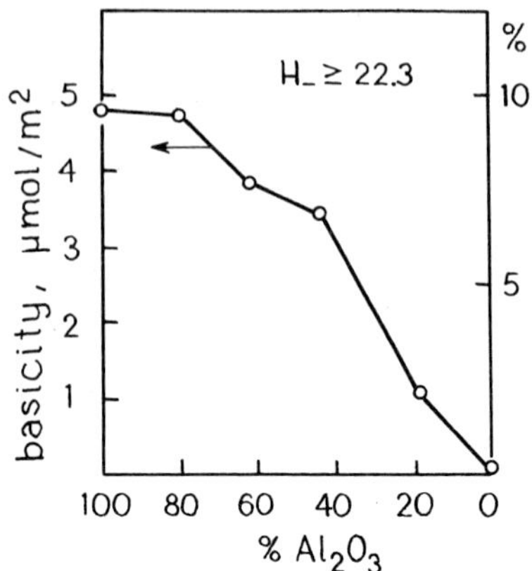

Figure 2. The basicity (Hammett scale) of silica-alumina

TABLE 1. Hammett factors of Substituents (Ref. 16)

$p-CH_3$	$m-CH_3$	H	p-Cl	$m-NO_2$	$p-NO_2$
-0.17	-0.07	0	0.23	0.71	1.24

3.2. EPR Measurements

A homogeneous yellow or red benzene solution of the $CpTiCl_2OPhX$ compounds changed colour to brown-black after addition of the reductant. The EPR spectra have wide singlets characteristic of anisotropic, axial symmetry of titanium(III) complexes (Ref. 13). The values of $g_\perp$ and $g_\parallel$ constants are presented in Table 2.

TABLE 2. $g_\perp$ and $g_\parallel$ constants of the EPR spectra of the $CpTiCl_2OPhX$ + BuLi system

X	$p-CH_3$	$m-CH_3$	H	p-Cl	$m-NO_2$	$p-NO_2$
$g_\perp$	1.991	1.984	1.981	1.992	2.033	2.007
$g_\parallel$	1.975	1.976	1.978	1.984	2.002	2.003

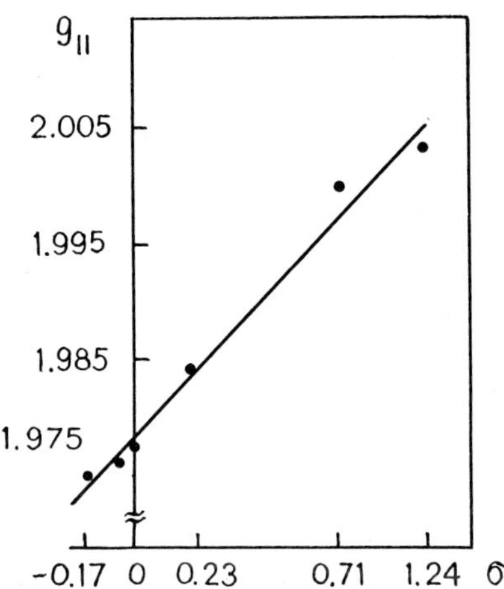

Figure 3. Relationship between the values of σ and $g_{\parallel}$.

The $g_{\perp}$ constants are larger than the $g_{\parallel}$ constants for all titanium(III) complexes and are very close to the g_e value (free electron) of 2.0023. These facts suggest a tetrahedral symmetry for the titanium(III) complexes obtained and unpaired electron localization on the $d_{x^2-y^2}$ orbital (Ref. 14).

The $g_{\parallel}$ constant values are very sensitive to the ligand-metal ion interaction within the complex (Ref. 15).

The $g_{\parallel}$ constant values of the titanium(III) complexes show dramatic changes depending on the kind of the substituent in the phenoxy ring and they correlate directly with the Hammett factors (see Fig. 3).
They produce also anisotropic EPR singlets characteristic of titanium(III) ions located in the axial symmetry of the ligands. The $g_{\perp}$ and $g_{\parallel}$ constants of the EPR signals obtained are presented in Table 3.

This proves the existence of the phenoxy rings in the coordination sphere of the titanium(III) ions in our systems and a strong influence of the donor-acceptor properties of the phenoxy ring on the unpaired electron density around the titanium(III) ions. This influence

TABLE 3. $g_{\perp}$ and $g_{\parallel}$ constants of the EPR spectra of the heterogenized system

Al_2O_3%	100	80	64	46	18	0
$g_{\perp}$	1.988	1.990	1.992	1.999	-	1.985
$g_{\parallel}$	1.962	1.959	1.982	1.983	-	1.979

can occur only when the titanium $d_{x^2-y^2}$ orbital is involved in a direct bond with the phenoxy ring ($d_{x^2-y^2}$ orbital should form π bond with π orbitals of the phenoxy ring) and the most probable structure of the titanium(III) complexes under investigation should be dimeric:

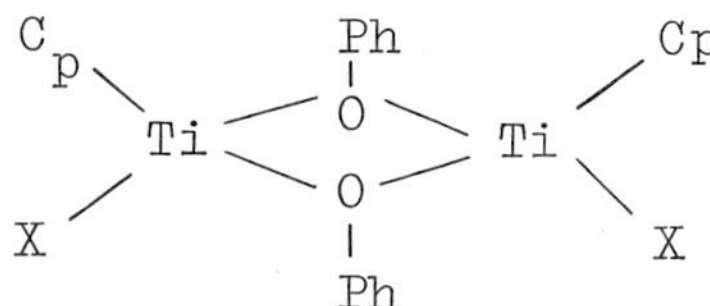

where X = alkyl, Cl or H

The heterogenized yellow or orange titanium(IV) surface complexes after reduction with BuLi gave brown or black grains. Part of titanium complexes was transferred from the grain surfaces to the solution as a result of the Ti-O bond rupture where besides alkylation of the Ti-Cl bonds, alkylation of the Ti-O bonds also occurred. After washing grains with hexane on the gel surfaces only titanium complexes strongly bonded to the surfaces remained on the grains. They should have the titanium-surface lattice oxygen bonds.

The $g_{\parallel}$ constants (which, as it was mentioned above, are very sensitive to the ligand-metal ion interaction) of the obtained signals largely depend on the gel composition, and as it is shown in Figure 4, the $g_{\parallel}$ values decrease with increasing basic (donor) properties of the supports.

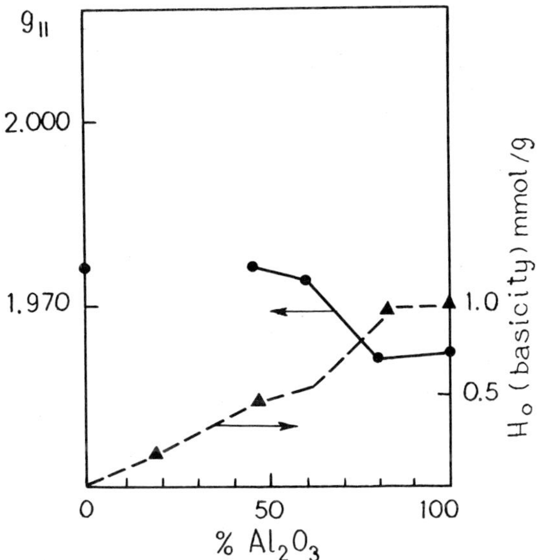

Figure 4. Alumina-silica basicity; $g_{\parallel}$ factors as function of alumina-silica composition.

For both homogeneous and heterogenized catalysts the increase in ligand donor properties causes the decrease in $g_{\parallel}$ constants of the titanium(III) complexes.

If the $g_{\parallel}$ constants for both systems are the same it would be possible formally to relate the determined alumina-silica gel composition with the determined σ Hammett factors of the phenoxy ring substituents.

The $g_{\parallel}$ constants of the EPR signals of the homogeneous complexes are in the range from 1.976 (para-CH_3) to 2.002 (para-NO_3).

The $g_{\parallel}$ constants of the EPR signals of the heterogenized complexes are in the range of 1.979 for least basic SiO_2 and 1.962 for the most basic Al_2O_3.

Since the ligand donor properties were responsible for the decrease in $g_{\parallel}$ constants in our systems, generally it may be stated that all alumina-silica gels have a stronger donor influence on titanium(III) complexes than the substituted phenoxy rings used. Only the most donor phenoxy ring (para-CH_3) characterized by σ = -0.27 may correspond to the least donor gel - SiO_2.

3.3. Olefin Hydrogenation

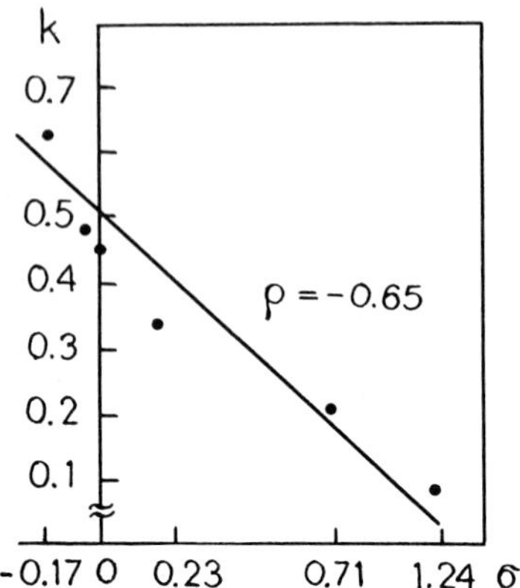

Figure 5. Relationship between the values of σ and k for cyclopentene hydrogenation

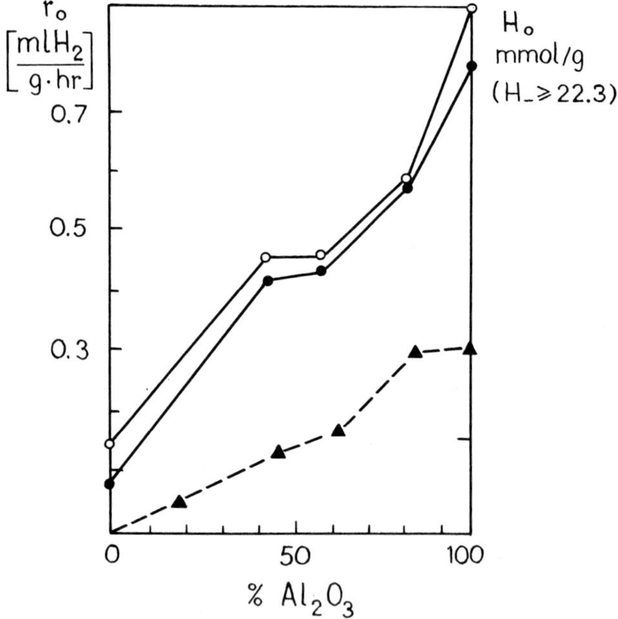

Figure 6. Alumina-silica basicity; ethylene (o) and propylene (●) hydrogenation over supported titanium(III) complexes.

The both kinds of the systems under consideration (heterogenized in propylene and ethylene and homogeneous in cyclopentene) were tested in hydrogenation.

The activity of both systems calculated as H_2 consumption per gram of catalyst per hour as a function of the σ Hammett factors for the homogeneous system (Figure 5) and as function of the support composition for the heterogenized system (Figure 6) is directly dependent on the donor properties of the ligands. The stronger donor properties of the ligand, the more efficient is the catalyst.

REFERENCES

1. B. Bogdanović, B. Henc, H.G. Karmann, H.G. Nossel, D. Walter and G. Wilke, Ind. Eng. Chem., 62, 34 (1970).
2. G. Henrici-Olive and S. Olive, Advan. Polym. Sci., 6, 421 (1969).
3. W. Skupiński and A. Wasilewski, J. Organometal. Chem., 282, 69 (1985).
4. W. Skupiński and S. Malinowski, J. Molec. Catal., 4, 95 (1978).
5. W. Skupiński, M. Zawartko and S. Malinowski, React. Kinet. Catal. Lett., 13, 319 (1980).
6. W. Skupiński and I. Cieslowska-Glinska, J. Organometal. Chem., 269, 29 (1984).
7. W. Skupiński and I. Cieslowska-Glinska, React. Kinet. Catal. Lett., 23, 349 (1983).
8. W. Skupiński and A. Wasilewski, J. Organometal. Chem., 220, 39 (1981).
9. H.A. Benesi, J. Am. Chem. Soc., 78, 5490 (1956).
10. K. Tanabe and H. Hattori, Chem. Lett., 625 (1976).
11. E. Fedoryńska, T. Woźniewski and S. Malinowski et.al., J. Colloid. Interf. Sci., 69, 469 (1979).
12. D.O. Scokart, F.D. Declerk, R.E. Sempels and P.G. Rouxet, J. Chem. Soc. Faraday, 73, 359 (1977).
13. J. Skupińska and W. Skupiński, React. Kinet. Catal. Lett., 16, 297 (1981).
14. J.E. Wertz, and J.R. Bolton, in: Electron Spin Resonance, Mc Graw Hill, N.Y. (1972).
15. I.N. Marow, N.A. Kostromina, EPR i NMR w Khimii Koordicionnyh Soedynenij, Nauka, Moskva (1979).
16. J. Shorter, Analysis in Organic Chemistry. An Introduction to Linear Free Energy Relationships, Oxford University Press, Oxford (1973).

OLIGOMERIZATION AND POLYMERIZATION OF MONOOLEFINS VIA HOMOGENEOUS AND HETEROGENIZED NICKEL CATALYSTS

W. Keim

Institut für Technische Chemie und Petrolchemie der Rheinisch-Westfälischen Technischen Hochschule Aachen, Worringer Weg 1, D-5100 Aachen, F.R.G.

ABSTRACT - The three chelate nickel complexes 1-3 will be discussed as catalyst precursors - in solution and heterogenized on organic and inorganic supports - for the oligomerization and polymerization of ethene and higher monoolefins.

1. INTRODUCTION

The nickel catalysed oligomerization of olefins is of considerable academic and industrial interest for the synthesis of linear and branched higher monoolefins. The linear olefins are key intermediates for detergents, plasticizers, and a variety of fine chemicals. The branched ones represent potential candidates for improving octance numbers. Of great industrial importance is the oligomerization of ethene to: α-olefins (SHOP-Process, 2 plants of about 400.000 t/a), the dimerization of propene (Dimersol-Process, 16 plants of 3 mio t/a), and the codimerization of propene with butene (2 plants).

Since many years we are working on the oligomerization of olefins. For our catalyst selection based on nickel the model shown in Fig. 1 was chosen

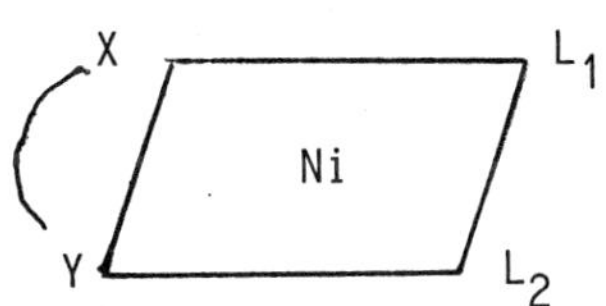

X⌒Y donor and acceptor
L_1 and L_2 must be able to dissociate easily

Figure 1. Model for olefin oligomerization

Square planar nickel complexes provide appropriate orbitals to interact with incoming olefins. Chelate ligands favour square planar structures. In addition, the chelate should minimize the chance of coordinating olefins formed during the catalytic cycle. X and Y were chosen among soft and hard ligands. The ligands L_1 and L_2 should easily dissociate, thus providing empty coordination sites for the olefins to be oligomerized.

2. CHELATE NICKEL COMPLEXES

The complexes 1-3 shown in Figure 2 fullfill the requirements of the proposed model. They are square planar as known from their x-ray structures. They also contain ligands which can dissociate easily.

Figure 2. Chelate complexes for olefin oligomerization

Interestingly, the complex 1 dissolved in toluene yields at 70-90°C and an ethene pressure of 30-50 bar α-olefins of > 99 % linearity and α-olefin content of > 95 %. Complex 1 is also active in the SHOP-Process (Ref. 1), but is practically inactive for other monoolefins. However, the complex 2 is active for the oligomerization of ethene and higher monoolefins (Ref. 2). Table 1 lists some results obtained with propene applying 2 in homogeneous phase.

TABLE 1. Oligomerization of propen with 2 (75 % conversion, 80°C)

Oligomers	w %	Linearity %
C_6	72.1	75
C_9	23.2	56
$C_{12}+$	4.7	33

The complex 3 in solution is inactive for oligomerising ethene or α-olefins. However, mixtures of bis(cyclooctadiene-1,5)nickel with aminophosphoranes yield active catalysts or the polymerization of ethene (Ref. 3).

A great disadvantage in olefin oligomerizations is the difficulty to separate the catalyst from the reaction products thus enabling catalyst recycle. In the SHOP-Process recycle was worked out operating in two-phase-systems (Ref. 4). Often such a recycle is impossible in polar solvents and a more general procedure is needed. Therefore, we have heterogenized complexes of type 1 to 3 applying organic and inorganic supports.

2.1 Ethene oligomerization using organic supports

Attempts to heterogenize 1 on polystyrene is outlined in Figure 3.

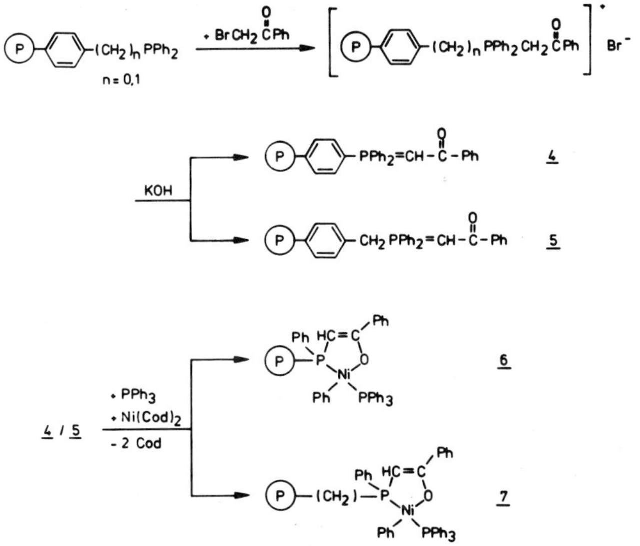

Figure 3. Heterogenization of complex 1 on polystyrene

The heterogenized systems 6 and 7 in toluene suspension yielded with ethene α-olefins (Ref. 5). In conformity with homogeneous 1, the oligomers produced are 96-100 % linear. The α-olefin content, however, varied over all experiments, indicates that the heterogenized systems isomerize the product olefins. The chain-growth factor ß lies in the same range as with catalyst 1. In long time experiments, the higher olefins (> C_{30}), formed, because of the geometric distribution of the product olefins, block the pores thus shortening catalyst life to an unbearable level.

By reacting 1,3-polydentates

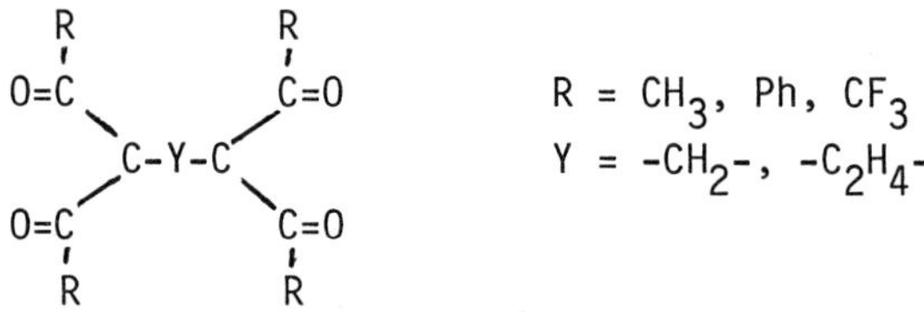

with bis(cyclooctadiene-1,5)nickel insoluble systems were obtained, which show only poor activity (< 5 %) for ethene, but the dimers are highly linear (> 70 %) (Ref. 6). These results are indicative for the importance of the chelate acetylacetonate of complex type 2 in oligomerization reactions.

2.2 Ethene oligomerization using inorganic supports

Impregnation of silica and silica-alumina with the nickel complexes 1 and 2 yields active supported solid-phase catalysts (SSPC). The results of the ethylene oligomerization experiments are summarized in Table 2.

The results can be interpreted that upon heterogenizing the unique porperties of the P⌒O and O⌒O chelate ligand of 1 and 2 prevail. (The supports alone or supports impregnated with a Ni(II) salt showed no activity at all under the same reaction conditions.) Catalyst 1/Al_2O_3-SiO_2 showed only poor linearity. This can be understood by assuming that insertion of primary product molecules such as butene or n-hexenes into the growing chain leads to branched oligomers, thus lowering the linearity of the product olefins.

TABLE 2. Ethylene oligomerization with SSP-catalysts in fixed-bed continous flow reactor

Catalyst	T (°C)	Turnover number (s^{-1})	Linearity (%)	ß-Factor (C_6-C_{16})
1/Al_2O_3-SiO_2	50	0.2	26	1.8
1/SiO_2	50	0.05	80	-
2/Al_2O_3-SiO_2	50	0.1	72	1.8
2/Al_2O_3-SiO_2	50	0.7	80	3.1
2/Al_2O_3-SiO_2	75	0.1	92	1.4
2/SiO_2	50	0.003	87	0.9

Support of 1 on pure SiO_2 gave highly linear products, whereas 1/ Al_2O_3-SiO_2 leads to branching. Using acetylacetonate as chelate (complex 2) didn't show a support dependence. All products using 1 or 2 are highly isomerized showing that the supported complexes are isomerization catalysts for the products olefins.

To controll the ß-factor by cutting off the higher molecular fractions, experiments were carried out synthesizing 1 and 2 in zeolithes. Indeed, by using faujasite a certain control of the ß-factor could be obtained via 2 as is evident from Figure 4.

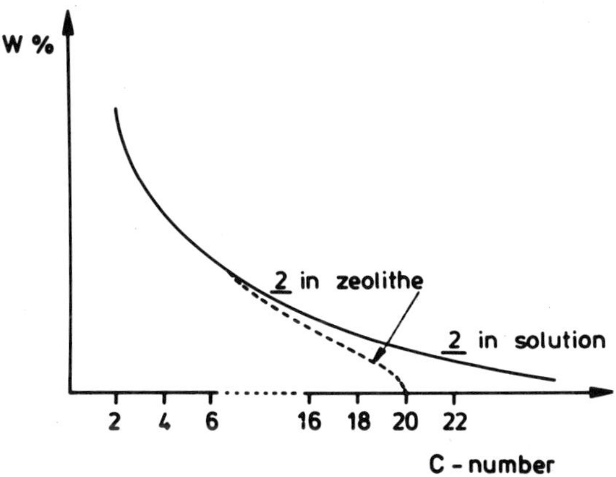

Figure 4. Product distribution of 2 in faujasite

Interestingly, by supporting complex 3 on SiO_2 the reaction with ethenes yields polyethylene PE II having a structure which lies between HDPE and LDPE as indicated in Figure 5.

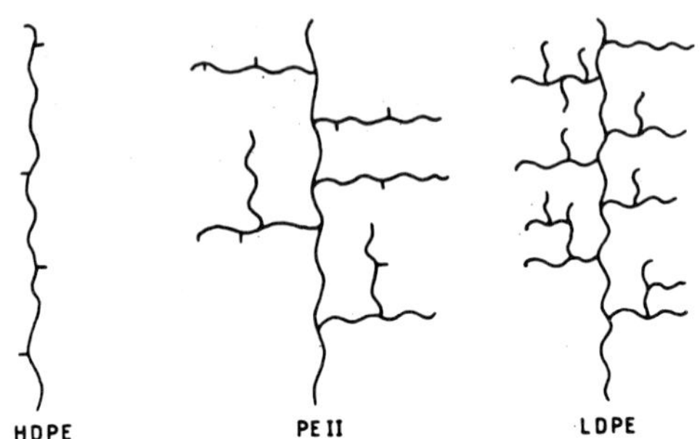

Figure 5. Polyethylene obtained by supporting complex 3 on SiO_2

2.3 Propene/butene oligomerization using inorganic supports

For industrial applications the linear oligomerization of C_3-C_5 monoolefins is of great interest. Therefore, propene and butene were included in our investigations. Only the supported complex 2 showed activity for propene/butenes. Figure 6 summarizes the results obtained with propene.

The use of Al_2O_3 lowers the linearity of the C_6 dimers dramatically, whereas when applying SiO_2 as support the linearity is maintained. Again, the ligand properties, namely the influence of the chelate, is not lossed upon supporting 2 when comparing homogeneous and heterogenized systems.

In Figure 7 we have compared three catalyst system: a) a Ziegler type system $Ni(hfacac)_2/AlR_3$ (Ref. 7); b) complex 2; c) complex 2/SiO_2.

It is evident from the dimer fraction, that in all three systems the ligand properties of hexafluoroacetylacetone prevail.

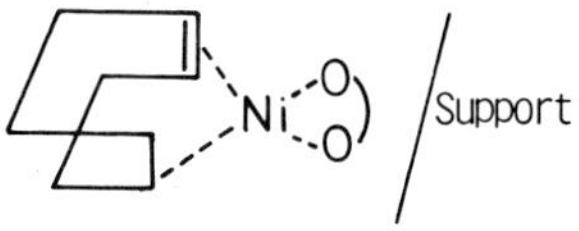

Ni /Support

SiO_2 = A ; Al_2O_3 = B

	A		B	
	%	Linearity %	%	Linearity %
C_6	72,1	78	68,1	32
C_9	23,2	48	25,2	31
C_{12}^+	4,7	-	6,7	-
Conversion	30	-	38	-

O O = hexafluoroacetyl acetone

Figure 6. Propene oligomerization with 2/SiO_2 and 2/Al_2O_3

Catalyst	$Ni(hfacac)_2$ i-Bu_3Al	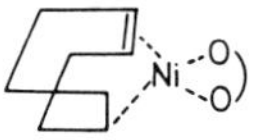 Ni	SiO_2/ Ni 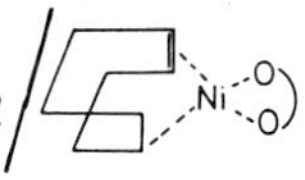
Components	%	%	%
Hexene-1	4,0	4,3	4,1
Hexene-2	59,5	56,0	58,0
Hexene-3	14,3	14,8	15,9
Methylpentenes	21,5	23,7	20,4
2,3-Dimethylbutenes	0,7	1,2	1,6

O O = hexafluoroacetylacetone

Figure 7. Comparison of three catalysts in propen oligomerization

2.4 Mechanistic considertations

Various mechanisms for the oligomerization of monooleins are discussed (Ref. 8 & 9). The ß-hydrogen elimination mechanism is the most generally accepted one. The first step comprises the insertion of an olefin into a metal-hydrogen bond. Further insertion of olefins into the metal-alkyl bond leads to growth products (propagation step), in which the hydrocarbon chain is bound via a metal carbon bond. A ß-hydrogen elimination step yields the oligomers under return to the metal hydride, which reenters into the catalytic cycle. A proposed reaction mechanism for complex 2 in solution is shown in Figure 8.

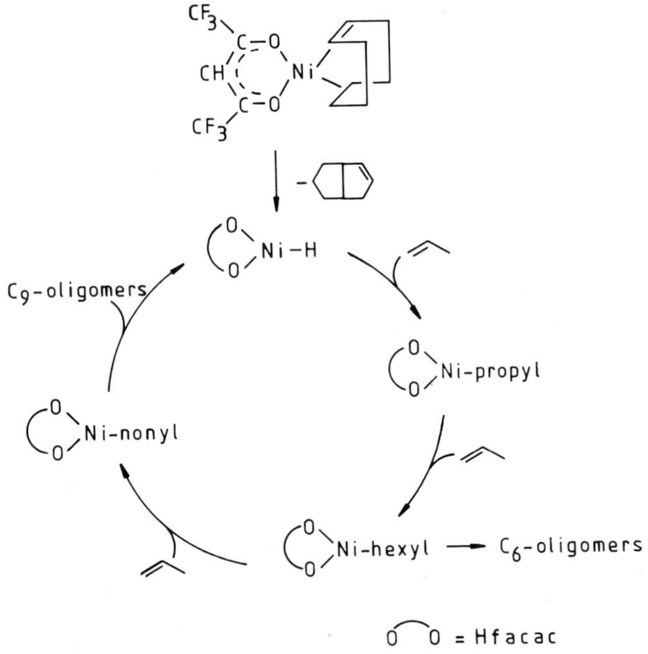

Figure 8. Proposed oligomerization mechanism for 2

Assuming a cyclooctadiene ligand dissociation, a nickel hydride complex will be formed, which follows the selfexplanatory path of Figure 8. Complexes 1-3 by itself are no catalysts, they serve only as precursors. Indeed, reaction of complex 2 with $P(C_6H_{11})_3$ gave

$[P(C_6H_{11})_3]_2(CF_3COCHCOCF_3)NiH$ which upon reaction with $NaBH_4$ could be converted into the known complex $[(C_6H_5)_3P]_2NiH_3BH_2$ (Ref. 10). In situ ^1H-NMR measurements of 2 with ethene under oligomerization conditions gave a signal at -16,3 ppm characteristic of a Ni-H bond. It can be assumed that a similar mechanism operates in the heterogenized systems.

REFERENCES

1. W. Keim, F.H.Kowaldt, R. Goddard and C. Krüger, Angew. Chem. Ind. Ed. Engl. 17, 466 (1978).
2. W. Keim, B. Hoffmann, R. Lodewick, G. Schmitt, J. Fleischhauer and U. Meier, J. Mol. Catal. 6, 79 (1979.
3. W. Keim, R. Appel, A. Storeck, C. Krüger and R. Goddard, Angew. Chem. Int. Ed. Engl. 20, 116 (1981).
4. W. Keim, Chem.-Ing. Techn. 56, 11, 850 (1984).
5. M. Peuckert and W. Keim, J. of Molecular Catalysis, 22, 289 (1984).
6. G. Kraus, Dissertation, RWTH Aachen, FRG, 1982.
7. Zhou Keyan, Gao Zhanxion, Jiang Jingyang, Wang Guozhi and W. Keim, J. of Dalian Inst. of Technology, Vol. 24, 2, 1-6 (1985).
8. W. Keim, A. Behr and M. Röper, "Comprehensive Organometallic Chemistry", Vol. 5, 371-462 (1982).
9. W. Keim, Annals of the N.Y. Academy of Sciences, Vol. 415, 191-201 (1983).
10. B. Despeyroux, Dissertation, RWTH Aachen, FRG (1981).

HYDROGENATION CATALYSTS BOUND TO FUNCTIONALIZED SILICA AND THEIR LEACHING

M.Čapka
Institute of Chemical Process Fundamentals, Czechoslovak Academy of Sciences, 165 02 Prague 6-Suchdol, Czechoslovakia

ABSTRACT - With the aim of supressing the metal leaching from heterogenized catalysts, novel types of transition metal complexes bonded to functionalized silica have been synthesized: a) neutral rhodium(I) complexes immobilized via bidentate phosphines, b) cationic rhodium(I) complexes bonded to silica by means of monodentate phosphines, c) titanium(IV) complexes bonded to immobilized h^5-cyclopentadienyl ligand. Hydrogenation activity, selectivity and stability of the above complexes is discussed in relation to their structure and the method of their preparation.

INTRODUCTION

Two decades ago, homogeneous and heterogeneous catalysis was studied separately. In the last decade, however, ever increasing attention has been paid to the so called heterogenized catalysts, which were expected to solve some technological problems connected with the application of active and selective but usually expensive soluble transition metal complexes (Refs. 1-3). In spite of the considerable effort, heterogenized catalysts, except polymerization ones, have not so far been applied in technical practice, the main disadvantage of them being rather labile bond between transition metal and support, resulting in metal leaching on reusing the catalyst. Therefore, we have centered on seeking for such types of heterogenized catalysts for which the process of metal leaching is supressed as much as possible.

In general, immobilized complex can be solubilized by three ways, the examples of which can be found in literature

- by the cleavage of the metal-ligand coordination bond
- by the cleavage of the ligand from the support surface
- by support degradation.

As in recent years we have studied systematically in our Laboratory the catalytic activity of transition metal complexes coordinately bound to inorganic supports modified with alkoxysilylalkyl substituted ligands (Refs. 4-8), we have been interested in the way of metal leaching in our catalysts. When studying the catalytic behaviour of neutral rhodium(I) complex of Wilkinson type immobilized via chiral propylphenylmenthylphosphine ligand, we have found that metal leaching results from the cleavage of the rhodium-phosphine coordination bond (Ref. 9).

It is not, of course, unexpected, since these neutral complexes are bonded to the support by one (I) or two (II) ligands and thus to realize the catalytic cycle, at least one of the metal-ligands bonds must dissociate to form vacant site for substrate coordination (Ref. 10).

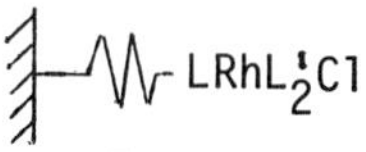

$LRhL'_2Cl$ (I) $RhL'Cl$ (II)

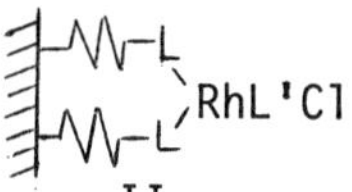

If such a site is created by the cleavage of the ligand bonded to the support, the complex is inevitably solubilized and that easier in the case of complexes of the type I than those of type II. This was confirmed by us both for achiral (Ref. 8) and chiral (Ref. 7) monodentate ligands. It is evident that such a cleavage would be less probably when the complex is bound via bidentate ligand (Refs. 2, 11-13).

RESULTS AND DISCUSSION

In our Laboratory the synthesis of silyl substituted bidentate ligands was made (Ref. 14), under conditions similar to those of analogous reaction of triethoxysilylalkyl halogenides with sodium diphenylphosphide (Ref. 15), (Scheme 1).

$$Ph_2P\frown P(Ph)H \xrightarrow{K} Ph_2P\frown P(Ph)K$$

$$\downarrow (EtO)_3Si(CH_2)_nX \longrightarrow (EtO)_3Si(CH_2)_nP(Ph)\frown PPh_2$$

$$Ph_2P\frown PPh_2 \xrightarrow{Li} Ph_2P\frown P(Ph)Li$$

n = 1, X = Cl

n = 3, X = Br

Scheme 1

The method based on the reaction of potassium with 1-(diphenylphosphino)-2-(phenylphosphino)ethane is advantageous in that dipotassium salt is not formed. On the other hand, the procedure is time-consuming and difficult (Ref. 16). 1-(Diphenylphosphino)-2--(phenylphosphino)ethane undergoes decomposition and can be purified only with difficulty. For that reason, we have tested also the reaction of lithium with readily accessible 1,2-bis(diphenylphosphino)ethane. In this case, formation of the dilithium salt is not avoided. However, we have found that this does not affect unfavourably the catalytic properties of the anchored complex.

The silylphosphines so prepared were used to functionalize silica. By this way, novel types of heterogenized rhodium(I) complexes were obtained, being coordinated to the support via bidentate ligand (Scheme 2).

$$SiO_2 \,|\!-OH + (EtO)_3Si\text{-}\wedge\!\!\wedge\text{-}PPh\frown PPh_2 \longrightarrow SiO_2 \,|\!-O-Si\text{-}\wedge\!\!\wedge\text{-}PPh\frown PPh_2$$

$$\downarrow Rh_2Cl_2(C_8H_{14})_4$$

$$SiO_2 \,|\!-O-Si\text{-}\wedge\!\!\wedge\text{-}PPh\frown PPh_2 \; (\cdots Rh^I \cdots)$$

$\text{-}\wedge\!\!\wedge\text{-} = -CH_2-, -(CH_2)_3-$

Scheme 2

As it is seen from the results summarized in Table I, the complexes are efficient catalysts for hydrogenation of unsaturated compounds under mild conditions (Ref. 12). In the case of 1,3-cyclooctadiene, the hydrogenation is nonselective, giving cyclooctane as the final product. The lower activity of the catalyst prepared with the use of the potassium salt is rather surprising. In contrast to it, the complexes immobilized via the ligand prepared by the reaction of lithium with 1,2-bis(diphenylphosphino)ethane are highly active. This is favourable in view of the easy accessibility of the ligands.

As both homogeneous and heterogenized catalysts show similar activity and selectivity, one can conclude that in this case the support plays less important role compared to the cooordination ability of the bidentate ligand and the way of its synthesis. The above catalysts thus extend rather limited series of complexes immobilized via bidentate ligands to organic (Refs. 2,11) or inorganic (Refs. 12,13) supports. Compared to the known procedures, the synthesis using lithium diphosphides represents the easier route to heterogenized catalysts.

On the other hand, one must admit that these complexes are active catalysts for rather limited series of unsaturated substrates. Therefore, we have been interested in cationic complexes of the type $/L_2Rh(cyclooctadiene)/^+A^-$ for which it is presumed that vacant coordination site is formed by removal of cyclooctadiene from the coordination sphere by its hydrogenation and phosphine ligands remain firmly bonded to transition metal during whole catalytic cycle (Refs. 17,18).

Although cationic rhodium(I) complexes have been immobilized on organic polymers, these catalysts were found to undergo decomposition to metal (Refs. 19-22). We have assumed that the reason of the low activity and instability of the above cationic complexes is that in these works analogs of homogeneous catalysts were not in fact introduced on support surface. Otherwise speaking have presumed that rhodium was bonded to the surface only by one polymer-bound ligand. Therefore we have studied (Ref. 23) immobilization

TABLE 1. Hydrogenation of 1,3-cyclooctadiene (7 mol) catalysed by neutral Rh(I) complexes (0.028 mmol) coordinated via heterogenized bidentate ligands; $P(H_2)$ = 176 kPa, temperature 60^0C, 5 ml of toluene, 5 ml of ethanol

Ligand	Prepared via	r $\mu mol\ min^{-1}$	Products, %		
			1,3-cyclo-octadiene	cyclo-octene	cyclo-octane
‖-CH_2PPh PPh_2	Li	229	7	29	64
‖-CH_2PPh PPh_2	K	13	61	27	12
‖-$(CH_2)_3PPh$ PPh_2	Li	361	9	78	13
‖-$(CH_2)_3PPh$ PPh_2	K	11	56	30	14
$(EtO)_3SiCH_2PPh$ PPh_2	Li	251	11	33	56
$(EtO)_3SiCH_2PPh$ PPh_2	K	19	58	27	15

of these complexes using silyl substituted alkyldiphenylphosphines and silica as support. This enabled us to anchor rhodium complexes by two ways (Scheme 3).

$$(EtO)_3Si(CH_2)_nPPh_2 + SiO_2 \text{-}OH \xrightarrow{\text{route A}} SiO_2\text{-}O\text{-}Si(CH_2)_nPPh_2$$

$$\downarrow /Rh(COD)(acac)/,\ THF,\ H^+A^-$$

$$(SiO_2\text{-}O\text{-}Si(CH_2)_nPPh_2)_xRh(COD)^+A^-$$

$$\uparrow SiO_2\text{-}OH$$

$$(EtO)_3Si(CH_2)_nPPh_2 \xrightarrow[\text{THF, } H^+A^-]{\text{route B; } Rh(COD)(acac)} |/(EtO)_3Si(CH_2)_nPPh_2/_xRh(COD)|^+A^-$$

n = 1.3 HA = $CH_3C_6H_4SO_3H$, acac = 2,4-pentadionate,
COD = 1,5-cyclooctadiene

Scheme 3

Procedure A is widely used and involves the synthesis of phosphinated support (by reaction of silica with trialkoxysilylalkyldiphenylphosphine in toluene) and the coordination of (1,5-cyclooctadiene)rhodium(I) 2,4-pentandionate in the presence of a strong acid. In the light of interesting results obtained recently by Reiss and Hetflejš (Refs. 24,25) in the synthesis of rhodium(I) arenesulphonate complexes, we have utilized p-toluenesulphonic acid in place of the strong inorganic acids commonly used for preparing cationic complexes. The immobilization of Rh(I) complex was found to be practically instantaneous. The number of coordinated phosphine ligands is determined by steric effects of ligands and by density of the phosphine groups in the vicinity of a given rhodium atom and not by the total P:Rh mol. ratio. One can assume that this ratio equals to 1 rather than to 2, which results in nonuniformity of the active sites.

By contrast to procedure A, procedure B provides conditions for structurally uniform active species. It involves preformation of soluble cationic rhodium(I) complex containing ω-triethoxysilylalkyldiphenylphosphines as ligands, followed by anchoring the so prepared complex to unmodified silica.

As demonstrated by the results given in Table 2, the cationic complexes catalyze hydrogenation of alkenes, alkynes and dienes. In the case of alkenes catalytic activity is comparable to that of the complexes reported by Reiss and Hetflejš (Ref. 24). The product distribution (TABLE 2) shows that dienes and alkynes are hydrogenated preferentially when compared to monoenes, in accordance (Refs. 26,27) to other Rh complexes of the type $/Rh(diene)L_2/^+A^-$.

TABLE 2. Hydrogenation of unsaturated substrates (7 mmol) catalysed by $|(SiO_2 \ \text{-O-SiCH}_2PPh_2)_x \cdot Rh(COD)|^+CH_3C_6H_4SO_3^-$ (0.028 mmol); p_{H_2} = 176 kPa, temperature 40°C, 5 ml of ethanol, 5 ml of toluene

Substrate	Initial reaction rate, μmol min^{-1}	Product distribution, %		
		diene	alkene	alkane
1,3-Cyclooctadiene	39	81	16	3
1,5-Cyclooctadiene	29	87	12	1
1-Heptine	58	67[a]	32	1
1-Heptene	145	-	51	49

[a] 1-Heptine

As cationic rhodium(I) complexes possesing silyl substituted phosphine ligands of the above type have not been reported, it was necesssary to pay attention also to the properties of their soluble analogues. In this case several unexpected results have been obtained.

Thus it was shown earlier by other authors (Ref. 24) that cationic rhodium(I) complexes possesing monodentate tertiary phosphines as ligands do not catalyze hydrogenation of Z-α-acetamidocinnamic acid. By contrast the use of ω-triethoxysilylalkyldiphenylphosphines leads to efficient hydrogenation catalysts (TABLE 3). Their activity depends strongly on the ligand to metal mol. ratio, the most efficient being those corresponding to $/Rh(diene)L_2/^+$. The ligand to metal ratio affects also the stability of the catalysts. The complexes with Rh:P mol.ratio = 1:1 are extremely sensitive to traces of oxygen and undergo easily decomposition to metallic rhodium.

TABLE 3. Hydrogenation of Z-α-acetamidocinnamic acid (7 mmol) catalysed by the soluble complexes $[(P\text{-ligand})_x Rh(COD)]^+ CH_3C_6H_4SO_3^-$ (0.028 mmol); reaction conditions see TABLE 2

Rh : P mol. ratio	k /min^{-1}/	
	$(C_2H_5O)_3SiCH_2P(C_6H_5)_2$	$(C_2H_5O)_3Si(CH_2)_3P(C_6H_5)_2$
1 : 1	54	74
1 : 1.5	101	122
1 : 2	141	124
1 : 3	2	0
1 : 4	1	0

From this aspect, also the results obtained in hydrogenation of Z-α-acetamidocinnamic acid catalysed by heterogenized catalysts (TABLE 4) deserve analysis. The catalysts prepared by procedure A

TABLE 4. Hydrogenation of Z-α-acetamidocinnamic acid (7 mmol) catalysed by the heterogenized cationic complexes $[(SiO_2\text{-ligand})_x.$ $.Rh(COD)]^+ CH_3C_6H_4SO_3^-$ (0.028 mmol), reaction conditions see TABLE 1

Procedure	k /min^{-1}/	
	}-O-$SiCH_2PPh_2$	}-O-$Si(CH_2)_3PPh_2$
A	82 (45[a], 9[b], 49[c])	13
B	424	544

[a]Second run, [b]third run, [c]the catalyst exposed to air for 20 h

(which does not ensure Rh:P mol.ratio = 1:2) are less efficient than homogeneous analogues. Although they could be exposed to air, their activity decreased rapidly on recycling due to decomposition to metallic rhodium, in analogy to the reported catalysts (Refs. 20-22,28).

However, dramatic change has been observed when procedure B was used. It involves preformation of the complex in solution and its subsequent anchoring to silica, ensuring thus the excess of the phosphine ligand for each molecule of the immobilized complex. In this case the catalysts exhibited the activity which was three to four times that was found for homogeneous analogue, in spite of unfavourable effect of transport phenomena on the catalytic behaviour of heterogenized catalysts.

This speaks for our assumption that the desired stoichiometry for each molecule of the immobilized complex is ensured. These results are not in discrepancy with our earlier finding (Ref. 29) than the catalysts with similar ligands prepared by procedure B are less efficient than those obtained by procedure A. In that case we immobilized neutral rhodium(I) complexes which are known to undergo dimerization and deactivation already in the course of their anchoring. Such a situation does not occur with cationic rhodium(I) complexes and thus the preformed soluble complex can be used with advantage. The higher catalytic activity of the complexes immobilized via longer chain spacer (TABLE 4) is in harmony with the assumption (Ref. 30) that the longer spacer ensures formation of surface complexes of the desired stoichiometry as well as with our recent results (Ref. 7) obtained in hydrogenation of Z-α-acetamidocinnamic acid catalysed by rhodium(I) complexes supported on silica via alkyldimenthyl ligands.

Summarizing, cationic rhodium(I) complexes bound to inorganic supports via alkoxysilyl substituted phosphines and prepared by the procedure ensuring the desired stoichiometry of catalyst sites exert considerably higher activity than their soluble analogues and could be thus used also in the study of chiral anchoring ligands.

Although promissing, the above results were obtained with the catalysts based on very expensive metal. Therefore, we were interested in replacing it by the cheaper metal. It is known that reduced titanium complexes (Ref. 31) efficient hydrogenation rate among catalysts. This led us to study these complexes since it is known that coordination via h^5-cyclopentadienyl ligand leads to relatively strong metal-ligand bond. One can thus assume that the metal immobilized via this ligand will not be solubilized. Only a few studies (Refs. 32,33) on heterogenized cyclopentadienyl-metal complexes bonded to inorganic supports reported so for might result from the apprehension that free surface hydroxyl groups remaining after functionalization of the support can exert unfavourable effect on catalytic properties of the resulting complexes. For that reason, in the case of functionalization of silica by cyclopentadienyl ligands followed by anchoring titanium(IV) complex, we have centered on single steps of immobilization, changes in the support texture and in the number of free surface hydroxyl groups and further on the structure of immobilized complexes.

For heterogenizing h^5-cyclopentadienyltitanium complexes, we have chosen stepwise formation of the complex on the support (Scheme 4), i.e. functionalization of silica by organosilicon reagent (i), trimethylsilylation of free surface hydroxyl groups (ii), transformation of the cyclopentadienyl group to the cyclopentadienyl anion (iii), coordination of the titanium complex (iv), and its reduction (v). As in the case of inorganic supports, such a procedure of immobilization has not yet been applied, we examined in detail single steps of immobilization and studied the structure of anchored complex and the accessibility of reaction centers (Ref. 34). Our success in preparing triethoxysilylalkylcyclopentadienes (Ref. 35) made it possible to use - in addition to the functionalizing agents containing cyclopentadiene-silicon bond also the silanes containing methylene or propylene spacer between silicon and cyclopentadienyl group.

As free hydroxyl groups would interfere in the subsequent transformation of cyclopentadienyl groups to the corresponding anions, the modified silica was further trimethylsilylated by the less ste-

$$\text{-OH, -OH, -OH} + R_xSi(CH_3)_{3-x}(CH_2)_n\text{-}C_5H_5 \xrightarrow{\text{i}} \text{-OH; -O, -O} \; Si(CH_3)_{3-x}(CH_2)_n\text{-}C_5H_5$$

ii: $(CH_3)_3SiCl$ + $/(CH_3)_3Si/_2NH$

$$\text{-OSi}(CH_3)_3\text{; -O, -O} \; Si(CH_3)_{3-x}(CH_2)_n\text{-}C_5H_5$$

iii: C_4H_9Li

$$\text{-OSi}(CH_3)_3\text{; -O, -O} \; Si(CH_3)_{3-x}(CH_2)_n\text{-}C_5H_4^{\ominus} \xrightarrow[+C_5H_5TiCl_3]{\text{iv}} \text{-OSi}(CH_3)_3\text{; -O, -O} \; Si(CH_3)_{3-x}(CH_2)_n\text{-}C_5H_4TiCl_2C_5H_5$$

v: C_4H_9Li → catalyst

$x = 1 - 3$; $n = 0, 1, 3$

Scheme 4

rically demanding and more effective trimethylsilylation agents, i.e. by a mixture of trimethylchlorosilane and hexamethyldisilazane (reaction (ii)), after which reaction the number of surface hydroxyl groups decreased to one third up to one fourth of the original value. We have not succeeded in decreasing further this value. Therefore, the reducing agent was used in excess in the reaction iii (Scheme 4), and the so prepared surface cyclopentadienyl anion was then used to coordinate (h^5-cyclopentadienyl)trichlorotitanium, giving the heterogenized analogue of titanocene (reaction (iv)).

The fact that titanium is bonded to the surface in the form of bis(h^5-cyclopentadienyl)dichlorotitanium complexes was confirmed by ESCA spectroscopy. The measurements were made with homogeneous $(C_5H_5)_2TiCl_2$ and its heterogenized analogues. It can be stated that the binding energies of titanium 2 $p_{3/2}$ electrons for all the samples differ from each other only within experimental errors (E_B = = 457.5 eV) and correspond to titanium in the oxidation state between +2 and +3. This formal oxidation state agrees with the assumption that only chloride anions withdraw electrons from titanium, while - similarly as in ferrocene - the back donation of electrons from the ligand to the central atom is very important in the case of h^5-cyclopentadienyl ligands. Data on core-level binding energies are summarized in TABLE 5.

TABLE 5. Core-level binding (eV) for titanium complexes (accuracy ±0.2 eV)

Complex	Line			
	Cl(2p)	C(1s)	Si(2p[a])	O(1s)
$(C_5H_5)_2TiCl_2$	198.8	285.0	-	532.7
⫲O-Si$(CH_2)_3C_5H_4TiCl_2(C_5H_5)$	199.1	284.8	103.4	532.8
⫲O-Si$CH_2C_5H_4TiCl_2(C_5H_5)$	199.1	284.7	103.4	532.6

[a]Used as internal standard for static charge referencing

As the other binding energies for crystalline $(C_5H_5)_2TiCl_2$ differ only slightly from those for heterogenized analogues, the structure of titanium complex on silica surface can be considered as very similar to the structure of the pure crystalline bis(h^5-cyclopentadienyl)dichlorotitanium. Therefore, free surface hydroxyl groups apparently do not enter the coordination sphere of titanium and will not thus affect the catalytic activity of the complex.

The presence of low valent titanium complexes on the support after reduction of heterogenized titanocene by butyllithium (step v) was proved by ESR spectroscopy. ESR spectra of the immobilized titanocene are depending on the excess of the reducing agent the greater excess of butyllithium, the more intense is ESR signal. The spectra were recorded with the use of a special sealed cell which made it possible to separate solid and liquid phase and to record

ESR spectra of both phases separately. The ESR signal was found only in the case of the solid phase. This proves that also after reduction with butyllithium, the titanium complex remains firmly bonded to the surface. Only after verification of these facts, the catalytic activity of the reduced titanium complexes has been kinetically studied (Refs. 36,37).

To get deeper insight into the action of the heterogenized catalyst, the course of the reaction in homogeneous phase should first be known. The reduced Cp_2TiCl_2 (Cp = cyclopentadienyl) was found to catalyze hydrogenation of 1-alkenes at atmospheric hydrogen pressure and at elevated temperatures (40-60°C). The kinetic study has shown that the reaction is zeroth order in hydrogen, the first order in the alkene and, unexpectedly, the half order in the catalyst, it has low activation enthalpy and negative activation entropy (ΔH^{*} = = 24 kJ.mol^{-1} and ΔS^{*} = -128 J.mol^{-1}K^{-1}). These results do not correspond to any of the so far reported mechanisms, we believe that they can be explained by the mechanism depicted in Scheme 5. We assume that the starting titanium(IV) complex is reduced to a dimeric complex of low valent titanium, which is likely the source of the highly active monomeric species which coordinate alkene in the rate determining step (the rate of hydrogenation depends on the structure of alkene). As no isomerisation of the unreacted alkene has been observed, one can assume that the coordination of alkene is followed by fast attack of hydrogen to form finally alkane which leaves coordination sphere, regenerating thus catalytically active complex which is then closed. This assumption is also supported by the independence of the reaction rate on hydrogen concentration.

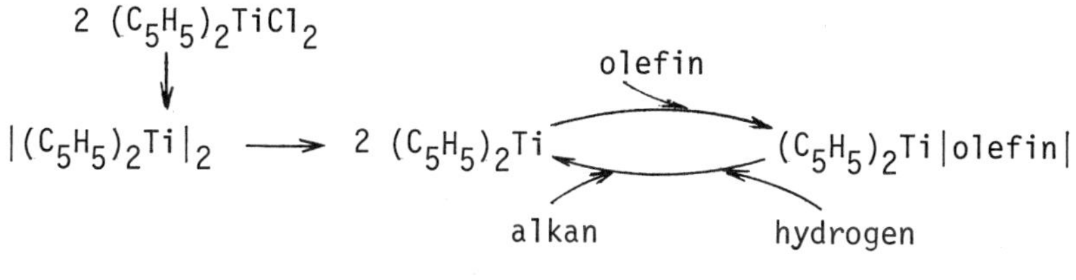

Scheme 5

On using the heterogenized catalysts, nearly the same activation parameters and identical orders in hydrogen and in the alkene were obtained in hydrogenation of 1-octene, while the r^0 vs. catalyst amount plot was linear (TABLE 6). This indicates that the reaction

TABLE 6. Hydrogenation of 1-octene catalysed by anchored catalysts

Complex	k ($l.mol^{-1}.s^{-1}$)	Reaction order olefin	hydrogen
SiO_2 -O-$Si(CH_2)_3$-Cp-$TiCl_2$-Cp	0.7	1.07	0
SiO_2 -O-Si-CH_2-Cp-$TiCl_2$-Cp	2.9	1.00	0
SiO_2 -O-Si-Cp-$TiCl_2$-Cp	3.3	0.98	0

rate does not depend on the catalyst concentration but only on its amount, which is typical of heterogeneous catalysts. This provides further support for the assumption that the catalytically active species remains bonded to the support in the course of the reaction. The dependence of the activity of the titanium catalyst on the length of the spacer separating the cyclopentadienyl group from the support surface is also of interest. It is seen that the most efficient catalyst is that containing the ligand with the shortest spacer and vice versa. This can be ascribed to the least mobile shortest spacer, inducing maximal restriction of mutual interaction of two catalytic sites anchored to the surface by preventing thus the deactivation of the sites by dimerization.

Compared to the homogeneous complex, the anchored catalysts have shown lower activity but they could be reused several times. This demonstrates again that the catalysts remain bonded to the support during the reaction, which makes them of interest for application in a continuous process (TABLE 7).

TABLE 7. Hydrogenation of 1-octene catalysed by heterogenized catalysts

Run	r^0 (mol.s^{-1})	Conversion after 15 min
1st	5.2	78
2nd	5.0	75
3rd	5.0	69
4th	3.4	53

Summarizing, the reported titanium catalysts are the actual heterogenized catalysts that remain attached to the support surface during the reaction and subsequent handling. Their activity (per metal atom) attains that of the more expensive rhodium complexes. The only disadvantage of these catalysts is their deactivation on repeated use, which is most likely caused by catalyst poisoning due to impurities present in the reaction mixture.

ACKNOWLEDGEMENTS. The author thanks all colleagues for stimulating discussions and assistance, especially Drs J. Hetflejš, A. Reissová, R. Řeřicha, J. Schraml, Mrs M. Czaková and Mr. I. Zbirovský (this Institute), Drs I. Bastl and K. Mach (J. Heyrovský Institute of Physical Chemistry and Electrochemistry, Czech. Acad. Sci., Prague) and Dr. K. Kreuzfeld (Zentral Institute für Organische Chemie, Bereich Komplexkatalyse, Rostock, DDR).

REFERENCES

1. Y.I. Yermakov, B.N. Zakharov and V.A. Kuznetzov, Catalysis by Supported Complexes. Elsevier, Amsterdam (1981).
2. C.U. Pittman, Jr., in Comprehensive Organometallic Chemistry, Vol. 8, p. 553. Pergamon Press, New York (1982).
3. D.D. Whitehurst, Chem. Technol. (1980), 44.
4. M. Czaková and M. Čapka, J. Mol. Catal. 11, 313 (1981).

5. Z.M. Michalska and M. Čapka, J. Mol. Catal. 11, 323 (1981).
6. Z. Broučková, M. Czaková and M. Čapka, J. Mol. Catal. 30, 241 (1985).
7. A. Kinting, M. Čapka and H. Krause, J. Mol. Catal. 33, 215 (1985).
8. M. Čapka, J. Hetflejš, V.M. Vdovin, V.E. Fedorov, N.A. Pritula and G.K. Fedorova, React. Kinet. Catal. Lett., in press.
9. M. Čapka, Collect. Czech. Chem. Commun. 42, 3410 (1977).
10. J.A. Osborn, F.H. Jardine, J.F. Young and G. Wilkinson, J. Chem. Soc. A (1966), 1711.
11. H.J. Kreuzfeld, Z. Chem. 18, 231 (1978).
12. M. Bartholin, Ch. Graillat, A. Guyot, G. Condurier, J. Bandiera and C. Naccache, J. Mol. Catal. 3, 17 (1977/78).
13. I. Kolb, M. Černý and J. Hetflejš, React. Kinet. Catal. Lett. 7, 199 (1977).
14. V. Zbirovský, H.J. Kreuzfeld and M. Čapka, React. Kin. Catal. Lett. 29, 243 (1985).
15. M. Čapka, Synth. React. Inorg. Met-Org. Chem. 7, 347 (1977).
16. K. Issleib and H. Weichmann, Z. Chem. 11, 188 (1971).
17. R.R. Schrock and J.A. Osborn, J. Amer. Chem. Soc. 98, 2134 (1976).
18. M.A. Garralda and L.A. Oro, Trans. Met. Chem. 5, 65 (1980).
19. G. Strucul, M. Bonivento, M. Graziani, E. Cernia and N. Palladino, Inorg. Chim. Acta 12, 15 (1975).
20. P. Pinna, M. Bonivento, G. Strukul, M. Graziani, E. Cernia and N. Palladino, J. Mol. Catal. 1, 309 (1975/76).
21. G. Strukul, P. D'Olimpio, M. Bonivento, P. Pinna and M. Graziani, J. Mol. Catal. 2, 179 (1977).
22. P. Pinna, C. Candilera, G. Strukul, M. Bonivento and M. Graziani, J. Organometal. Chem. 159, 91 (1978).
23. V. Zbirovský and M. Čapka, Collect. Czech. Chem. Commun., in press.
24. J. Reiss and J. Hetflejš, Collect. Czech. Chem. Commun., in press.
25. J. Reiss and J. Hetflejš, React. Kinet. Catal. Lett., in press.
26. R.R. Schrock and J.A. Osborn, J. Amer. Chem. Soc. 98, 2143 (1976).
27. R.R. Schrock and J.A. Osborn, J. Amer. Chem. Soc. 98, 4450 (1976).
28. L.A. Oro and R. Sariego, React. Kinet. Catal. Lett. 21, 445 (1982).
29. Z. Kozák and M. Čapka, Collect. Czech. Chem. Commun. 44, 2624 (1979).
30. J.M. Brown and H. Molinari, Tetrahedron Lett. (1979), 2933.
31. D.W. Nacomber, W.P. Hart and M.D. Rausch, Adv. Organometal. Chem. 21, 1 (1982).
32. F.R.W.P. Wild, G. Gubitosa and H.H. Britzinger, J. Organometal. Chem. 148, 73 (1978).
33. R. Jackson, J. Rudelsen, D.J. Thompson and R. Whelan, J. Organometal. Chem. 125, 57 (1977).
34. A. Reissová, Z. Bastl and M. Čapka, Collect. Czech. Chem. Commun., in press.

35. A. Reissová and M. Čapka, Synth. React. Inorg. Met. Org. Chem., in press.
36. A. Reissová, Thesis, Czechoslovak Acad. Sci., Prague (1984).
37. A. Reissová and M. Čapka, Proceedings of Vth Czech.-Ital. Symp. on Catalysis, Bechyně, Czechoslovakia, p. 99 (1985).
38. G.H. Brubaker, R.H. Grubs and E.S. Kavan, J. Organometal. Chem. 120, 49 (1976).
39. H.H. Brintzinger, J.E. Bercaro, L.H. Marvisch and L.G. Bell, J. Am. Chem. Soc. 94, 1219 (1972).

TRANSITION METAL COMPLEXES WITH POLYSTYRENE-ATTACHED AMINE-PHOSPHINE LIGANDS AS HYDROSILYLATION CATALYSTS

Z.M. Michalska
Institute of Polymers, Technical University,
ul. Zwirki 36, 90-924 Lódź, Poland

ABSTRACT - Complexes of Pt(II), Ru(II) and Ir(I) with chelating amine-phosphine ligands $PPh_2PC_6H_4CH_2NMe_2$(PCN) and $PPh_2C_6H_4NMe_2$(PN) were used as new and effective catalysts for the hydrosilylation of C=C and C=O bonds in homogeneous and heterogenized systems. High activity was found for the platinum(II) complex attached to macrorecticular polystyrene resins, whereas the nonsupported analogue could not act as catalyst under comparable conditions due to its insolubility. Homogeneous and heterogenized complexes of Ru(II) with PN ligands were also active, especially in hydrosilylation of ketones. High yields were obtained in the hydrosilylation of hexene-1 only when monohydrosilanes with electronegative substituents were used in the absence of solvents. The catalytic activity of iridium(I) complex, IrCl(CO)PCN, in hydrosilylation of C=C bonds was also observed provided that undesirable side reactions, e.g., isomerization, were eliminated.

1. INTRODUCTION

Phosphine complexes of group VIII elements are known to be good catalysts for hydrosilylation of olefins though their activity strongly depends on the P/M molar ratio. It was observed that in many cases (Refs. 1-3) the catalytic system reached the highest activity when P/M was 1. Therefore the complexes with amine-phosphine ligands such as $Ph_2PC_6H_4CH_2NMe_2$(PCN) and $Ph_2PC_6H_4NMe_2$

(PN) seemed to be attractive, new catalysts that satisfy this demand. However, a disadvantage of these complexes is their poor solubility in suitable solvents for homogeneous catalysis. That is why in cases, where the metal complexes were not soluble, they were used in their heterogenized form.

An additional advantage in using this type of ligands is their chelate character. A serious problem encountered in catalysis by supported complexes is metal leaching from the support into solution. Bi- or polydentate binding of the metal center to a support usually improves the stability of the polymer catalyst (Refs.4-7). An additional reason for using chelates with mixed donor groups is that the phosphine end will stabilize the metal at low oxidation state and firmly bind it to a support, while the weakly coordinated amine will be a labile group that easily undergoes dissociation.

In this report, the study of the reactivity of the complexes of platinum(II), ruthenium(II) and iridium(I) with amine-phosphine chelating ligands in the hydrosilylation reactions will be presented as an extention of the earlier work on the analogous system of rhodium(I) (Ref. 8).

2. RESULTS AND DISCUSSION

Homogeneous complexes of Pt(II), Ru(II) and Ir(I) with amine-phosphine chelating ligands were prepared as described in (Refs. 9,10). The polymer-attached counterparts were prepared by the ligand exchange between homogeneous complex and PCN or PN functionalized polymers or by synthesis in situ by treating polymeric ligands with the appropriate metal compound as shown in Scheme. Catalytic activities of these materials were evaluated in hydrosilylation of alkenes and ketones.

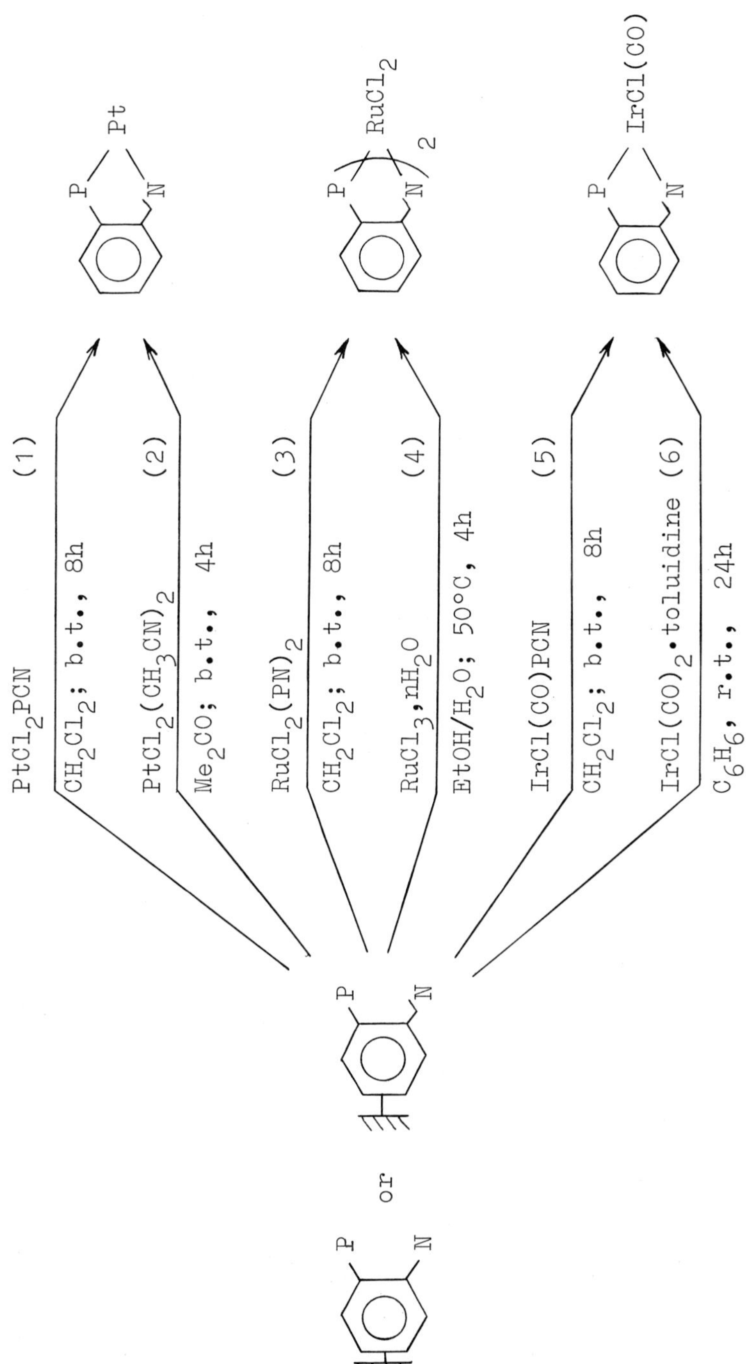
P
N
or
P
N
$PtCl_2PCN$ (1)
CH_2Cl_2; b.t., 8h
$PtCl_2(CH_3CN)_2$ (2)
Me_2CO; b.t., 4h
$RuCl_2(PN)_2$ (3)
CH_2Cl_2; b.t., 8h
$RuCl_3,nH_2O$ (4)
$EtOH/H_2O$; 50°C, 4h
IrCl(CO)PCN (5)
CH_2Cl_2; b.t., 8h
$IrCl(CO)_2$·toluidine (6)
C_6H_6, r.t., 24h
Pt
$RuCl_2$
2
IrCl(CO)

Scheme

2.1. Platinum Catalysts

Catalytic properties of the $PtCl_2PCN$ complex were investigated in homogeneous and heterogenized systems. $PtCl_2PCN$ appeared to be insoluble in benzene -- the usual reaction medium. Attempts to hydrosilylate hexene-1 in the CH_2Cl_2/benzene solution in which platinum complex was dissolved were unsuccessful. A high activity was found for the heterogenized analogue produced in the reaction between precursor $PtCl_2(CH_2CN)_2$ and PCN-functionalized polystyrene resins (reaction (2) in Scheme). As shown in Table 1, the addition of hydrosilylanes to hexene-1 was completed in 15 min. at 30°C. The activity remained unchanged after five consecutive runs. A lower activity was shown by the polymer-attached complex prepared in a ligand exchange reaction (1) in Scheme. The reason for this difference might be the different mechanism of heterogenization. It seems likely that similarly to the analogous rhodium system (Ref.8) some of the platinum atoms of the homogeneous complex are linked to the polymer support not by both chelate ligands but by the phosphine group being the stronger donor which replaces weakly coordinated amine. In such a situation the platinum catalytic center can be surrounded by the two phosphine ligands: one derived from the parent chelate and the other from the polymer-attached one. The increase in P/Pt ratio in the hydrosilylation reaction usually involves induction periods or inhibition of the reaction. To confirm this hypothesis, $PtCl_2(CH_3CN)_2$ was supported on the phosphinated and aminated polymers. No activity was found for the former system, whereas the latter showed a high activity.

These results indicate that reaction (2) is the method of choice of the preparation of the highly active heterogenized platinum(II) catalyst.

TABLE 1. Hydrosilylation of hexene-1 catalyzed by Pt(II) complex

$R_3SiH\ CH_2 = CH(CH_2)_3CH_3 \longrightarrow R_3Si(CH_2)_5CH_3$

Catalyst	Silane	Solvent	Temp./Time °C h	Yield[a)]
$PtCl_2PCN$	Me_2PrSiH	CH_2Cl_2/C_6H_6	20 - 50/0 - 72	0
	Et_3SiH	"	"	0
	Me_2PhSiH	"	"	0
Ⓟ— $PCNPtCl_2$[b)]	Me_2PrSiH	C_6H_6	30/1,5	93
	Et_3SiH	"	30/1,5	90
	Me_2PhSiH	"	30/1,5	95
Ⓟ— $PCNPtCl_2$[c)]	Me_2PrSiH	"	30/0,25	100[d)]
	Et_3SiH	"	30/0,25	100[d)]
	Me_2PhSiH	"	30/0,25	100[d)]
Ⓟ— $PPh_2/_2PtCl_2$[c)]	Me_2PhSiH	"	30/24	0
Ⓟ— $NMe_2/_2PtCl_2$[c)]	Me_2PhSiH	"	30/0,25	100

a) based on hexene-1; calculated by GLC; polymer: 8% DVB; 20% CMS; 0,72% P;
b) prepared in (Ref. 1);
c) prepared in (Ref. 2); [≡SiH]: [hexene-1] : [catalyst] = 1 : 1 : 2,5 x 10^{-5};
d) recycled 5 times

2.2. Ruthenium Catalysts

Complexes of Ru(0), Ru(I) to Ru(III) with phosphine ligands were used as homogeneous catalysts for the hydrosilylation of C=O bonds (Refs. 11-13). Their application to the addition of hydrosilanes to olefins did not give positive results. Only recently (Ref. 14) suitable conditions for the application of effective catalysts in hydrosilylation of alkenes-1 were found.

The results of the present study showed that the ruthenium(II) complexes with amino-phosphine ligands behaved as homogeneous catalysts in a similar way as those described above. As seen from Table 2, both homogeneous and polymer-supported $RuCl_2(PN)_2$ complexes were active in O-silylation of ketones. For dihydrosilanes the reaction rates were higher than for monohydrosilanes, and cyclic ketones were silylated more easily than dialkyl ones.

In the hydrosilylation of hexene-1 (Table 3) the best results were obtained with silanes with electron-withdrawing substituents in the absence of a solvent when the reaction was conducted in air. This was consistent with the earlier findings for Ru(II) phosphine complexes (Ref. 14). The addition of dihydrosilanes was considerably slower and gave monosubstituted silanes similarly as in catalysis by the rhodium system (Ref. 15). Longer reaction times were needed for the polymer-attached system.

2.3. Iridium Catalyst

In contrast to rhodium compounds, the corresponding low spin d^8 complexes of iridium, e.g., $IrCl(CO)(PPh_3)_2$ and $IrH(CO)(PPh_3)_3$, do not catalyze hydrosilylation of alkenes. This is due to the fact that they undergo oxi-

TABLE 2. Hydrosilylation of ketones catalyzed by Ru(II) complex at 120°

Catalyst	Ketone	Silane	Time h	Yield[a)] %
$RuCl_2(PN)_2$	MeCOMe	$MePhSiH_2$	2	92
		Me_2PhSiH	4,5	81
	MeCOEt	Et_2SiH_2	1	95
	[cyclopentanone]=O	$MePhSiH_2$	0,5	100
	[cyclopentanone]=O	MePhSiH	2	89
	[cyclohexanone]=O	$MePhSiH_2$	0,5	100
Ⓟ-PN/$_2$RuCl$_2$[b)]	MeCOMe	$MePhSiH_2$	3,5	98
	MeCOEt	Pr_2SiH_2	3,5	91
	[cyclopentanone]=O	$MePhSiH_2$	0,5	100

a) based on silane; calculated by GCL.
b) polymer: 8% DVB; 20% CMS; 0,72% P; prepared in (Ref. 3)

[≡ SiH] : [ketone] : [catalyst] = 1 : 1 : 1 x 10^{-3}.

TABLE 3. Hydrosilylation of hexene-1 catalyzed by Ru(II) complex at 120°

Catalyst	Silane	Time	Yield[a)]
$RuCl_2(PN)_2$	$(EtO)_3SiH$	3	95
	$(MeO)_3SiH$	3	98
	$MePhSiH_2$	48	78
	Pr_2SiH_2	48	83
Ⓟ-$PN/_2RuCl_2$[b)]	$(EtO)_3SiH$	6	47
	$(MeO)_3SiH$	6	52
	$MePhSiH_2$	72	38
	Ph_2SiH_2	72	40

a) based on hexene-1; calculated by GLC

b) polymer: 8% DVB; 20% CMS; 0,72% P; prepared in (Ref. 3)

[≡SiH]: [hexene-1] : [catalyst] = 1 : 1 : 1×10^{-3}.

dative addition with silanes to form catalytically inactive bisphosphine adducts of the type $IrHCl(CO)(PPh_3)_2SiR_3$ having a stable Ir-P bond (Ref.16). It was therefore believed that the iridium(I) complex with the amine-phosphine ligand, in which amino end is a weaker donor, would not form a stable intermediate with silanes.

The results summarized in Table 4 show that IrCl(CO)PCN is an active catalyst for the addition of the silanes to C=C double bonds, provided that the interfering side reactions are eliminated. These include an extensive isomerization of alkene which gives internal isomers which are inactive with regard to hydrosilylation. To exclude this reaction, the compounds which could not isomerize were used, namely, ethylene and trimethylvinylsilane. This latter, on the other hand, easily undergoes disproportionation with trialkylsilanes at high reaction temperatures. As can be seen from Table 4, the best results were obtained when silanes with bulky substituents were used.

Various solvents were tested for a homogeneous reaction because of poor solubility of IrCl(CO)PCN. The highest yields were achieved in benzene, which dissolves the iridium(I) complex at a reflux temperature. This problem is avoided when heterogenized metal complex is used. Its activity is similar to that of the homogeneous catalyst.

3. CONCLUSIONS

It has been shown that catalytic systems based on the complexes of platinum(II), ruthenium(II) and iridium(I) with chelate amine-phosphine ligands were active with regard to the hydrosilylation of C=C and C=O bonds. The use of the heterogenized form of the platinum(II) complex allowed us to discover the highly active catal-

TABLE 4. Hydrosilylation of C=C bonds catalyzed by Ir(I) complex

Catalyst	Substrates	Solvent	T/Time °C/ h	Yield[a)]
IrCl(CO)PCN	$Me_2PrSiH + CH_2 = CH_2$	CH_2Cl_2/C_6H_6	50/24	91
Ⓟ-PCNIrCl(CO)[b)]	$MePr_2SiH + CH_2 = CH_2$	C_6H_6	50/24	88
IrCl(CO)PCN	$MePr_2SiH + CH_2 = CHSiMe_3$	CH_2Cl_2	50/24	51
	$MePr_2SiH + CH_2 = CHSiMe_3$	THF	50/24	22
	$Ph_3SiH + CH_2 = CHSiMe_3$	CH_2Cl_2/C_6H_6	120/11	95
	$Ph_3SiH + CH_2 = CHSiMe_3$	C_6H_6	120/75	100
Ⓟ-PCNIrCl(CO)[c)]	$Ph_3SiH + CH_2 = CHSiMe_3$	C_6H_6	120/12	85
	$MePh_2SiH + CH_2 = CHSiMe_3$	"	120/3.5	91
	$NphPhMeSiH + CH_2 = CHSiMe_3$	"	120/4	94

a) based on vinylsilane; calculated by GLC
b) polymer support: 8% DVB; 20% CMS; 1,62% P; prepared in (Ref. 6)
c) polymer: 8% DVB; 10% CMS; 0,83% P; prepared in (Ref. 6)

[≡ SiH] : [C=C] : [catalyst] = 1 : 1 : 2×10^3.

yst. In contrast, its nonsupported analogue could not act as catalyst due to solubility problems. Thus the benefit of heterogenized catalysis was demonstrated. The results also indicate that the amine-phosphine bidentate ligands provide a suitable environment of the iridium atom promoting its activity in hydrosilylation of double bonds. However, clean reaction conditions have to be maintained. A study of the ruthenium(II) complexes with amine-phosphine ligands has shown that these are promising catalysts for the addition of hydrosilanes to ketones and alkenes.

The supported and homogeneous catalysts exhibited similar behaviour suggesting that the active species in the two systems were similar.

4. EXPERIMENTAL

4.1. Materials

The complexes $PtCl_2PCN$ (Ref. 9), $RuCl_2(PN)_2$ (Ref. 9), IrCl(CO)PCN (Ref. 10) and $IrCl(CO)_2$(p-toluidine) (Ref. 17) were prepared by the known methods. The compounds o-(diphenylphosphine)-N,N-dimethylbenzylamine (PCN) and o-(diphenylphosphine)-N,N-dimethylaniline (PN) were prepared as described in (Refs. 18 and 10). Hexene-1, ketones and the solvents were dried and distilled before use. The silanes were prepared by published methods (Ref. 19).

4.1.1. Synthesis and Functionalization of Polystyrene Resins. Copolymers of styrene, divinylbenzene and chloromethylstyrene were prepared in suspension at 88°C for 8h using benzoyl peroxide as an initiator and heptane as a pore-forming agent. The detailed procedure is given in (Ref. 8). Functional groups were attached to the selected polymers in the reaction between chloromethyl groups of the resin and the excess lithiated ligand Li-PCN or Li-PN, as described elsewhere (Ref. 8).

4.1.2. Preparation of Polymer-Supported Metal Complexes in the Ligand Exchange Reaction (Reactions (1), (3) and (5) in Scheme). In a typical experiment, the appropriate metal compound ($PtCl_2PCN$, $RuCl_2(PN)_2$ and $IrCl(CO)_2PCN$) (0,1 mmol) in a CH_2Cl_2 solution was added at stirring to the polymer-attached PCN or PN ligand (1 g) suspended in CH_2Cl_2 under nitrogen. The mixture was refluxed for 6-8h, then allowed to stand at room temperature overnight. The solids were filtered off and extracted with boiling CH_2Cl_2 in a Soxhlet apparatus for 2h and dried under vacuum.

4.1.3. Preparation of Polymer-Supported Metal Complexes in the Synthesis in situ (Reactions (2), (4) and (6) in Scheme). Preparation of Ⓟ-$PCNPtCl_2$. A solution of $PtCl(CH_3CN)_2$ complex (0,2 mmol) in acetone was added to the suspension of the polymer-attached PCN ligand (1 g) in acetone. The mixture was refluxed for 4h at stirring, then the solid was filtered off, Soxhlet-extracted with acetone and dried under vacuum.

4.1.4. Preparation of Ⓟ-$PN/_2RuCl_2$. $RuCl_3 \cdot 3H_2O$ (0,2 mmol) was dissolved in a $EtOH/H_2O$ solution and added dropwise to the polymer-attached PN ligand (1 g) in ethanol. The mixture was stirred at 50°C until the colour of the solution disappeared, then repeatedly washed with ethanol and dried under vacuum.

4.1.5. Preparation of Ⓟ-PCNIrCl(CO). 1 g of polymer-PCN ligand swelled in benzene was treated with a benzene solution of $IrCl(CO)_2$ (p-toluidine) complex

(0,1 mmol) at ambient temperature for 24h. A colourless solution was withdrawn with a syringe, the solid was extracted with benzene and dried under vacuum.

4.1.6. Catalytic Hydrosilylation. General procedure. The catalyst, solvent and silane were placed in the reactor equipped with a magnetic stirrer which was then purged with nitrogen and the temperature was adjusted to that required. The appropriate alkene or ketone was then injected. Reactions catalyzed by the ruthenium system were conducted in the atmosphere of air without solvent.

The GLC analyses of reaction products were carried out on a Chromatron GCHF 18,3 (Berlin) gas chromatograph, using a 2 m column of 10% OV-101 on a 80-100 mesh Chromosorb WAW equipped with flameionization detectors.

REFERENCES

1. H.M. Dickers, R.N. Haszeldine, L.S. Malkin, A.P. Mather and R.V. Parish, J.Chem.Soc.Dalton, 308 (1980).
2. D.C. Apple, K.A. Brady, J.M. Chance, N.E. Heard and T.A. Nile, J.Mol.Catal., 29, 55 (1985).
3. R.A. Faltynek, Inorg.Chem., 20, 1357 (1981).
4. C.U. Pittman, Jr., Quock Ng, A. Hirao, W. Honnick and R. Hanes, in Relation entre catalyse homogène et catalyse hétérogène, Ed. CNRS, Paris (1978).
5. R.J. Card and D.C. Neckers, Inorg.Chem., 17, 2345 (1978).
6. N.L. Holy, in Fundamental Res.Homog.Catal., M. Tsutsui (Ed.), vol. 3, Plenum, N.Y. (1979), 691 p.
7. M. Berglund, C. Anderson and R. Larsson, J.Organomet.Chem., 258, 195 (1983).
8. Z.M. Michalska, J.Mol.Catal., 19, 345 (1983).
9. T.B. Rauchfuss, F.T. Ratino and D.M. Roundhill, Inorg. Chem., 14, 652 (1975).
10. T.B. Rauchfuss, J.L. Clements, G.F. Angew and D.M. Roundhill, Inorg.Chem., 16, 775 (1977).
11. R.J.P. Corriu and J.J.E. Moreau, J.Chem.Soc.Chem. Commun., 38 (1973).
12. C. Eaborn, K. Odell and A. Pidcock, J.Organometal. Chem., 63, 93 (1973).

13. P. Svoboda, R. Rericha and J. Helflejs, Coll.Czech. Chem.Commun., 39, 1324 (1974).
14. B. Marciniec and J. Guliński, J.Organomet.Chem., 253, 349 (1983).
15. I. Ojima, T. Kogure, I. Nihanyanagi and Y. Nagai, Bull.Chem.Soc. Japan, 45, 3506 (1972).
16. I. Ojima, K. Yamamoto and M. Kumada, in Aspects of Homogeneous Catalysis, Reidel, Dordrecht, vol. 3, 186 p. and refs.therein.
17. U. Klabunde, Inorg.Synth., 15, 82 (1974).
18. T.B. Rauchfuss and D.M. Roundhill, J.Am.Chem.Soc., 96, 3098 (1974).
19. V. Bazant, V. Chvalovsky, J. Rathousky, in Organosilicon Compounds, vol. 4(1), Publ. of the Inst. Chem.Process Fund., Czech.Acad.Sciences, Prague, (1973).

HYDROSILYLATION BY FIXED COMPLEXES. INFLUENCE OF THE SURFACE ON CATALYSIS OF THE REACTION OF HYDROSILYLATION BY FIXED COMPLEXES OF GROUP VIII METALS

V.O. Reikhsfeld, N.K. Skvortsov and V.S. Brovko
Lensovet Technological Institute, Leningrad 198013
USSR

ABSTRACT - The peculiarities of the catalytic action of transition metal complexes supported on either organic polymer or modified inorganic matrices have been examined in catalytic hydrosilylation. Although the mechanisms of action of supported metal complexes and their soluble analogs are essentially similar, there are some specific effects caused by the influence of the surface. These are the dependences of the catalytic activity and selectivity on the nature of the matrix, on the degree of saturation of the surface by metal complexes and on the structure of surface onium groups. The supporting of complexes also influences regio- and stereoselectivity of the reaction. The use of supported transition metal complexes allows one to control the efficiency of the hydrosilylation process.

1. INTRODUCTION

The development of new types of catalysts based on transition metal complexes fixed on supports by different methods is a promising approach to the improvement of catalytic processes of practical importance. The new catalysts must show the activity and selectivity comparable to homogeneous catalysts. Besides, they must be stable and capable of multiple use. The supports used must have little effect on the fixed complex and on the components of the reacting media. So, the evaluation of the effects of fixation on the properties of heterogenized complexes is necessary.

We have studied the influence of the surface using Group VIII metal complexes fixed onto organic or inorganic supports by ionic bonds according to the scheme:

$$\left\{\overset{+}{E}R_3\,[ML]^- \qquad \text{or} \qquad \left\{O\text{-}Si(CH_2)_n\overset{+}{E}R_3\,[ML]^-,\right.\right.$$

where E = N, P; L = ligand.

The advantages of heterogenization of metal complexes by ion-exchange was discussed recently (Ref. 1).

The hydrosilylation is one of the most convenient reactions for the study of the effects of the metal complex fixation. In the present paper we report the data on hydrosilylation of olefins with some organosilanes.

The reaction occurs via two routes:

$$RCH{=}CH_2 + R_3SiH \longrightarrow \begin{cases} RCH_2CH_2SiR_3 & \beta\text{-addition} \\ CH_3CH(R)SiR_3 & \alpha\text{-addition.} \end{cases}$$

So it is possible to evaluate not only the activity of catalysts, but also the regioselectivity (the relative yield of side products formed in the reaction of disproportionation, e.g., $RCH_2CH_2SiHR_2$) and enantioselectivity when studying hydrosilylation by means of organosilanes having the asymmetric silicon atoms.

2. CATALYSIS BY COMPLEXES SUPPORTED ON ION-EXCHANGE RESINS

We prepared the series of catalysts by fixation of complexes of platinum and other transition metals (Ref.2) on ion-exchange resins such as AV-17 and functionalized silica supports and found them to possess as high activity as that of similar homogeneous catalysts (Ref. 3). The catalyst stability was characterized by its ability to remain unaffected during many catalytic cycles or many hours of a continuous process.

A detailed kinetic investigation of the regularities of reactions that occur in the presence of these catalysts and study of the isotope effect in the deuterosilylation with dimethylphenyl deuterosilane, as well as of the influence of the reagent and the solvent structures indicate the great similarity between the mechanisms of action of the fixed metal complexes and their homogeneous analogs (Ref. 4). At the same time a number of peculiar features in the behaviour of catalysts containing surface complexes were observed. The ion-exchange for immobilization of complexes of Pt(II), Pt(IV), Rh(III), Rh(IV), Ir(III), etc., has been studied extensively in (Ref. 2). To put Pt complexes onto high-basic anionites, a great amount of platinum is required, up to 20% by mass, at a maximum saturation, but as was convincingly shown in (Refs. 5,6), the reaction proceeds with a high rate only within the surface layer of a granule. It is impossible to increase the specific activity by lowering degree of deposition, because in the case of a uniform distribution over the granule volume the concentration of surface active centers is decreased too.

X-ray data indicates that during fixation, platinum is uniformly distributed over the diameter of a catalyst granule regardless of the degree of surface saturation (Fig.1). This shows that application of technical anionites as polymer matrices to fix acidic complexes of Pt and other transition metals is not efficient.

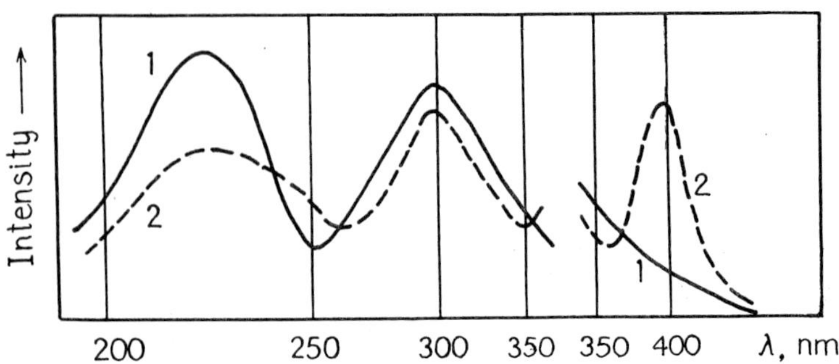

Figure 1. X-ray spectra for 1 - 3%Pt, 2 - 9%Pt, 3 - 16%Pt.

3. CATALYSIS BY COMPLEXES SUPPORTED ON FUNCTIONALIZED INORGANIC SUPPORTS

We have to obtain efficient catalysts with the use of silica supports (silica gel, macroporous glass) functionalized by organosilicon compounds containing quaternary ammonium or phosphonium ligands (Refs. 3,4). The specific activity of these catalysts is quite comparable to that of similar homogeneous model compounds (Table 1).

The peculiar feature of the catalysts is that the lowering of activity associated with the saturation state of the support surface with the metal complex is not proportional to the lowering of the concentration of the metal. This means that the unreacted quaternary group of the surface exerts additional negative effect on the reaction rate (Table 2).

For a more detailed study of the hydrosilylation reaction in the presence of these catalysts, we have examined the influence of various factors on the kinetics of the reaction. First, we have found the dependence of the duration of an induction period and the rate of addition on the nature of the surface onium group (Table 3).

TABLE 1. Specific activity of fixed platinum complexes $[\{\text{-OSi(CH}_2)_3\text{ER}_3]_2\text{PtCl}_6$ (Ref. 1)

Catalyst Functionalized SiO_2		Specific activity mole/(g-atom of Pt)·sec	
E	R	1-heptene DMFS[a]	styrene + MDCS[b]
N	Me	0.7	0.7
N	Et	2.4	0.8
N	Bu	3.9	1.2
P	Bu	1.7	2.5
P	Ph	1.7	6.9
Ion-exchange resin		$7.2\ 10^3$	-
$(C_{12}H_{25}PPh_3)_2PtCl_4$		-	9.4[c]

a) DMFS - dimethylphenylsilane, b) MDCS - methyldichlorosilane, c) with 1-hexene.

TABLE 2. Relative rate of hydrosilylation in the presence of fixed complexes at different surface saturation (F)

F = Pt/ER_3 g-atom/mole	1-heptene - DMFS[a]		styrene - MDCS[b]	
	I[c]	W[d]	I[c]	W[d]
1.0	-	-	1.0	1.0
0.55	1.0	1.0	1.6	0.24
0.25	2.0	0.85	11.4	0.13
0.09	3.1	0.21	-	-
0.03	4.5	0.08	16.8	0.08

a), b) - see Table 1, c) - $I = \frac{\tau_{sat.}}{\tau_F}$, d) - $W = \frac{K_{2sat.}}{K_{2F}}$

TABLE 3. Influence of onium groups on the reaction rate

Catalysts (1)		1-hexene - MDCS				styrene -MDCS			
E	R	k_1 10^3	k_2 10^3	ΔH_2	ΔS_2	k_1 10^3	k_2 10^4	ΔH_2	ΔS_2
N	Me	0.50	0.60	71	-65	0.72	0.33	30	-224
N	Et	-	-	74	-58	0.83	0.36	-	-
P	Bu	1.00	0.69	-	-	5.56	1.19	48	-154
P	Ph	1.40	0.76	83	-27	8.33	3.57	47	-145

Dimensions: k_1, sec^{-1}, k_2, mole/mole·sec; ΔH_2, kJ/mole, ΔS_2, $J \cdot mole^{-1}K^{-1}$

It was shown that in the case of addition of dichloromethylsilane to olefins, the catalytic activity increases in the order: $\sim\overset{+}{N}Me_3 < \sim\overset{+}{N}Et_3 < \sim\overset{+}{N}Bu_3 < \sim\overset{+}{P}Bu_3 < \sim\overset{+}{P}Ph_3$. This fact is quite unexpected, because according to the existing mechanisms the reaction proceeds in the inner coordination sphere of platinum (Ref. 7). In our view, the structure of the counter-ion affects the sorptive power with regard to the hydride silane, which contributes to the interaction of the latter with the surface platinum complexes. This view keeps in with the change in the reaction kinetic parameters, which indicates that the difference in reaction rates for hydrosilylation catalyst with different onium groups is determined mainly by the change in entropy factor (Table 3).

It should be noted that under the conditions studied, further reduction to Pt(0) during the reaction never occurs, although this often occurs in catalysis by H_2PtCl_6 (Ref. 8). UV data obtained by us confirm the PT(IV)→Pt(I transition under the action of the hydride silane or of the reaction mixture, both for suspensions of catalyst (1) in octadecane and for model soluble salts $(C_{16}H_{33}NMe_3)2PtCl_6$ (Fig. 2). In all cases, the presence of a maximum in the region 340-400 nm indicates a d - d transition, characteristic of Pt(II) (Ref. 9). A XPS analysis of the hydrosilylation catalysts isolated after the reaction shows that the binding energy of the 4f 7/2 electrons changes from 74.8 eV Pt(IV) to 73.1 eV, which also is in good accord with the Pt(IV)→Pt(II) transition.

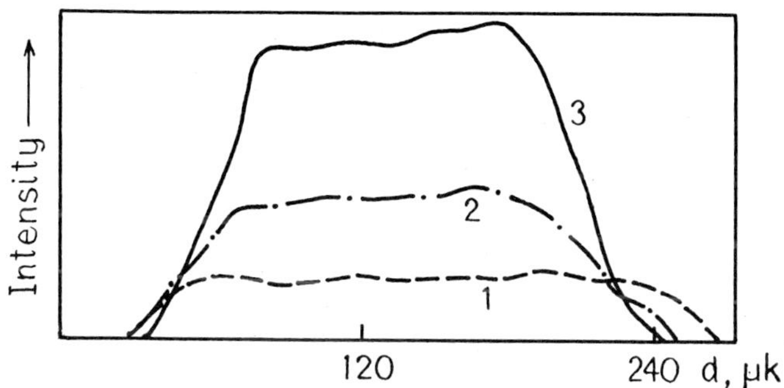

Figure 2. UV spectra of the HMC before (1) and after (2) the reaction.

Fixation of transition metal complexes on the modified carrier enhances the regioselectivity of the addition of hydride silanes to olefins in the direction of the formation of the end product (Table 4).

TABLE 4. Hydrosilylation of substituted styrenes p-$XC_6H_4CH=CH_2$

X	Relative amounts of isomers α/β, % mole			
	Cl_3SiH	Cl_2MeSiH	$ClMe_2SiH$	Et_3SiH
Cl	0/100	5/95 (31/69)[a]	10/90	-
H	0/100	0/100 (31/69)[a]	0/100	5/95
Me	-	5/95 (36/64)[a]	10/90	-

[a] In the presence of Speier's catalyst

The increase in the selectivity of the reaction can be attributed to the stability of the structure of the catalytically active centers, which is provided by the surface onium groups ER_3^+ and the rigidity of the inorganic carrier. These advantages are not fully manifested for catalysts on polymer matrices, possibly because of the nonrigidity of the macromolecular chains. For this reason, a certain amount of the α-isomer is observed in the hydrosilylation of styrene in the presence of an AV-17-8 anion-exchanger. We have also found that the surface has great effect on the direction of the catalytic hydrosilylation of olefins with optically active organosilanes. The effect is quite opposite to that observed in the course of the homogeneous reaction (Ref. 10). While the latter leads to the products with retention of the configuration (Ref. 11), the reaction in the presence of the platinum catalyst fixed on the ion-exchange resin results in the inversion of configuration of optically active organosilanes. It is clear that the direction of the possible interaction between the organosilan and the olefin is under great influence of the surface on which the catalytic complex is fixed.

In conclusion, we would like to state that fixation of complexes on the surface of suitable supports provide for the possibility of increasing efficiency and selectivity of catalysts and also altering these parameters by changing the nature of the surface.

REFERENCES

1. V.O. Reikhsfeld, N.K. Skvortsov and V.S. Brovko, I-s Soviet-Indian Seminar on Catalysis, Institute of Catalysis, Novosibirsk, 52-57 (1984).
2. A.V. Nikitin and V.O. Reikhsfeld, Zh.Obsch.Khim.,55 2079-2084 (1985).
3. V.S. Brovko, N.K. Skvortsov and V.O. Reikhsfeld, Zh. Obsch.Khim.,51, 467 (1981).
4. N.K. Skvortsov, V.S. Brovko and V.O. Reikhsfeld, Zh.

Obsch.Khim.,54, 2310-2316 (1984).
5. N.A. Filippov, V.O. Reikhsfeld and T.N. Zaslavskaja, Zh.Obsch.Khim.,47, 1374 (1977).
6. V.O. Reikhsfeld, N.A. Filippov and A.V. Nikitin, Zh. Obsch.Khim.,47, 959 (1977).
7. V.O. Reikhsfeld, Silicon Organic Compounds and Materials on Their Basis (in Russian), 5-13, Nauka, Leningrad (1984).
8. V.B. Pukhnarevich, B.A. Trofimov, L.I. Kopylova and M.G. Voronkov, Silicon Organic Compounds and Materials on Their Basis (in Russian),43, 2691-2694 (1973).
9. J. Chatt, G.A. Gamlen and L.E. Orgel, J.Chem.Soc.Am., 486-496 (1985).
10. L.H. Sommer, J.E. Lyons and H. Fujimoto, J.Am.Chem. Soc., 91, 7051 (1969).
11. V.O. Reikhsfeld, N.I. Flerova, N.A. Filippov and T.N. Zaslavskaja, Zh.Obsch.Khim., 50, 2017 (1980).

COMPARISON OF THE CATALYTIC PROPERTIES OF HOMOGENEOUS AND HETEROGENIZED CARBONYLPHOSPHINE-COBALT COMPLEXES IN THE OLEFIN HYDROFORMYLATION REACTION

G.A. Korneeva, O.L. Butkova, L.I. Zvezdkina,
N.A. Pritula, S.M. Loktev and Yu.B. Kagan
A.V. Topchiev Institute of Petrochemical Synthesis of the Academy of Sciences of the USSR, Moscow 117912, USSR

ABSTRACT - To elucidate the possibility of formation of strongly fixed on silica cobalt hydrocarbonyl complexes, modified by phosphine ligands, whose catalytic properties in olefin hydroformylation should not differ from those observed for their homogeneous analogs, the regularities of $HCo(CO)_4$ modification by mono- and bidentate phosphine and silylphosphine ligands were studied. To obtain a heterogeneous catalyst, which is active, selective and stable in the gas-phase propylene hydroformylation, the ligands were chosen by taking into account the results obtained. The surface ligand concentrations were determined provided that the ligand environment of the fixed active center did not differ from that in the homogeneous system.

1. INTRODUCTION

Heterogenization of metal complexes via their supporting on inorganic carriers is a way to obtain solid catalysts with high stability and efficiency, and with selectivity comparable with that of their homogeneous analogs.

One may expect that the active center, supported on solid surface, will preserve the catalytic activity and the selectivity typical for them in solution, if after heterogenization it retains ligand environment and structural features of the organometallic complex.

In the present work we have performed immobilization of cobalt hydrocarbonyl complexes modified by tri-n-butylphosphine (PBu_3) by exchanging one butyl ligand substituent by the silyl, having functional groups which can easily be bound with the surface of the inorganic oxide carrier.

Surface functionalization was done with the use of $(CH_2)_nR_2SiOR'$, which (as a result of its reaction with the surface silica hydroxyl groups) binds with the surface through the fragment (Si) -O-Si-C, stable under the hydroformylation conditions ((Si) - is the silica surface). Organoalkoxysylil substituent has to meet the following requirements: a) due to the inert hydrocarbon fragment it has to preserve modified properties of the bound ligand, since substitution of one substituent at phosphor atom is known to result in a substantial change in catalytic activity and selectivity of the complex formed (Ref. 1); b) the length of hydrocarbon chain has to provide a sufficient distance between the metal atom and the surface; c) each cobalt atom has to be available for bond formation with two ligands anchored on the surface. In view of these requirements we chose the following values: n=2,3, $R=CH_3$, $R'=OCH_3$, OC_2H_5.

To elucidate the possibility of formation of the stable complex whose catalytic properties should not differ from those of its homogeneous analogs, we have studied the regularities of $HCo(CO)_4$ modification in the hydroformylation conditions in mono-, bidentate- and silylphosphine ligand solutions. We chose the types of ligands suitable for obtaining active, selective and stable heterogenized catalysts and determined the surface ligand concentrations, which provide the same ligand environment of the supported active center as in the homogeneous system (methods of modification, identification of complexes and parameters of hydroformylation performance (Refs. 3-5); methods of synthesis of diphosphine (Ref. 4), silylphosphine (Ref. 6) and sylyldiphosphine ligands (Ref. 7) were used as described earlier).

2. REGULARITIES OF MODIFICATION. CARBONYL GROUP SUBSTITUTION IN $HCo(CO)_4$ BY PHOSPHINE LIGANDS

Homogeneous selective catalysts for olefin hydroformylation are known to be formed "in situ" in hydrocarbons from $HCo(CO)_4$ and PBu_3 as a result of the CO substitution by the ligand (Ref. 3). Substitution of CO in $HCo(CO)_4$ by a tri-n-butylphosphine ligand is also known to proceed as consecutive reversible reactions:

$$HCo(CO)_4 \underset{(-1)}{\overset{(1)}{\rightleftharpoons}} HCo(CO)_3(PBu_3) \underset{(-2)}{\overset{(2)}{\rightleftharpoons}} HCo(CO)_2(PBu_3)_2 \underset{(-3)}{\overset{(3)}{\rightleftharpoons}} HCoCO(PBu_3)_3.$$

The rate constants could be estimated as $k_2 \gg k_{-2} < k_1$.

A degree of CO substitution increases in hydroformylation conditions with increasing ligand concentration; it is accompanied by the decrease in catalytic activity of the complex formed and the increase in selectivity with respect to the normal structure products. Figure 1 represents variations in activity and selectivity of the homogeneous catalyst in propylene hydroformylation with increasing ligand concentration. The compositions of the catalytic system formed under these conditions at different PBu_3 concentrations have been determined by IR-spectroscopy method (Table 1).

At P:Co = 15-20, $HCo(CO)_2(PBu_3)_2$ is the main product formed in reaction (its yield is about 98-99%, IA, Table 2), i.e. at hydrocarbonyl modification in the ligand solution the formation of $HCo(CO)_4$ is practically suppressed. Suppression of reactions (-2) and (-1) seems

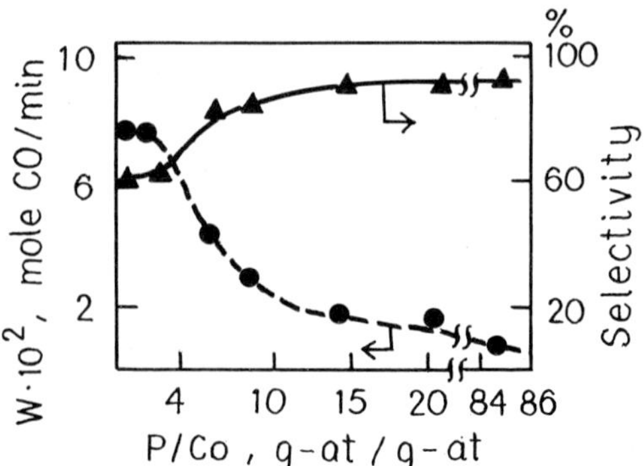

Figure 1. Propylene hydroformylation in the presence of cobalt hydrocarbonyls modified to different extents by butylphosphine. Solvent - n-heptane (80 ml); concentration of Co - 3 mmol/l; 180 °C; $CO:H_2$= 1:1; 10 MPa.

TABLE 1. Composition of hydrocarbonylphosphine cobalt complexes under hydroformylation conditions at different tri-butylphosphine concentrations. Reaction parameters are the same as in Fig. 1)

P:Co, mole ratio	Content of the catalytic system ($L=PBu_3$)	P:Co mole ratio	Content of the catalytic system ($L=PBu_3$)
0-2	$HCo(CO)_4$	8-12	$HCo(CO)_3L > HCo(CO)_2L_2$
2-3	$HCo(CO)_4 > HCo(CO)_3L$	>15	$HCo(CO)_2L_2 \gg HCo(CO)L_3$*
3-5	$HCo(CO)_3L > HCo(CO)_2L_2$		$HCo(CO)_3L$
5-8	$HCo(CO)_3L \gg HCo(CO)_4$, $HCo(CO)_2L_2$	>80	$HCo(CO)_2L_2 \geq HCo(CO)L_3^*$

* $(HCo(CO)(PBu_3)_3$ is catalytically inactive

to be the necessary condition for the stability of the catalyst supported with the help of the bound ligand,

TABLE 2. The composition of disubstituted cobalt hydrocarbonyl complexes used for hydroformylation

Index	Complex composition	Index	R	Characteristic bands, cm^{-1}
I	$HCo(CO)_2(PBu_2R)_2$	IA	Bu	1970 s. 1945 s. 1895 v.s.
		IB	$-(CH_2)_3-Si(Me)(Me)-OEt$	1968 s. 1942 s. 1985 v.s.
II	$H(CO)_2Co$ chelate: $P(Bu)(Bu)-(CH_2)_{12}-P(Bu)(R)$, n=2,4	IIA	Bu	1910 s. 1928 v.s. 1990 v.s.
		IIB	$-(CH_2)_3Si(Me)(Me)-OEt$ (≯ 2)	1905 m. 1920 v.s. 1892 s.
III	$H(CO)_2Co$ chelate: $P(Bu)(Bu)-R-P(Bu)(Bu)$	IIIA	$-(CH_2)_n-$ n=2,4 (⩾ 2)	1910 m. 1928 v.s. 1990 v.s.
		IIIB	$-(CH_2)_2-Si(Me)(OMe)-(CH_2)_2$ (⩾ 2)	1905 s. 1920 s. 1982 v.s.
IV(1)	$HCo(CO)_2(PBu_2-R-PBu_2)_2Co(CO)_2H$	IVA	$-(CH_2)_4-$	1970 s. 1945 s. 1895 v.s.
IV(2)	$[HCO(CO)_2(PBu_2-R-PBu_2)_2]_x$	IVB	$-(CH_2)_2-Si(Me)(OMe)-(CH_2)_2$	1978 s. 1942 s. 1889 m.

Seven-membered chelated IIA, IIB and IIIB hydrocarbonyl species (n=4) are stable at 20-80 °C; under hydroformylation conditions at a ligand concentration corresponding to P:Co = 15:20 they are converted species IV. This proceeds during an induction period of the propylene hydroformylation in the presence of III (n=4) at P:Co = 15:20.

since only in this case cobalt is not leached from the surface. One can expect that such effect will be achieved at high surface ligand concentrations. When the increase in surface ligand concentration is limited by the content of the silica surface OH-groups, modification by bound bidentate phosphine ligands would made it possible to increase this concentration, and thus to provide the high content of the supported metal and the stability of surface complexes.

However, under conditions of homogeneous propylene hydroformylation when bidentate $Bu_2P(CH_2)_4PBu_2$ was used as ligand at $P:Co \approx 15-20$, the disubstituted hydrocarbonyl species with the open (non-chelate) ligand structure (IV, Table 2) was formed (Refs. 3,4). As follows from IR spectra (in region of vibration of CO group and H-Co bond) and catalytic properties in hydroformylation, their properties are similar to those of $HCo(CO)_2(PBu)_2$ (see Table 2). The formation of a stable five-membered chelate ring in the disubstituted cobalt hydrocarbonyl species was observed in the case of $Bu_2P(CH_2)_2PBu_2$ even at very low (P:Co = 2) ligand concentrations. Such species were formed in the case of $Bu_2(P(CH_2)_4PBu_2$, only at relatively low temperatures (20-60 °C), which appears to be due to the decrease in metallocycle stability as the number of members in the chelate ring increases (Ref. 4).

However, all chelate disubstituted hydride structures studied are completely inactive under the hydroformylation conditions employed (Table 2).

3. SURFACE FUNCTIONALIZATION BY MODIFYING LIGAND

Mono- and bidentate ligands (IB, IIB, IIIB, IVB, Tables 2 and 3) on the silica surface are characterized by the modifying properties similar to those observed when no silyl fragment was the molecule (IA, IIA, IIIA, IVA). This refers both to the regularities of modification and conversion of the complexes obtained and their

TABLE 3. The catalytic properties of disubstituted cobalt hydrocarbonyl complexes modified by phosphine and silylphosphine ligands (parameters of hydroformylation are the same as in Figure 1).

Index	P:Co mole ratio	Catalytic properties	
		Rate at P:Co= 20,mol CO/min	Selectivity, n-products, mol%
IA	15÷80	0.025	93-94
IB	15÷80	0.025	93-94
IIA	2÷20	inactive	
IIB	2÷20	inactive	
IIIA	2÷20	inactive	
IIIB	2÷20	inactive	
IVA	⩾2	0.025	94-96
IVB	⩾2	0.025	94-96

properties in propylene hydroformylation (Ref. 5). A similarity of the corresponding spectra in the range of 1600-2200 cm^{-1} indicates the similar complex structure and ligand metal environment in the case of phosphine and the corresponding silylphosphine ligands. Both in the presence and absence of a functional group in the substituent at the P atom, bidentate ligand complexes are active only in the case of the open (non-chelate) ligand structures. Note that regularities of $HCo(CO)_4$ modification studied both with mono- and bidentate silylphosphine ligands that form open structures do not differ much from those observed for PBu_3 (see Fig. 1) (Ref. 3-5).

The fact that the functional groups (silyl fragments) in a ligand have no effect on their modifying properties makes it possible to consider them to be suitable for obtaining bound ligands. The regularities of catalyst modification by bidentate ligand show the ways to obtain higher concentrations of the surface phosphine groups.

4. DETERMINATION OF THE SURFACE LIGAND CONCENTRATIONS EFFECTIVE FOR OBTAINING ACTIVE AND SELECTIVE HETEROGENIZED CATALYSTS

The carbonyl cobalt complex was heterogenized on silica by interacting cobalt hydrocarbonyl with a preliminary anchored ligand (such procedure provides a uniform distribution of Co over the surface (Ref. 8)):

1. (Si) -OH + R'O-Si~~~P $\xrightarrow{-R'OH}$ (Si) -O-Si~~~P

2. (Si) -O-Si~~~P $\xrightarrow[-2\ CO]{HCo(CO)_4}$ ((Si)-O-Si~~~P$)_2$Co(CO)$_2$H

(R'O-Si~~~P : silylphosphine ligand).

Liquid-phase propylene hydroformylation in the presence of $HCo(CO)_4$ and excess of PBu_3 in solution when SiO_2 was introduced into a reactor simultaneously made it possible to estimate the influence of the silica surface on the catalytic properties of the system studied. The results are given in Table 4. A silica was used with average pore radius ~130 Å, BET surface 185 m^2/g, OH surface group content 4-5 by 100 $Å^2$). The experiments were conducted in the presence of $HCo(CO)_4$ + PBu_3 syste (P:Co = 5 and 10).

A comparison of the results of runs 2 and 3 shows that SiO_2 having surface hydroxyls interacts with $HCo(CO)_4$. At this interaction cobalt from solution is bound on the surface in inactive form. This is evidenced by decrease of the reaction rate and the cobalt concentration in solution (run 3). Simultaneously the catalytic selectivity decreases as low as the values characteristic of nonmodified catalysts (60% products of the normal structure). This decrease may indicate that the ligand also interacts with silica.

TABLE 4. Effect of silica on the catalytic properties of cobalt carbonyl complexes modified by tributylphosphine in propylene hydroformylation (reaction parameters are the same as indicated in Figure 1).

Catalytic system	P:Co mole ratio	Hydroformylation rate, mole CO/min	n-product yield, mole%	Co amount in solution [b]
$HCo(CO)_4$	-	0.078	60	100
$HCo(CO)_4$, PBu_3	5	0.050	75	100
$HCo(CO)_4$, PBu_3, (Si)-OH	5	0.036	61	32
$HCo(CO)_4$, PBu_3, (Si)-$OSiMe_3$ [a]	5	0.048	75	100
$HCo(CO)_4$, PBu_3, (Si)-OH	10	0.049	76	98
$HCo(CO)_4$, PBu_3	10	0.035	87	100

[a] Si-OH was eliminated by reaction with trimethyletoxysilane according to (Ref. 9);
[b] in % of initial content.

When surface hydroxyl groups are substituted by a trimethylsilyl fragment (run 4) there is no effect of the surface on the catalytic properties of the modified cobalt carbonyl. As follows from the comparison of runs 2 and 4, $HCo(CO)_4$ and PBu_3 do not interact with dehydroxylated silica surface.

With increasing ligand concentration (run 5) up to P:Co = 10 in the presence of (Si)-OH the reaction rate increased as compared to run 3, and practically all cobalt was found in the solution, i.e. cobalt did not react with the silica surface. In this case probably the ligand is adsorbed on the surface. This is suggested by the increased selectivity (as compared to the experiment without silica (run 6)).

So the heterogenization of the carbonyl phosphine me-

TABLE 5. Gas-phase propylene hydroformylation in the presence of heterogenized disubstituted hydrocarbonylphosphine cobalt complexes (180 °C; 10 MPa; C_3H_6:Co:H_2=1:4.5:4.5; space velocity 1500 h^{-1})

Index (see Table 2)	P:Co mole ratio	Productivity at stable activity, g/l·h	Period of stable activity, h	Selectivity of n-products mol.%
IB	15	100	20	92-93
	22[a)]	120	500	92-93
IIB, IIIB	15	inactive		
	22	inactive		
IVB	15	150	25	93-95
	22[a)]	400	300	93-95

[a)] During the test about 10% of cobalt of the initial loading was removed from the catalyst.

tal complexes may be accompanied by the partial inactivation of the phosphine ligands due to their interaction with the surface. An irreversible adsorption of a metal also is possible with the formation of inactive surface compounds.

The results of testing of heterogenized cobalt complexes in hydroformylation, as well as the coincidence of IR spectra of the surface complexes formed with those of the homogeneous systems allow us to conclude that the surface catalytic complexes can retain the properties characteristic of the complexes in solution.

The results of continuous tests of the complexes heterogenized on silica at P:Co = 15 and 22 in the gas-phase propylene hydroformylation are given in Table 5. IR spectra of these complexes (measured in Nujol) were identical with those observed for the corresponding individual silylphosphine complexes (Table 2).

In the presence of the heterogenized catalyst IB

the turnover number was about 100 min^{-1}, to be compared to 115 min^{-1} for its homogeneous analog.

REFERENCES

1. F.E. Paulik, Catal. Rev., 6, 49 (1972).
2. V.M. Vdovin, V.E. Fedorov, N.A. Pritula and G.K. Fedorova, Izv. Akad. Nauk SSSR, ser. khim., 1, 181 (1982).
3. Yu.B. Kagan, G.A. Korneeva, O.L. Butkova, M.V. Shishkina and A.N. Bashkirov, Neftekhimia, 20 (6), 869 (1980).
4. Yu.B. Kagan, G.A. Korneeva, O.L. Butkova and M.V. Shishkina, Neftekhimia, 23 (1), 77 (1983).
5. Yu.B. Kagan, O.L. Butkova, G.A. Korneeva and S.M. Loktev, Izv. Akad. Nauk SSSR, ser. khim., 8, 1714 (1985).
6. L.I. Zvezdkina, N.A. Pritula, L.V. Karlova, V.M. Vdovin, Yu.B. Kagan and A.N. Bashkirov, Neftekhimia, 17 (6), 873 (1977).
7. O.L. Butkova, L.I. Zvezdkina and N.A. Pritula, Izv. Akad. Nauk SSSR, ser. khim., 10, 2390 (1980).
8. Yu.B. Kagan, G.A. Korneeva, L.V. Karlova, N.A. Pritula, V.M. Vdovin and A.N. Bashkirov, in: Catalysts containing the anchoring complexes, Symposium Proceedings, Yu.I. Yermakov (Ed.), Novosibirsk, p. 11 and 154 (1980).
9. M. Čapka and J. Hetflejs, Coll. Czech. Chem. Commun., 39, 154 (1974).

SUPPORTED LIQUID PHASE HYDROFORMYLATION CATALYSTS CONTAINING RHODIUM-DIBENZOPHOSPHOLE COMPLEXES

Jes Hjortkjær,
Technical University of Denmark, Building 227, 2800 Lyngby, Denmark.
Penka Toromanova-Petrova,
Higher Institute of Chemical Technology, Burgas 8010, Bulgaria.

ABSTRACT

Catalysts obtained by supporting solutions of $HRh(CO)(PPh_3)_3$ in phenyl-dibenzophosphole (DBP) on silica catalyze the hydroformylation of propene. The modifications observed with respect to activity and regioselectivity relative to similar SLP-catalysts in which the liquid phase is PPh_3, is ascribed to interaction between the phosphorous lone pair and the π-electrons of the benzophosphole ring system, and to lowered configurational constraints against anti-Markownikoff insertion of propene into the metalhydrogen bond.

A phenomenological examination of the hydroformylation kinetics disclose reaction orders of 0, 0 and 1 with respect to carbon monoxide, hydrogen and propene. The activation energies for the formation of the two aldehydes were essentially the same, implying that the regioselectivity is independent of temperature.

INTRODUCTION

Recently several industrial propene hydroformylation plants based on the use of Rh-complexes as catalysts have been established (1,2). In these plants the product (butanal), is also the reaction medium in which the catalyst precursor, hydridocarbonyltris(triphenylphosphine)-rhodium(I) is dissolved together with a great excess of phosphine, because of the favourable influence on regioselectivity. When this approach is carried to the extreme by using molten triphenylphosphine as reaction medium, a rapid and highly selective conversion of propene to n-butanal is observed (3). A "heterogeneous" version of such a catalytic system may be established by supporting the liquid reaction medium (in which the catalytic complex is dissolved) on the surface of a solid support. Such SLP-catalysts exhibit very high selectivity, regioselectivity and activity in propene hydroformylation, and may be a realistic alternative to the usual industrial liquid phase hydroformylation (4,5,6). A further improvement of these catalysts may be established by replacing the triphenylphosphine ligand by 5-phenyl-5H-dibenzophole (DBP), because the benzophosphole ring system may reasonably be expected to exert a greater electronic "elasticity", leading to an improved ability to counterbalance the charge fluctuations on the catalyst metal center, induced by oxidative addition and reductive elimination processes during the catalytic cycle. For olefin hydrogenation an essentially improved activity has been demonstrated with rhodium-DBP complexes (7,8) and for "homogeneous" styrene hydroformylation enlarged activity has also been reported (9). Especially in relation to SLP-catalysts it is also important that phenyldibenzophosphole under reaction conditions might reasonably be expected to have a lower vapor pressure compared to triphenylphosphine (10).

We now present data concerning the effect on catalytic behavior in relation to propene hydroformylation of SLP-catalysts in which the liquid phase is DBP. At the same time a phenomenological examination of the reaction kinetics is reported.

EXPERIMENTAL

5-Phenyl-5H-dibenzophosphole (DBP) was prepared from tetraphenylphosphoniumbromide using methods previously reported (11), and hydridocarbonyltris(triphenylphosphine)rhodium(I) was prepared from rhodium trichloride trihydrate (Engelhard Industries) using the method of Ahmad et al. (12).

The support was a porous silica (Grace) consisting of 100-140 mesh particles with a total surface area of 400 m^2/g, a mean pore diameter of 10 nm and pore volume 1 $cm^3 \cdot g^{-1}$.

The catalysts were prepared by physical mixing of the three components: Support, liquid phase component (DBP) and $HRh(CO)(PPh_3)_3$. At room temperature all components are solid, but when the catalyst is heated to reaction temperature the DBP will melt (m.p. 94°C) and form a liquid phase distributed on the surface and/or in the pores of the solid support. In this liquid phase $HRh(CO)(PPh_3)_3$ will be dissolved and the equilibria:

$$HRh(CO)(PPh_3)_3 \rightleftharpoons HRh(CO)(PPh_3)_2(DBP) \rightleftharpoons HRh(CO)(PPh_3)(DBP)_2 \rightleftharpoons HRh(CO)(DBP)_3$$

will be established. The DBP-concentration is very high relative to the PPh_3-concentration, because PPh_3 can only come from dissociation of $HRh(CO)(PPh_3)_3$, which is added corresponding to a $HRh(CO)(PPh_3)_3$-concentration of only $4{,}0 \cdot 10^{-2}$ mol complex/mol DBP (see below). So, after complete dissociation of the complex the maximum PPh_3-concentration in the liquid phase is 12%. Besides, the DBP-complex is known to dissociate far less than the phosphine complex (13), so that the equilibria under the conditions described must be assumed to be shifted to the right, leaving essentially all complex in the form of $HRh(CO)(DBP)_3$.

In these catalysts the liquid-loading (mass of liquid present per unit of support) was 0,6 (g DBP)$\cdot$(g support)$^{-1}$ and the concentration af complex in the liquid phase was app. $4{,}0 \cdot 10^{-2}$ (mol complex)$\cdot$(mol DBP)$^{-1}$, which corresponds to a P/Rh ratio of app. 25.

The hydroformylation reactions were studied using a microcatalytic flow system. A known quantity of each catalyst (typically 0,5g) was placed in a stainless steel reactor where it formed a short bed supported on a porous, stainless steel plug sealed across the reactor tube. The reactor was placed in a thermostatted oil bath. As outlined earlier (4) for differential conversion of propene in the reactor the rate of hydroformylation r, (mol$\cdot$sec$^{-1}\cdot$(g Rh)$^{-1}$) is defined as:

$$r_1 = \frac{F \cdot x}{W_{Rh}}$$

where x is the fractional conversion of propene in the reactor for a flow rate with respect to propene of F, and W_{Rh} represents the weight of rhodium present in the reactor and introduced into the silica support as $HRh(CO)(PPh_3)_3$. If r_2 and r_3 are the rates of formation of n-butanal and iso-butanal respectively, the regioselectivity, R, is defined by the value of the ratio r_2/r_3. In this work no formation of side products was seen and so $r_1 = r_2 + r_3$. Mixtures of the reactants, and in some cases nitrogen, having various compositions were prepared such that the partial pressure of each reactive component could be independently varied. Flow rate control was by means of a precision fine metering valve. After reduction to atmospheric pressure, the exit gas was periodically analyzed by injection of samples

($0.5\ cm^3$) into a gas chromatograph. Analysis of the components was achieved using a SCOT column of squalane (50 ft•0.02 in.) (Perkin Elmer F11). Suitable calibration procedures were carried out using mixtures of known composition.

RESULTS AND DISCUSSION

As explained earlier there are two important factors, which underlie our wish to examine the catalytic performance of the present dibenzophosphole-complex-SLP catalysts:

First because the dibenzophosphole ring system is expected to induce a higher "electroncapacity" into the catalytic complex, compared to similar complexes with phosphine ligands. This will probably facilitate the oxidative addition and reductive elimination reactions in the catalytic cycle. Secondly because SLP-catalysts based on dibenzophole rather than phosphine are expected to be more stable, simply because of lowered vapor pressure of the ligand, which of course would lead to less sublimation of ligand, which is also a problem in industrial hydroformylation. In order to evaluate the present findings it is therefore helpful to consider them in close comparison with results of previous work on hydroformylation with dissolved dibenzophosphole complexes, and with SLP-hydroformylation catalysts containing triphenylphosphine as liquid reaction medium.

Liquid phase hydroformylation with phosphole complexes has only been reported in two cases (9,14). According to Hayashi et al. (9) Rh-dibenzophosphole complexes catalyze the hydroformylation of styren with an activity, which is slightly higher than with the correspondin phosphine complexes (relative rates 1,1:1) and also regioselectivity is approximately the same with the two kinds of ligands. This is in striking contrast to our findings for the SLP-catalysts, because we observe a decline in activity to approximately 20% of the activity found for similar phosphine based SLP-catalysts (compare fig. 1 with fig. 4 in ref. 4). It is difficult to account for this lowered activity if it can not be related to the electronic changes induced by the DBP-group, which as previously mentioned are expected to have the opposite effect because of higher electroncapacity. On the other

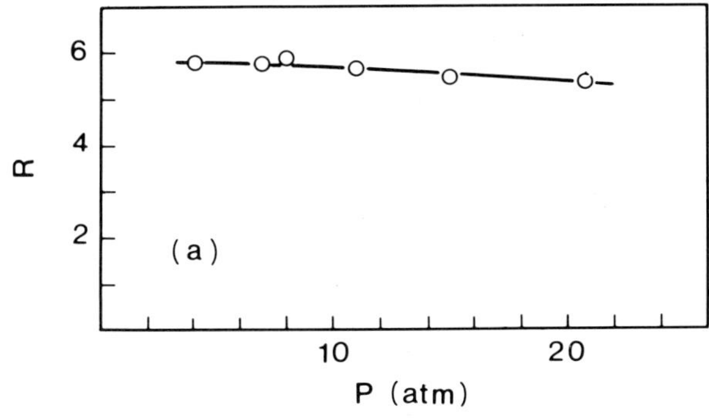

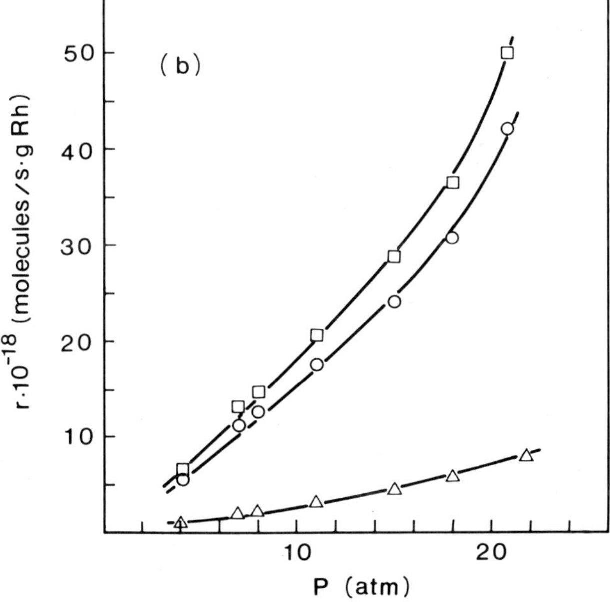

Figure 1. Regioselectivity (a) and hydroformylation rate (b) (total rate (□), n-butanal formation rate (o) and isobutanal formation rate (Δ)) as a function of total pressure with $P_{C_3H_6}:P_{CO}:P_{H_2}=1:1:1$ and temperature = 100°C.

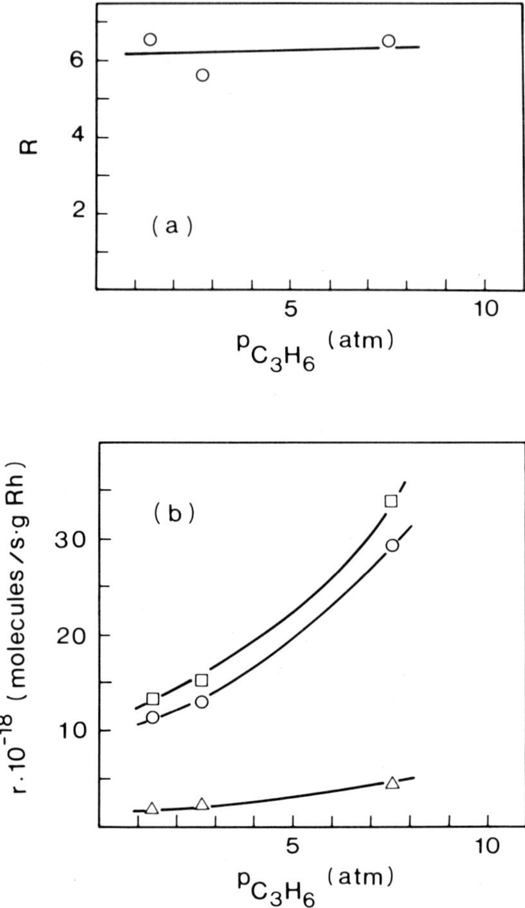

Figure 2. Regioselectivity (a) and hydroformylation rate (b) (total rate (□), n-butanal formation rate (o) and isobutanal formation rate (Δ)) as a function of propene partial pressure with $P_{CO}=P_{H_2}=2{,}74$ atm and temperature = 100°C.

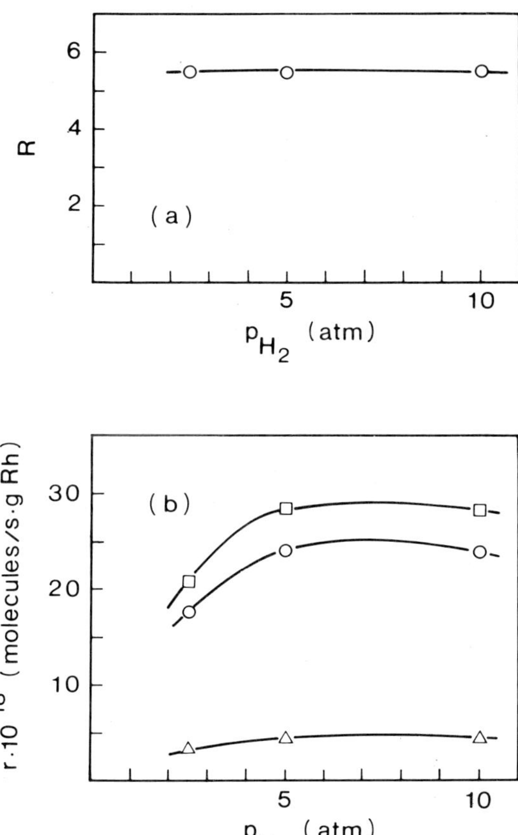

Figure 3. Regioselectivity (a) and hydroformylation rate (b) (total rate (□), n-butanal formation rate (o) and isobutanal formation rate (Δ)) as a function of hydrogen partial pressure with $P_{C_3H_6}=P_{CO}=5$ atm and temperature = 100°C.

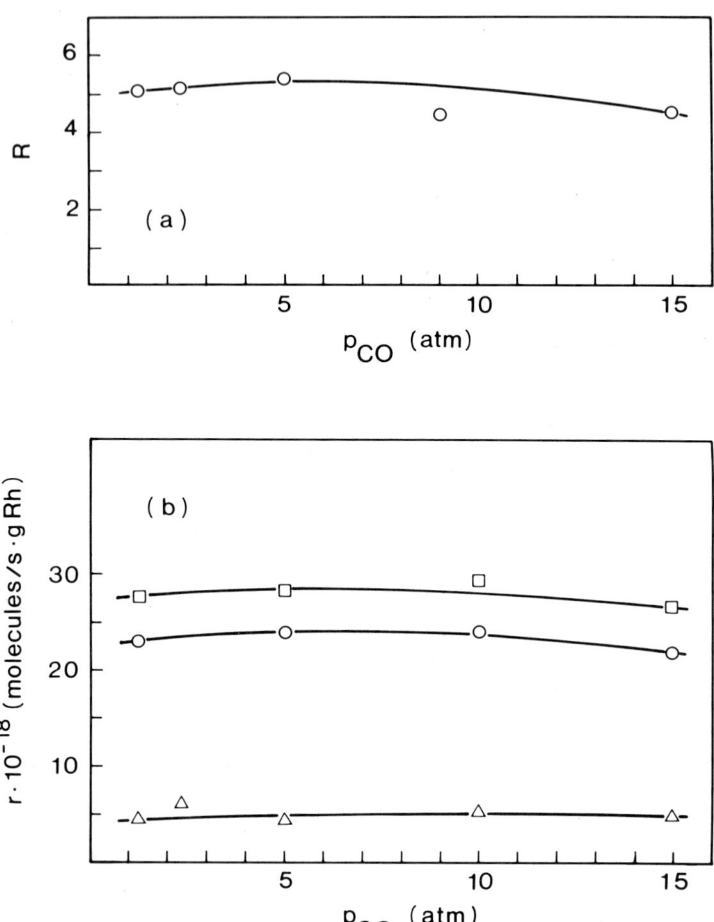

Figure 4. Regioselectivity (a) and hydroformylation rate (b) (total rate (□), n-butanal formation rate (o) and isobutanal formation rate (Δ)) as a function of carbon monoxide partial pressure with $P_{H_2} P_{C_3H_6}$ =5 atm and temperature 100°C.

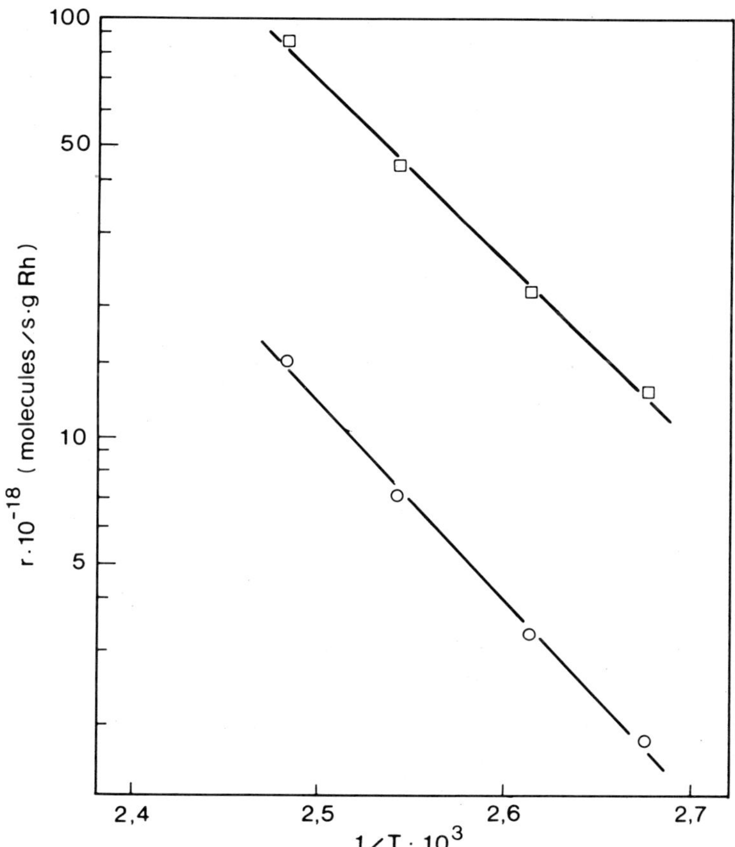

Figure 5. Arrhenius plot with $P_{C_3H_6}:P_{CO}:P_{H_2}=1:1:1$ and a total pressure of 10 atm.(□) = n-butanal formation rate and (o) = iso-butanal formation rate.

hand, this enlarged electron capacity of the phosphole ring system may result in a decreasing electron density on the rhodium atom due to interaction between the π-system and the phosphorous lone pair. But also the increased rigidity of the DBP ligand relative to PPh_3 could possibly have a decreasing effect on catalytic activity, although it seems rather unlikely that this steric effect is so important relative to the electronic effect expected from the DBP group. However, another remarkable dissimilarity between the two kinds of SLP-catalysts is with respect to regioselectivity as a function of carbon monoxide pressure. While the regioselectivity of the PPh_3-catalysts varies from app. 25 at low carbon monoxide to app. 6 at carbonmonoxide pressures higher than 5 atm (4,6), the regioselectivity of the DBP-catalysts is almost independent of carbon monoxide pressure, and stays constant at app. 5 (fig. 4), i.e. very close to the value found at higher carbon monoxide pressures with the PPh_3-catalysts.

The high sensitivity of the regioselectivity for the PPh_3-catalysts has previously (6) been referred to a change in mechanism from the associative to a mechanism involving the square trans phosphine, $HRh(CO)(PPh_3)_2$ (3). This shift, which was observed in low pressure liquid phase hydroformylation of hex-1-ene with $HRh(CO)(PPh_3)_3$ as catalyst precursor, was dependent on the carbon monoxide pressure and on the extent to which excess PPh_3 was present (15), so that with the very high PPh_3 concentration in the SLP-catalysts the shift is expected to take place at a carbon monoxide of app. 1 atm. (fig. 11 in ref. 6). Since this mechanism involves only mono-carbonyls it is of course a condition that the mono-carbonyl can coordinate the olefin involved, but for DBP-Rh complexes (and contrary to similar PPh_3-Rh complexes) it has been shown (16) that in a solution containing both $HRh(CO)(DBP)_3$ and $HRh(CO)_2(DBP)_2$ only the di-carbonyl coordinates the olefin ("the olefin is selectively trapped") (16). This means that only the di-carbonyl DBP-complexes are active in the catalytic reaction, and a lowered carbon monoxide pressure can not be expected to lead to an increased regioselectivity due to increased configurational constraints - as seen for the PPh_3-complexes (15) - against anti-Markownikoff addition of the olefin to the DBP-Rh complex. In accordance with these assumptions we find for the DBP-catalysts a carbon monoxide independent regioselectivity of app. 5, i.e. very close to the value found with the PPh_3-catalysts at high carbon monoxide pressure (4,6). Another consequence of the inability of $HRh(CO)(DBP)_3$ to coordinate olefins would be that this complex should not be able to catalyze olefin hydrogenation. Other DBP-Rh complexes have been reported to be active in olefin hydrogenation (7,8,17) but there are no reports on $HRh(CO)(DBP)_3$ being active, however, this remains to be investigated, but we have seen no trace of propane formation even at high hydrogen partial pressure and/or low carbon monoxide partial pressure.

An increase in total pressure of the reactant mixture ($P_{C_3H_6}:P_{H_2}:P_{CO}=1:1:1$) results in an increase in rates of formation of both n- and iso-butanal (fig. 1). For the reaction conditions employed the rates seem to be a little more than 1. order in total pressure, and at the same time a slight increase in regioselectivity from 5 to 6 is observed (fig. 1). The rate dependencies on propen partial pressure, hydrogen partial pressure and carbon monoxide partial pressure are given in figures 2, 3 and 4, respectively. For carbon monoxide partial pressures of more than app. 1 atm the hydroformylation rates and the regioselectivity are essentially independent (fig. 4), and the same is the case for hydrogen partial pressures of more than 5 atm (fig. 3), so that all the influence of total pressure may be referred to the influence of propene partial pressure, in agreement with our results reported in fig. 2. This behavior is not unusual for liquid phase hydroformylation catalysts, accordingly a near zero order dependence with respect to carbon monoxide has previously been found (4,5,18,19), and a first order dependence with respect to alkene is also reported for widely varying conditions (3,5,18). For hydrogen it is usual to find first order dependence (3,18,20), but with SLP-catalysts orders of 0,09 (5) and 0,4 (6) have been reported.

An increase in temperature results in an increase in hydroformylation rate, but at high temperature the Arrhenius plot (fig. 5) was not linear, probably due to pore diffusion limitations and/or partly decomposition of the catalytic complex. From the linear part of the Arrhenius plot the activation energies for the formation of the two aldehydes were calculated. For n-butal E = 81,5 kJ/mol and for iso-butanal E = 89,8 kJ/mol leading to a slight decrease of regioselectivity with increasing temepratur. These activation energies are higher than found for the analogous PPh_3-catalysts (6), and as the DBP-catalysts are also more stable they can be used at higher temperatures than the PPh_3-catalysts. Accordingly, at 130°C with a 1:1:1 reactant mixture at a total pressure of 10 atm. no deactivation could be observed after 240 h. with the DBP-catalysts, and the total hydroformylation rate was $1,7 \cdot 10^{-4}$ $mol \cdot sec^{-1} \cdot (gRh)^{-1}$ which compares favourably with app. $1,3 \cdot 10^{-4}$ $mol \cdot sec^{-1} \cdot (gRh)^{-1}$ at 100°C with PPh_3-SLP-catalysts (4,6).

The DBP-catalysts are very sensitive to oxygen. Even traces of oxygen in the reactant mixtures leads to a rapid, but temporary increase in activity and decrease in regioselectivity, and this change in regioselectivity is considered to reflect a modification in nature of active species rather than a change in the number of catalytically active centres. The regioselectivity falls to app. 1,5-2, which is close to the value previously reported for hydroformylation catalysts without "excess" ligand (4), and is therefore considered to be due to the formation Rh-carbonyls without any DBP-ligand.

ACKNOWLEDGEMENT

We wish to thank Peter Simonsen for his very competent assistance with the experimental work, especially the purification and mixing of reactant gasses. Most of the hydroformylation experiments were carried out by Niels Munk and Ulla Brandstrup.

REFERENCES

1. E.A.V. Brewster, Chemical Engineering, Nov. 1976, 90.
2. R. Fowler, Hydrocarbon Processing, Sept. 1976, 247.
3. C.K. Brown, G. Wilkinson, J. Chem. Soc. A (1970) 2753.
4. J. Hjortkjær, M.S. Scurrell, P. Simonsen, J. Mol. Cat. 6/6 (1979) 405.
5. L.A. Gerritsen, Thesis, Delft University Press, 1979.
6. J. Hjortkjær, M.S. Scurrell, P. Simonsen, H. Svendsen, J.Mol. Cat. 12/2 (1981) 179.
7. D.G. Holah, A.N. Hughes, B.C. Hui, Can. J. Chem. 50 (1972) 3714.

8. D.E. Budd, D.G. Holah, A.N. Hughes, B.C. Hui, Can. J. Chem. 52 (1974) 775.
9. T. Hayashi, M. Tanaka, I. Ogata, J. Mol. Cat. 6 (1979) 1.
10. M. V. Forward, S.T. Bowden, W.J. Jones, J. Chem. Soc., supplement (1949) 121.
11. H. Hoffmann, Chem. Ber., 95 (1962) 2563.
12. N. Ahmad, S.D. Robinson, M.F. Uttley, J. Chem. Soc., Dalton Trans., (1972) 843.
13. D.G. Holah, A.N. Hughes, B.C. Hui, Can. J. Chem. 53 (1973) 3674.
14. Shell International Research Maatschoppy N.V., NL 6, 516, 164 (1966), Chem. Abstr., 66 (1967) 28392r.
15. J. Hjortkjær, J. Mol. Cat., 5 (1979) 377.
16. J.M. Brown, L.R. Canning, A.G. Kent, P.J. Sidebottom, J. Chem. Soc., Chem. Commun., (1982) 721.
17. D.G. Holah, I.M. Hoedless, A.N. Hughes, B.C. Hughes, B.C. Hui, D. Martin, Can. J. Chem. 52 (1974) 3758.
18. K.L. Olivier, F.B. Booth, Am. Chem. Soc. Div. Pet. Chem. Prepr., General Papers, 14 (1969) A7.
19. P. Cavalieri d'Oro, L. Raimondi, G. Pagani, G. Montrasi, G. Gregorio, G.F. Oliveri del Castillo, A. Andreetta, Symp. on Rhodium in Homogeneous Catalysis, Veszprem, Hungary 1978.
20. D. Evens, J.A. Osborn, G. Wilkinson, J. Chem. Soc. A, (1968) 3133.

VAPOR PHASE CARBONYLATION OF ORGANIC COMPOUNDS WITH SUPPORTED GROUP VIII METAL CATALYSTS.

K. Fujimoto, K. Omata, T. Shikada, and H. Tominaga
Department of Synthetic Chemistry Faculty of Engineering
The University of Tokyo Hongo 7-3-1, Bunkyo-ku, Tokyo 113, JAPAN.

ABSTRACT - The essential of the catalysis of supported group VIII metals for methanol carbonylation was formation of the surface carbonyls and nondissociative adsorption of carbon monoxide together with dissociative adsorption of a halide promoter onto the carbonyl.

1.INTRODUCTION

Rhodium has been known to be highly active as catalysts for methanol carbonylation in both complex and metal forms (Refs. 1-2). Nickel tetracarbonyl is also excellent for the reaction (Ref. 3). The authors have found that nickel metal supported on active carbon shows outstanding activity for the vapor phase carbonylation of organic compounds (Ref. 4 & 5). In the present study the authors have studied the details of the catalysis of Ni and other group VIII metals supported on active carbon for methanol carbonylation.

2.EXPERIMENTAL

2.1. Catalyst.

Catalysts were prepared by impregnating a commercially available active carbon (AC,Takeda Shirasagi C, 1200m^2/g, wood char, steam activation) with metal salts from their aqueous solutions followed by reducing in flowing hydrogen at 400°C for 2h. The preparation method has been described in detail elsewhere (Ref. 4). Metal salts used were : $RhCl_3 \cdot 3H_2O$, $IrCl_4 \cdot H_2O$, $PdCl_2$, $RuCl_3$, $Fe(NO_3)_2 \cdot 9H_2O$, $Co(NO_3)_2 \cdot 6H_2O$, $Ni(AcO)_2 \cdot H_2O$. The metal loading was 2.5wt%/AC as metal.

2.2. Apparatus and Procedure.

The activity test was conducted in a stainless steel flow type reactor under pressurized conditions. The conditions were 11atm, 180-350°C, W/F=5g of catalyst·h/mol, CO/methanol/methyl iodide feed ratio=100/19/1. A vaporizer was located at the top of the reactor for the purpose of mixing and heating the reactants. Liquid reactants such as methanol, methyl iodide were fed with syringe type micro-feeder. Products and reactants were collected respectively by a heated sampling bulb and analyzed by gas chromatographs.

Adsorption of CO and methyl iodide (MeI) was measured using a volumetric technique in a flow cell connected to a conventional volumetric vacuum line. Methyl iodide was purified by a repeated frozen-pump-thew cycle. The catalyst was reduced in flowing hydrogen at 400°C for 2h and then evacuated at the same temperature. Adsorption isotherm was determined by measuring before and after evacuating for 0.5h at room temperature (for CO) and 200°C (for MeI), and the amount of irreversible adsorption was calculated as the difference of the two isotherms at 200 Torr. Temperature programmed reaction (TPR) and desorption (TPD) were performed according to a similar procedure as that of the previous study (ref. 9). The samples, on which adsorbed species were hold, were heated in flowing H_2 (TPR) or He (TPD) at a constant rate (400°C/h) and the desorbed products were detected by a gas chromatograph.

3. RESULTS AND DISCUSSION

3.1. Reaction Pathways in Methanol Carbonylation.

We have reported that the methanol carbonylation on the Ni/AC catalyst proceeds along the reaction pathways as described in equation(1) (Ref. 5).

$$\begin{array}{ccccc} \mathrm{MeOH} & \longrightarrow & \mathrm{AcOME} & \rightleftarrows & \mathrm{AcOH} \\ & \searrow\ \nearrow & & \searrow\ \nearrow & \\ & \mathrm{DME} & & (\mathrm{Ac_2O}) & \end{array} \quad (1)$$

It has been demonstrated by the thermodynamic calculation that lower temperature and higher CO partial pressure is favorable for the selective synthesis of acetic acid. Under reaction conditions adopted in this study, the formation of acetic acid is substantially free from the equilibrium limitation.

3.2. Catalytic Features of Ni/AC Catalyst.

The activities of Ni on three types of support are shown in TABLE 1 in terms of MeOH conversion. Ni shows an excellent activity for the carbonylation only when it is supported on AC.

TABLE 1 Adsorption and MeOH Carbonylation Activity

Catalyst	Activity[a] (%)				H_2 uptake (µmoles/g)	CO uptake (µmoles/g)	CO/H[b]	CO/Ni
	AcOMe	AcOH	DME	CH_4				
Ni/SiO_2	1.1	0	0.4	0.4	37	206	2.8	0.48
Ni/Al_2O_3[c]	2.2	0	96.2	0.7	7.2	237	16.5	0.56
Ni/AC	21.6	69.1	2.1	3.0	6.7	677[d]	50.5	1.59

a MeOH conversion to at 290 °C (300 °C on Ni/AC), 10Kg/cm^2-G, W/F=10 g·h/mol, CO/MeOH/MeI=20/19/1 molar ratio with 10wt%Ni/SiO_2 or Al_2O_3 and 2.5wt%Ni/AC.

b Adsorbed CO molecules par adsorbed hydrogen atom.

c Adsorption was measured at room temperature.

d CO uptake at 80 Torr.

It is apparent that AC exerts a special influence on the supported Ni. The results of H_2 adsorption and CO adsorption are also summarized in TABLE 1. The amount of chemisorbed H_2 was far less than that of CO for every sample and thus the ratio of chemisorbed CO to that of H was exceedingly higher than 1.0. It should be also noted that the amount of chemisorbed CO is greater than total Ni loaded on AC, which indicates that more than one molecule of CO is adsorbed on a surface Ni atom.

TPR of adsorbed CO with hydrogen has been known to be useful for investigating the character of chemisorbed CO on supported metal catalysts (Ref. 8). Figure 1 shows the results of TPR of CO which is adsorbed on Ni supported on Al_2O_3, SiO_2, and AC, respectively.

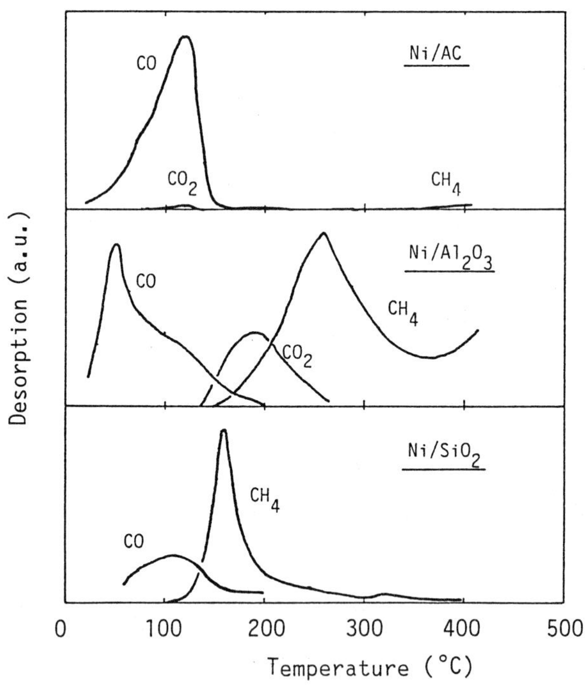

Figure 1 TPR Profile of Adsorbed CO

The reactivity of CO on Ni/AC is markedly different from that on other supports. In the case of the Ni/Al_2O_3 or the Ni/SiO_2, CO is desorbed only at low temperature and then CH_4 (and CO_2) are formed at higher temperature. In the case of Ni/AC, however, almost all of the adsorbed CO were desorbed at temperature below 120°C. It has been generally accepted that the first step of CH_4 formation by hydrogenation of carbon monoxide is the dissociation of C-O bond (equation 2, Ref. 9). Then the resultant $C_{(a)}$ and $O_{(a)}$ would react with either hydrogen or CO as shown in equations (2) to (5).

$$CO_{(a)} \longrightarrow C_{(a)} + O_{(a)} \qquad (2)$$

$$C_{(a)} + 4H_{(a)} \longrightarrow CH_4 \qquad (3)$$

$$O_{(a)} + CO_{(a)} \longrightarrow CO_2 \qquad (4)$$

$$O_{(a)} + 2H_{(a)} \longrightarrow H_2O \qquad (5)$$

Thus, it is clear that on the Ni/Al_2O_3 catalyst or the Ni/SiO_2 catalyst large part of adsorbed CO is in the dissociated or nearly dissociated form. On the Ni/AC catalyst CO is adsorbed in the non-dissociative way and the bond between nickel and CO is weak as suggested from the low temperature desorption spectrum. In contrast to the Ni/Al_2O_3 and the Ni/SiO_2, the C-O bond dissociation (reaction 2) hardly proceeds on the Ni/AC and therefore the formation of CH_4 (reaction 3) or CO_2 (reaction 4)is small.

Thus the characters of adsorbed CO on Ni/AC differes significantly for that on the Al_2O_3 or on the SiO_2. That is, the suppression of H_2 adsorption on Ni/AC suggests there exists a unique interaction with AC such as new kind of SMSI, and the large uptake of CO and the TPR spectrum indicate the formation of subcarbonyl. The molecular CO bonded weakly on Ni should play an important role in the carbonylation of MeOH. The authors have already reported (ref. 11) that the reaction order of the MeOH carbonylation with respect to the partial pressure of MeI, which is essential for the reaction, is 0.1 on the Ni/AC. This suggests that MeI is adsorbed strongly on the Ni under reaction conditions and thus it is expected to be suitable as probe molecule for the determination of surface area of the dispersed nickel. Chemisorption of MeI was measured at 250 °C to estimate the number of active site of the Ni catalyst under reaction temperature. Dispersion of Ni on AC was calculated from the uptake of chemisorbed MeI at 250°C by:

$$D_{Ni} = MeI_a / Ni_{total} \qquad (6)$$

where, D_{Ni} is dispersion of Ni, MeI_a is uptake of MeI in moles/g, and Ni_{total} is amount of supported Ni in moles/g. The dispersion level is illustrated in Figure. 2 as a function of the Ni loading. Apparently, the dispersion of Ni decreases with Ni loading, and thus the particle size of Ni is estimated to increase. The rate of carbonylation has a maximum at about 5% of Ni loading. However, the catalytic activity in terms of turn over number N_M in h^{-1} was independent of the metal dispersion as presented in Figure. 2. It is apparent that the carbonylation of MeOH over the Ni/AC catalyst is apparently a structure-insensitive reaction.

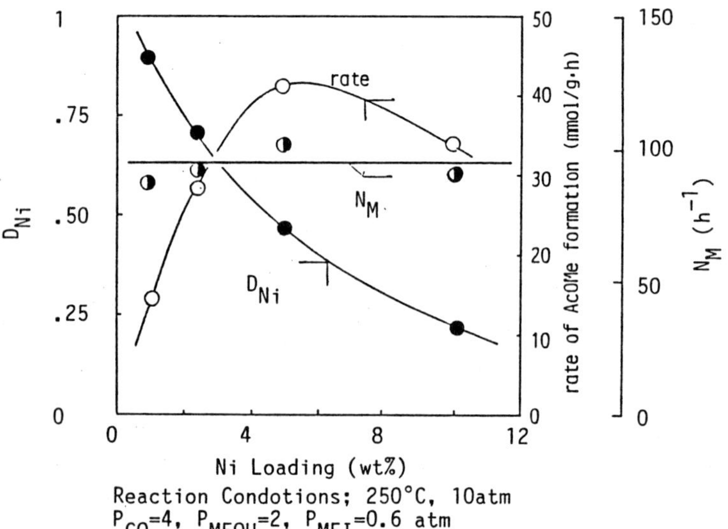

Figure 2 Effect of Ni Loading on D_{Ni} and N_M

3.3. Catalytic Activity of Group VIII Metal Catalysts.

Methanol conversions to methyl acetate and acetic acid on group VIII metals supported on active carbon are illustrated in Figure 3. The yields of methyl acetate are shown by dotted lines and those of acetic acid by full lines. On each catalyst acetic acid formation is predominant at higher temperature and that of methyl acetate has a maximum point.

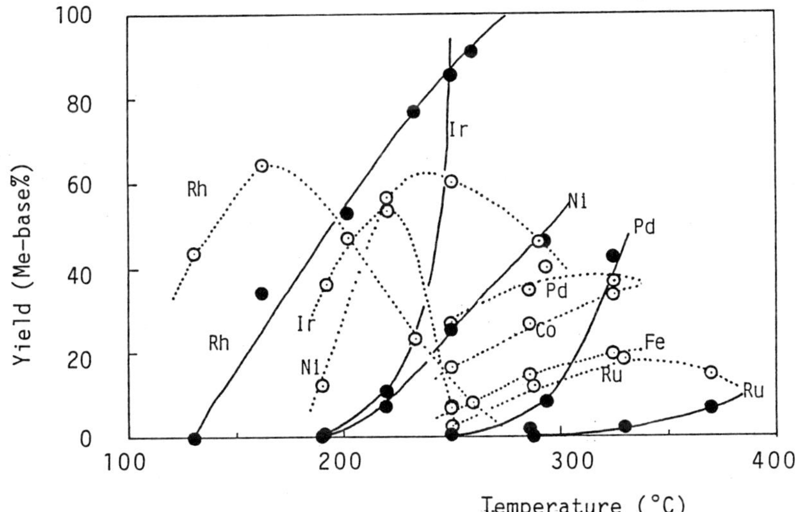

Figure 3 Activity of MeOH Carbonylation

The Turn Over Number (TON) and the activity of carbonylation were defined by equations(7) and (8), respectively:

$$\text{TON} = (Y_{AcOH} + Y_{AcOMe}/2) \times F_{MeOH} \;/\; M \;/\; 100 \;(h^{-1}) \quad (7)$$

$$\text{ACTIVITY} = T_{10}^{-1} \;(K^{-1}) \quad (8)$$

where Y_{AcOMe} and Y_{AcOH} are the yield of methyl acetate and acetic acid from methanol in mol% , respectively. F_{MeOH} is the feed rate of methanol (mmol/h/g of catalyst) and M is the amount of supported metal (mmol/g of catalyst). T_{10} indicates the temperature where TOF of the catalyst is 10 h^{-1}. The order of activity evaluated from equation(8) of the catalysts is as follows:

$$Rh > Ir > Ni > Pd > Co > Ru > Fe \qquad (9)$$

This order is in good agreement with that reported by Kawabata et al. (Ref. 7) for the carbonylation of methyl acetate in liquid phase.

Halide promoter such as methyl iodide is essential for the carbonylation of methanol for every catalyst in both liquid and vapor phase. The reaction order with methyl iodide is nearly 1 for Rh catalysts, whereas the value for the Ni/AC catalyst is close to 0. The difference is most likely attributed to the difference in their affinity for halide promoter. Probably the affinity is one of the main factors of the catalytic materials for good activity. In Figure 4 the activity defined by the equation(8) are plotted as a function of the affinity, where the

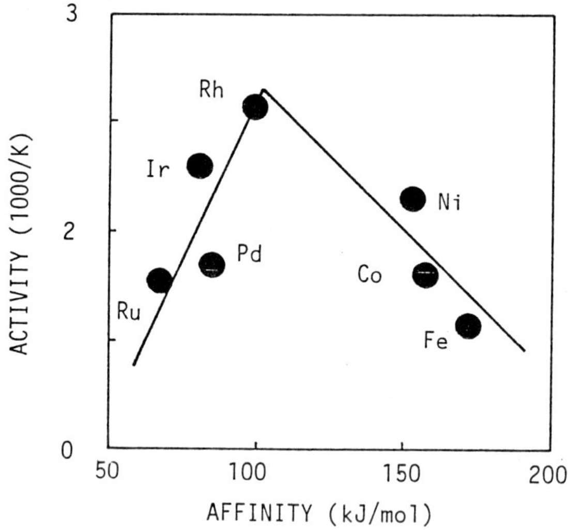

Figure 4 Activity as a Function of Affinity

affinity of the metal for halogen is defined by the equation(10):

$$\text{AFFINITY} = \Delta H_f^{o}(Cl)/ON \qquad (10)$$

where $H_f(Cl)$ is the standard enthalpy of formation of metal chloride and ON is the oxidation number of the compound. The value correspond to the heat of formation per one metal-chlorine bond.

The relation between the activity and the affinity represents a volcano shape. On the metal which locates at the left hand side of the peak, such as Ru, Pd, and Rh, the oxidative addition of MeI to the metal should be slow because of their poor affinity to halide ion and thus the step could determine the rate of methanol carbonylation. Therefore the stronger is the affinity, the more quickly the reaction proceeds. Although the affinity level is represented as the enthalpy of formation of metal chloride because of the lack in the data of iodide system, the TPD spectra of MeI proved to be well consistent with the concept in Fig. 4. Namely, the order of peak temperature of MeI desorption was Ru < Pd < Rh. On the contrary, although the rate of oxidative addition should be large enough on the metal located on the right hand side of the slope, which would have strong affinity to halide ion, the too strong adsorption of methyl iodide suppresses the adsorption of other reactants to retards other steps. Therefore, the more strongly methyl iodide is adsorbed, the lower the activity . The high affinity of metals should exhibits the low reaction order with respect to MeI, which has been confirmed by the fact that the observed value is 0.1 for the Ni/AC catalyst, whereas it is 0.7 for the Rh/AC catalyst (Ref. 10).

4. CONCLUSION

Thermodynamics indicates that methanol carbonylation is favored at higher partial pressure of carbon monoxide and lower temperature for the selective synthesis of acetic acid. CO was adsorbed weakly upon the Ni/AC catalyst in the associative form, while it was adsorbed upon the Ni/Al_2O_3 or the Ni/SiO_2 catalyst more strongly and in part dissociative form. It was indicated that MeOH carbonylation over Ni/AC is a structure-insensitive reaction. The catalytic activity of group VIII metals supported on active carbon was in the order as follows :

$Rh > Ir > Ni > Pd > Co > Ru > Fe$

The relation between the activity and the affinity of the metals for halide ion is presented as a volcano shape. The rate determining step of the reaction was discussed on the basis of the affinity and the reaction order with respect to methyl iodide.

REFERENCE

1. Roth J. F., Craddock J. H., Hershman A., and Paulik F. E., Chem. Technol, 23, 600(1971).
2. Robinson K. K., Hershman A., Craddock J. H., and Roth J. F., J. Catal., 27, 389(1972).
3. Reppe W., Ann. Chem., 58, 274(1953).
4. Fujimoto K., Shikada T. Omata K., and Tominaga H., Ind. Eng. Chem. Prod. Res. Dev., 21, 429(1982).
5. Omata K., Fujimoto, K., Shikada, T., and Tominaga, H., Ind. Eng. Chem. Prod. Res. Dev., 24, 234(1985).
6. Luft G., and Schrod M., J. Mol. Catal., 20, 175(1983).
7. Kawabata Y., Hayashi T., Mori K., and Ogata I., CSJ Annual Meeting, 1982, Tokyo, 45th, 4K12.
8. Fujimoto, K., Kameyama, M., Kunugi, T., J. Catal., 61, 7(1980).
9. Zagli, A. E., J. Catal., 56, 453(1979).
10. Fujimoto K., Omata K., Shikada T., and Tominaga H., Ind. Eng. Chem. Prod. Res. Dev. 22, 436(1983).

HIGHLY ACTIVE HOMOGENEOUS AND SUPPORTED PALLADIUM AND PLATINUM COMPLEX CATALYSTS WITH NITROGEN CONTAINING LIGANDS IN HYDROGENATION OF UNSATURATED HYDROCARBONS

V.M. Frolov, O.P. Parenago, L.P. Shuikina,
A.V. Novikova, A.I. Elnatanova, G.M. Cherkashin,
E.G. Kliger and E.Ya. Mirskaya
A.V. Topchiev Institute of Petrochemical Synthesis of the USSR Academy of Sciences, Moscow 117912, USSR

ABSTRACT - Novel homogeneous and mineral-supported nitrogen-containing complexes of palladium and platinum as hydrogenation catalysts were obtained. Catalysts based on palladium complexes are highly active in selective hydrogenation of dienes and alkynes to alkenes, the selectivity of the reaction being 98-99% at an almost complete conversion of the substrate. Catalysts based on platinum complexes are active in hydrogenation of olefins, dienes and alkynes to the corresponding alkanes. A molar activity of the supported metal complexes in the hydrogenation is more than 20 000 and 100 000 molen substrate/(g-atom metal·h) for platinum and palladium complex catalysts, respectively, at a metal content on the support surface of 0.03-0.05 mass%, 20 °C and a hydrogen pressure of 0.1 MPa.

I. INTRODUCTION

The hydrogenation of unsaturated compounds, which is widely used in industry for production of a great number of valuable chemical products, is catalyzed by metal complex systems based on Group VIII metals (Refs. 1-3). Most works on hydrogenation of unsaturated hydrocarbons deal with the catalysts containing organophosphorus ligands such as phosphines and phosphites. Meanwhile, a considerable theoretical and practical interest is shown in metal complexes with ligands having other heteroatoms, for example, nitrogen, in their structure.

The objective of this work was to synthesize novel platinum and palladium complex systems with nitrogen-containing ligands and to study their catalytic properties. Preparation of obtaining homogeneous catalysts was based on the use of ligands (higher aliphatic amines) with high affinity for the solvent; this ensured retention of platinum and palladium complexes in low oxidation states in the solution.

2. EXPERIMENTAL

Homogeneous metal complex catalysts were synthesized by interacting a metal compound with the corresponding amine in a water-free aromatic solvent (for example, toluene) at 50-70 °C in argon. The initial metal concentration in the solvent was 1×10^{-2} mol/l, the amine to metal molar ratio being equal to two. The Pd(II) or Pt(IV) complexes so obtained showed a high catalytic activity in hydrogenation due to their treatment with organoaluminium compounds (usually with diisobutylaluminum hydride). The treatment with a reactant was also performed in a toluene solution at a molar ratio Al : metal $\geq$ 4.

The use of some palladium compounds in combination with various amines (vide infra) led to the formation of catalytically active complexes without organoaluminum co-catalyst.

Supported metal complex catalysts were obtained by contacting toluene solutions of homogeneous catalysts with the surfaces of mineral supports preheated in vacuum at 300 °C. Aluminium, silicon and magnesium oxides, barium sulphate, calcium carbonate, zeolites as well as oxides of rare earth metals were used as supports. Heterogenation of the catalytically active metal complexes resulted in that the metal was put on the surface and remained strongly bound to the support during subsequent experimentations. An optimal metal content on the support surface was typically from 0.03 to 0.05 mass%.

Hydrogenation was carried out in a shaking ($\geqslant$ 300 shakes/min) quartz reactor at temperatures from 10 to 60 °C and hydrogen pressure from 0.02 to 0.1 MPa. The reaction kinetics was studied in a system providing an automative control of the hydrogen pressure in the reactor during experimentation. The reaction products were analyzed chromatographically in capillary coloumns.

3. RESULTS AND DISCUSSION

3.1. Homogeneous Hydrogenation Catalysts

Three types of palladium complex catalytic systems have been realized depending on the chemical character of an initial palladium compound and a nitrogen containing ligand.

The first catalyst type (A1) was prepared by treating palladium chloride and alkali metal palladate complexes with primary aliphatic amines having C_8-C_{13} alkyl groups (Ref. 4) with strong reducing agents (organoaluminium compounds, lithium alkyls). The A1 type catalysts revealed high activity in selective hydrogenation of conjugated dienes and alkynes into alkenes. In this case a preliminary promotion of initial or reduced complexes with water or molecular oxygen was necessary to obtain the catalytic activity (Refs. 5,6).

The second catalyst type (A2) was prepared as a result of interaction of palladium chloride or acetate either with tertiary aliphatic amines also having sufficiently long alkyl groups (e.g., trioctylamine, trinonylamine) or with alkyldisubstituted anilines (N,N-dipropylaniline, N,N-dioctylaniline).

Use of tertiary amines with short alkyl groups or triphenylamine did not lead to the formation of homogeneous palladium catalysts: a precipitate of the insoluble product was formed in the course of the interaction (Ref. 7).

IR and PMR spectroscopy studies of the product of the reaction between $PdCl_2$ and trinonylamine showed the splitting of Pd-Cl bonds, absorption bands at 320 and 235 cm^{-1}, which refer to the stretching vibrations of the terminal and bridged Pd-Cl bonds, respectively, were absent in the long-wave range of the spectrum. Besides that, the formation of a coordination bond between palladium and a quaternized nitrogen atom as well as a Pd--H bond was demonstrated. The latter one appeared simultaneously together with a C=C double bond in one of the hydrocarbon groups of the amine. The IR and PMR spectroscopy data confirmed the presence of an olefinic fragment in the amine molecule.

In this way a reduction of palladium occurs in the course of the interaction of the palladium salt with a tertiary amine. A hypothetical scheme can be represented as follows:

$$PdCl_2 + 2(C_9H_{19})_3N \rightarrow \underset{\substack{| \\ HC{=}{=}CH-(CH_2)_n \\ | \\ (CH_2)_{6-n} \\ | \\ CH_3}}{H-Pd-} \quad \overset{\substack{Cl^- \\ |}}{-N^+}\!\!\begin{array}{l} (CH_2)_8-CH_3 \\ (CH_2)_8-CH_3 \end{array} + [(C_9H_{19})_3NH]^+Cl^-$$

A bidentate coordination of the metal with a nitrogen atom and a double bond was shown in (Ref. 8) with platinum(II) complexes with o-vinylaniline or o-vinyl-N,N-dimethylaniline. A formation of the coordination bond between palladium and the double bond seems to be a factor which facilitates the formation and stabilization of reduced palladium complexes with nitrogen containing ligands. It is quite possible that the reduced complexes of palladium with triphenylamine or dimethyldodecylamine are unstable just because of the absence of this possibility.

The suggested structure of the product of the interaction between palladium chloride and trinonylamine is of course a hypothetical one. Real structures of this type must be much more complicated. One cannot exclude the formation of cluster type fragments with Pd-Pd bonds.

The third catalyst type (A3) was prepared using bis-π-allylpalladiumchloride as a starting compound. In this case the obtainment of active catalytic species was also possible without use of organoaluminium reducing agents. But unlike $PdCl_2$ systems, the reduction of π-allyl palladium complex proceeds easily with primary amines, e.g., octylamine. In the case of dioctylamine its interaction with allylpalladiumchloride is accompanied by a partial decomposition of the complex formed. The resulting solution contains 50 or 80% of the total palladium for the molar amine/palladium ratio being 2 or 4, respectively. A reaction between the allylpalladium complex and tertiary amines resulted in the formation of insoluble products.

Table 1 contains results of hydrogenation of isoprene as a model substrate in the presence of three types of palladium complex catalysts. As follows from the data, the catalysts are hyghly active.

Data for hydrogenation of various conjugated dienes and alkynes in the presence of a A1 type catalyst are given in Table 2.

TABLE 1. Hydrogenation of isoprene in the presence of palladium complex catalysts with nitrogen containing ligands; [Pd] = 5×10^{-4} M; solvent=toluene; amine:Pd=2, Al:Pd = 4

Starting palladium compound	Ligand	Reducing agent	Turnover number, h^{-1} (20 °C, 0.1 MPa)
$PdCl_2$	$C_8H_{17}NH_2$	$HAl(iso\text{-}Bu)_2$	64000
Na_2PdCl_4	$C_{11}H_{23}NH_2$	$HAl(iso\text{-}Bu)_2$	52000
$PdCl_2$	$(C_8H_{17})_3N$	-	32000
$Pd(CH_3COO)_2$	$(C_9H_{19})_3N$	-	43400
$PdCl_2$	$(C_8H_{17})_2NC_6H_5$	-	4500
$PdCl_2$	$(CH_3)_2C_{12}H_{25}N$	$HAl(iso\text{-}Bu)_2$	34700
$(\pi\text{-}C_3H_5PdCl)_2$	$C_8H_{17}NH_2$	-	33200
$(\pi\text{-}C_3H_5PdCl)_2$	$(C_8H_{17})_2NH$	-	2900

TABLE 2. Hydrogenation of conjugated dienes and alkynes in the presence of the $PdCl_2$-$C_8H_{17}NH_2$-$HAl(iso\text{-}Bu)_2$ catalyst system

Substrate	Hydrogenation products	Turnover number, h^{-1} (20 °C, 0.1 MPa)
Isoprene	Isopentenes	64000
Cyclopentadiene	Cyclopentene	30100
Pentadiene-1,3	Pentenes	65000
Cyclooctadiene-1,3	Cyclooctene	23800
Diphenylacetylene	cis-Stilbene	18100
Heptyne-3	cis-Heptene-3	25000
Octyne-4	cis-Octene-4	17100

According to a chromatographic analysis the selectivity of hydrogenation to olefins attains 98-99% at conversions as high as 95%. The reaction proceeds stereo-

TABLE 3. Hydrogenation of unsaturated hydrocarbons in the presence of the H_2PtCl_6-amine-$HAl(iso\text{-}Bu)_2$ system

Nitrogen containing ligand	Substrate	Hydrogenation product	Turnover number, h^{-1} (20 °C, 0.1 MPa)
$C_8H_{17}NH_2$	Hexene-1	Hexane	4100
$(C_8H_{17})_2NH$	Hexene-1	Hexane	5900
$(C_9H_{19})_3N$	Hexene-1	Hexane	17000
$(C_9H_{19})_3N$	Heptyne-3	Heptane	12000
$(C_9H_{19})_3N$	Pentadiene-1,3	Pentane	10400

selectively in the case of disubstituted alkynes: disubstituted olefins formed contain 95-98% of cis-configuration.

An interaction of H_2PtCl_6 with octyl-, dioctyl- or trioctylamine in toluene resulted in the formation of platinum (IV) complex irrespective of the nature of the nitrogen containing ligand. The complexes revealed their catalytic activity only after treatment with organoaluminium compounds; additives of water or alcohol were necessary to promote the reduced complexes (Table 3).

As seen from the data, platinum complex catalysts provided hydrogenation of olefins, dienes and alkynes into the corresponding alkanes. The maximum rate was observed for terminal olefins.

3.2. Supported Hydrogenation Catalysts

Heterogenized catalysts were obtained by supporting reduced palladium and platinum complexes on the surface of mineral carriers (Ref. 9).

Catalytic activity data for supported catalysts are given in Table 4.

It should be noted that supporting led to an increase

TABLE 4. Hydrogenation of olefins, dienes and alkynes in the presence of supported catalysts (0.03-0.05% Pd or Pt by weight). Solvent = toluene. The A2/support catalysts were activated with oxygen at 0.1 MPa

Catalyst type	Support	Substrate	Turnover number, h^{-1} (20 °C, 0.1 MPa)
A1/support	Al_2O_3	Pentadiene-1,3	68100
A1/support	SiO_2	Diphenylacetylene	26500
A1/support	Nd_2O_3	Isoprene	73700
A1/support	Al_2O_3	2-Methylhexene-5-yn-3-ol-2	35400
A1/support	NaX zeolite	Diphenylacetylene	27500
A2/support	NaX zeolite	Heptyne-3	25800
A2/support	Al_2O_3	Pentadiene-1,3	62500
A2/support	Al_2O_3	Isoprene	121000[a)]
A2/support	NaX zeolite	1-(tetrahydrofuryl-2-heptyn-2-ol-1	13100
A3/support	Al_2O_3	Octyne-4	32100
A3/support	Al_2O_3	Diphenylacetylene	21100
Pt/support	Al_2O_3	Hexene-1	20500
Pt/support	Al_2O_3	Pentadiene-1,3	18000

a) Solvent = Hexane

in activity as compared to the homogeneous catalysts. Aside from that, unlike homogeneous catalysts the supported ones were stable towards hydroxyl containing compounds which allowed the hydrogenation of substrates with hydroxyl groups (Table 4).

An interesting feature of supported A2 type palladium containing catalysts was their ability to be strongly promoted by molecular oxygen. In this case the hydro-

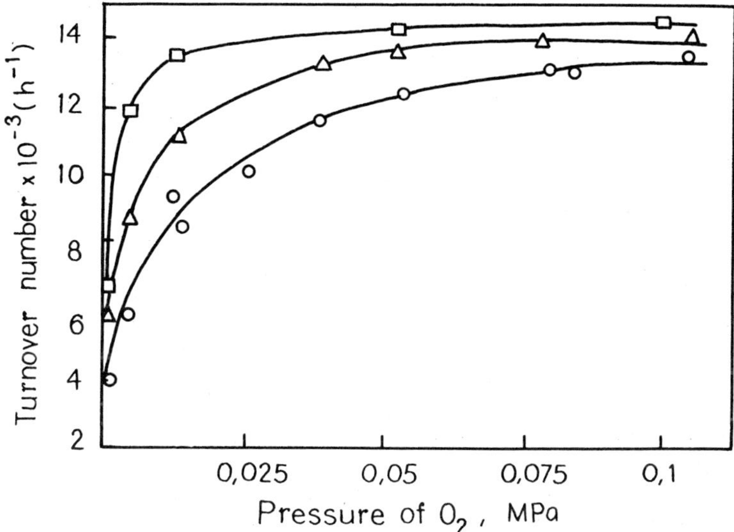

Figure 1. Dependence of activity in hydrogenation of diphenylacetylene on oxygen pressure. Catalyst: $PdCl_2$-$(C_9H_{19})_3N$/support (0.07% Pd by weight). Exposure time - 18 h, 20 °C, 0.07 MPa H_2

genation rate increased with increasing oxygen pressure during activation up to 0.1 MPa if the exposure time was sufficiently high (a few hours). The activation by oxygen was most efficient in the case of hydrogenation of acetylenic compounds (Fig. 1).

To explain this relationship, a formal scheme was put forward in which two types of active sites are involved. The first type has the activity C_1 without oxygen treatment and chemisorbs oxygen reversibly until the equilibrium state is attained. A fast reaction of the oxygen-covered sites with hydrogen provides the second type sites having the activity C_2:

$$Pd_n + O_2 \underset{}{\overset{K}{\rightleftharpoons}} Pd_n \cdot O_2 \quad (1)$$

$$Pd_n \cdot O_2 + H_2 \longrightarrow \text{Active site II} \quad (2)$$

where Pd_n stands for the supported complex.

In this way the total activity of the A2 type supported catalysts is a function of C_1, C_2 and the specific coverage of oxygen, Θ. In terms of Langmuir's isotherm the total activity can be expressed as

$$C = C_1(1 - \Theta) + C_2\Theta \quad (3)$$

where

$$\Theta = \frac{\alpha P}{1 + \alpha P} \quad (4)$$

(α and P are the adsorption coefficient of O_2 and its pressure, respectively).

A multiple regression treatment allowed separate calculation of C_1, C_2 and α (Table 5).

TABLE 5. Values of C_1, C_2, α and squared correlation coefficient, R^2, for a catalyst: A2/support. Substrate= =diphenylacetylene

Support	α, MPa^{-1}	C_1, h^{-1}	C_2, h^{-1}	R^2
NaX	24.6	4800	16850	0.90
Al_2O_3	44.7	6340	16550	0.83
SiO_2	29.4	5700	15100	0.93

As follows from the Table, high values of R^2 evidence in favour of the proposed scheme.

The A1/support catalysts were practically insensitive towards oxygen, while the activity of A3/support catalysts decreased after a contact with oxygen. Supported platinum containing catalysts were insensitive with respect to oxygen.

Palladium-based catalysts of the type A2/support were shown to be quite stable to some catalytic poisons (water, thiophene, triphenylphosphine) and to preserve their activity and selectivity in the absence of the solvent upon heating up to 200 °C (Ref. 10).

As shown by special experiments on reaction products removal and introduction of a new portion of the substrate, it is possible to carry out a series (more than 10) of consecutive cycles of hydrogenation on one catalyst sample at an almost constant reaction rate. Thus, all types of the supported metal-based catalysts are characterized by a high stability of their catalytic action.

Using the reaction of isoprene hydrogenation as an example, we have compared the activity and selectivity of supported catalysts of the A2/support type with those of the catalysts of selective hydrogenation of dienes and alkynes currently used in practice. The results obtained indicate that the palladium-containing catalysts proposed by us are 30-40 times more selective and active than the Lindlar palladium-based catalyst.

4. KINETICS OF HYDROGENATION

Kinetics of hydrogenation of conjugated dienes and acetylenes in the presence of both homogeneous and supported Pd-catalysts is characterized by the concentration independence of the reaction rate which results from the complete coverage of active sites by the substrate (Fig. 2). This accounts for the high selectivity of hydrogenation to olefins.

Using the method of dosed introduction of an inhibi-

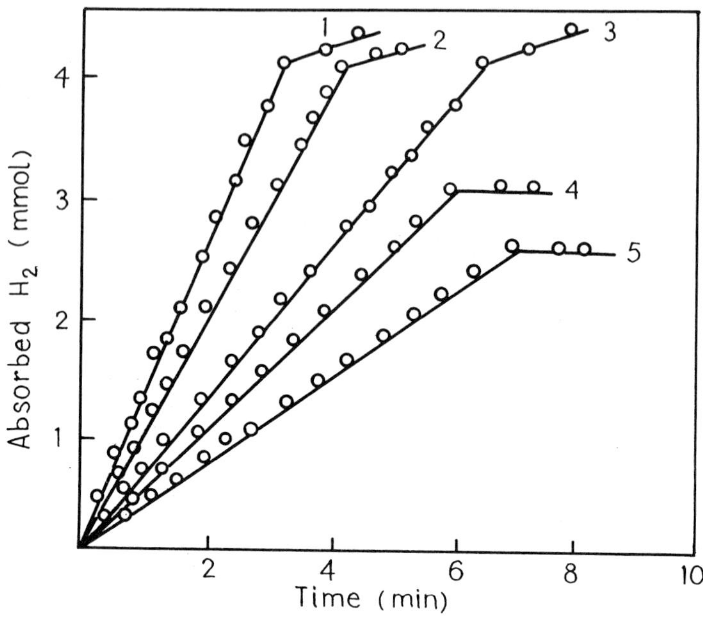

Figure 2. Kinetics of hydrogenation of dienes and alkynes: isoprene (1), pentadiene-1,3 (2), cyclopentadiene (3), heptyne-3 (4), octyne-4 (5); 0.05% Pd by weight, 20 °C, 0.5 MPa H_2

tor (carbon monoxide) the upper limit of the number of active sites was determined. It proved to be 7-12% based on total Pd atoms for the A2/support system which evidences in favour of the cluster character of this catalyst.

It may be concluded that a new method of preparation of homogeneous and supported catalysts allows high rates of hydrogenation of unsaturated hydrocarbons and their derivatives. The method provides wide opportunities for the synthesis of highly active homogeneous and heterogenized catalysts on the basis of the platinum group metals.

REFERENCES

1. Aspects of Homogeneous Catalysis, R. Ugo (Ed.), vol. 1, Carlo Manfredi, Milano (1970).
2. B.R. James, Homogeneous Hydrogenation, A Wiley-Interscience Publication John Wiley & Sons, New York, London, Sydney, Toronto (1973).
3. F.J.McQuillin, Homogeneous Hydrogenation in Organic Chemistry, D. Reidei Publishing Company, Dordrecht Holland, Boston, USA (1977).
4. V.M. Frolov, O.P. Parenago, L.P. Shuikina, B.A. Dolgoplosk, G.A. Stepanov, V.Sh. Feldblum and T.V. Konovalova, USSR Patent No 725375, Bull. Izobr., No 25 (1981).
5. L.P. Shuikina, G.M. Cherkashin, O.P. Parenago and V.M. Frolov, Dokl. Akad. Nauk SSSR, 257, No 3, 655 (1981).
6. V.M. Frolov, O.P. Parenago, L.P. Shuikina and G.M. Cherkashin, React. Kinet. Catal. Lett., 16, 115 (1981).
7. O.P. Parenago, G.M. Cherkashin and L.P. Shuikina, Neftekhim., 25, 589 (1985).
8. M.K. Cooper and D.W. Yaniuk, J. Organomet. Chem., 221, 231 (1981).
9. V.M. Frolov, O.P. Parenago, L.S. Kovaleva, A.V. Novikova, A.I. Elnatanova, E.I. Mirskaya and E.G. Kliger, USSR Patent No 1066975, Bull. Izobr., No 2 (1984).
0. V.M. Frolov, O.P. Parenago, L.S. Kovaleva, E.I. Mirskaya and A.I. Elnatanova, Neftekhim., 24, 783 (1984).

HYDROGENATION OF SOME ORGANIC COMPOUNDS CATALYZED BY PLATINUM AND PALLADIUM COMPLEXES

D.M. Shopov and S.K. Rakovsky
Institute of Kinetics and Catalysis, Bulgarian Academy of Sciences, Sofia 1113, Bulgaria

ABSTRACT - The synthesis of platinum and palladium complexes with organic dyes immobilized on some organic and inorganic supports is discussed. The catalytic activity of these complexes was studied in the hydrogenation of nitro-groups, conjugated double bonds and acetylenic bond.

1. INTRODUCTION

In recent years a constant increase is observed in the use of homogeneous catalysts in chemical industry. Probably, the role of heterogeneous catalysts will be predominant in the future, compared with that of homogeneous catalysts, but these latter are very important for explanation of the catalytic process mechanisms. They also make a significant contribution to the overall amount of chemical production. According to the calculations that have been done, there exist in the world 23 industrial processes, in which soluble, including immobilized, metal compounds are used as catalysts. The estimates given above are in good agreement with the data by Waddams (Ref. 1). At present the world production of chemical industry with the use of homogeneous catalysts is about 16 millions of tons and it is expected that by the year 2000 it will be doubled. This prognosis shows that obviously the chemistry and technology of homogeneous processes will be developed at accelerated rates.

Further on we will use the term "homogeneous catalysts", having in mind immobilized ones too. Our investigations (Refs. 3,4), and also studies by other authors (Refs. 5-7) show that the mechanism of the catalytic action does not change significantly after the catalysts are immobilized.

In the present paper we consider the synthesis and catalytic properties of supported complexes of platinum and palladium. As a basis for the consideration of this question serves our own, although modest, contribution to the methods of the synthesis of homogeneous catalysts and immobilized complexes of platinum and their use.

In the present work we have used the excellent monographs and reviews (Refs. 8-23) and extremely valuable works by Halpern (Ref. 24), James (Ref. 25), Yermakov (Ref. 26), Moiseev (Ref. 27), Shilov (Ref. 28), Khidekel (Refs. 29,30), Ugo (Ref. 31), Maitlis (Ref. 32) and Sokolskii (Ref. 33).

It should be noticed that our results have been obtained in active collaboration with the groups of Professor M.L. Khidekel and Professor Yu.I. Yermakov.

2. PLATINUM COMPLEXES

2.1. Synthesis

Special attention is paid to the synthesis of the platinum and palladium complexes having conjugated organic dyes as ligands. Earlier it has been shown (Refs. 29 & 30) that ligands of this type take part in the catalytic stage via an electron transfer. On the basis of these theoretical premises, for the first time by us (Ref. 34) and in the group of Professor M.L. Khidekel (Ref. 35) the complex of platinum with 1-phenyl-azo-2-naphtol (PAN) was synthesized. This complex turned out to be a highly effective homogeneous catalyst in the reaction of nitrobenzene (NB) hydrogenation. In order to obtain better technological characteristics of the complex, it was supported on different carriers, for example, on alumina, silica, charcoal and ceramics. The most convenient support turned out to be γ-Al_2O_3 with a total specific surface of 120 m^2/g, which is used as a reforming catalyst. The synthesis of the complex is made according to Scheme 1.

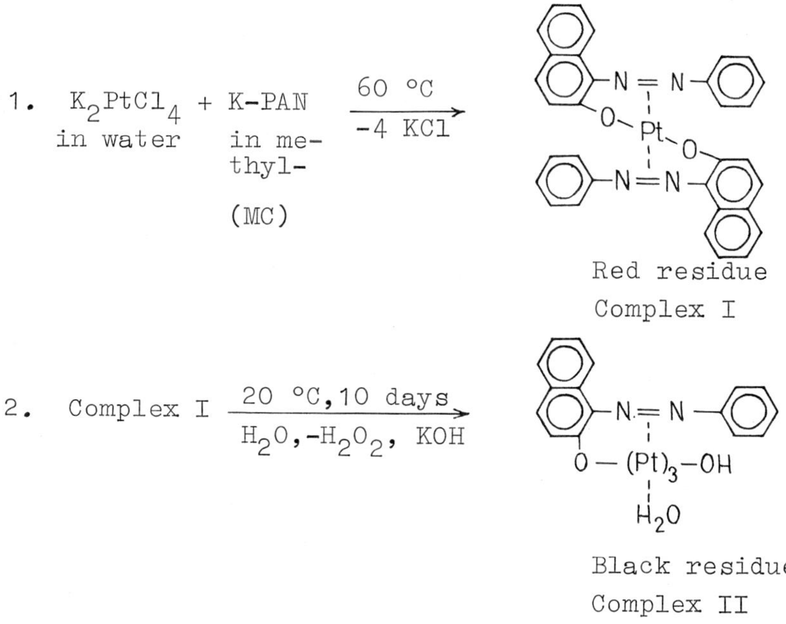

1. K_2PtCl_4 in water + K-PAN in methyl- (MC) $\xrightarrow[-4\ KCl]{60\ °C}$ Red residue Complex I

2. Complex I $\xrightarrow[H_2O, -H_2O_2,\ KOH]{20\ °C, 10\ days}$ Black residue Complex II

Scheme 1

A water solution of K_2PtCl_4 was mixed with a solution of K-PAN in MC and a red precipitate was formed (Complex I). Later on, the precipitate was stirred in air for 10 days, and a dark, almost black, residue was formed (Complex II). Our suggestion of the structure of Complex I is based on the UV-VIS and IR spectroscopic data and on the analogy in the synthesis of similar complexes of other metals, such as Co, Ni and Fe. A more detailed analysis of this synthesis shows that if the initial compounds are dissolved in a mixed solvent H_2O : MC = 1 : 10 with the subsequent stirring of the solution no precipitates are formed even after 10 days. Therefore, for the formation of Complex I it is necessary that K-PAN be precipitated when the solutions are mixed. Then the reaction proceeds on the surface of Complex I. We suggested that the reaction of the formation of Complex II proceeds through several stages according to Scheme 2:

1. $-OK + PtCl_4^{-2} \longrightarrow -OPtCl$

2. $PtCl_4^{-2} + KOH \longrightarrow Pt(OH)_2Cl_2^{-2}$

3. $-OPtCl + KOH \longrightarrow -PtOH$

4. $-OPtOH + Pt(OH)_2Cl_2^{-2} \longrightarrow -OPtPtOH + H_2O_2$

5. $-OPtPtOH + Pt(OH)_2Cl_2^{-2} \longrightarrow -OPtPtPtOH$

Scheme 2

The data on the composition and the structure of Complex II were obtained by elementary analysis, UV-VIS, IR, XPS and mass-spectrometry methods. Table 1 gives the data on the elemental composition from which the overall formula of Complex II was derived:

$$Pt_3C_{16}H_{15}N_2O_4$$

It describes exactly the composition of Complex II given above as a cluster compound. According to the RPES (Fig. 1), the platinum in Complex II has an electron state different from that of the metal. UV-VIS-reflectance spectroscopy and IR-spectra showed that Complex II has ligands with the structure of PAN and a Pt-O bond.

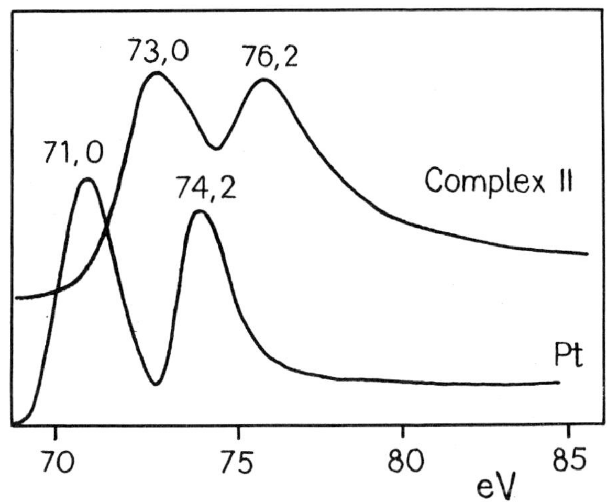

Figure 1. X-ray photoelectron spectra in the region Pt 4f

TABLE 1. Elemental analysis of Complex II

Element	Weight contents Exper., %	Weight contents Theor., %	Molecular mass Exper., %	Molecular mass Theor., %
Platinum	62-65	66.2		
Carbon	21.7	21.7		
Hydrogen	1.7	1.7	880-886	884
Nitrogen	3.2	3.2		
Oxygen	-	7.2		
Chlorine	traces	-		
Potassium	traces	-		

The molecular mass of the complex was measured by the Knauer method and was in the interval 880-886 (Table 1). It corresponds exactly to the overall formula given above. All data we have obtained support our suggestion that the formation of Complex II proceeds according to Scheme 2. Besides that, Complex II turned out to be an amourphous substance according to the X-ray analysis. A small amount of it dissolves in organic solvents (for example, in DMF - to 1% and in ethanol and hydrocarbons significantly to a lesser extent than 0.5%). Even in those solvents we did not obtain real solutions.

2.2. Hydrogenation of Nitrobenzene

The catalytic properties of Complex II were studied after its immobilization on Al_2O_3, the content of platinum in the species being 0.5 mass %.

The NB hydrogenation reaction was carried out in the temperature range from 15 to 70 °C at a partial pressure of hydrogen from 200 to 712 mm Hg, a total pressure in the system of 712 mm Hg maintained by argon. The volume of the reactor supplied with a magnetic stirrer was 8 ml. The amount of the absorbed hydrogen was registered automatically (Ref. 36). With the help of GLC and HPLC it was found that the hydrogenation of NB in DMF produces phenylhydroxylamine (PHA) with a 97% yield, and in ethanol - aniline with a 96% yield. The hydrogen absorption rate increases with increasing catalyst amount and hydrogen pressure. The dependence of the hydrogenation rate on the NB concentration (Fig. 2) shows that the reaction order on NB changes from 1 to 0. The kinetic curve becomes a straight line in coordinates 1/W : 1/NB (line 2), which corresponds to the kinetics of the Wilkinson's catalyst, $RhCl(PPh_3)_3$ or, in other words, to the Lainwiver-Berk dependence. During the investigations it was found that the catalyst (0.4 g) absorbed hydrogen (5 ml) and after that the hydrogenation process began.

The activation energy of the catalyst modification equals 3.5 kcal/mol, and the activation energy of the hydrogenation process is 9 kcal/mol. The kinetic equa-

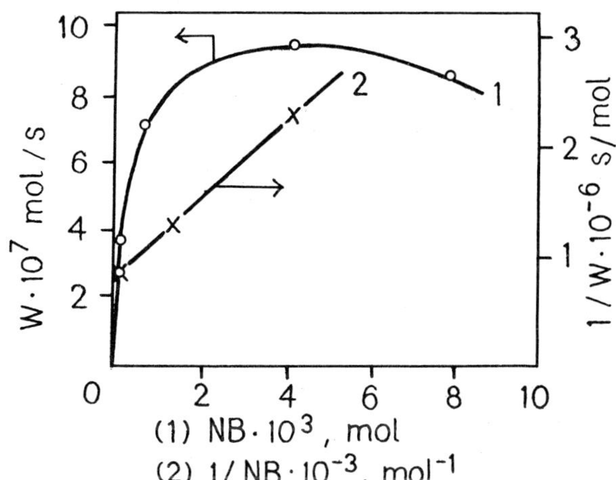

Figure 2. Dependence of the hydrogen absorption rate on the concentration of NB and its transformation in reciprocal coordinates

tion of the hydrogenation rate can be expressed as follows:

$$W = \frac{k_1 K_1 [H_2][Cat][NB]}{1 + K_1 [H_2] + K_2 [NB]}$$

The rate equation corresponds to the following scheme:

$$\begin{array}{ccc} Cat + H_2 & \xrightleftharpoons{K_1} & [Cat \ldots H_2] \\ NB \downarrow\uparrow K_2 & & NB \downarrow k_1 \\ [Cat \ldots H_2] & \overset{H_2}{\not\rightleftharpoons} & Cat + PHA \end{array}$$

As can be seen, the scheme includes several equilibrium stages: substrate coordination, hydrogen transfer and catalyst regeneration. However, it cannot explain the solvent effect. Probably, the solvent takes part in the coordination to an intermediate catalytic complex together with PHA, but in order to be able to compete with PHA it should have a good complexing ability. We proved that fact by carrying out the hydrogenation in a mixed solvent DMF, ethanol. During the transition from DMF to ethanol a constant increase of the concentration of the formed aniline was observed. Therefore, in the above scheme the reaction k_1 is a complex multistage process.

Our conclusion is that Complex II immobilized on γ-Al_2O_3 is a highly effective and selective catalyst for hydrogenation of NB.

2.3. Hydrogenation of Olefins and Acetylenes

The activity of the immobilized complex was studied in the hydrogenation reaction of cyclohexene, 1-heptene, and 1-decene. The reaction conditions were the same as in the NB hydrogenation.

However, the hydrogen absorption rate was to 1-2 orders lower than in the case of NB hydrogenation. A conclusion can be made that the hydrogenated group or part of the molecule connected with it are taking part in the complexformation with the catalyst. An isolated double bond and alkyl groups do not have such properties. The latter conclusion made us try to hydrogenate a dienic bond in the presence of an olefinic bond. As a substrate the acetyl derivative of the natural steroid was chosen:

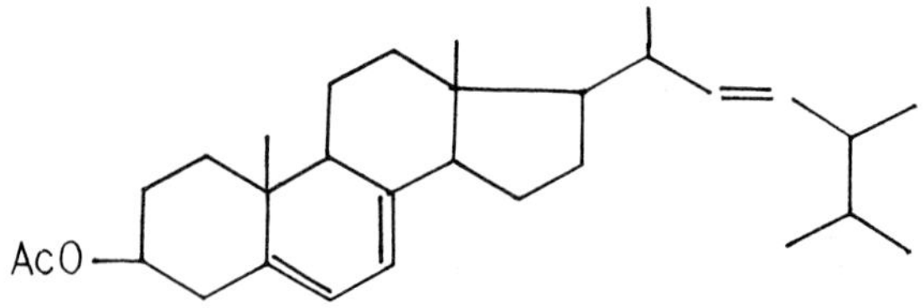

AcO

It turned out that the steroid is hydrogenated at a

moderate rate in the cycle B, and the double bond in the side chain remains unchanged. Comparing the effectivity and the stereoselectivity of the catalyst Pd/$BaSO_4$, partly poisoned with quinoline, suggests that the natural compound, containing an acetylenic bond:

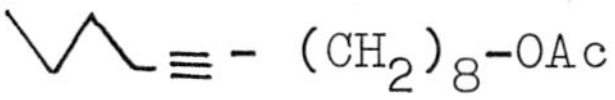

$$\equiv -(CH_2)_8-OAc$$

is hydrogenated by our catalyst at a lesser rate, but with a greater stereoselectivity towards formation of cis-olefins.

Recently, using the Pt-PAN catalyst we have succeeded in conducting a selective hydrogenation of the conjugated double bond of the methyl ester of the ergosterol:

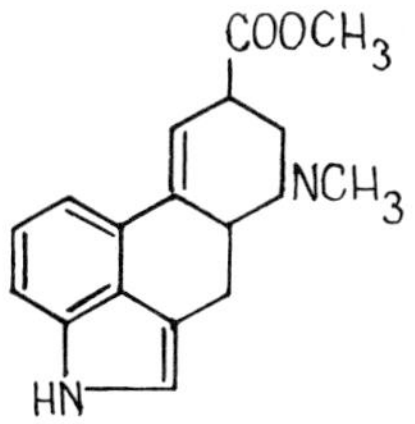

$COOCH_3$ NCH_3 HN

3. PALLADIUM COMPLEXES

3.1. Synthesis

Another direction of our investigations is synthesis of palladium complexes with dyes (Refs. 37-40). Especially good hydrogenation catalyst turned out to be the Pd(II) complex with 1,2,4-trihydroxy antraquinone (purpurine) and its modification with different mono- and bidendate ligands, such as 8-hydroxyquinoline, dimethylglioxime, chloranilic acid and 1-phenyl-azo-2-naphthol.

It is not possible to obtain the complex of Pd(II) with purpurine by direct interaction of the components in the solution, because it is unstable. The supported analogue of the latter is synthesized by a preliminary immobilization of the purpurine on γ-Al_2O_3 followed by the treatment of the immobilized dye with a solution of K_2PdCl_4 or $Pd(OAc)_2$. On the basis of the reflectance UV-VIS and IR spectroscopies we have suggested that as a result of such "consecutive synthesis", on the surface of the carrier, palladium complexes, containing as ligands purpurine and the above modifying additives, are formed. The most probable structures of these complexes are as follows:

Catalyst A

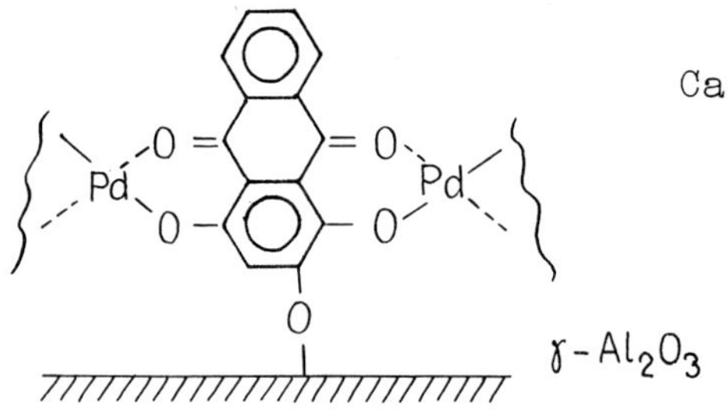

γ-Al_2O_3

Catalyst B

γ-Al_2O_3

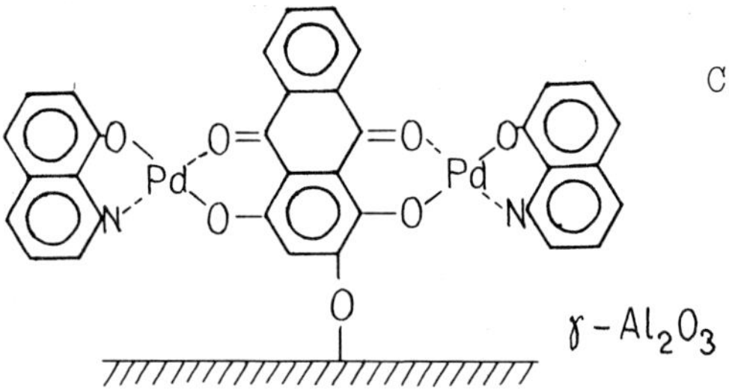

Catalyst C

γ-Al_2O_3

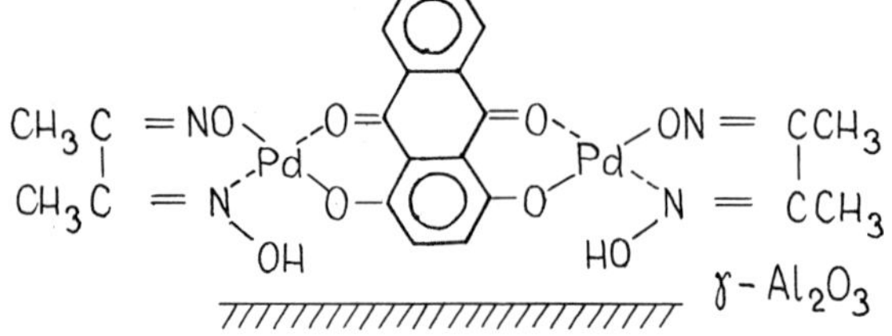

Catalyst D

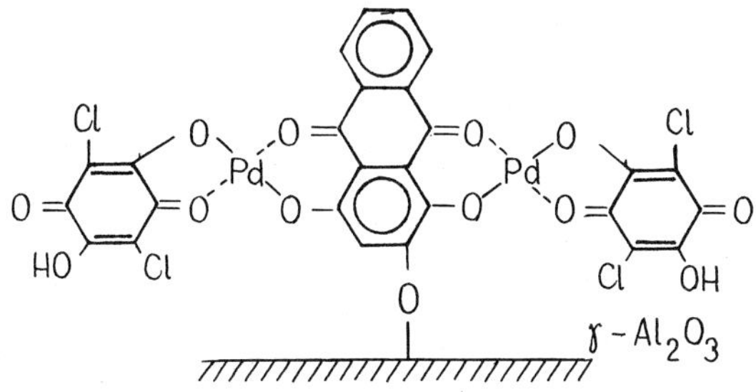

Catalyst E

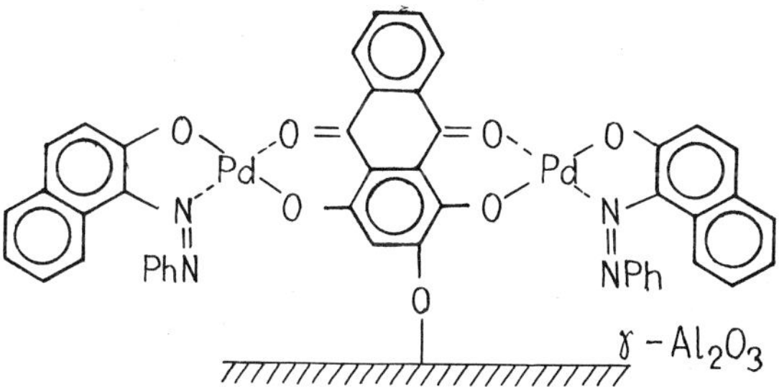

The amount of palladium per gram of the catalyst is 0.62% mass. The modified catalysts obtained by the treatment of Catalyst A, have the above structures.

It should be noted that independently of the very strong complexing ability of some ligands, including 8-oxiquinoline, "separation" of palladium from the purpurine was not observed. Along with alumina, charcoal and hitozane were also used as carriers:

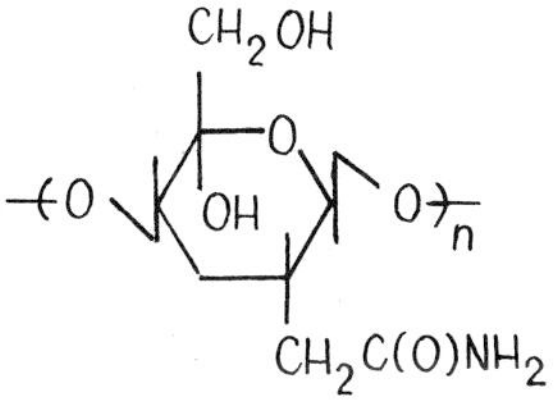

3.2. Nitrobenzene Hydrogenation

As we have stated above, the catalysts were investigated in the hydrogenation of a series of substituted nitrobenzenes. The corresponding results are presented in Table 2.

TABLE 2. Results of hydrogenation of a series of substituted nitrobenzenes with different palladium catalysts

Substrate	T, °C	Solvent	W, min^{-1}	Products	Yield
			A		
1,3,5-trinitrobenzene	67	DMF	20	3,5-diaminonitrobenzene	97
2,4-dinitrophenol	20	EtOH	15	2-amino-4-nitrophenol : 2-nitro-4-aminophenol= 1:1	98
As 2.	67	As 2.	44	As 2.	97
2,4-dinitroanisole	67	n-Octane	43	2-amino-4-nitroanisole:2-nitro-4-aminoanisole= 1 : 1	97
			B		
NB	67	DMF	60	Aniline	97.5
			C		
NB	67	DMF	22	Aniline	98.5
			D		
p-Nitrotoluene	67	DMF	28	p-Aminotoluene	97
			E		
As 7.	67	DMF	21	As 7.	97.5

Catalyst A, supported on charcoal and hitozane in the conditions of example 2 (Table 2), showed hydrogenation rates 44 and 44 min^{-1}, respectively, with the yield of 97%. The activity of the other catalysts decreased, when the support was changed by charcoal and hitozane.

The results in Table 2 show that the catalysts are effective in the reactions of nitrobenzene hydrogenation and it proceeds to the formation of aminoderivatives, but the stereospecificity is low, when o- and p-isomers should be obtained. The hydrogenation rate when our catalysts are used is greater than that of the classical palladium catalysts supported on charcoal under the same experimental conditions. For instance, the rate of 2,4-dinitroanisole hydrogenation is 43 min^{-1}, while the classical catalyst gives only 8 min^{-1}. Catalysts B, C, D and E hydrogenate better NB and its monosubstituted derivatives, in comparison with catalyst A.

When the effectivity of Catalyst A is tested in the hydrogenation reaction of the natural steroid which is mentioned above, it is found that the double bond in the hydropyridine ring is hydrogenated predominantly.

3.3. Hydrogenation of Other Compounds

Palladium catalysts were used during the hydrogenation of some simple olefins such as styrene, vinilacetate, stylbene, cianostylbene, maleic and fumaric acids. In all the cases the reaction rates were rather low; we consider that these catalysts are not suitable for hydrogenation of simple olefins, but they proved to be efficient in the selective hydrogenation of NB, mononitrobenzenes, polynitrobenzenes, natural products, etc.

Finally, the general conclusion can be expressed in the following way:

- a method has been found for the synthesis of catalysts on the basis of a cluster compound of platinum and the mechanism of its action in the hydrogenation reactions of some organic compounds and natural products has been studied;
- a series of catalysts have been obtained on the basis

of the palladium complex with purpurine. The studies showed that these catalysts are specifically active in the hydrogenation of polynitrobenzene.

REFERENCES

1. A.D. Waddams, Chemicals from Petroleum (4th Edition) J. Murray, London (1978).
2. A. Nakamura and M. Tsutsui, Principles and Applications of Homogeneous Catalysis, J. Wiley & Sons, N.Y. (1980).
3. V.V. Ivanova, S.K. Rakovsky and D.M. Shopov, 22nd ICCC, v. 2, p. 714, August 1982, Budapest, Hungary.
4. V.V. Ivanova, S.K. Rakovsky, V.V. Georgieva and D.M. Shopov, 4th ISHC, v. IV, p. 62, September 1984, Leningrad, USSR.
5. A.D. Pomogailo and M.V. Klynev (USSR)., Dep. Doc., VINITI, 414-84, 1984.
6. L. Červeny and V. Ruzichka, Chem. Listy, 79, 296 (1985).
7. P.M. Henry, Palladium-Catalyzed Organic Reactions, Adv. Organom. Chem., 13, 363 (1975).
8. E.L. Muetterties, Science, 196, 839 (1977).
9. P.M. Maitlies, The Organic Chemistry of Palladium, v. 1 and v. 2, Acad. Press, N.Y., London (1971).
10. B.R. James, Homogeneous Hydrogenation, J. Wiley & Sons, N.Y. (1973).
11. P. Bonchev, Komplexoobrazovanie i Kataliticheskaya Aktivnost (in Russian), Mir, Moskva (1975).
12. G. Henrici-Olivé, S. Olivé, Coordination and Catalysis, Verlag Chemie-Weinhein, N.Y. (1977).
13. F.J. McQuillin, Homogeneous Hydrogenation in Organic Chemistry, Reidel Publ. Comp., Boston, USA (1975).
14. Aspects of Homogeneous Catalysis, R. Ugo (Ed.), v. 1, Carlo Manfredi, Editore-Milano (1970).
15. E.A. Karakhanov, A.G. Dedov and A.S. Loktev, Usp. Khim., 54, 289 (1985).
16. M.L. Khidekel, Sov. Sci. Rev., B, Chem. Rev., 4, 284 (1982).
17. Metal Complexes of Platinum Group in Synthesis (in Russian), Vses. Seminar, published in Chernogolovka (1983).
18. P.M. Maitlies, Pure and Appl. Chem., 33, 489 (1973).
19. M.J. Bruce, J. Organom. Chem., 53, 141 (1973).
20. J.L. Garnett, Catal. Rev., 5, 229 (1971).
21. C. Masters, Homogeneous Transition-Metal Catalysis, Chapman and Hall, London, New York, 1981.
22. I.I. Moiseev, Zh. Vsesouz. Khim. Obschest. im D.I. Mendeleeva, 22, 30 (1977).

23. P.R. Hartley, The Chemistry of Platinum and Palladium, Appl. Sci. Publ., London (1973).
24. J. Halpern, Acc. Chem. Res., 3, 386 (1970).
25. B.R. James, Adv. Organomet. Chem., 17, 319 (1979).
26. Yu.I. Yermakov, Izv. Sib. Otd. Akad. Nauk SSSR, ser. Khim., 58 (1976).
27. I.I. Moiseev, Zh. Vsesouz. Khim. Obschest. im D.I. Mendeleeva, 27, 309 (1982).
28. A.E. Shilov and A.A. Steinman, Coord. Chem. Rev., 27, 97 (1977).
29. E.G. Chepaikin and M.L. Khidekel, Koord. Khim., 4, 643 (1978).
30. I.P. Lavrenev, M.L. Khidekel, Usp. Khim,, 52, 596 (1983).
31. R. Ugo, Catal. Rev. Sci. Eng., 11, 225 (1975).
32. P.M. Maitlies, Chem. Soc. Rev., 10, 1 (1981).
33. D.V. Sokolskii, Hydrogenation in Solutions (in Russian), Nauka, Alma-Ata (1979).
34. S.K. Rakovsky, V.V. Ivanova, D.M. Shopov, A.P. Shepelkin and Yu.I. Yermakov, 4th ISHC, 1984, v. 1, p. 164, Leningrad.
35. E.G. Chepaikin and M.L. Khidekel, J. Molec. Catal., 4, 103 (1978).
36. Bulgar. Patent No 36 152.
37. A.V. Bulatov, S.K. Rakovsky, V.V. Georgieva, V.V. Ivanova, D.M. Shopov and M.L. Khidekel, 4th ISHC, 1984, v. IV, p. 62, Leningrad.
38. S.K. Rakovsky, A.V. Bulatov, V.V. Ivanova, V.V. Georgieva, L.K. Nenchev and D.M. Shopov, 1st Nation. Conf., 1985, v. 1, p. 95, Bulgaria.
39. Bulgar. Patent No 66 693 (1984).
40. Bulgar. Patent No 66 694 (1984).

SYNTHESIS AND CATALYTIC PROPERTIES OF METAL-CONTAINING MACROLIGANDS ON INORGANIC CARRIERS

V.R. Mkrtychan, A.F. Lunin, V.V. Farafonov,
R.A. Karakhanov and A.L. Lapidus
I.M. Gubkin Institute of Petrochemical and Gas Industry, 117917 Moscow, USSR

ABSTRACT - Heterogenized metal-complex catalysts (HMCs) have been developed in which conjugated double bond polymers on mineral supports act as macroligands for palladium, copper, iron, chromium, vanadium and other metal ions. HMCs, due to their high thermal and chemical stability in chemically active media, can be used in high temperature catalytic processes. The catalytic activity of HMCs has been studied in carbonylation, hydration and oxidation reactions.

1. INTRODUCTION

The central problem the developers of heterogenized metal-complex catalysts (HMCs) on mineral supports have to face is the low stability of HMCs at elevated temperatures and in chemically active environment which considerably restricts their use in practice.

Presently there are a large number of minitely elaborated techniques for attaching metal complexes on available supports which differ either in the type of the carrier used (polymeric, mineral) or in the method of attachment. Not intending to cover all known carrier types, we would only mention the most widely used ones:
a) ion-exchange polymers;
b) organic polymers, both inert (e.g., styrene/divinyl benzene copolymers) and chemically modified (e.g., chelate-forming);

c) inorganic materials (silica, aluminosilicate and glass). Inorganic carriers, containing hydroxyls as surface functional groups are known to be only slightly active, and the possibility of developing stable HMCs is very limited. To overcome this disadvantage, there have been proposed methods for modifying the carrier surface by amination, sulfochlorination, addition of groups with active chlorine, etc. Using these methods it is possible to prepare solid carriers featuring good complexing ability.

A broad class of HMCs have been prepared on the basis of the carriers which exhibit a high catalytic activity in organic synthesis. However, apart from unquestionable advantages typical of any metal-complex catalyst on a solid support, such HMCs also have certain shortcomings. For instance, the HMCs on ion-exchange polymers or cross-linked copolymers containing chelate ligands have a technologically feasible granulated particle shape but low chemical, thermal and swelling stability. The HMCs on modified linear or graft copolymers also have a relatively low chemical and thermal stability and, besides, an irregular particle shape. Inorganic supports modified by organic ligands do have appropriate particle shapes, but can react with organic media. Under the effect of heat, thermal degradation may take place in organic fragments of the organo-mineral carrier.

An attempt to obtain HMCs by anchoring the metal on thermally stable polymers, for example, the molybdenum or tungsten complex on oligophenolene (Ref. 7), though interesting by itself, can hardly be regarded as promising because granulated HMCs cannot be prepared by this method.

To surmount these difficulties, we have developed a process for preparing HMCs on polymer-mineral carriers. The feature that distinguishes this process from the prior ones is polymerization of monomers directly on the inorganic carrier producing a polymer with conjugated double bonds, characterized by a high chemical and thermal stability, comparable with or superior to that of mineral carriers (Ref. 3). The process provides for obtaining

a) polymer-mineral supports with a "ready-made" ligand group and

b) polymer-mineral supports containing active chlorine which may be substituted by the desired structures capable of complexing reactions.

The first method is simpler, but restricts the range of polymer-mineral substrates suitable for complexing.

The second method may be used to obtain a broad variety of fixed ligand structures.

2. EXPERIMENTAL

The mineral carrier was impregnated with a monomer solution, polymerized under atmospheric pressure at 400-450°C or in an autoclave. The mineral material was silica with a 160 m^2/g specific surface area and a 0.83 cm^3/g total pore volume, and aluminosilicate with a specific surface of 210 m^2/g. Prior to impregnation, the mineral carriers were heated for 4 h at 400°C under 10^{-2} Torr.

The monomer polymerizable to macroligand was 2,4,6-tri-(phenylamino)-sym-triazine (Triphenylmelamine).

The monomer used for the synthesis of polymer-mineral carriers with an active chlorine atom was 2,4,6-trichloro-sym-triazine (Cyanurchloride). The polymerization occurs via the scheme:

1.

$$n\ \text{(2,4,6-tris(NHAr)-1,3,5-triazine)} \longrightarrow \left[\begin{matrix}\text{NHAr}\\ |\\ \text{C}\end{matrix}=\text{N}\right]_{3n}$$

Poly(phenyl cyanamide)

2. n (cyanuric chloride: triazine ring $C_3N_3Cl_3$) $\longrightarrow$ $\left[\begin{array}{c} Cl \\ | \\ C=N \end{array}\right]_{3n}$ (Polychlorocyane)

Polychlorocyane was further modified. Amino-,hydroxy- and carboxy-containing compounds different in structure were used as modifiers (Table).

The HMC was prepared by treating the macroligands, usually at room temperature, with solutions of transition metal compounds.

The product was in the form of black-colored pellets whose shape and size were controlled by the particular mineral material used. As a rule, the HMC particles are spherical with diameters ranging between 0.1-0.5 and 2-5 mm.

3. RESULTS AND DISCUSSION

3.1. Structure of HMC

The polymerization, modification and complexing processes were monitored with the help of IR-spectroscopy, atomic absorption analysis and X-ray microanalysis.

According to elemental analysis, the amount of poly-(phenyl cyanamide) on the support (sample No.7, Table 1) is below 5.5% on silica and 7.0% on aluminosilicate. The quantity of polychlorcyane (PCC) on aluminosilicate is usually 3 to 6% and depends on the technique used to attach cyanurchloride to the mineral carrier.

The polyconjugated polymer structure is verified by the typical IR band at 1630 cm^{-1} of macroligand samples.

The percentage of metal in the HMC was determined by the atomic absorption technique according to the metal content of the acid solution following decomposition of of the metal complex. This method of determining metal in the sorbent is more accurate than its determination according to the difference before and after the complexing. The complexing ability data for different HMCs are given in the Table.

The X-ray microanalysis (RMA) technique was resorted to for a quantitative determination of the concentration of the chlorine in polychlorcyane and on the carrier and that of palladium in the HMC as well as for a rough estimation of element distribution in the sample bulk. The analysis was carried out with the aid of a JSM-U3 scanning microscope and a KEVEX energy-dispersion spectrometer under accelerating voltage of 15 kV and currents of 10^{-9}-10^{-10} A. As established earlier (Ref. 4), such conditions are optimum considering the thermal and radiation stability of polymers. The concentrations of Cl and Pd were determined relative to the external reference. The samples were broken in halves, a protective ca. 200 Å thick carbon layer was sprayed on the surfaces, the samples placed in the microscope column and analyzed according to the K_{α}-line of the characteristic radiation of chlorine and palladium. The sampling was done for 100 s over 100 sq.microns of surface; the broken HMC pellet surfaces were scanned edge-to-edge with averaging over 10 samples. Under these conditions the relative element determination error was 2-3%.

Fig. 1 shows the X-ray spectra of aluminosilicate samples impregnated with cyanurchloride (1) and PCC on aluminosilicate (Fig. 1 a). The content of chlorine in the monomer was 4%, in the polymer 2%, indicating a 50% loss of chlorine during polymerization at high temperatures. A comparison of the chlorine concentration data for PCC samples collected from different surface regions (edges, middle) shows the polymer to be uniformly distributed over the bulk of the carrier (Fig. 1 b).

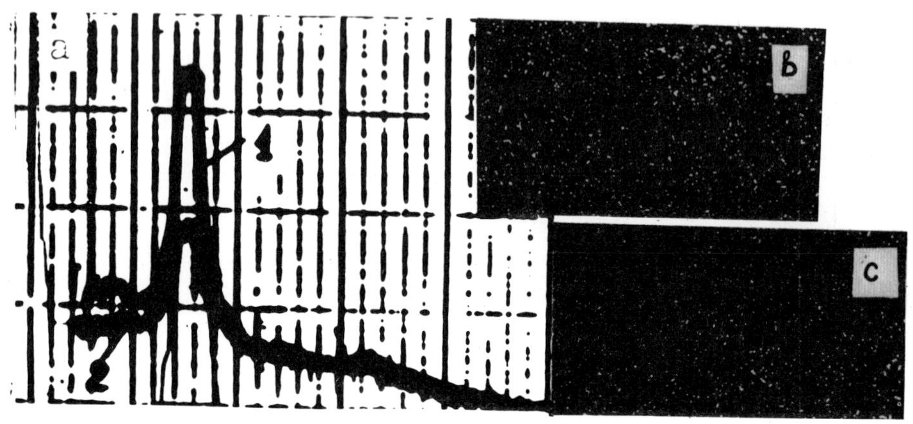

Fig. 1. X-ray K_α-Cl spectra of aluminosilicate samples with cyanurchloride (1) and polychlorocyane (2) - a; X-ray K_α-Cl pictures of PCC/aluminosilicate samples taken at different surface sites: edge - b, middle - c, x 10^3.

Figs. 1 b and 1 c show micrographs of broken HMC pellet surfaces taken in the characteristic X-ray band of chlorine. One may see fluctuations of chlorine distribution.

Fig. 2 presents microphotographs of the surfaces of the Pd containig HMC (sample 1, Table 1) taken in the characteristic X-ray band of palladium. According to the atomic absorption analysis, the palladium content in this sample is 1.4% and becomes 1.0% after heating at 150°C. The Pd concentration determined by the XDM technique is 1.2%; the microphotographs taken in the characteristic radiation of palladium show its sufficiently uniform distribution over the pellet surface (Figs. 2 a, 2 b).

The HMCs prepared by the method described above feature a high thermal and chemical stability. After boiling the samples in water for 12 h the relative stability of the complex, determined as a ratio of the sorption capacity after exposure to the initial sorption capacity, was 0.85 to 1.0, and 0.8 to 0.95 in acetone. The pictures of Pd distribution patterns over the bulk of sample 1 prior to (Fig. 2a) and after the thermochemical treatment (Fig. 2b) clearly illustrate this.

TABLE 1. Complexing ability of macroligands

Sample No	Monomer	Modifier	Ligand	Metal	Metal content in HMC mg-eq./g (%)
1.	Cyanur chloride	4-amino-1,2,4-triazole	N-N-NH- (triazole ring)	Pd	0.23 (1.2)
2.	cyanur-chloride	benzoimidazole	(benzimidazole ring)	Pd	0.13((0.7)
3.	cyanur-chloride	glycine	$-NH-CH_2-COOH$	Cu	0.46 (1.5)
4.	cyanur-chloride	diethylamine	$-N(C_2H_5)_2$	Cu Fe	0.18 (0.6) 0.21 (0.6)
5.	cyanur-chloride	triethylamine	$-N(C_2H_5)_3$	Cu Fe	0.15 (0.5) 0.07 (0.2)
6.	cyanur-chloride	monoethanolamine	$-NHC_2H_4OH$	Cu Fe Co	0.31 (1.0) 0.28 (0.8) 0.23 (0.7)
7.	triphenylmelamine	monoethanolamine	-C=N-NHAr	Pd Co Ni Cu V	0.20 (1.1) 0.44 (1.3) 0.29 (1.0) 0.28 (0.9) 0.15 (0.4)

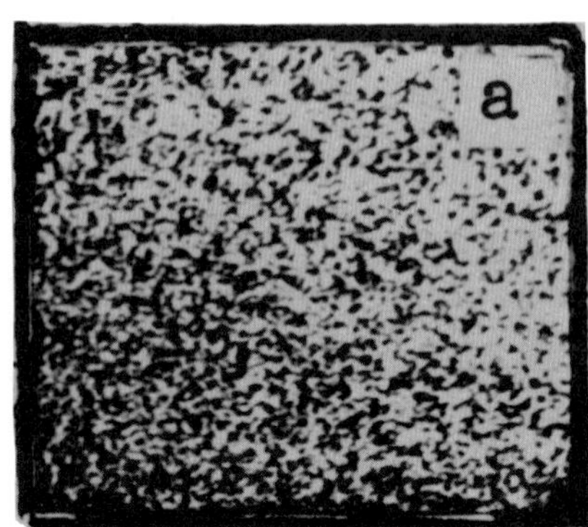

Fig. 2. X-ray K_α-Pd pictures of Pd-HMC (sample 1) prior to exposure (edge, middle) - a; after exposure (edge, middle) -mb, x 10^3.

The Pd-containing HMCs are thermally stable up to 150-200°C, and the Pd-PPCA-aluminosilicate sample (sample 7) up to 250°C.

3.2. Catalytic Properties of HMCs

The catalytic activity of Pd-containing HMCs was studied in nitro-compound carbonylation reactions. The HMCs showed a high capacity and provided the yield of aromatic and aliphatic carbamates of 50-92%. The selectivity was as high as 95%.

Copper-containing HMCs (samples 3-6) were active in selective hydration reactions.

The most difficult problem is the choice of HMC for oxidation process. Special operation conditions, elevated temperatures and strong oxidants (such as oxygen and hydroperoxides) make special demands on a catalyst to be used. Despite a relatively high thermooxidation stability of conjugated double bonds of polymers, the only substance that successfully stands the effect of oxygen for a long time is polyphenylcyanamide (PPCA, sample 7). The reaction of liquid phase isopropylbenzene oxidation with oxygen was chosen as a model to study the catalytic activity of HMCs. The study was carried out on a standard equipment consisting of 3 cubic cm. glass reactor provided with a stirrer.

Active centers on a catalyst surface and the surface itself play a substantial role in oxidation processes and in those associated with hydroperoxide accumulation and decomposition, in particular. As is known, developed surface with a large number of narrow pores may cause uncontrollable hydroperoxide decomposition, decay of active particles and appearance of undesirable products. The surface area of the carrier with attached polymer was 4 m^2/g, that is, too small to considerably affect the oxidation process. In the presence of mineral supports, PPCA, silica and aluminosilicate have been reported not to affect cumene oxidation rate. Both pumice and polymer without a mineral support decrease the rate of oxygen adsorption in comparison with thermal autooxidation process, a starting rate of the oxidation being decreased with a simultaneous increase in induction period. Further decrease of oxygen adsorption and increase in induction period on polymer-mineral matrices show the catalyst deactivation. It is associated presumably with a change in the number of active centers on the polymer surface due to its interaction with the mineral support. Pumice, oppositely, increasing surface of the polymer 10- to 20-fold, multiplies the number of surface active centers as well. This results in the reduction of the induction period and in the increase of both starting and maximum reaction rates. The amount of the polymer in the reaction medium enhances the oxidation rate, but to a limited extent. As soon as the polymer content exceeds 20%, the oxidation rate slows down. This phenomenon is due to the change of the rate ratios of radical chain oxidation elementary stages. In order to investigate the mechanism of the oxidation and choose a catalyst, we have studied the hydroperoxide decomposition process. Substances leading cumene hydroperoxide decomposition to 1-methyl-1-phenylethanol (MPE) may be considered to be most promising oxidation catalysts. A large content of other products, such as acetophenone (AP), phenol (Ph), α-methylstyrene (α-MS) rules out the possibility of a chain process. The data in Table 2 show that only a pure polymer and PPCA on pumice are appropriate for future studies of cumene oxidation to hydroperoxide. However, this process can be carried out when the rate of hydroperoxide decomposition is not very high and is much less than the rate of the oxidation. The effect of the coordinating metal of the hydroperoxide decomposition process is shown in Table 3.

TABLE 2. Effect of the mineral support on the rate of hydroperoxide decomposition and on the content of products (120°C, catalyst content 4%, hydroperoxide content 0.5 mol/l)

Catalyst	W_{ROOH} x 10^{-4} mol/l·sec.	Content of cumene hydroperoxide decomposition products, % mass.			
		AP	MPE	Ph	α-MS
alumosilicate	24.2	15.1	-	84.9	-
silica gel		no decomposition			
pumice		no decomposition			
PPCA	0.7	9.8	90.2	-	-
PPCA/AS	20.6	26.1	-	74.9	-
PPCA/silica	12.6	30.5	65.0	-	4.5
PPCA/pumice	4.1	4.5	95.5	-	-

TABLE 3. Effect of the nature of the complexing agent on the rate of the HMC-catalyzed hydroperoxide decomposition and on the content of products (120°C, hydroperoxide content 0.58 mol/l, catalyst content 4%, PPCA/pumice).

Metal	W_{ROOH} x 10^{-4}, mol/l·sec.	Content of decomposition products, %				
		AP	MPE	Ph	α-MS	dicumene peroxide
Ni	5.98	8.9	91.1	-	-	-
Co	8.56	16.5	83.5	-	-	-
Cr	8.75	5.8	1.2	58.2	-	34.8
Fe	9.17	25.3	30.0	44.7	traces	

As follows from Tables 2 and 3, the hydroperoxide decomposition rate increases 1.5- to 2-fold after the formation of Ni and Co HMC; Ni and Co being the only ions which do not change the route of the decomposition.

A study of kinetics in different solvents (ethylbenzene, chlorbenzene, decane) has shown that hydroperoxide decomposition goes through the formation of the complex of ROOH with a catalyst. The dissociation constant of the complex has been determined. Most catalytic reactions going through the formation of an intermediate complex have one common feature, that is, a limiting hydroperoxide concentration upon reaching which the decomposition rate steadily rises. For instance, the Ni-HMC shows a linear growth of the decomposition rate up to 0.8 mol/l and then, as the hydroperoxide concentration rises up to 1.2 mol/l, the decomposition rate decreases dramatically to show a zero growth. The change of catalyst content affects the hydroperoxide decomposition rate in a similar way. A supplementary study is needed to explain these irregularities. Table 4 presents data on the catalytic activity of HMCs in cumene oxidation.

TABLE 4. Effect of the nature of the metal ion on oxygen absorption rate and cumene oxidation products distribution (PPCA/pumice reaction time 3 h, 120°C, catalyst content 4%).

Metal ion	W x 10^{-4}, mol/l·sec.	Cumene conversion,%	Content of oxidation products			
			hydroperoxide	AP	MPE	Ph
	1.26	5.2	71.2	5.8	23.0	-
Ni	1.86	6.3	79.1	1.7	19.2	-
Co	1.19	4.9	-	16.3	83.7	
Cr	1.23	5.1	-	9.8	39.2	29.4
Fe	1.14	4.7	-	14.9	31.9	30.9
Mn	0.17	0.7	-	42.9	57.1	-

All metals except for Ni are found to reduce polymer carrier ability to accumulate hydroperoxide. The Ni-HMC increases W_{max} of the oxidation.

The increase of the catalyst content from 4% to 16% leads to the growth of cumene conversion from 6.3% to 19.3% with a slight fall of the selectivity.

The growth of the content of Ni^{+2} ions on the catalyst surface from 0.1% to 1% decreases cumene conversion 1,3--fold and the oxidation rate 2-fold. This can be attributed to a decrease in the number of centers responsible for initiation as a result of complexing. If the suggestion that nitrogen atoms are responsible for metal coordination is correct, it is them (or at least one of them) to determine the initiating ability of both the HMC and its matrix. The addition of the catalyst to a hydroperoxide-initiated cumene oxidation slows down the oxidation. The rate of pure homogeneous process is $1.105 \cdot 10^{-5}$ mol/l·sec., the catalyst oxidation rate is $0.126 \cdot 10^{-5}$ mol/l·sec. and that of a total process is $0.256 \cdot 10^{-5}$ mol/l·sec. . The radical process chain length has been determined from the kinetic data (14.5 for the initiated homogeneous oxidation and 2.17 for the heterogeneous HMC process). Data obtained as well as oxygen and hydroperoxide adsorption studies allow one to suggest that cumene oxidation is of pure heterogeneous nature on polymers and polymer-based HMCs with oxygen activation on active centers. The metal of HMC plays a substantial role in all elementary process stages strengthening the effect of the polymer mineral support. There are certain prospects for a practical use of the catalyst developed because a large amount of hydroperoxide (up to 45-50%) can be selectively accumulated (70-75% selectivity) in the cumene oxidation process.

A number of catalysts for the cumene conjugated oxidation with olefins (dodecene) and aldehydes (benzaldehyde) have been obtained by varying mineral supports and by coordinating metals or their mixtures. Despite the relatively low yields of the main products (olefin oxide and acid), the high selectivity of the process (up to 90%) gives the hope for a practical application of the experimental results obtained.

REFERENCES

1. G.V. Lisichkin, A.Yu. Yuffa, Heterogeneous metal-complex catalysts (in Russian), pp.50-134, Khimia, Moscow (1981).
2. USA Patent No 4200052 (1981).
3. G. Göbel, Z. Stornick, Z.Angew.Makromol.Chem.,71, 221 (1978).
4. E.V. Egorov, P.F. Novikov,,D.R. Ragozin, V.L. Zeitlin, Dokl.Akad.Nauk SSSR, 146, 6, 1360 (1962).
5. A. Guyot, M. Bartholin, Int.Chim., 166, 221 (1977).
6. M. Bartholin et al., J.Molec.Catal., 2, 307 (1977).
7. V.A. Sergeev, L.I. Vdovina, in Catalysts containing supported complexes, Yu.I.Yermakov (Ed.), Novosibirsk, Institute of Catalysis,1, 159 (1980).
8. USSR Authorship Certificate No 952.316 (1979).
9. A.E. Chalykh, I.N. Sapozhnikova, A.D. Aliev, Vysokomol.Soed., A21, 1664 (1979).
10. Yu.I. Yermakov, G.K. Boreskov et al., Dokl.Akad. Nauk SSSR, 214, 146 (1974).

EFFECTS OF BASIC MACROMOLECULES ON THE COBALT(II)PHTHALOCYANINE-CATALYZED AUTOXIDATION OF HYDROPHILIC AND AMPHIPHILIC THIOLS

W.M. Brouwer, A.J.M.S. Robeerst, P. Piet and A.L. German
Laboratory of Polymer Chemistry
Eindhoven University of Technology
P.O. Box 513
5600 MB EINDHOVEN
The Netherlands

SYNOPSIS

Polymeric catalysts consisting of cobalt(II)phthalocyanine-tetra-sodium-sulfonate ($CoPc(NaSO_3)_4$) bound to basic polymers exhibit a high activity in the oxidation of water soluble thiols to disulfides with molecular oxygen in aqueous media. The used polymeric ligands are water soluble polyamines and poly(quaternary ammonium hydroxides) (ionenes). Mainly three parameters determine the catalytic effect: the linear charge density on the polymer, the amount of basic groups in the polymer domain and the ionic strength in the solution. Isolation of the Co-sites by the polymer coils highly contributes to the rate accelerations observed. The effect on the oxidation rate of the hydrophobicity of the thiols, 2-mercapto-ethanol and dodecylmercaptan, is studied with $CoPc(NaSO_3)_4$ in the presence of cationic soaps or cationic polymers as a catalyst. The polar 2-mercaptoethanol is oxidized most easily with hydrophilic, highly charged polycations; introduction of hydrophobic groups in the polymers results in a decreased catalytic activity. The oxidation rate of dodecylmercaptan is considerably enhanced, not only in the case of conventional or polymer soap micelles, but surprisingly also when highly charged hydrophilic 2-4 - and 2-6 -ionenes are used. This latter result can not be interpreted in terms of hydrophobic interactions between substrate and catalyst. Complex formation between the dodecylmercaptide anions and the positively charged ionenes has to be considered, probably analogous to the complex formation between polyelectrolytes and oppositely charged surfactants.

INTRODUCTION

Cobalt(II)phthalocyanine is a well known catalyst for the oxidation of thiols to disulfides [1]:

$$4\ RSH + O_2 \longrightarrow 2\ RSSR + 2\ H_2O \qquad (1)$$

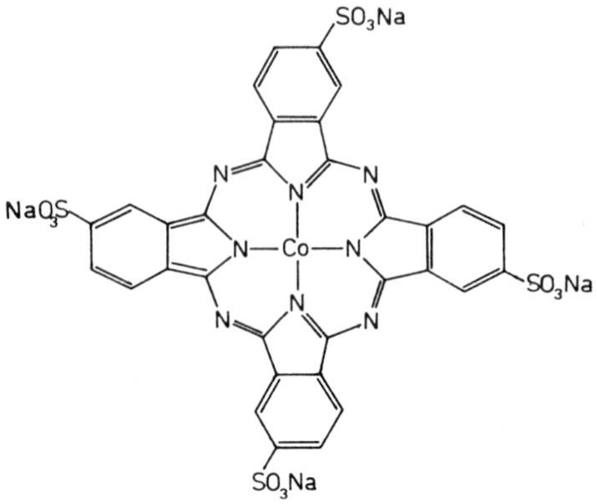

Figure 1. Chemical structure of cobalt(II)phthalocyanine-tetrasodium sulfonate (CoPc)

Attachment of the water soluble cobalt(II)phthalocyanine-tetrasodium sulfonate ($CoPc(NaSO_3)_4$; Fig. 1) to poly(vinylamine) (PVAm) caused a large enhancement in the rate in comparison to the monomeric analog 1,3 propanediamine [2]. For this particular polymeric catalyst system the effect of pH [3], ionic strength, substrate concentration, oxygen pressure, temperature [4] and molecular weight of the polymeric ligand [2] on the oxidation rate of 2-mercaptoethanol (RSH) to 2,2 dithiodiethanol (RSSR) was investigated. The polymeric catalyst exhibited an enzyme-like behavior, e.g. Michaelis-Menten kinetics (Scheme 1) with respect to both thiol and oxygen and a bell shaped pH dependence.

Scheme 1

$$E + S + O_2 \overset{K_s}{\rightleftharpoons} O_2\text{-}E\text{-}S \overset{k_2}{\longrightarrow} E + P$$

(E stands for polymer-bound $CoPc(NaSO_3)_4$; S for substrate (thiol) and P for products)

The enhanced polymeric catalyst activity was partially attributed to prevention of the dimerization of $CoPc(NaSO_3)_4$ by the basic polymer coils and to enrichment of these coils with thiol anions, the reactive species. The attachment of $CoPc(SO_3)_4^{4-}$ to the polymer was attributed to coordinative and Coulombic interaction [5,6], but the nature of the interaction in terms of chemical structures has not as yet been elucidated.

From catalytic experiments using copolymers of vinylamine and vinylalcohol as polymeric ligands, it could be deduced that reaction rate is linearly dependent on the positive charge density on these ligands [7].

We recently reported a cationic micellar effect on the oxidation of the sparingly water soluble, long chain substrate dodecylmercaptan (DM) by molecular oxygen with cobalt(II)phthalocyanine-tetra-sodium sulfonate ($CoPc(NaSO_3)_4$) as a catalyst [8], see eq. 1 (with RSH = DM). The oxidation rate of a polar substrate, 2-mercaptoethanol (ME), however, could not be accelerated under these conditions. It was shown that in the conventional micellar catalysis of DM both electrostatic and hydrophobic interactions occurred between substrate and micelle, but that in the polyelectrolyte catalysis of ME primarily electrostatic forces were important.

Polysoaps (polymer micelles) may provide the structural characteristics of both conventional micelles and polyelectrolytes. The catalytic activity of conventional micelles and polyelectrolytes has been extensively studied in the past three decades, but efforts to understand the catalytic efficiency of polymer micelles mainly stem from recent years [9-14].

In polyelectrolyte catalysis the electrostatic field of the macro-ion contributes to an enhanced activity either by accumulation of substrate in the polymer domain or by desolvation of the activated complex [15]. Sakurada [16] proposed that, apart from electrostatic interactions, also hydrophobic interactions play an important role in the polyelectrolyte catalysis. Specially when substrates with hydrophobic moieties are used, the hydrophobicity of the polymers largely appears to determine rate, due to concentration effects, environmental effects or both. Moreover, polymers capable of forming intra-molecular side-chain aggregates often exhibit rates much higher than those observed for similar polymers which are not able to form such micellar domains [10,12,14].

In order to gain more insight into the structural characteristics needed to develop efficient catalysts capable of oxidizing either DM or ME, the rate accelerations of both DM and ME oxidation with various polymer catalysts, polysoaps (polymer micelles) as well as polyelectrolytes, have been investigated.

EXPERIMENTAL

The polymers used in this investigation and some of their physical properties are shown in Table 1.

The linear ionene bromides were prepared as described by Rembaum [20]. The 3-3/C_{12} ionene polysoap was synthesized according to the method of Sonnessa et al. [21]. Poly(4-vinylpyridine) (Polysciences Inc.) was quaternized with n-butylbromide (Fluka) in methanol at 60°C for 70 hr. The degree of quaternization was derived from the bromide content of the dried polymer as determined by argentometric titration (Volhardt). Poly(vinylaminehydrochloride) was synthesized according to the method of Hart as described by Hughes and St. Pierre [22].

The carbon, nitrogen and hydrogen analysis results and the I.R. spectra indicated that the products have the desired composition. All polymers appeared to be water soluble.

Hexadecyltrimethylammoniumbromide (Fluka, analytical grade) was used without further purification. Dodecylmercaptan (DM) and 2-mercaptoethanol (ME) were (vacuum) distilled and their purity (> 99%) was checked before use by iodometric titration. The polymer solutions were eluted through an anion exchange column (Amberlite IRA 401), with the exception of the 3-3/C_{12} ionene, in order to replace the halogenide by hydroxyl-ions. The 3-3/C_{12} ionene was used in the bromide form since this polysoap appeared to have a very long residence time in the ion exchange column.

Table 1. Polymers and surfactant used in the investigation

polymer or soap	abbreviation	molecular weight g mole^{-1}	average linear charge density parameter, λ [a)]
x-y/R ionenes [b)]:			
x=2, y=4, R=CH_3	2-4/C_1	$\bar{M}_w = 10^4$ [c)]	1.45
x=2, y=6, R=CH_3	2-6/C_1	$\bar{M}_w = 10^4$ [c)]	1.15
x=2, y=10, R=CH_3	2-10/C_1	$\bar{M}_w = 6.10^3$ [c)]	0.82
x=3, y=3, R=$C_{12}H_{25}$	3-3/C_{12}	$\bar{M}_w = 10^4$ [c)]	1.45
hexadecyltrimethyl-ammoniumbromide	CTAB	-	1.15 [d)]
poly(4-vinylpyridine), quaternized with n-butyl-bromide (83 mole%)	PVP/C_4(83)	$\bar{M}_w = 10^5$	2.49
poly(vinylamine)	PVAm	$\bar{M}_n = 5.10^4$	0.99 (pH=7.8) [e)] 0.0 (pH=13) [e)]

a) $\lambda = \frac{\sigma e^2}{4\pi\varepsilon bkT}$, where e = proton charge, σ = degree of ionization of the polymer, ε = dielectric constant, b = average linear distance between ionizable (ionized) groups, kT = Boltzmann term, λ taken from polyelectrolyte theory, see ref (17)

b) $\left(- \overset{\overset{C}{|}}{\underset{\underset{R}{|}}{N}}{}^{\oplus} - (CH_2)_x - \overset{\overset{C}{|}}{\underset{\underset{R}{|}}{N}}{}^{\oplus} - (CH_2)_y - \right)_n$ $Br^{\ominus}$ $Br^{\ominus}$

c) estimation from the intrinsic viscosity data in 0.4 M KBr, using the Mark-Houwink constants for 3-4/C_1 ionenebromide (18)

d) estimation for rod-like micellar structure. Ionic strength = 0.25 M. Linear charge density = (surface charge density)$^{\frac{1}{2}}$, calculated with data of ref (19)

e) determined by potentiometric titration

Polymer concentrations were in the range 4.10^{-4} - 10^{-3} M (repeating units). $CoPc(NaSO_3)_4$ concentration was in the range $(1.9-3.9).10^{-7}$ M. It was verified that in these ranges specific rate did not depend significantly on the $CoPc(NaSO_3)_4$ and polymer concentrations, with the exception of the $3-3/C_{12}$ ionene and the $PVP/C_4(83)$ polymers where rate appeared to be less than first order in the $CoPc(NaSO_3)_4$ concentration. The catalyst preparation and the catalytic measurements were performed as described previously [4].

The turbidity τ during turbidimetric titration of the polymeric catalyst with sodiumdodecylsulfate (SDS) in the presence of DM was obtained from transmission measurements using a Carl Zeiss PMQ II spectrophotometer operating at a wavelength of 600 nm with 1 cm cells.

RESULTS AND DISCUSSION

The rate accelerations observed for both DM and ME oxidation in several catalyst systems are presented in Table 2. It is clearly shown that introduction of hydrophobic moieties in the polymer (e.g. compare $3-3/C_{12}$ with $2-4/C_1$ ionene) dramatically lowers the catalytic activity for polar substrate (ME) oxidation. In addition, the ME oxidation rate decreases when polymer charge density decreases (e.g. within the series $2-4/C_1$, $2-6/C_1$ and $2-10/C_1$ ionene), as was shown earlier [23]. The presence of cationic charge is also necessary in the DM oxidation as may appear from the catalytic indifference of non-charged PVAm for DM oxidation at pH=13. The DM oxidation capabilities of the polysoap $3-3/C_{12}$ and of CTAB (conventional micelles) are relatively high. The hydrophilic $2-4/C_1$ ionene, however, exhibits catalytic activity not exclusively for the polar thiol ME, but <u>unexpectedly</u> also for the long chain thiol DM. DM oxidation activity is also exhibited by the $2-6/C_1$ ionene and to a lesser extent by the $2-10/C_1$ and $PVP/C_4(83)$ polymers, despite the inability of these short-branched polymers to form intramolecular micellar domains [12,14]

Table 2. Rate accelerations (v/v_0) and kinetic constants for the oxidation of dodecylmercaptan (DM) and 2-mercaptoethanol (ME) in various catalyst systems

catalyst	DM [a]			ME [b]		
	v/v_0	$10^{-2}k_2$ / s^{-1}	K_m^{-1} / $mole^{-1}dm^3$	v/v_o	$10^{-2}k_2$ / s^{-1}	K_m^{-1} / $mole^{-1}dm^3$
$3\text{-}3/C_{12}$	18	e)	e)	2.7	e)	e)
CTAB	28	3.9±0.5	17±3	0.3 c)	e)	e)
$2\text{-}4/C_1$	27	3.4±0.5	55±6	32	39±5	35±6
$2\text{-}6/C_1$	27	e)	e)	22	e)	e)
$2\text{-}10/C_1$	7 d)	0.9±0.1	63±7	3.1	e)	e)
$PVP/C_4(83)$	4	e)	e)	2.0	e)	e)
PVAm	1	e)	e)	32	45±6	10±1

a) pH = 13

b) pH = 8

c) pH = 10

d) pH = 11.5

e) not determined

v is the average initial rate (first minute of oxygen consumption)

$v_{o,DM}$ = 4 ml O_2/μmole Co, min in a polymer and surfactant free system, pH = 13

$v_{o,ME}$ = 38 ml O_2/μmole Co, min in a polymer and surfactant free system, pH = 10

DM and ME concentrations were 4.10^{-2} and 2.10^{-1} M, respectively

$CoPc(NaSO_3)_4$ concentration was 1.95 10^{-7} M

To our knowledge, this is the first time a non-micellar, hydrophilic polyelectrolyte is reported to exhibit a hydrophobic substrate conversion efficiency, comparable with that of conventional and polymer micelles.

In several papers dealing with the conversion of hydrophobic substrates by polymeric catalysts the importance of hydrophobic (side)groups in the polymers has been emphasized. Shinkai et al. [12], however, suggested an enhancement of the <u>local</u> hydrophobicity when strongly hydrophobic substrates bind to a hydrophobic polymer. They pointed out that this effect could become more influential than the <u>overall</u> polymer hydrophobicity.

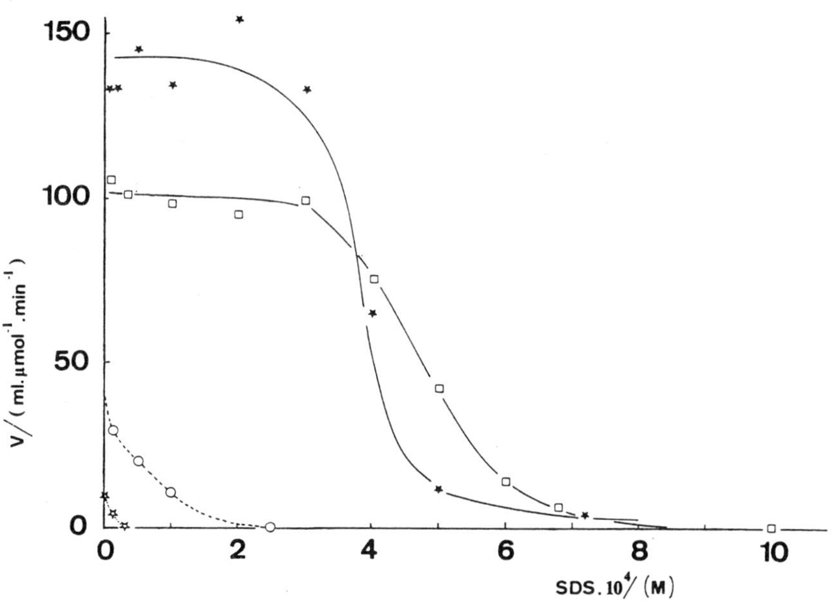

Figure 2. Effect of SDS on the oxidation rate of DM in the systems CoPc/PVAm at pH = 10.0 (open stars), CoPc/(2-4/C_1) at pH = 12.9 (full stars), CoPc/(2-6/C_1) at pH = 13.0 (squares), and CoPc/(2-10/C_1) at pH = 10.9 (circles). T = 25°C; polymer concentrations 4.10^{-4}M; [DM] = 4.2 x 10^{-2}M; [CoPc] = 3.9 x 10^{-7}M.

In the present case, dodecylmercaptide anions probably interact with the positively charged polymers, forming complexes analogous to those between polyelectrolytes and oppositely charged surfactants [24,25].

In order to provide supportive evidence to this hypothesis, the effect of increasing amounts of the anionic surfactant sodium dodecyl sulfate (SDS) on the reaction rates of DM oxidation with CoPc/ionenes was measured.

In Fig. 2 the initial oxidation rate is plotted versus the SDS concentration for various polymer ligands. As shown before, PVAm hardly shows any activity in DM oxidation and the relatively low activity of 2-10/C_1 rapidly decreases. The 2-4/C_1 and 2-6/C_1 ionenes, however, are showing a remarkable behavior in remaining practically constant at low SDS concentration, followed by a rapid decrease at [SDS] = 4×10^{-4} M. This concentration appears to be equal to the polymeric cation concentration, strongly suggesting that the positive charge on the polymer is shielded by stochiometric complex formation. The expected occurrence of strong interaction between SDS and ionene may lead to insoluble salt formation and as a consequence to turbidity phenomena.

Therefore, in separate experiments the polymeric catalyst solution was titrated turbidimetrically with SDS (Fig. 3) in the presence of DM. A maximum turbidity was observed at an [SDS]/[2-6/C_1] ratio close to unity. These findings are leading to the following model for the interaction between the anionic surfactant SDS and the cationic ionenes (Fig. 4).

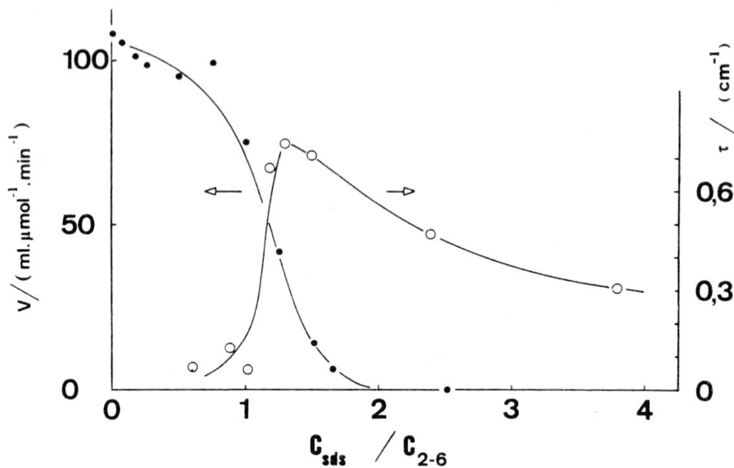

Figure 3. Effect of the ratio [SDS]/[2-6/C$_1$] on the rate of oxidation and the turbidity (τ) in the system CoPc/(2-6/C_1). Conditions as in Figure 2.

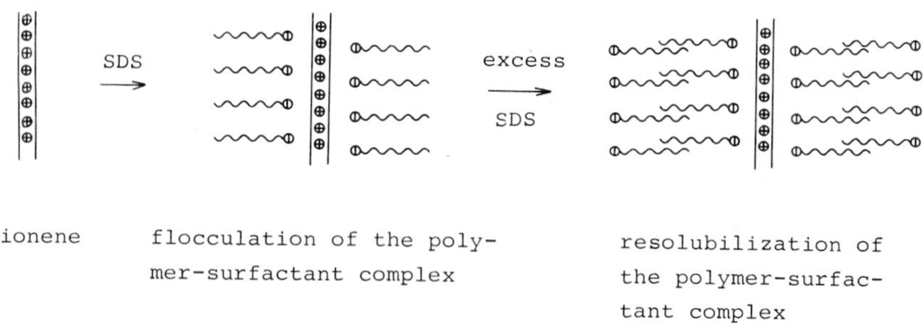

Figure 4. Schematic representation of the interaction between the anionic surfactant SDS and an ionene.

In the absence of SDS, 2-6/C_1 ionene dissolved readily in water as a result of its ionogenic character: $\tau = 0$ and reaction rates are high. The addition of SDS will cause adsorption of DS^- ions onto the positively charged polymer, leading to a polymer-surfactant complex with zero net charge (Fig. 4). This stochiometric complex formation will be favored by a concerted mechanism of ionic interaction between DS^- and the cations on the polymer backbone, and hydrophobic (polymer-micellar) domain formation involving SDS "side chains". The turbidity reaches a maximum value as a result of flocculation of the uncharged complex at an [SDS]/[2-6/C_1] ratio close to unity and close to the inflection point of the rate curve (Fig. 3). Obviously, the "stochiometric shielding" is much more effective than the small-ion (e.g. OH^-) shielding. Addition of excess SDS resolubilizes the polymer-surfactant complex and the polymer solution becomes transparent again (Figures 3 and 4). Catalytic activity is extremely low as a result of increased shielding and ionic substrate repulsion. From these observations it is conceivable that DM also strongly adsorbs to polyelectrolytes, creating a hydrophobic mantle around the polymer backbone, thus supplying its own polymer micelles required for catalysis of hydrophobic substrate oxidation.

In this train of thought it can also be understood why 2-10/C_1 ionene shows a rather low activity in DM oxidation (see Table 2 and Fig. 2). At first sight, the C_{10} apolar backbone sequences might seem favorable here. However, the C_{10} backbone sequences are too short to form polymer micelles with DM, since > C_{14} backbone sequences were reported [14] to be required in order to obtain polysoap behavior. On the other hand, the dodecylmercaptide ions, acting as C_{12} "side chains", are also too short in the case of the 2-10/C_1 ionene, since the length of the side chains required to form polymer micelles is related to and should amply exceed the length of the apolar backbone sequences [26].

Additional information can be derived from kinetic measurements. Most of the catalytic systems investigated so far exhibit Michaelis-Menten saturation kinetics toward thiol. The apparent binding constants (K_m^{-1}) and turn-over numbers (k_2) are listed in Table 2. It is shown that for $2\text{-}4/C_1$ and $2\text{-}10/C_1$ the binding of DM to the catalyst is even more efficient than in conventional micelles.

In Fig. 5 the effect of salt (NaCl) on the oxidation rate of DM and ME in the $2\text{-}4/C_1$ system is shown. In the DM oxidation (dots in Fig. 5) the addition of a simple salt hardly affects rate, which strongly indicates that hydrophobic polymer-substrate interactions within the micellar domains created by the long-chain substrate ions themselves, are playing an important role.

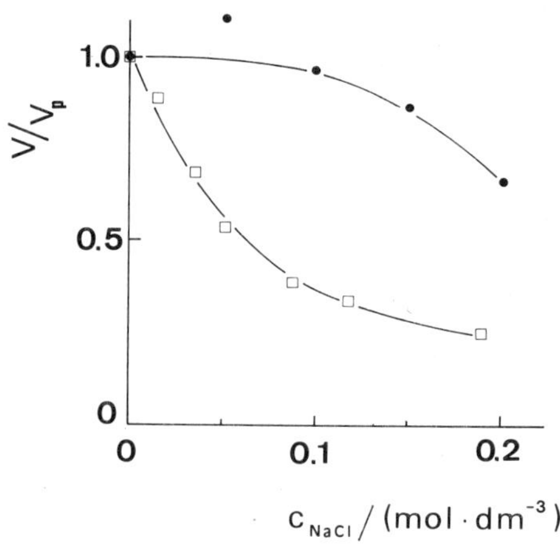

Figure 5. Effects of NaCl on the oxidation rate of DM (dots) and ME (squares) in the catalytic system $CoPc/(2\text{-}4/C_1)$; v and v_p are the reaction rates in the presence and absence of salt, successively.

In view of the pronounced effect of NaCl addition on the oxidation rate of ME (squares in Fig. 5), electrostatic polymer-substrate interactions seem to prevail in the 2-4/C_1 catalysis of ME as we pointed out earlier for ME oxidation with the $CoPc(NaSO_3)_4$/PVAm catalyst [8].

These findings seem to fit in completely with the "Hard and Soft Acids and Bases" concept [27]. Pearson [27], and Weber and Gokel [28] reported that quaternary cations (quats) tend to pair preferably with the "softest" ion available, i.e. in the present case the large lipophilic dodecylmercaptide anion. In this line of thought, for the same quat substantial competition may be expected when Cl^- ions are added to a system containing "harder" substrate anions, viz. hydroxyethylmercaptide anions.

In conclusion, we have shown that hydrophobic moieties in polyelectrolytes have a rate retarding effect on the oxidation of ME. Hydrophilic polyelectrolytes with a high polymer charge density exhibit the best performance for this hydrophilic substrate.

On the other hand, we found that not only amphiphilic but also hydrophilic ionenes exhibit high activity in the oxidation of DM in water, comparable with the activity of conventional CTAB micelles. From the apparent binding constant (K_m^{-1} = 55 M^{-1}) it appeared that DM is effectively bound to the non-micellar hydrophilic polymer catalyst 2-4/C_1, probably since the adsorbed substrate creates a hydrophobic mantle around the polymer chain, leading to a polymer micellar structure as schematically indicated in Fig. 6a.
This is reminiscent of the complex formation between polyelectrolytes and oppositely charged surfactants. The comparably high activities observed with the polysoap 3-3/C_{12} and the conventional CTAB micelles, will be the result of the occurrence of micellar structures as depicted in Fig. 6b and c. Further detailed studies on these polymer-substrate interactions are required.

The catalytic effects of other kinds of polymer micelles and of immobilized polymeric DM and ME oxidation catalysts are now being investigated.

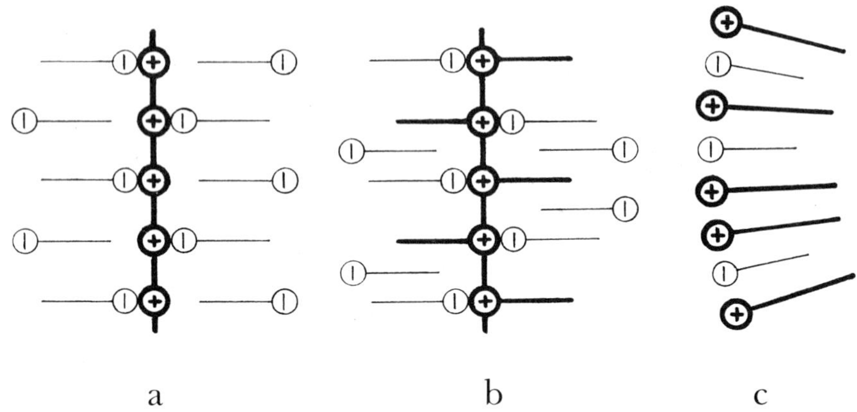

Figure 6. Schematic depiction of the proposed interaction between dodecylmercaptide anions (———⊖) and the polyelectrolyte 2-4/C_1 (a), the polysoap 3-3/C_{12} (b), and the conventional cationic soap CTAB (c).

REFERENCES

1. N.N. Kundo and N.P. Keier, Russ. J. Phys. Chem., 42 (1968), 707
2. W.M. Brouwer, P. Piet and A.L. German, Makromol. Chem., 185 (1984), 363
3. W.M. Brouwer, P. Piet and A.L. German, Polym. Bull., 8 (1982), 245
4. W.M. Brouwer, P. Piet and A.L. German, J. Mol. Catal., 22 (1984), 297
5. J.H. Schutten and J. Zwart, J. Mol. Catal., 5 (1979), 109
6. W.M. Brouwer, P. Piet and A.L. German, Polym. Commun., 24 (1983), 216
7. W.M. Brouwer, P. Piet and A.L. German, J. Mol. Catal., 29 (1985), 335
8. W.M. Brouwer, P. Piet and A.L. German, J. Mol. Catal., 29 (1985), 347
9. C.G. Overberger, T.W. Smith and K.W. Dixon, J. Polym. Sci. Symp., 50 (1975), 1
10. T. Kunitake, S. Shinkai and S. Hirotsu, Biopolymers 15 (1976), 1143
11. T. Ishiwatari, T. Okubo and N. Ise, J. Polym. Sci. Polym. Chem. Ed., 18 (1980), 1807

12. S. Shinkai, S. Hirakawa, M. Shimomura and T. Kunitake, J. Org. Chem., 46 (1981), 868
13. E.J. Delaney, L.E. Wood and I.M. Klotz, J. Am. Chem. Soc., 104 (1982), 799
14. V.E. Rodriguez Nieves, E.J. Ribaldo, R. Baroud and F.H. Quina, J. Polym. Sci. Polym. Lett. Ed., 20 (1982), 433
15. N. Ise in A. Rembaum and E. Sélégny (eds): "Polyelectrolytes and their applications", Reidel, Dordrecht (1975), 71
16. I. Sakurada, J. Pure Appl. Chem., 16 (1968), 236
17. A. Katchalsky, Pure Appl. Chem., 26 (1971), 327
18. A. Rembaum and H. Noguchi, Macromolecules, 5 (1972), 261
19. P. Debye and E.W. Anacker, J. Phys. Chem., 55 (1951), 644
20. A. Rembaum, W. Baumgartner and A. Eisenberg, J. Polym. Sci. Polym. Lett. Ed., 6 (1968), 159
21. A.J. Sonnessa, W. Cullen and P. Ander, Macromolecules, 13 (1980), 195
22. A.R. Hughes and T.St. Pierre, J. Macromol. Synthesis 6 (1977), 31
23. W.M. Brouwer, P. Piet and A.L. German, J. Mol. Catal., 31 (1985), 169
24. I.A. Robb in E.H. Lucassen-Reynders (ed): "Anionic Surfactants", surfactant science series, vol. 11, Marcel Dekker, New York (1981), ch. 3
25. A. Harada and S.-i. Nozakura, Polym. Bull., 11 (1984), 175
26. E.G. Knapick, J.A. Hirsch and P. Ander, Macromolecules, 18 (1985), 1015
27. R.G. Pearson: "Hard and Soft Acids and Bases", Dowden, Hutchinson and Ross, Inc., Stroudsburg, Pa, 1973
28. W.P. Weber and G.W. Gokel: "Phase Transfer Catalysis in Organic Chemistry, Springer Verlag, Berlin, 1977, 7

STUDY ON THE INTERACTION OF SURFACE ALKYL AND HYDRIDE COMPOUNDS OF Ti, Zr AND Hf WITH CARBON MONOXIDE

G.A. Nesterov, V.A. Zakharov, S.A. Vasnetsov and E.A. Paukshtis

Institute of Catalysis, Novosibirsk, 630090, USSR

ABSTRACT - The interaction of carbon monoxide with surface compounds with the composition $(Si\text{-}O)_nMR_{4-n}$ (M-Ti,Zr,Hf; R-CH_2Ph,H; n=1÷3) has been studied using IR-spectroscopy. The interaction of CO with surface benzyl and hydride compounds of Ti,Zr and Hf follows a two-stage mechanism. At low temperatures (170-250 K) a reversible adsorption of CO on a transition metal ion is observed. As the temperature of CO adsorption is raised above 293 K, carbon monoxide is inserted into the metal-hydrogen and metal-alkyl bonds to give surface formyl and acyl compounds.

1. INTRODUCTION

In recent years, a number of works on the study of individual organic and hydride compounds of Group IV transition metals with carbon monoxide have been published (Refs. 1-8). Interest shown in these reactions is due to that insertion of carbon monoxide into the metal-hydrogen and metal-alkyl bonds is thought to be a step of many catalytic reactions (Ref. 1). The insertion of carbon monoxide into the metal-alkyl bond with the formation of acyl complexes was observed for $Cp_2MR(R')$ (M-Ti,Zr, Hf; R-CH_3, CH_2Ph, C_3H_7, I, Cl; R'-CH_3, C_6H_5 (Refs. 2,3) $Ti(CH_2Ph)_4$ (Ref.4) and Cp_2TiR(R-C_6H_5, o-$CH_3C_6H_4$) (Ref.5). For the case of $Cp_2Zr(CH_3)_2$, $Cp_2Zr(CH_2Ph)_2$, $Ti(CH_2Ph)_4$ this reaction is reversible. For $Cp_2Ti(CH_2Ph)_2$ the interaction with CO leads to displacement of benzyl ligands with the formation of $Cp_2Ti(CO)_2$ and $(PhCH_2)_2CO$ (Ref. 2). X-ray studies of reaction products of alkyl Ti, Zr, Hf compounds with carbon monoxide have revealed that the resulted acyl complexes have similar metal-carbon and metal-oxygen distances (Ref. 3). According to IR data, these compounds have considerably lower frequencies, ν_{CO}, than the carbonyl groups of acyl compounds (Refs. 2-4). From these facts it is concluded that the acyl group in these compounds acts as a bidentant three-electron ligand.

The interaction of Ti, Zr and Hf hydride compounds with CO also includes the step of CO insertion into the metal-hydrogen bond, but leads to the formation of complex dimer products (Refs. 6-8). This step is preceded by a reversible adsorption of CO on a metal ion, which occurs below 208 K, as ascertained by IR data (Ref. 8).

We have studied the interaction of carbon monoxide with benzyl and hydride complexes of Ti, Zr and Hf, anchored on silica. The data obtained are compared with the results known for the homogeneous analogs of these compounds.

2. EXPERIMENTAL

$M(CH_2Ph)_4$ (M - Ti, Zr, Hf) was synthesized as described in (Ref. 9). Silica with a BET surface of 260 m^2/g and a pore volume of 1.6 cm^3/g was used as a support. Catalysts were prepared by interacting benzene solutions of $M(CH_2Ph)_4$ with silica, predehydroxylated at 693 K. The catalysts were washed with pure benzene and dried in vacuum at 293 K. Surface Ti, Zr and Hf hydride compounds were prepared by treating the corresponding surface benzyl compounds with hydrogen at 393-453 K. IR spectra were registered using a Specord-75IR.

3. RESULTS AND DISCUSSION

3.1. Interaction of Carbon Monoxide With Surface Benzyl Compounds of Ti, Zr and Hf

It has been found (Ref. 11) that when supported on silica, tetrabenzyltitanium interacts with the surface hydroxyl groups by reaction (1)

$$(Si\text{-}OH)_n + Ti(CH_2Ph)_4 \rightarrow (Si\text{-}O)_n Ti(CH_2Ph)_{4-n} + (4-n)CH_3Ph \qquad n = 1 \div 3 \qquad (1)$$

Tetrabenzylzirconium and tetrabezylhafnium interact with silica in the same way. IR spectra of surface compounds of $(Si\text{-}O)_n M(CH_2C_6H_5)_{4-n}$ have absorption bands from benzyl ligands, that may be ascribed, according to (Refs. 10,11), to δ_{C-H} at 1385 cm^{-1}; to skeleton vibrations of the benzene ring at 1445, 1480 and 1585 cm^{-1} and to ν_{C-H} of the benzene ring at 3020, 3065, 3075 and 3090 cm^{-1} (Table 1). Adsorption of CO at 100 K on surface benzyl compounds brings about new absorption bands at 2135, 2155, 2170 and 2175 cm^{-1} (Fig. 1, tables 1,2). The absorption bands from benzyl ligands remain unchanged. Evacuation of the samples at 100 K leads to recovery of the initial spectra. The 2135 and 2155 cm^{-1} bands characterize stretching vibrations of CO, adsorbed on the surface centers of silica. This is supported by the experiments on CO adsorption on silica, dehydroxylated at 693 K. We ascribe the 2170 and 2175 cm^{-1} bands (Tables 1,2) to ν_{CO}, coordinated on Ti, Zr and Hf ions. Note that CO adsorption on Ti, Zr and Hf ions is stronger than on the surface centers of silica. When CO is adsorbed in amounts much smaller than a monolayer, the 2170 and 2175 cm^{-1} bands appear first. Upon evacuation of the samples containing the adsorbed carbon monoxide the 2135 and 2155

TABLE 1. IR data on the interaction of CO with surface benzyl and hydride compounds of Ti, Zr and Hf

System	An active group in a surface compound and its absorption bands in IR spectra	ν_{CO} (cm^{-1}) at various temperatures of adsorption		Absorption bands of active groups in surface compounds upon the interaction with CO at various temperatures (cm^{-1})	
		100-170 K	293 K	100-170 K	293-393 K
$Ti(CH_2Ph)_4/SiO_2$	Ti - CH_2Ph 1440, 1475, 1585 3020, 3065, 3075, 3095	2135, 2155, 2170	1490 1635	unchanged	1450, 1485, 1595, 3025, 3070, 3085, 3115
$Zr(CH_2Ph)_4/SiO_2$	Zr - CH_2Ph 1440, 1475, 1585 3020, 3065, 3075, 3095	2135, 2155, 2170	1510 1620	unchanged	1455, 1485, 1595, 3035, 3070, 3085, 3115
$Hf(CH_2Ph)_4/SiO_2$	Hf - CH_2Ph 1445, 1480, 1590, 3020, 3065, 3075, 3090	2135, 2155, 2175	1510 1625	unchanged	1460, 1490, 1600, 3030, 3070, 3085, 3115

$Ti(CH_2Ph)_4/SiO_2/H_2$	Ti - H 1560	1970, 2040, 2060, 2080, 2110, 2135, 2155, 2170	1535	1530	
$Zr(CH_2Ph)_4/SiO_2/H_2$	Zr - H 1625	2020, 2050, 2075, 2090, 2135, 2155, 2170	1535	1520, 1550	
$Zr(BH_4)_4/SiO_2$ *)	Zr - H 1665	2040, 2135, 2155	1540	1560	
$Hf(CH_2Ph)_4/SiO_2/H_2$	Hf - H 1670	2035, 2065, 2080, 2135, 2155, 2175	1490, 1540	1525, 1635	

*) data of this work (Ref. 14).

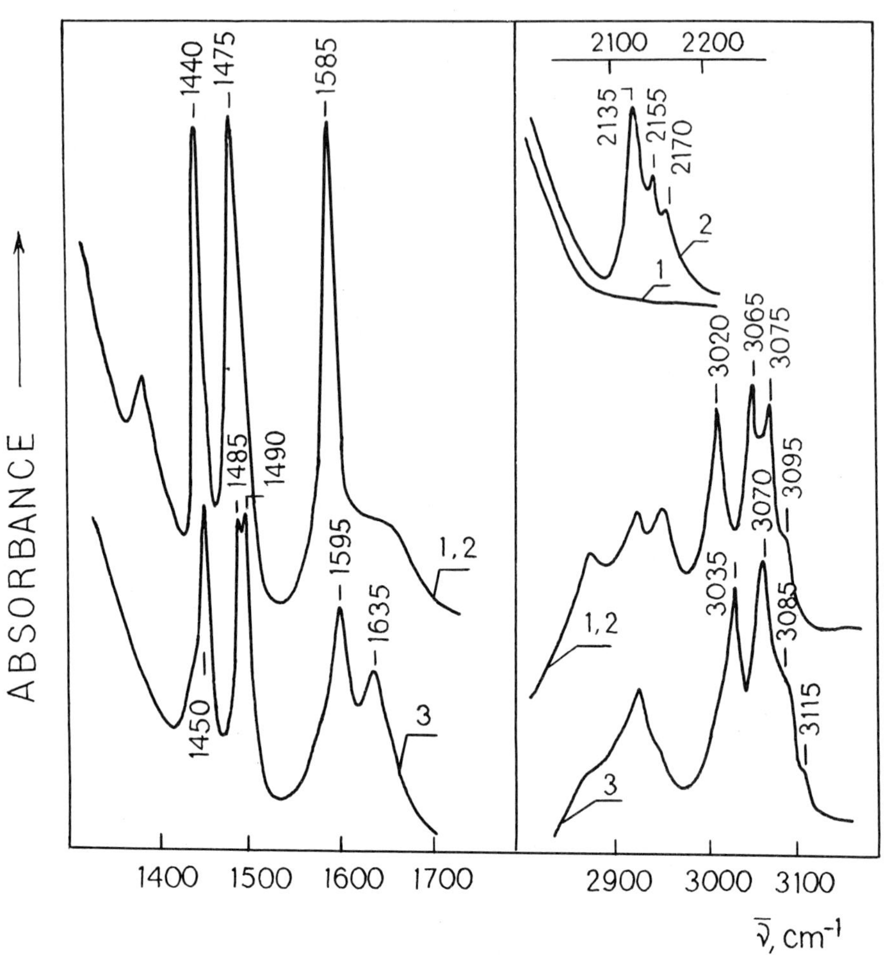

Fig. 1. IR spectra of the $Ti(CH_2Ph)_4/SiO_2$ sample (1) after the adsorption of carbon monoxide (5 Torr) at temperatures: (2) 100 K, and (3) 393 K.

cm^{-1} bands disappear first.

A revesible adsorption of carbon monoxide is observed also during the interaction of CO with some individual complexes at low temperatures. Thus, $Cp_2Ti(CH_2Ph)_4$ and $Ti(CH_2Ph)_4$ reversibly interact with one and two CO molecules, respectively, at temperatures below 220 K

(ν_{CO}= 1867, 1952 cm^{-1}) (Refs. 2,4).

The adsorption of CO on $M(CH_2Ph)_4/SiO_2$ systems at $T > 293$ K causes irreversible changes in IR spectra (Fig. 1, Tables 1,2). The absorption bands from benzyl ligands lose their intensity and shift by 5-15 cm^{-1} towards higher frequencies. Besides, for all systems new absorption bands appear within 1490-1635 cm^{-1} (Tables 1,2). We attribute such changes to the insertion of CO into the M-C bonds to give surface acyl compounds of Ti, Zr and Hf (scheme 2). Such interpretation is confirmed by the literature data. For example, the interaction of CO with $Ti(CH_2Ph)_4$ at temperatures above 250 K produces a $Ti(CH_2Ph)_2(COCH_2Ph)_2$ compound, with ν_{CO} at 1635 cm^{-1} (Ref. 4). The same value of ν_{CO} has been registered by us for $Ti(CH_2Ph)_4/SiO_2$ (Fig. 1, curve 3 and Table 2). Also, during the interaction of CO with $Cp_2M(CH_2Ph)_2$ (M - Ti, Zr) compounds at 293 K CO is inserted into the M-C bonds to form $Cp_2M(COCH_2Ph)_2$ compounds , wherein (ν_{CO}) CO frequencies are within 1520-1540 cm^{-1} (Ref. 3).

The close values of ν_{CO} in acyl compounds for homogeneous (Refs. 2-8) and supported systems suggest that the acyl group in supported systems acts as a bidentate three-electron ligand.

Thus, the interaction of CO with benzyl complexes of Ti, Zr and Hf, anchored on silica, obeys a two-stage mechanism. At low temperatures (100-170 K) a reversible adsorption of CO is observed. On raising the interaction temperature above 293 K the insertion of CO into the $M-CH_2Ph$ bonds takes place:

$$L_xM-CH_2Ph \overset{CO}{\rightleftharpoons} L_x\overset{\swarrow CO}{M} - CH_2Ph \rightarrow L_x\overset{\swarrow}{M}-\overset{O}{\overset{\|}{C}}-CH_2Ph . \qquad (2)$$

3.2. The Interaction of Carbon Monoxide With Surface Hydride Compounds of Ti, Zr and Hf

Surface hydride compounds of Ti, Zr and Hf were prepared by the treatment of surface benzyl compounds of the same metals with hydrogen using the technique proposed in (Refs. 11, 12).

$$(Si\text{-}O)_{n} M\text{-}(CH_2Ph)_{4-n} \xrightarrow{H_2} (Si\text{-}O)_{n} M(H)_{4-n} + (4-n)CH_3Ph \qquad n = 1 \div 3 \qquad (3)$$

Surface hydrides of Ti, Zr and Hf have respectively the absorption bands at 1560, 1625 and 1670 cm^{-1} (Table 1).

Upon the adsorption of CO at 100-170 K on the $(Si\text{-}O)_{n} M(H)_{4-n}$ systems the bands from Ti, Zr and Hf hydrides are abolished , while some new bands appear (Fig.2 and Tables 1,2). On evacuating the samples at the same temperature the initial IR spectra are recovered. The new absorption bands within 1970-2200 cm^{-1} may be assigned to carbon monoxide, adsorbed on a transition metal ion. During CO interaction with $(C_5Me_5)_2ZrH_2$ and $(C_5Me_5)_2HfH_2$ at 200 K a reversible adsorption of CO occurs (Ref. 8) and is characterized by the absorption bands at 2044 and 2036 cm^{-1} respectively.

Upon the interaction of CO with surface hydrides of Ti, Zr and Hf, in contrast to homogeneous hydride complexes, numerous absorption bands from the adsorbed carbon monoxide are observed (Tables 1,2). This may be due to inhomogeneity of the composition and coordination state of surface compounds, as well as to adsorption of several CO molecules in the metal coordination sphere.

Appearance of new absorption bands at 1500-1650 cm^{-1} after CO adsorption at 100-170 K may be associated with the change in the frequency of vibrations of the metal-hydrogen bonds ($\nu_{M-H} \rightarrow \nu'_{M-H}$). This transition is rever-

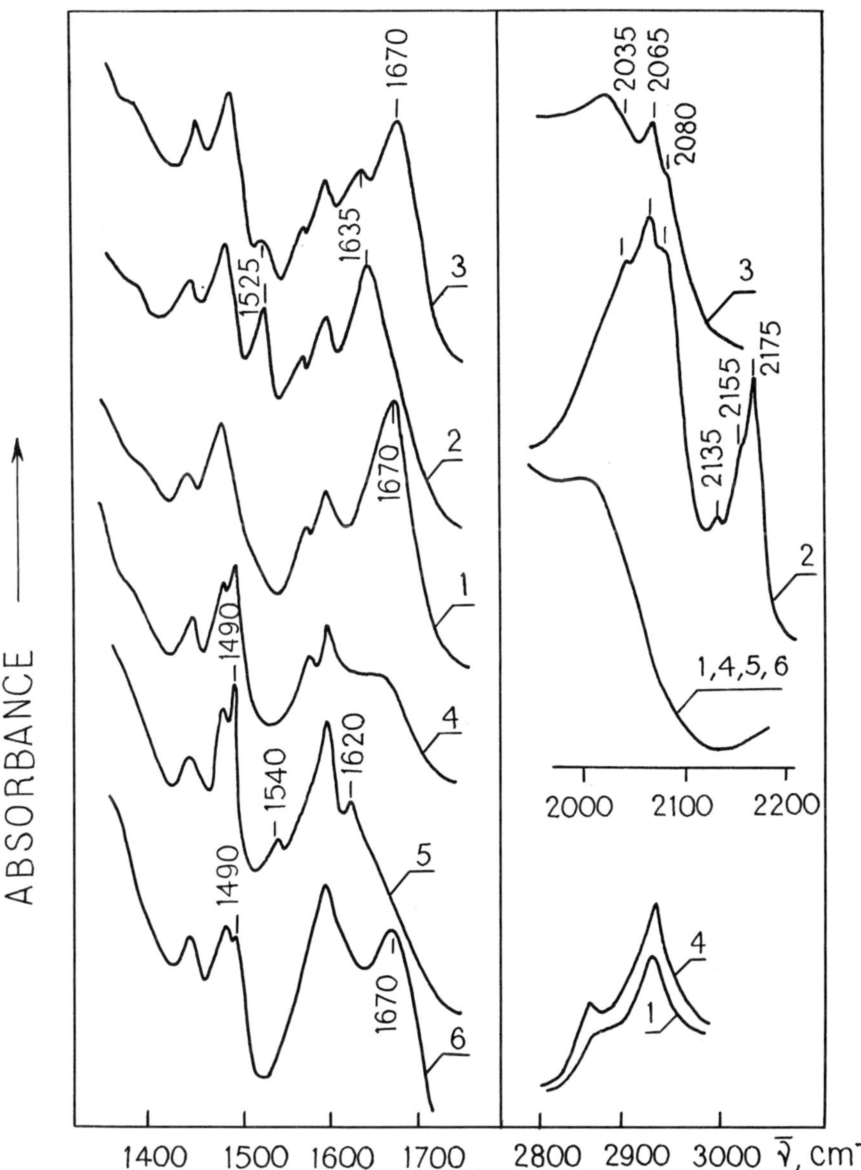

Fig. 2. IR spectra of the $Hf(CH_2Ph)_4/SiO_2/H_2$ system (1) after CO adsorption at various temperatures: (2) 100 K, the ratio CO/Hf ≈ 3; (3) 100 K, the ratio CO/Hf ≈ 0.5; (4) 293 K; (5) 330 K; (6) sample (5) is treated with hydrogen at 453 K.

sible; during the desorption of carbon monoxide (evacuation at 100-170 K) a back transition $\nu_{M-H'} \rightarrow \nu_{M-H}$ occurs, e.g. ν_{Hf-H} (1670 cm^{-1}) $\rightleftarrows$ $\nu_{Hf-H'}$ (1525,1635 cm^{-1}). Note also that the adsorption of small amounts of CO (less than a monolayer) runs first on Ti, Zr and Hf

TABLE 2. IR data on the states of CO during its interaction with benzyl and hydride compounds of Ti,Zr and Hf

No	Active group in a surface compound	Absorption bands of CO (cm^{-1})	
		coordination	insertion
1	Ti - CH_2Ph	2170	1490, 1635
2	Zr - CH_2Ph	2170	1510, 1620
3	Hf - CH_2Ph	2175	1510, 1625
4	Ti - H	1970, 2040, 2060 2080, 2110	1535
5	Zr - H	2020, 2050, 2075, 2090	1535
6	Hf - H	2035, 2065, 2080	1490, 1540, 1620

hydrides (adsorption bands within 2000-2100 cm^{-1}) and then on surface centers of silica (2135, 2155 cm^{-1}). When the samples are gradually evacuated, the absorption bands disappear in a reverse order. This indicates that the interaction of CO with the surface Ti, Zr and Hf hydrides is stronger than with the surface centers of silica. New bands in the region 1500-1650 cm^{-1} cannot be assigned to ν_{CO} in the surface formyl compounds of Ti, Zr and Hf. It will be shown below that these compounds are formed only at higher temperatures of the interaction and have different absorption bands.

Thus, during the interaction of CO with surface hydrides of Ti, Zr and Hf at 100-170 K a reversible adsorption of CO occurs on the surface hydrides; as a result, the absorption bands shift from ν_{M-H} towards lower frequencies.

A 5-10 min treatment of the $(Si-O)_{n}M(H)_{4-n}$ systems with carbon monoxide, at 293-393 K followed by evacuation of the samples at the same temperatures, leads to the disappearance of absorption bands, ν_{M-H}, and appearance of new bands in the range 1490-1620 cm^{-1} (Fig. 2, Tables 1,2). The absorption bands from the adsorbed carbon monoxide at 2000-2200 cm^{-1} are also abolished. Such changes in the IR spectra suggest that carbon monoxide is inserted into the M-H bonds to give surface formyl compounds, characterized by the absorption bands in the range 1490-1620 cm^{-1}. This supposition is supported by the growth of intensities of ν_{C-H} in the range 2800-3000 cm^{-1} (Fig. 2), which indeed must take place during CO insertion into the M-H bonds accompanied by the formation of new C-H bonds.

For soluble organometallic complexes of Zr and Hf the insertion of CO into the M-H bonds is observed also at temperatures above 220 K (Refs. 7,8). In this case either a complex dimer product is formed

$$(C_5Me_5)_2ZrH_2 + 3CO \rightarrow (C_5Me_5)_2Zr\begin{matrix}\diagup O \diagdown \\ \diagdown H \quad H \diagup\end{matrix}\overset{H}{\overset{|}{C}}=C-O-\overset{H}{\overset{|}{Zr}}(C_5Me_5)_2 \qquad (4)$$

or disproportionation of the reaction products takes place

$$2(C_5Me_5)_2ZrH_2 + 3CO \rightarrow (C_5Me_5)_2Zr(H)(OCH_3) + (C_5Me_5)_2Zr(CO)_2. \qquad (5)$$

Thus, the interaction of CO with surface Ti, Zr and Hf hydride compounds like with surface Ti, Zr and

Hf alkyl compounds, proceeds by a two-stage mechanism. At low temperatures a reversible adsorption of CO on the Ti, Zr and Hf ions is observed. On raising the temperature ($\geqslant$ 293 K) CO is inserted into the M-H bonds

$$L_xM\text{-}H \xrightleftharpoons{CO,\ 100\text{-}170\ K} L_xM\begin{matrix}\diagup CO \\ \diagdown H\end{matrix} \xrightarrow[>293\ K]{} L_xM\begin{matrix}\diagup C\text{-}H \\ \nwarrow \| \\ O\end{matrix} \quad . \qquad (6)$$

Hydrolysis of the surface formyl compounds leads to methanol formation.

When treating surface formyl compounds, formed in the system $Hf(CH_2Ph)_4/SiO_2/H_2/CO$, with H_2 at 453 K surface hydrides of hafnium are regenerated (Fig. 2, 6). In this case the absorption bands 1490, 1540 and 1620 cm^{-1} assigned to surface formyl compounds of Hf decrease, and the 1670 cm^{-1} absorption band from ν_{Hf-H} appeares. These data indicate the possibility of occurrence of reaction (7).

$$L_xHf\begin{matrix}\diagup C \diagdown \\ \| \\ O\end{matrix} H \xrightarrow[453\ K]{H_2} L_xHf\text{-}H \qquad (7)$$

The evacuation of the sample at 453 K after CO adsorption (without treatment with hydrogen) does not lead to reappearance of surface hydrides of hafnium.

A comparison of our data on the interaction of CO with hydride and alkyl compounds of Ti, Zr and Hf, anchored on silica, with the data for their individual organometallic analogs (Refs. 1-8) leads us to the following conclusions.

In both cases the interaction of CO with hydride and alkyl complexes proceeds by a two-stage mechanism through the step of adsorption, followed by the insertion of carbon monoxide into the M-C and M-H bonds, individual and anchored Ti, Zr and Hf complexes having similar reactivities with respect to CO.

The differences between individual and anchored complexes of Ti, Zr and Hf during their interaction with CO are related to that the complexes in solution may form dimers due to interaction of initial complexes with products of their reaction with CO. These reactions are improbable for anchored complexes because of their rigid fixation on the silica surface. Unlike systems in solution, anchored complexes have several states on the support surface that differ in ligand environment. As a result, the interaction of anchored complexes, especially of hydride compounds, with carbon monoxide produces several forms of adsorbed carbon monoxide.

ACKNOWLEDGEMENT. The authors are thankful to Prof. Thiele who has kindly donated the $M(CH_2C_6H_5)_4$ samples.

REFERENCES

1. C. Masters, Adv.Organomet.Catal., 17, 61 (1978).
2. G. Fachinetti, C. Floriani, J.Organomet.Chem., 71, C5 (1974).
3. G. Fachinetti, G. Fochi, C. Floriani, J.Chem.Soc. Dalton Trans., 1946 (1977).
4. A. Roder, K.-H. Thiele, G. Palyi, L. Marko, J.Organomet.Chem., 199, C31 (1980).
5. G. Fachinetti, C. Floriani, A. Posell, S. Pucci, J. Chem.Soc.Chem.Commun., 269 (1978).
6. J.H. Tueben, E.J.M. De Boer, A.H. Klazinga, E. Klei, J.Molec.Cat., 13, 107,(1981).
7. J. Manriquez, D. McAlister, R. Sanner, J. Bercaw, J. Am.Chem.Soc., 100, 2716 (1978).
8. J. Marsella, C. Cartis, J. Bercaw, K. Calton, J.Am. Chem.Soc., 102, 7244 (1980).
9. V. Giannini, U. Zucchini, E. Albizatti, J.Polym.Sci. Polym.Lett., 8, 405 (1970).
10. W. Bruser, K.-H. Thiele, P. Zdunneck, F. Brune, J. Organomet.Chem., 32, 335 (1971).
11. G.A. Nesterov, V.A. Zakharov, E.A. Paukshtis, E.N. Yurchenko, Kinet.Katal., 20, 429 (1979).
12. V.A. Zakharov, V.K. Dudchenko, E.A. Paukshtis, L.G. Karakchiev, Yu.I. Yermakov, J.Molec.Catal., 2, 421, (1977).
13. M.A. Makarova, G.A. Nesterov, V.A. Zakharov, V.V. Volkov, K.G. Myakishev, Vysokomol.Soed., 25B, 485 (1983).

14. G.A. Nesterov, E.A. Paukshtis, V.A. Zakharov, React. Kinet.Catal.Lett., 26, 357 (1984).

RECENT STUDIES OF OXIDE-BOUND ORGANORHODIUM COMPLEXES

Jeffrey Schwartz
Department of Chemistry, Princeton University
Princeton, New Jersey 08544 USA

ABSTRACT - Oxide-bound organorhodium complexes were prepared by reaction between tris(allyl)rhodium and hydroxyl group-containing metal oxides. Reactions between these oxide-supported organorhodium complexes and hydrogen have been discovered and have been described as processes involving heterolytic activation of that small molecule. In this concept, coordinatively unsaturated rhodium binds the dihydrogen molecule as a ligand which, in turn, labilizes the hydrogen as protons. Migration of a proton from a rhodium-hydride bond to another ligand in the coordination sphere of the metal gives rise to the observed hydrogenolysis products. Alkanes and hydrogen are conceptual analogs in that both classes of compounds can interact with a coordinatively unsaturated metal in the same way. Accordingly, we have found that alkanes are readily activated by coordinatively unsaturated rhodium complexes. In fact, two mechanisms exist for alkane activation, one of which directly parallels that which has been noted for hydrogen, and the other involves acidification of vicinal C-H bonds which occurs on electrophilic attack upon the alkane. Routes for the conversion of methane to other organic products using these rhodium complexes are described.

INTRODUCTION

"Homogeneous" catalysis, or catalysis by discrete, soluble transition metal complexes, is dominated by studies of the chemistry of such species in "soft"[1] ligand environments (e.g. consisting of phosphines, sulfides or carbonyls), and fascinating chemical processes, many of practical significance, have been developed through an understanding of their reactivity patterns.[2] Historically, one focus of research in "homogeneous" catalysis has been the elucidation of changes in reactivity of a complex which are effected by subtle modification of the ligand environment of the metallic center. It is of interest to us to determine how a gross change in this ligand environment affects the reactivity of a metal system bound to it; specifically, we aim to elucidate changes in rules of molecular reactivity which occur when the "soft" ligand environment of a transition metal complex is replaced by a "hard", oxygen-based one.[1] In this context, solid metallic oxides were chosen to provide this oxygen-based ligation.

The chemistry of oxide-bound metal complexes of the "early" transition metals has been extensively studied by several groups[3] and has been developed industrially, for example, to yield new types of olefin polymerization[3e] or oxidation[3c] catalysts; no such comparable body of work exists regarding the "late" metals. In that context, most work has focused on attachment of metal carbonyl clusters to oxide supports[4] and on the creation of particulate metals attached to these oxides.[5]

THE CHEMISTRY OF OXIDE-BOUND RHODIUM(ALLYL)COMPLEXES

We chose to focus our attention on the chemistry of oxide-ligated rhodium complexes, given the many interesting and important reactions which exist for this metal in "classical", "soft-ligand"-based homogeneous catalysis. In all of the reactions described below, we have used a variety of chemical procedures to characterize our complexes stoichiometrically to determine the nature of non-oxide ligation about the metal. We have also used several spectroscopic techniques to further characterize these species.

Tris(allyl)rhodium reacts[6] with hydroxyl groups of silica with evolution of 1 equiv of propylene to give Rh(III) species [SiO]Rh(allyl)$_2$, **1**. (We use the terminology "[SiO]" simply to indicate covalent bonding between the oxide and the metal, but details of this interaction are not yet known. We do not know, for example, if more than one oxygen atom interacts with the rhodium center at any time.) Chemical degradation of **1** yields 2 equiv of propylene, confirming the stoichiometric assignment as shown.

Hydrogen reacts slowly with **1** at room temperature. During this time 1 eq of propane is evolved. Hydrogen uptake measurements correlated with the amount of propane thus obtained and these data are consistent, stoichiometrically, with the formation of [SiO]Rh(allyl)H, **2**. The infrared spectrum of **2** shows strong absorptions attributable to the allyl and hydride ligands (ν_{Rh-H}

[Scheme 1]

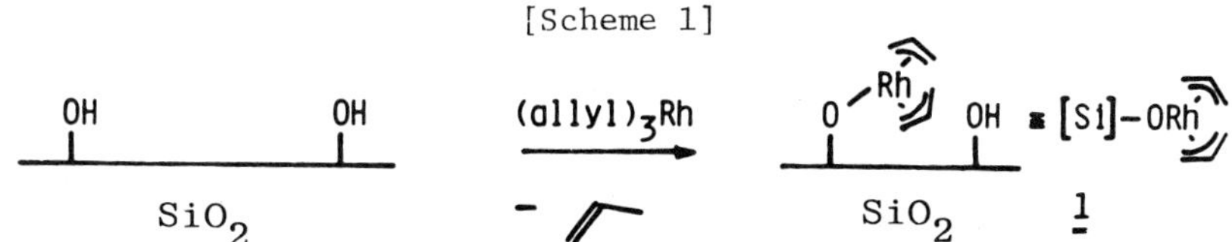

= 2010 cm^{-1}; weaker absorption at 1800 cm^{-1}). Both of these hydride ligand bands are observable in vacuo; the latter band is quite broad, and it is possible that a fraction of the rhodium centers in **2** are bridged by hydride ligands. The allyl group of **2** is not removed by hydrogenation (1 atm, room temp). The stoichiometry of **2** can be determined by chemical degradation using methanol (which liberates ca. 1 eq of propylene) and by treatment with methyl iodide (which liberates ca. 1 eq of methane). We note[7] that if hydrogenation is attempted before chemical deposition has occurred (i.e. before formation of 1 is complete) tris(allyl)rhodium adsorbed on the oxide rapidly reacts to give a black material which shows no infrared transmission, and which may be rhodium metal.

A series of XPS experiments was performed on **2** to corroborate its assignment as one of rhodium in an oxidized state relative to Rh metal. A single Rh species was found with an oxidation potential at 308.0 eV (Rh 3d 5/2 vs Si 2p). Thermolysis of this material at 400° followed by XPS analysis revealed that rhodium metal had been formed (oxidation potential at 307.25 eV).[9a] XPS data obtained for **2** falls outside a range of values (308.4 eV to 311.3 eV)[9b,c] which has been determined for Rh(III) salts, a range which overlaps with that for typical Rh(I) complexes (307.6 eV to 309.6 eV).[9b] XPS data for covalent organometallic complexes, however, must be interpreted with caution:[9c,9d] although **2** is formally a complex of Rh(III), actual positive charge build-up on the metal in this hydride-ligated species may be quite low.

Treatment[6] of **2** with excess gaseous HCl slowly leads to nearly complete removal of the allylic group, and infrared analysis of this material shows that a new rhodium hydride species (ν_{Rh-H} = 2110 cm^{-1}) has been obtained. This material reacts with cyclohexane followed by hydrogen to yield cyclohexyl chloride, and we propose this compound to be [SiO]Rh(H)Cl, 3.

Treatment[6] of 3 with hydrogen results in its conversion to a new species containing IR absorption bands at 2080 and 2020 cm^{-1}. This process can be followed readily by IR as can the reversion of 4 to 3 in the presence of gaseous HCl. Dihydrido species 4 reacts in the presence of excess olefin by uptake of 1 eq of the alkene to yield a new species [SiO]Rh(alkyl)H, 5, assigned as such on the basis of IR (ν_{Rh-H} = 2032 cm^{-1}) and by subsequent hydrogenolysis of 5 to 4 with concomitant liberation of 1 eq of the corresponding alkane.

[Scheme 2]

1 —H_2→ [Si]-ORh(allyl)H (2)

2 —olefin⇌(H_2; -alkane)— [Si]-ORh(allyl)-alkyl

2 —PMe_3→ [Si]-O-Rh(allyl)(PMe_3)$_2$H

2 —HCl→ [Si]—O-Rh(Cl)H (3)

3 —1) cyclohexene, 2) H_2→ 4 + cyclohexyl chloride

3 —PMe_3→ [Si]-ORh(Cl)(PMe_3)$_2$H

3 ⇌ 4 (H_2; HCl)

[Si]—ORh(alkyl)H (5) ⇌ (olefin; H_2; -alkane) [Si]—ORh(H)H (4) —PMe_3→ [Si]—O-$RhH_2(PMe_3)_3$ (6)

<u>A Unified Mechanistic Concept for Reactions of Supported Organorhodium Complexes with Hydrogen</u>

The simple conversion of 1 to 2 using H_2 is an example of activation of hydrogen by a complex of formally Rh(III). Although analogous phenomena have been noted in aqueous medium,[10] activation of hydrogen by Rh(III) species is not a common occurrence in the "soft" ligand chemistry of this metal in which hydrogen is activated by an "oxidative" process.[11] A similar process of hydrogen activation occurs in conversions of either 5 or 3 to give 4. The transformations, 1 → 2, 2 → 3, and 4 → 5 are also interesting from the point of view of the nature of the products which are formed. For example, until recently organorhodium(III) hydride species had been considered unstable with regard to rapid reductive elimination; there now exist many examples of such species[12] as isolated entities (and of organoiridium (III) congeners[13]). Another interesting observation is that complex 3 can be prepared in the presence of excess HCl; typically, metal hydride ligands are labile to acids, such as HCl, to yield dihydrogen and salts.

The stability of 1, 2, 3, 4 and 5 to "reductive elimination" may be due to the presence of oxide ligation stablization of Rh(III) relative to Rh(I) through stronger O-Rh bonding in the high-valent form. As a solid support, the oxide could also stabilize coordinatively unsaturated species such as 2, 3 and 4 against clusterification. That these rhodium species are unsaturated (or are highly labile toward deligation of μ-oxo centers of the oxide) can be demonstrated by noting their reactions with simple Lewis bases: 2, 3 and 4 react with small cone angle phosphines with stoichiometries predictable by normal rules of coordinative saturation. In other words, in these oxide-bound materials, the existence and number of "vacant" coordination sites can be determined or controlled in a predictable way as can be done in "homogeneous" organometallic chemistry (see Scheme 2).

The combination of high oxidation state and coordinative unsaturation suggests that these rhodium complexes could be electrophilic. We believe that this postulated electrophilic nature of coordinatively unsaturated Rh(III) species 2, 3 and 5 is responsible for their activation of hydrogen in ligand hydrogenolysis steps. Activation of H_2 has been observed by aqueous Rh(III)[10] and by numerous metal oxides.[17] In these cases, the concept of "heterolytic" activation of hydrogen has been developed: the coordinatively unsaturated metal center acts as a sink for "H^-", and a base in the environment of the metal center stabilizes the released proton. A comparable scheme has been suggested[18] to explain hydrogenolysis of $(\eta^5\text{-}C_5H_5)_2Zr(R)X$ species in which the 16 electron Zr(IV) complex is proposed to form a these systems cannot be explained convincingly. Complex **2** also catalyzes rapid (< 1 min) equilibration of a 50:50 H_2-D_2 mixture at room temperature.[20]

Activation of Alkanes

Simple alkanes are conceptual analogs of dihydrogen in that they both should be able to form a two center, three electron bond with an unsaturated metal center utilizing electron pairs in bonding σ orbitals; analogous intramolecular C-H bond coordination is now well established experimentally[21] and has been calculated to be facile intermolecularaly.[22] To probe interactions between alkanes and **2**, we studied their reaction in the presence of D_2 gas. This led to H-D exchange in the alkane, and after long reaction times an equilibrium mixture of

[Figure 1]

deuterated species was obtained.[20] It is well-known that many particulate metals will catalyze exchange between alkanes and deuterium gas;[21] for rhodium, this process usually results in an initially observable polyexchange of C-H bonds for C-D ones. In contrast, we note[7] sequential H-D exchange occurred when **2** was used to catalyze this process between ethane, for example, and D_2 (See Figure 1, entry 1). Decomposition of **2** (heating at 400° under one atmosphere of hydrogen for 4 hrs) gave a black species which showed no infrared absorptions attributable to Rh-H stretching modes. Not surprisingly this species showed catalytic activity for H-D exchange which was different from that of **2**. Noteworthy is the observation that whereas **2** gives rise to a distribution of deuterated ethanes relatively high in d_1-and low in d_6-ethane, this other black substance gives rise to a bimodal pattern showing a high degree of d_6-ethane relative to d_1 (see entry 2). We had noted that simple thermolysis of **2** yields rhodium metal (by XPS analysis) and, therefore, we compared this black material with rhodium on silica obtained by conventional molecular hydrogen complex followed by transfer of a "proton" from a Zr-H bond to the Zr-carbon bond. (A bona fide molecular hydrogen complex has now been prepared[19] by reaction between a coordinatively unsaturated Mo species and H_2). For hydrogenolysis of **2**, **3**, or **5**, coordination of molecular hydrogen followed by proton transfer to a ligand in the coordination sphere of the rhodium would yield the reaction products.

[Scheme 3]

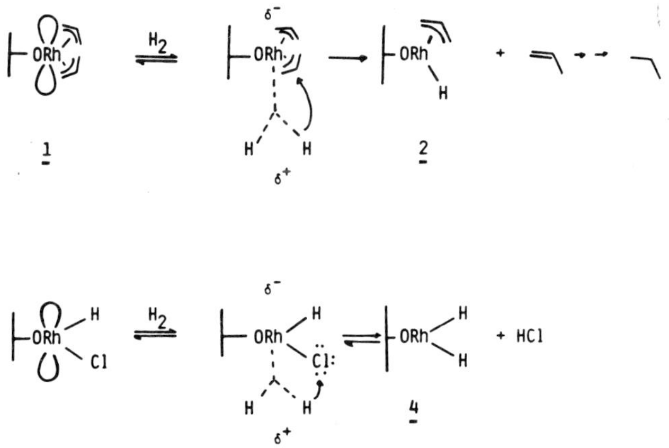

When **2** is exposed to D_2, H-D exchange between the atmosphere and residual hydroxyl groups of the silica support is observed.[20] An infrared spectrum of **2** taken after a 10 min exposure to deuterium gas (1 atm, 20°C) showed strong very broad absorption centered at 2600 cm^{-1} assigned to -OD vibrations consistent with the notion that in the presence of **2** H_2 or D_2 act as a source of H^+ or D^+. In contrast to rapid exchange of protonic OH groups with D_2, only slow exchange of the Rh-H group occurs with the

[Scheme 4]

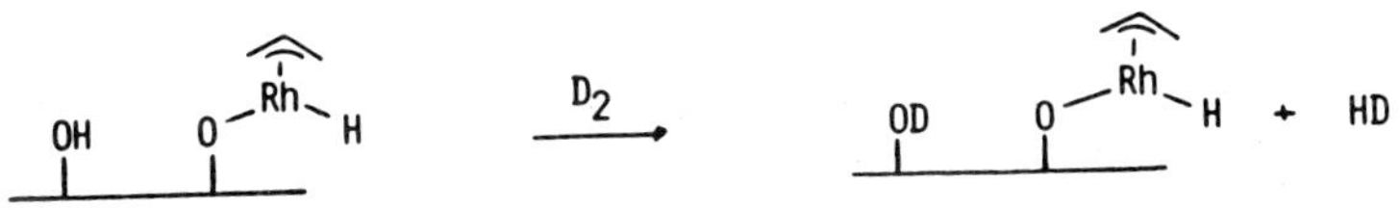

atmosphere. Until detailed structural information for **2** can be ascertained, the origin of slow exchange of the hydride ligand in

methods. Interestingly, rhodium on silica thus prepared behaves in a fashion similar to that noted for the pyrolysate (see entries 3 and 4). As well, we note that overloading a silica sample with tris(allyl)Rh relative to acidic -OH groups also gave a black material on treatment with hydrogen which catalyzed H-D exchange with ethane in a bimodal fashion analogous to either the pyrolysate or rhodium on silica prepared by conventional mechanisms; so did $Rh(allyl)_3/KBr$ (no chemical deposition, see entry 5). We believe, therefore, that <u>in the absence of chemical anchoring</u> to the oxide, hydrogenation of the allyl rhodium complex will give rhodium metal;[24] the bimodal distribution noted for H-D exchange catalyzed by rhodium metal obtained in the variety of ways described above may be related to the microscopic nature of active sites in particulate rhodium metal (particle size, number and nature of crystal faces, etc.), but in each case ethane-d_6 either predominates or is formed as a major component of the mixture (at low conversion) in contrast to that distribution resulting from use of **2**.

Figure 2. Deuterium incorporation into ethane at low conversion.

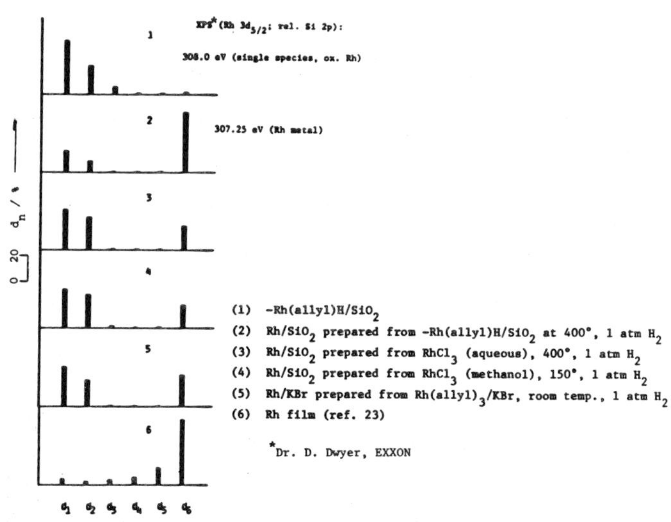

Mechanisms for H/D Exchange

The notion that **2** might act as an electrophilic site for activation of an alkane (or for dihydrogen) suggests that one possibility for H/D exchange involves hydride abstraction from the alkane. To probe this concept, we examined[7] rates for H/D exchange of a series of alkanes catalyzed by **2**. As shown in Table 1, steric factors are important in the H/D exchange process; however, alkanes amenable to activation by a hydride abstraction-based route (branched species containing tertiary C-H bonds) reacted more slowly in the H/D exchange process than did their linear analogs. A radical-based H/D exchange reaction should also be favored by the presence of tertiary C-H bonds. Therefore, it seems that both hydride and H• abstraction pathways can be ruled out based on these relative rate measurements. As shown in Table 1, rates for H/D exchange of methane and neopentane are remarkably slow compared with those noted for the series of linear or branched hydrocarbons. In fact, it seems that alkanes demonstrating rapid H/D exchange are those capable of forming olefins; those alkanes examined which are incapable of olefin formation undergo slow H/D exchange. To understand mechanisms whereby complex **2** catalyzes H/D exchange, we performed a series of stoichiometric reactions. Deuterated species [Si(OD)~SiO]Rh(allyl)D•(**7**) was obtained when **1** was treated with 1 atm D_2. Species **7** reacts slowly with methane to give CH_3D and

Table 1. Rates for D_2/alkane exchange catalyzed by (SiO)-Rh(allyl)H.

Alkane	Type*	Turnover (mol/h-mol Rh)
CH_4	α,β	~0.0008
C_2H_6	α,β	1.2
C_3H_8	α,β	5.9
n-C_4H_{10}	α,β	28.0
i-C_4H_{10}	α,β	6.8
n-C_5H_{12}	α,β	179.0
i-C_5H_{12}	α,β	37.0
neo-C_5H_{12}	α	~0.0008

Catalyst amount=50 mg (Rh=0.012 mol); alkane=2.0 mol; D_2=4.0 mol; reaction temperature=100°C. NOTE: Chain branching slows the reaction:

(1) —C⊕ by "H⁻" abstraction;

(2) —C⊙ by "H·" abstraction are NOT important processes here.

*Upon attack at a C_α- H bond, is there a C_β- H bond present?

[SiO]Rh(allyl)H. The slow rate of this stoichiometric reaction is consistent with slow catalytic H-D exchange rates noted in Table I for methane, but it is not adequate to explain those higher rates noted for most other alkanes. When this same material (containing 0.08 mmol Rh-D, 0.43 mmol O-D) was heated with ethane under similar conditions, 0.35 mmol deuterium incorporation, close to the equilibrium amount expected with regard to total deuterium content of 7 (deuteride and -OD sites) was detected in the ethane, as was a band in the infrared spectrum attributed to rhodium-hydride stretches. Given that exchange between ethane C-H bonds and protonic -OD centers is noted, we propose that two pathways exist for H-D exchange in ethane, but that only a single one exists for H-D exchange in methane. We have called these pathways "hydridic" and "protonic" and they are shown in Scheme 5. According to this scheme, activation of methane can be followed only by intramolecular ligand rearrangement to the hydridic center, analogous to our proposed mechanism for H_2-cleavage of Rh-allyl units. Coordination of an alkane such as ethane, however, provides another pathway for activation: electrophilic attack on an α C-H bond acidifies β C-H units and, therefore, deprotonation of the coordinated alkane can occur.

[Scheme 5]

TWO PATHWAYS EXIST FOR H/D EXCHANGE
("Protonic", fast; "Hydridic", slow)

"Protonic"

$7 \xrightarrow{CH_3CH_3}$ ... → alkane-d_n

"Hydridic"

$7 \xrightarrow{CH_3CH_3}$ ... → alkane-d_n

Activation of Methane

The notion that methane can be activated by an intramolecular ligand rearrangement route from a two-electron, three-centered intermediate analogous to that one proposed for activation of H_2 was utilized to develop a pathway for a chemically interesting conversion of that species to an organic product, and we have demonstrated that methane can replace hydrogen in both of the hydrogen activation steps shown in Scheme 2. For example, when **1** is treated with H_2 **2** results; similarly we found that when **1** was treated with methane (two days, 100°), a mixture of hydride complexes **2** and **8** were obtained. Propylene and a small amount of butane and butene were also detected. Methanolysis of an aliquot of the resulting organorhodium species gave propane, butene and a small amount of hexane. Since protonolysis of **1** is known to yield propane and hexane, the reactions shown in Scheme 6 can

[Scheme 6]

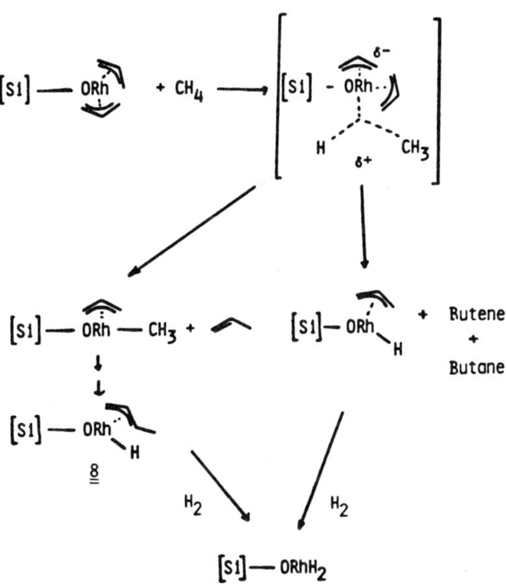

account for our data (methane activation was proven by use of $^{13}CH_4$). Hydrogen also reacts with hydridochloride complex 3 to give dihydride 4, and therefore 3 was treated with methane. This results in the formation of methyl chloride (0.85 equiv); activation of methane was confirmed by use of $^{13}CH_4$. Here, infrared analysis performed on rhodium-containing materials showed the presence of dihydride **4**, and broadened absorption centered at 2040 cm^{-1} suggested that **4** could be contaminated with another hydride species, perhaps a (methyl)rhodium(hydride) complex (**9**; for the analogous[SiO]Rh(H)Bu, ν_{Rh-H} = 2010 cm^{-1}). Indeed when this material (**9**) was treated with chlorine, methyl chloride was obtained (0.2 equiv). These observations can be explained by the sequences shown in Scheme 7.

[Scheme 7]

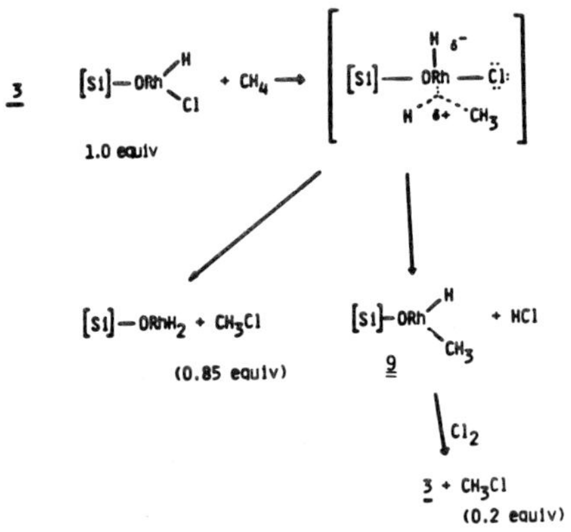

3 [Si]—ORh(H)(Cl) + CH_4 → [[Si]—ORh(H)(:Cl:)(H···CH_3)]

1.0 equiv

[Si]—$ORhH_2$ + CH_3Cl (0.85 equiv)

[Si]—ORh(H)(CH_3) + HCl

9

Cl_2

3 + CH_3Cl (0.2 equiv)

The stoichiometric rections in Scheme 6 suggested the possibility for a catalytic cycle for chlorination of methane using **3**. Indeed we found this to be the case.[26] Complex **2** also served as the precursor for a catalyst for chlorination of methane (trichloropropane was also obtained). Neither Rh metal prepared from thermolyzed **2** nor conventionally prepared rhodium/silica (from $RhCl_3 \cdot 3H_2O$) showed activity for methane chlorination. Although the mechanism for the metal-catalyzed reaction has not yet been proven, we believe that it parallels the steps noted in the stoichiometric cases shown in Scheme 6, and involves the metal acting as a Lewis acid. As shown in Scheme 6, the Lewis acidic metal center coordinates a C-H bond and competitive migration of either proton or the alkyl fragment to a ligand in the coordination sphere of metal generates the mixture of products obtained. (Although free radical-based chlorination of methane at low conversions also gives a distribution which is high in methyl chloride and low in carbon tetrachloride, relative rates for chlorination of C_1 compounds are different from those reported[27] for the free-radical case. For example, using 3 relative rates[25] for chlorination of CH_4, CH_3Cl, CH_2Cl_2, and CH_3Cl are in the order 1.0, 1.6, 0.5, 0.4. For free-radical chlorination, these relative rates are reported[27] to be 1.0, 1.8, 1.3, 0.6. In both cases, therefore, methyl chloride is more reactive than is methane; for the free radical route, so is methylene chloride. In contrast, in our case, although methyl chloride is more reactive than is methane, methylene chloride is less reactive than is CH_4.) Using 3 as a catalyst, methane conversion to chlorinated products is found to be first-order with respect to methane zero-order with regard to chlorine pressure, consistent with a mechanism in which alkane activation (either in the coordination step or in the intramolecular rearrangement step) is rate-determining, and in which chlorine attack upon rhodium complexes is relatively fast.

The mechanism for CH_4 activation shown in Scheme 7 does _not_ involve oxidation state changes for the metal in the course of the chlorination sequence. Rather, it simply involves the metal acting as a Lewis acid. Therefore, it seemed likely that other metallic species capable of acting as Lewis acids should also be able to catalyze methane chlorination; rates would vary according to the electrophilicity of the metal and the ability of the "methane complex" to undergo intramolecular rearrangement. In fact, platinum species in acidic media are well-known to activate alkanes,[28] and we found[26] that a variety of silica-bound halometal complexes could catalyze methane chlorination. The most active of these used molybdenum; zirconium(IV) complexes (d^0), and even an aluminum one ($[SiO]AlCl_2$) could suffice.

Oxide-Dependent Deposition Aggregation

Although we had noted that a variety of oxides could be used for chemical attachment of organorhodium complexes, we had not originally studied how variation in the oxide itself affects the structure or reactivity of complexes bound thereto, and deposition control was attained only by adjusting _total_ content of

TABLE 2. METAL-CATALYZED CHLORINATION OF METHANE

Catalyst Precursor	Metal Amount (mmol)	Time (h)	CH_3Cl (%)	CH_2Cl_2 (%)	$CHCl_3$ (%)	CCl_4 (%)	Total Conversion	Turnover Number $\times 10^4$ (mol/mol-metal/sec)
SiO_2	-	11	2.5	trace	trace	0	2.5	-
[Si]ORhHCl	0.024	11	12.6	1.1	trace	trace	13.7	1.7
[Si]OAlMe$_2$	0.065	11	5.3	0.8	trace	0.4	8.1	0.38
[Si]O$_2$Zr(allyl)$_2$	0.024	12	4.7	0.4	0.1	trace	5.2	0.60
[Si]O$_2$Zr(CHCMe$_3$)$_2$	0.033	11	5.0	trace	0.6	trace	6.2	0.57
*[Si]O$_2$Zr(CHCMe$_3$)$_2$	0.033	11	5.7	0.6	0.6	trace	6.9	0.63
[Si]O$_2$Mo(allyl)$_2$	0.029	10	20.9	8.2	2.3	trace	12.2	3.7
*[Si]O$_2$Mo(allyl)$_2$	0.029	10	23.3	17.6	12.5	2.3	55.7	6.4
[Si]O$_2$MoCl$_2$	0.029	12	13.8	8.8	trace	trace	15.6	1.5
[Si]O$_2$MoO$_2$	0.029	12	23.7	9.6	2.4	trace	35.7	3.4

* Catalyst treated with Cl_2 before adding CH_4/Cl_2 mixture.

Cl_2 = 4 mmol; CH_4 = 1.2 mmol; at 100°C

Catalyst amount = 100 mg

hydroxyl groups on the oxide. The more subtle notion was not considered that relative locations might vary of reactive hydroxyl groups on oxides of different structure, although this distribution of reactive sites would control deposition of one equivalent of an organometallic with regard to another. Silica and alumina are among the most commonly used oxides, in general, for chemical attachment to reactive organometallic complexes, and we found that an interesting oxide-dependent deposition phenomenon distinguishes these two materials: although for both we utilize samples containing a large excess of acidic hydroxyl groups relative to total rhodium deposited, we note that for silica, deposition of rhodium complexes occurs selectively to generate dimers; on alumina (of comparable total acidity) monomers are formed.

$Rh(C_3H_5)(CO)_2$ and SiO_2. Treatment of dried SiO_2 with a pentane solution of $Rh(C_3H_5)(CO)_2$ for 24h resulted in an off-white solid showing i.r. absorptions at 2107(mw), 2091(s), and 2033s cm^{-1}. The similarity of this spectrum to that observed for $[Rh(CO)_2Cl]_2$ suggested a possible dimeric structure. A sample of ^{13}CO enriched $[Rh(CO)_2Cl]_2$ was prepared, examined by i.r. and converted into $Rh(C_3H_5)(CO)_2$. This sample was used to prepare the SiO_2 supported carbonyl **10** (Fig. 3). Comparison of the supported sample with that of ^{13}CO-enriched $[Rh(CO)_2Cl]_2$ and $Rh(C_3H_5)(CO)_2$ clearly shows the monomeric nature nature of the (allyl)dicarbonyl complex (Fig. 4) and the dimeric nature of sup-

[Figure 4]

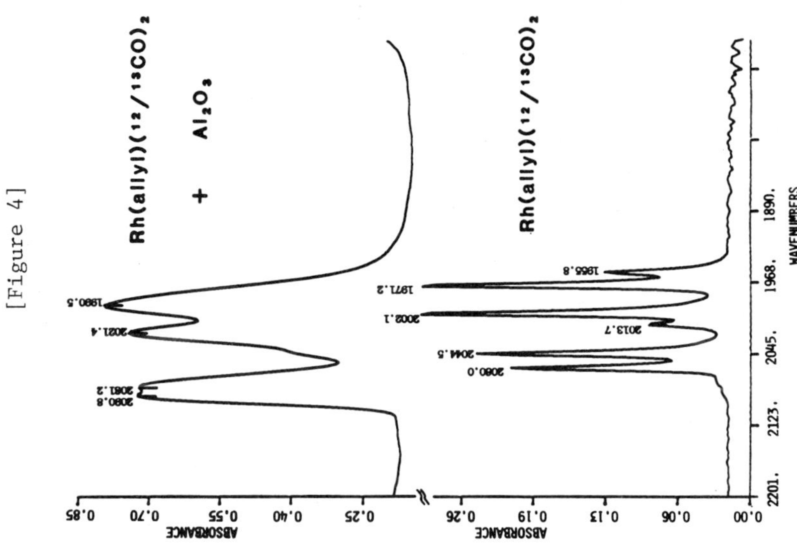

[Figure 3]

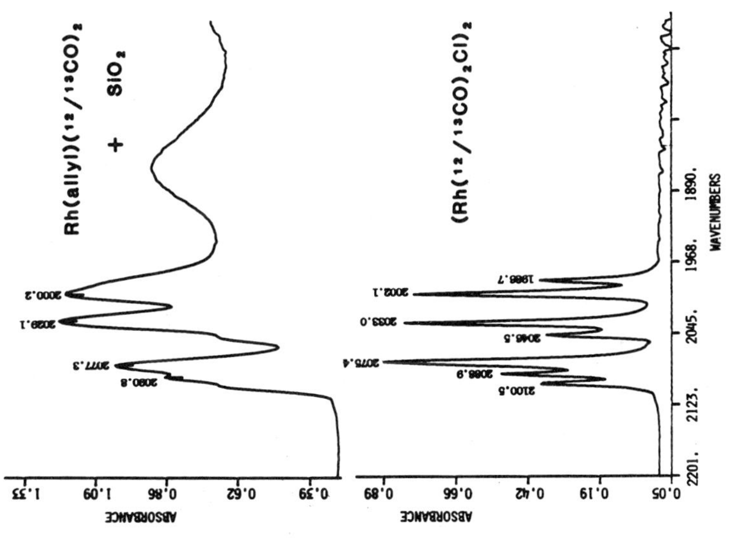

ported carbonyl complex **10**. The good agreement of the Fig. 3 spectra indicates that the surface species is a close structural analogue to $[Rh(CO)_2Cl]_2$.

$Rh(allyl)_2$-SiO_2 has been shown to react with CO generating a supported carbonyl complex and 1,5 hexadiene. For comparison purposes a sample of $Rh(allyl)_2$-SiO_2 was treated with the same $^{12}CO/^{13}CO$ mixture; here, too, a <u>dimeric</u> structure was indicated for **10** (Fig. 5).

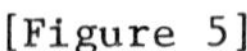
[Figure 5]

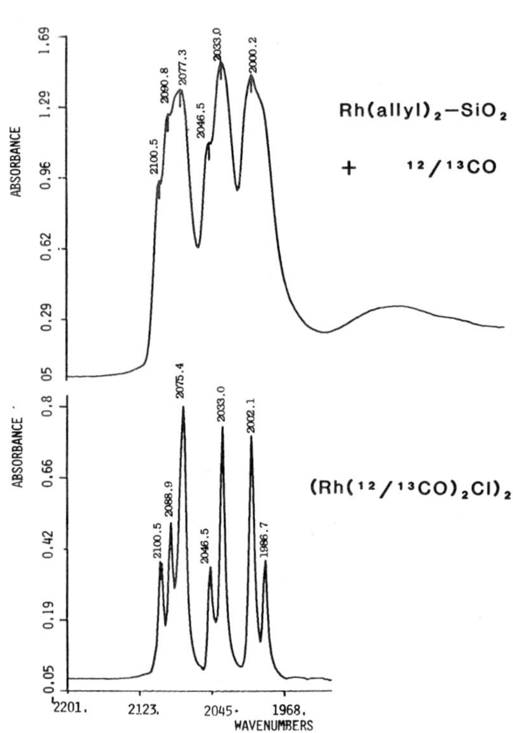

<u>$Rh(C_3H_5)(CO)_2$ and Al_2O_3</u>. Treatment of dried Al_2O_3 with a pentane solution of $Rh(C_3H_5)(CO)_2$ for 24h resulted in a pale yellow solid showing i.r. absorptions at 2091s and 2021s cm^{-1}. The same ^{13}CO enriched sample of $Rh(C_3H_5)(CO)_2$ used for deposition onto SiO_2 gave an alumina-supported product showing a relative intensity pattern analogous to that of its <u>monomeric</u> (allyl)dicarbonyl precursor (Fig. 4). An overall shift to higher frequency due to

the basic nature of the support was detected. The observation of a monomeric dicarbonyl on this support is in agreement with the product derived from rhodium carbonyl cluster precursors.

HOMOGENEOUS OR HETEROGENEOUS?

Throughout this discussion little attention has been paid to the relegation of our oxide-bound complexes to either the classical "homogeneous" or "heterogeneous" categories. Conventionally, the criterion of solubility is that which is used to categorize classical "homogenous" and "heterogenous" systems, and these classical contexts in fact refer to reactions from the perspective of the chemist, i.e., the presence of catalyst and substrate in one or more phases. Another perspective on the reaction is that one from the point of view of the substrate molecule: does it encounter one type of active site, or does it encounter many? A better consideration for defining a catalyst as "homogeneous" therefore may be that one relating to its possessing only a single type of active site, and in fact, we prefer to describe these species as "homogeneous, yet insoluble" entities.

A "Marriage of Convenience?"

Does the use of an oxide support simply constitute a "marriage of convenience?" Such "marriages" exist when catalyst species are reversibly bound to an insoluble support: the catalyst species perform their role in solution and are removed from the reaction medium upon reattachment to that support.[29] The oxide-bound complexes described herein do not fit this "definition" of a "marriage of convenience" in that they perform their role chemically bound to the oxide; they are complexes in an oxygen-based ligand environment:

- This environment apparently stabilizes the metal in its "harder" or higher-valent form.
- The oxide ligand can immobilize bound complexes (and facilitate recovery from a reaction medium) and inhibit clusterification; this can be used to advantage, for example, for olefin[6,30] or, especially, arene[30,31] hydrogenation.
- A high-valent complex in a coordinatively unsaturated

environment is apparently electrophilic; this, in turn, can be put to advantage in the context of alkane activation.

To create an "ensemble catalyst" by attaching an active metal complex to a support which itself is structurally modifiable is an especially appealing concept. As in any good "marriage" synergy, here the interplay between reactivity patterns for the bound metal complex and physical properties of the support, can be exploited to maximize reactivity or selectivity for the couple in ways not possible to accomplish for the partners acting alone.[32] It is perhaps this aspect of oxide-bound metal complexes which holds the greater allure for future research endeavors in this field.

ACKNOWLEDGEMENTS. The author wishes to acknowledge the efforts of Michael Ward, Tai-Nang Huang, Nobu Kitajima, Greg McNulty, Barry Posin, Kevin Cannon, and Eric Fisher for elucidating the chemistry of oxide-bound rhodium complexes. He also wishes to thank the National Science Foundation, Celanese Corporation, ARCO and Chevron Oil for financial support of research in this area, Dr. Dan Dwyer of Exxon for performing XPS analyses, and Professor Kurt Mislow for the use of his NMR Line Shape Program.

REFERENCES AND NOTES

(1) For a review of "hard" and "soft" character see: R. G. Pearson, "Hard and Soft Acids and Bases"; Dowden, Hutchinson and Ross: Stroudsburg, Pa., 1973.

(2) For example, see: G. W. Parshall, "Homogeneous Catalysis", Wiley: New York, 1980.

(3) (a) D. G. H. Ballard, Adv. Catal. **1973**, 23, 263; (b) V. A. Zakharov, V. K. Dudchenko, E. A. Paukshtis, L. G. Karakchiev and Yu.I. Yermakov, J. Mol. Cat. **1977**, 2, 421; (c) Yu.I. Yermakov and B. N. Kuznetsov, J. Mol. Catal. **1980**, 9, 13; (d) Yu.I. Yermakov, B. N. Kuznetsov and V. Zakharov, "Catalysis by Supported Complexes", Elsevier: Amsterdam, 1981; (e) Y. Iwasawa, M. Yamagishi and S. Ogasawara, Chem. Commun. **1980**, 871; (f) J. P. Candlin and H. Thomas, Adv. Chem. Ser. **1974**, 132, 212; (g) F. J. Karol, G. L. Karapinka, C. Wu, A. W. Dow, R. N. Johnson and W. L. Carrick, J. Poly. Sci. **1972**, A1, 10, 2621; G. L. Karapinka, US Pat. 3709853.

(4) For example, see (a) J. Evans and G. S. McNulty, J. Chem. Soc. Dalton, **1984**, 1123; (b) R. Psaro, R. Ugo, G. M. Zanderighi, B. Besson, A. K. Smith and J. M. Basset, J. Organomet. Chem. **1981**, 213, 215.

(5) Y. Ryndin, R. Hicks, A. Bell and Yu. I. Yermakov, J. Catal. **1981**, 70, 287.

(6) M. D. Ward and J. Schwartz, J. Mol. Catal. **1981**, 11, 397.
(7) N. Kitajima and J. Schwartz, J. Mol. Catal. **1984**, 26, 165.
(8) D. Dwyer, personal communication.
(9) (a) This value is within the range reported for Rh metal of varying particle size: T. Huizinga, H. F. J. van't Blick, J. C. Vis and R. Prins, Surf. Sci. **1983**, 135, 580; (b) For a range of values for oxidized Rh species, as well as for Rh metal, see S. L. T. Anderson, K. L. Watters and R. F. Howe, J. Catal. **1979**, 59, 340; **1981**, 69, 212; (c) A. D. Hamer, D. G. Tisley and R. A. Walton, J. Chem. Soc., Dalton Trans. **1973**, 116; (d) J. A. Conner, "Handbook of X-Ray and UV Photoelectron Spectroscopy", Heyden: Philadelphia, 1977, p. 185.
(10) (a) J. Halpern and B. R. James, Can. J. Chem. **1966**, 44, 671; (b) B. R. James and G. L. Rempel, Can. J. Chem. **1966**, 44, 233.
(11) For example, see References 2, ch. 3.
(12) (a) W. D. Jones and F. J. Feher, Organometallics, **1983**, 2, 562; (b) R. A. Perlana and R. G. Bergman, Organometallics, **1984**, 3, 508.
(13) (a) A. H. Janowicz and R. G. Bergman, J. Am. Chem. Soc. **1982**, 104, 352; (b) J. K. Hoyano and W. A. G. Graham, J. Am. Chem. Soc. **1982**, 104, 3723.
(14) F. R. Jensen and B. Rickborn, "Electrophilic Substitution of Organomercurials:, McGraw-Hill: New York, N.Y., 1968.
(15) L. F. McGahey and F. R. Jensen, J. Am. Chem. Soc. **1979**, 101, 4397.
(16) J. A. Labinger, D. W. Hart, W. E. Siebert III and J. Schwartz, J. Am. Chem. Soc. **1975**, 97, 3851.
(17) (a) A. Ozaki, "Isotopic Studies of Heterogeneous Catalysis"; Kodansha: Tokyo, 1977; pp 13-15; (b) R. J. Kokes, A. L. Dent, C. C. Chang, L. T. Dixon, J. Am. Chem. Soc. **1972**, 94, 4429; (c) R. L. Burwell, Jr., G. I. Haller, K. C. Taylor, J. F. Read, Adv. Catal. **1970**, 20, 1.
(18) K. I. Gell, B. Posin, J. Schwartz and G. M. Williams, J. Am. Chem. Soc. **1982**, 104, 1846.
(19) G. J. Kubas, R. R. Ryan, B. I. Swanson, P. J. Vergamini and H. J. Wasserman, J. Am. Chem. Soc. **1984**, 106, 451.
(20) M. D. Ward and J. Schwartz, Organometallics, **1982**, 1, 1030.
(21) M. Brookhart and M. L. H. Green, J. Organometal. Chem. **1983**, 150, 395.
(22) N. Koga, S. Obara and K. Morokuma, J. Am. Chem. Soc. **1984**, 106, 4625.
(23) C. Kemball, Adv. Catal. **1959**, 11, 233.
(24) The reaction between a silica/(allyl)rhodium species and H_2 has been reported to give silica/metallic rhodium: H. C. Foley, S. J. DeCanio, K. D. Tau, K. J. Chao, J. H. Onuferko, C. Dybowski and B. C. Gates, J. Am. Chem. Soc. **1983**, 105, 3074. In striking contrast to **2** (which has a strong IR absorption at 2010 cm^{-1} [ν(Rh-H)], this other material shows no IR absorption in the region 2000-1050 cm^{-1} in the absence of excess H_2; therefore this material is fundamentally different from **2**, and conclusions of that study are not germane to a discussion of the chemistry of **2**. Formation of rhodium metal in this other case may, in fact, be due to incomplete chemical deposition of the organometallic onto the

oxide, or to overloading of the support relative to available OH groups. Deposition stoichiometry was not reported in this case.
(25) N. Kitajima and J. Schwartz, J. Am. Chem. Soc. **1984**, 106, 2220.
(26) N. Kitajima and J. Schwartz, Proc. VIII ICCC, **1984**, **V**, 453.
(27) P. Goldfinger, G. Huybrechts and G. Martens, Trans. Faraday Soc. **1961**, 57, 210.
(28) N. F. Gol'dschleger, V. V. Eskova, A. E. Shilov and A. A. Shteinman, Russ. J. Phys. Chem. **1972**, 46, 785.
(29) For example, see B. C. Gates, J. Catal. **1975**, 40, 255.
(30) T.-N. Huang and J. Schwartz, J. Am. Chem. Soc. **1982**, 104, 5244.
(31) M. D. Ward and J. Schwartz, J. Am. Chem. Soc. **1981**, 103, 5253.
(32) For one such example, in the case of zeolite-encapsulated catalyst complexes, see reference 30 and T.-N. Huang, J. Schwartz and N. Kitajima, J. Mol. Cat. **1984**, 22, 389.

MECHANISM OF ANCHORING Pt(II) AND Pt(IV) ACIDOCOMPLEXES ON γ-Al_2O_3 MODIFIED BY MINERAL ACIDS

T.V. Zymbal, L.Ya. Alt, V.P. Doronin and V.K. Duplyakin
Institute of Catalysis, Omsk 664029, USSR

ABSTRACT - Adsorption mechanism of Pt(II) and Pt(IV) chloride complexes on γ-Al_2O_3 modified by mineral acids is studied. As shown by ^{195}Pt NMR and IR spectroscopies, the adsorption on supports pretreated with HBr or HNCS follows predominantly the mechanism of ligand substitution. In this case the support surface has "bridging" ligands that are coordinatively bound to Al atoms. The adsorption on γ-Al_2O_3 modified by $HClO_4$ or HF occurs by the mechanism of ion exchange.

1. INTRODUCTION

Supported catalysts are widely used in industry. At present considerable interest is being shown in processes, which occur at anchoring the precursor of the active components on the support and at further thermal treatments because these steps are crucial for the formation of the active component and, hence, for the catalytic properties.

Strong adsorption of metal complexes on the surfaces of all supports usually occurs at the step of impregnation during preparation of supported catalysts, a typical example of which is Pt/Al_2O_3. The literature contains contradictory opinions concerning the mechanism of interaction between platinum complexes and alumina surface modified by mineral acids. Two main mechanisms (i) ion exchange between metal-containing compounds and surface groups of the support (Refs. 1,2) and (ii) ligand substitution in the inner coordination sphere of the metal complex (Ref. 3) - are usually considered. The hypothesis that dissolved aluminium ions of the support participate in adsorption is also discussed (Ref. 4).

Investigations devoted to H_2PtCl_6 adsorption on γ-Al_2O_3 do not contain direct experimental data on the composition of complexes in adsorption layers. Conclusions about the reaction mechanisms are made in these works on the basis of indirect evidence such as variations in pH of impregnating solution and in concentrations of noncoordinated ligands measured during adsorption.

We have made an attempt to elucidate the main mechanisms of interaction of Pt(II) and Pt(IV) complexes with the surface of γ-Al_2O_3 pretreated with mineral acids.

2. EXPERIMENTAL

γ-Al_2O_3 with a specific surface area of 240 m^2/g and a total pore volume of 0.65 cm^3/g was used as support. It was modified by 1 N aqueous solutions of HBr, HF, $HClO_4$ and a 0.45 N solution of HSCN. Platinum was supported from aqueous solutions of H_2PtCl_6 and H_2PtCl_4 by gradually adding them to the suspended alumina pretreated with mineral acids (support particle diameter ranged 0.25 to 0.5 mm). After supporting had been completed, the suspension was filtered; the precipitate was immediately washed with a small amount of water, and surface platinum complexes were desorbed using a 8 N solution of $HClO_4$.

The mechanism of the interaction of platinum complexes with ligands fixed on the surface was judged from the composition of complexes desorbed from the support. ^{195}Pt NMR was employed to analyze solutions obtained. Spectra were taken on a CXP-300 Bruker spectrometer at room temperature. For some samples diffuse reflectance spectra (M-40 spectrometer) and IR-spectra (IR-75 spectrometer Karl Zeiss Jena) were also registered.

An analysis of the initial H_2PtCl_6 solution has shown that about 85% mol. platinum is in the form of $[PtCl_6]^{2-}$ and 15% mol., in the form of $[PtCl_5(H_2O)]^-$.

3. RESULTS AND DISCUSSION

3.1. Interaction of Pt(IV) and Pt(II) Complexes With the Surface of γ-Al_2O_3 Treated With HBr

Preliminary halogenation of γ-Al_2O_3 is commonly used in preparation of Pt/Al_2O_3 catalysts to provide a uniform distribution of an active component over the support particle. As a result, the support adsorptive capacity can change substantially. This data is in conformity with the well-known idea (Refs. 1,5) that halogen acids compete with chloroplatinate ions for adsorption sites and, consequently, decrease the number of these sites for platinum, the mechanism of adsorption remaining unchanged.

Results of the study of the composition of platinum complexes in solution after desorption from the support surface containing 3.6 wt% Br obtained by NMR are listed in Table 1. In this experiment the support was kept in a H_2PtCl_6 impregnation solution (concentration 0.23 mol/l) for a period from 1 to 30 min. Formation of mixed chlorine-bromine complexes of platinum is evidence that the reaction of ligand substitution occurs on the support. The number of coordinated surface bromine ligands ranges from 1 to 3. The resulting mixed chlorine-bromine complexes desorbed from the support surface after a 1- and a 5-min contacting of the impregnation solution with the support have only a cis-structure (Ref. 6). It can be seen in Table 1 that the reaction of ligand substitution, $[PtCl_6]^{2-} + Br^-$, on the alumina sup-

TABLE 1. Content of platinum complexes in solution after desorption and in a residual solution.[a)]

	Adsorption time, min	Content of complexes, mol%				
		$[PtCl_6]^{2-}$	$[PtCl_5Br]^{2-}$	$[PtCl_4Br_2]^{2-}$	$[PtCl_3Br_3]^{2-}$	$[PtCl_2Br_4]^{2-}$
Solution after desorption	1	33	33	23	11	-
	5	32	36	23	9	-
	30	25	37	27	11	-
Residual solution	30	22	34	26	11	7

a) Amount of adsorbed Pt was 2.4% by weight.

port occurs during 1 min, and further the composition of mixed complexes changes slightly. A similar reaction in a homogeneous phase occurs rather slowly (for several days at the concentrations employed by us (Ref. 7)). An increase in reactivity of a bromine ion fixed on γ-Al_2O_3 by a factor of 10^4 may be caused by its polarization in the adsorption layer of the support.

After 30 min of contacting the impregnation solution with the support, mixed chlorine-bromine complexes appear in an equilibrium solution. This fact indicates a reversible adsorption. In this case, the complexes are in the form of both cis- and trans-isomers.

It was of interest to follow changes in the ligand-substitution process in the inner coordination sphere of the complex on gradual (dosage for 30 min) addition of H_2PtCl_6 to the impregnation solution (see Table 2).

As follows from Table 2 (sample 1), a decrease in the current concentration of the metal-complexon during desorption results in that no chloroplatinate ion is in the adsorption phase, in other words, all of the platinum is fixed on the surface by a coordination mechanism, an average statistical adsorption site having four bromine ions (in contrast to 1.7 at one-time addition of H_2PtCl_6).

On decreasing the content of bromine on the surface of the support after a 3 h calcination of the sample containing 3.5% Br in air at 400°C a chloroplatinate ion appears in the adsorption phase, and an average degree of ligand substitution falls to 1.6 (sample 2). The presence of a $[PtCl_6]^{2-}$ anion in the solution after desorption indicates that some of the platinum is fixed by the mechanism of ion exchange with the surface bromine ions.

The presence of only cis-isomers permits one to suggest the so-called "bridging" surface halides, which are located in the inner coordination sphere of platinum and also are coordinatively bound to the aluminium of the

TABLE 2. Content of platinum complexes in solution after desorption (regulated addition of H_2PtCl_6).

Sample	Content of Br, wt%	Content of adsorbed Pt, wt%	Content of Pt complexes, mol%					
			$[PtCl_6]^{2-}$	$[PtCl_5Br]^{2-}$	$[PtCl_4Br_2]^{2-}$	$[PtCl_3Br_3]^{2-}$	$[PtCl_2Br_4]^{2-}$	$[PtClBr_5]^{2-}$
1	3.5	2.0	-	-	-	35	38	27
2	1.6	1.9	16	32	31	21	-	-

support. When adsorbed on the support, these halides cannot participate in a cis-trans-isomerization because of inhibition of an intersphere exchange, occurrence of which has been proved by our experiments. On heating samples containing mixed chlorine-bromine complexes anchored on the support, trans-isomers appear in the adsorption phase after 2 h and 1 h, respectively.

From all of the above, one may conclude that the main (in some cases the only) mechanism of platinum fixation on halogenated supports is that of ligand substitution. Formation of complexes in the adsorbent phase is characterized by a high degree and a high rate of ligand substitution and is determined by a concentration ratio of the fixed ligand to the metal complex in solution.

It seemed interesting to compare the mechanism of adsorption of an octahedral complex, $[PtCl_6]^{2-}$, with that of interaction between a flat complex, $[PtCl_4]^{2-}$, and a halogenated γ-Al_2O_3 surface. For this purpose we used the above described method.

During H_2PtCl_4 adsorption onto the support, containing 3.6 wt% bromine, the ligands in the inner coordination sphere of the metal are substituted by surface bromide ions. Support for this comes from ^{195}Pt NMR, which ascertains the formation of mixed $[PtCl_{4-n}Br_n]^{2-}$ complexes (n=1÷4) by desorption in solution (see Table 3).

TABLE 3. Content of Pt(II) complexes after desorption (regulated addition of H_2PtCl_6; 2.2% Pt).

Content of Pt(II) complexes, mol.%				
$[PtCl_4]^{2-}$	$[PtCl_3Br]^{2-}$	$[PtCl_2Br_2]^{2-}$	$[PtClBr_3]^{2-}$	$[PtBr_4]^{2-}$
-	14	27	32	27

As can be seen in Table 3, all of the platinum is fixed on the surface by the mechanism of ligand substitution. The $[PtCl_2Br_2]^{2-}$ complex in the adsorbent phase is in the form of cis- and trans-isomers. This fact suggests that a flat $[PtCl_4]^{2-}$ complex can adsorb onto the support surface both by one edge of the square and by the whole plane of the square. The ratio of the isomers formed (trans:cis=1:1) differs significantly from an equilibrium ratio in the homogeneous phase (trans:cis= =1:2).

Thus, halogen acids do not compete with platinum complexes for adsorption sites. On the contrary, preliminary halogenation (HCl, HBr) produces adsorption sites of a new origin and, thus, provides platinum fixation predominantly via a coordination mechanism.

Additional support for the ligand-substitution mechanism is provided by diffuse reflectance spectra, which show bands at 280-300 nm corresponding to charge transfer and characteristic of mixed chlorine-bromine complexes of Pt(IV) (Ref. 8).

3.2. Ligand Substitution in the System "H_2PtF_6 - γ-Al_2O_3 Pretreated With HCl".

As is known, reactions of $[PtF_6]^{2-}$ anions with halides in solutions are thermodynamically allowed, but are very slow. The equilibrium of the reaction

$$[PtF_6]^{2-} + 6Cl^- \rightleftharpoons [PtCl_6]^{2-} + 6F^-$$

is strongly shifted to the right. Since substitution of the first ligand in $[PtF_6]^{2-}$ is the rate-determining step, formation of mixed fluorine-chlorine complexes of Pt(IV) in solution is impossible (Ref. 9). An UV-spectroscopy analysis of the solutions after desorption from the support has confirmed the absence of mixed complexes in the adsorbent phase, most of platinum being in the form of $[PtCl_6]^{2-}$. A noticeable acceleration of ligand substitution in the inner sphere of the complex ($10^5 \div 10^6$ times in comparison with the reaction in solution) was observed.

Evidently, both on the support and in the solution the rate-determining step is that of the substitution of the first ligand F^- by Cl^-, which is followed by a rapid substitution of the remainder F ligands due to diffusion of adsorbed complexes and ligands on the surface or, possibly, due to noncoordinated Cl^- ions emerging in the solution. As a result, $[PtCl_6]^{2-}$ is formed.

3.3. Interaction of H_2PtCl_6 With Thiocyanate Ions Fixed on γ-Al_2O_3

IR spectra of the samples prepared by impregnating γ-Al_2O_3 with HNCS or NH_4NCS solutions have absorption bands, ν_{CN}, at 2053 and 2100 cm^{-1}. The band at 2053 cm^{-1} is characteristic of the solvated anion $(NCS)^-$ (Ref. 10) that has an ionic bond with γ-Al_2O_3. A 2100 cm^{-1} band is assigned by us to a thiocyanate ion covalently bound to an aluminium atom of the support via a nitrogen atom. The literature has no evidence on the type of coordination of thiocyanate groups with aluminium atoms, nor it reports on the existence of any coordination bonding. Ascription of the 2100 cm^{-1} band to thiocyanate groups covalently bound to aluminium is in conformity with the ability of a "hard" acid (Al^{3+}) to coordinate to a "hard" base (thiocyanate group coordinated via a nitrogen atom) (Ref. 11).

IR data on the samples with thiocyanate ions fixed on γ-Al_2O_3 are shown in Table 4. During desorption from the acid (HNCS), a proportion of the thiocyanate groups coordinatively bound to the surface aluminium atoms is markedly greater than their proportion on the support treated with a solution of $NH_4(NCS)$ salt.

TABLE 4. IR data for thiocyanate ions fixed on γ-Al_2O_3

Sample	Compounds to be supported	Concentration of NCS on the surface, $\frac{mg\text{-}eq}{g}$	Frequencies, cm^{-1}	Intensity ratio of the bands D_{2100}/D_{2053}
1	H(NCS)	0.48	2108	1.0
			2053	
2	$NH_4(NCS)$	0.10	2100	0.6
			2053	

The interaction of H_2PtCl_6 with fixed thiocyanate ions gives rise to absorption bands at 2175-2180 (ν_1), 2120-2125 (ν_2) and 2050 cm^{-1} (ν_3). Following the literature data (Ref. 12), we have attributed absorption band ν_1 to bridging thiocyanate groups of the type $\geqslant$Al-N-C-S-Pt and ν_2 to terminal thiocyanate groups bound to the central platinum atom through a sulfur atom. The absorption band that corresponds to the terminal group of thiocyanate seems to be due to the interaction of $[PtCl_6]^{2-}$ with a solvated anion $(NCS)^-$, which produced ν_{CN}= 2050 cm^{-1} before the interaction with a platinum atom.

IR data for supports with various contents of platinum are shown in Table 5. At low platinum contents, bridging structures do not form on the support; the spectra show only the absorption bands corresponding to the terminal thiocyanate groups.

^{195}Pt NMR study of the products of desorption of surface complexes has shown that both $[PtCl_6]^{2-}$ and mixed complexes with the composition $[PtCl_{6-n}(NCS)_n]^{2-}$ are present in the adsorption phase, n = 1÷6 changing as a function of the surface concentration of fixed thiocyanate ions.

TABLE 5. IR data for anchored platinum complexes

Sample	Content of Pt, wt%	Absorption frequencies, ν_{CN}, cm^{-1}			Intensity ratio	
		ν_1	ν_2	ν_3	D_1/D_2	D_2/D_3
1	2.3	2176	2125	-	0.64÷0.75	-
2	1.34	-	2118	2055	-	2.82
3	0.41	-	2110	2049	-	0.89

Our results show that during $[PtCl_6]^{2-}$ adsorption on the surface of γ-Al_2O_3 modified by (NCS) both ligand substitution and ion exchange mechanisms can occur. Their contributions are determined by the concentrations of the fixed thiocyanate and the supported platinum. Note that an IR spectroscopic study of the $[PtCl_6]^{2-}$ + +NCS/ γ-Al_2O_3 system has provided direct evidence that "bridging" ligands exist during anchoring metal complexes on modified supports.

3.4. Adsorption of H_2PtCl_6 on Supports Pretreated With $HClO_4$ or HF

The use of perchloric or hydrofluoric acid to modify γ-Al_2O_3 brings about the situation where supported ligands cannot substitute chloride ions in the inner coordination sphere of the $[PtCl_6]^{2-}$ complex because of steric or thermodynamic factors.

When H_2PtCl_6 was adsorbed onto the support pretreated with ClO_4^- or F^-, the composition of complexes in the adsorption phase corresponded to that of the initial solution of chloroplatinic acid: 85% $[PtCl_6]^{2-}$and 15%

$[PtCl_5(H_2O)]^-$. From this it follows that anchoring occurs by an ion exchange mechanism. A gravimetric analysis with the use of diantipyrinepropyl methane has made it possible to determine the concentration of ClO_4^- ions that emerge in the solution during the course of platinum supporting. The concentration found suggests that the ion exchange on the surface occurs stoichiometrically with the substitution of two ClO_4^- ions by a $[PtCl_6]^{2-}$ anion.

Thus, in the case of fixation of Pt(II) and Pt(IV) complexes on the surface of γ-Al_2O_3 modified by mineral acids, the process follows either the mechanism of ligand substitution or that of ion exchange. The contribution of each mechanism can be controlled by selecting initial metal complexes and methods of support modification. Based on the knowledge of adsorption mechanisms, it is possible to carry out intentional synthesis of surface complexes at the step of catalyst preparation, which is important in practice. We feel that the principles of complex formation with the participation of surface ligands established in this work are of general character and can be used for purposeful synthesis of coordination sites of various nature and composition.

REFERENCES

1. B.B. Zharkov, G.N. Maslyanskii, N.E. Buyanova et al., Zh. Prikl. Khim., 45, 1434 (1972).
2. R.W. Maatman, P. Manaffy, Ph. Hoekstra and C. Addink, J. Catal., 23, 105 (1971).
3. J.C. Summers and S.A. Ausen, J. Catal., 52, 445 (1978).
4. E. Santacesaria, S. Carra and J. Adami, Ind. Eng. Chem. Prod. Res. Develop., 16, No 1, 41 (1977).
5. S. Sivasanker, A.V. Ramaswamy and P. Ratnasamy, in Preparation of Catalysis, II, Elsevier, Amsterdam (1978).
6. V.K. Duplyakin, V.P. Doronin, T.V. Zymbal, L.Ya. Alt and A.S. Belyi, Dokl. Akad. Nauk SSSR, 281, 89 (1985).
7. B.N. Peschevitskii, A.B. Belyaev, R.L. Schekotikhina

and G.D. Malchikov, in Kinetics of Ligand Substitution Reactions in Complexes of Platinum Metal Compounds with Gold, Reference Book (in Russian), Nauka, Novosibirsk (1974).
8. I.B. Lipnitskii and D.S. Umreiko, Zh. Prikl. Spektrosk., 11, 670 (1969).
9. V.A. Schipachev, Thesis, Institute of Neorg. Chemistry, Novosibirsk (1980).
10. Yu.A. Kharitonov, in Vibrational Spectra in Inorganic Chemistry (in Russian), Nauka, Moscow (1971).
11. M.G. Day and J. Selbin, in Theoretical Inorganic Chemistry, Reinhold Book Corp., N.Y.-Amsterdam-London (1971).
12. J. Chatt and L.A. Duncanson, Nature, 178, 997 (1956).

NATURE OF SURFACE COMPOUNDS OBTAINED BY INTERACTION OF RHENIUM AND MOLYBDENUM COMPLEXES WITH γ-ALUMINA AND THEIR ROLE IN CATALYSIS

F.K. Schmidt

A.A. Zhdanov State University, Irkutsk 664003, USSR

ABSTRACT - Experimental data on the composition and structure of surface compounds formed at supporting rhenium and molybdenum carbonyls and ammonium paramolybdate on γ-alumina, and on their transformations at thermal activation are reported. The role of OH-groups of γ-Al_2O_3 in the formation of the surface species is discussed. The data on the composition of the surface "subcarbonyls" were obtained.

1. INTRODUCTION

In recent years an ever increasing interest has been shown in the interaction of transition metal compounds with the support with a view to elucidate the mechanism of formation as well as the composition and the structure of catalytically active sites. This field of heterogeneous catalysis was stimulated, first of all, by the progress in catalysis by both homogeneous and supported metal complexes which has led to a deeper insight into the mechanism of catalysis (Ref. 1).

This contribution deals with the nature of surface compounds formed by the interaction of rhenium and molybdenum carbonyl complexes as well as ammonium paramolybdate with γ-alumina and their role in some catalytic reactions.

Naturally, the choice of these systems was not a

mere chance. Alumina-supported molybdenum catalysts belong to polyfunctional systems, therefore, regulation of their properties, first of all selectivity, requires knowledge of the composition and structure of surface compounds. Besides, alumina-molybdenum systems form the basis for preparing a number of industrial catalysts to be used in large-scale processes, such as hydrotreating of fuels.

The possibility of using different physical methods to study surface composition is an advantage of the catalytic systems based on metal carbonyls on mineral supports (Refs. 1,2). This class of catalysts is suitable for investigation of the nature of surface compounds and the influence of the state of the support hydroxyl groups on the transformations of surface complexes during thermal activation of catalysts.

2. $Re_2(CO)_1/Al_2O_3$ SYSTEM

2.1. Reaction of Rhenium Carbonyl Complexes with γ-Alumina (Refs. 2,3)

Formation of catalysts of olefin disproportionation during thermal activation of rhenium carbonyl complexes on mineral supports is followed by the evolution of carbon monoxide and hydrogen. Method of thermal desorption with a GLC control allowed us to establish that, unlike individual $Re_2(CO)_{10}$ which decomposes with a simultaneous loss of all carbonyl ligands, this compound on γ--alumina loses only three CO groups per one rhenium atom at 453 K. No elimination of carbon monoxide occurs on further heating up to 600 K; above this temperature all CO ligands are lost.

Similar investigations have been performed using the following supported complexes: $H_3Re_3(CO)_{12}$, $BrRe(CO)_5$, $PhSn[Re(CO)_5]_3$, $Ph_3SnRe(CO)_5$.

The temperatures of the beginning of CO evolution are close for three latter compounds containing a $Re(CO)_5$ fragment, being a little higher for the brominated complex. On heating $H_3Re_3(CO)_{12}/\gamma-Al_2O_3$ the elimination of carbon monoxide starts below 370 K, while at 450 K the complex loses two CO groups per each Re atom. Therefore, it is evident that whatever the initial structure and the composition of rhenium carbonyl complexes and their concentration on the surface, a stable compound with molecular formula $Re(CO)_2$ is formed on the surface.

At 450 K $Re_2(CO)_{10}/\gamma-Al_2O_3$ system provides evolution of gaseous hydrogen in an amount of ca. 1.07 H per each atom of Re. In this case, hydrogen evolution can only be due to the reaction of the complex with the hydroxyl groups of the support surface.

Trinuclear supported complex $H_3Re_3(CO)_{12}$ gives one hydrogen molecule per each atom of the rhenium supported at 450 K. This value is 1.6 to 1.7 H atoms per Re atom with the tin-containing complexes, while with $BrRe(CO)_5$ no hydrogen evolution is observed up to 450 K.

Mössbauer spectroscopy has been used to study the decomposition of $Ph_xSn[Re(CO)_5]_{4-x}$ on the surface of Al_2O_3. It has been established, that thermal treatment of these complexes results in a complete release of organyl substituents from the tin atom and the formation of its oxides (SnO_2 and SnO in a molar ratio of ca. 4:1). The formation of tin oxides is not caused by the oxidation of the complex with a molecular oxygen which can be presented in an inert gas as an impurity. This has been proved by the coincidence of Mössbauer spectra of the samples activated in helium atmosphere and in high vacuum.

Data on quantitative analysis of the products obtained on heating the supported rhenium complexes to 450 K are compiled in Table 1.

TABLE 1.

Initial complex	Mole ratios			Surface species
	CO:Re	H:Re	C_6H_6:Re	
$Re_2(CO)_{10}$	3	1	-	2 [$Re(CO)_2$ ads]
$H_3Re_3(CO)_{12}$	2	2	-	3 [$Re(CO)_2$ ads]
$BrRe(CO)_5$	3	-	-	[$Re(CO)_2$ ads]
$Ph_3SnRe(CO)_5$	3	1.6-1.7	3	[$Re(CO)_2$ ads], SnO_2 ; SnO

Based on these data the reactions proceeding on the support surface can be represented as follows:

$$Re_2(CO)_{10} + 2\ (-OH) \longrightarrow 2\ [-O-Re(CO)_2] + H_2 + 6\ CO$$

$$H_3Re_3(CO)_{12} + 3\ (-OH) \longrightarrow 3\ [-O-Re(CO)_2]\ 3\ H_2 + 6\ CO$$

$$BrRe(CO)_5 \longrightarrow [BrRe(CO)_2] + 3\ CO$$

$$Ph_3SnRe(CO)_5 \begin{cases} \xrightarrow{+5\ -OH} H_2 + [-O-Re(CO)_2] + SnO_2 + 3PhH + 3CO \\ \xrightarrow{+3\ -OH} [-O-Re(CO)_2] + SnO + 3PhH + 3CO \end{cases}$$

where (-OH) is a surface hydroxyl group.

An assumption of the existence of the second route on $Ph_3SnRe(CO)_5$ decomposition, which occurs on its adsorp tion on less hydroxylated parts of the surface, allows us to explain the reason of bivalent tin oxide formation as well as to understand non-stoichiometric amount of the hydrogen released, since no hydrogen is formed via this route.

2.2. Spectroscopic Study of $Re_2(CO)_{10}$ Interaction with the γ-Al_2O_3 Surface (Ref. 3)

IR-spectrum of the supported $Re_2(CO)_{10}$, recorded in the carbonyl ligand absorption region, is identical with that for $Re_2(CO)_{10}$ in paraffin solutions, which indicates the absence of chemical interaction between the carbonyl complex and the support at room temperature. On heating this system in inert atmosphere or in high vacuum above 400 K an absorption band (1986 cm^{-1}) of the initial complex is abolished, and two new bands appear at 2025 and 1905 cm^{-1}. This process is completed at 450 K to give an intermediate structure displaying two bands in the carbonyl region of the spectrum, which are stable up to 580 K. Further heating leads to the disappearance of 2025 and 1905 cm^{-1} bands, which proves that they belong to the same surface compound. Based on a quantitative analysis of the amount of CO formed, the structure ascribed to this form is $Re(CO)_2$. Activation of $Re_2(CO)_{10}$ supported on γ-alumina containing different amounts of fluorine and showing widely different acidity allows one to observe a shift of carbonyl bands in the spectrum towards higher frequencies with increasing acidity. This shift indicates formation of the chemical bond between rhenium and the support:

$$\underset{\diagup\ \ \ \diagdown}{\overset{F^-}{\overset{|}{Al}}}\overset{}{\underset{O}{}}\underset{\diagup\ \ \ \diagdown}{\overset{O-H^+}{\overset{|}{Al}}} + Re(CO)_5 \longrightarrow \underset{\diagup\ \ \ \diagdown}{\overset{F^-}{\overset{|}{Al}}}\underset{O}{}\underset{\diagup\ \ \ \diagdown}{\overset{O-Re(CO)_2}{\overset{|}{Al}}} + 3CO + H_2 .$$

An analysis of IR spectra of the alumina OH-groups before and after activation of the supported rhenium carbonyl does not prove the existence of a donor-acceptor interaction between the Re atom and hydroxyl groups (the formation of these linkages is followed by the shift of absorption bands of OH groups towards lower frequencies and/or their broadening). On the other hand, some decrease in the OH absorption band intensities has been registered after heating the $Re_2(CO)_{10}$ supported on γ-Al_2O_3 to 453 K. This observation is consistent with the hydrogen formation due to decomposition of hydroxyl groups according to the scheme suggested for the interaction of the complex with the support.

A study of the activation of the supported rhenium carbonyl by photoelectron spectroscopy has revealed a positive shift of the Re 4f 7/2 lines by 0.8 eV. Taking into account the value for the metallic rhenium (40.6 eV) as well as for Re_2O_7 (46.5 eV) a change of the formal charge on the rhenium atom by ca. 1 has been established. Most rhenium atoms have the same oxidation state, which is in good agreement with all previous conclusions about the formation of one "subcarbonyl" surface form.

IR spectroscopy investigation of the surface -O--$Re(CO)_2$ form has shown that:

(1) this form does not react with carbon monoxide in the temperature range from 270 to 570 K with the formation of species with higher content of CO ligands;

(2) it is inert with respect to oxygen up to 570 K, an irreversible disappearance of the bands at 2025 cm^{-1} and 1905 cm^{-1} being observed above this temperature;

(3) an exposure of the sample towards propylene for a long time does not lead to any noticeable changes in the spectrum.

Stability of the surface -O-$Re(CO)_2$ form and its inability to coordinate CO and C_3H_6 molecules have been explained by the formation of a trinuclear surface cluster

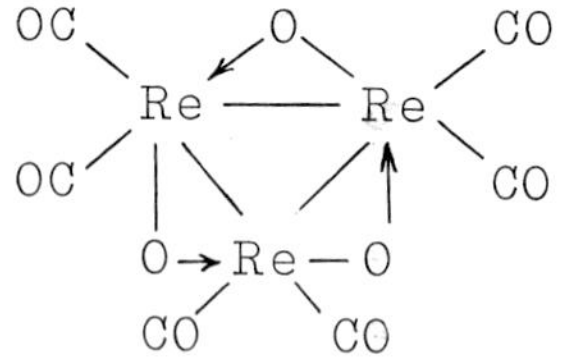

```
OC        O         CO
  \     ↙   \      /
   Re ——————— Re
  /  | \    / ↑    \
OC   |  \  /  |     CO
     O → Re — O
        /   \
      CO     CO
```

Support for this idea comes from the complete coincidence of IR spectra of the activated $Re_2(CO)_{10}/\gamma-Al_2O_3$

and $H_3Re_3(CO)_{12}/\gamma-Al_2O_3$, where no Re-Re bond rupture occurs for the latter complex.

The amount of propylene desorbed from the activated $Re_2(CO)_{10}/\gamma-Al_2O_3$ surface has been estimated, the propylene being displaced from the metal coordination sphere by carbon monoxide. Only small part (ca. 1% of the Re atoms) is able to coordinate C_3H_6. Mononuclear rhenium species may be responsible for the catalytic activity, though they could not be detected by physical methods because of their negligible content (less than 1%).

The kinetics of reaction was followed spectroscopically at different temperatures

$$Re_2(CO)_{10} \longrightarrow 2\,[-O-Re(CO)]_2 + 6\ CO.$$

The concentration of the starting material has been judged from optical density of the 2070 cm^{-1} band which is characteristic of physically adsorbed carbonyl. The first order has been established, the effective activation energy being 100.4 $\pm$ 4.4 kJ/mol.

3. $Mo(CO)_6/\gamma-Al_2O_3$ SYSTEM

As was found by different authors (Howe et al. (Refs. 5-7), Burwell and Brenner (Refs. 8,9), Kadushin and Matyshak (Refs. 10,11), thermal activation of $Mo(CO)_6/\gamma-Al_2O_3$ is followed by elimination of carbonyl ligands with the formation of the subcarbonyl molybdenum complexes on the surface. However, the results obtained by these authors, are contradictory in some aspects.

The reactivity of the γ-alumina surface is determined by the conditions of the pretreatment of the support (Ref. 12). In this connection, transformations of $Mo(CO)_6$ on $\gamma-Al_2O_3$ have been studied by IR and UV spectroscopies after a one-hour exposure of the samples in vacuum at

373, 473, 573 673 and 777 K. The low temperature treatment (up to 623 K) of the support has been found to provide a transformation of the supported $Mo(CO)_6$ having one strong absorption band at 1986 cm^{-1} to a stable surface subcarbonyl complex characterized by bands at 1725, 1790, 1935 and 2020 cm^{-1}. Howe et al. observed similar changes of IR spectra in somewhat different experimental conditions (Ref. 13). After Howe, we denote this substance as $Mo(CO)_x$. No bands that might be characteristic of the compounds intermediate between $Mo(CO)_6$ and $Mo(CO)_x$ have been observed.

At increase of the temperature of pretreatment of the support above 663 K a quite different spectral picture was observed. Supporting $Mo(CO)_6$ by sublimation in vacuum at 253 K results in the increase of absorbance at 1750-1810 cm^{-1} and the appearance of the new bands at 2000, 2030 and 2115 cm^{-1}; these bands gradually disappear as the temperature achieves 273 K. No evolution of carbon monoxide in appreciable amounts was detected. Later on, the increase in intensities of the bands at 1898 (weak), 1938 and 2070 cm^{-1} as well as a shoulder at 1870 cm^{-1} were observed. A slow heating to 333 K is followed by the growth of the shoulder at 1870 cm^{-1} and diminution of the bands at 1938 and 2070 cm^{-1}. The transformation of the surface species is completed at 373 K giving the substance characterized by three absorption bands at 1870, 1980 and 1598 cm^{-1} (weak), which is stable on heating to 463 K. Above this temperature all carbonyl bands simultaneously decrease and eventually vanish, which indicates that they belong to the same subcarbonyl structure.

Volumetric determinations of the amount of carbon monoxide eliminated allowed us to establish that by the moment of formation of a stable species the initial complex $Mo(CO)_6$ had lost about half of its carbonyl ligands, so the probable species formed is $Mo(CO)_3$.

Addition of oxygen to a cell containing the sample preheated to 373 K at $2.66 \cdot 10^3$ Pa at room temperature leads to the formation of a new species characterized by a weak band at 2022 cm^{-1}, which possibly, is due to the oxidation of molybdenum. A higher pressure of oxygen in the cell leads to the disappearance of the bands characteristic of the surface carbonyl.

On addition of carbon monoxide to the sample of $Mo(CO)_3 \cdot$ads at $6.65 \cdot 10^3$ Pa the absorption bands in IR spectrum of the latter slowly disappear; simultaneously new bands at 1938 and 2070 cm^{-1} as well as a shoulder at 1898 cm^{-1} appear. Such spectrum can be attributed to the intermediate surface species formed by addition of CO ligands to $Mo(CO)_3 \cdot$ ads. Weak bands at 2000, 2030 and 2115 cm^{-1} appear as the carbon monoxide pressure is raised to $4 \cdot 10^4$ Pa at room temperature. The process is completed after several hours with the formation of $Mo(CO)_6$ which is characterized by one absorption band in the carbonyl region, identical with that of the initial complex.

Taking into account the sequence of appearance of new bands in IR spectra on heating γ-alumina-supported $Mo(CO)_6$ as well as on formation of the initial $Mo(CO)_6$ from the surface $Mo(CO)_3 \cdot$ads species by CO chemisorption, the spectral picture observed can be interpreted as follows.

At the first stage, a low temperature transformation of $Mo(CO)_6$ from physically adsorbed to chemisorbed state seems to result due to donor-acceptor interaction of one carbonyl ligand with the Lewis acid centre of the surface, followed by distortion of the regular octahedron symmetry. In this case the adduct is formed via coordination of the lone pair of oxygen in the ligand, which leads to a significant decrease of stretching vibration frequencies for the coordinated CO group and increase of those for the ligands unbound with aprotic acids. Absorption bands at 2000, 2030 and 2115 cm^{-1} in $Mo(CO)_6$ chemically adsorbed on the acide centre are assigned to free CO ligands, while those in the 1750-1810 cm^{-1} range to CO groups coordinated to acid centres of different strength. This is supported by the fact that no carbon monoxide is evolved in noticeable amounts below 273 K.

Absorption bands at 1898, 1938 and 2070 cm^{-1} are characteristic of the molybdenum pentacarbonyl structure coordinated with an oxygen atom of the surface hydroxyl. A convincing evidence is the coincidence of these frequencies with those at 1895, 1940 and 2080 cm^{-1} reported for the complex with a similar structure - $Mo(CO)_5 \cdot$ hexanol (Ref. 5) which also has the C_4-symmetry ($2A_1 + E$). Because of the low stability of an intermediate molybdenum structure with four ligands its absorption band in IR spectrum could not be observed. As stated above, the bands at 1980, 1870 and 1598 cm^{-1} are characteristic of the final stable product - $Mo(CO)_3 \cdot$ ads. Its structure can be discussed based on the IR spectrum obtained.

As was stated by Kazusaca and Howe (Ref. 7), a low-frequency band of the activated supported carbonyl complex at 1600 cm^{-1} does not belong to carbonate-carboxylate structures. Therefore, a very strong decrease of the frequency can be explained assuming that part of surface complexes has bridging ligands attached to aprotic centres (I)

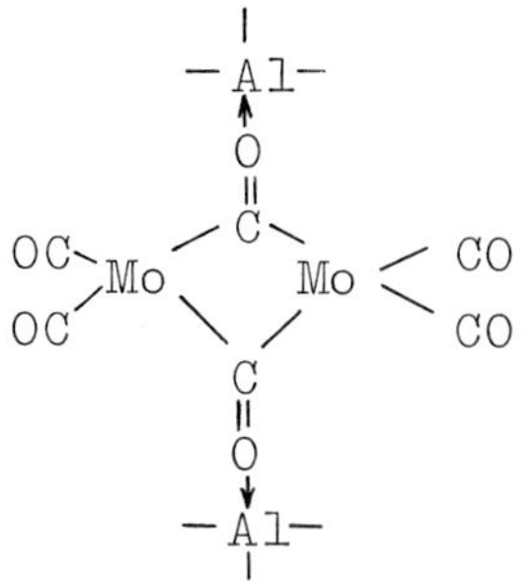

```
              |
            -Al-
              ↑
              O
              ‖
             C
OC        /     \         CO
   >Mo            Mo <
OC        \     /         CO
             C
              ‖
              O
              ↓
            -Al-
              |
```

(I)

The intensity of the 1598 cm^{-1} band in most spectra obtained is, however, rather small. So, a fraction of these structures on the surface can be estimated not more than a few percents of the total amount of "sub-carbonyl" forms of the transition metal.

Most surface $Mo(CO)_3 \cdot$ ads forms have, probably, three terminal CO ligands (II)

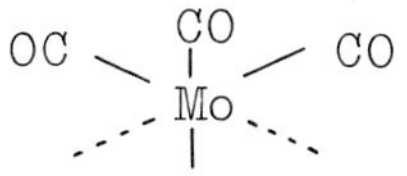

$$\begin{array}{ccccc} OC & \diagdown & \overset{CO}{\underset{Mo}{|}} & \diagup & CO \\ & \cdot\cdot\cdot & | & \cdot\cdot\cdot & \end{array} \qquad (II)$$

and are characterized by very intense bands at 1870 and 1980 cm^{-1}.

The results obtained lead us to conclude that transformation of $Mo(CO)_6$ on γ-alumina surface heated to or above 663 K is completed giving the $Mo(CO)_3 \cdot$ ads form at 353 K which can exist in the form of at least two different structural isomers (I and II) whose relative amount is determined by acidic properties of the support. On addition of even small amounts of propylene all bands ascribed to carbonyl ligands disappear. Hence, one may conclude that molybdenum species obtained on the support pretreated at high temperatures in vacuum are inactive in the reaction under investigation. However, it cannot be excluded that they serve as "precursors" of active centres, since passing helium through the sample for 30 min at room temperature gave an active catalyst.

CO addition at low pressure to the $Mo(CO)_3 \cdot$ ads sample prepared in high vacuum leads to formation of the initial $Mo(CO)_6$. However, at a subsequent evacuation, transformation of the carbonyl complex occurs via a different route to give a stable species with absorption bands at 1725, 1790, 1935 and 2020 cm^{-1}. Such species have also been obtained on the supports treated at temperatures below 623 K and were referred to as $Mo(CO)_x$.

The existence of the two routes of transformation

of the supported $Mo(CO)_6$ leading to $Mo(CO)_3$ or $Mo(CO)_x$ is caused by the difference in properties of the support surface. This conclusion has been made from the following experiment. γ-Alumina samples prepared at 750 K were treated for some minutes with carbon monoxide under $5 \cdot 10^4$ Pa at room temperature. $Mo(CO)_6$ was supported after evacuation, and its transformations during heating were followed using IR-spectroscopy. The transformation of the complex occured via the second route. Adsorption of CO is, probably, followed by blocking Lewis acid centres thus preventing formation of the Mo hexacarbonyl forms chemically adsorbed on aprotic centres at the first stage of the transformation. This effect causes the proceeding of the reaction via the second route yielding $Mo(CO)_x$.

It is known that conditions of the alumina thermal treatment determine the state of its surface and affect the amount and strength of acid centres. To clarify the effect of Lewis centres on formation of the $Mo(CO)_x$, the rate of this process has been measured on alumina samples pretreated at different temperatures (not higher than 623 K). The rate of the $Mo(CO)_x$ formation is found to be independent of temperature of pretreatment of the support.

A study of IR spectra with a synchronous pressure control in the spectrometer cell suggests that the $Mo(CO)_x$ structure proposed earlier corresponds also to the form $Mo(CO)_3$. The compound appears to have the following structure:

```
                 O
                 ||
OC             C             CO
   \         /   \         /
    Mo <           > Mo <                 (III)
   /         \   /         \
OC             C             CO
                 ||
                 O
```

When supporting $Mo(CO)_6$ on three γ-alumina samples with different concentrations of OH-groups, the highest activity in propylene metathesis has been shown by the sample with a partly hydroxylated surface, while no catalysis has been observed on the γ-alumina with a large surface concentration of hydroxyl groups (Refs. 8,14). Decomposition of $Mo(CO)_6$ on the fully dehydroxylated γ-alumina surface proceeds similarly to that of supported $Mo(CO)_6$, which is additional evidence that hydroxyl groups participate in stabilization of "subcarbonyl" forms. CO evolution from the $Mo(CO)_3 \cdot ads$ complex occurs at temperatures above 423 K and is followed by the formation of H_2 and small amounts of CH_4. Hydrogen formation indicates the oxidation of molybdenum, its main proportion being in the oxidation state from +2 at 573 K to +6 at 773 K (Refs. 8,14).

4. MoO_3/ γ-Al_2O_3 SYSTEM

This system is known as polyfunctional catalyst. Up to now, many questions concerning the nature of surface active centres, the mechanism of their formation and their role in particular reactions remain unclear. Suffice it to note that about ten models of the state of molybdenum oxide on alumina have been suggested in recent years. As follows from the analysis of literature data, the existing variety of views is the following:

(i) at a content of Mo of about $2-4 \cdot 10^{18}$ atoms/m^2 the surface compounds exist in a non-crystallized state and, therefore, the methods used do not provide unequivocal information about the molybdenum state on the surface;

(ii) the results of investigation are often described within the framework of the model of the uniform state of Mo in MoO_3/γ-Al_2O_3, while facts of nonuniformity of

Mo surface compounds being ignored; and

(iii) experimental data on the role of alumina in the catalyst structure formation are insufficient.

One of the approaches to study $MoO_3/\gamma-Al_2O_3$ system is based on the dissolving of Mo compounds from the support surface. By correctly selecting conditions of dissolving, it is possible to determine the content and the nature of various Mo compounds in $MoO_3/\gamma-Al_2O_3$ as well as the role of hydroxyl groups of Al_2O_3 in the formation of molybdenum surface compounds.

4.1. Solubility of Molybdenum Compounds of Molybdenum-Alumina Catalysts and their Identification (Ref. 15)

Let us consider the data on solubility of molybdenum compounds of molybdenum-alumina catalysts in H_2O and in 10 N aqueous ammonia. The solubility of $MoO_3/\gamma-Al_2O_3$ in water depends on experimental conditions. In static conditions the process of dissolving is complicated by the side-processes of redistribution and readsorption of the dissolved molybdenum anions. It was found that under flow conditions the dissolving of two types of Mo surface compounds does occur, the rate of dissolving of easily soluble W_s-form being twenty times as large as that of poorly soluble W_d-form.

The amount of Mo compounds in $MoO_3/\gamma-Al_2O_3$, which are insoluble in 10 N NH_4OH (S-form), increases with increasing the temperature of calcination, Mo concentration and the surface of the support. The samples for further investigation have been chosen based on the solubility data listed in Table 2.

An analysis of UV and IR spectra of A, B, C and D samples as well as of $\gamma-Al_2O_3$ indicates that molybdenum compounds of octahedral configuration (Mo(o)) are

TABLE 2. Characteristics of $MoO_3/\gamma-Al_2O_3$ complexes and methods of their preparation

Sample	MoO_3 contents, %	BET surface m^2/g	Amount of F^- adsorbed, mmol/g catalyst[a)]	Method of preparation
A (initial, W_s+W_d+M+ $+S$ forms of Mo)	10.7	200	-	Impregnation with ammonium paramolybdate solution, drying and calcination at 873 K, 3 h
B (W_d+M+S forms)	9.9	-	-	W_s-form extracted from sample A with further drying
C ($W_{d/2}+M+S$ forms)	9.5	-	-	W_d-form partly extracted from sample B with further drying
D (S-form)	2.6	-	-	A with W_s- and W_d forms extracted by 10 N NH_4OH. Dried.
	6.7	200	0.40	Supporting of MoO_4^{2-} in flow conditions (unreacted MoO_4^{2-} washed off from Al_2O_3 by flowing water)
	5.7	130	0.30	
	4.3	100	0.25	
	2.9	60	0.15	

a) The amount of basic OH groups on $\gamma-Al_2O_3$ as determined by the fluorine ion adsorption from solution.

soluble in water. Molybdenum compounds in both octahedral and tetrahedral (Mo(t)) configurations can dissolve in ammonia. A comparative analysis of Mo-O stretching vibrations in IR spectra has been performed for the samples A, B and C and model compounds such as MoO_3, $Al_2(MoO_4)_3$, K_2MoO_4, $K_2Mo_2O_7$, $K_2Mo_3O_{10}$; Al_2O_3 impregnated with ammonium salts of iso- and heteropolymolybdenum acids (30% MoO_3). The W_s-form of Mo has been found to be the isopolycompound of the octahedral Mo (945 cm^{-1} (Mo=O), 915 cm^{-1} (Mo-O-Mo), 667 cm^{-1} (Mo-O)). The spectrum of B_1 is characterized by absorption bands at 945-900, 850-800, 750-700 and 520 cm^{-1} which correspond to those in IR spectra of the individual $K_2Mo_2O_7$ and $K_2Mo_3O_{10}$. A comparison of B_1 and C_1 IR spectra suugests that the W_d-form can be related to polymeric Mo(o), that is to a part of molybdenum compounds of the type of alkali metal di- and trimolybdates. The intensity of the absorption bands at 945 and 900 cm^{-1} in IR spectrum of C_1 tends to fall with extraction of the W_d-form.

It was found that formation of the Mo(t) surface ions is due to the alumina OH-groups with ν = 3780 and 3730 cm^{-1}, while that of the Mo(o) surface ions is due to those with ν = 3780 cm^{-1}. OH groups are not involved in the formation of the Mo S-form.

4.2. The Principal Stages of the Formation of Molybdenum-Alumina Catalysts (Refs. 16,17)

For the impregnation stage the mechanism of adsorption of molybdate ions on the support surface is most important. As is seen from Table 2, the quantity of OH groups measured on various alumina samples from the experiments on F-exchange is proportional and approximately equal to the amount of the molybdate ions interacting with alumina during impregnation. According to literature data, at room temperature the fluorine ions exchange with the alumina OH groups which are characterized by high frequency bands in IR spectrum (Ref. 18). Thus, the following ion-exchange reaction proceeds during impregnation:

$$\overset{\mid}{\underset{\mid}{Al}} -OH + MoO_3OH^- \longrightarrow \overset{\mid}{\underset{\mid}{Al}} -O-MoO_2H + OH^-.$$

The results on the selective dissolving of molybdenum compounds from the calcined $MoO_3/\gamma-Al_2O_3$ show that the formation of the Mo(t) moiety is due to the OH-groups with $\nu = 3780$ and $3730\ cm^{-1}$. Therefore, the OH-groups with $\nu = 3730\ cm^{-1}$ become active during the calcination and are involved in the following reaction:

$$\overset{\mid}{\underset{\mid}{Al}} -O-MoO_2OH + HO-\overset{\mid}{\underset{\mid}{Al}} \xrightarrow{\text{calcination}} \overset{\mid}{\underset{\mid}{Al}} -O-\underset{O\ \ \ O}{\overset{}{Mo}}-O-\overset{\mid}{\underset{\mid}{Al}}$$

In the γ-alumina model suggested by Lippens, the OH groups with $\nu = 3730$ and $3780\ cm^{-1}$ are located close to each other thus forming chains. Such arrangement of OH groups on γ-alumina surface predetermines the formation of the molybdenum compounds with the structure similar to that of di- or trimolybdate of alkali metals, i.e. to the chain structure consisting of Mo(o) and Mo(t) or MoO_5. Proceeding from these premises a suggestion has been made that the formation of the molybdenum compounds with the structure of di- or trimolybdate of alkali metal includes the participation of Mo(t) surface ions.

The molybdenum isopolycompound (W_s-form) is formed on the γ-alumina acid-base pair whose existence is due to the OH-groups with $\nu = 3780$ and $3700\ cm^{-1}$. An assumption has been made that the ion-exchange adsorption of molybdate ions takes place during the impregnation. At the calcination stage, the OH groups act as acid thus providing the association reaction of the molybdate ions:

$$Mo_7O_{24}^{6-} + 4\ H_2O \rightleftharpoons 7\ MoO_4^{2-} + 8\ H^+.$$

Taking into account both literature and our data the

scheme of the formation of molybdenum-alumina catalysts has been suggested (see Fig. 1).

The OH groups with $\nu = 3780\ cm^{-1}$ are seen to be of exceptional importance in the $MoO_3/\gamma-Al_2O_3$ structure formation, the generation of the Mo surface compounds being dependent not only on the number of these OH groups but also on their arrangement on γ-alumina surface. According to literature data, an absorption band at $3780\ cm^{-1}$ is that of OH groups involved in the interaction with tetracoordinated aluminium atoms in $\gamma-Al_2O_3$ (Ref. 19).

The degree of disorder of Al_2O_3 is known to be determined primarily by the disorder in the arrangement of Al atoms in tetrahedral cavities of the γ-alumina lattice. The disagreements on the $MoO_3/\gamma-Al_2O_3$ structure are, probably, due to that the investigators employed different alumina samples for catalyst preparation.

4.3. Role of Molybdenum Compounds in the Formation of the Catalytically Active Sites (Ref. 20)

The influence of the W_s-form of molybdenum on the catalytic activity in cyclohexane dehydrogenation, heptane hydrocracking, thiophene hydrogenolysis, butadiene hydrogenation and propene metathesis has been studied.

The cyclohexane conversion has been found to decrease with decreasing W_s-form of Mo in the sample, being almost zero with a sample containing no W_s-form. Thus, molybdenum isopolycompounds form active sites of cyclohexane dehydrogenation. Under experimental conditions (the temperature of the reduction and of the reaction is 673 K) the Mo ions in the W_s-form and W_d-form are in the same oxidation states; however, the W_d-form is practically inactive. Based on these facts it has been concluded that the active sites for the cyclohexane dehydrogenation are formed from the Mo W_s-form, the activity of the catalyst being dependent both on the oxida-

1. IMPREGNATION

In solution:

$$Mo_7O_{24}^{6-} + 4\ H_2O \rightleftharpoons 7\ MoO_4^{2-} + 8\ H^+$$

$$MoO_4^{2-} + H^+ \rightleftharpoons MoO_3OH^-$$

On the support:

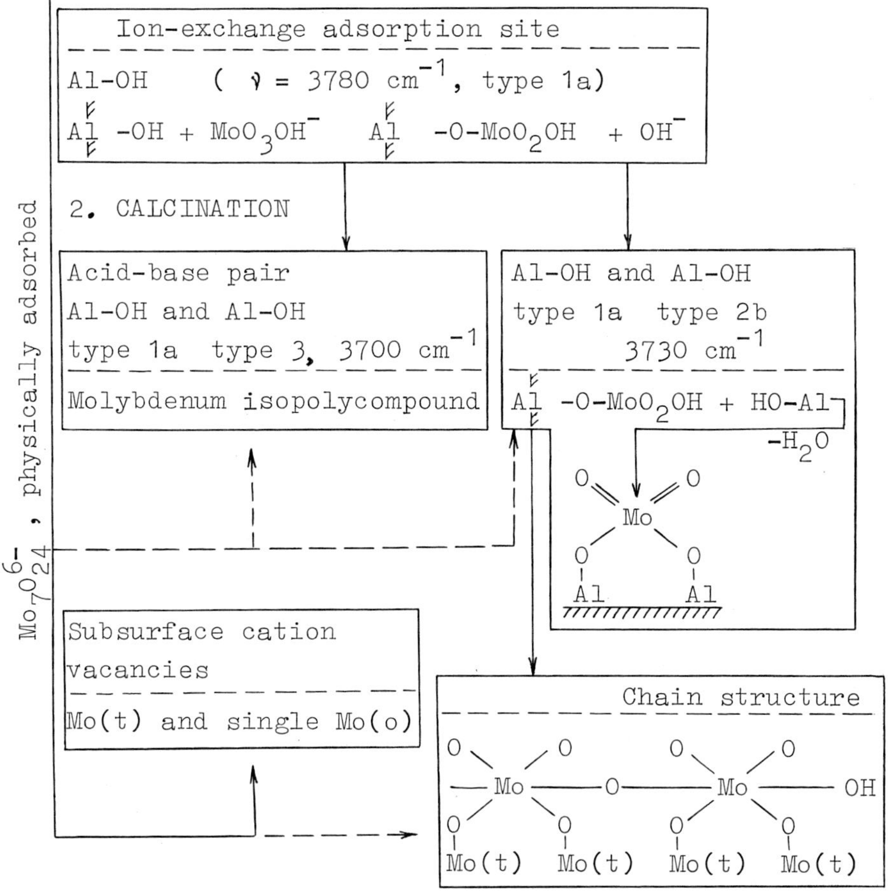

Ion-exchange adsorption site

Al-OH (ν = 3780 cm^{-1}, type 1a)

Al -OH + MoO_3OH^- Al -O-MoO_2OH + OH^-

2. CALCINATION

Acid-base pair
Al-OH and Al-OH
type 1a type 3, 3700 cm^{-1}

Molybdenum isopolycompound

Al-OH and Al-OH
type 1a type 2b
3730 cm^{-1}

Al -O-MoO_2OH + HO-Al

$-H_2O$

O O Mo O O Al Al

Subsurface cation vacancies

Mo(t) and single Mo(o)

Chain structure

O O O O
Mo —O— Mo — OH
O O O O
Mo(t) Mo(t) Mo(t) Mo(t)

$Mo_7O_{24}^{6-}$, physically adsorbed

Figure 1. Scheme of the formation of noncrystalline molybdenum compounds in MoO_3/γ-Al_2O_3 (Mo content in MoO_3/γ-Al_2O_3 = 2-4·10^{18} atoms/m^2)
— Established ways of formation; - - possible ways of formation (molybdate ions supplier)

tion state of the Mo ions and on their disposition in active sites.

There is a linear dependence between the amount of the oxygen adsorbed and the catalyst activity in thiophene hydrogenolysis, while in cyclohexane dehydrogenation this relation corresponds to an S-shape curve.

The oxygen absorption at room temperature is mainly due to the molybdenum compounds with the chain structure.

Therefore, the catalytically active sites for heptane hydrocracking and thiophene hydrogenolysis are formed from the molybdenum compounds with the chain structure. A linear relationship in the thiophene hydrogenolysis indicates that the other molybdenum compounds are also active in this reaction. The experiments showed the Mo S-form to be catalytically inactive. "Single" Mo(t) should also be considered as inactive because they do not undergo reduction under experimental conditions. Thus, the active sites in the thiophene hydrogenolysis are formed from the Mo(t) as a fragment of the chain structure.

The Mo W_s-form is a "poison" for the propene metathesi The active sites for this reaction are formed from the molybdenum compounds with the chain structure. One of the reasons for catalyst deactivation is a side reaction of polymerization.

REFERENCES

1. Mechanism of Catalysis. G.K. Boreskov and T.V. Andrushkevitch (Eds.), pp. 1,11, Nauka, Novosibirsk (1984).
2. O.L. Kosynskii, Thesis, Irkutsk State University, Irkutsk (1984).
3. O.L. Kosynskii, A.A. Kadushin, O.V. Krylov, F.K. Schmidt and N.E. Kolobova, Heterogeneous Catalysis, Proc. 4th Intern. Symp., part 1, pp. 403-409, Varna (1979).
4. A.A. Kadushin, O.L. Kosynskii and F.K. Schmidt, in Proc. 4th Soviet-French Seminar on Catalysis (1980).

5. R.T. Howe, D.E. Davidson and D.A. Whan, J. Chem. Soc., Faraday Trans. I, 68, 2266-2274 (1972).
6. R.T. Howe and J.K. Leith, J. Chem. Soc. Faraday Trans. I, 69, 1167-1177 (1973).
7. A. Kazusaca and R.T. Howe, J. Mol. Catal., 9, 183-198 (1980).
8. R.L. Burwell and A. Brenner, J. Mol. Catal., 1, 77-84 (1976).
9. A. Brenner and R.L. Burwell, J. Catal., 58, 353-363 (1978).
10. A.A. Kadushin, V.A. Matyshak, N.A. Kutyreva, A.V. Sklyarov and A.G. Vlasenko, Kinet. Katal., 23, 1167-1171 (1982).
11. V.A. Matyshak, A.A. Kadyshin, A.V. Kurman, N.A. Kutyreva and V.A. Matyshak, Kinet. Katal., 24, 1255-1258 (1983).
12. O.L. Kosynskii, in Hydrodynamics and Transfer Phenomena in Two-Phase Disperse Systems, pp. 122-127, Irkutsk (1982).
13. R.T. Howe, Inorgan. Chem., 15, 486-487 (1976).
14. R.L. Burwell, J. Catal., 86, 301-314 (1984).
15. F.K. Schmidt, Kim En Khva, K.P. Zhdanova, B.V. Timashkova and N.V. Alekseeva, in Hydrodynamics and Transfer Phenomena in Two-Phase Disperse Systems, pp. 113-119, Irkutsk (1982).
16. F.K. Schmidt, Kim En Khva, K.P. Zhdanova, B.V. Timashkova and N.V. Alekseeva, Kinet. Katal., 24, 1691 (1983).
17. F.K. Schmidt, Kim En Khva, K.P. Zhdanova, in Catalysis and Progress in Chemical Technology, pp. 189-195, Nauka, Novosibirsk (1984).
18. G.D. Chukin, Zh. Strukt. Khim., 17, 122-128 (1976).
19. H.-P. Boehm and H. Knözinger, in Catalysis - Science and Technology, Berlin, Akademie-Verlag, v. 40, pp. 40-189 (1983).
20. F.K. Schmidt and Kim En Khva, React. Kinet. Catal. Lett., 14, 283-288 (1980).

LIQUID PHASE OXIDATION OF ETHYLBENZENE IN THE PRESENCE OF METAL-POLYMER COMPLEX CATALYSTS WITH SPECIALLY PREARRANGED STRUCTURES OF ACTIVE CENTRES

A.A. Efendiev, D.D. Orujev, T.N. Shakhtakhtinsky and V.A. Kabanov

Institute of Theoretical Problems of Chemical Technology, Baku 370143, USSR, State University, Moscow, USSR.

ABSTRACT - Complexes of cobalt with a copolymer of diethyl ester of vinylphosphonic acid and acrylic acid specially prearranged for ethylbenzene and α-phenylethyl peroxide have been obtained. The prearrangement consists of formation of the complexes in the presence of the mentioned hydrocarbon substrate, subsequent fixation of the formed structure by intermolecular crosslinking and removal of the template substrate from the crosslinked complexes. It was shown that such prearrangement resulted in a considerable increase in activity and selectivity of the catalysts of liquid phase oxidation of ethylbenzene.

Interest in polymer-supported metal complexes as catalysts has largely grown in the last few years (Ref. 1). Such catalysts are of particular interest in a new and quickly developing field of catalysis by solid supported metal complexes (Ref. 2). The use of polymer complexons as solid supports opens up new possibilities to vary ligand surrounding and control catalytic properties of complexes. These possibilities are provided by the macromolecular nature of the polymer complexons, to be more precise, by the flexibility of macromolecules, their capability of acquiring different conformations and possibilities of varying dislocations of functional complexing groups both in a macromolecular chain and in a three-dimensional polymer matrix. It is important that in this case the complexes are directly formed from transition metals and polymeric ligands.

The objective of the present work was to prepare metal-polymer complexes with structures of active centers specially prearranged for certain hydrocarbon substrates to increase the catalytic activity.

The possibility of such prearrangement does not seem unrealistic if we consider some literature data on specific silica gels which are highly selective with respect to previously chosen compounds (Refs. 3,4). Adsorbents were prepared by forming gels in the presence of the compounds to be adsorbed. The adsorption mechanism was explained in terms of "traces" imprinted by the template compounds.

Whitmore (Ref. 5) and Balandin (Ref. 6) have shown the increase in the rate of hydrogenation of alkyl oranges when using silica-supported catalysts formed in the presence of the same dyes. Balandin et al. have synthesized silica gels in the presence of polycondensation products of D,L-alanine esters and found that they might be used for directed polycondensation of aminoacids (Ref. 7). However, the lack of reproducible methods of preparation of specific silica gels was mentioned in Reference 8. Moreover, the adsorption specificity of silica gels is easily lost during subsequent treatments and even during storage (Ref. 9), apparently, due to the fact that the so-called "traces" or "imprints" are not chemically fixed.

We have developed a new principle of preparation of metal-polymer complexes. The principle is based on the use of the "memory" of the polymer composition and consists of the conformational prearrangement of macromolecules of non-crosslinked polymer complexons to a position favourable for complexing the metal, and the subsequent fixation of conformations, optimum for complex formation, by means of intermolecular crosslinking (Refs. 10-14). Crosslinked macromolecules can "keep in mind" their conformational prehistory, that is, conformations advantageous for complexing the metal, which markedly enhances the complex-forming capacity of the polymer.

One of the examples is a copolymer of diethyl ester of vinylphosphonic acid (DEVPA) and acrylic acid (AA), the macromolecules of which were arranged for complexing with cobalt (Ref. 15). It was shown that complex-forming capacity of the polymer with respect to cobalt increased more than thrice due to prearrangement.

Liquid phase oxidation of alkylaromatic hydrocarbons, such as xylenes and ethylbenzene, carried out in the presence of prearranged complexes has shown that the latter ones are much more active than non-prearranged complexes (Refs. 16-18).

Further investigations have indicated that improvement of catalytic properties of the prearranged complexes was not only due to a higher content of a transition metal but also to essential structural changes of the complexes (Ref. 19). It has been shown that the prearrangement of the polymer leads to the formation of uniform complexes with the structure identical with that of their low molecular weight models in solution.

Meanwhile, in the case of the non-prearranged polymer the formation of various complexes with structures essentially different from their low molecular weight models was observed. For this reason the prearranged complexes are more active and selective than non-prearranged ones even with the same metal content in both specimens. Such effect of the prearrangement is a good premise for the formation of uniform active centres which, in turn, is an important condition for the preparation of catalysts with high activity and selectivity.

An attempt has been made to increase further the activity of metal-polymer complex catalysts by means of formation and subsequent fixation of the active centre structure specially prearranged for the hydrocarbon substrate.

The approach proposed is based on the principle suggested by one of us (Ref. 20) for preparation of synthetic models of enzymes with a tertiary structure of macromolecules optimized for sorption of the low molecular weight substrate.

The general point of our method is the interaction of noncrosslinked polymer complexes with the hydrocarbon substrate (or an intermediate product of its chemical transformation) in solution, i.e. under conditions when macromolecules are still sufficiently mobile. The next step is fixation of the structure of active centres by crosslinking and removal of the template substrate from the crosslinked catalyst.

Complexes of cobalt with the above mentioned copolymer of DEVPA and AA were chosen for such prearrangement, and the reaction of liquid phase oxidation was carried out in the presence of the prearranged complexes. According to (Ref. 17) the structure of the complexes can be described by the following scheme:

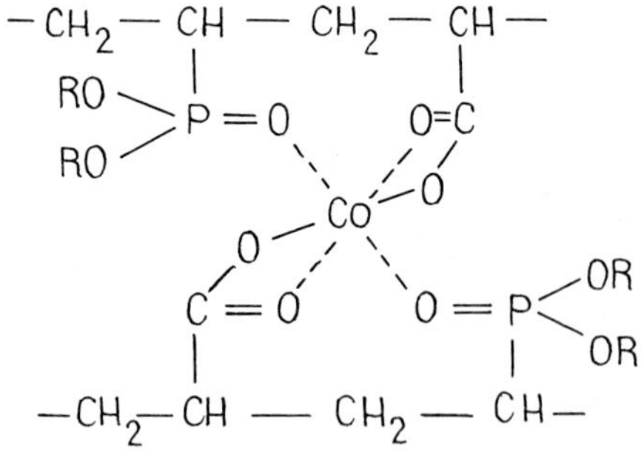

```
 — CH2 — CH — CH2 — CH —
         |            |
   RO \               |
        > P = O      O=C
   RO /        \    / /
                \  / O
                 Co
             O —/ \ \
            /  /    \
           C = O     O = P < OR
           |             |   OR
  — CH2 — CH — CH2 — CH —
```

The copolymer of DEVPA and AA with a number-average molecular mass of 160000 contained 9.2% of phosphorus. Ethanol solution of $CoCl_2$ was added dropwise to ethanol solution of the copolymer until a cobalt content of 4.2 meq/g was achieved. The resulting solution was mixed with an excess ethylbenzene and kept for 8-16 hours. The mixture was frozen with liquid nitrogen; freeze drying was carried out at 10^{-4} - 10^{-5} torr and -30 °C for 2-4 days to remove the solvents. At the end of the drying period the cooling of the specimen was first terminated under vacuum. Then after room temperature had been achieved, the vacuum treatment was stopped.

Dry metal-polymer complexes were mixed with N,N' -methylenediacrylamide. Tablets, prepared from the mixture, were vacuum treated at 10^{-4} - 10^{-5} torr and heated at 140 °C for 3 hours. The crosslinked complexes were ground in a vibratory mill and a fine fraction with a particle size ranging from 0.06 to 0.09 mm was used as a catalyst.

The liquid phase oxidation of ethylbenzene was carried out at 90 °C and atmospheric pressure in a glass reactor fitted with inlet and outlet pipes, a condenser and a thermostat. Gaseous oxygen was used as an oxidant. Gas chromatographic analyses of the reaction products were performed with LHM-8 Model 6. A 0.3 m column of 15% dinonylphtalate on chromatron N-AW (0.2-0.3 mm) was used with nitrogen carrier gas.

For comparison, an identical fraction of non-prearranged complexes was also used. As found by low temperature nitrogen adsorption, specific surfaces of both specimens are practically identical and range from 0.2 to 0.3 m^2/g.

Kinetic curves of ethylbenzene oxidation in the presence of prearranged and non-prearranged complexes are presented in Figure 1. It is seen that the rate of the reaction in the presence of prearranged complexes is much greater than that in the presence of non-prearranged complexes. Kinetic curves of formation of the oxidation reaction

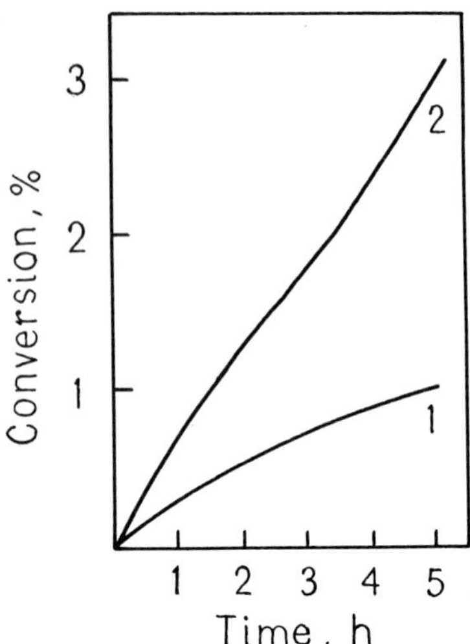

Figure 1. Kinetic curves of ethylbenzene oxidation in the presence of complexes crosslinked without prearrangement (1) and after prearrangement for ethylbenzene (2).

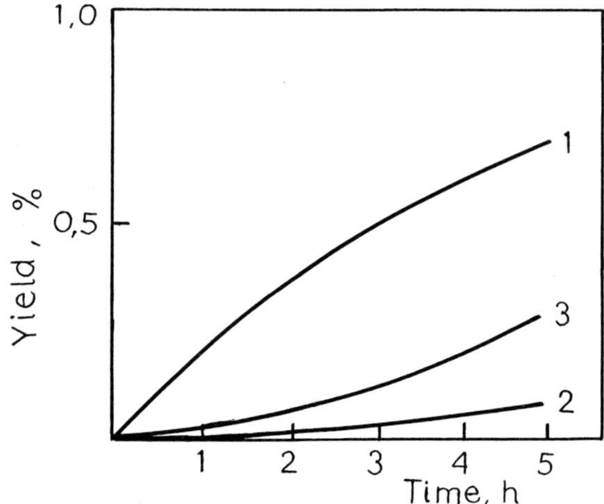

Figure 2. Kinetic curves of formation of α-phenylethyl hydroperoxide (1), methylphenylcarbinol (2) and acetophenone (3) in the presence of non-prearranged complexes.

products, namely, α-phenylethyl hydroperoxide, methylphenylcarbinol and acetophenone in the presence of both

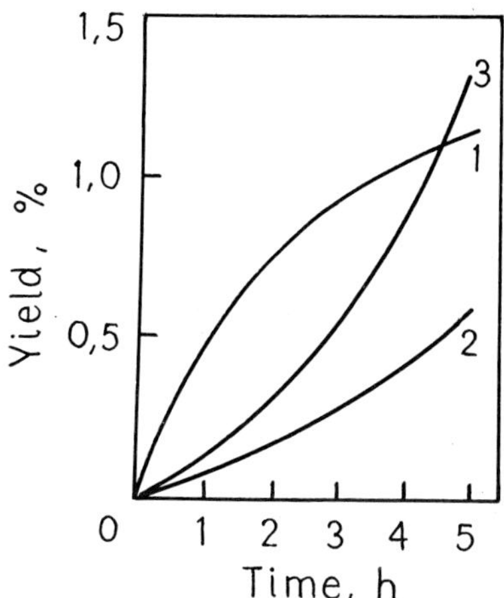

Figure 3. Kinetic curves of formation of α-phenylethyl hydroperoxide (1), methylphenylcarbinol (2) and acetophenone (3) in the presence of complexes prearranged for ethylbenzene

types of complexes are given in Figures 2 and 3. It can be seen that the prearrangement of the complexes results in a considerable increase in the formation rates of all three products, the initial rate of hydroperoxide formation being much more increased than the rate of its decomposition into methylphenylcarbinol and acetophenone.

The next step was to examine the specificity of the prearrangement. With this in mind, complexes prearranged for α-phenylethyl hydroperoxide were prepared using the same procedure.

Kinetic curves of formation of the oxidation reaction products in the presence of the complexes prearranged for hydroperoxide are presented in Figure 4. A comparison of the data presented in Figures 3 and 4 shows that the prearrangement of complexes for α-phenylethyl hydroperoxide increases significantly the activity and the selectivity of the catalyst.

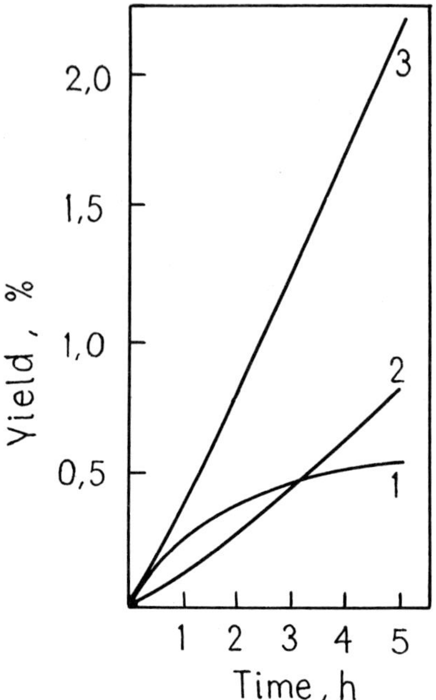

Figure 4. Kinetic curves of formation of α-phenylethyl hydroperoxide (1), methylphenylcarbinol (2) and acetophenone (3) in the presence of complexes prearranged for α-phenylethyl hydroperoxide

The results obtained show that the structural prearrangement of active centres of metal-polymer complexes for the hydrocarbon substrate leads to a pronounced increase in activity and selectivity of the catalyst. The prearrangement can also be used as an additional instrument for mechanistic studies of catalytic processes using metal-polymer complex catalysts.

REFERENCES

1. Polymer-Supported Reactions in Organic Synthesis, P. Hodge and D. Sherrington (Eds.), Moscow, Mir (1983)
2. Yu.I. Yermakov, V.A. Zakharov and B.N. Kuznetsov, Oxide-Supported Complexes in Catalysis, Moscow, Nauka (1980).
3. M.V. Polyakov, Zh. Fiz. Khim., 2, No 6, 799-805 (1931)
4. F.H. Dickey, J. Phys. Chem., 59, No 8, 695-707 (1955)

5. W.F. Whitmore and A. Revukas, J. Am. Chem. Soc., 59, 1500-1502 (1937).
6. A.A. Balandin, V.V. Patrikeev, A.F. Sholin, L.Ya. Bichenkova and A.M. Taber, Dokl. Akad. Nauk SSSR, 169, No 6, 1362-1365 (1966).
7. V.V. Patrikeev, T.D. Kozarenko and A.A. Balandin, Izv. Akad. Nauk SSSR, Otd. Khim. Nauk, 170-171 (1962).
8. V.V. Patrikeev, A.F. Sholin and I.A. Nikiforova, Izv. Akad. Nauk SSSR, Otd. Khim. Nauk, 1031-1035 (1963).
9. F.A. Yampolskaya, Yu.I. Polischuk and M.V. Polyakov, Dokl. Akad. Nauk SSSR, 173, No 4, 870-872 (1967).
10. V.A. Kabanov, A.A. Efendiev and D.D. Orujev, USSR Pat. No 502907, Bull. Izobr., No 6, p. 58 (1976).
11. V.A. Kabanov, A.A. Efendiev, D.D. Orujev and N.M. Samedova, Dokl. Akad. Nauk SSSR, 238, No 2, 356-358 (1978).
12. V.A. Kabanov, A.A. Efendiev and D.D. Orujev, J. Appl. Pol. Sci., 24, 259-267 (1979).
13. V.A. Kabanov, A.A. Efendiev and D.D. Orujev, Vysokomol. Soed., 21A, No 3, 589-595 (1979).
14. A.A. Efendiev and V.A. Kabanov, Pure & Appl. Chem., 54, No 11, 2077-2092 (1982).
15. A.A. Efendiev, D.D. Orujev and V.A. Kabanov, Dokl. Akad. Nauk SSSR, 255, No 6, 1393-1395 (1980).
16. A.A. Efendiev, H.L. Shick, T.S. Murtuzayeva, D.D. Orujev and T.N. Shakhtakhtinsky in: Synthesis and Catalytic Properties of Supported Complexes of Transition Metals, p. 211-214, Novosibirsk (1977).
17. A.A. Efendiev, T.N. Shakhtakhtinsky, L.F. Mustafaeva and H.L. Shick, Ind. Eng. Chem. Res. Prod. Dev., 19, No 1, 75-76 (1980).
18. A.A. Efendiev, H.L. Shick, L.F. Vekilova and T.N. Shakhtakhtinsky, in: Catalysts Containing Supported Complexes, p. 179-182, part 2, Novosibirsk (1980).
19. V.A. Kabanov, L.S. Molochnikov, S.A. Ilyichev, O.N. Babkin, Yu.M. Sultanov, D.D. Orujev and A.A. Efendiev, Azerb. Khim. Zh., No 4, 92-94 (1985).
20. V.A. Kabanov, On a Problem of Artificial Enzyme. In: Advances in the Chemistry and Physics of Polymers, Moscow, Khimiya (1973).

CATALYSTS PREPARED VIA HETEROGENIZATION OF METAL HALOGENIDES IN HYDROCARBON OXIDATION

A.Ya. Yuffa, V.V. Berentsweig, N.V. Vorontsova and G.P. Matsenko
State University, Tyumen 625003; State University, Moscow 117234, USSR

ABSTRACT - Data on the structure and properties of catalysts prepared via heterogenization of transition metal halogenides on silica are reported. It has been established that anomalously coordinated species are formed during the process of anchoring transition metal halogenides on silica modified by lithium. A tendency to the "island"-type distribution of active component on the support has been revealed during catalyst activity tests in hydrocarbon oxidation reactions. It has been found that modification of anchored halogenides by ligands leads to formation of surface complexes. A kinetic model, which takes into account the metal ion participation in the steps of initiation and chain termination, has been suggested. A correlation between redox potentials of homogeneous analogues of heterogenized complexes and rate constants of initiation in the hydrocarbon radical oxidation has been found.

1. INTRODUCTION

A further development and a wide industrial application of undoubtedly promising heterogeneous metal complex catalysis is hampered by a number of factors: i) insufficient understanding of the mechanism of the active surface formation in the process of heterogenization of transition metal complexes; ii) scarce information about optimal surface concentrations of the active component in heterogeneous metal complex catalysts (HMCC) and

iii) inadequate scientific research work on oxidation catalysis in the presence of the HMCC. This latter remark is important for two reasons:

(1) Greater part of oxidation reactions in industry is carried out either under conditions of homogeneous catalysis characterized by low processibility or under hard conditions of traditional heterogeneous catalysis (Ref.1 Naturally, HMCCs would help to eliminate the above mentioned shortcomings while preserving the advantages of homogeneous and heterogeneous catalysts (Ref. 2).

(2) Oxidation-type reactions are sometimes a much more sensitive test for active surface structures than the most sophisticated physico-chemical techniques used for catalysts characterization (Ref. 3). We may hope that combination of physico-chemical and catalytic methods of HMCC testing will provide a more reliable information about the structure of the catalysts based on heterogenized metal complexes as well as the reaction mechanism in their presence.

These problems are discussed in the present paper.

2. STUDIES ON FORMATION OF HMCCs PREPARED VIA ANCHORING TRANSITION METAL HALOGENIDES

Catalysts were prepared in accordance with the well-developed techniques (Refs. 4-6):

$$\text{⫻}\!-\!OH \xrightarrow[-RH]{RLi} \text{⫻}\!-\!OLi \xrightarrow{MCl_n} \text{⫻}\!-\!O)_xMCl_{n-x} \xrightarrow{mL} \text{⫻}\!-\!O)_xML_mCl_{n-x}\,,$$

where M=Co, Cu, Ni, Mo, Fe, Mn; L=phen, py, EDTA, hin, bipy, 8-oxhin.

The metal content control in the samples was performed by the photometric and atomic absorption methods.

ESR spectra were taken using a reflactance-type instrument with $\lambda \simeq 3.2$ cm. The XPS-measurements were carried out with an ES-200B spectrometer. The vacuum in the measurement chamber was from 10^{-8} to 10^{-9} Torr. The non-monochromatizing emission of an aluminium anode $h\nu$ = 1486.6 eV was used to produce exciting radiation. Electron diffuse reflactance spectra were obtained with a Pay Unicam SP-8000 instrument.

Studies of the copper-containing samples indicate a considerable difference between the state of copper ions in the samples obtained by impregnation and those obtained by chemical anchoring (Cu content varied from 0.09 to 1.73 weight %). Thus, the ESR spectrum of the impregnated $CuCl_2/SiO_2$ sample had either an anisotropic signal typical of the massive $CuCl_2$, or a symmetrical singlet with the g-factor 2.18 and ΔH = 80 Öe. The anchored $CuCl_2$ samples produced only weak signals from Cu^{2+} ions in coordination of tetragonal pyramid with the signal intensity exceeding 1% of the number of Cu cations introduced. At the same time the XPS-measurements exposed a photoelectron peak with E_b = 931.5 eV, typical of the $Cu2p_{3/2}$ level of Cu^{2+} ions.

It should be noted that no ESR signal was observed both in the samples, obtained by anchoring from alcoholic solutions and in the catalysts obtained by heterogenization of $CuCl_2$ from aqueous and acetonitrile solutions. This fact precludes us from making the conclusion that copper ions anchored on the surface can be reduced with alcoholic solutions. The above-mentioned fact can be explained only by the formation of strongly interacting Cu^{2+} cations on the surface.

To give a comprehensive picture, we used ESR method to study the samples, obtained by heterogenization of $CuCl_2$ (1). All samples obtained by anchoring $CuCl_2$ from solutions in alcohol and acetonitrile, in argon and in air, displayed weak signals from isolated Cu^{2+} cations in the tetragonal pyramid coordination, whereas the samples obtained by impregnating silica with CuCl from alcoholic solutions produced no such signal.

The results obtained allow us to suggest a model of formation of the catalyst active surface. Even small concentrations of Cu^{2+} cations anchored on silica (0.1--0.5 nm^{-2}) do not disperse over the surface and do not form isolated cations. Apparently, modification of the silica surface by lithium changes its energy and makes the surface considerably more homogeneous. Cu^{2+} cations migrate easily when they interact with the surface and form the "island"-type structures with a strong ion-ion interaction. So the ESR signals are hampered. It is possible that the initially chemisorbed copper ions form a sort of energy traps and that further anchoring of the copper halogenide molecules takes place in the environment of these cations. A similar "island"-type chemisorption of organic molecules was discovered also in (Ref. 7). An "island"-type chemisorption of metal halogenides on the silica surface modified by lithium is of general character, this is proved by the absence of an ESR signal from Mn^{2+} ions anchored on silica.

When discussing the driving forces of the "island"-type anchoring of the transition metal halogenides it should be mentioned that low metal concentrations lead to the formation of abnormally coordinated surface structures (Figure 1).

The effect of these structures is crucial when anchoring is carried out at low temperatures. Note that the above mentioned structures are formed from the single-site chemisorption centers, the proportion of which, as stated in (Ref. 8), is the largest at low surface concentrations of anchored halogenides.

It can be assumed that one of the reasons of the "island"-type chemisorption of halogenides on the silica surface modified by lithium is the interaction of the

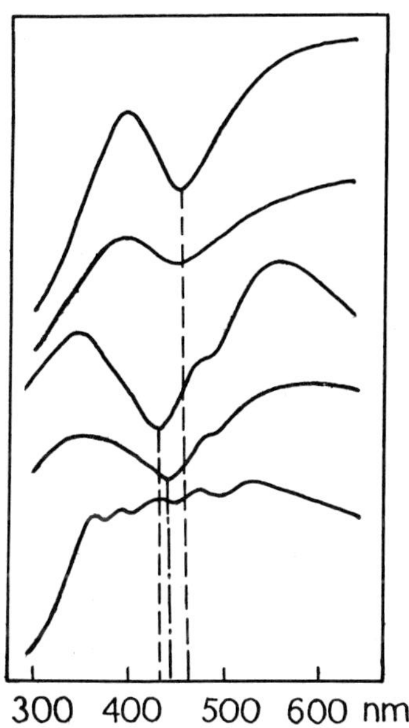

Figure 1. Absorption spectra of anchored (1,2,3) and impregnated (4,5) catalysts $NiCl_2/SiO_2$: 1 - 4.6; 2,5 - 1.3; 3 - 0.8; 4 - 4.8 weight % Ni. Temperature of anchoring: 1, 2 - 30 °C, 3 - 10 °C.

heterogenized molecules with coordinatively unsaturated surface complexes. The same approach can be applied to the results of the work (Ref. 3).

Anchored halogenides can serve as matrix for synthesis of the surface complexes. We mentioned earlier that $CoCl_2$ anchored catalysts, modified by different ligands, can vary the selectivity of the liquid-phase oxidation of tetralin over a wide range (Ref. 6). In this connection examination of the complex formation products on the surface was of great interest. Table 1 shows the results obtained by XPS and diffuse reflactance spectroscopy.

Our results differ from those in paper (Ref. 9). We have stated that the complex composition is defined by

TABLE 1. XPS & diffuse reflectance spectroscopy of $CoCl_2/SiO_2$ samples modified by different ligands

	Ligand modifier	ΔE_b line, eV	E_{sat}, eV	I_{sat}/I_{main}	Range (nm) λ_{max} 400 - 800
1	-	781.0	4.7	0.45	590 630 710
2	pyridine	782.0	2.7	0.44	530 590 660
3	αα'-bipyridyl	782.3	3.5	0.46	
4	EDTA	780.9	3.7	0.33	465 510 560
5	8-hydroxyquinoline				575
6	phenanthroline	780.8	4.8	0.30	

the ratio L:Co. Absorption bands in the range of 13000-18000 cm^{-1} allow us to speak of a predominant pseudo-tetrahedral Co^{2+} ion coordination in the $CoCl_2/SiO_2$ anchored sample and of the surface pyridine complexes (Refs. 10 & 11). Changes in spectra parameters with increasing the Py:Co ratio indicates the complex formation with 2 pyridine molecules. The surface complexes with bidental ligands are characterized by a prevalent non-tetrahedral cobalt-ion coordination. Similar picture was previously observed for synthesis of surface acetylacetone complexes (Ref. 8).

The difference in Co^{2+} states also follows from XPS data (Table 1). Thus, the high values of $E_bCo2p_{3/2}$ in surface pyridine and bipyridile complexes shows that the value of an effective positive charge on the transition metal ion is of great importance. Taking into account small values of ΔE ("main line-satellite") which characterize the higher degree of the Co-Cl ion bond in cobalt complexes anchored on the silica surface, it is possible to suggest that chlorine anions are in the outsphere environment of the transition metal ion. This supposition is confirmed by that the lowest values of Cl2p line bond energy can be observed in the XP-spectra of the complexes under study. Such low values testify to a higher electron density on the chlorine ion.

3. CATALYTIC PROPERTIES OF HETEROGENIZED COMPLEXES

The catalytic properties of anchored transition metal halogenides and of their complexes were explored in liquid-phase oxidation reactions of cyclohexene, tetralene and cumene in a static system over the temperature range from 60 to 100 °C. As was found, an optimum activity depends on the weight of catalyst and transition metal concentration on silica surface. Thus, with increasing catalyst concentration from 1 to 50 g/l (1.73 weight % Cu) the initial oxidation rate increases, but the specific catalytic activity decreases (Figure 2).

It can be mentioned that at the same Cu content in the samples the specific catalytic activity falls sharply with the rise of the metal surface concentration, whereas this change is not so obvious when the catalyst concentration increases (Figure 2). It should be considered that in the chain-radical reaction

$$W = \frac{k_p}{k_t} [RH] \cdot \sqrt{W_i} , \qquad (1)$$

where $W_i = k_i [C_M][RH][O_2]$; k_p - rate constant of chain growth; k_t - rate constant of chain termination.

An increase in concentrations of the active centers C_M, leads to a greater reaction rate according to

$$W = k_o \cdot [C_M] , \qquad (2)$$

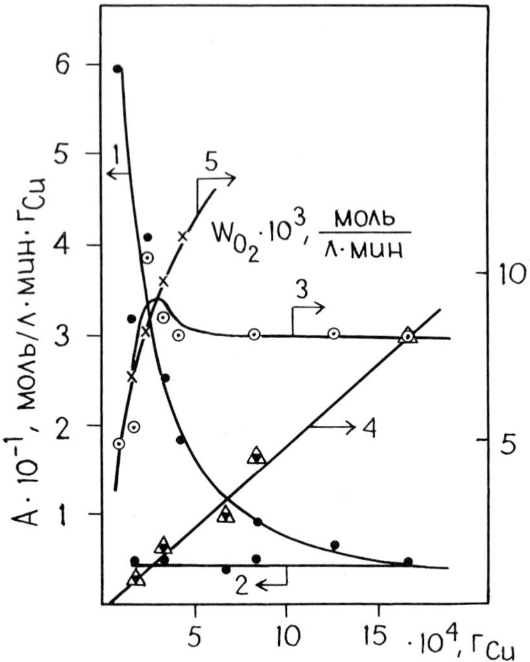

Figure 2. Specific catalytic activity (per 1 g Cu, A) and initial reaction rate (W) vs Cu content in samples: 1,3 - at variation of surface coverage; 2,4 - at various catalyst concentrations; 5 - rated curve 3. Conditions: 75 °C, 10 ml of cumene, 01 g of catalyst for the study of Cu surface coverage.

where $k_o = \frac{k_p}{k_t} [RH]^{3/2} \cdot k_i^{1/2} \cdot O_2^{1/2}$.

In our case this equation harmonizes only the initial part of the reaction rate dependence on surface metal concentrations (Figure 2, curves 3 and 5). This fact confirms our idea of an "island"-type mechanism of anchoring the transition metal halogenides on silica surface: a greater "island" size at high metal concentrations does not generate a greater number of active centers, thus the total substrate conversion during the catalytic reaction remains limited. The similar dependence was discovered earlier in (Ref. 6).

The adequate model which describes the specificity of kinetics of the oxidation process in the presence of different cobalt complexes was obtained. The model, along with the stages typical of the non-catalytic cyclohexene, includes also stages responsible for the catalyst effect on these reactions:

1. initiation due to radical bimolecular decomposition of hydroperoxide:

$$[Co(II)]_s + 2\,ROOH \xrightarrow{k_i'} [Co(II)]_s + RO_2^{\bullet} + RO^{\bullet} + H_2O; \quad (3)$$

2. initiation due to activation of molecular oxygen:

$$[Co(II)]_s + O_2 + RH \xrightarrow{k_i''} [Co(II)]_s + HO_2^{\circ} + R^{\bullet}; \quad (4)$$

3. stages of quadratic and linear chain termination reactions:

$$[Co(II)]_s + RO_2^{\bullet} \rightleftarrows [Co(II)\cdot RO_2^{\bullet}]_s \quad (5)$$

$$[Co(II)\cdot RO_2^{\circ}]_s + RO^{\bullet} \xrightarrow{k_t''} [Co(II)]_s + R' = O + ROH + O_2 \quad (6)$$

$$[Co(II)\cdot RO_2^{\circ}]_s \xrightarrow{k_t'} [Co(III)]_s + R' = O + OH^- . \quad (7)$$

It was found that there is a linear correlation between the logarithms of constants k_i' and k_i'' and redox potentials of the pair Co(III)/Co(II) in these complexes (Figure 3).

This correlation can indicate that stage (3), e.g., is a consecutive elementary interaction of two hydroperoxide molecules with heterovalent cobalt ions in heterogenized complex:

$$[Co(II)]_s + ROOH \longrightarrow [Co(III)OH]_s + RO^{\bullet} \quad (8)$$

$$[Co(III)OH]_s + ROOH \longrightarrow [Co(II)]_s + RO_2^{\bullet} + H_2O. \quad (9)$$

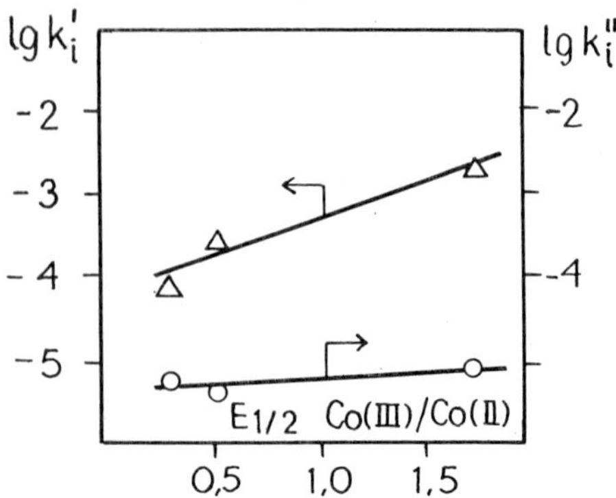

Figure 3. Dependence of initiation rate constants (k_i' and k_i'') of the liquid phase cyclohexene oxidation in the environment of heterogenized cobalt complexes on redox potentials of Co(III)/Co(II). Conditions: 75 °C, 2 ml of cyclohexene (without solvent), catalyst concentration 25 g/l.

Taking into account that the maximum value of k_i corresponds to a "non-complexated" cobalt ion, we can suppose that Co(III) in Co(II) reduction stage limits this process. Ligands while stabilizing the metal ion in the Co(III) state, reduce the rate at this stage.

The best way to achieve maximum conversions of various hydrocarbons is to apply cobalt complexes with pyridine and bipyridyl ligands (Table 2). It can reasonably be supposed that an optimal correlation between redox potentials and kinetic parameters of oxidation is realized in these cases.

Anchored metal complex catalysts have, along with a high activity, a noticeable effect on the selectivity of liquid phase oxidation. It should be noted that preserving high catalytic activity and specificity of homogeneous contacts, these catalysts acquire in some cases specific features of heterogeneous catalysts. For example heterogenized chlorides of Mn(II) and particularly of Cu(II) in liquid phase oxidation of cumene and cyclohe-

TABLE 2. Catalytic properties of anchored cobalt chloride complexes in hydrocarbon oxidation

Ligand-modifier	Oxidation of cyclohexene			Oxidation of cumene
	$W_{O_2} \cdot 10^3$, $\frac{mole}{l \cdot s \cdot g}$	Selectivity ROOH, %	$\frac{R'=O}{ROH}$	$W_{O_2} \cdot 10^3$, $\frac{mole}{l \cdot s \cdot g}$
1 -	4.3	10.5	2.5	1.38
2 Pyridine*	5.0	11.2	2.0	0.94
3 Pyridine**	8.0	11.2	2.5	0.79
4 Bipyridyl*				0.85
5 Bipyridyl**	5.0			0.64
6 EDTA*	1.7	10.0	2.4	
7 EDTA**	0.3			
8 Phenanthroline*	1.2	22.6	5.8	0.21
9 Phenanthroline**	0.3			0.12
10 8-Hydroxyquinoline*	3.7	14.9	3.9	
11 8-Hydroxyquinoline**	0.3	17.0	3.4	

* Ratio L:Co = 1; ** Ratio L:Co = 2.

xene with high (80-90%) selectivity provide the corresponding hydroperoxides. Such selectivity is more characteristic for heterogeneous catalysts, which catalyze the process by activation of molecular oxygen (Ref. 12), and less characteristic for homogeneous ones, which decompose efficiently hydroperoxides (Ref. 13).

It is obvious now that high selectivity of hydroperoxide formation is due not only to activation of molecular oxygen but also to the specificity of the heterogeneous catalysts. This specificity may be attributed to the formation of surface "island"-type structures with a low activity in hydroperoxide decomposition. The specificity of "island"-type structures causes changes in oxidation selectivity of tetraline (at constant general conversion and high cobalt surface concentrations) (Ref. 6). The specific activity of a heterogeneous catalyst may be due to the prevalent formation of active polymolecular structures in the adsorption layer (Ref.14

In conclusion, it should be noted that the proposed "island"-type mechanism of heterogenization of transition metal halogenides on silica surface has a considerable effect on catalytic reaction processes in the presence of the above mentioned catalysts.

The "island"-type chemisorption prevents to obtain high dispersion of metallic catalysts, obtained by heterogenized complex reduction (Ref. 15).

Moreover, it can be expected that an irregular "island"-type distribution of metal ions on the support surface will be quite useful in preparing catalysts for consecutive reactions, in the first place when equilibrium yields of intermediate products are low. The "island"-type active centers show synergetic effects during catalytic reactions in the presence of anchored transition metal halogenides. Tentative results of our investigation of these catalysts make us feel optimistic.

ACKNOWLEDGEMENTS. The authors wish to thank Prof. A. Slinkin, Dr. E. Shpiro, Dr. A. Kucherov and Dr. A. Stakheev for their assistance in carrying out the XPS and ESR measurements as well as for their valuable remarks and helpful discussions.

REFERENCES

1. J.E. Lyons, Appl. Industr. Catal., 3, 131-214 (1984).

2. G.V. Lisichkin and A.Ya. Yuffa, in Heterogeneous Metal Complex Catalysts, Khimiya, Moskva (9181).
3. R. Grago and D. Pribich, Inorg. Chem., 24, 1983-1985 (1985).
4. A.Ya. Yuffa, L.M. Ryazanova, G.M. Gvozdeva and M.Yu. Turova, Kinet. Katal., 22, 1465-1468 (1981).
5. A.Ya. Yuffa, V.V. Berentsweig, G.V. Kudryavtsev and G.V. Lisichkin, Kinet. Katal., 22, 1469-1474 (1981).
6. V.V. Berentsweig, A.Ya. Yuffa and L.V. Bajitova, Neftekhim., 21, 254-258 (1981).
7. C. Lochmüller, A.S. Colorn and M.L. Hunnicutt, Analit. Chem., 55, 1344-1348 (1983).
8. A.Ya. Yuffa, G.V. Lisichkin in Catalysts Containing Supported Complexes, Yu.I. Yermakov (Ed.), Nauka, Novosibirsk (1980).
9. B.I. Lobov, N.I. Shutaya and V.V. Berkman, Zh. Neorg. Khim., 29, 1758-1762 (1984).
10. L. Sestili and C. Furlani, J. Inorg. Nucl. Chem., 32, 1997-2008 (1970).
11. L. Banci, A. Bencini, C. Benelli, D. Gatteschi and C. Zanchini, Structure and Bonding, 52, 37 (1982).
12. L.Ya. Margolis, in Hydrocarbon Oxidation on Heterogeneous Catalysts, Khimiya, Moscow, 1977.
13. I.P. Skibida, Usp. Khim., 46, 1729-1751 (1975).
14. Yu.D. Norikov, E.A. Blumberg, M.D. Irmatov and Sh.K. Bogfishvilli, Dokl. Akad. Nauk SSSR, 223, 1187-1189 (1975).
15. A.Yu. Stakheev, A.Ya. Yuffa, E.S. Shpiro, G.V. Antoshin, A.A.Slinkin, E.A. Fedorovskaya and Kh.M. Minachev, Kinet. Katal., 25, 750-753 (1984).

H_2-INDUCED ELECTRON EXCHANGE IN TiO_2-SUPPORTED $RhCl(PPh_3)_3$

G. Munuera[1], V. Rives-Arnau[1*], A. R. González-Elipe[1], J. Soria[2], J.C. Conesa[2]

[1]Dpto. Química General, Universidad de Sevilla, Seville (Spain).
[2]Ins. Catálisis y Petroleoquímica, C.S.I.C., Madrid (Spain).

ABSTRACT - TiO_2(P-25, Degussa)-supported Wilkinson's complex (5% w/w) has been studied. X.p.s. and i.r. spectra show that ca. 50% of the complex is oxidized to Rh^{III} and $OPPh_3$, and TG/MS analysis indicate that the Cl ligand is lost in vacuo at ca. 423 K, when adsorption of CO and H_2 only occurs on the unoxidized Rh^{I} complex giving bis-carbonyl and Rh^{III}-hydride species. Adsorption of H_2, but not of CO, produces a transmittance drop at $\leq 2000\ cm^{-1}$ which disappears by outgassing at $T \gg 423$ K, and intense e.p.r. lines due to Ti^{3+} ions, in equilibrium with H_2 in the gas phase at $T \gg 423$ K. A mechanism is proposed for the reversible electron exchange between the Rh^{III}-hydride and the TiO_2 support on the basis of previous results with $RhCl_3/TiO_2$.

1. INTRODUCTION

The catalytic properties of Rh^{I} complexes heterogeneized on suitable supports in reactions involving olefines, CO and/or H_2 are widely studied. It is generally accepted (Ref. 1) that complexes having a labile ligand ($Rh(\eta^3-C_3H_5)(CO)(PPh_3)_2$, $Rh(\eta^3-C_3H_5)_3$, etc.) readily produce 'low coordinated', highly reactive catalytic species, while other results (Ref. 2) suggest that an excess of melt PPh_3 on the support surface modifies the properties of Rh catalysts and controls the growth and physical properties of the metal particles (Ref. 3 & 4).

Results are here reported for the system $RhCl(PPh_3)_3/TiO_2$ in comparison with previous studies on $RhCl_3/TiO_2$ (Ref. 5), in order to study such surface 'low coordinated' species. TiO_2 was chosen as support, instead of more conventional SiO_2 or Al_2O_3, since it is now well known that this oxide has a strong capacity to interact with noble metals (Ref. 6 & 7) modifying their catalytic activity and selectivity, as well as their capacity to chemisorb H_2 and CO (SMSI phenomena).

2. EXPERIMENTAL

The specimen was prepared by impregnating TiO_2 (P-25, Degussa) with a

*Present address: Facultad de Farmacia, Salamanca (Spain).

benzene solution of $RhCl(PPh_3)_3$ (Fluka); the detailed method has been described elsewhere (Ref. 8). Amounts of the complex and the oxide were chosen to yield 5% w/w, so that the calculated molecular surface, assuming 100% dispersion of the complex, was close to that obtained from loading (i.e., 1.65 vs. 1.54 nm^{-2}).

Adsorption isotherms of N_2 at 77 K were measured in a conventional system. TG/DTG analysis was carried out in a Cahn electrobalance; e.p.r. (at 77 and 300 K), i.r. and x.p.s. (at 300 K) spectra were recorded respectively with JEOL (mod. JES-3PX, X-band), Perkin-Elmer (mod. 577) and Leybold-Heraeus (mod. LS10, excitation source MgKα). Further experimental details have been given elsewhere (Ref. 5 and 9 to 11).

3. RESULTS

3.1. X.p.s. Analysis and Thermal Evolution of $RhCl(PPh_3)_3/TiO_2$

X.p.s. of the sample at 300 K, Fig. 1, shows a partial oxidation of the Wilkinson's complex upon deposition on TiO_2 and exposure to the air, as detected by the broad Rh(3d) peaks, and intense O(1s) signals at 532.4, 531.4 and 529.7 eV, present in the spectrum of parent TiO_2. From the spectra, the P/Rh, Cl/Rh and O/Ti ratios were 2.9, 1.4 and 2.7 (O/Ti =2.3 for TiO_2), indicating a larger oxygen content at the surface. Deconvolution of the Rh(3d) peaks indicates a 50% oxidation of Rh^{I} to Rh^{III}.

Adsorption isotherms of N_2 at 77 K show an increase in the adsorption capacity for $P/P_0 > 0.3$ (mesopore region), but without the appearance of hysteresis, after incorporation of the complex; the BET surface area was close to that of the TiO_2 support ($49 \pm 1\ m^2\ g^{-1}$). TG/DTG data suggest a two-stages thermal decomposition of the supported complex, with maxima at 475 and 650 K. Total weight loss was close to the expected one.

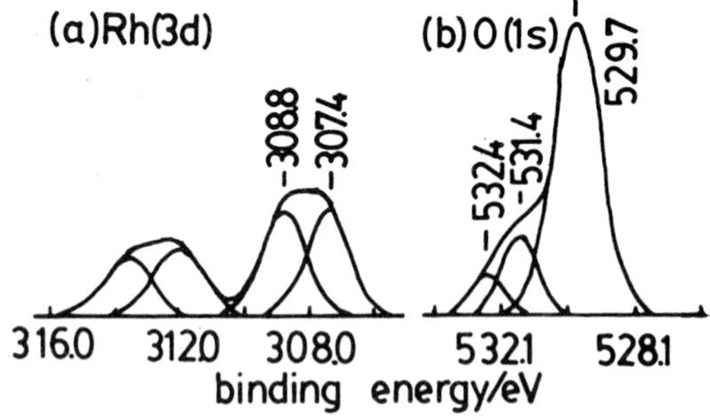

Figure 1. X.p.s. of $RhCl(PPh_3)_3/TiO_2$. (Ref. C(1s) at 286.4 eV).

I.r. assessment of the decomposition in vacuo ($P<10^{-4}\ Nm^{-2}$) of the supported complex using 50 K step-by-step heating is summarized in Fig. 2.A. After outgassing at $T\leqslant423$ K, the spectrum shows four small bands in the range 2000–1660 cm^{-1} (out-of-plane ν_{C-H} overtones of Ph groups), 2960, 2920 and 2850 (organic contaminants on the support), 1590, 1475 and 1435 (PPh_3 ligands), 3060 (aromatic ν_{C-H}) and a small shoulder at 2000 cm^{-1}; two bands (not shown in the figure) at 1120 and 1090 cm^{-1} suggest the presence of $OPPh_3$. Outgassing at 523 K removes the band at 1475 cm^{-1} and a medium-width band at ca. 2000 cm^{-1} was now clearly observed. At $T\geqslant573$ K the spectrum shows a band at 3720 cm^{-1} (isolated OH groups at the support), while that at 2000 cm^{-1} has strenghted. Outgassing at 673 K removes this band, but not those at 1590 and 1435 cm^{-1}, reducing the OH bands and leading to a greyish-blue coloured sample. A yellowish-white substance condensed on the cool parts of the i.r. cell at $T\approx450$ K was identified as $OPPh_3$ by MS and i.r. measurements.

MS analysis of the evolved light species shows peaks due to H_2O and HCl (this from the loss of the Cl ligand) at $T\leqslant450$ K, while at higher temperatures the cracking pattern of the phosphine was observed with increasing intensity. X.p.s. of the solid residue after decomposition in vacuo at 673 K shows the lack of P and Cl, and a strong signal of C, confirming the partial cracking of the phosphine ligands.

The e.p.r. spectrum of the sample outgassed at 300 K shows a signal with $g\approx g_e$ due to C impurities on the support (Ref. 8), Fig. 2.B, that narrows as the outgassing temperature increases, until a very narrow

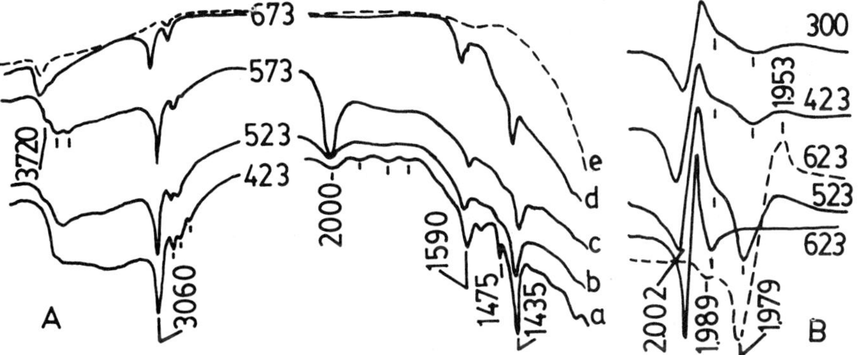

Figure 2. (A) I.r. and (B) e.p.r. spectra of the sample and the support (dotted line) outgassed 1 h at the indicated temperatures (K).

shape ($\Delta H \simeq 6$ G) is reached at 623 K. Oxygen at 300 K removes this signal, but it is recovered by outgassing at 300 K for several hours or at 623 K for a few minutes. For TiO_2, this signal could be sometimes detected under similar conditions, but was irreversibly removed by oxygen at 623 K. In addition, two Ti^{3+} signals ($g_{\perp}=1.989$, $g_{\parallel}=1.979$) develop with the outgassing treatment. The intensity of the latter signal passes through a maximum and then decreases for higher outgassing temperatures; this signal is very much intense than that observed in similarly treated TiO_2, where it continuously grows with increasing outgassing temperatures; this indicates that Ti^{3+} ions are produced by reduction of the support during outgassing and decomposition of the complex, but they are lost through reduction of the Rhodium ions.

3.2. Adsorption of CO and H_2 on $RhCl(PPh_3)_3/TiO_2$

These processes were studied to characterize the surface Rh^I species in samples milded decomposed at $T \leqslant 450$ K, after removal of the Cl ligand and before cracking/removal of the $PPh_3/OPPh_3$ ligands. On samples outgassed at 373 K, adsorption of CO takes place at 373 K, but not at 300 K, indicating that this process is somewhat activated. No further adsorption of CO occurs at 373 K after outgassing at 300 K, i.e., preadsorbed CO is not removed at 300 K; however, new adsorption occurs, but only ca. 50% of the original value, after outgassing at 423 K. Monolayer capacities, assuming Langmuir-type isotherms, were ca. 1.0 and 0.5 $ml_{STP}\ g^{-1}$, respectively, the former corresponding to $CO{:}Rh^I=2{:}1$, on the basis of the amount of Rh^I present in the sample (50%, x.p.s.).

A strong uptake of H_2 also occurs at 423 K, but not at 300 K, after outgassing at 423 K; further adsorption of H_2 is only observed after outgassing at 423 K, but in a lower extent (25%) than in the fresh sample outgassed at this temperature, indicating that some of the preadsorbed H_2 is not removed even in these conditions. The ratio between the monolayer capacities for CO and H_2 was close to 2:1, i.e., Rh^I species having two unsaturated coordination sites should be involved in the adsorption of both CO and H_2 in these mildly decomposed samples.

3.3. I.r. and e.p.r. assessment of the adsorbed species

Adsorption of CO at 300 K on a sample outgassed at 373 K only produces

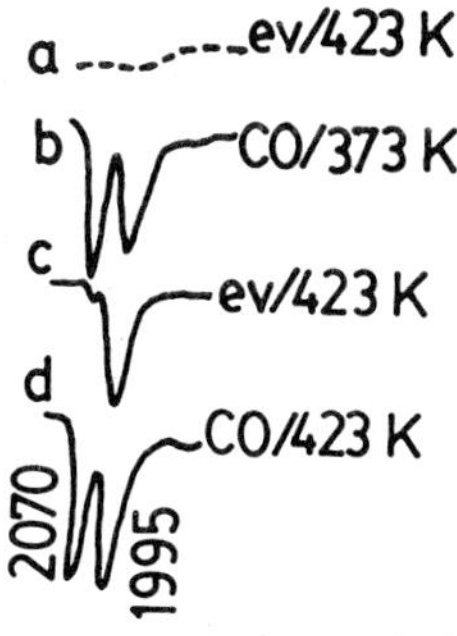

Figure 3. I.r. spectra of the sample and the support (dotted line) and their interaction with CO in the indicated conditions.

very weak i.r. bands at 2070 and 1995 cm^{-1} which strength at 373 K, Fig. 3.b. The band at 2070 cm^{-1} is slightly removed on outgassing at 373 K, but completely at 423 K, and restored by CO readsorption (500 $N\,m^{-2}$) at 423 K, but not at 300 K. Absorbance ratio for the two bands was 1 ± 0.1, thus suggesting species such as:

$$\square\!\!-\!Rh^{I}\!-\!\square\,(P)_3 \underset{-\,CO}{\overset{+\,CO}{\rightleftharpoons}} OC\!-\!Rh^{I}\!-\!\square\,(P)_3 \underset{-\,CO}{\overset{+\,CO}{\rightleftharpoons}} OC\!-\!Rh^{I}\!-\!CO\,(P)_3$$

The reversible adsorption/desorption of the CO ligands leading to the band at 2070 cm^{-1} corresponds (see volumetric data) to ca. 50% of the total adsorbed CO. The remaining adsorbed CO, responsible for the band at 1995 cm^{-1}, was more strongly held and can only be removed after outgassing at T>523 K, this treatment irreversibly leading to partial cracking and loss of the $PPh_3/OPPh_3$ ligands, as discussed above. E.p.r. spectra of the sample does not show any change upon CO adsorption.

On a sample outgassed at 423 K, changes were not observed upon adsorption of H_2 (500 $N\,m^{-2}$) at 300 K, but at 423 K the transmittance of the sample greatly drops for $\nu<2000$ cm^{-1}, Fig. 4.A.b, only a small, broad band at 2000 cm^{-1} being detected. This phenomenon was almost completely reversible, since removal of the gas phase at 300 K restores the original spectrum, Fig. 4.A.c, except for a strong, sharp, unresolved band at 2000-1975 cm^{-1}. Outgassing at 423 K only slightly decreases the intensity of the shoulder at 1975 cm^{-1}, and the whole process is reversible on adsorption of H_2 at 423 K.

When a sample outgassed at 300 K was treated with H_2 at 300 K, the e.p.r. spectrum at 77 K shows a slight increase of the Ti^{3+} signal, the original spectrum being recovered by outgassing at 300 K, Fig. 4.B.a. However, adsorption of H_2 at 423 K strengths this signal at g=1.972,

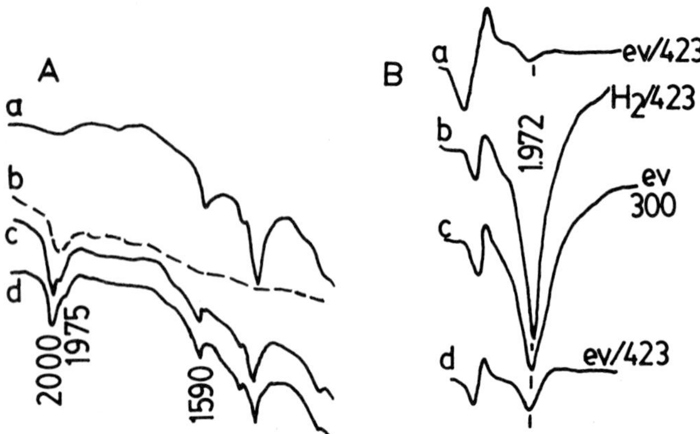

Figure 4. (A) I.r. and (B) e.p.r. assessment of the interaction of H_2 with the specimen (for conditions, see text).

Fig. 4.B.b. The new Ti^{3+} signal was partially removed by outgassing at 423 K for a few minutes, Fig. 4.B.d; similar experiments using the TiO_2 support did not show any Ti^{3+} signal in these conditions. This behaviour indicates that the Ti^{3+} ions are not formed by the hydrogen reduction of the support in these conditions (423 K), but by some sort of reversible electron exchange between the Rh^{III}-hydride and TiO_2.

It is worthy of note that under the above conditions, signals due to paramagnetic Rh species (e.g., Rh^{II}, d^7) were not observed either at 77 or 300 K, suggesting that these species if formed, must be rather unstable on the TiO_2 surface.

4. DISCUSSION

X.p.s. of the samples, Fig.1, indicate that ca. 50% of the Rh^I species are oxidized in air to Rh^{III}, and the intensified O(1s) peak at 532.4 eV and the i.r. bands at 1120 and 1090 cm^{-1} clearly show the formation of $OPPh_3$, suggesting a reaction such as eqn. 1. TG analysis also con-

$$Rh^{I}Cl(PPh_3)_3 + O_2 \longrightarrow Rh^{III}Cl(O)(OPPh_3) \qquad (1)$$

firms that some of the PPh_3 ligands are oxidized, leading to a slightly higher weight loss. From x.p.s., P/Rh=2.9 (expected value 3), indicating that the PPh_3 ligands are not lost on supporting the complex.

Nitrogen adsorption at 77 K suggests that the TiO_2 surface remains exposed to the gas phase, confirmed as well by the absence of hysteresis in the isotherms; the observed changes in mesoporosity ($P/P_o >$ 0.3) should be then ascribed to the presence of the bulky $PPh_3/OPPh_3$ ligands on the TiO_2 surface. According to these results, we may assume that the complex was originally well dispersed (though 50% oxidized), probably forming a monolayer on the support.

The thermal treatment in vacuo leads to the loss of the Cl ligand (as HCl) at $T \leqslant 423$ K and to removal of the $PPh_3/OPPh_3$ ones in the range 623-673 K, involving a reduction of the support by reacting with the PPh_3 ligands (eqn. 2), so that electrons (Ti^{3+} ions observed by e.p.r.)

$$PPh_3 + O^{=} \longrightarrow OPPh_3 + 2\,e^- \;(Ti^{3+}) \qquad (2)$$

are progressively formed and react with the Rh^I species according to eqn. 3. Reduction of all the Rh^I ions to metallic Rh would require

$$Ti^{3+} + Rh^I \longrightarrow Ti^{4+} + Rh^o \qquad (3)$$

oxidation of ca. 20% of the PPh_3 ligands according to eqn. 2. Moreover, the loss of the $PPh_3/OPPh_3$ ligands in the narrow range 623-673 K shifts eqn. 3 to the right, decreasing the Ti^{3+} e.p.r. signal and stabilizing the Rh^I species by the PPh_3 ligands at the TiO_2 support. In fact, the phosphinated ligands-assisted dispersion of the metallic ions has been claimed (Ref. 12) to explain the effect of an excess of phosphine on the increased activity and regio-selectivity of supported $Rh(H)(CO)(PPh_3)_3$ catalysts.

Adsorption of CO on mildly (423 K) outgassed samples occurs only on the Rh^I species forming bis-carbonyl species (sharp i.r. bands at 2070 and 1995 cm^{-1}), while no band due to Rh^{III}-carbonyl species was observed. This behaviour differs from that of $RhCl_3/TiO_2$ samples previously studied by us (Ref. 5), where adsorption of CO, even at 300 K, immediately produces Rh^{III} mono-carbonyl species characterized by a sharp i.r. band at 2138 cm^{-1}, although readily and quantitetively reduced in excess of CO according to eqn. 4, the new $Rh^I(CO)_2$ moiety being charac-

$$Rh^{III}Cl_n(CO) + 2\,CO + OH^- \longrightarrow Rh^ICl_n(CO)_2^+ + CO_2\;(g) \qquad (4)$$

terized by i.r. bands at 2088 and 2020 cm^{-1}. The band at 2070 cm^{-1} in our sample can be fully removed, without disturbing that at 1995 cm^{-1}, while for the bis-carbonyl species from $RhCl_3/TiO_2$ heating in vacuo removes both bands almost simultaneously, though that at 2088 cm^{-1} is now slightly more stable. These facts, together with the ca. 20 cm^{-1} shift, suggest that phosphinated $Rh^{I}(CO)_2^+$ moieties on the TiO_2 surface are more highly coordinated than the Rh^{I} species formed by CO reduction of $RhCl_3/TiO_2$, and so the existence of species of the type $Rh(CO)_2(PPh_3)_3^+$ (EAN=18), easily transformed according to eqn. 5, may be assumed. In addition, contrary to the behaviour observed for $RhCl_3/TiO_2$ samples outgassed at 423 K, CO cannot react under the mild con-

$$Rh(CO)_2(PPh_3)_3^+ \underset{+\,CO}{\overset{-\,CO}{\rightleftharpoons}} Rh(CO)(PPh_3)_3^+ \qquad (5)$$

ditions used in this work, indicating that the Rh^{III} species present in these samples should be fully coordinated with $O^{=}_{latt}$ and/or $OPPh_3$, probably forming mono- or poly-nuclear complexes.

The presence of Rh^{I}-phosphinated species having two unsaturated coordination sites is further confirmed by the strong band at ca. 2000 cm^{-1} after heating in vacuo at 523-573 K, before the loss of the PPh_3/ $OPPh_3$ ligands occurs. This band could be ascribed to Rh-H bonds formed by a side-reaction on the low-coordinated Rh^{I} species (EAN=14), which may react intramolecularly (ortho-metallation), forming the monohydride $(Ph_3P)_2\overline{Rh(H)(PPh_2C}_6H_4)$, which becomes stabilized at the TiO_2 surface, a process already observed in similar systems (Ref. 12). Formation of a Rh-C bond, followed by extensive cracking of the PPh_3 ligands at higher temperatures, might account for the generation of the sharp carbon e.p.r. signal (g=2.002) observed in these conditions, and that finally becomes incorporated into the metallic clusters upon thermal decomposition (Ref. 8).

The reversible formation of Ti^{3+} ions at 423 K at the TiO_2 support during H_2 adsorption seems to be related to hydrogen still bonded after outgassing at 300 K, but removed at 423 K, while the loss of transmittance below 2000 cm^{-1} may be due to the small amounts of H_2 reversibly removed at 300 K. The measured CO/H_2 ratio of 2 suggests that most of the adsorbed H_2 is on the same Rh^{I} species adsorbing CO to produce the bis-carbonyl leading to Rh^{III} species (EAN=16), eqn. 6, still coordinatively unsaturated and characterized by stable i.r. bands at 2000-1975 cm^{-1} as observed for $RhCl_3/TiO_2$ samples (Ref. 11). While for the bis-carbonyl (EAN=18) any further expansion of the coordina-

$$L_3Rh\genfrac{}{}{0pt}{}{\square}{\square} + H_2 \longrightarrow L_3Rh\genfrac{}{}{0pt}{}{H}{H} \qquad (L=PPh_3/OPPh_3) \qquad (6)$$

tion sphere is prevented, here basic ligands can be accepted to yield fully coordinated Rh^{III} species. In oxide-supported Rh complexes, coordinatively unsaturated Rh^{III} ions and basic ($OPPh_3$, $O^=$) species may facilitate the heterolytic activation of H_2. A general reactivity pattern has been proposed by Ward et al. (Ref. 13), where Rh^{III} species act as active Lewis acids. According to these authors, eqn. 7 can be written. The initially formed Rh^{II}-trishydride (EAN=19) would be a highly reducing species, so that electron transfer to the support would

$$L_nRh^{III}(H)_2 + H_2\,(g) \longrightarrow \left[L_n - \overset{H \cdots\cdots H}{Rh^{III}(H)_2} \longrightarrow L_n - Rh^{II}(H)_3 + H^+_{TiO_2} \right] \longrightarrow$$

$$\longrightarrow L_nRh^{III}(H)_3 + e^-_{TiO_2} \qquad (7)$$

generate Ti^{3+} species (detected by e.p.r.), and a fully coordinated Rh^{III} complex (EAN=18). Since the amount of Ti^{3+} species only represents a small fraction of the total Rh loading, eqn. 7 seems to occur only on a small number of sites. A similar phenomenon has been previously reported by us (Ref. 9) using M/TiO_2 (M=Pt, Rh, Pd, Ru), though in this case the e.p.r. signal due to Ti^{3+} was reversibly removed by outgassing even at 300 K, indicating a much weaker form of adsorbed H_2.

Meanwhile, the transmittance drop, also observed in these systems, has been related to the transfer (spill-over) of the protons to the TiO_2 surface, where they may be stabilized in a highly polarizable structure, as discussed elsewhere (Ref. 14). Outgassing at 300 K will produce reactions as those in eqns. 8 and 9, so that several types of

$$2\,H^+_{TiO_2} + O^=_{latt}/OH^-_{ads} \longrightarrow H_2O\,(g) + V_{O\,(TiO_2)} \qquad (8)$$

$$L_nRh^{III}(H)_3 + H^+ \longrightarrow H_2\,(g) + L_nRh^{III}(H)_2 \qquad (8bis)$$

$$L_nRh^{III}(H)_3 + 2\,Ti^{3+}_{TiO_2} \longrightarrow \tfrac{1}{2}H_2\,(g) + n\,L\,(g) + Rh^o + 2\,Ti^{4+}_{TiO_2} \qquad (9)$$

Rh^{III}-hydrides, besides the main Rh^{III} monohydride characterized by the sharp i.r. band at ca. 2000 cm^{-1}, could be present. This would explain the recovery of the i.r. transmittance at 300 K (removal of H^+ according to step 8), and the loss of the Ti^{3+} e.p.r. signal at slightly higher temperatures (step 8bis), being completely irreversible once the $PPh_3/OPPh_3$ ligands are removed (step 9).

Our results indicate that an easy spill-over of hydrogen via electrons and protons transfer occurs from the Rh^{III}-hydride complexes to the support, the process probably being determined by the formation of highly reducing Rh^{II}-polyhydride complexes (EAN=19) that transfer electrons to the semiconducting support. This type of spill-over cannot be observed on insulating supports (SiO_2, Al_2O_3) where the electronic structure does not allow the complex-semiconductor electron transfer.

ACKNOWLEDGMENT. Thanks are due to CAICYT and Fundación Ramón Areces for finantial support.

REFERENCES

1. Yu. I. Yermakov; Catal. Rev. 13, 77 (1976); J. Mol. Cat. 9, 13 (1980).
2. Th. G. Spek, J. J. Scholten; J. Mol. Cat. 3, 81 (1977/78).
3. G. Cocco, S. Enzo, F. Pina, G. Strukul; J. Cat. 82, 160 (1983).
4. F. Pina, M. Gonizzi, G. Strukul, G. Cocco, S. Enzo; J. Cat. 82, 171 (1983
5. J. C. Conesa, M. T. Sainz, J. Soria, G. Munuera, V. Rives-Arnau, A. Muñoz; J. Mol. Cat. 17, 231 (1982).
6. S. J. Tauster, S. C. Fung, R. L. Garten; J. Am. Chem. Soc. 100, 170 (1978
7. G. C. Bond, S. Burch in Specialist Periodical Reports, Catalysis, vol. 6, p. 27, The Royal Society of Chemistry, London (1983).
8. V. Rives-Arnau, G. Munuera; Appl. Surf. Sci. 6, 122 (1980).
9. J. C. Conesa, J. Soria; J. Phys. Chem. 86, 1392 (1982).
10. J. C. Conesa, P. Malet, A. Muñoz, G. Munuera, M. T. Sainz, J. Sanz, J. Soria, in Int. Congr. Catal., 8th (1984), vol. 5, p. 217.
11. J. C. Conesa, G. Munuera, A. Muñoz, V. Rives, J. Sanz, J. Soria; Stud. Surf. Sci. Cat. 17, 149 (1983).
12. J. Hjortjaer, M. S. Scurrell, P. Simonsen; J. Mol. Cat. 6, 405 (1979); ibid. 10, 127 (1981).
13. M. D. Ward, J. Schwartz; Organometallics 1, 1030 (1982).
14. H. Noller, B. Mayerböck, G. Zundel; Surf. Sci. 33, 82 (1972).

PHOTOCATALYTIC REDOX REACTIONS SENSIBILIZED BY TRIS-BIPYRIDINERUTHENIUM(II) COMPLEX IN MICROHETEROGENEOUS SYSTEMS

I.M. Tsvetkov, O.V. Gerasimov, S.V. Lymar and V.N. Parmon

Institute of Catalysis, Novosibirsk 630090, USSR

ABSTRACT - Photocatalytic oxidation of H_2O to molecular oxygen in "sacrificial" systems containing: a photocatalyst ($Ru(bipy)_3^{2+}$), a catalyst for H_2O oxidation (Co compounds) and an irreversible electron acceptor (either $S_2O_8^{2-}$ or Co(III) ammonia complexes) has been studied in microheterogeneous systems of three different types: micellar solutions, microemulsions and suspensions of lipid vesicles. It has been shown that photocatalyst adsorption on sodium dodecylsulfate micelle surface can lead to a more than 10-fold increase in the selectivity of H_2O oxidation compared to homogeneous solution. A probable mechanism for this phenomenon is suggested. The possibility of creation of isolated photocatalytic reaction-center for oxygen evolution from water in water-in-oil microemulsions and internal cavities of vesicles has been demonstrated. The problem of conjugation of such reaction-center inside vesicle with the electron phototransfer to the acceptor localized in an aqueous phase outside vesicle is discussed.

The development of molecular photocatalytic systems for photoinduced cleavage of H_2O into H_2 and O_2 is of interest for modelling photosynthesis of plants (see for review Refs. 1-3). The greatest progress in this direction has been achieved by using microheterogeneous systems that make possible to obtain spatial separation of photogenerated starting reactants for water cleavage, namely strong one-electron oxidant and reductant (see review of Ref. 4 and literature therein). Some difficulties can arise when using these reactants in subsequent "dark" catalytic reactions of H_2O oxidation to O_2 and its reduction to H_2 due to the fact that microheterogeneous systems have developed water-organic interfaces. We believe it is particularly difficult to carry out a four-electron oxidation of H_2O. The very first attempt to catalyze this reaction by homogeneous Co catalysts in solutions of lipid vesicles (Ref. 5) did reveal significant changes in the activity and selectivity of these catalysts compared to homogeneous solutions.

In order to choose the appropriate catalysts for the photocatalytic oxidation of H_2O to O_2 and to clarify the specific effect of system heterogeneity on this reaction, we have examined microheterogeneous systems of three different types: micellar solutions, microemulsions and suspensions of lipid vesicles. Photocatalytic oxidation of H_2O was studied in so-called "sacrificial" processes when H_2O oxidation is accompanied by the simultaneous reduction of the specially introduced irreversible electron acceptors ("sacrifice"(A)):

$$A + 2H_2O \xrightarrow[\text{PhC, Cat}]{h\nu} A_{red} + 4H^+ + O_2\uparrow , \qquad (1)$$

where PhC, Cat and A_{red} are the photocatalyst, the catalyst for H_2O oxidation and the acceptor reduction product, respectively.

EXPERIMENTAL

Chemical or pure grade reactants $CoCl_2 \cdot 6H_2O$, $Co(NH_3)_5Cl_3$, $K_2S_2O_8$, $NaIO_4$, disodium salt of ethylenediaminetetraacetic acid (Na_2EDTA) and organic solvents: pure grade p-xylene, I-pentanole and chloroform were used without further purification. Sodium dodecylsulfate

(SDS) was recrystallized from a water-ethanole solution. Tris-bipyridineruthenium ($Ru(bipy)_3^{2+}$), dichloride and diperchlorate, cetylviologen (CV^{2+}), dibromide and $[Co(NH_3)_4CO_3]Cl$ complex were synthesized according to the literature (Refs. 6-9). "Water-in-oil" microemulsions and suspensions of egg lecithin vesicles were prepared according to (Refs. 10,11). Oxygen and CO_2 formed during irradiation of 8 ml solution degassed by evacuation were recorded by a chromatographic technique (Ref. 5). For irradiation, we applied a 1000 W xenon lamp equipped with either a ZhS-12 glass-filter ($\lambda > 420$ nm) for $S_2O_8^{2-}$ -containing solutions, or a combination of SS-5+SZS-9 glass-filters ($400 < \lambda < 500$ nm) for solutions that contained $[Co(NH_3)_5Cl]^{2+}$ and $[Co(NH_3)_5CO_3]^+$. Rate constants for quenching of the $^*Ru(bipy)_3^{2+}$ were determined from the lifetime measurements by using an ORTEC 9200 nanosecond fluorometer.

RESULTS AND DISCUSSION

Since strong oxidants formed in photocatalytic systems can be consumed not only for H_2O oxidation but also for the side-reactions of oxidation of some other (e.g. organic) compounds that are present in these systems, the major parameter to characterize the behaviour of these systems is selectivity. For "sucrificial" photocatalytic system the selectivity towards H_2O oxidation to O_2 can be determined as the ratio of the actually evolved O_2 to its stoichiometric quantity in reaction (1). This ratio measured by the moment of the complete consumption of the acceptor was taken as H_2O oxidation selectivity (β) in experiments described below.

In micellar solutions we have studied the formation of O_2 in a "sucrificial" system that contains $Ru(bipy)_3^{2+}$ (a photocatalyst), $CoCl_2$ (a catalyst for H_2O oxidation), and either $[Co(NH_3)_5Cl]^{2+}$ or $S_2O_8^{2-}$ (irreversible electron acceptors). To determine the O_2 formation stoichiometry, we have taken into account that the persulphate-anion is the acceptor of two electrons, whereas the Co complex is a one-electron acceptor (see below). As is seen from table, the presence of SDS micelles practically does not influence the β value, if $[Co(NH_3)_5Cl]^{2+}$ is used as an acceptor. On the contrary, the addition of SDS to persulphate-containing solutions increases the amount of O_2 evolved (Fig. 1). At [SDS] = 0.3 M it corresponds to the more than 10 fold increase of β (Table).

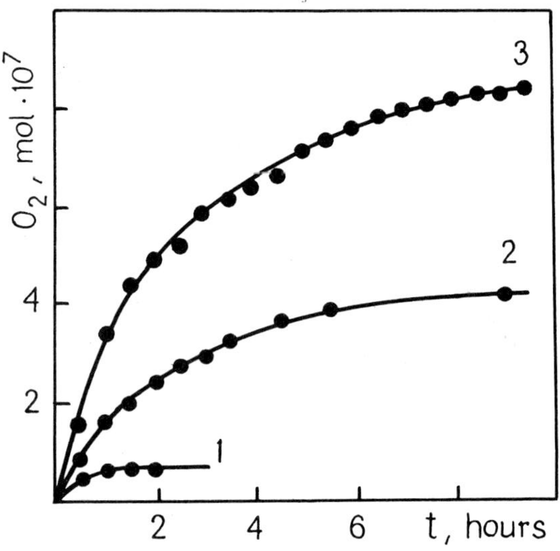

Fig. 1. Kinetics of photocatalytic evolution of O_2(with $S_2O_8^{2-}$ as an acceptor): 1 - homogeneous solution; 2-3 - micellar solution; [SDS] = 0.02 (2), and 0.3 M (3). (For concentration of the rest components refer to table).

Oxygen evolution stops after the complete consumption of persulphate. Addition of another portion of $K_2S_2O_8$ into the reaction mixture restores the photoinduced oxygen evolution. It should be noted that under irradiation of all the solutions listed in table, noticeable (~10^{-6} mol) amounts of CO_2 evolve into the gas phase. It indicates that the most important side-processes competing with reaction (1) are the reactions of deep oxidation of organic compounds.

TABLE. Selectivities of H_2O (β) and rate constants for quenching (k_q) of $^*Ru(bipy)_3^{2+}$ by an acceptor in homogeneous and micellar solutions

Acceptor	β, %, (k_q, $10^8 M^{-1} s^{-1}$)	
	Homogeneous solution	Micellar solution
$[Co(NH_3)_5Cl]^{2+}$	0.4 (16)	0.3 (57)
$S_2O_8^{2-}$	1.7 (6.2)	22 (0.73)

Acetate buffer 0.1 M, pH 4.8; $[Ru(bipy)_3^{2+}] = 2\ 10^{-4}$M, $[Co^{2+}] = 10^{-3}$M; [acceptor] = 10^{-3}M. In micellar solution [SDS] = 0.3 M when determining β and $2 \cdot 10^{-2}$M in k_q measurements.

During photocatalytic reactions in persulphate-containing solutions partial destruction of photo-catalysts takes place that is recorded from the intensity decrease and the broadening of the light-absorption band of $Ru(bipy)_3^{2+}$ at $\lambda \simeq 450$ nm. In micellar solutions the destruction is much weaker.

Thus SDS micelles distinctly promote the photocatalytic evolution of O_2 in persulphate-containing solutions. Some possible reasons for this phenomenon can be clarified on the basis of the known data concerning the reaction mechanism in homogeneous solutions. It is suggested (Refs. 12,13) that the process under discussion includes the following reactions:

$$Ru(bipy)_3^{2+} \xrightarrow{h\nu} {}^*Ru(bipy)_3^{2+} \quad (2)$$

$$^*Ru(bipy)_3^{2+} + S_2O_8^{2-} \longrightarrow Ru(bipy)_3^{3+} + SO_4^{2-} + SO_4^{\dot{-}} \quad (3)$$

$$Ru(bipy)_3^{2+} + SO_4^{\dot{-}} \longrightarrow Ru(bipy)_3^{3+} + SO_4^{2-} \quad (4)$$

$$Ru(bipy)_3^{3+} + 1/2H_2O \xrightarrow{Cat} Ru(bipy)_3^{2+} + H^+ + 1/4O_2\uparrow \quad (5)$$

The latter reaction takes place only in the presence of catalyst (Cat) which in our case is Co(II) salt introduced into solution. In principle, H_2O can also be oxidized by the anion radical $SO_4^{\dot{-}}$:

$$SO_4^{\dot{-}} + 1/2H_2O \xrightarrow{Cat} SO_4^{2-} + 1/4O_2 + H^+. \quad (6)$$

Discrepancy between the amounts of O_2 evolved in homogeneous persulphate-containing solution and the stoichiometry of reaction (1) is due to the oxidant consumption to oxidize photocatalyst ligands:

$$Ru(bipy)_3^{3+} + Ru(bipy)_3^{2+} \longrightarrow Ru(bipy)_3^{2+} + P \quad (7)$$

$$Ru(bipy)_3^{2+} + SO_4^{\dot{-}} \longrightarrow SO_4^{2-} + P \quad (8)$$

$$Ru(bipy)_3^{3+} + P \text{ (or } P') \longrightarrow Ru(bipy)_3^{2+} + P' \quad (9)$$

$$SO_4^{\dot{-}} + P \text{ (or } P') \longrightarrow SO_4^{2-} + P', \text{ etc.} \quad (10)$$

Here P denotes the primary product of photo-catalyst destruction which is the Ru(II) complex of $(bipy)_2Ru(bipy')$ type with a partially oxidized ligand (bipy'). Further oxidation of P yields deeper destruction products until the detected CO_2 is formed. Apparently, P and P' products are oxidized much easily compared to the relatively stable $Ru(bipy)_3^{2+}$, hence one can suggest that the most essential routes of side-processes are reactions (9) and (10) leading to the useless consumption of oxidants. The contribution of reactions (7) and (8) to the side-processes seems to be significantly lower, but their effect on the H_2O oxidation selectivity is very high, since they account for the formation rate of easily oxidized products - P.

When using a $[Co(NH_3)_5Cl]^{2+}$ cation as an electron acceptor, its irreversible destruction

$$^{*}Ru(bipy)_3^{2+} + [Co(NH_3)_5Cl]^{2+} \longrightarrow Ru(bipy)_3^{3+} + Co^{2+} + 5\ NH_3 + Cl^- \qquad (3')$$

is followed by reactions of types (5), (7) and (9) resulting in the evolution of O_2 and CO_2.

Due to the strong adsorption of $Ru(bipy)_3^{2+}$ and $Ru(bipy)_3^{3+}$ cations on negatively charged surfaces of micelles in a solution of SDS (Refs. 14,15) the reactions involving the participation of the above complexes take place directly on water-micellar interface. There are no grounds to suggest that the presence of micelles changes significantly the above mechanism of H_2O photocatalytic oxidation. It is more likely that the addition of SDS alters only the rates of some reactions involving the participation of $Ru(bipy)_3^{2+}$ and $Ru(bipy)_3^{3+}$ adsorbed on micelles. Here one can suggest that the rates of reactions (5) and (7) are weakly dependent on the presence of SDS micelles in solution. Actually, these reactions also take place if $[Co(NH_3)_5Cl]^{2+}$ is used as an acceptor. In this case, however, the introduction of SDS into solution exerts practically no effect on O_2 photogeneration. Therefore one can suggest that the specificity of persulphate-containing systems manifests itself primarily in reactions with the participation of $S_2O_8^{2-}$ and $SO_4^{\dot{-}}$ anions. Thus, in micellar solutions the rate of reaction (3) decreases significantly, which is directly evidenced by the decrease of the observed rate constant of $^{*}Ru(bipy)_3^{2+}$ quenching by $S_2O_8^{2-}$ anion (Table). This effect can readily be interpreted by the strong Coulombic repulsion of $S_2O_8^{2-}$ anions from the negatively charged micellar surfaces with $Ru(bipy)_3^{2+}$ cations adsorbed on them into the bulk solution. On the contrary, electrostatic attraction of the $[Co(NH_3)_5Cl]^{2+}$ cation to the surfaces of micelles increases the rate constant of reaction (3'). This electrostatic effect of charged micelles on the reaction rate between anions and cations is well known for the quenching of excited states (Ref. 4). By analogy with reaction (3), one can also suggest a significant rate decrease of reactions (4) and (8) in micellar solutions. It is evident that the retardation of reactions (3) and (4) cannot affect the selectivity of H_2O oxidation but leads only to the observed decrease in the overall rate of $S_2O_8^{2-}$ acceptor consumption.

To give reasons for the increase of O_2 yield in the presence of micelles, we believe the most important is to take into account the rate decrease of reaction (8) that is the primary destruction of photocatalyst and the formation of P. Actually, the rates of side-reactions (9) and (10) that are responsible for the useless consumption of photogenerated oxidant are directly dependent on the P (P') concentration. Therefore even insignificant increase in the primary destruction of photocatalyst can considerably enhance the contribution of secondary reactions of bipyridyl ligand oxidation, which leads to the decrease in H_2O oxidation selectivity.

Thus photocatalyst adsorption on the interface of microheterogeneous micellar systems is in principle capable of promoting the photocatalytic oxidation of H_2O to molecular oxygen. In the system examined the mechanism of such an effect seems to be the protective action of the interface with respect to oxidative destruction of photocatalyst ligands.

Suspensions of lipid vesicles and microemulsions are essentially more complicated in structure than micellar solutions. But at the same time they are considered nowadays to be the most promising objects for accomplishing a closed-loop process of H_2O photocatalytic cleavage into H_2 and O_2 (Refs. 3,4).

Lipid vesicles are microscopic (about 300 Å in diameter) spherical species formed by a bilayer lipid (lecithin in the experiments described in what follows) membrane, whose inner and outer volumes could have unmiscible aqueous solutions of different composition. Reliable spatial separation between oxidant and reductant photogenerated in vesicle suspensions can be obtained by the electron transfer through the membrane by using an electron-carrier embedded into membrane. From the viewpoint of the H_2O photocatalytic cleavage, the promising systems are those (Refs. 16,17) wherein the electron transfer through a membrane is accompanied by the formation of strong oxidants $Ru(bipy)_3^{3+}$ in the internal cavities of vesicles. These systems permit us to attempt to congugate the electron phototransfer across a membrane with the catalytic oxidation of H_2O inside vesicles. The main difficulty arised here is due to an extremely small size (ca.100 Å) of the inner volume of the vesicle. Evidently, in this case one could try either to use the reported homogeneous catalysts of oxygen evolution, which should be introduced into the cavity of lipid vesicle during the process of its formation,or to organize assembling the water-insoluble catalyst from the components introduced inside the vesicles.

We have started studying H_2O photocatalytic oxidation inside small isolated water volumes by using a "water-in-oil" microemulsion consisting of p-xylene, 1-pentanol, SDS and water (Ref. 10). Aqueous droplets of this microemulsion are convenient models for the internal cavities of vesicles and at the same time they are convenient for standard physico-chemical studies. Photocatalytic evolution of O_2 in such a microemulsion has been carried out by introducing $Ru(bipy)_3^{2+}$, and an irreversible acceptor of the electron – $[Co(NH_3)_4CO_3]^+$ into the dispersed aqueous phase. Due to the electron phototransfer from $Ru(bipy)_3^{2+}$ to the acceptor, a strong oxidant – $Ru(bipy)_3^{3+}$ is formed, while the reduced cobalt complex decomposes to produce Co(II) similarly to reaction (3'). The catalyst of H_2O oxidation via reaction (4) seems to be formed inside the droplets from the accumulated Co(II).

The diameter of a water droplet of the microemulsion (~150 Å) is quite close to the inner cavity of lecithin vesicle. This enabled us to use the discussed results to create a reaction centre to the photocatalytic oxygen evolution in the inner volumes of lipid vesicles by introducing there $Ru(bipy)_3^{2+}$ and $[Co(NH_3)_4CO_3]^+$. The sequence of reactions leading to O_2 production under visible light illumination ($400 < \lambda < 500$ nm) seems to be the same for both microemulsion and vesicle suspension studied. A conceptual sequence of these reactions for the vesicle is given in Fig. 2.

Measured selectivities of H_2O oxidation in microemulsions and vesicle suspensions were β = 5 and 2%, respectively. It is slightly lower than in the respective homogeneous aqueous solution where β = 9.5%. Decrease of the β value in microheterogeneous systems is, apparently, due to the consumption of photogenerated $Ru(bipy)_3^{3+}$ for the oxidation of organic compounds that form microemulsions and vesicles. This suggestion agrees with the β decrease observed previously (Ref. 5) upon adding vesicles into the aforesaid "sacrificial" $Ru(bipy)_3^{2+}$-$CoCl_2$ - $[Co(NH_3)_5Cl]^{2+}$ system.

The evolution of O_2 from the inner cavities of the vesicles is proved by the fact that a great excess of Na_2EDTA in the outer volume of vesicles had no effect on the yield of oxygen. Indeed, Na_2EDTA is known to reduce efficiently the $Ru(bipy)_3^{3+}$ particles, and introduction of Na_2EDTA into any available homogeneous system for the photocatalytic O_2 evolution inevitably leads to the complete suppression of water oxidation. Thus the discussed experiments demonstrate a possibility to design the photochemical reaction centre for water oxidation to oxygen in the inner volumes of lipid vesicles even in the presence of the reducing media (needed to hydrogen evolution) outside vesicles and, therefore, to create on this basis a system with the complete cycle of water cleavage.

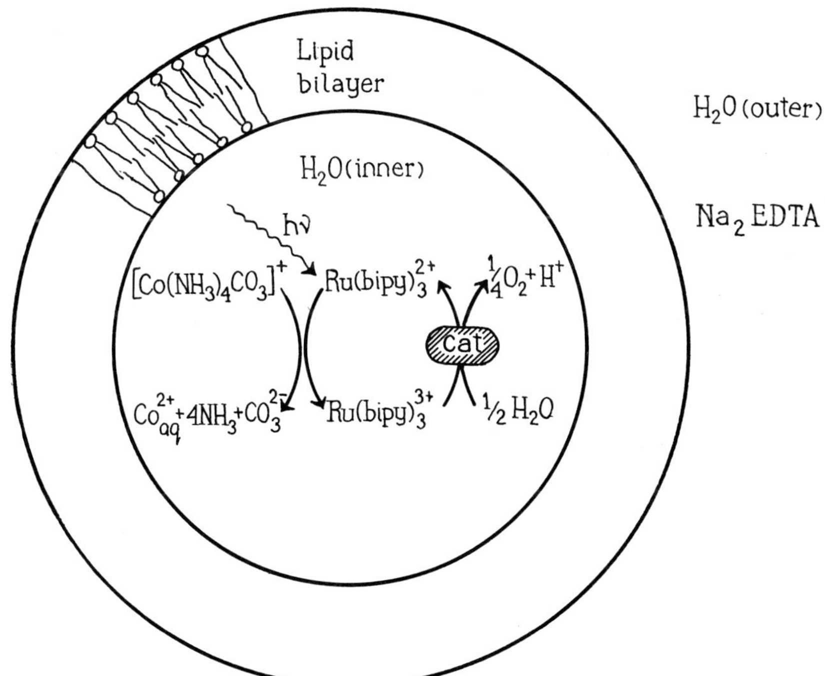

Fig. 2. Suggested scheme for functioning of the reaction centre for O_2 evolution in the internal cavity of lipid vesicle. Cat. is the catalyst for H_2O oxidation based on Co compounds (Refer to the text).

For the creation of such system in the suspension of vesicles it is required that water oxidation inside vesicles be conjugated with the electron transfer across the vesicle membranes. To accomplish this conjugation, we have attempted to use vesicles with an electron-carrier (cetylviologen) embedded into membranes. It has been shown previously (Refs. 16,17) that in such vesicles electron phototransfer is possible across a membrane from $Ru(bipy)_3^{2+}$ localized inside a vesicle to the acceptor outside vesicles. This transfer leads to the formation of $Ru(bipy)_3^{3+}$ inside vesicles. It could be expected that the introduction of the appropriate catalyst into the internal cavities of vesicles would lead to the photocatalytic oxidation of water inside these cavities conjugated with the acceptor reduction outside the vesicles. For a realization of this possibility, we have modified the system shown in Fig. 2 by introducing an electron-carrier (cetylviologen) into the vesicle membrane and a strong irreversible electron acceptor - $NaIO_4$, instead of Na_2EDTA, into the outer aqueous solution. Such modification resulted in increase of the maximum total yield of O_2 obtained under illumination by more than a factor of 40 in comparison with the system without $NaIO_4$. Unfortunately, in spite of unambiguous evidences in favour of an extremely important role of cetylviologen in the process under discussion, we cannot state with certainty that O_2 evolution in this system is caused by the electron phototransfer across the membrane onto IO_4^- but not by IO_4^- penetration into the inner cavity of the vesicle. The study to answer this question is underway. At the same time, the selective introduction of Na_2EDTA into either the inner or outer volume of the vesicles have provided a reliable evidence in favour of the fact that the O_2 photogeneration in the $NaIO_4$-containing system does occur in the inner cavity of the vesicle. Here it is essential, that taking into account the necessity to use four ions of a one-electron oxidant to form one O_2 molecule via reaction (5), the turn-over number of photocatalyst in the course of O_2 evolution in the $NaIO_4$-containing system is not lower than four even if one does not take into account a much higher consumption of photogenerated $Ru(bipy)_3^{3+}$ in side-processes of oxidation of organic compounds and its own ligands. It indicates the catalytic character of functioning of the reaction centre for H_2O photooxidation in the internal cavity of vesicle.

Thus the results indicate that it is possible to create model artificial reaction centre for the photocatalytic oxidation of H_2O to O_2 in microheterogeneous systems. By developing routes to conjugate such centres with spatially-separated reaction centres for H_2O reduction to H_2, one can hope to create a model photocatalytic system with a closed cycle of H_2O cleavage.

REFERENCES

1. Energy Resources Through Photochemistry and Catalysis, Ed. M. Gratzel. - N.-Y.: Acad.Press, 1983.
2. Photochemical Conversion and Storage of Solar Energy, Ed. J.S. Connoly. - N.-Y.: Acad.Press, 1981.
3. Photocatalytic Conversion of Solar Energy, part 1,2, Ed.by K.I. Zamaraev, Nauka, Novosibirsk, 1985.
4. S.V. Lymar, V.N. Parmon, Ibid, part 2, pp. 186-244.
5. I.M. Tsvetkov, S.V. Lymar, V.N. Parmon, K.I. Zamaraev, Kinet.Katal., 27, 109-116 (1986).
6. R.H. Paleer, T.S. Piper, Inorg.Chem., 5, 5, 864-878, (1966).
7. C.F. Lin, N.C. Lin, J.C. Bajlar, Inorg.Chem., 3, 8, 1085-1087 (1964).
8. H.T. van Dam, J.J. Ponjee, J.Electrochem.Soc., 121, 1555 (1974).
9. F.B. Glinka, N.G. Klyuchnikov, Chemistry of Complex Compounds, Prosveshchenie, Moskva, 1967.
10. S.V. Lymar, M.I. Khramov, V.N. Parmon, K.I. Zamaraev, Khim.Fiz., 2, 550-553 (1983).
11. S.V. Lymar, I.M.Tsvetkov, V.N. Parmon, K.I. Zamaraev, Khim.Fiz., 1, 405-410 (1982).
12. N.K. Khannanov, V.Ya. Shafirovich, Dokl.Akad.Nauk SSSR, 260, 1418-1421 (1981).
13. M. Neumann-Spallart et al., Helv.Chim.Acta, 63, 1111 (1980).
14. D. Meisel, M.S. Matheson, J. Rabani, J.Am.Chem.Soc., 100, 1, 117-122 (1978).
15. O.V. Gerasimov, V.N. Parmon, Zh.Fiz.Khim., 2, 111 (1986).
16. V.N. Parmon et al., React.Kinet.Catal.Lett., 23, 409 (1983).
17. E.E. Yablonskaya, V.Ya. Shafirovich, Nouv.J.Chim., 8, 2, 117-122 (1984).

HYDROGEN-TRANSFER REACTIONS CATALYZED BY HETEROGENIZED RUTHENIUM(II) COMPLEXES BOUND TO POLYCARBOXYLATE MATRICES

G. Valentini, A. Cecchi, C. Di Bugno, G. Braca and G. Sbrana

Dipartimento di Chimica e Chimica Industriale, Università di Pisa, 56100 Pisa, Italy

ABSTRACT - $RuH_2(PPh_3)_4$ has been attached to different polycarboxylate matrices (carboxylic acid resins, polyacrylic acid, eventually post-crosslinked with hexamethylene diamine) to give heterogenized carboxylato-derivatives.
The polymer-supported catalysts, as well as polymer-bound ruthenium systems obtained by terpolymerization of bis(acrylo)bis(triphenylphosphine) complexes, have been tested as catalysts in the hydrogen-transfer from alcohols to aldehydes.
The information obtained by IR spectroscopy on the nature of the coordinated metal species and correlations with the catalytic properties of the systems are discussed.
The possibility to isolate reactive reaction intermediates by using polymeric catalysts is also evidenced.

1. INTRODUCTION

Soluble carboxylato(triphenylphosphine)ruthenium(II) complexes have been recently shown to be among the more active ruthenium catalysts in the hydrogen transfer from alcohols to aldehydes (Ref. 1).

Mechanistic investigation on the modifications of the catalytic precursors under reaction conditions revealed that bis(carboxylato)-derivatives retain one carboxylate ligand in the coordination sphere of the metal, whereas mono(carboxylato)-derivatives undergo a complete release of the ligand (Ref. 1).

The immobilization of ruthenium complexes onto polymeric matrices containing carboxylate pendents with formation of heterogeneous bis(carboxylato)-derivatives was therefore of interest in order to get active and easily reusable catalysts.

2. RESULTS AND DISCUSSION

The preparation of heterogenized carboxylate systems was approached following different strategies based either on the use of polymeric supports as carboxylic resins and linear polyacrylic acid (PAA), or on the polymerization of performed bis(acrylato)ruthenium(II) complexes (Ref. 2).

The ruthenium precursor chosen for the anchoring reaction was $Ru_2(PPh_3)_4$ which can react with the carboxylic groups of the matrices giving either mono(carboxylato)- or bis(carboxylato)-species (Equations 1 and 2).

$$\text{Ⓟ-COOH} + RuH_2(PPh_3)_4 \rightarrow \text{Ⓟ-COORuH}(PPh_3)_3 + H_2 + PPh_3 \quad (1)$$

$$2\ \text{Ⓟ-COOH} + RuH_2(PPh_3)_4 \rightarrow \text{Ⓟ-COORu}(PPh_3)_2\text{OOC-Ⓟ} + 2H_2 + 2PPh_3 \quad (2)$$

The formation of hydrido(carboxylato)-derivatives generally occurs with weak carboxylic acids (pKa > ca.4.2), whereas bis(carboxylato)-species are formed with stronger acid groups (Ref. 3).

The acid strength of carboxylic groups in polymeric systems cannot be predicted since apparent pKa values higher (Ref. 4) or lower (Ref. 5) than the corresponding low-molecular-weight analogs have been observed. Bis(carboxylato)ruthenium species may be expected to be formed, however, under conditions of appropriate chains flexibility and availability of the pendent functionalities.

2.1. Ruthenium Complexes Anchored to Carboxylate Resins

The use of functionalized resins as supports is a widely employed technique to obtain insolubilized metal derivatives (Refs. 6,7). The main drawbacks generally concerned are related to difficulties in achieving high metal loadings, due to the limited availability of the reactive functionalities, and in the characterization of the anchored metal species.

As far as the former aspect is concerned, the porosity and swellability of the resin, and bulkiness of the metal precursor play key roles.

In this view, Amberlite IRC50 was chosen as support on account of high exchange power (8.0 meq/g dry) and large pores size (800 Å), using tetrahydrofuran (THF) as reaction medium since it provides both a high swelling of the resin and a complete solubility of the metal precursor.

The metal loading of a sample obtained using an initial COO/Ru molar ratio of 0.8 (AMB1 in Table 1) was limited (Ru%, 2.38) and it could be increased only very slightly working with a lower COO/Ru molar ratio (AMB2 in Table 1).

In this case the limiting factor seems to be the accessibility of the metal precursor to the carboxylate functions of the resin. Accordingly, the use of a microporous resin (Kastel C101, average pores size 30 Å) was completely unsuccessful for the preparation of anchored catalysts.

2.2. Ruthenium Complexes with Polyacrylic Acid (PAA)

The use of a soluble poly-carboxylate ligand as polyacrylic acid, in which the carboxylate pendents are more accessible, allowed to obtain higher metal loadings and also modulate them by varying the reaction conditions.

The metal content of the polymer-bound systems (PAA1, PAA2 and PAA3 in Table 1) increased, albeit non linearity, upon decreasing the COO/Ru molar ratio used in the preparation, whereas the P/Ru ratio of the coordinated species decreased correspondingly, due to the increased difficulty of incoming metal species in achieving coordination onto the polymer. These P/Ru ratios are considerably lower than those found with the rigid support Amberlite, and confirm that flexible chains cause a significant displacement of the bulky triphenylphosphine ligands from the metal precursor (Ref. 8).

The systems with the higher metal loadings (PAA2 and PAA3) showed however a marked metal leaching under catalytic conditions (see later). Therefore, in order to suppress or reduce this phenomenon, the preparations of these catalysts were replicated adding hexamethylenediamine (HMD) to produce cross-linking through the free carboxylic groups (Ref. 9).

The characteristics of the products obtained (PAA2H and PAA3H in Table 1) are different from those of the above described resin- and linear PAA-derivatives since HMD introduces rigidity in the system but also competes with metal precursor in coordination.

2.3. Polyacrylato Ruthenium(II) Complexes Obtained by Polymerization of Organometallic Monomers

The preparation of new bis(acrylato)bis(triphenylphosphine)ruthenium(II) complexes (I, II and III), and their co-polymerization with MMA and 1,4-butanediol-dimethacry-

TABLE 1. Analytical data for ruthenium-carboxylate complexes immobilized on different polymeric supports

Complex	Metal precursor	Elemental analysis					P/Ru	COO/Ru		COO/N		Cross-linking %
		Ru, %	P, %	C, %	H, %	N, %		initial	final	initial	final	
AMB1	a	2.38	1.93	53.23	6.65	-	2.60	8	28.6	-	-	-
AMB2	a	2.65	2.08	53.54	6.60	-	2.56	4	25.7	-	-	-
PAA1	a	4.58	1.40	51.46	5.81	-	1.0	20	25.6	-	-	-
PAA2	a	6.93	2.05	54.04	6.02	-	0.96	10	15.3	-	-	-
PAA3	a	10.92	2.75	56.86	4.94	-	0.82	5	8.5	-	-	-
PAA2H	a	7.98	2.68	54.15	5.88	2.32	1.09	10	10.9	5	6.8	14.7
PAA3H	a	8.62	3.70	56.30	5.19	0.78	1.40	5	9.1	5	11.4	8.8
TER	b	1.19	0.66	61.48	8.12	-	1.80	-	-	-	-	11
TER/CO	c	1.15	0.67	63.37	7.82	-	1.90	-	-	-	-	30

[a] $RuH_2(PPh_3)_4$; [b] $Ru(OCOCH=CH_2)_2(PPh_3)_2$; [c] $Ru(CO)_2(OCOCH=CH_2)_2(PPh_3)_2$

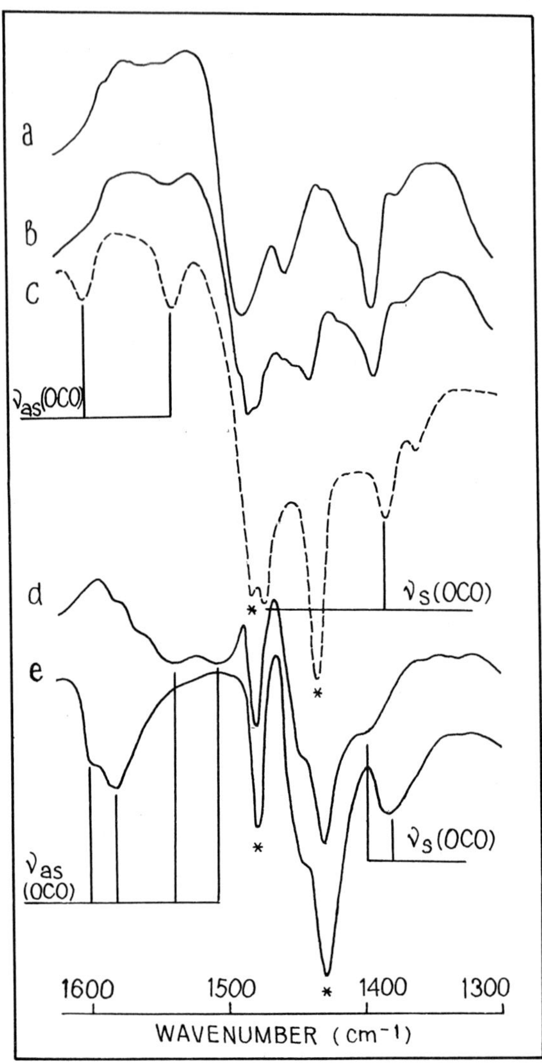

Figure 1. IR spectra of heterogenized ruthenium complexe in the region between 1600 and 1300 cm^{-1}.
(a) Amberlite IRC50; (b) resin-bound complex AMB1; (c) expanded computer-aided difference between spectra b and a; (d) PAA-bound complex PAA3; (e) the same complex after treatment in the solid state with CO.
* Absorptions due to PPh_3.

late (BDDMA) has been recently reported by us (Ref. 2).

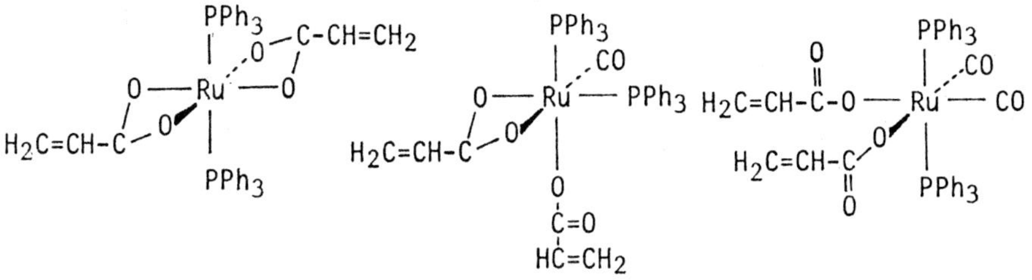

(I) (II) (III)

The procedure represents a very reliable, even though time-consuming, strategy for the obtainment of heterogenized complexes where metal species are all in the same, predetermined coordinative situation, the structure of the metal precursor being retained during the polymerization (Ref. 2).

Cross-linked ter-polymers of (I) and (III) (TER and TER/CO, respectively, in Table 1) were preferred as catalysts for their higher insolubility.

2.4. Characterization of the Polymer-Bound Metal Species

The major information on the nature of the coordinated metal species was given by IR spectroscopy.

In the case of the resin-bound complexes (AMB1 and AMB2) the IR absorptions of the carboxylate ligands coordinated to the metal were very weak and partially blurred by strong absorptions of the support (Figure 1, spectra a and b). A computer-aided subtraction of the spectrum of the resin allowed however to reveal the absorptions due to the $\nu_{as}(OCO)$ and $\nu_s(OCO)$ modes together with bands of PPh_3 (Figure 1, spectrum c).

The absorptions at 1600 and 1382 cm^{-1} ($\Delta\nu = 208\ cm^{-1}$) and those at 1537 and 1471 cm^{-1} ($\Delta\nu = 66\ cm^{-1}$) suggest the presence of both mono- and bi-dentate carboxylate ligands bound to the metal centres (Refs. 3,10).

The systems bound to linear PAA contain the carboxylate ligands predominantly with a bi-dentate coordination (Figure 1, spectrum d).

No useful information on the type of coordination of the carboxylate ligands could be obtained, on the contrary, from the IR spectra of cross-linked PAA-derivatives of strong and broad bands of the ammonium-carboxylate groups (Ref. 9).

On the basis of the IR spectra it was not possible however to decide whether one or more carboxylate groups were present in the coordination sphere of the metal. A possible way to circumvent this problem was suggested by the observed reactivity of low-molecular-weight carboxylato-models towards carbon monoxide.

Preliminary experiments on $Ru(OCOCH_3)_2(PPh_3)_2$(IV), $Ru(CO)(OCOCH_3)_2(PPh_3)_2$ (V) [ν(CO) 1954 cm^{-1}] and $RuH(OCOCH_3)(PPh_3)_3$ (VI) showed in fact that the treatment in the solid state with carbon monoxide produces the corresponding carbonyl derivatives by simple change of the coordination of the carboxylate ligand(s) from bi-dentate to mono-dentate. So a mixture of the mono(carbonyl) complex (V) and bis(carbonyl) complex $Ru(CO)_2(OCOCH_3)_2(PPh_3)_2$ (VII) [ν(CO) 2032 and 1978 cm^{-1}] was obtained when (IV) was reacted with CO at 50 atm for 2 h, whereas the bis(carbonyl) complex (VII) was the unique product for longer reaction times (12 h), or starting from (V). Analogously, the hydrido complex (VI) gave $RuH(CO)(OCOCH_3)(PPh_3)_2$ (VIII) [ν(CO) 1930 cm^{-1}].

On this ground, the ν (CO) frequencies of the generated carbonyl derivatives are of interest since they can provide correlations with the number of coordinated carboxylate ligands.

The IR spectra of the heterogenized complexes after reaction in the solid state with carbon monoxide at 50 atm and room temperature for 12 h indicate the change of the coordination of the carboxylate ligand from bi- to mono-dentate (Figure 1, spectra d and e) and the formation of the almost two main carbonyl species A and B (Figure 2, solid-line spectra). Species A, having a pair of ν(CO) frequencies at 2050 and 1996 cm^{-1} comparable with the ν(CO) absorptions of (VII), are probably bis(carbonyl)bis(carboxylato)-species. The species B, having a unique ν(CO) absorption which falls in the range 1956-1952 cm^{-1}, may correspond either to a mono(carbonyl)bis(carboxylato)-species analogous to (v), or, more likely, to a mono(carbonyl)mono(carboxylato)-species analogous to (VIII) with a lower content of PPh_3.

The relative content of the mono(carbonyl)mono(carboxylato)-species B increases with increasing the metal loading in the linear PAA derivatives (about 14% in PAA1 and 46% in PAA3 (Ref. 11)), as far as in the crosslinked systems.

The IR spectra of the TER and TER/CO systems have been already reported (Ref. 2). It is important to outline here that the ν (CO) frequencies of the TER/CO system (2055 and 1922 cm^{-1}) were slightly shifted towards higher wavenumbers with respect to those of the starting monomer (III) (2032 and 1978 cm^{-1}), analogously to what observed for poly-carboxylato-derivatives with respect to homogeneous models.

This seems to suggest that poly-carboxylate matrices behave as stronger acids than the corresponding low-molecular-weight analogs.

The treatment of the non(carbonyl)TER system with CO under pressure resulted in the formation of a carbonyl

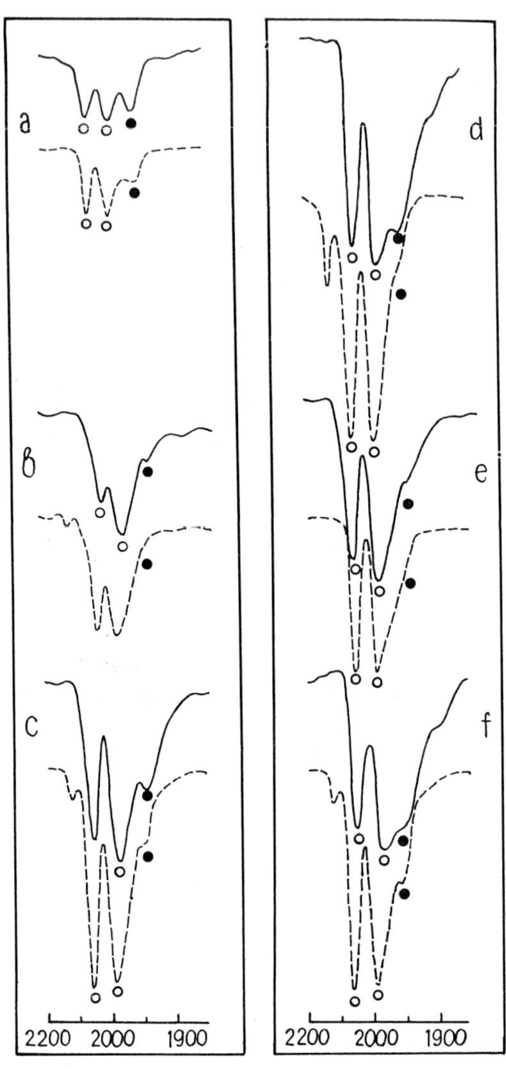

Figure 2. IR spectra of heterogenized ruthenium complexes in the region between 2200 and 1900 cm^{-1} after treatment in the solid state with CO (solid-line spectra) and recovered after catalytic runs (dotted-line spectra). (a) Resin-bound complex AMB1; (b) PAA-bound complex PAA1; (c) PAA-bound complex PAA2; (d) PAA-bound complex PAA3; (e) PAA-bound and postcrosslinked complex PAA2H; (f) PAA-bound and postcrosslinked complex PAA3H. o Species "A". ● Species "B".

derivative analogous to TER/CO having ν (CO) frequencies at 2053 and 1988 cm^{-1} (Figure 4, spectrum a).

These evidences indicate that the behaviour of polymer-bound carboxylate species strictly parallel that of the acetato models.

2.5. Catalytic Properties of the Heterogenized Systems

The above described heterogenized systems were tested as catalysts in the hydrogen transfer from cyclohexanol to n.hexaldehyde (Eq. 3) at 100°C using toluene as solvent.

$$n.C_6H_{11}CHO + \text{C}_6\text{H}_{11}\text{-OH} \longrightarrow n.C_6H_{11}CH\ OH + \text{C}_6\text{H}_{10}\text{=O} \qquad (3)$$

Activity, selectivity and metal leaching of the systems were evaluated in successive cycles (Figure 3). The activity is expressed as the maximum number of turnovers performed by each ruthenium atom per hour (TN_{max}), whereas the selectivity is calculated as (moles of n.hexanol formed/moles of n.hexaldehyde converted) x 100, being the aldehyde substrate possibly consumed by acid-catalyzed aldol condensation reactions.

The metal leaching is the percent of the metal charged in each catalytic run which was found in solution at the end of the experiment.

According to the generally observed lower activity of heterogenized derivatives with respect to "homogeneous" models (Refs. 6,12), all the poly(carboxylato)-systems exhibited TN_{max} lower than (VII), which is the best ruthenium catalyst so far reported for the above reaction (TN_{max}: 16, selectivity: 80%) (Ref. 1).

As a common feature, the polymer-bound complexes showed also in the first catalytic cycle a selectivity lower

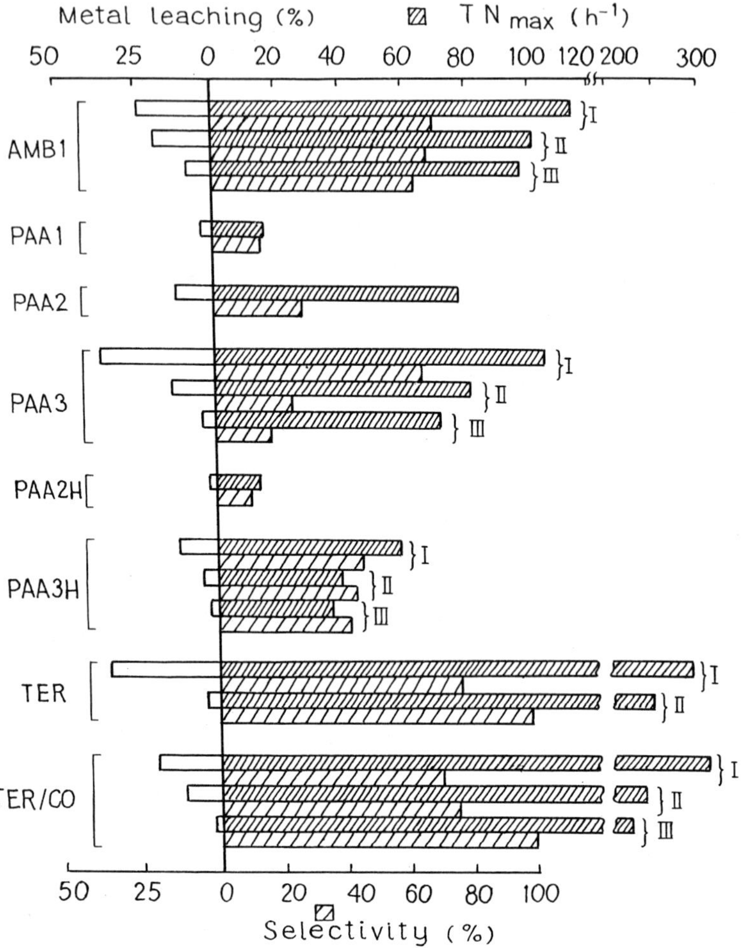

Figure 3. TN_{max}(h^{-1}), selectivity and metal leaching of the heterogenized ruthenium catalysts in the hydrogen-transfer from cyclohexanol to n.hexaldehyde at 100°C in toluene as solvent.
Reaction conditions. Cyclohexanol:2M; n.hexaldehyde:1M; Molar ratio Ru/n.hexaldehyde: 1/50.

than the model complex (VII).

As far as these aspects are concerned, it is necessary to remind here that the hydrogen-transfer reaction has been demonstrated for soluble carboxylato-derivatives to proceed through an initial dissociation of one carboxylate group, whereas the second carboxylate ligand, where present, always remains in the coordination sphere of the metal (Ref. 1).

The free carboxylic acid groups present in the polymeric matrices, or produced during the activation of the catalysts, can be involved in some acid-catalyzed reactions of the aldehyde substrate and are responsible for the observed lower than 100% selectivity in the hydrogen transfer reaction.

The metal leaching of the polymer catalysts was different for the different systems and always more pronounced in the first catalytic cycle. Except for TER and TER/CO complexes, the extent of metal leaching can be related to the amount of mono(carboxylato)ruthenium species present in the fresh catalysts (see Figure 2), which undergo dissociation under reaction conditions.

Postcrosslinking of the PAA derivatives resulted in a more limited metal leaching with respect to "linear" derivatives.

The resin- and PAA-bound complexes showed a comparable behaviour upon recycling in that activity, selectivity and metal leaching progressively decreased.

The catalysts bound to linear PAA showed poor catalytic properties when the metal loading was low (PAA1) and improved performances, but lower heterogeneity, upon increasing the metal content (PAA2 and PAA3). It is interesting to note that the catalytic behaviour of the PAA3 system in the second cycle strictly parallels that of the PAA2 system; upon computing the amount of the metal lost by the former complex in the first cycle, the two systems had a comparable content of metal. This seems to indicate that the catalytic properties of these systems mainly depend on the COOH/Ru ratio of the working system

The decreased selectivity of these catalysts upon recycling can be attributed to the previously mentioned increase of the free carboxylic groups in the polymeric system and their easier accessibility due to release of metal species into solution.

The heterogenized derivatives obtained by polymerization of the bis(acrylato)-precursors (TER and TER/CO) showed the best catalytic properties as expected on account of their strict similarity with low-molecular-weig bis(carboxylato)-analogs and absence of free carboxylic groups in the starting polymeric catalyst.

These systems showed in fact a selectivity in the first cycle only a little bit lower than the homogeneous model (VII), whereas the catalytic activity was the highest among the heterogenized catalysts and about 5 times lower than that of the reference complex.

The rather high metal leaching of these systems in the first catalytic run, which however significantly dropped out upon recycling, was due to the partial solubilization of oligomeric fractions of the products still containing carboxylato-metal species, as indicated by IR spectra of solids recovered by evaporation of the reaction solvent.

It is interesting to note a very striking feature of these complexes upon recycling relative to the progressive increase of the selectivity, which reached one hundred per cent in the second or third cycle. This peculiar behaviour, related to the progressive consumption of the acid groups generated in the activation of the catalysts with formation of the corresponding esters, offers interesting advantages on the homogeneous analogs and overcompensate for their higher activity.

2.6. Reaction Mechanism

In order to get information on the nature of the polymer-bound metal species involved in the catalysis, an IR study was carried out on the heterogenized complexes recovered after catalytic experiments.

All the insoluble catalysts contained carbonyl(carboxylato)ruthenium species due to the occurrence of a stoichiometric decarbonylation of the aldehyde substrate under reaction conditions, analogously to what found for soluble carboxylate derivatives (Ref. 1).

The catalysts bound to Amberlite and linear or postcrosslinked PAA showed ν(CO) absorptions in the IR spectrum at frequencies very similar to those shown by the fresh samples after treatment with CO (cfr. dotted-line spectra with solid-line spectra in Figure 2). The absorptions at lower wavenumbers, designated above as species B, were however as more reduced in intensity as the metal leaching of the complex under catalytic conditions was higher (cfr. Figures 2 and 3).

This observation and the supposed analogies between the mechanisms of the homogeneous and heterogenized (carboxylato)-species strongly support the contention that species B are generated by mono(carboxylato)ruthenium species.

An extra absorption at _ca_. 2120 cm^{-1} was also present in all the investigated spectra, probably due to ν (Ru-H) of hydrido species involved in the catalytic process. A treatment of the recovered catalysts with CO under pressure did not produce changes in the IR spectra of the products.

A rather different behaviour was observed in the case of catalysts TER and TER/CC. Both systems showed identical IR spectra inthe ν(CO) region after catalysis (Figure 4, spectrum b), consistent with the presence of

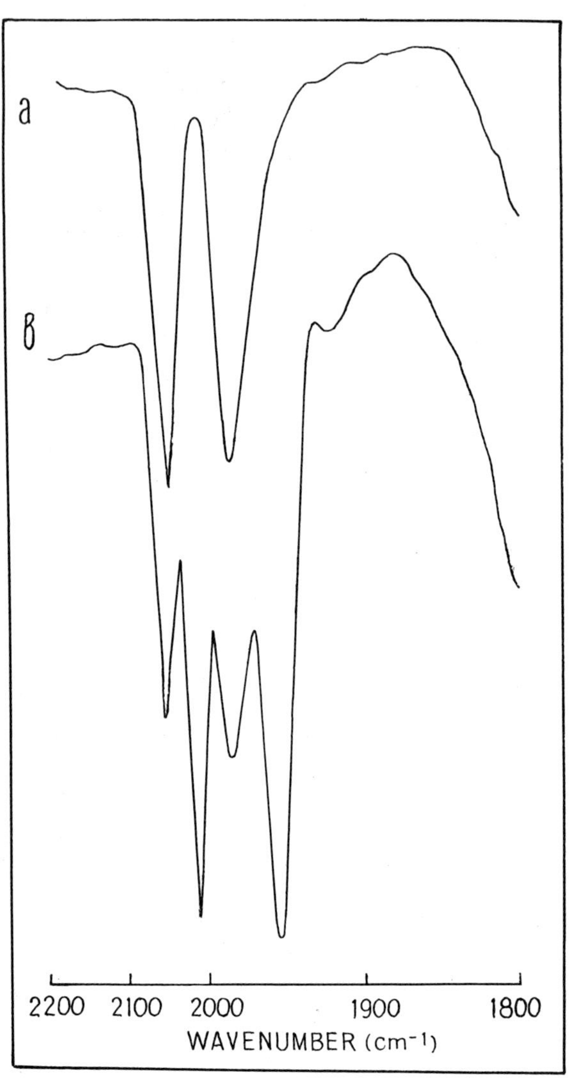

Figure 4. IR spectra in the region between 2200 and 1800 cm^{-1} of heterogenized ruthenium complexes obtained by ter-polymerization of bis(acrylato) complexes.
(a) Complex TER after treatment in the solid state with CO;
(b) complex TER/CO recovered after catalytic runs.

two different bis(carbonyl) species having absorptions at 2055-1988, and 2012-1954 cm^{-1}, respectively.

The former species corresponds to the bis(carbony)bis (carboxylato)-species initially present in the TER/CO system by decarbonylation of the aldehyde substrate. The latter species, for which the shift of the ν(CO) frequencies towards lower wavenumbers is indicative of a higher electron density around the metal centre, is probably an alkoxo-species where a carboxylate ligand has been replaced by an alkoxide group. The formation of an analogous alkoxo-intermediate has been demonstrated in solution for the model complex (VII), but the isolation of the product was impossible (Ref. 1).

The yields of formation of alkoxide derivatives in heterogenized or homogeneous systems, calculated on IR data, were 65% and 40%, respectively. Because it is known (Ref. 13) that carboxylic groups of weaker acids are more reluctant to undergo dissociation, this evidence suggests that the acrylate groups in the polymer system behave as stronger acid groups than acetic acid.

When the TER/CO complex was allowed to react with cyclohexanol at 100°C for 2 h, the IR spectrum of the recovered product showed only the ν(CO) frequencies of the alkoxo-species.

Upon treating the system with an excess of boiling acetic acid, the bis(carbonyl)bis(acetato) complex (VII) was obtained quantitatively together with 0.87 moles of cyclohexyl acetate per mole of ruthenium, this confirming the existence of the alcohol in the coordination sphere of the metal.

The possibility of isolating a reactive catalytic intermediate bound to a macromolecular system may be attributed to the peculiar stabilizing effect of the polymeric ligand and accounts also, together with availability problems, for the lower activity of the heterogenized catalysts with respect to homogeneous models.

3. CONCLUSIONS

The use of pre-formed polymeric supports for the obtainment of heterogenized catalysts may show some complicances due to the difficulty of the anchoring of meta moieties having only a pre-determined structure and chemical environment.

These drawbacks can be conveniently overcome by the preparation and polymerization of tailor-made organometallic monomers.

Our results suggest, indeed, new and very stimulatin potentialities concerned with the use of polymer-bound metal complexes for the isolation of reaction intermediates and a better understanding of catalytic processes.

4. EXPERIMENTAL

IR spectra were taken for samples as KBr disks on Pe kin-Elmer 283 grating spectrometer equipped with a 3600 Data Station and suitable softwares for spectra subtraction and areas calculations. The apparatus used for hydrogen-transfer experiments and GLC analysis of the reaction products were as previously described (Ref. 1).

Solid catalysts were recycled by siphonying off the decanted solution followed by drying under a stream of N_2.

The determination of P was carried out by X-ray fluorescence (Ref. 14). Ruthenium was determined spectrophotometrically (Ref. 15).

$RuH_2(PPh_3)_4$ (Ref. 16), $Ru(OCOCH_3)_2(PPh_3)_2$ (Ref. 17), $Ru(CO)(OCOCH_3)_2(PPh_3)_2$ (Ref. 18), $Ru(CO)_2(OCOCH_2)_2(PPh_3$ (Ref. 19), $Ru(OCOCH{=}CH_2)_2(PPh_3)_3$ (Ref. 2), and $Ru(CO)_2(OCOCH{=}CH_2)_2(PPh_3)_2$ (Ref. 2) were prepared accoring to literature methods.

Amberlite IRC50 (Rohm and Haas) was dried under vacuum at 50°C after extraction with anhydrous ethanol for 4 h in a Soxlet apparatus. Hexamethylenediamine (Fluka AG, 97%) was used as purchased.

All solvents were thoroughly degassed before use and all all operations were carried out in a nitrogen atmosphere. Tetrahydrofuran was distilled prior to use.

For the treatments with CO (Johnson and Matthey,99.8%), the solid complexes contained in glass vials were charged under nitrogen in a stainless-steel autoclave which, after evacuation, was pressurized at 50 atm.

5. PREPARATION OF THE POLYCARBOXYLATE RUTHENIUM COMPLEXES

5.1. Amberlite IRC50 Ruthenium Complexes

Dried Amberlite IRC50 (1.14 gr) was allowed to swell for 15 min in refluxing THF. A solution of $RuH_2(PPh_3)_4$ (1.31 gr for AMB1 and 2.62 gr for AMB2) in the same solvent (50 ml) was then added and the reaction mixture was refluxed for 12 h. After filtration of the solvent, the pale-brown resin beads were washed with THF (100 ml), n.hexane (20 ml) and dried (AMB: 1.2 gr, Ru yield 25%; AMB2: 1.3 gr, Ru yield 15%). Analytical data are collected in Table 1.

5.2. Polyacrylato-Ruthenium Complexes

Polyacrylic acid (300 mg) was dissolved in refluxing THF (30 ml) in the presence of few drops of water. An appropriate amount of $RuH_2(PPh_3)_4$ in order to achieve COO/Ru molar ratios of 20, 10 or 5 in THF (50 ml) was then added drop-wise and the solution was refluxed for 24 h. No solid product separated spontaneously from the red-brown solutions of preparations with COO/Ru ratios of 20 and 10. On evaporation to drieness, pale-yellow powders were obtained which were washed with the reaction solvent (100 ml) and dried (PAA1: 0.26 gr, Ru yield 89.6%; PAA2: 0.40 gr, Ru yield 78.9%). In the preparation carried out with a COO/Ru ratio of 5 a deep-violet solid separated spontaneously and was filtered, washed as above and dried (PAA3: 0.46 gr, Ru yield 55.7%). Analytical data are listed in Table 1.

5.3. Postcrosslinked Polyacrylato-Ruthenium Complexes

The procedures followed for the preparations of the postcrosslinked PAA derivatives were analogous to those described above for the obtainment of the PAA2 and PAA3 complexes, except that hexamethylenediamine (HMD) (0.049 gr, COO/N = 5) in THF (10 ml) was added drop-wise to the reaction mixture. In the former case the addition was made at the end of the addition of the precursor ruthenium complex, whereas in the latter one it was started when one half of the ruthenium solution has been added, and then the feeding of the two solutions proceeded parallelly.

The solid products recovered by filtration of the solvent, washed and dried were, respectively, PAA2H (pale-brown, 0.43 gr, Ru yield 75.8%) and PAA3H (deep-green, 0.44 gr, Ru yield 46.4%). Analytical data are reported in Table 1.

ACKNOWLEDGEMENTS. The present work was carried out under the research program Progetto Finalizzato "Chimica Fine e Secondaria" of the C.N.R., Roma (Section Ab).

REFERENCES

1. G. Valentini, A. Cecchi, C.Di Bugno, G. Sbrana and G. Braca, submitted for publication to J.Organomet. Chem.
2. G. Valentini, A. Cecchi, G. Sbrana and G. Braca, submitted for publication to J.Chem.Soc.Dalton.
3. S.D. Robinson and M.F. Uttley, J.Chem.Soc.Dalton, 1912 (1973).
4. A. Katchalsky and J. Gilles, Rec.Trav.Chim.Pays-Bas, 68, 879 (1949).
5. A.J. Cornish-Bowden and J.R. Kowles, Biochem., 113, 353 (1969).
6. C.U. Pittman, Jr., in Comprehensive Organometallic Chemistry, G. Wilkinson, F.Gordon and E.W.Abel (Eds) Vol.8, p. 553, Pergamon, Oxford (1982).
7. Yu.I. Yermakov, B.N. Kuznetsov and V.A. Zakharov,in Catalysis by Supported Complexes, p. 9, Elsevier, Amsterdam (1981).
8. G. Valentini, G. Sbrana and G. Braca, J.Mol.Catal., 11, 383 (1981).
9. G. Valentini, G. Sbrana, G. Braca and P.Da Prato, J.Catal.,96 (1985).
10. N.F. Curtis, J.Chem.Soc.(A), 1579 (1968).
11. Calculations were made on the assumption that the for the two carbonyl species is equivalent. The areas of the bis(carbonyl) species were computed as one half.
12. F. Ciardelli, G. Braca, C. Carlini, G. Sbrana and G. Valentini, J.Mol.Catal.,14 , 1 (1982).
13. A. Dobson and S.D. Robinson, Inorg.Chem., 16, 137 (1977).
14. E.D. Marshall and R.R. Richard, Anal.Chem.,22, 795 (1950).
15. L. Leoni, F. Marchetti, G. Sbrana, G. Braca and G. Valentini, Ann.Chim.,69, 39 (1979).
16. J.J. Levinson and S.D. Robinson, J.Chem.Soc.(A), 2947 (1970).
17. J.D. Gilbert and G. Wilkinson, J.Chem.Soc.(A),1749 (1969).
18. A. Spencer and G. Wilkinson, J.Chem.Soc.Dalton,786, (1974).
19. B.F.G. Johnson, R.D. Johnston, J. Lewis and I.G. Williams, J.Chem.Soc.(A), 689 (1971).

Part 3

Catalysis by Metal Clusters and Dispersed Metal Particles

SURFACE ANALOGUES OF METAL CARBONYL ADDUCTS AND THEIR POSSIBLE ROLE IN CO HYDROGENATION - A REVIEW

H. Knözinger

Institut für Physikalische Chemie, Universität München, Sophienstraße 11, 8000 München 2, W.-Germany

ABSTRACT - The structures of metal carbonyls adsorbed on oxide surfaces show clear analogies with those of molecular C- and O-bonded adducts. Also, CO chemisorbed on suitable sites of clean, promoted and supported metal surfaces undergo interactions which can be reasonably compared to those in molecular Σ- or Π-CO adducts. Similarities in reactivities between molecular adducts and corresponding surface species can be found. Lewis acid assistance in surface adducts may well be involved in C-O bond scission and in the migratory CO insertion. This latter step can be considered as the termination step in the sequence of elementary steps in the CO + H_2 reaction leading to oxygen-containing products with two or more C-atoms. Control of the rate of this step by modifiers and additives which are capable of adduct formation, should permit control over the catalyst selectivity.

1. INTRODUCTION

The catalytic hydrogenation of carbon monoxide on an industrial scale is almost exclusively carried out by means of heterogeneous supported metal catalysts. The possible mechanistic features of this reaction towards hydrocarbon (Fischer-Tropsch) products or to oxygenates have been discussed by Muetterties and Stein (Ref. 1). A crucial step in the sequence of elementary reactions of CO hydrogenation is the activation of the CO molecule. Although some authors (Ref. 2) favour formyl species as initial intermediates in the formation of hydrocarbons on heterogeneous catalysts there is a large body of experimental evidence available today (Ref. 3-7) which conclusively demonstrates that dissociative chemisorption of CO on metal surfaces is the initiating step in the synthesis of hydrocarbons. On the other hand, CO reduction to CH_4 in homogeneous systems can occur without cleavage of the C-O bond and the most likely intermediate in that case is a formyl species. A tentative reaction sequence, as suggested by Ugo and Psaro (Ref. 8), is shown in Scheme 1,

$$M-CO \xrightarrow{M^*-H} M-C{\overset{H}{\lessgtr}}_{O} \xrightarrow{M^*-H} M-CH_2OH$$

$$M-CH_2OH \underset{OH^-}{\overset{H^+}{\rightleftharpoons}} M=CH_2^+$$

$$CH_4 \xleftarrow{M^*-H} M-CH_3 \xleftarrow{M^*-H} M=CH_2^+$$

Scheme 1

where M* - H indicates a strongly reducing hydride which may be located on the metal skeleton of a polynuclear comlex or may be an independent molecular entity. Alternatively, the formyl species can also lead directly to methanol or via the M-CH_3 intermediate to ethanol by CO-insertion (Ref. 8).

In heterogeneously catalyzed hydrogenation of CO, metal formyl species have been considered as intermediates for methanol synthesis (Refs. 1,4,7). Although the metal formyl could also play a role in the formation of ethanol and higher alcohols, there is strong evidence for an independent reaction sequence (Refs. 9,10). CO insertion into a metal alkyl intermediate as a termination step of the chain growth is a plausible route (Refs. 1,9). Insertion reactions of CO into metal-alkyl bonds are known to be promoted by oxophilic metals, i.e. Lewis acids (Ref. 11). In the presence of such metals, the carbonyl ligand can be C- and O-bonded: M-CO-L (L is a Lewis acid or cation), and its reactivity is therefore greatly enhanced (Ref. 11).

Support (Refs. 12-14) and additive effects (Refs. 14-22) in heterogeneous catalysis are well known to dramatically influence the selectivity of the CO + H_2 reaction towards hydrocarbons, C_1-, C_2-, and higher oxygenates.

In addition, C- and O-bonded carbonyl species may also play a role in methanation and higher hydrocarbon formation. Catalytic hydrogenation of CO was reported to occur in $AlCl_3$ melts in the presence of $Ir_4(CO)_{12}$ (Ref. 23) and in BBr_3 melts containing $Os_3(CO)_{12}$ (Ref. 24). Although the actual catalytic species could not be identified, intermediate adducts of the form Ir-CO-Al and Os-CO-B were inferred based on the observation of very low carbonyl stretching frequencies.

Methanation activity has also been observed with supported osmium catalysts when Os^{2+} surface species were detected (Refs. 25-28), although the complete absence of very small metal aggregates could not be evidenced with certainty. Enhanced activity of these catalysts was demonstrated when the Al_2O_3 support contained chlorine and hence, developed increased Lewis acidity (Ref. 27). It was therefore suggested (Refs. 27, 28), that this unusual kind of CO reduction was possibly assisted by Lewis acid sites (cus Al^{3+}) of the support surface via C- and O-bonding: Os^{2+}-CO-Al^{3+}_{cus}.

Support effects in hydrocarbon formation have been reported by several groups (Refs. 29-31). Bracey and Burch (Ref. 31) have demonstrated for the Pd/TiO_2 system that enhanced activity for the CO + H_2 reaction is found, irrespective of whether or not strong metal-support interactions (SMSI) are operative. They have therefore concluded that the role of the support is to create new active sites at the interface between metal aggregate or particle and the support. This type of site is considered to permit chemisorption of CO with C- and O-bonding. Similar suggestions have been made by others (Refs. 28, 32-34).

The present literature thus provides some indication of the possible importance of C- and O-bonded carbon monoxide in catalytic CO hydrogenation on heterogeneous catalysts. It is the purpose of the present contribution to compare structures and properties of molecular compounds with those found in adsorption systems on solid surfaces, and to ultimately search for correlations with the activity and selectivity of heterogeneous catalysts including support and modifier effects.

2. MOLECULAR C- AND O-BONDED METAL CARBONYLS

The formation and structures of C- and O-bonded metal carbonyls have recently been reviewed by Horwitz and Shriver (Ref. 11). Only a few representative examples will therefore be mentioned here. So-called Σ-CO- and Π-CO-compounds are distinguished (Ref.11). The former involve simple Lewis acid - base interactions between an acceptor atom L and the basic carbonyl oxygen: M-CO-L. Both terminal and bridging CO ligands can form such Lewis acid adducts (Ref. 11), but due to the higher basicity of bridging CO ligands (Ref. 35), the Σ-CO chemistry for polynuclear carbonyls is much richer than that for mononuclear carbonyls. In contrast, the Π-CO bonded compounds generally involve less polar bonding and are known primarily in dinuclear metal-metal bonded complexes, but a few examples have been reported for polynuclear carbonyls. Π-CO compounds which can be depicted as

[1]

may be considered as molecular analogues of precursors for dissociative chemisorption on metal surfaces (Refs. 1, 34). The examples to be mentioned here have been chosen according to the relevance of the acceptor element as a modifier in heterogeneous catalysts or as constituent of the metal oxide support.

2.1 Σ-CO Compounds

When $Na[Mn(CO)_5]$ is dissolved in THF two species can be identified, one of which is to be described as a contact on pair (Ref. 36):

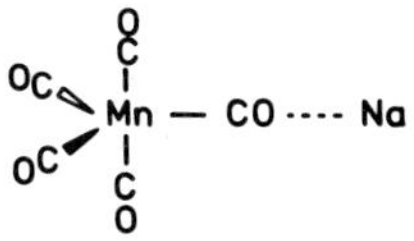

[2]

otassium is a frequently used promoter in heterogeneous atalysis and is an effective acceptor for adduct formaion. In $K_2[Fe(CO)_4]$ closest interactions with the K^+ ion re K-O distances of 0.271 nm indicating a Σ-CO interacion (Ref. 37). Ion pairing with Σ-CO interaction was alo reported for $K_2[H_2Ru_4(CO)_{12}]$ (Ref. 38). In this case n edge-bridging carbonyl ligand is involved in the aduct formation. Mg^{2+} can also act as the acceptor for Σ -CO bonded carbonyls (Refs. 39, 40). An example is gien in Fig. 1. The short Mg-O bond is evident as well as he shortening of the Mo-C distance and the lengthening f the C-O distance. It should be noted too that the Mo-C-rray is nearly linear whereas the Mg-O-C array is bent.

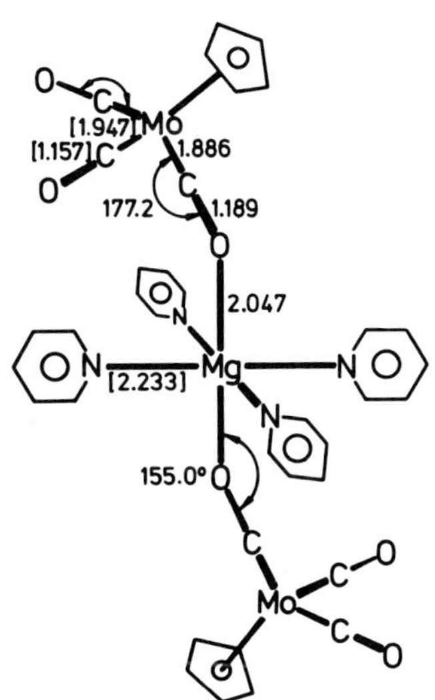

ig.1: Structure of $[(C_5H_5)Mo(CO)_3]_2Mg(C_5H_5N)_4$. ond lengths are given in Ångstrom units [from Ref. 39].

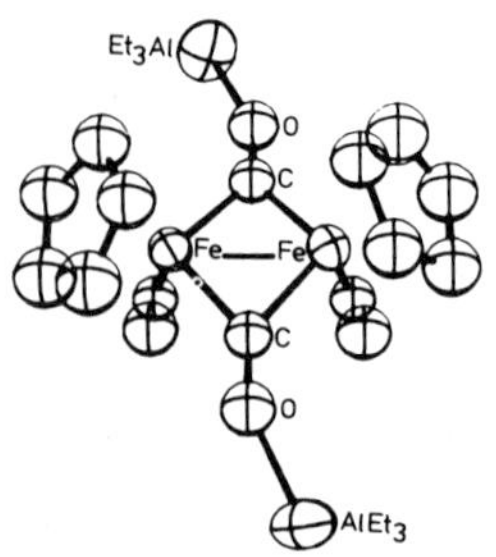

Fig.2: X-ray structure of the adduct $[C_5H_5]_2Fe_2\ [CO]_4\cdot 2Al(C_2H_5)_3$. [from Ref. 41].

A large variety of carbonyl compounds with Σ-CO-inter action with group III acceptors have been described (Ref 11). Fig.2 shows the X-ray structure of $[Cp_2Fe_2(CO)_4]\cdot[2\ AlEt_3]$ (Ref. 41). The figure clearly indicates the ben Al-O-C coordination of the bridging carbonyls.

It is interesting to note that Σ-CO interactions ca induce structural changes as, e.g. in $Ru_3(CO)_{12}$. This clus ter contains only terminal carbonyl ligands in solutio and in the solid state. In interaction with $AlBr_3$ in to luene solution, however, a CO bridged adduct is forme (Ref. 42).

The early transition metals in high oxidation state (e.g. Ti^{4+}, Zr^{4+}, Hf^{4+}) are relatively strong Lewis acid and they do indeed form Σ-CO adducts (Ref. 11). Simpl Lewis acid adducts are also formed between metal carbo nylates and various Mn^{2+} complexes (Ref. 43). The struc tures of these latter compounds are very similar to tho se shown in Fig. 1 for Σ-CO compounds with Mg^{2+}.

The Lewis acid-base interaction in these adducts lead to a significant weakening of the C-O bond which document itself in the sometimes extremely low carbonyl stretchin frequencies (Refs. 11, 44). Some characteristic data ar summarized in Table 1.

TABLE 1. Carbonyl Stretching Frequencies of Typical Σ-CO-Compounds.

Compound	Σ-CO-Ligand[a)]	ν(Σ-CO) cm^{-1}	Ref.
$[(Cp)Mo(CO)_2CO]_2Mg(py)_4$	T	1667	(39)
$[(Cp)Cr(CO)_2CO]_2Mn(py)_4$	T	1652	(43)
$[(Cp)Mo(CO)_2CO]_2Mn(py)_4$	T	1650	(43)
$(Cp)(CO)_2Mo(CO)Ti(thf)(Cp)_2$	T	1650	(126)
$(Cp)(CO)_2Mo(CO)Zr(CH_3)(Co)_2$	T	1545	(127, 128)
$(Cp)_2Fe_2(CO)_4$ BCl_3	E	1463	(42)
$Fe_2(CO)_9$ $AlBr_3$	E	1557	(42)
$Fe_3(CO)_{12}$ $AlBr_3$	E	1548	(42)
$Ru_3(CO)_{12}$ $AlBr_3$	E	1535	(42)
$Co_3(CO)_9COTi(Cl)(Cp)_2$	F	1401	(129)
$Co_3(CO)_9COZr(Cl)(Cp)_2$	F	1373	(130)
$Co_3(CO)_9COHf(Cl)(Cp)_2$	F	1373	(130)
$[Co_3(CO)_9CO]_2TiCpCo(CO)_4$	F	1268, 1233	(131)

a) T = terminal, E = edge-bridging; F = face bridging

2.2 Π-CO Compounds

The clusters

$[HFe_4(CO)_{12}(\mu_3\text{-}\eta^2\text{-}CO)]^-$ (Ref. 45)

and $[(\eta^5\text{-}C_5H_5)_3Nb_3(CO)_6(\mu_3\text{-}\eta^2\text{-}CO)]$ (Ref. 46) both contain a highly activated carbonyl ligand that can be considered as a model for the side-on adsor-

bed precursor state for C-O bond scission on metal sur faces. In the Fe_4 cluster anion, the activated CO ligan bridges four iron atoms and this resembles sites at th intersection of two close-packed face-centered cubic me tal planes, such as (111) and ($\bar{1}$11). In contrast, the lig and μ_4-η^2-CO in the cluster
$[Co_2Mo_2(\mu_2\text{-}CO)_3(\mu_4\text{-}\eta^2\text{-}CO)(\eta\text{-}C_5H_5)_2(\eta\text{-}C_5Me_5)_2]$
(Ref. 47) resembles adsorption on coordinatively unsatu rated sites on stepped metal surfaces (e.g., at intersec tions of (111) and (001) planes of face-centered cubi metals). The structure of this cluster is schematicall shown in Fig. 3.

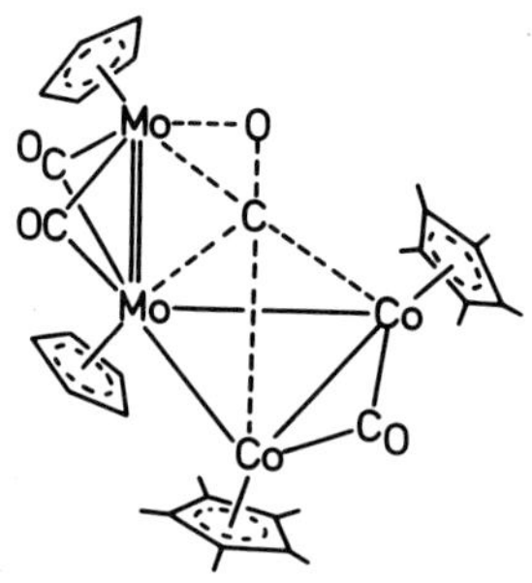

Fig.3: Structure of $[Co_2Mo_2(\mu_2\text{-}CO)_3(\mu_4\text{-}\eta^2\text{-}CO)(\eta\text{-}C_5H_5)_2(\eta\text{-}C_5Me_5)_2]$. [from Ref. 47].

As in the case of Σ-CO compounds, Π-CO interaction al so leads to significant weakening of the C-O carbony bond. Some typical infrared carbonyl stretching frequen cies for Σ-CO bonded complexes are summarized in Table 2.

2.3 Reactivity of C- and O-bonded carbonyls

As shown above, the C-O bond is significantly lengthene and weakened in Σ-CO and in Π-CO compounds. The CO-bor must therefore be strongly activated in these compounds so that its cleavage and hydrogenation of the CO molecul le is expected to be facilitated. In fact, there are se veral reports describing CO cleavage and hydrocarbon for

TABLE 2. Carbonyl Stretching Frequencies of Typical Π-CO Compounds.

Compound	$\nu(\Pi\text{-CO})\ cm^{-1}$	Ref.
$(Cp)Co(CO)_2Zr(Cp^*)_2$	1683	(132)
$(Cp)_3NbMo(CO)_3$	1560	(133)
$(Cp)_3Nb_3(CO)_7$	1330	(134)
$(Cp)_2(Cp^*)_2Mo_2Co_2(CO)_4$	1633	(47)

ation in the presence of strong acids (Refs. 23, 24, 48 50). Perhaps the nicest example of cluster chemistry to imic the heterogeneously catalyzed hydrogenation of CO, s provided by the $[Fe_4(CO)_{13}]^{2-}$ cluster anion (Refs. 51, 2), for which Π-CO interaction has been shown to occur n the course of methane formation. CO insertion reactions R = alkyl)

$$L_nM{<}^{R}_{CO} \rightleftharpoons \left[L_nM-C{<}^{R}_{O}\right] \xrightarrow{L'} L_nM{<}^{L'}_{C(=O)-R} \qquad [3]$$

ith formation of acyl ligands are known to be assisted y electron acceptors. It is likely that the correspon- ing rate enhancement proceeds via initial attack of the lectron acceptor on the carbonyl oxygen of the CO ligand nvolved in the insertion reaction (Ref. 11). The elec- ror acceptor function may be played by protons or main roup cations, but some examples have also been reported or transition metal ions.

The nature of the cation strongly influences the rate f alkyl migration. Rate enhancements for the insertion eaction in $[Fe(CO)_4R]^-$ by a factor of the order of mag- itude of 10^2 have been reported (Ref. 53) as compared o a large noncoordinating cation. An increase of the ra- e of methyl migration by a factor of 10^8 was reported or $Mn(CO)_5(CH_3)$ in the presence of Lewis acids such as

$AlCl_3$ or $AlBr_3$ (Ref. 54). It is interesting to note that not only is there a kinetic enhancement of the insertion reaction, but also a stabilization of the insertion product by formation of acyl adducts in the presence of Lewis acids (Ref. 55).

While the CO insertion in a metal alkyl bond appears to be approximately thermoneutral, the CO insertion in a metal-hydrogen bond has an endothermicity of about 125 $\pm$ 20 kJ mol^{-1} (Ref. 56). In fact, metal formyl complexes are only formed by external hydride transfer via nucleophilic attack by a strong classical hydride on a coordinated CO (Ref. 57). $[\eta^5-C_5(CH_3)_5]_2ZrH_2$ has been shown to form formyl complexes with CO (Ref. 58). Titanium (Ref. 59) and niobium (Ref. 60) hydrido complexes may also develop favourable hydridic character.

It is important in relation to an understanding of factors determining selectivity, that neither alkali metal ions nor Lewis acids have been found to promote hydride migration to form metal formyl complexes in a manner analogous to cation assisted alkyl migration (Ref. 11).

3. SURFACE ANALOGUES OF C- AND O-BONDED CARBONYLS

3.1 Adsorption of metal carbonyls on metal oxide surfaces

When $Mo(CO)_6$ is contacted with partially dehydroxylated Al_2O_3, partial decarbonylation occurs and adsorbed Mo(CO)- species are formed at about 370K (Refs. 61, 62). Brown (Ref. 32) reinterpreted the original assignments of infrared bands and gave convincing arguments in favour of a C- and O-bonded species:

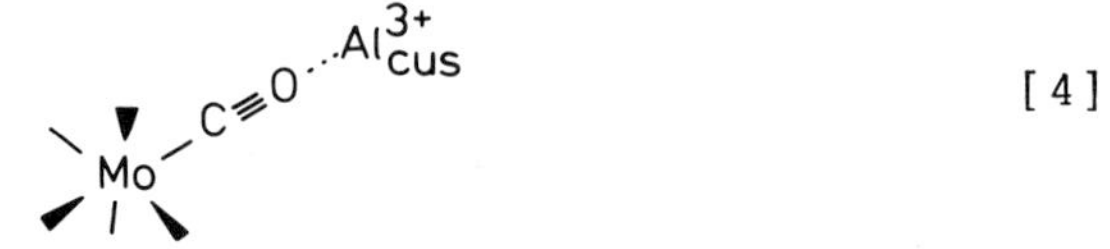

[4]

where the acceptor is considered to be a coordinatively unsaturated Al^{3+}_{cus} site of the support. The interaction of $Cr(CO)_6$ with a reduced chromia/silica also led to the formation of surface adducts (Ref. 63):

$$(CO)_5Cr —— CO\cdots Cr^{2+}_{cus}. \qquad [5]$$

Similar interactions have been observed for $Cr(CO)_6$ (Ref. 64) and for $Ni(CO)_4$ and $Fe(CO)_5$ (Ref. 65) adsorbed on MgO.

Several examples have been described for the adsorption of carbonyl clusters with formation of Σ-CO species on oxides. Interactions of bridging carbonyl ligands with hydroxyl protons probably occurred when $Fe_3CO)_{12}$ was adsorbed on HY zeolite (Refs. 66, 67) and silica (Ref. 68). When adsorbed on alumina or magnesia, $Fe_3(CO)_{12}$ forms an anionic hydride $[HFe_3(CO)_{11}]^-$ (Refs. 69, 70). This species gives a broad band at 1598 cm^{-1}. The corresponding band of the molecular analogue $[N(C_2H_5)_4]$ $[HFe_3(CO)_{11}]$ is found at 1709 cm^{-1}. The frequency shift of 111 cm^{-1} was considered characteristic of adduct formation with Σ-CO interaction, and the following structure was suggested for the surface-bound cluster:

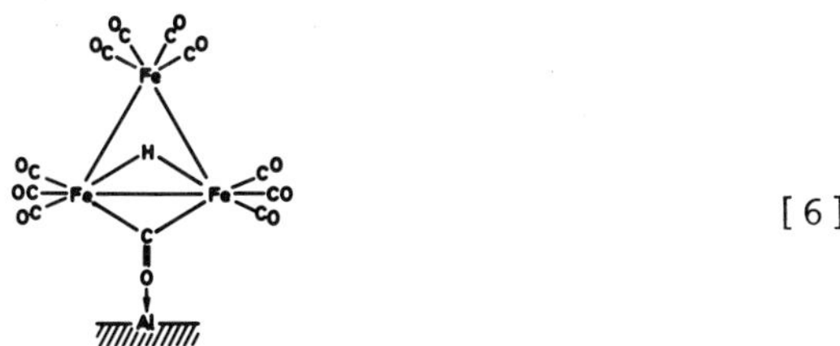

[6]

A similar surface chemistry was observed with $Ru_3(CO)_{12}$ adsorbed on both Al_2O_3 and MgO (Ref. 70). Still larger red-shifts for face- and edge-bridging CO ligands were reported by Tessier-Youngs et al. (Ref. 71) when $(\pi-C_5H_5)_4Fe_4(CO)_4$, $(\pi-C_5H_5)_3Ni_3(CO)_2$, $(\pi-C_5H_5)_2Ni_2(CO)_2$ were contacted with dehydroxylated $\gamma-Al_2O_3$. The observed frequency shifts fell between those measured for the molecular 1:1 adducts with $Al(C_2H_5)_3$ and $AlBr_3$ (see Table 1). It was therefore concluded that the clusters formed surface Σ-CO compounds involving Al^{3+}_{cus} cations as acceptor sites:

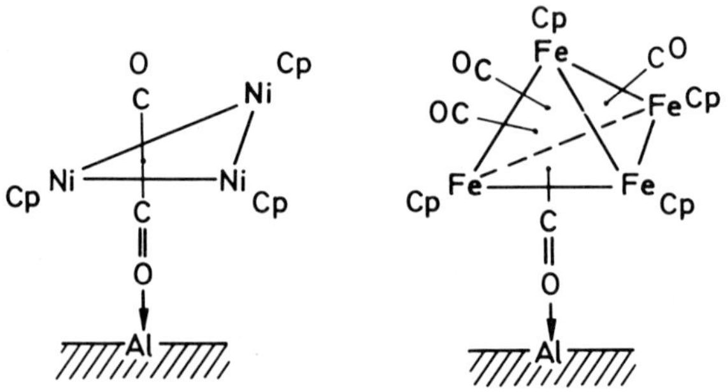

[7]

The examples mentioned above demonstrate that acceptor sites on oxide surfaces are capable of forming surface Σ-CO compounds with metal carbonyls in much the same manner as in molecular adducts.

It has also been shown (Refs. 33, 72) for at least two examples that such Σ-CO interactions can induce migratory insertion in close analogy to these reactions in homogeneous systems. When $Mn(CO)_5(CH_3)$ was contacted with partially or fully dehydroxylated γ-Al_2O_3 surfaces in pentane solution a very fast methyl migration was observed (Ref. 72) with formation of the surface species [8A]

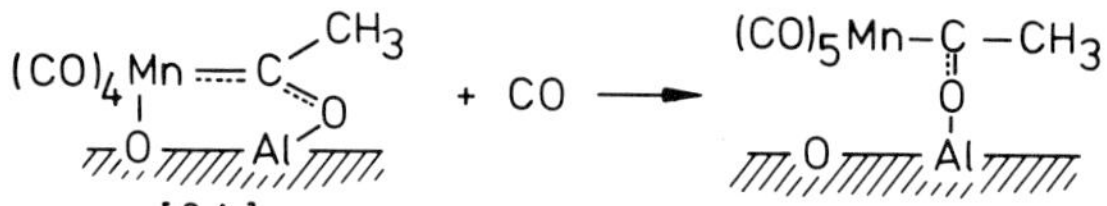

$$(CO)_4Mn{=}C(CH_3){-}O{\to}Al \;[8A] + CO \longrightarrow (CO)_5Mn{-}C(CH_3){=}O{\to}Al$$ [8]

in close analogy to the $AlBr_3$ induced migratory insertion reaction (Ref. 54) mentioned in section 2.3. The analogous reaction occurred with $(\eta^5$-$C_5H_5)Fe(CH_3)(CO)_2$. The methyl migration reaction so promoted by Lewis acid and base sites of γ-Al_2O_3 was at least 10^3 times faster for $Mn(CO)_5(CH_3)$ than the reaction in the presence of CO alone and 10^6 times faster for $(\eta^5$-$C_5H_5)Fe(CO)_2(CH_3)$. CO addition to the surface species [8A] occurred with formation of a Lewis acid bound acyl complex though this transformation was found to be much slower than for the molecular analogues.

In conclusion, surface Lewis acid sites can strongly enhance the rate of migratory insertion reactions in adsorbed metal carbonyls.

3.2 CO-Adsorption on Clean and Alkali-Modified Metal Surfaces

CO chemisorption on clean metal surfaces has been studied extensively by electron energy loss spectroscopy (EELS). Extremely low carbonyl stretching frequencies with simultaneously increased M-C frequencies have been reported for several systems. CO stretching frequencies at 1412 and 1557 cm^{-1} have been attributed to CO bound to kink sites of a stepped Pt surface (Ref. 73), while a band at 1520 cm^{-1} on a Ni surface was considered to be due to CO bound to a step site (Ref. 74). Three CO chemisorption states were identified on the Fe(111) surface (Ref. 75) which are related to the topography of this crystal face (Fig. 4): (i) CO on "deep hollow" site with ν(C-O) = 1530 cm^{-1}; (ii) CO on "shallow hollow" site with ν(C-O) = 1850 cm^{-1}; (iii) CO on "on-top" site with ν(C-O) = 2000 cm^{-1}. The lowest CO stretching frequencies ever found were reported for CO adsorbed on a Cr(110) surface in the range 1150-1330 cm^{-1} (Ref. 76). Electron stimulated desorption ion angular distribution (ESDIAD) suggested a strongly inclined molecular axis of the CO molecule relative to the macroscopic surface. These experimental results can probably best be explained by assuming Π-CO interactions analogous to those described in section 2.2 for molecular Π-CO compounds (see Table 2).

As pointed out by Eberhardt et al. (Ref. 77), the large shift in CO stretching frequency which cannot reasonably be explained in the Blyholder model (Ref. 78), can be considered to be a result of π-bonding, where in a simple molecular orbital picture this involves both metal donation to the CO 2π level and direct 1π interaction with

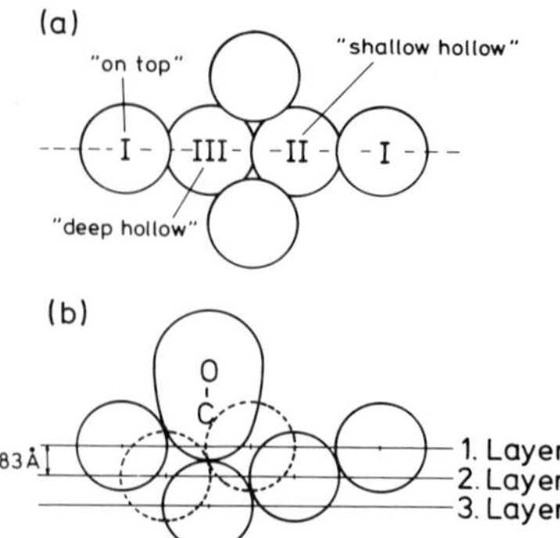

Fig. 4: Surface topography of Fe(111) surface.
(a) Top view: Roman numbers I, II, and III indicate atoms in the first , second and third layer (see part b).
(b) Cross section along the broken line in (a). A CO molecule adsorbed in the "deep hollow" site is shown (see Ref. 75).

the acceptor metal site. This view is supported by the red-shift of only about 300 cm^{-1} relative to the gas phase frequency for CO^- (with one electron in the 2π orbital) (Ref. 79) as compared to shifts of 600 to 1000 cm^{-1} for the CO chemisorption species mentioned above.

Alkali coadsorption on metals leads to enhanced rates of CO dissociation (Ref. 80-83). CO/alkali/metal systems have therefore attracted increasing interest as model systems for elucidating the promoter effect of alkali modifiers in CO hydrogenation catalysts. All experimental results consistently indicate a weakening of the C-O bond in the presence of alkali coadsorbed on the metal surface. However, some discrepancy still exists regarding the question of whether the undisputed lateral interaction between CO and the alkali additive (Na, K) is a "direct" or "through metal" interaction. Several groups favour indirect "through metal" interactions (Refs. 84-87) while others provide evidence for "direct" interaction. Although this question is certainly not clarified with certainty, strong arguments seem to exist in favour of direct interactions.

Strong red-shifts of the carbonyl frequencies are induced by K coadsorption as shown by EELS. E.g., bands at 1350 and 1490 cm^{-1} have been reported for (CO+K)/Ru(001) (Ref. 88). Similarly large red-shifts (Refs. 89, 90) and an increase in vibrational overtone anharmonicity (Ref. 90) have been reported by others. The same arguments as mentioned above for clean metal surfaces, would also support π-bonding as the origin of the large frequency shift (Ref. 77). In addition, the increased anharmonicity of 70 cm^{-1} observed for (CO+K)/Ru(0001) (Ref. 90) as compared to only 40 cm^{-1} for CO^- (Ref. 79), points towards the fact that enhanced π-back donation induced "through metal" by the potassium cannot reasonably be assumed. Also noticeable is the absence of any larger frequency shifts for increasing CO exposures in the case of (CO+K)/Ru(001) at a potassium coverage of 0.33 (Ref. 88). This certainly indicates the absence of strong competition for back-donation electrons among the CO molecules as would be expected if "through-metal" interactions were operative.

Thermal desorption spectroscopy (TDS) indicated strong interactions between CO and the modifier Na (Ref. 91) or K (Refs. 87, 88, 92, 93) since CO and Na or K, respectively, desorbed at the same temperature. ESDIAD data for low coverages of CO+K (Ref. 92) and of CO+Na (Ref. 91) on Ru(001) were consistent with a model in which the CO axis is strongly inclined to the surface. For (CO+K)/ Ru (001) (Ref. 92), TDS provided evidence for strong interaction between CO and K, and additional information from ESDIAD and LEED measurements suggested the formation of ordered species having $K_x(CO)_{2x}$ stoichiometry.

A detailed study of the system (CO+K)/Fe(111) using combined EELS and TDS techniques demonstrated the complexity of the system, but also supported the interpretation that there is a direct interaction between CO and potassium (Ref. 93).

Angle-resolved photoelectron spectroscopy (ARUPS) studies of the systems (CO+K)/Cu(100) and (CO+K)/ Ru(0001) were supportive of the π-bonding configuration of the CO molecule, since it provided direct experimental evidence for the involvement of the 1π and 2π molecular orbitals of CO in the bonding (Ref. 77). Metastable quenching spectroscopy MQS (the surface is probed with metastable He(2^1S) atoms, the MQ spectrum being a plot of the ejected electron current as a function of the electron kinetic energy) of the (CO+K)/Ni(111) system (Ref. 94) suggested that the CO molecules (at low CO and K coverages) were adsorbed in close proximity to the K atoms and were standing upright at 118K, but were very much tilted away from the normal or even lie flat on the surface after warming the system to 298K.

In conclusion, although there may still be alternative interpretations, strong arguments exist in favour of direct interaction between CO and additive atoms which closely resemble those in molecular Π-CO compounds. The corresponding surface structures may, hence, be depicted schematically (Ref. 102) as follows (P indicates a promoter atom):

[9]

3.3 CO Adsorption on Supported Metal Catalysts

Only few infrared spectroscopic data are available indicating the presence of C- and O- bonded CO molecules on supported metal catalysts. Kellner and Bell (Ref. 95) observed weak and broad bands in the 1700-1500 cm^{-1} region for CO chemisorbed on Ru/Al_2O_3 which they attributed to μ-bonded carbonyls interacting with Al^{3+}_{cus} sites on the Al_2O_3 support surface. Vannice and Sudhakar (Ref. 96) have recently suggested the direct attractive interaction of the carbonyl oxygen of chemisorbed CO with Ti^{3+} and/or oxygen vacancies generated at the Pt/TiO_2 interface. These interactions are analogous to those mentioned in the introduction (Refs. 27-34).

Ichikawa et al. (Refs. 22, 97) reported carbonyl infrared bands at 1520 cm^{-1} for a $Rh\text{-}Mn/SiO_2$ catalyst, which was built up at the expense of the characteristic band at 1880 cm^{-1} of the bridging CO adsorbed on unpromoted Rh/SiO_2 catalysts. Ti and Zr additives gave similar but smaller frequency shifts. It could be shown that the additives were present in ionic form as Mn^{2+}, Ti^{4+}/Ti^{3+}, and Zr^{4+}/Zr^{3+}. Hence, the infrared evidence clearly suggests an adduct formation involving attractive interaction between carbonyl oxygen and promoter ion much in the same manner as in molecular Σ-CO or Π-CO compounds. The additive ions can be located either on the support surface close to the metal/support interface or on the surface of the metal particle itself:

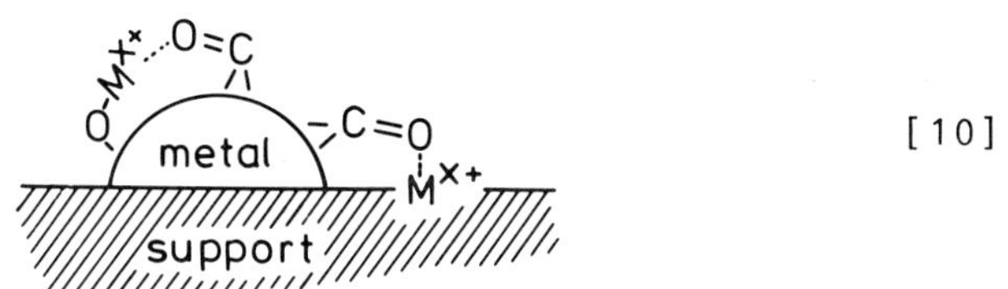

[10]

Ichikawa et al. (Ref. 22) also observed the formation of acetyl species under high pressure working conditions with these catalysts, which is an indication of CO insertion.

4. RELATIONS BETWEEN ADDUCT FORMATION AND CATALYTIC CHEMISTRY OF HETEROGENEOUS CO HYDROGENATION

Hydrocarbons, methanol and higher oxygenates are the products of the heterogeneously catalyzed carbon monoxide hydrogenation. The selectivities are strongly dependent on reaction conditions and the nature of the catalyst. Some metals (typically Ni, Co, Fe) exclusively produce hydrocarbons, while others (e.g. Cu/ZnO, Pd) give high methanol yields. Still other transition metals appear to be extremely flexible such as Rh, which gives hydrocarbons only as a pure metal or supported on SiO_2, but which can be made to produce methanol and/or C_2- and C_2^+- oxygenates in rather high yields when supported on certain oxides and by addition of certain promoters.

As regards the mechanism of the CO hydrogenation reactions, it is now generally agreed that (i) hydrocarbon formation proceeds through dissociative chemisorption of CO (Refs. 3-8); (ii) methanol formation requires ionic species (Refs. 15, 18, 98-100) and proceeds through a formyl intermediate (Ref. 101); (iii) C_2 and higher oxygenates are formed through a route independent of the methanol formation (Refs. 9, 10). These products are considered to be formed from alkyl chains growing on a metal particle by CO insertion as the likely termination step (Refs. 9, 18, 22). A (certainly oversimplified) reaction model for the heterogeneously catalyzed CO hydrogenation is given in Scheme 2:

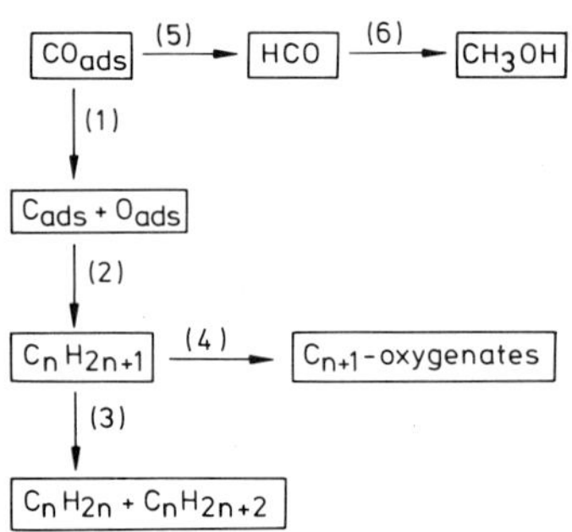

Scheme 2

Selectivity control requires a delicate balance of rate coefficients for the competing reaction steps and of the surface concentrations of intermediates involved. These parameters are determined by the reaction conditions and by the nature of the catalyst used. The present discussion will be confined to properties of the catalysts only and will particularly concentrate on the question of whether C- and O-bonded carbonyls might be involved in some of the selectivity-determining steps.

4.1 Carbon Monoxide Dissociation

CO dissociation is involved in both, hydrocarbon and C_2 and higher oxygenate formation as the initial step. For dissociation to occur on a metal surface (despite the fact that the heat of dissociative adsorption must be favourable as compared to molecular adsorption (Ref. 9)), C and O atoms of the adsorbed CO molecule must interact with surface atoms simultaneously in the precursor state. This is possible e.g. on the "deep" and "shallow hollow" sites on a Fe(111) surface (Ref. 75) or on step and kink sites. Π-CO bonded species can be envisioned on such sites (see section 3.2). Hence, supported metal particles should provide a high density of sites of this type to be effective in CO dissociation.

Alternatively, dissociation rates may be enhanced by promoters, or supports. As discussed in section 3.2, alkali metal coadsorption on clean transition metal surfaces leads to increased rates of CO dissociation (Refs. 80-83) which can be interpreted as being due to a tilted CO configuration (Π-CO bonded species). Significant rate enhancements for hydrocarbon formation were observed on Rh/SiO_2 catalysts when they were promoted by MnO and MoO_2 (Ref. 18). Sachtler et al. (Ref. 102) have recently emphasized the possibility of promoter action by means of stabilization of C- and O-bonded precursor states for CO dissociation.

A suitable choice of the support can have equally favourable influence. Vannice and Garten (Ref. 103) observed that Ni/TiO_2 was about ten times more active than Ni/SiO_2. It has been shown that under reducing conditions Ti^{x+} and O^{2-}-ions are transported onto the surface of transition metal particles supported on TiO_2 (Refs. 104-108). This fact removes some of the geometric limitations for the formation of Π- or Σ-CO bonded surface species, as pointed out by Sachtler et al. (Ref. 102). In addition, adduct-like interactions could also be possible along the interface between metal particle and support according to [10] even if the support constituents are not diffusing onto the metal surface (Refs. 29-34, 96). It should be noted that the C-O-L array is not linear in molecular Σ-CO adducts (see section 2.1). This fact may be favourable for the corresponding interaction on a solid surface.

Chuang et al. (Ref. 109) have recently reported that alkali metal additives (Li, Na, K) reduced the activity for hydrocarbon formation on Rh/TiO_2. This result is considered not to be in contradiction to the trends mentioned above. As TiO_2 was the support in this study, a promoting effect by Ti^{x+} may already have existed in the unmodified catalyst and the alkali additive may have blocked the interface sites responsible for CO activation, so that a potential promoting effect of the alkali was obscured.

4.2 Carbon Monoxide Insertion

According to Sachtler (Refs. 7, 18), formation of C_2 and higher oxygenates is best described by assuming a dual site model. In this model, chain propagation is considered to occur on metal sites (ensembles of metal atoms) as in the case of hydrocarbon formation. The termination step, however, is believed to occur on ionic metal sites most likely by CO insertion leading to acyl groups. The necessity of ionic species for oxygenate production has already been emphasized (Ref. 110) and CO insertion as a likely termination step for the formation of primary alcohols, aldehydes etc. has been demonstrated experimentally (Ref. 9).

In the Sachtler model (Refs. 7, 18) it is assumed that alkyl groups migrate or spill over towards ionic sites on which CO-insertion is more likely to occur due to the reduced back-donation and an increased positive charge on the carbonyl carbon atom. Therefore the latter becomes more susceptible to nucleophilic attack by an alkyl group. Ionic species may be stabilized by certain (basic) additives or may even be formed in CO atmosphere via disintegration of small metal aggregates (Refs. 111-114). On the other hand, Ponec (Ref. 115) has shown that a correlation between concentration of Rh^{x+} ions and C_2^+ oxygenate yields did not exist for Rh catalysts.

For Rh/SiO_2 catalysts it has been shown that additives such as Mn, Ti, Zr, Hf, Th, V, Ta, and Nb significantly increased the activity for C_2^+ oxygenate formation (Ref. 20-22, 116-118) and Ichikawa et al. (Refs. 22, 97) demonstrated the presence of C- and O-bonded adsorbed CO on promoted Rh/SiO_2 catalysts. Van den Berg et al. (Ref. 18) provided experimental evidence for close contact on an atomic scale between the promoter ions and rhodium. The suggestion of Ichikawa et al. (Ref. 97) depicted in structure [10] that the promoter ions may be involved as acceptors in these C- and O-bonded species looks therefore rather logical. As pointed out in section 2.3, migratory CO insertion is significantly promoted in C- and O- bonded compounds, and Ichikawa et al. (Ref. 22) have indeed observed the formation of acetyl species on promoted Rh/SiO_2 catalysts.

Increased yields of oxygenates with simultaneously increased hydrocarbon yields on promoted catalysts may simply be a consequence of the enhanced rate of C-O bond scission. Selectivity towards C_2^+ oxygenates, however, is controlled by the relative rates of the termination steps (3) and (4) according to Scheme [2]. Normally, surface concentrations of CO_{ads} are very high but the rate constants for CO-insertion appear to be low (Ref. 7). It is these rate constants which should be increased relative to the rate constants of chain growth termination by ß-H abstraction or α-H addition if high C_2^+ oxygenate selectivities are required.

Metals with a strong tendency to adsorb CO in a bridging configuration (like Rh) which is favourable for the formation of Σ-CO (or Π-CO) species and the addition of promoters which are known to act as acceptor sites, should fulfill essential requirements for high C_2^+ selectivities. Interestingly, most of the effective modifiers mentioned above are known to form C- and O-bonded metal carbonyls (Ref. 11). This model for C_2^+ oxygenate formation does not obviously require ionic species of the active metal (e.g. Rh^{x+}) unless these act as acceptors themselves. However, the model requires multiple metal sites for CO dissociation and adsorption of CO in a bridging configuration. These multiple metal sites may also be necessary for ß-H abstraction, but proper promotion can still be expected to have a benefitial effect on C_2^+ oxygenate selectivity since enhancement factors of several orders of magnitude on the rate of migratory CO insertion have been reported for homogeneous reactions (see section 2.3). It must be emphasized, that suitable promoter ions may be provided by the support itself (e.g. Rh/V_2O_5 (Refs. 9, 115, 135) if close contact with the metal on an atomic scale is possible (transport of support ions on the metal surface, interface area).

4.3 Formyl Species

There is convincing evidence that formyl intermediates play a decisive role in CH_3OH synthesis. In fact, Kuznetsov et al. (Ref. 10) have detected formyl groups by ^{13}C--NMR on Rh/La_2O_3 by ^{13}C-NMR and Lavalley et al. have detected HCO species by FT infrared spectroscopy on ZnO (Refs. 119 -121) and on Cu-ZnO (Ref. 119). The most convincing evidence has been provided by the chemical trapping method (Refs. 101, 121, 122). It is also agreed that ionic metal species, e.g. Cu^{+} (Refs. 98, 99, 122), Pd^{x+} (Refs. 15, 100), Rh^{x+} (Refs. 18, 100) are essential for methanol synthesis. If one assumes the formyl to be the initial intermediate, it must be formed by nucleophilic attack of H on the carbonyl carbon. Therefore, coordination of CO onto ionic sites with reduced π-back donation may be favourable. It is also known that formyl species coordinated to ionic sites are energetically more favourable than coordination on zero-valent metal sites (Ref. 123). However, as mentioned in section 2.3, the preparation of metal formyl complexes by hydrogen migration from the metal is not known, although it has been suggested that the process might, in fact, be operable (Refs. 124, 125). Also, Lewis acid assistance has not been reported up to now (see section 2.3), but a possible role of C- and O-bonded precursors on solid catalyst surfaces can certainly not be completely ruled out.

5. CONCLUSIONS

The species formed when metal carbonyls are brought in contact with oxide surfaces provide structural analogies with molecular C- and O-bonded adducts. Also, CO chemisorbed on metal surfaces - unsupported or supported - may undergo interactions which can reasonably be described as surface analogues of Σ- or Π-CO compounds, in which certain additives (modifiers or promoters) or constituents of the support surface may play the role of the Lewis acid. For few cases it has been shown that the reactivities of these surface adducts are comparable to those of their molecular counterparts. Thus, the Lewis acid assistance of migratory CO insertion has been verified on Al_2O_3-surfaces.

Although the significance of these surface adducts in heterogeneously catalyzed reactions involving CO as a reactant, namely in CO hydrogenation, are not proven, there is a number of characteristics which can be reasonably explained when Σ- or Π-CO surface species are inferred as intermediates. Thus Π-CO adducts may well be considered as precursors for the dissociative chemisorption of CO. The Lewis-acid assistance in Σ-CO species may significantly enhance the rate of insertion into metal-alkyl bonds to form acyl intermediates. Considering this step as the termination step in the CO + H_2 reaction which leads to oxygen - containing (C_2^+) products, catalyst selectivity should be significantly influenced by promoters which are suitable for Lewis-acid assistance of this step. It is interesting to note that effective modifier elements are known to provide strong tendency towards formation of molecular adducts. Hence, it is suggested that some of the ideas discussed in this review could well help to design catalysts with qualitatively predictable properties, namely selectivities, for the CO + H_2 reaction.

ACKNOWLEDGEMENTS

The financial support for work done in this laboratory by the Deutsche Forschungsgemeinschaft and by the Fonds der Chemischen Industrie is gratefully acknowledged, as well as the extremely helpful discussions with Prof. B. C. Gates, Prof. C. Kappenstein, and Prof. V. Ponec.

REFERENCES

1. E.L. Muetterties and J. Stein, Chem. Rev., 79, 479 (1979).
2. G. Henrici-Olivé and S. Olivé, Angew. Chem. Int. Ed. Engl., 15, 136 (1976).
3. V. Ponec and W. A. van Barneveld, J. & E. C. Product Research & Development, 18, 268 (1979).
4. A. T. Bell, Catal. Rev.-Sci. Eng., 23, 203 (1981).
5. P. Biloen and W.M.H. Sachtler, Adv. Catal., 30, 165 (1981).
6. W.M.H. Sachtler, Chem.-Ing.-Tech. 54, 901 (1982).

7. W.M.H. Sachtler, Proc. 8th Intern. Congr. Catal., W.-Berlin, 1984, Dechema, Frankfurt, and Verlag Chemie Weinheim, 1984, Vol. 1, p. 151
8. R. Ugo and R. Psaro, J. Mol. Catal., 20, 53 (1983).
9. T. L. F. Favre, G. van der Lee and V. Ponec, J. Chem. Soc. Chem. Commun., 230 (1985).
10. V. L. Kuznetsov, A. V. Romanenko, I. L. Mudrakovskii, V. M. Mastikhin and Yu. I. Yermakov, Proc. 8th Intern. Congr. Catal., W.-Berlin, 1984, Dechema, Frankfurt, and Verlag Chemie, Weinheim, 1984, Vol. V., p. 3
11. C. P. Horwitz and D. F. Shriver, Adv. Organomet. Chem. 23, 219 (1984).
12. F. Fajula, R. G. Anthony and J. H. Lunsford, J. Catal. 73, 237 (1982).
13. R. F. Hicks and A. T. Bell, J. Catal., 90, 205 (1984).
14. H. Orita, S. Naito and K. Tamaru, J. Chem. Soc. Chem. Commun., 993 (1983).
15. J. M. Driessen, E. K. Poels, J. P. Hindermann and V. Ponec, J. Catal., 82, 26 (1983).
16. Y. Kikuzono, S. Kagami, S. Naito, T. Snishi, and K. Tamaru, Faraday Disc. Chem. Soc., 72, 135 (1982).
17. H. A. Dierkse, P. W. Lednor and P. C. Versloot, J. Chem. Soc. Chem. Commun., 814 (1982).
18. F. G. A. van den Berg, J. H. E. Glezer and W.M.H. Sachtler, J. Catal., 93, 340 (1985).
19. S. C. Chuang, J. G. Goodwin, Jr., and I Wender, J. Catal., 92, 416 (1985).
20. T. P. Wilson, P. H. Kasai and P. C. Ellgen, J. Catal., 69, 193 (1981).
21. M. Ichikawa, Chemtech., 674 (1982).
22. M. Ichikawa, T. Fukushima and K. Shikakura, Proc. 8th Intern. Congr. Catal., W.-Berlin, 1984, Dechema, Frankfurt, and Verlag Chemie, Weinheim, 1984, Vol. II, p. 69
23. Hsu-Kun Wang, H. W. Choi and E. L. Muetterties, Inorg. Chem. 20, 2661 (1981).
24. H. W. Choi and E. L. Muetterties, Inorg. Chem., 20 2664 (1981).
25. R. Psaro, R. Ugo, G. M. Zanderighi, B. Besson, A. K. Smith and J. M. Basset, J. Organomet. Chem., 215, 215 (1981).
26. M. Deeba, J. P. Scott, R. Barth and B. C. Gates, J. Catal., 71, 373 (1981).
27. H. Knözinger, Y. Zhao, B. Tesche, R. Barth, P. Epstein, B. C. Gates and J. P. Scott, Faraday Disc. Chem. Soc., 72, 53 (1982).
28. E. O. Odebunmi, Y. Zhao, H. Knözinger, B. Tesche, W. H. Manogue, B. C. Gates and J. Hulse, J. Catal., 86, 95 (1984).
29. G. B. McVicker and M. A. Vannice, J. Catal., 63, 25 (1980).

30. R. Burch and A. R. Flambard, J. Chem. Soc. Chem. Commun., 123 (1981); and React. Kinet. Catal. Lett., 17, 23 (1981).
31. J. D. Bracey and R. Burch, J. Catal., 86, 384 (1984).
32. T. L. Brown, J. Mol. Catal., 12, 41 (1981).
33. D. F. Shriver, in: Catalytic Activation of Carbon Monoxide, P. C. Ford (Ed.) ACS Symp. Series 152, 1 (1981).
34. B. C. Gates and H. Knözinger, in: Metal Clusters in Catalysis, B. C. Gates, L. Guczi and H. Knözinger (Eds.), Elsevier, Amsterdam 1986, in print
35. D. F. Shriver, J. Organomet. Chem., 94, 259 (1975).
36. Ref. (11), p. 228.
37. R. G. Teller, R. G. Finke, J. P. Collman, H. B. Chin and R. Ban, J. Am. Chem. Soc., 99, 1104 (1977).
38. K. E. Inkritt and S. G. Shore, Inorg. Chem., 18, 2817 (1979).
39. S. W. Ulmer, P. M. Skarstad, J. M. Burlich and R. E. Hughes, J. Am. Chem. Soc., 95, 4469 (1973).
40. G. B. McVicker, Inorg. Chem., 14, 2087 (1975).
41. N. J. Nelson, N. Kime and D. F. Shriver, J. Am. Chem. Soc., 91, 5173 (1969).
42. J. S. Kristoff and D. F. Shriver, Inorg. Chem., 81, 499 (1974).
43. T. Blackmore and J. M. Burlitch, J. Chem. Soc. Chem. Commun., 405 (1973).
44. H. Knözinger, in: Metal Clusters in Catalysis, B. C. Gates, L. Guczi and H. Knözinger (Eds.), Elsevier, Amsterdam, 1986, in print
45. M. Manassero, M. Sandoni and G. Longoni, J. Chem. Soc. Chem. Commun., 919 (1976).
46. W. A. Herrmann, H. B. Biersack, M. L. Ziegler, K. Weidenhammer, R. Siegel and D. Rehder, J. Am. Chem. Soc., 103, 1692 (1981).
47. P. Brun, G. M. Dawkins, M. Green, A. D. Miles, A. G. Orpen and F. G. A. Stone, J. Chem. Soc. Chem. Commun., 926 (1982).
48. R. E. Stimson and D. F. Shriver, Organometallics, 1, 787 (1982).
49. L. I. Shoer and J. Schwartz, J. Am. Chem. Soc., 99, 5831 (1977).
50. C. Masters, C. van der Woude and J. A. van Doorn, J. Am. Chem. Soc., 101, 1633 (1979).
51. K. H. Whitmire and D. F. Shriver, J. Am. Chem. Soc., 102, 1456 (1980).
52. M. A. Drezdon and D. F. Shriver, J. Mol. Catal., 21, 81 (1983).
53. J. P. Collman, R. G. Finke, J. N. Cawse and J.I. Brauman, J. Am. Chem. Soc., 100, 4766 (1978).
54. T. G. Richmond, F. Basolo and D. F. Shriver, Inorg. Chem., 21, 1272 (1982).
55. S. B. Butts, S. H. Strauss, E. M. Holt, R. E. Stimson, N. W. Alcock and D. F. Shriver, J. Am. Chem.

Soc., 102, 5093 (1980).
56. C. Masters, Homogeneous Transition Metal Catalysis - a Gentle Art, Chapman & Hall, London, New York, 1981, p. 227
57. C. P. Casey and S. M. Neumann, J. Am. Chem. Soc., 98, 3595 (1976).
58. J. M. Manriquez, D. R. McAlister, R. D. Scanner and J. E. Bercaw, J. Am. Chem. Soc., 100, 2716 (1978).
59. L. J. Guggenberger and F. N. Tebbe, J. Am. Chem. Soc., 98, 4137 (1976).
60. J. A. Labinger, Kwai Sam Wong and W. R. Scheidt, J. Am. Chem. Soc., 100, 3254 (1978).
61. A. Brenner and R. L. Burwell, Jr., J. Catal., 52, 353 (1978).
62. A. Kazusaka and R. F. Howe, J. Mol. Catal., 9, 183 (1980).
63. E. Guglielminotti, J. Mol. Catal., 13, 207 (1981).
64. E. Guglielminotti and A. Zecchina, J. Chim. Phys., 78, 891 (1981).
65. E. Guglielminotti, A. Zecchina, F. Boccuzzi and E. Borello, in: Growth and Properties of Metal Clusters, J. Bourdon (Ed.). Elsevier, Amsterdam, 1980, p. 165
66. D. Ballivet-Tkatchenko and G. Condurier, Inorg. Chem., 18, 558 (1979).
67. J. Evans, Chem. Soc. Rev., 10, 159 (1981).
68. K. Lázár, K. Matusek, J. Mink, S. Dobos, L. Guczi, A. Vizi-Orosz, L. Markó and W. M. Reiff, J. Catal., 87, 163 (1984).
69. F. Hughues, J. M. Basset, Y. Ben Taarit, A. Choplin, M. Primet, D. Rojas and A. K. Smith, J. Am. Chem. Soc., 104, 7020 (1982).
70. J. M. Basset and A. Choplin, J. Mol. Catal., 21, 95 (1983).
71. C. Tessier-Youngs, F. Correa, D. Pioch, R. L. Burwell, Jr., and D. F. Shriver, Organometallics, 2, 898 (1983).
72. F. Correa, R. Nakamura, R. E. Stimson, R. L. Burwell and D. F. Shriver, J. Am. Chem. Soc., 102, 5112 (1980).
73. H. Hopster and H. Ibach, Surface Sci., 77, 109 (1978).
74. W. Erley, H. Ibach, S. Lehwald and K. Wagner, Surface Sci., 83, 585 (1979). 75.
U. Seip, M.-C. Tsai, K. Christmann, J. Küppers and G. Ertl, Surface Sci., 139, 29 (1984).
76. N. D. Shinn and T. E. Madey, Phys. Rev. Lett., 53, 2481 (1984).
77. W. Eberhardt, F. M. Hoffmann, R. de Paola, D. Heskett, J. Strathy, E. W. Plummer and H. R. Moser, Phys. Rev. Lett., 54, 1856 (1985).
78. G. Blyholder, J. Phys. Chem., 68, 2772 (1964).
79. H. Ehrhardt, L. Langhans, F. Linder and K. S. Taylor, Phys. Rev., 173, 222 (1969).

80. G. Brodén, G. Gafner and H. P. Bonzel, Surface Sci., 84, 295 (1974).
81. M. P. Kiskinova, G. Pirug and H. P. Bonzel, Surface Sci., 133, 321 (1983).
82. S. R. Kelemen, A. Kaldar and D. J. Dwyer, Surface Sci., 121, 45 (1982).
83. C. T. Campbell and D. W. Goodman, Surface Sci., 123, 413 (1982)
84. M. E. Dry, T. Shingles, L. F. Boshoff and G. F. Oosthuizen, J. Catal., 15, 190 (1969).
85. H. P. Bonzel, J. Cac. Sci. Technol., A2, 866 (1984).
86. J. J. Weimer and E. Umbach, Phys. Rev. B30, 4863 (1984).
87. J. J. Weimer, E. Umbach and D. Menzel, Surface Sci., 155, 132 (1985).
88. F. M. Hoffmann, J. Hrbek and R. A. de Paola, Chem. Phys. Lett. 106, 83 (1984).
89. J. E. Crowell, E. L. Garfunkel and G. A. Somorjai, Surface Sci. 121, 303 (1982).
90. F. M. Hoffmann and R. A. de Paola, Phys. Rev. Lett., 52, 1697 (1984); J. Vac. Sci. Technol., A2, 1339 (1984).
91. F. P. Netzer, D. L. Doering and T. E. Madey, Surface Sci., 143, L363 (1984).
92. T. E. Madey, C. Benndorf, N. D. Shinn, R. L. Kurtz and R. Stockbauer, Proc. Symp. Surface Science, Obertraun, Austria, 1985, p. 10
93. U. Seip, I. C. Bassignana, J. Küppers and G. Ertl, Surface Sci., 160, 400 (1985).
94. J. Arias, J. Lee, J. Dunaway, R. M. Martin and H. Metiu, Surface Sci., 159, L433 (1985).
95. C. S. Kellner and A. T. Bell, J. Catal., 71, 296 (1981).
96. M. A. Vannice and C. Sudhakar, J. Phys. Chem., 88, 2429 (1984).
97. M. Ichikawa and T. Fukushima, J. Phys. Chem., 89, 1564 (1985).
98. R. G. Herman, K. Klier, G. W. Simmons, B. P. Finn and J. B. Bulko, J. Catal., 56, 407 (1979).
99. K. Klier, Adv. Catal., 31, 243 (1982).
100. E. K. Poels, P. J. Mangnus, J. V. Welzen and V. Ponec, Proc. 8th Intern. Congr. Catal., W.-Berlin, 1984, Dechema, Frankfurt, and Verlag Chemie, Weinheim, 1984, Vol. II, p. 59
101. J. P. Hindermann, A. Kiennemann, A. Chakor-Alami and R. Kieffer, Proc. 8th Intern. Congr. Catal., W.-Berlin, 1984, Dechema, Frankfurt, and Verlag Chemie, Weinheim, 1984, Vol. II, p. 163
102. W. M. H. Sachtler, D. F. Shriver, W. B. Hollenberg and A. F. Lang, J. Catal., 92, 429 (1985).
103. M. A. Vannice and R. L. Garten, J. Catal., 56, 236 (1979).
104. D. E. Resasco and G. L. Haller, J. Catal., 82, 279

(1983).
105. D. N. Belton, Y.-M. Sun and J. M. White, J. Am. Chem. Soc., 106, 3059 (1984).
106. H. R. Sadeghi and V. E. Henrich, J. Catal., 87, 279 (1984).
107. X.-Z. Jiang, T. F. Hayden and J. A. Dumesic, J. Catal., 83, 168 (1983).
108. S. Takatani and Y.-W. Chung, J. Catal., 90, 75 (1984).
109. S. C. Chuang, J. G. Goodwin, Jr., and J. Wender, J. Catal., 92, 416 (1985).
110. P. R. Watson and G. A. Somorjai, J. Catal., 74, 282 (1982).
111. H. F. J. Van't Blik, J. B. A. D. van Zon, T. Huizinga, J. C. Vis, D. C. Koningsberger and R. Prins., J. Phys. Chem., 87, 2264 (1983).
112. M. Primet, J. Chem. Soc. Faraday I, 74, 2570 (1978).
113. M. Zaki, G. Kunzmann, , B. C. Gates and H. Knözinger, to be published
114. B. Tesche, E. Alizo de Delgado and H. Knözinger, in preparation
115. V. Ponec, data presented at Spring Meeting of EUROCAT Group, Nijmegen, March 1985
116. P. C. Ellgen, W. J. Barley, M. M. Bhasin and T. P. Wilson, Adv. Chem. Ser., 178, 147 (1979).
117. M. Ichikawa, Bull. Chem. Soc. Japan, 51, 2273 (1982).
118. M. Ichikawa, K. Shikakura and M. Kawai, Proc. China-Japan-USA Symp. Heterogeneous Catalysis, Dalian, China, 1982, paper A 07 J.
119. J. Saussey, J.-C. Lavalley, J. Lamotte and T. Rais, J. Chem. Soc. Chem. Commun., 278 (1982).
120. J.-C. Lavalley, J. Saussey and T. Rais, J. Mol. Catal., 17, 289 (1982).
121. J. Saussey, J.-C. Lavalley, T. Rais, A. Chakor-Alami, J. P. Hindermann and A. Kiennemann, J. Mol. Catal., 26, 159 (1984).
122. G. A. Vedage, R. Pitchai, R. G. Herman and K. Klier, Proc. 8th Intern. Congr. Catal., W.-Berlin, 1984, Dechema, Frankfurt, and Verlag Chemie, Weinheim, 1984, Vol. II., p. 47
123. N. A. Anikin, A. A. Bagaturyants, G. M. Zhidomirov and V. B. Kazansky, Russ. J. Phys. Chem. Eng. Transl., 57, 393 (1983).
124. B. H. Byers and T. L. Brown, J. Organomet. Chem., 127, 181 (1977).
125. W. J. Miles and R. J. Clark, J. Organomet. Chem., 131, 93 (1977).
126. J. S. Merola, R. A. Gentile, G. B. Ansell, M. A. Modrick and S. Zentz, Organometallics, 1, 1731 (1982).
127. J. A. Marsella, J. C. Huffman, K. G. Caulton, B. Longato and J. R. Norton, J. Am. Chem. Soc., 104, 6360 (1982).
128. B. Longato, J. R. Norton, J. C. Huffman, J. A. Mar-

sella and K. G. Caulton, J. Am. Chem. Soc., 103, 209 (1981).
129. G. Schmid, V. Bätzel and B. Stutte, J. Organomet. Chem. 113, 67 (1976).
130. B. Stutte, V. Bätzel, R. Boese and G. Schmid, Chem. Ber., 111, 1603 (1978).
131. G. Schmid, B. Stutte and R. Boese, Chem. Ber., 111, 1239 (1978).
132. P. T. Barger and J. E. Bercaw, J. Organomet. Chem., 201, C39 (1980).
133. A. A. Pasynski, Yu. V. Skripkin, I. L. Eremenko, V. T. Kalinnikov, G. G. Aleksandrov, V. G. Andrianov and Yu. T. Struchkov, J. Organomet. Chem., 165, 49 (1979).
134. W. A. Herrmann, M. L. Ziegler, K. Weidenhammer and H. Biersack, Angew. Chem. Int. Ed. Engl., 18, 960 (1979).
135. J. Kowalski, G. v. d. Lee and V. Ponec, Appl. Catal., 19, 423 (1985).

EXAFS, HREM AND MÖSSBAUER CHARACTERIZATION OF Rh BIMETAL CLUSTERS ON OXIDES OR CARBON AND THEIR CATALYTIC BEHAVIORS IN CO BASED REACTIONS.

Masaru Ichikawa

Research Institute for Catalysis, Hokkaido University, Sapporo, 060 Japan

ABSTRACT - Well defined Rh-Co and Rh-Fe bimetal catalysts were prepared from the bimetal carbonyl clusters such as $Rh_{4-x}Co_x(CO)_{12}$(x=2,3) and $[FeRh_4(CO)_{15}][NMe_4]_2$ supported on metal oxides and an amorphous carbon. Much higher activities for the gas phase hydroformylation of ethylene and propene were achieved on carbon-supported Rh-Co bimetal clusters, rather than on Rh-only and Co-only catalysts. Selectivity to normal-aldehyde was also enhanced by increasing Co content in the Rh-Co bimetal cluster catalysts supported on carbon are explained by their reduced states of Co (hardly retained on oxide supports e.g., Al_2O_3 or ZnO). EXAFS and Mössbauer studies suggest that Co in Rh-Co and Fe in Rh Fe bimetal clusters impregnated on SiO_2 and Al_2O_3 exist in higher oxidation states even after a strong H_2 reduction(400°C), and they are coordinated with Rh atoms having a direct Rh-Co-O and possibly Rh-Fe-O bondings. This is reflected in an anchoring role for preventing the Rh aggregates under the prevailing conditions, and acting a breacker to devide the Rh ensemble as well as an electronic donor-lignad to stabilize the acyl intermediates for production of aldehyde in the hydroformylation catalysed by an isolated Rh atom.

1. INTRODUCTION

Activities and selectivities of the supported metal catalysts in CO based reactions, e.g., CO hydrogenation and olefin hydroformylation, are strongly dependent on the states of metal dispersion and of the supporting materials(metal oxides or carbon)(ref. 1-5, 21). Moreover, the supported bimetal catalysts exhibit a promotion of the catalytic activity and selectivity, additionally improving the stability of the catalyst life, which are markedly different from those of the catalysts of the constituent metals. Their catalytic, electronic and structual properties are associated with the metal compositions of surface bimetal species and geometric situation of metal ensembles, which are influenced with the metal-support interaction at the metal interface. Sinfelt et al(ref. 6-7) have investigated by EXAFS spectroscopy the surface compositions of metal crystallites consisting of Ru-Cu, Pt-Ir and Rh-Ir(1:1) bimetallics dispersed on alumina or silica and reported that the catalysts conventionally prepared gave the heterogeneous compositions near the surface of an alloy particle, being generally different from the bulk region. We have also reported(ref.8) by the EXAFS studies on Rh-Fe bimetal catalysts, combined with Mössbauer measurement(ref.9,10) that Fe atoms are mostly in the state of Fe^{3+}, and are located in the metal-support interface forming bonds with Rh atoms and with oxygen atoms of silica support. For higher Fe/Rh ratio, the excess Fe atoms are in the state of Fe^{0} and are on the surface of Rh particles. The structual models of the Rh-Fe catalysts well explain the catalytic properties(ref.11,12) in $CO+H_2$ reaction as well as the role of $Fe^{3+}$to anchor the Rh particles on the SiO_2 support.

Since 1974, numbers of the pioneering works have been done (ref.3-4, 13-17) to support the idea that molecular cluster compounds are used as precursors to produce highly dispersed metal/bimetal aggregates (less than 10Å in size) supported on the different metal oxides. Some recent reviews(ref.16-20) claim that cluster-derived catalysts provide differences/advantages over the conventional metal catalysts. This may be associated with their unique structures and homogeneous metal compositions of the supported clusters which are strongly bound with the oxide supports. We have previously reported(ref.4,21)that Rh-Co carbonyl clusters supported on suitable basic oxides such as ZnO, MgO and La_2O_3 catalyse the olefin hydroformylation in high efficiencies under the mild reaction conditions. The rates and linear-aldehyde selectivities depend not only on the kinds of precursor carbonyl clusters but also of the metal oxides as supporting materials Geoffroy and Vannice(ref.22) reported the active Ru-Fe bimetal catalysts from mixed metal carbonyl clusters supported on an amorphous carbon. They suggested that a Ru-Fe/carbon is in raft-like structure and keeps the lower oxidation state, hardly on other oxide supports. Accordingly, we use here an amorphous carbon as well as the different metal oxides such as ZnO and SiO_2 to disperse bimetallics e.g., $Rh_{4-x}Co_x(CO)_{12}$(x=0,2,3,4) and $[FeRh_4(CO)_{15}][NMe_4]$ and studied on thier morphologies and surface metal compositions of supported bimetals by means of EXAFS, Mössbauer and HREM. Higher activities of olefin hydroformylation catalysed by Rh-Co and Rh-Fe bimetal clusters on an amorphous carbon are discussed in terms of the ensemble-size and of the elecronic ligand effect by Co or Fe atoms bound with Rh atoms.

We have found that Co or Fe atoms in the Rh bimetal ensembles exhibit the marked promotion of hydroformylation activities over 10 times and improve the selectivities to linear-aldehyde formation. Varying the size of metal clusters, surface metal compositions and kind of support materials with different metal-support interaction, will be reflected in the activity and product distributions in other CO based reactions such as CO hydrogenation into C_1-C_2 oxygenates as previously reported (ref. 1,3,12,21). Some related works in using the bimetal carbonyl clusters in catalyst preparation are exemplified in Table 1.

TABLE 1. Preparation of Bimetal Cluster-Derived Catalysts

bimetal clusters	support	preparation	test reactions	ref.
$Rh_{4-x}Co_x(CO)_{12}$	ZnO, MgO	decarbonylation (evac) H_2-red.	hydroformylation, CO+H_2	(1,4)
	Al_2O_3	pyrolysis	reforming	(13)
$H_2RhOs_3(CO)_{10}$			olefin	(33)
$ClAuOs_3(CO)_{10}$ $HAuOs_3(CO)_{10}$	P-polymer		hydrogenation isomerization	(23)
$H_2FeRu_3(CO)_{12}$ $Fe_2Ru(CO)_{12}$ $FeRu_2(CO)_{12}$	Carbon Al_2O_3, SiO_2	pyrolysis H_2-red.	CO+H_2	(22,24, 25)
$[Fe_2Mn(CO)_{12}][NEt_4]_2$	Al_2O_3, SiO_2	H_2-red.	CO+H_2	(26)
$[H_3RuOs_3(CO)_{12}]^-$	Al_2O_3	H_2-red.	hydrogenation isomerization	(28)
$CpWIr_3(CO)_{11}$, $Cp_2W_2Ir_2(CO)_{10}$	Al_2O_3	H_2-red	CO+H_2	(27)

2. EXPERIMENTAL

2.1. Catalyst Preparation

Carbonyl cluster compounds, $Rh_2Co_2(CO)_{12}$, $RhCo_3(CO)_{12}$ and $[FeRh_4(CO)_{15}][NMe_4]_2$ were synthesized after Martinengo and Longoni(ref.29,30). $Rh_4(CO)_{12}$, $Co_4(CO)_{12}$ and $Fe_3(CO)_{12}$ were purchased from Strem Chemicals Inc., and purified by recrystallization from pentane solution. Each carbonyl cluster was impregnated from suitable organic solutions; $Rh_{4-x}Co_x(CO)_{12}$ in pentane and $[FeRh_4(CO)_{15}][NMe_4]_2$ in THF onto an amorphous carbon(KISHIDA Chem. Co., S.A.=500 m^2/g; 20 mesh), SiO_2 gel (Davison #57, S.A.=280 m^2/g;10-20 mesh) and γ-Al_2O_3(NISHIOKOGYO A-11, 280 m^2/g; 60 mesh). The organic solvent was carefully removed by a vacuum distillation. The resulting materials(1-4 wt% metal loading) were charged in a reactor, and subject to heattreatment(20-200°C), followed with H_2-admission at the different temperatures(200-400°C). Removal of CO from each supported carbonyl clusters proceeded almost completely by the heattreatment at 150°C in vacuo for 1 h, which is confirmed by the change of carbonyl bands in the ir spetroscopic observation with a Hitachi EPI-59A grating spectrometer. The entire procedure was performed in a vacuum or under inert atmosphere of N_2, since most of the carbonyl clusters are comparatively air-sensitive in a solution or upon impregnation.

2.3. Apparatus and Procedures

2.3.1. Hydroformylation of ethylene and propene

The kinetic measurements of the atmospheric pressure hydroformylation (in gas phase) were carried out at 50-180°C, using an open flow-mode reactor(pyrex-glass tubing; id 18X500 mm long), as described elsewhere (ref. 1,3,21). A mixture gas of C_2H_4(C_3H_6), CO and H_2(total pressure 1 atm) was passed through the catalyst bed(1-5 g cat) at the rates of 40-120 ml/min). The hydroformylation product(as well as those in a CO+H_2 reaction) such as aldehyde and alcohols were dissolved in a water condensor(50 ml H_2O) by bubbling the effluent gas. The analysis of the oxygenated products were conducted by FID(flame ionization detector) gc with Porapak Q 4m at 180°C and Chromosorb 101 3m at 135° C. Acetone or ethanol is used as an internal standard. The effluent gas analysis was performed by TCD gc with Al_2O_3-DMF for C_3H_6/C_3H_8 and porapak Q 3m at 70°C for C_2H_4/C_2H_6. The gases had a purity higher than 99.9%v/v and pretreated with active carbon and MS-5A column to eliminate a moisture and possible carbonyls such as $Fe(CO)_5$ and $Ni(CO)_4$ from the gas cylinders.

2.3.2. EXAFS, Mössbauer and HREM measurements

All the x-ray absorption measurements were performed at room temp. by use of the EXAFS(Extended x-ray Absorption Fine Structure) spectrometer at Beam Line 10B in Photon Factory of National Laboratory for

High Energy Physics(KEK-PF)(ref.34).
The catalyst samples packed in the specially designed cell with a 500 μ Kapton film window was sealed with He or N_2, and were free from exposure to air or moisture during the measurements. The data for Rh K edge and Co K edge EXAFS were taken, compared with those of a starting carbonyl cluster, Rh, Co, Rh_2O_3, Co_2O_3 powders as a reference. Mössbauer spectra were recorded at room temperature with a constant acceleration spectrometer[^{57}Co-Rh] under H_2 atmosphere, using a glass cell with Kepton film windows. Mössbauer spectra were fitted by comp-utor with calculated spectra consisting of Lorenstian shaped lines having the different Mössbauer parameters. Isomer shift is given relative to α-Fe. High resolution electron microscopy(HREM) was carried out on the sample of $Rh_4(CO)_{12}$ and $Rh_2Co_2(CO)_{12}$ deposited on Al_2O_3 and carbon, followed with the heattreatment in vacuo. The size of the metal particles in the resulting catalysts was measured in the range of less than 10Å.

3. RESULTS AND DISCUSSIONS

3.1. Hydroformylation of Ethylene and Propene Catalysed on Carbon-Supported Rh-Co Bimetal Clusters.

3.1.1. Activity and Selectivity

The hydroformylation of ethylene, giving C_2H_5CHO and C_2H_6 as a simple hydrogenation product proceeded at 50-180°C catalytically on $Rh_{4-x}Co_x(CO)_{12}$(x=0,2,3,4) supported on an amorphous carbon. C_3H_7OH as a further hydrogenated product was also obtained with a small yield on the Co-rich Rh bimetal catalysts and $Co_4(CO)_{12}$-derived catalyst. The results are summerized in Table 2, where the rates of C_2H_5CHO(or C_3H_7CHO) formation in used cluster basis are much enhanced with increasing the Co content in the Rh-Co bimetal catalysts, compared with those on the catalysts from $Rh_4(CO)_{12}$ or $Co_4(CO)_{12}$. On the other hand, in the propene hydroformylation it was found that the selectivities to linear aldehyde were increased from 69% to 73%, 78% on the catalysts from $Rh_4(CO)_{12}$, $Rh_2Co_2(CO)_{12}$ and $RhCo_3(CO)_{12}$, respectively, as shown in Fig. 1 and Fig. 2. Although the $Co_4(CO)_{12}$-derived catalysts gave near 90% normal isomer selectivity in propene hydroformylation, its specific rates per Co atom were considerably lower:100-200 times smaller than those on the $Rh_2Co_2(CO)_{12}$ and $RhCo_3(CO)_{12}$-derived

Table 2. Catalytic performance of Rh, Rh-Co bimetal and Co cluster-derived catalysts supported on carbon in ethylene and propene hydroformylation.

catalyst cluster on carbon	$C_2H_4 + CO + H_2$ at 112°C			
	C_2H_5CHO	C_3H_7OH	C_2H_6	(mmole/mmole cluster/min)
$Rh_4(CO)_{12}$	0.042	tr	0.034	
$Rh_2Co_2(CO)_{12}$	0.462	tr	0.111	
$RhCo_3(CO)_{12}$	0.578	tr	0.114	
$Co_4(CO)_{12}$	0.013	0.003	0.008	
	$C_3H_6 + CO + H_2$ at 135°C			
	C_3H_7CHO	C_4H_9OH	n-isomer select.(%)	C_3H_8
$Rh_4(CO)_{12}$	0.007	tr	69	0.005
$Rh_2Co_2(CO)_{12}$	0.075	tr	73	0.014
$RhCo_3(CO)_{12}$	0.167	0.004	78	0.025
$Co_4(CO)_{12}$	0.002	0.001	84	0.001

C_2H_4: CO: H_2 = 40:40:40 ml/min, C_3H_6:CO:H_2 = 20:40:40 ml/min

Fig.1. Temperature dependency of rates and normal isomer selectivity for ethylene and propene hydroformylation on $Rh_2Co_2(CO)_{12}$/Carbon (0.53 mmole) C_2H_4:CO:H_2=40:40:40 ml/min, C_3H_6:CO:H_2=20:40:40 ml/min

Fig. 2 Ethylene and propene hydroformylation on $RhCo_3(CO)_{12}$/Carbon (0.51 mmole)

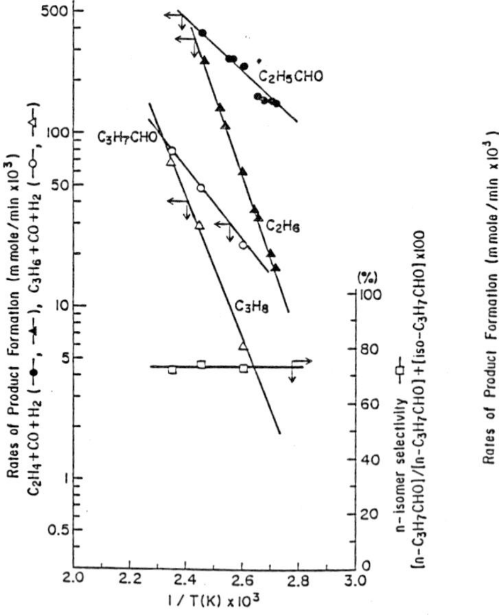

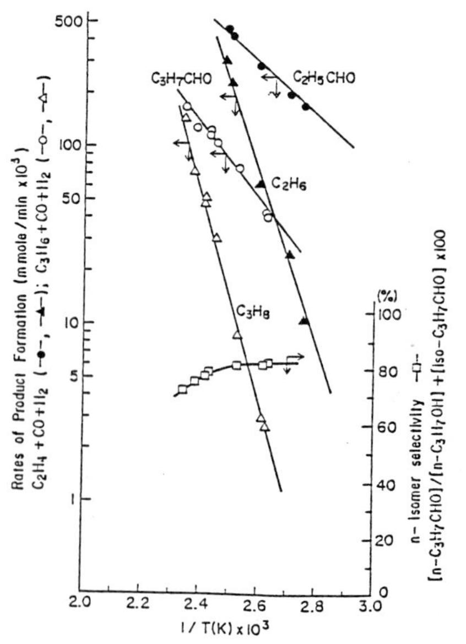

catalysts. Accordingly, such higher rates and higher selectivities to linear aldehyde on the Rh-Co bimetal catalysts could not be explained simply by the additional formation of aldehyde with the individual Rh and Co atoms in the catalysts. In fact, if $Rh_4(CO)_{12}$ and $Co_4(CO)_{12}$ were codeposited(1:1 and 1:3 molar ratios) onto the amorphous carbon from the physical mixture solution, the resulting $Rh_4(CO)_{12}$+$Co_4(CO)_{12}$ /carbon catalysts provided the rates and normal-isomer selectivities were almost same as those on the $Rh_4(CO)_{12}$/carbon, indicating no enhancement for the reaction by the presence of Co atoms in the catalyst. Additionally, it is of interest to note here that the Rh-Co catalysts prepared by the conventional coimpregnation of $RhCl_3$ and $CoCl_2$(1:1 and 1:3 Rh/Co on the carbon gave no appreciable effect on the rates and linear aldehyde selectivities in propene hydroformylation in comparison with those of the Rh-only catalysts.

3.1.2. Dependency of the cluster loading upon activity and selectivity The variation in the rates of propene hydroformylation on $Rh_2Co_2(CO)_{12}$ /carbon catalyst was studied by changing the amount of cluster loaded, the result is shown in Fig. 3. Rates of C_3H_7CHO formation at 152 and 132^oC were propotional to the amount of $Rh_2Co_2(CO)_{12}$, in keeping the normal isomer selectivities around 73%. For higher loading of the cluster above 3 mmole per 10 g of carbon, the rates of aldehyde form-

Fig.3 Dependency of amount of $Rh_2Co_2(CO)_{12}$ upon activity and selectivity in propene hydroformylation on the $Rh_2Co_2(CO)_{12}$/carbon.

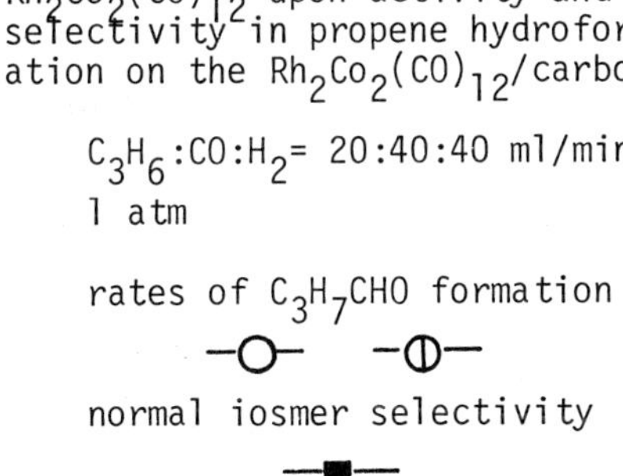

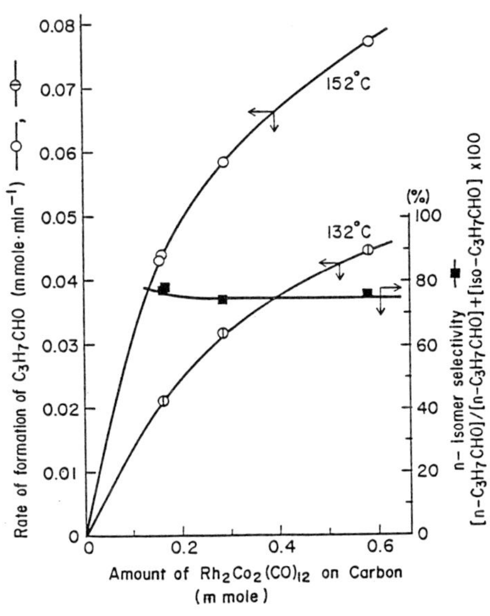

ation were comparatively suppressed from those estimated by the linearity relationship as shown in Fig. 3. This is consistent with our interpretation that the Rh_2Co_2 unit might be responsible for the active site of olefin hydroformylation and present in a uniform dispersion on an amorphous carbon. Vannice et al(ref.22) suggested the bimetal crystallites derived from $FeRu_2(CO)_{12}$ are trapped in the small-pore structure of the carbon.

3.1.3. Pressure Dependency of Reactant Gases

Kinetic measurements of the C_3H_7CHO as the result of CO insertion and C_3H_8 in a simple hydrogenation of propene were studied on $RhCo_3(CO)_{12}$ /carbon catalyst. For the hydroformylation rates, as shown in Fig.4 and Fig. 2, lower activation energies around 13 kcal/mole and higher order with respect to H_2 and propene pressure and lower negative order to CO pressure. In contrast with these, Rates of C_3H_8 formation had higher activation energies around 34 kcal/mole and the pressure dependencies of first order to H_2 and propene and 1.5 minus order to CO. The obtained kinetic parameters for the product formation were basically consistent with the results measured on $Rh_4(CO)_{12}/ZnO$ as previously reported(ref. 4). This suggests that the olefin hydroformylation reaction proceeds similarly on all the Rh cluster-derived catalysts supported on the different supports materials. The isolated Rh in the Rh bimetal ensembles may be active for the reaction.

3.1.4. Promotion Effects of Co and Fe in Rh bimetal clusters for the Olefin Hydroformylation.

It is markedly interesting to find that Co atoms in the Rh-Co bimetal catalysts supported on an amorphous carbon enhance the rates of aldehyde production in one or two order of the magnitude(in Rh atom basis) over those on the $Rh_4(CO)_{12}$-derived catalyst, as shown in Fig. 5. The selectivities to linear aldehyde were moderately increased by increasing Co content, similarly as those on $Rh_{4-x}Co_x(CO)_{12}/ZnO$. Comparing with the promotion effects by Co atoms in the Rh-Co bimetal catalysts on ZnO, much higher enhancements for the rates of aldehyde formation were achieved on the catalysts supported on carbon. The differences in the activities would be associated with the different oxidation of Co atoms in the Rh-Co ensembles supported on ZnO and

Fig. 4 Pressure dependencies of rates of C_3H_7CHO and C_3H_8 formation over $RhCo_3(CO)_{12}$/carbon with varying the partial pressure of reactant gases. 140°C, SV=288 h^{-1}

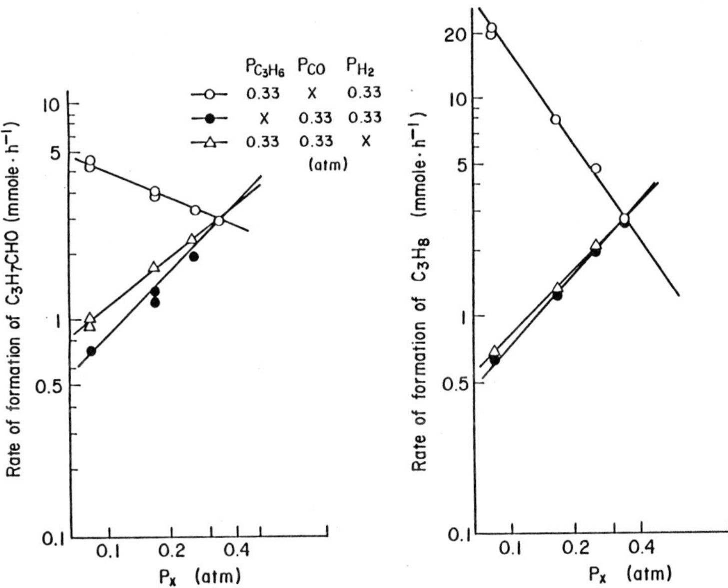

carbon, on which Co atoms easily keep their lower oxidation state as suggested by Vannice et al.(ref. 22), rather than other oxide supports. Furthermore, it is obvious in Fig. 1 and 2 that a simple hydrogenation of propene to propane is much effectively suppressed on the Rh_2Co_2 and $RhCo_3$ supported catalysts, compared with that on Rh_4 catalyst. It seems that Co atoms act as a inert breaker to devide the Rh ensemble, being active for CO insertion and to depress H_2 oxidative adsorption, similarly in the case of Rh-Zn/SiO_2(ref.35, 36). On the other hand, Co atoms in a Rh-Co bimetal play a role as an electronic donor ligand to stabilize the linear aceyl intermediate in aldehyde formation, possibly due to the bi-site interaction with acyl oxygen as depicted in the following:

CO, C_3H_7 (δ−) / Rh — Co ⟶ C_3H_7–C=O···Co / Rh — Co ⟶(H_2) n-C_3H_7CHO ↗ ; Rh —H— Co

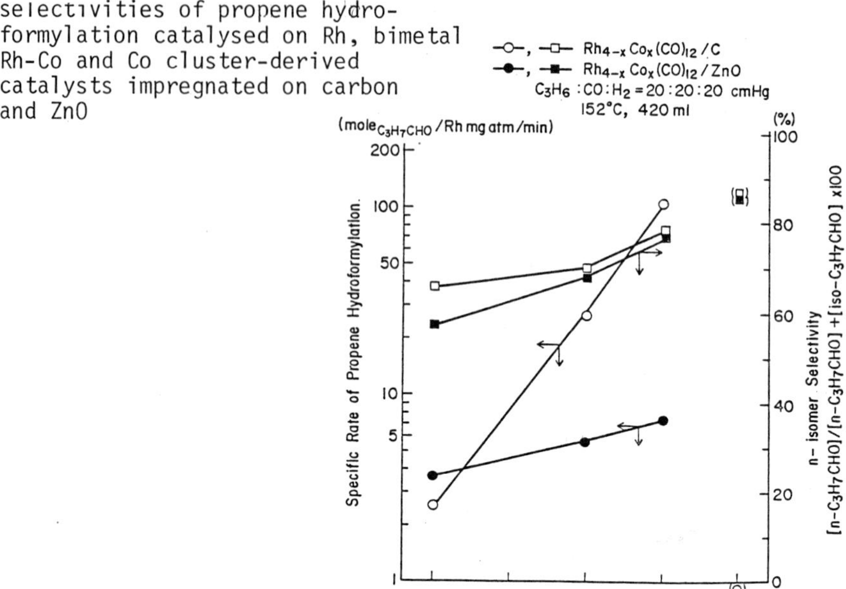

Fig. 5 Activities and n-isomer selectivities of propene hydroformylation catalysed on Rh, bimetal Rh-Co and Co cluster-derived catalysts impregnated on carbon and ZnO

In this connection, when Fe-Rh bimetal catalysts were prepared from $[FeRh_4(CO)_{15}][NMe_4]_2$ highly dispersed on silica gel, the resultant catalysts provided higher activities for ethylene hydroformylation over 50 times larger than on the $Rh_4(CO)_{12}/SiO_2$ catalyst. SiO_2-supported Fe catalyst from $Fe_3(CO)_{12}$ after H_2-reduction was inactive for the hydroformylation of ethylene. Fe atoms like Co atoms in Rh-Co bimetal catalysts act as an electronic ligand for acid-promoted CO insertion in the hydroformylation,as suggested by Shriver(ref.37).

3.2. Characterization of Supported Rh bimetal Clusters

3.2.1. EXAFS Studies on Al_2O_3- and Carbon-Supported Rh-Co bimetal Clusters.

EXAFS spectra were recorded for catalysts derived from $Rh_4(CO)_{12}$, $Rh_2Co_2(CO)_{12}$, $RhCo_3(CO)_{12}$ and $Co_4(CO)_{12}$ impregnated on γ-Al_2O_3 and an amorphous carbon. Carbonyls of the impregnated clusters are eliminated by a mild oxidation at room temperature, followed with H_2 reduction at 400°C or evacuation at 150°C in a vacuum. Preliminary result has been recently reported in ref. 31. Examples of the Fourier transforms of $k^3\chi(k)$, which were calculated by use of the Rh-K EXAFS and Co-K EXAFS data are presented in Fig. 6 and 7. After the reverse Fourier transformation for the region of the peaks corresponding to Rh-Rh, Rh-Co, Rh-O and Co-O, Co-Co bonds, the curve fitting analysis were carried out by using the emprical parameters derived from the EXAFS data of $Rh_4(CO)_{12}$, $Co_4(CO)_{12}$, $RhCo_3(CO)_{12}$, Rh_2O_3 and Co_2O_3 as the references. The results of the curve-fitting analysis on all the catalysts derived from the carbonyl clusters and metal salts on Al_2O_3 are given in Table 3. When the dehydrate γ-Al_2O_3 was impregnated with $Rh_2Co_2(CO)_{12}$, the coordination numbers(C.N.) and bond lengths(R) are nearly the same as those of the original carbonyl clusters. This indicates that the Rh-Co bimetal cluster skelton is retained even in the pyrolyzed/ H_2-reduced catalysts, and the bimetal cluster units are highly and uniformaly dispersed on oxide surface. Addionally, the situation is the same also in the catalysts derived from $RhCo_3(CO)_{12}$, as shown in Table 3. In the similar procedures we have observed the Rh-K and Co-K edge EXAFS on the catalysts of $Rh_2Co_2(CO)_{12}$ and $RhCo_3(CO)_{12}$ impregnated on carbon. In the Fourier transform for the Rh-K

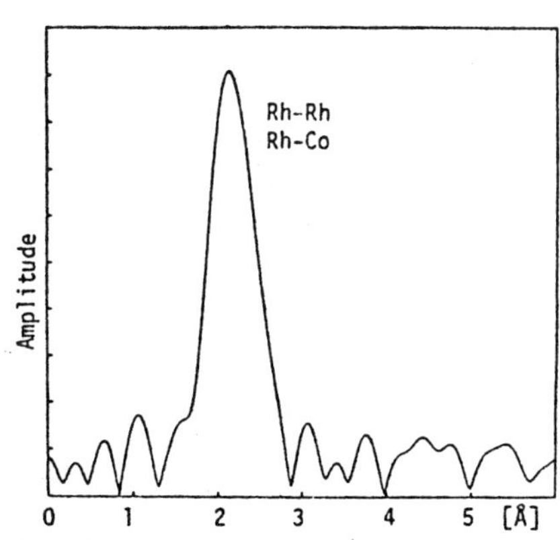

Fig.6. Fourier transform of $k^3 \cdot \chi(k)$ (Rh-K) for $Rh_2Co_2(CO)_{12}/Al_2O_3$ reduced with H_2 at 400°C

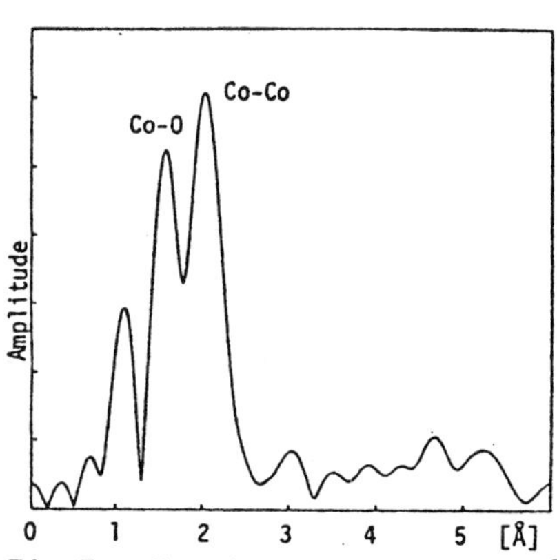

Fig.7. Fourier transform of $k^3 \cdot \chi(k)$ (Co-K) for $RhCo_3(CO)_{12}/Al_2O_3$ reduced with H_2 at 400°C

Table 3. Curve-fitting results for Rh-Co bimetal catalysts prepared from carbonyl clusters supported on γ-Al_2O_3.

samples	Rh-O		Rh-Co		Rh-Rh		Co-O		Co-Co		Co-Rh	
(metal wt %)	C.N.	R	C.N.	R	C.N.	R	C.N.	R	C.N.	R	C.N.	R
$Rh_2Co_2(CO)_{12}$			2	2.63	1	2.73			1	2.55	2	2.63
2) pyrolysis (2 wt %)	1-2	2.14	-	-	1	2.64	2-3	1.96	-	-	-	-
3 H_2 red. (4 wt %)	1-2	2.20	1-2	2.49	2-3	2.64	2	1.97	1-2	2.49	1-2	2.53
$RhCo_3(CO)_{12}$			3	2.60					2	2.53	1	2.61
7) pyrolysis (4 wt %)	1	2.22	1-2	2.56	-	-	1-2	1.99	1-2	2.43	-	-
8) H_2 red. (4 wt %)	2	2.18	3	2.58	-	-	1-2	1.98	1-2	2.43	-	-
9) $Rh_4(CO)_{12}$ + $Co_4(CO)_{12}$ H_2 red. (4 wt %)	2	2.18	-	-	4-6	2.66	2	2.02	1	2.45	-	-
10) Rh:Co = 1:1 salt (4 wt %)	-	-	4-6	2.58	7-8	2.67	1-2	2.05	3-4	2.47	1-2	2.62
11) Rh:Co = 1:3 salt (4 wt %)	-	-	4-6	2.59	7-8	2.66	-	-	10	2.48	-	-

and Co-K EXAFS data, peaks corresponding to Rh-O and Co-O bonds are observed both $Rh_4(CO)_{12}$ and Rh-Co bimetal cluster catalysts. The results suggest that Co atoms rather than Rh in the dispersed Rh-Co bimetals are attached with the surface oxygen atoms of the oxide supports in forming the Rh-Co-O bonding. On the other hand,when the physical mixture of $Rh_4(CO)_{12}$ and $Co_4(CO)_{12}$ was used (sample 9), the resulting catalyst sample had no Rh-Co bonds. This is well explained by the model location(Fig. 8) that Rh ensembles and Co ensembles are separated from each other without no scrambling among Rh and Co clusters during the catalyst preparation. In contrast to these, when the catalyst was prepared by impregnating γ-Al_2O_3 or carbon with a solution of $RhCl_3$ and $CoCl_2$ and then reduced with hydrogen at 400°C (sample 10 and 11 in Table 3), the total coordination number around Rh is nearly 12, but that around Co was rather small(C.N.=6), indicating that relatively large metal particles with Co-rich surface compositions are formed. The particle surface is covered almost with Co atoms, which give lower activities for olefin hydroformylation as mentioned above.

Fig. 8 Proposed Model for the Rh-Co Bimetal Catalysts

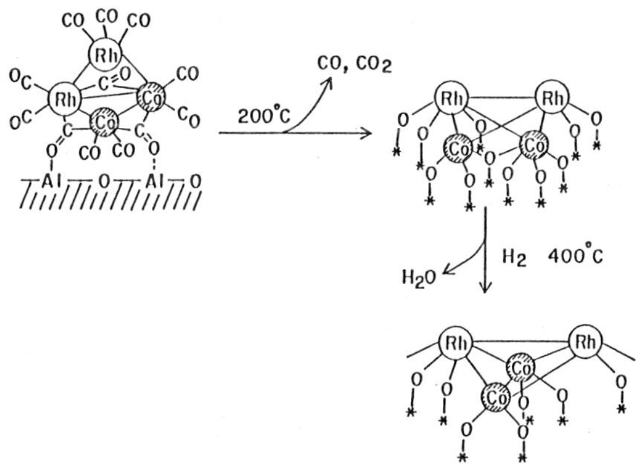

3.2.2. Mössbauer studies on SiO_2-supported Rh-Fe bimetal Cluster.

The valence states of Fe in the bimetal catalyst prepared from $[FeRh_4(CO)_{15}][NMe_4]_2$ impregnated on SiO_2 were observed by in situ Mössbauer spectroscopy. The observed Mössbauer parameters (I.S.=0.42 mm/sec and Q.S.=0.87 mm/sec, relative to α-Fe for the catalyst(Fig. 9) are reasonably consistent with those in SiO_2-supported Rh-Fe prepared from the conventional impregnation of $RhCl_3$ and $FeCl_3$, as previously reported(ref.10). As shown in Fig.9, Fe atom is present in the state of Fe^{3+}in the Rh_4Fe cluster-derived catalyst, even after the strong H_2-reduction and in the presence of Rh, which is active for hydrogenation. Fe^{3+} does not form as a bulk iron oxide, but is highly dispersed. Although there is a limited information on the possible location of Fe in the Rh-Fe bimetal catalyst, it is reasonable to postulate that Fe^{3+}ions are bound with not only Rh atoms but

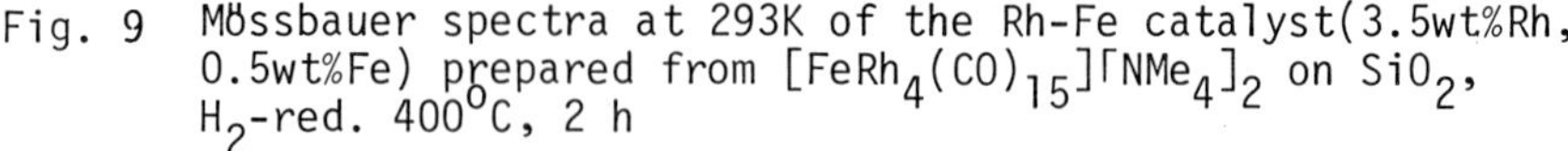
Fig. 9 Mössbauer spectra at 293K of the Rh-Fe catalyst(3.5wt%Rh, 0.5wt%Fe) prepared from $[FeRh_4(CO)_{15}][NMe_4]_2$ on SiO_2, H_2-red. 400^oC, 2 h

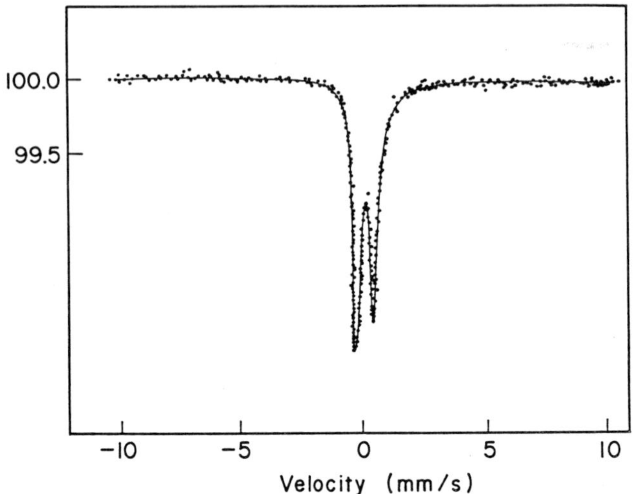

with oxygen atoms of silica support in forming a Rh-Fe-O bonding. It is probably accepted by the analogy in the case of Rh-Fe/SiO_2 by the EXAFS studies(ref.10). Fe atoms in a Rh_4Fe bimetal cluster bound with SiO_2 seem to play a role for anchoring the Rh aggregates to prevent a sintering, and also for modifying to enhance the olefin hydroformylation catalysed by Rh atoms.

4. CONCLUSIONS

Rh-Co and Rh-Fe bimetal catalysts are prepared from the metal carbonyl clusters such as $Rh_{4-x}Co_x(CO)_{12}$(x=2,3) and $[Rh_4Fe(CO)_{15}][NMe]_2$ which are impregnated on an amorphous or metal oxides such as ZnO and SiO_2. They exhibited unique catalytic activities and selectivities for the olefin hydroformylation, in depending on the metal compositions of the original clusters used and the kind of supports. We found that much higher activities in the hydroformylation of ethylene and propene were achieved on carbon-supported Rh-Co and SiO_2-supported Rh-Fe bimetal clusters, compared with those on the Rh-, Co- and Fe-only catalysts as well as the conventionally prepared catalysts from the metal salts. The difference due to the cluster impregnation is associated with the high dispersion and uniform metal composition of a discrete bimetal cluster formed on the support surfaces. Co atoms in the Rh-Co and Fe in the Rh-Fe bimetal catalysts promote the rates of hydroformylation catalysed by Rh and improve the linear aldehyde selectivities. In order to characterize the resulting Rh-Co and Rh-Fe bimetal catalysts from the carbonyl clusters, we have conducted in situ EXAFS, Mössbauer and HREM studies, and elucidated the formation of a direct bimetal bonding such as Rh-Co(or Rh-Fe), which is also bound with the oxygen of oxide support. In contrast, the EXAFS data revealed that the conventional Rh-Co catalysts from metal salts had a heterogeneous metal composition: Co atoms cover the bulk Rh particle in the phase segregation. Higher activities of carbon-supported Rh-Co bimetal clusters are explained by their reduced states of Co(hardly retained on metal oxide supports) in a Rh-Co ensemble on carbon. It is also found that Co in Rh-Co and Fe in the Rh-Fe clusters on Al_2O_3 or SiO_2 are present in higer oxidation states even after a strong H_2 reduction(400^oC), and they are coordinated with Rh atoms, which implies that Co and Fe ions at the metal/support interface. This is associated with an anchoring effect for preventing the sintering of the Rh aggregates and bimetal ensembles and to keep the catalytic activities and activities at the prevailing reaction conditions. They also affect the catalytic performance in enhancing the rates of olefin hydroformylation and relatively increasing the linear aldehyde selectivities, compared with those on the Rh-only catalysts. The promotion effect by Co in the Rh-Co bimetal catalysts could be explained in terms of an electronic ligand effect to stabilize the acyl intermediate by the interaction between acyl-oxygen and Co ion/atoms. Co and Fe atoms/ions act also as a inert breaker to devide the Rh ensemble into an isolated Rh atom, which would be active for the olefin hydroformylation, and a simple hydrogenation of olefin is suppressed at the same time.

ACKNOWLEDGEMENT

The author is much indebted to T.Yokoyama, Dr. N. Kosugi and Prof. H. Kuroda of Tokyo University for the EXAFS measurements and analysis of Rh-Co bimetal cluster-derived catalysts. I also express my thanks to Prof. W.M.H. Sachtler and D.F. Shriver, Northwestern University by their helpful discussion and interest in the supported bimetal cluster-derived catalysts in studying the olefin hydroformylation , and to Dr. S. Iijima, Meijo University in cooperation for the measurement of an atomic image of supported bimetal cluster by high resolution electron microscopy.

REFERENCE

1. M. Ichikawa, Chemtech, 674 (1982); M. Ichikawa and K. Shikakura, Stud. Surf. Sci. Catal., 7, 925 (1981)
2. V.L. Kuznetsov, A.T. Bell and Y.I. Yermakov, J. Catal., 65, 374 (1980); Yu. I. Yermakov, and V.A. Zakharov, Adv. Catal., 24, 173 (1980)
3. M. Ichikawa, Bull. Chem. Soc., Japan, 51, 2268, 2273(1978): J.C.S Chem. Commun., 566 (1978)
4. M. Ichikawa, J. Catal., 59, 67(1979); 56, 127(1978)
5. A.K. Smith, A. Theorier, J.M. Basset, R. Ugo, D. Commereuc and Y. Chauvin, J. Am. Chem. Soc.,100, 2590 (1978); F. Hugues, B. Besson, P. Bussiere, J.A. Delmon, M. Leconte and J.M. Basset, ACS sym. ser., 192, 255 (1982)
6. J.M. Sinfelt, Acc. Chem. Res., 10, 15 (1977); R.L. Garten and J.H. Sinfelt, J. Catal., 62, 127 (1982)
7. G.H. Via, G. Meitzner, F.W. Lytle and J.H. Sinfelt, J. Chem.Phys. 79, 1527 (1983): J.H. Sinfelt, G.H. Via, F.W. Lytle, J. Chem. Phys., 72, 4832 (1980), 76, 2779 (1982)
8. M. Ichikawa, T. Fukushima, T. Yokoyama, N. Kosugi, and H. Kuroda, J. Phys. Chem., in press (1986)
9. J.W. Niemantsverdriet, A.M. van der Kaan, J.J. van Loef and W.N. Delgass, J. Phys. Chem., 87, 1292 (1983), J. Catal., 89, 138 (1984)
10 Y. Minai, T. Fukushima, M. Ichikawa, T. Tominaga, J. Radioanal. Nucl. Chem. Lett., 87, 189 (1984)
11. M.M. Bhasin, W.J. Bartley, P.C. Ellgen and T.P. Wilson, J. Catal 54, 120 (1978)
12. M. Ichikawa, T. Fukushima and K. Shikakura, Proceedings 8th Intnl Congr. Catal.(Berlin) vol-II, p69-80 (1984)
13. J.R. Anderson and D.M. Mainwaring, J. Catal., 35, 162 (1974)
14. A.K. Smith and J.M. Basset, J. Mol. Catal., 2, 229 (1977)
15. M. Ichikawa, J.C.S. Chem. Commun., 11, 26 (1976) ; Shokubai,17, 164 (1975); Chem Lett., 335 (1976)
16. B.C. Gates and J. Lieto, Chemtech, 195 (1980), J. Catal., 67, 303 (1981); M. Deeba, J.P. Scott, R. Barth and B.C. Gates, J. Catal., 71, 373 (1981)
17. A. Brenner and D.A. Hucul, Inorg. Chem., 18, 2836 (1979)
18. M. Ichikawa, "Tailored Metal Catalysts" ed. Iwasawa, p183-263 (1985) D. Reidel Pub.
19. J. Zwart and R. Snel, J. Mol. Chem., 30, 305 (1985)
20. J. Phillips and J.A. Dumesic, Applied Catal., 9, 1 (1984)
21. M. Ichikawa, K. Sekizawa, K. Shikakura and M. Kawai, J. Mol. Chem., 11, 167 (1981)
22. M. Kaminsky, Ki.J. Yoon, G.L. Geoffroy and M.A. Vannice, J. Catal., 91, 338 (1985)

23. R. Pierantozzi, K.J. McQuade, and B.C. Gates, Stud. Surf. Sci. Catal., 7, 941 (1981)
24. Y. Iwasawa and M. Yamada, J.C.S., Chem. Commun., 675 (1985)
25. L. Guczi, Z. Schay, K. Matsusek, I. Boyayad and G. Steffler, Stud. Surf. Sci. Catal., 7, 211 (1981)
26. V.L. Kuznetsov, A.F. Danilynk, I.E. Kolosova, and Yu.I. Yermakov, React. Kinet. Catal.,Lett., 21, 249 (1982)
27. J.R. Shapley, S.J. Hardwick, D.S. Foose, G.D. Stucky, M.R. Churchill. C. Bueno and J.P. Hutchinson, J. Am. Chem. Soc., 103, 7383 (1981)
28. J.R. Budge, J.P. Scott and B.C. Gates, J.C.S., Chem. Commun., 342 (1983)
29. S. Martinengo, P. Chini, V.G. Albano, and F. Cariati, J. Organomet. Chem., 59, 379 (1973)
30. A. Ceriotti, G. Longoni, R. Della Pergola, B.T. Heaton, and D.O. Smith, J. C.S. Dalton Trans., 1433 (1983)
31. T. Yokoyama, K. Yamazaki, N. Kosugi, H. Kuroda, M. Ichikawa and T. Fukushima, J.C.S. Chem. Commun.,962 (1984)
32. S. Iijima and M. Ichikawa, J. Catal., 94, 313 (1985)
33. J. Lieto, M. Walf, B.A. Matrana, M. Prochazka, B. Tesche, H. Knözinger, and B.C. Gates, J. Phys. Chem., 89, 991 (1985)
34. H. Oyanagi, T. Matsushita, M. Itoand H. Kuroda, KEK Report 83/30 (1984); N. Kosugi and H. Kuroda, Program EXAFS-1, Research Center for Spectroscopy, The University of Tokyo (1985)
35. M. Ichikawa, A.J. Lang, D.F. Shriver and W.M.H. Sachtler, J. Am. Chem. Soc., 107, 7216 (1985)
36. W.M.H. Sachtler and M. Ichikawa, J. Phys. Chem., in press(1986)
37. D.F. Shriver, ACS sym. ser., 152, 1 (1981); C.P. Horwitz and D.F. Shriver, Adv. Organomet. Chem., 23, 219 (1984)

TRACING THE PATHWAYS OF REACTIVE INTERMEDIATES ON METAL CLUSTER COMPLEXES

H.D. Kaesz, W. Krone-Schmidt, W.J. Sieber, N.M. Boag and C.B. Knobler

Department of Chemistry & Biochemistry,

University of California, Los Angeles, CA 90024, USA

ABSTRACT - The edge double-bridged complexes $Ru_3\{\mu\text{-}H,\mu\text{-}O{=}CR\}(CO)_{10}$ (R= Me, Et, Ph, or NMe_2) have been reacted with dimethylacetylene or some diaryl acetylenes at ambient pressure and temperature. A variety of products including trinuclear μ_3-π-allyl and di-nuclear σ-π-vinyl complexes are obtained. The dinuclear {μ-σ-π-vinyl, μ-carboxamido} complexes of ruthenium are seen to be fluxional, by contrast to dinuclear bis-{μ-acyl} derivatives, which are geometrically rigid both on the NMR and on the laboratory time frames.

1. INTRODUCTION

Earlier, we reported the reaction of $Ru_3\{\mu\text{-}H,\mu\text{-}O{=}CR\}$-$(CO)_{10}$ (R= Me, Et, Ph, or NMe_2) with ethylene under mild conditions, see Scheme 1, top part.[1] This leads to dinuclear _gem_-_bis_-acyl derivatives with no trace of the isomeric _vic_ derivatives. The _vic_ isomers are obtained by double alkylation of $Ru_3(CO)_{12}$ with RLi, followed by oxidation with $[R'_3O][BF_4]$, middle part of Scheme 1.[2] No _gem_ derivatives are observed under these conditions, nor are the _vic_ derivatives observed to isomerize into the _gem_ form.

By contrast, the reaction of ethylene with $Os_3\{u\text{-}H,u\text{-}O{=}CR\}(CO)_{10}$, requires elevated temperature and pressure, see lower part of Scheme 1.[2] This leads to _tri_-nuclear _bis_-acyl derivatives that are obtained in an equilibrium mixture of _vic_ and (predominantly) _gem_ isomers. The _tri_-nuclear _vic_ isomer is obtained exclusively in the low

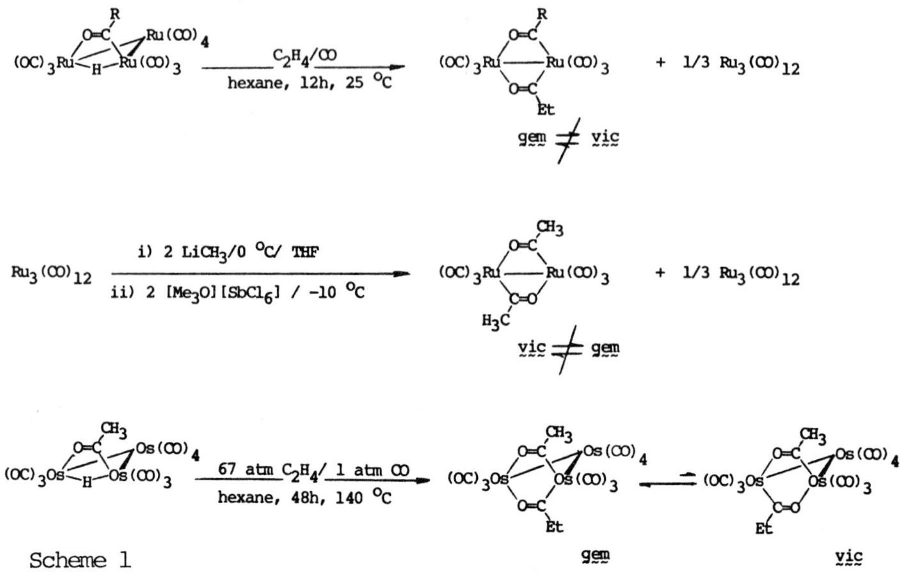

Scheme 1

temperature di-alkylation of $Os_3(CO)_{12}$ followed by oxidation, and is seen to equilibrate at room temperature and in polar solvents to the same product ratio obtained from the ethylene reaction, namely, a mixture of 15% *vic*- and 85% *gem*-$Os_3\{\mu\text{-}O{=}CR\}_2(CO)_{10}$.[2]

Interconversion of the *bis*-acyl isomers requires the acyl group to pass through a σ,π-intermediate or transition state, Scheme 2.

Scheme 2

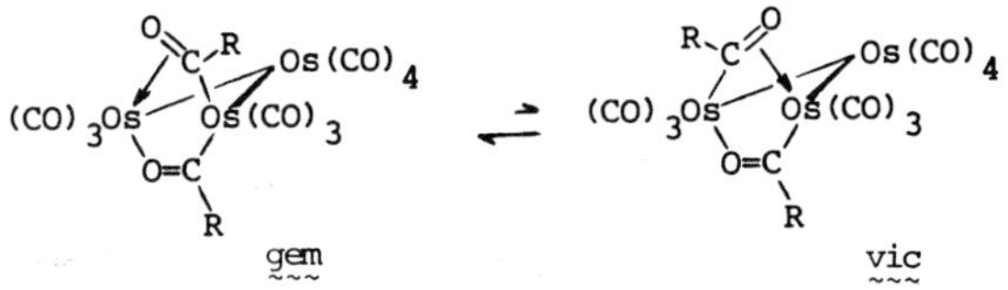

In a triosmium derivative whose structure has been determined,[3] a long (non-bonding) separation of 3.67(1) Å is seen between the *bis*-acyl bridged osmium atoms.[3] This could accommodate σ,π-bonding for the acyl group. In the diruthenium complexes on the other hand, we find a strong *bis*-acyl bridged metal-metal bond of 2.686(1) Å).[1] Thus, the acyl groups may be "locked" into the σ,σ-bonding mode affording a higher barrier to isomerization.

We were thus interested to obtain a mixed edge double-bridged complex containing the μ-σ,π-vinyl group accompanying a μ-acyl or a μ-carboxamido group, to examine the effects of such a combination on the metal-metal bond length and on the fluxional properties in such a derivative. The μ-σ,π-vinyl group can be obtained by reaction of a hydrido-metal cluster complex with a suitable acetylene.[4-11]

2. RESULTS

Our studies of the reaction of $M_3\{\mu\text{-H},\mu\text{-O=CNMe}_2\}(CO)_{10}$ (M= Ru, **1a**, or M= Os, **1b**)[12] with two types of acetylenes are shown in two Schemes which follow. The reactions with MeC≡CMe, **2**, are shown in Scheme 3. From **1a**, three isolatable new products are obtained namely, $Ru_3\{\eta^3\text{-}CH_2CHC(H)CH_3\}\{\mu\text{-}(CO),\mu\text{-O=CNMe}_2\}(CO)_8$, **4a**, $Ru_2\{\mu\text{-}\sigma,\pi\text{-}C(Me)\text{=CHMe},\mu\text{-O=CNMe}_2\}(CO)_6$, **5a**, and $Ru_2\{\mu\text{-O=C-CMe=CMe-}(\eta_2\text{-CMe=CHMe}\}\{\mu\text{-O=CNMe}_2)(CO)_5$, **6a**. The structures for **4a** and **6a** have been determined and are presented in Figures 1 and 2 below.

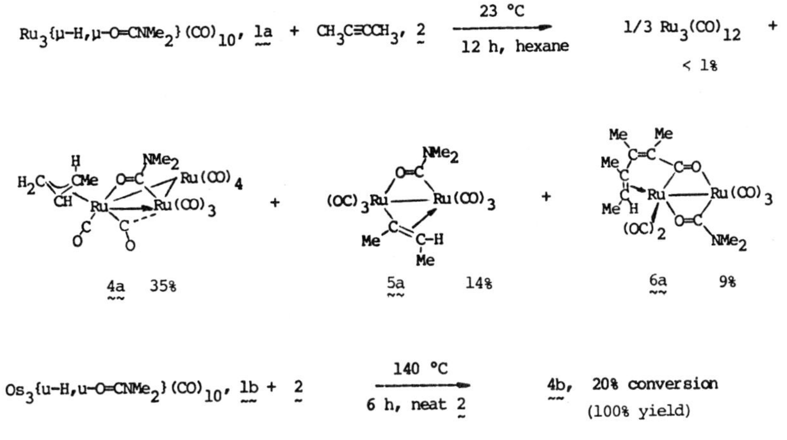

$Ru_3\{\mu\text{-H},\mu\text{-O=CNMe}_2\}(CO)_{10}$, **1a** + $CH_3C{\equiv}CCH_3$, **2** —(23 °C, 12 h, hexane)→ 1/3 $Ru_3(CO)_{12}$ (< 1%) + **4a** 35% + **5a** 14% + **6a** 9%

$Os_3\{\mu\text{-H},\mu\text{-O=CNMe}_2\}(CO)_{10}$, **1b** + **2** —(140 °C, 6 h, neat **2**)→ **4b**, 20% conversion (100% yield)

Scheme 3

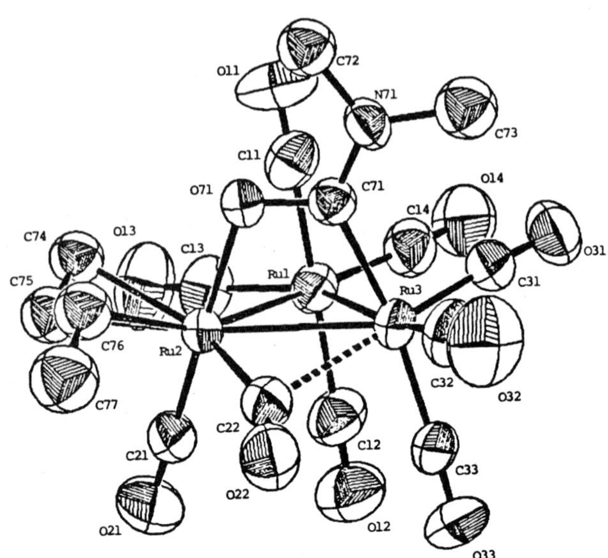

Figure 1. ORTEP of Ru_3{n^3-$CH_2CHC(H)CH_3$}{u-(CO),u-O=CNMe$_2$}$(CO)_8$, 4a; Å: Ru2-Ru2= 2.867(1); Ru1-Ru3= 2.829(1); Ru2-Ru3= 2.783(1).

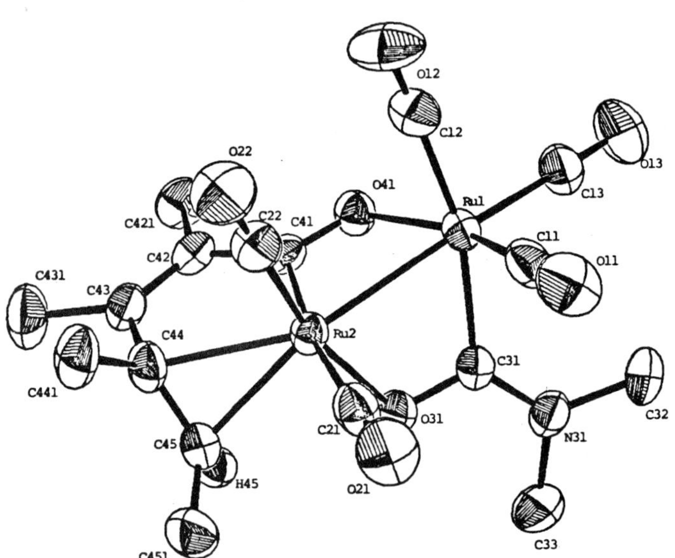

Figure 2. ORTEP of Ru_2{u-O=C-CMe=CMe-(n^2-CMe=CHMe}{u-O=CNMe$_2$)$(CO)_5$, 6a; Ru1-Ru2= 2.681(1) Å.

The metal-metal separation in 6a is 2.681(1) Å, close to that observed for the dinuclear *bis*-acyl ruthenium complex mentioned in the introduction.

For the tri-osmium complex 1b, more vigorous conditions are required for reaction using excess 2 as its own solvent under autoclave conditions, lower part, Scheme 3. Only the η^3-allyl derivative 4b is obtained. The position of the allyl group in complexes 4a or 4b, as well as the attachment of the {μ-O=C-CMe=CMe-(η^2-CMe=C(H)Me} group in 6a suggest that initial substitution of acetylene occurrs on the metal to which the oxygen of the {μ-O=CNMe$_2$} group is attached. This follows the path of PR_3 or ^{13}C substitution observed for these and related edge double-bridged derivatives.[13,14]

In the first-formed complexes of MeC≡CMe, a 1,2-hydrogen shift on the acetylene accompanies a *beta*-shift of hydrogen from the metal framework giving an η^3-allyl complex. This either predominates, as in 4a, or is the exclusive product, as in 4b. In product 6a a new μ-acyl group is formed in a vicinal arrangement relative to the μ-carboxamido group. The new organic group involves a *cis,cis*-oligomerization sequence of two acetylenes before the CO insertion giving the metalocyclic construction: Ru-Ru-C(O)-C(Me)=C(Me)-C(Me)=CH(Me). This latter contrasts with an interpolated CO insertion in an analogous iron system, accompanied by one *cis*- and one *trans*- substituted acetylene, *i.e.* Fe-Fe-C(R)=C(R)-C(O)-C(Ph)=C(H)Ph.[10]

To avoid formation of the π-allyl derivative, we carried out the reaction with aryl acetylenes, see Scheme 4 and Figure 3, next page.

$Ru_3\{\mu\text{-}H,\mu\text{-}O{=}CNMe_2\}(CO)_{10}$, **1a** + ArC≡CAr $\xrightarrow[\text{12 h, hexane}]{\text{23 °C}}$ 1/3 $Ru_3(CO)_{12}$ (10%) +

Ar = Ph, $p\text{-}CH_3C_6H_4$, (p-tol), or $p\text{-}(CH_3O)C_6H_4$, (p-anisyl)

$(OC)_3Ru\{O{=}C(NMe_2)\}\{C(Ar){=}C(H)Ar\}Ru(CO)_3$ ⇌ $(OC)_3Ru\{O{=}C(NMe_2)\}\{C(Ar){=}C(H)Ar\}Ru(CO)_3$ [80 - 46%] + $Ru(CO)_3$ / $Ru(CO)_3$ (**A**) + $Ru_2(CO)_6(Ar_2C_2)_2$, isomer of **A** [18 - 23%]

Scheme 4

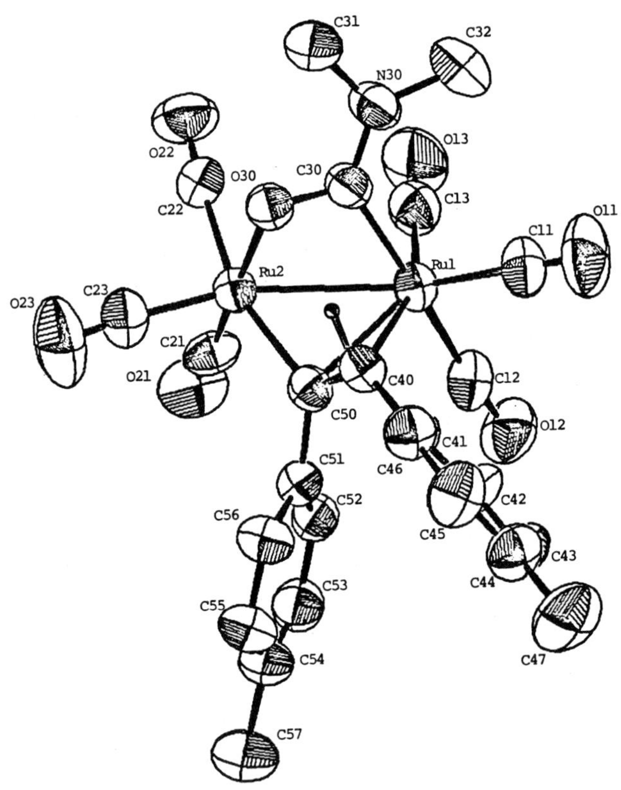

Figure 3. ORTEP of $Ru_2\{\mu\text{-}\sigma,\pi\text{-}C(p\text{-}tol){=}C(H)p\text{-}tol,\mu\text{-}O{=}CNMe_2\}(CO)_6$, **7b**; Ru1-Ru2= 2.739(1) Å.

^{13}C NMR of the dinuclear products $Ru_2\{\mu\text{-}\sigma,\pi\text{-}C(Ar){=}C(H)Ar, \mu\text{-}O{=}CNMe_2\}(CO)_6$, Ar= Ph, 7a, ($p\text{-}CH_3C_6H_4$), 7b, or ($p\text{-}CH_3O\text{-}C_6H_4$), 7c, indicate the existence of geminal and vicinal isomers in solution. The spectrum for 7a at −50 °C is shown as Figure 4. An unequal population of the two isomers is observed; if the predominant species in solution is that which is observed in the structure determination (of 7b, Fig. 3), then it would be the vicinal isomer. The resonances assigned as C^2 (one for each isomer, of unequal intensity) show a splitting of 157 Hz in the H-coupled spectrum. The signals of the two isomers merge into a single set of resonances at +30 °C, giving evidence for rapid interchange.

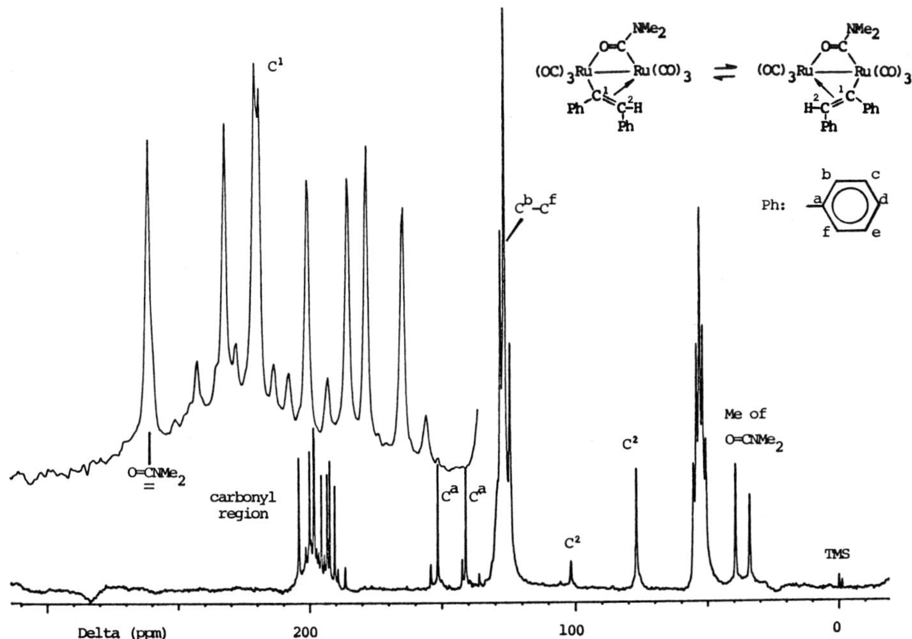

Figure 4. $\{^1H\}^{13}C$ NMR of $Ru_2\{\mu\text{-}\sigma,\pi\text{-}C(Ph){=}CHPh, \mu\text{-}O{=}CNMe_2\}(CO)_6$, 7a, CH_2Cl_2 sol'n., −50 °C.

The dinuclear ruthenium complexes containing a μ-σ,π-vinyl group show a metal-metal separation *above* 2.7 Å and are fluxional, by contrast to the di-nuclear *bis*-μ-acyl complexes in which the metal-metal separation is *below* 2.7 Å and whose geometry is rigid. The *vicinal*-{μ-σ,π-vinyl,μ-carboxamido} complex is favored in the solid (structure for complex **7b**, Fig. 3) and may also be favored in solution as the active isomer for acetylene homologation, (structure of complex **6a**, Fig. 2).

ACKNOWLEDGMENT. This work was supported by a Grant from the National Science Foundation to H.D.K., a NATO-SERC Fellowship to N.M.B., and a NATO Fellowship from the DAAD (German Academic Exchange Service) to W.J.S.

REFERENCES

1. C.E. Kampe, N.M. Boag and H.D.Kaesz, J. Mol. Catal., **21**, (1983) [Proceedings of the 4th International Symposium on the Relations Between Homogeneous and Heterogeneous Catalysis] pp. 279-291.
2. C.M. Jensen, Y.-J. Chen and H.D. Kaesz, J. Am. Chem. Soc., **106**, 4046-4047 (1984).
3. Y.-J. Chen, C.B. Knobler and H.D. Kaesz, unpublished results.
4. A.J. Deeming, S. Hasso and M. Underhill, J. Chem. Soc. Dalton Trans., 1614 (1975).
5. A.D. Clauss, M. Tachikawa, J.R. Shapley and C.G. Pierpont, Inorg. Chem., **20**, 1528-1533 (1981).
6. (a) W.G.Jackson, B.F.G. Johnson, J.W. Kelland, J. Lewis and K.T. Schorpp, J. Organomet. Chem., **87**, C27-C30 (1975).
(b) J.J. Guy, B.E. Reichert and G.M. Sheldrick, Acta Cryst., **B32**, 3319-3320 (1976).
7. A.G. Orpen, A.V. Rivera, E.G. Bryan, D. Pippard and G.M. Sheldrick, J. Chem. Soc. Chem. Commun., 723-724 (1978).
8. S. Aime, D. Osella, L. Milone, A.M. Manotti-Lanfredi and A. Tiripicchio, Inorg. Chim. Acta, **71**, 141-147 (1983).
9. E. Sappa, A. Tiripicchio and P. Braunstein, Chem. Revs., **83**, 203-239 (1983).
10. J. Ros, X. Solans, M. Font-Altaba and R. Mathieu, Organometallics, **3**, 1014-1020 (1984).
11. M. Lourdichi and R. Mathieu, Nouv. J. Chim., **6**, 231 (1982).

12. (a) A. Mayr, Y.-C. Lin, N.M. Boag and H.D. Kaesz, Inorg. Chem., 21, 1704 (1982).
(b) R. Szostak, C.E. Strouse and H.D. Kaesz, J. Organomet. Chem., 191, 243-260 (1980).
(c) Y.-C. Lin, A. Mayr, C.M. Knobler and H.D. Kaesz, J. Organomet. Chem., 272, 207-229 (1984).
13. A. Mayr, Y.-C. Lin, N.M. Boag, C.E. Kampe, C.B. Knobler and H.D. Kaesz, Inorg. Chem., 23, 4640-4645 (1984).
14. C.E. Kampe and H.D. Kaesz, Inorg. Chem., 23, 4646-4653 (1984).

RECENT STUDIES ON THE ORGANIC CHEMISTRY OF POLYNUCLEAR METAL CENTRES

S.A.R. Knox

Department of Inorganic Chemistry, The University,
Bristol, BS8 1TS, U.K.

ABSTRACT - The tri-hydride $[Ru_3(\mu\text{-}H)_3(CO)_3(\eta\text{-}C_5H_5)_3]$ provides an excellent entry into the organic chemistry of the tri-ruthenium centre. Reactions with a variety of unsaturated hydrocarbons produce complexes whose nature and reactivity are discussed in relation to processes occurring on a metal surface. Transformations of hydrocarbons through hydrogen loss and addition and through carbon-carbon bond cleavage and formation are induced thermally and photochemically, exemplified by studies on the conversions of ethylene and ethyne to μ_3-ethylidyne, shown to follow different pathways, and their 'dimerisation' to give μ_3-C(Me)CHCH and μ_3-CHC(Me)CH. Activation of co-ordinated hydrocarbons is also achieved through oxidation, assisted by the $\eta\text{-}C_5Me_5$ ligand, when proton loss is promoted.

1. INTRODUCTION

In recent years we have described how, from the complexes $[Ru_2(CO)_4(\eta\text{-}C_5R_5)_2]$ (R=H or Me), a considerable organic chemistry of the diruthenium centre can be developed, providing insights into the behaviour of organic species at the dinuclear sites present on a metal surface. In this lecture related organic chemistry of the triruthenium centre is outlined, built upon the reactive hydrido cluster $[Ru_3(\mu\text{-}H)_3(CO)_3(\eta\text{-}C_5H_5)_3]$. The different reactivity of dinuclear and trinuclear metal sites is an important aspect of heterogeneous catalysis and the contrasts between organo-diruthenium and organo-triruthenium chemistry can be viewed in this light.

2. TRIRUTHENIUM PRECURSORS

There is a variety of synthetic routes by which μ-alkylidene complexes of formula $[Ru_2(CO)_2(\mu\text{-}CHR)(\mu\text{-}CO)(\eta\text{-}C_5H_5)_2]$ (2;R=H, Me, Et, Ph, CO_2Et, etc.) may be obtained from the dimer $[Ru_2(CO)_4(\eta\text{-}C_5H_5)_2]$ (1) (Ref. 1-3). Upon heating in the solid state these all provide good yields of the μ_3-alkylidyne triruthenium complexes $[Ru_3(\mu_3\text{-}CR)(\mu\text{-}CO)_3(\eta\text{-}C_5H_5)_3]$ (3). Compared with diruthenium complexes the organic species appear to be very tightly bound in the latter and only a few reactions have been observed (see below). A much better entry into organo-triruthenium chemistry is

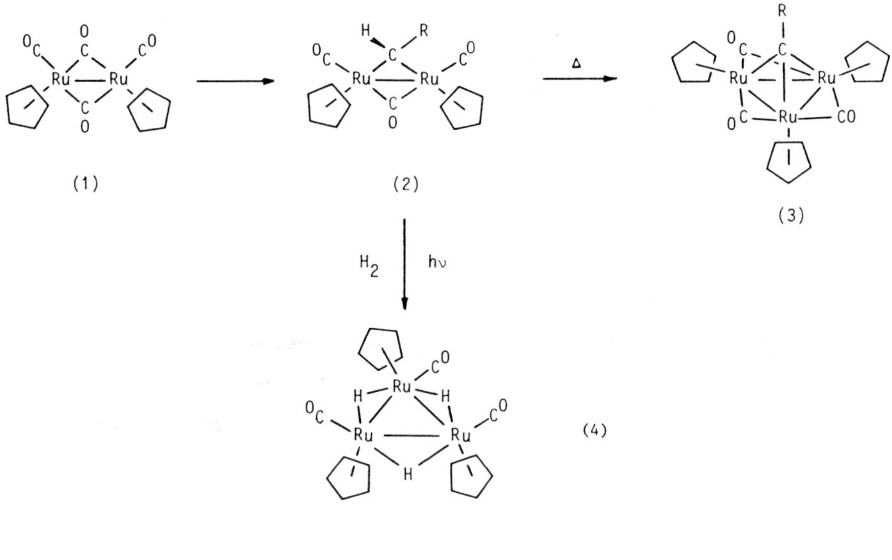

(1) (2) (3) (4)

provided by the blue trihydride $[Ru_3(\mu\text{-}H)_3(CO)_3(\eta\text{-}C_5H_5)_3]$ (4), prepared by bubbling hydrogen through a toluene solution of any μ-alkylidene complex (2) under u.v. irradiation (Ref. 4). From the μ-CH_2 complex a 76% yield is obtained. The solid state structure has been established by X-ray diffraction; this structure is maintained in solution at room temperature but on warming a fluxional process occurs on the n.m.r. time scale to average the environments of the cyclopentadienyl and hydride ligands.

The reactions of (4) with Ag^+ or H^+, deuterium, and CO are summarised below:

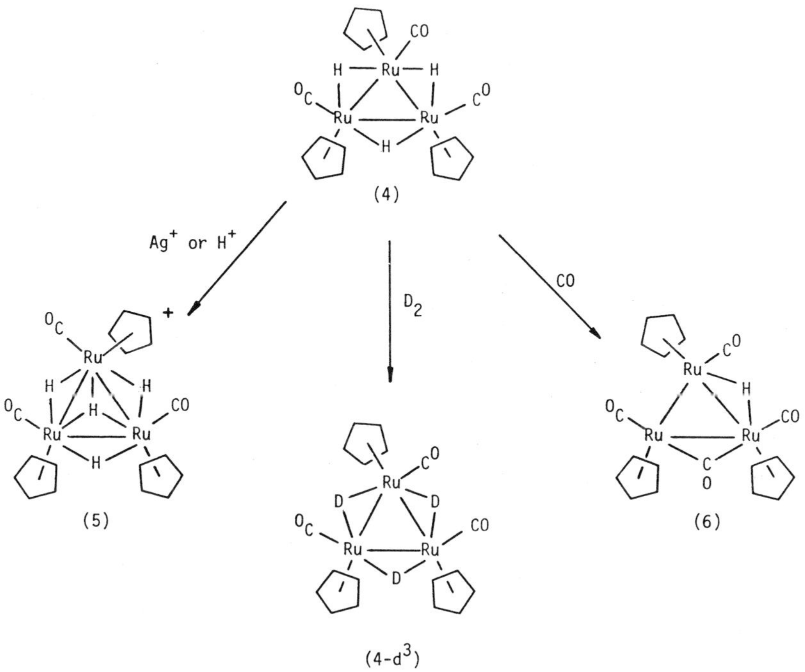

(4)

Ag^+ or H^+

D_2

CO

(5)

$(4\text{-}d^3)$

(6)

Cyclic voltammetry reveals that (4) undergoes a reversible one-election oxidation and when treated with Ag^+ the cation $[Ru_3(\mu\text{-}H)_3(\mu_3\text{-}H)(CO)_3(\eta\text{-}C_5H_5)_3]^+$ (5) is formed, presumably via the radical cation $(4)^{\ddagger}$ which abstracts a hydrogen radical from solvent. Cation (5) is also obtained directly by protonation of (4). A more useful reaction occurs when (4) is heated in toluene at 100°C under an atmosphere of deuterium, H/D exchange occurring rapidly to yield $[Ru_3(\mu\text{-}D)_3(CO)_3(\eta\text{-}C_5H_5)_3]$ $(4\text{-}d^3)$, a complex which has been valuable in probing the mechanisms of the reactions of (4) with unsaturated hydrocarbons. Treatment of (4) with CO effects the substitution of two hydrogens to give purple $[Ru_3(\mu\text{-}H)(\mu\text{-}CO)(CO)_3(\eta\text{-}C_5H_5)_3]$ (6), an unstable complex which does not generate any organo-triruthenium chemistry. The ready loss of two hydrogens from (4) does indicate, however, the likely first stage of the reactions of (4) with unsaturated hydrocarbons, which are now described.

3. REACTIONS WITH ALKENES

The trihydride $[Ru_3(\mu\text{-}H)_3(CO)_3(\eta\text{-}C_5H_5)_3]$ (4) reacts with a variety of alkenes in boiling toluene, with results exemplified in Fig. 1. In the reactions with ethylene and 1-alkenes $RCH{=}CH_2$ (R=Me, Et, or Ph) the three cluster-bound hydrogens and one alkene hydrogen are lost in a rearrangement which affords the μ_3-alkylidyne species (3). This thermal reaction of (4) with ethylene is complete within 4h and gives (3; R=Me) in 56% yield. A comparable yield can be achieved in a photochemically-induced reaction, but over about a day. However, there is an aditional low yield product, the C_4-ligand complex $[Ru_3(\mu\text{-}CO)(\mu_3\text{-}CO)(\mu_3\text{-}CHCHCMe)(\eta\text{-}C_5H_5)_3]$ (7) arising from two molecules of ethylene. More will be said later about such species, which evolve regularly from photochemical reactions of (3) and (4) with alkynes.

On the basis of the reaction of (4) with CO, outlined earlier, the initial product of the reaction with ethylene is likely to be $[Ru_3(\mu\text{-}H)(C_2H_4)(CO)_3(\eta\text{-}C_5H_5)_3]$. This could lose H_2 and rearrange to a μ_3-ethylidyne complex (3) in several ways, e.g. via alkenyl, vinylidene, or ethylidene intermediates. The reaction of (4) with ethyne sheds some light on these possibilities and discussion will therefore be held over until that is described. However, at this stage the involvement of metal-bound hydrogen in the organic transformation can be discounted; treating (4-d^3) with ethylene under both thermal and photochemical conditions gave the μ_3-CMe complex (3) containing no significant amount of deuterium.

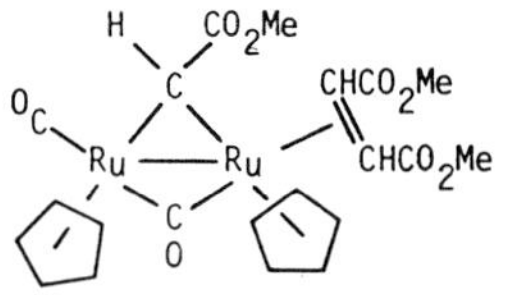

(9)

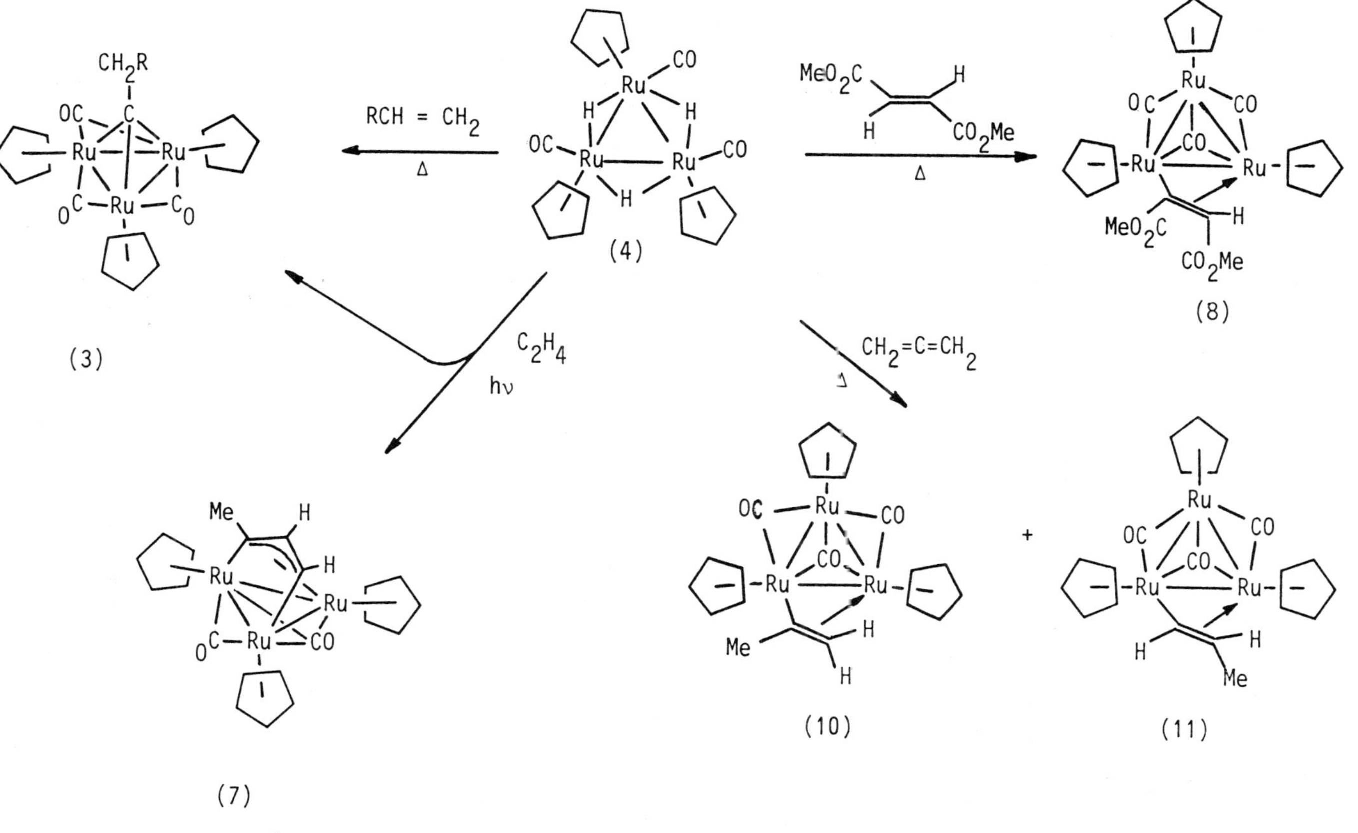

Fig. 1. Reactions of $[Ru_3(\mu\text{-}H)_3(CO)_3(\eta\text{-}C_5H_5)_3]$ with alkenes.

No alkene to alkylidyne rearrangement is possible for dimethyl fumarate, trans-$CH(CO_2Me)=CH(CO_2Me)$, and instead a μ-alkenyl complex $[Ru_3(\mu\text{-}CO)_2(\mu_3\text{-}CO)[\mu\text{-}C(CO_2Me)CH(CO_2Me)](\eta\text{-}C_5H_5)_3]$ (8) is formed from (4), in a reaction reminiscent of those of $[Os_3(\mu\text{-}H)_2(CO)_{10}]$ with alkenes (Ref. 5,6). A surprising co-product of this reaction is the diruthenium μ-alkylidene complex $[Ru_2(CO)[CH(CO_2Me)=CH(CO_2Me)](\mu\text{-}CHCO_2Me)(\mu\text{-}CO)(\eta\text{-}C_5H_5)_2]$ (9), derived from a cleavage of the double bond of the alkene.

Allene reacts rapidly (<1h) with (4) in boiling toluene to give as shown the two isomeric μ-propenyl complexes (10) and (11), in 12:1 ratio (Fig. 1). The μ-$C(Me)CH_2$ complex (10) is the expected product of hydrogen migration to a co-ordinated allene, with (11) being generated by a subsequent hydrogen shift. As will be described later, the complexes (10) and (11) are also obtained by the reaction of propyne with (4); in the same way (8) is formed by employing dimethyl acetylenedicarboxylate, $MeO_2CC{\equiv}CCO_2Me$.

Cluster breakdown occurs when (4) is heated with butadiene and only the diruthenium products $[Ru_2(CO)_2(\mu\text{-}C_4H_6)(\eta\text{-}C_5H_5)_2]$ (12) and $[Ru_2(CO)(\eta^2\text{-}C_4H_6)(\mu\text{-}CO)_2(\eta\text{-}C_5H_5)_2]$ (13) are isolated, the former being shown through X-ray diffraction to contain the hydrocarbon bridging in the S-trans configuration. As expected, photolysis

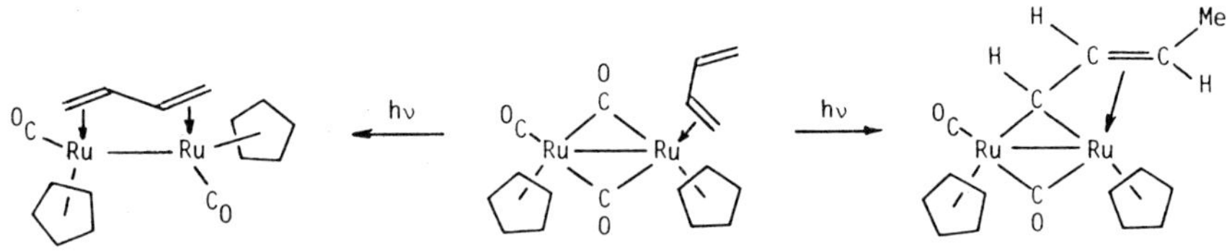

(12) (13) (14)

of the η^2-butadiene complex (13) yielded (12), but also $[Ru_2(CO)(\mu\text{-}CO)(\mu\text{-}CHCHCHMe)(\eta\text{-}C_5H_5)_2]$ (14), an isomer of (12) related by a formal 1,4 hydrogen shift.

4. REACTIONS WITH ALKYNES

When ethyne is bubbled through a boiling toluene solution of (4) the complex is consumed within 15 minutes, affording the μ_3-ethylidyne species (3) in 17% yield and the new μ-ethenyl complex $[Ru_3(\mu\text{-}CO)_2(\mu_3\text{-}CO)(\mu\text{-}CHCH_2)(\eta\text{-}C_5H_5)_3]$ (15) in 45% yield. The latter is evidently an intermediate in the formation of the former since under the same conditions (15) isomerised quantitatively to (3) over 2.5h. The same experiments performed with ($4\text{-}d^3$) gave (15) and (3) containing μ-CHCHD and μ-CH_2D ligands respectively, confirming the expected involvement of metal-bound H/D in the process. This first observation of an apparently direct transformation of $CHCH_2$ to CCH_3

(4) $\xrightarrow[h\nu]{HC_2H}$ (15) $\xrightarrow{\Delta}$ (3)

at a polynuclear metal centre contrasts with an osmium system where $[Os_3(\mu\text{-}H)(\mu\text{-}CHCH_2)(CO)_{10}]$ converts to $[Os_3(\mu\text{-}H)_2(\mu_3\text{-}CCH_2)(CO)_9]$ on heating, and treatment with hydrogen is then required to generate $[Os_3(\mu\text{-}H)_3(\mu_3\text{-}CMe)(CO)_9]$ (Ref. 6,7). The interconversion of μ-$CHCH_2$ and μ-CCH_3 is unknown at a dinuclear metal centre but alkyl substitution at the β-carbon reduces the barrier sufficiently that μ-CHCHR $\rightleftharpoons$ μ-CCH_2R isomerisation may be observed (Ref. 8,9).

No evidence of the μ-ethenyl complex (15) was obtained in the reaction of (4) with ethylene, yet it is sufficiently long-lived in boiling toluene that were it an intermediate in the formation of the μ-CCH_3 product (3) it would have been detected easily. It is therefore clear that the transformations of ethylene and ethyne to μ_3-CCH_3 at the triruthenium centre follow different paths. Some possibilities are summarised in Fig. 2. For ethylene, the ruling out of the μ-ethenyl intermediate leads to the possibility that

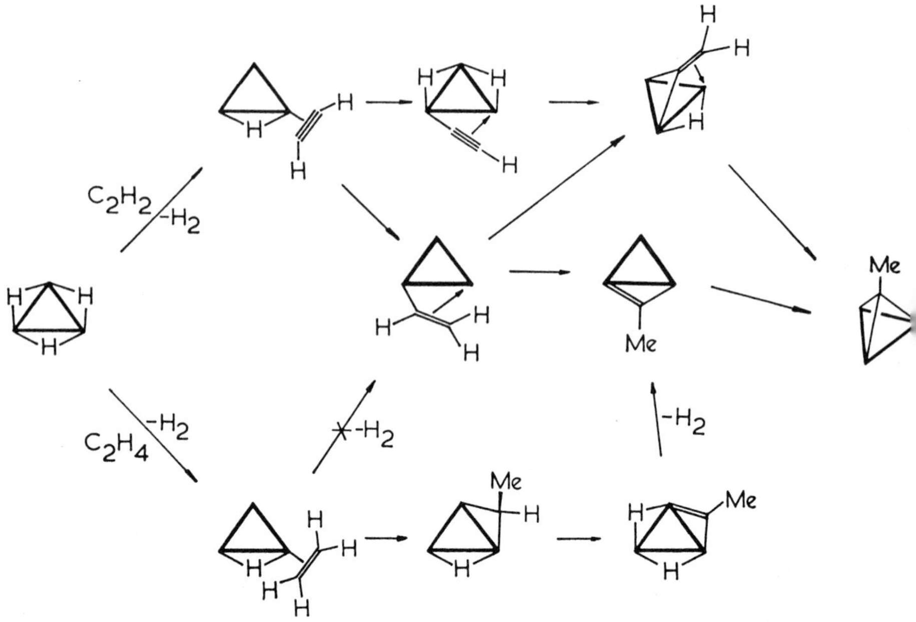

Fig. 2. Possible transformations of ethylene and ethyne at a triruthenium centre.

isomerisation to μ-ethylidene occurs, followed by hydrogen transfer to the cluster and H_2 loss. Both the CH_2=CH_2 to μ-$CHCH_3$ and μ-$CHCH_3$ to μ_3-CCH_3 steps have been observed separately in our previous work (Ref. 4,10). The intermediacy of a μ-ethenyl intermediate in the transformation of ethyne is established. However, the μ_3-CCH_3 species is formed in this reaction too rapidly to have all evolved via the slow isomerisation of μ-$CHCH_2$. An additional pathway is indicated, perhaps involving the μ-ethynyl species shown. The final coordination of an alkylidyne to the triruthenium unit is a very favourable one; only on heating above 400°C do these complexes release the organic species as hydrocarbons whereas the μ-alkylidene complexes (2) do so at about 200°C.

The Pt(111) surface has three-fold symmetry and is composed of triangular sites like that in (4). It is fascinating, therefore, to note that both ethylene and ethyne chemisorbed on this surface also rearrange to a μ_3-ethylidyne species (Ref. 11,12). This suggests that (4) may serve as a useful model for the organic rearrangements on such a surface and that the pathways outlined above may be relevant to the surface situation.

Complexes $[Ru_3(\mu\text{-}CO)_2(\mu_3\text{-}CO)(\mu\text{-}CRCHR)(\eta\text{-}C_5H_5)_3]$, analogous to (15), are formed in high yield when (4) is heated with alkynes RC_2R (R=Me, Ph, or CO_2Me). Migration or loss of an R group would now be required to generate a μ_3-alkylidyne complex but this does not occur; unlike (15) these complexes are thermally stable. The isomeric complexes $[Ru_3(\mu\text{-}CO)_2(\mu_3\text{-}CO)(\mu\text{-}CHCHR)(\eta\text{-}C_5H_5)_3]$ and $[Ru_3(\mu\text{-}CO)_2(\mu_3\text{-}CO)\text{-}(\mu\text{-}CRCH_2)(\eta\text{-}C_5H_5)_3]$, derived from (4) and alkynes RC_2H (R=Me or Ph), are however able to undergo hydrogen shifts to produce $\mu_3\text{-}CCH_2R$ species, and do so readily on heating.

All but two of the μ-alkenyl complexes of type (15) are fluxional, undergoing the characteristic oscillatory σ/π rearrangment of such a ligand.[7] The exceptions are the μ-CR=CHR (R=Ph or CO_2Me) complexes, whose strongly π-accepting double bonds would not easily oscillate between metal centres. Comparative X-ray diffraction studies on the μ-CR=CHR (R=Me or CO_2Me) complexes confirm the strengthened coordination induced by an electronegative substituent.

It was seen above that whereas the thermal reaction of ethylene with $[Ru_3(\mu\text{-}H)_3(CO)_3(\eta\text{-}C_5H_5)_3]$ (4) gave (3), derived from single molecule of the alkene, the photochemically-induced reaction also produced the C_4-ligand complex (7). A similar but stronger tendency towards carbon-carbon bond formation under u.v. irradiation is apparent in the reactions of (4) with alkynes. Under these conditions ethyne reacts with (4) over 16h to yield (3) but no μ-ethenyl complex (15), and has 70% of the yield located in three products each derived from two molecules of ethyne. These are the complexes (16) (50%), (17) (15%), and (18) (5%) illustrated. Complex (16), characterised by X-ray diffraction, has an unusual zwitterionic

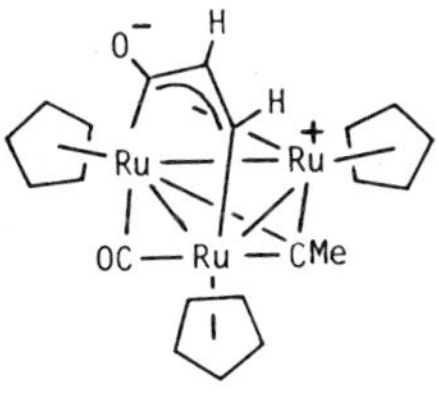

(16)

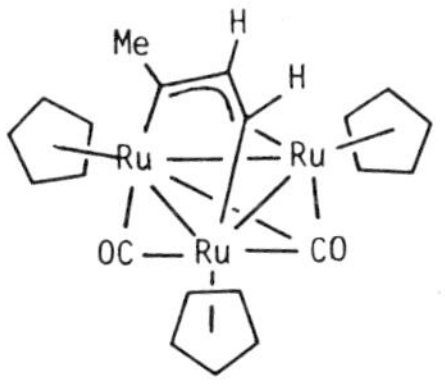

(17)

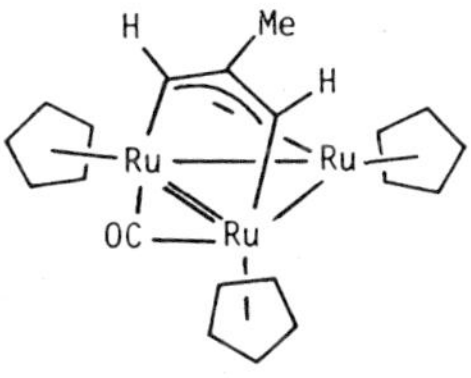

(18)

character and is seen to be structurally related to (17) and (18). Indeed, (16) and (17) are isomers, differing in whether ethylidyne or CO is linked with an ethyne unit. Complex (18), on the other hand, has one less CO than these two, providing an electronic deficiency which the cluster presumably overcomes through additional metal-metal bonding. Another feature of (18) is of interest; the C_4 ligand has a branched chain, requiring that although formed from two molecules of ethyne a carbon-carbon bond cleavage must also be involved. There is precedent for this in the work of Stone (Ref. 13) and Chisholm (Ref. 14), the latter observing a μ-C(Me)C(Me)C($SiMe_3$)⟶μ-C(Me)-C($SiMe_3$)C(Me) interconversion to occur on the n.m.r. time scale via a suggested μ-cyclopropenyl intermediate. However, neither u.v. irradiation nor heating would induce the rearrangement of (17) to (18). Likewise, the isomers (16) and (17) could not be interconverted.

The presence of an ethylidyne unit in each of (16), (17) and (18) suggests the possible intermediacy of complex (3) in their formation, and treatment of the latter with ethyne under u.v. irradiation does indeed give (16) (45%) and (17) (30%). It is likely that in this reaction an initial CO substitution occurs to give a complex $[Ru_3(\mu_3\text{-CMe})(HC_2H)(CO)_2(\eta\text{-}C_5H_5)_3]$ containing coordinated ethyne which subsequently links with either CO, giving (16), or ethylidyne, giving (17). However, if such an intermediate is involved in the formation of (16) and (17) from the trihydride (4) there must be another pathway to it, perhaps via the μ-ethenyl species, because (16) and (17) are formed much more rapidly from (4) than from (3) (16h vs. 4d). The linking of ethyne and ethylidyne to give a μ_3-CHCHCMe ligand at a triruthenium centre has been previously observed by Keister (Ref. 15).

In early experiments with ethyne the gas was 'purified' by passing through concentrated sulphuric acid, a precaution quickly dispensed with when the major product (60%) of the reaction between this 'ethyne' and (4) under u.v. irradiation proved to be an SO_2 complex

$[Ru_3(\mu\text{-}SO_2)(\mu\text{-}CO)_2(\mu_3\text{-}CMe)(\eta\text{-}C_5H_5)_3]$ (19), identified by X-ray diffraction. As above, the complex is produced in equally good yield, but more slowly (67 c.f. 19h.), from the μ_3-ethylidyne complex

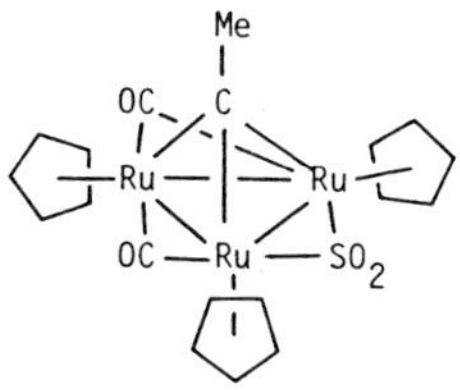

(19)

(3) and 'purified' ethyne. Strangely, it was not possible to prepare (19) from (3) and either SO_2 or SO_3 and its evolution remains obscure.

5. PENTAMETHYLCYCLOPENTADIENYL TRIRUTHENIUM CHEMISTRY

The change from an η-C_5H_5 to an η-C_5Me_5 ligand has a marked effect on the organic chemistry of the diruthenium centre and it was therefore of interest to attempt the synthesis of the η-C_5Me_5 analogue of (4). However, passing hydrogen through a solution of $[Ru_2(CO)_4(\eta\text{-}C_5Me_5)_2]$ under u.v. irradiation gave not $[Ru_3(\mu\text{-}H)_3(CO)_3(\eta\text{-}C_5Me_5)_3]$, a trimer of the $RuH(CO)(\eta\text{-}C_5Me_5)$ unit, but $[Ru_2(\mu\text{-}H)_2(CO)_2(\eta\text{-}C_5Me_5)_3]$ (20) (Ref. 10), the dimeric formulation presumably being favoured sterically. Recently it has been reported that photolysis of $[RuEt(CO)_2(\eta\text{-}C_5H_5)]$ in pentane at -30°C generates unstable $[Ru_2(\mu\text{-}H)_2(CO)_2(\eta\text{-}C_5H_5)_2]$ (Ref. 16). Only a 5% yield of (20) is

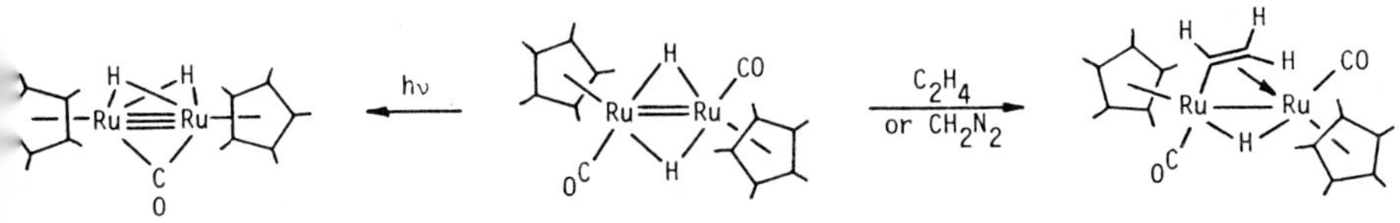

hν

C_2H_4 or CH_2N_2

(20)

obtained by the above route and the complex is better prepared (25%) by treating [$RuCl(CO)_2(\eta-C_5Me_5)$] with $LiAlH_4$. Some limited chemistry of (20) has been explored, as illustrated.

The major product of the photolysis of [$Ru_2(CO)_4(\eta-C_5Me_5)_2$] under hydrogen is an unusual triruthenium cluster radical [$Ru_3(\mu-CO)_3$-(μ_3-CO)($\eta-C_5Me_5)_3$] (21). This is one electron short of the 48 valence electrons present in isostructural [$Ru_3(\mu-CO)_3(\mu_3-CMe)(\eta-C_5Me_5)_3$] (22); i.e. two-electron CO replaces three-electron ethylidyne. The

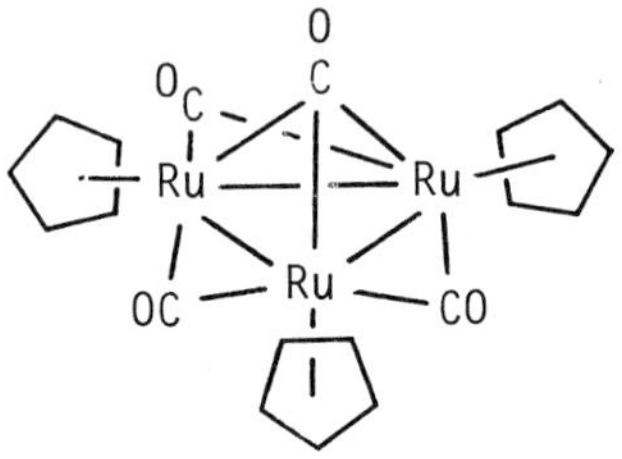

(21)

complex is paramagnetic with a magnetic moment of 1.8 B.M. but as yet no e.s.r. signal has been observed. The cyclic voltammogram of (21) shows three reversible waves at E° = -0.72, -0.10, and +0.85V corresponding to the existence of a 48-electron anion, a 46-electron cation, and a 45-electron dication. Work in progress suggests that the first two of these may be isolable. This, and other chemistry of (21) under investigation, will be described. As expected, (21) reacts rapidly with NO, a radical itself.

The stability of 47-electron (21) suggests that the μ_3-ethylidyne complex should be oxidised readily and this is so. The cyclic voltammogram has two reversible waves at E° = 0.17 and 0.86V and both the monocation (23) and di-cation (24) have been isolated (see Fig. 3). The former, isoelectronic with (21) and therefore also paramagnetic (μ 2.0 B.M.), is stable and its structure has been established by X-ray diffraction. The latter is less stable and slowly deprotonates to afford the μ_3-vinylidene cation [$Ru_3(\mu-CO)_3(\mu_3-CCH_2)(\eta-C_5Me_5)_3$]$^+$ (25), which may be treated with $NaBH_4$ to regenerate (22) or with MeLi to give the μ_3-propylidyne complex (26), completing the homologation of (22) (Ref. 17).

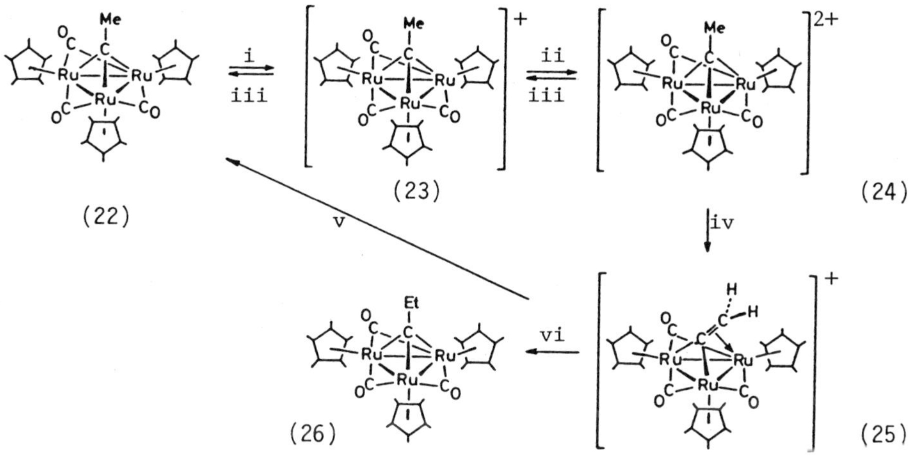

Fig. 3. Reagents: i, H^+ or Ag^+; ii, Ag^+; iii, $LiBHEt_3$; iv, $-H^+$; v, $NaBH_4$; vi, MeLi.

When $[Ru_2(CO)_4(\eta\text{-}C_5H_5)_2]$ is subjected to u.v. irradiation under hydrogen it does not produce complexes analogous to (20) and (21) because of the generation of $[Ru(CO)_2(\eta\text{-}C_5H_5)]$ radicals which readily attack cyclopentadienyl rings to yield the tetraruthenium μ_3-cyclopentadienylidene complex $[Ru_4(CO)_6(\mu_3\text{-}C_5H_4)_2(\eta\text{-}C_5H_5)_2]$ (Ref. 18). Thermolysis of $[Ru_2(CO)_4(\eta\text{-}C_5H_5)_2]$ in decalin proceeds differently, giving the first trigonal bipyramidal pentaruthenium cluster $[Ru_5(CO)_3(\mu_3\text{-}CO)_3(\eta\text{-}C_5H_5)_4]$ (27). A co-product is

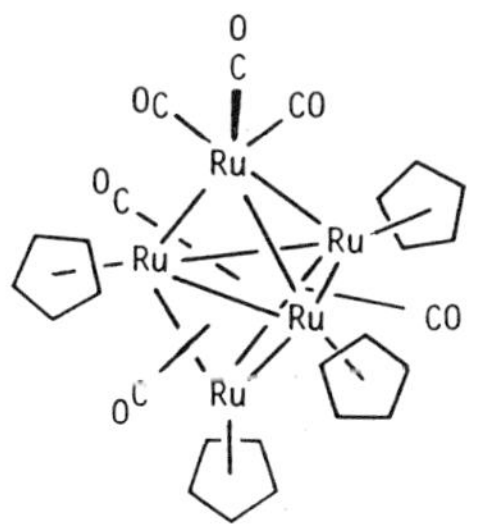

(27)

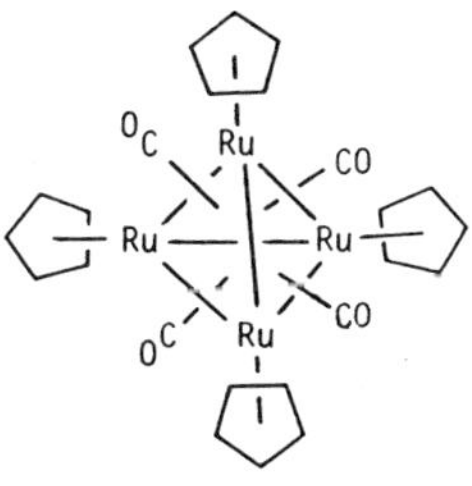

(28)

$[Ru_4(\mu_3\text{-}CO)_4(\eta\text{-}C_5H_5)_4]$ (28), to which (27) is related through the replacement of a μ_3-CO ligand by an isolobal capping $Ru(CO)_3$ group. Only the analogue of (28) is formed when $[Ru_2(CO)_4(\eta\text{-}C_5Me_5)_2]$ is heated in decalin, but an unprecedented abstraction of hydrocarbon fragments occurs in boiling ether solvents, yielding μ_3-alkylidyne complexes analogous to (22). In tetrahydrofuran, which contains a C_4 unit, all four C_1 - C_4 complexes $[Ru_3(\mu\text{-}CO)_3(\mu_3\text{-}CR)(\eta\text{-}C_5Me_5)_3]$ (R=H, Me, Et, and Pr) are produced in a combined yield of about 20%, whereas diglyme, containing only C_1 and C_2 units, affords the μ_3-CH and μ_3-CMe species alone in a similar yield.

6. CATALYTIC ACTIVITY OF $[Ru(\mu\text{-}H)(CO)(\eta\text{-}C_5R_5)]_n$ (n = 2 or 3)

Both the hydrides $[Ru_3(\mu\text{-}H)_3(CO)_3(\eta\text{-}C_5H_5)_3]$ (4) and $[Ru_2(\mu\text{-}H)_2(CO)_2\text{-}(\eta\text{-}C_5Me_5)_2]$ (20) are catalysts, or catalyst precursors, for the isomerisation and hydrogenation of 1-alkenes, but are only moderately active. Isomerisation of 2- and 3-alkenes is catalysed very slowly. There are major differences between the Ru_2- and Ru_3-based systems. Whereas the presence of hydrogen gas improves the isomerisation activity of (4), with only slight hydrogenation induced, the gas is required to achieve any significant isomerisation at all by (20), and then at the expense of major competitive hydrogenation. Since the first step in catalytic isomerisation by (4) is likely to be loss of H_2 the beneficial effect of its presence is presumably due to its stabilising the system. During catalysis the triruthenium system remains blue [the colour of (4)] and (4) has been subsequently isolated nearly quantitatively. This and the chemistry of the complex, which was seen above to normally involve retention of the Ru_3 cluster, leads to the conclusion that the catalytic cycle is based on this unit. A possible scheme for the isomerisation and hydrogenation of 1-hexene by (4) is illustrated in Fig. 4. Complex (4) also slowly catalyses the hydrogenation of cyclohexa-1,3-diene to cyclohexene specifically and of 1,3-butadiene to 1- and 2-butenes and butane.

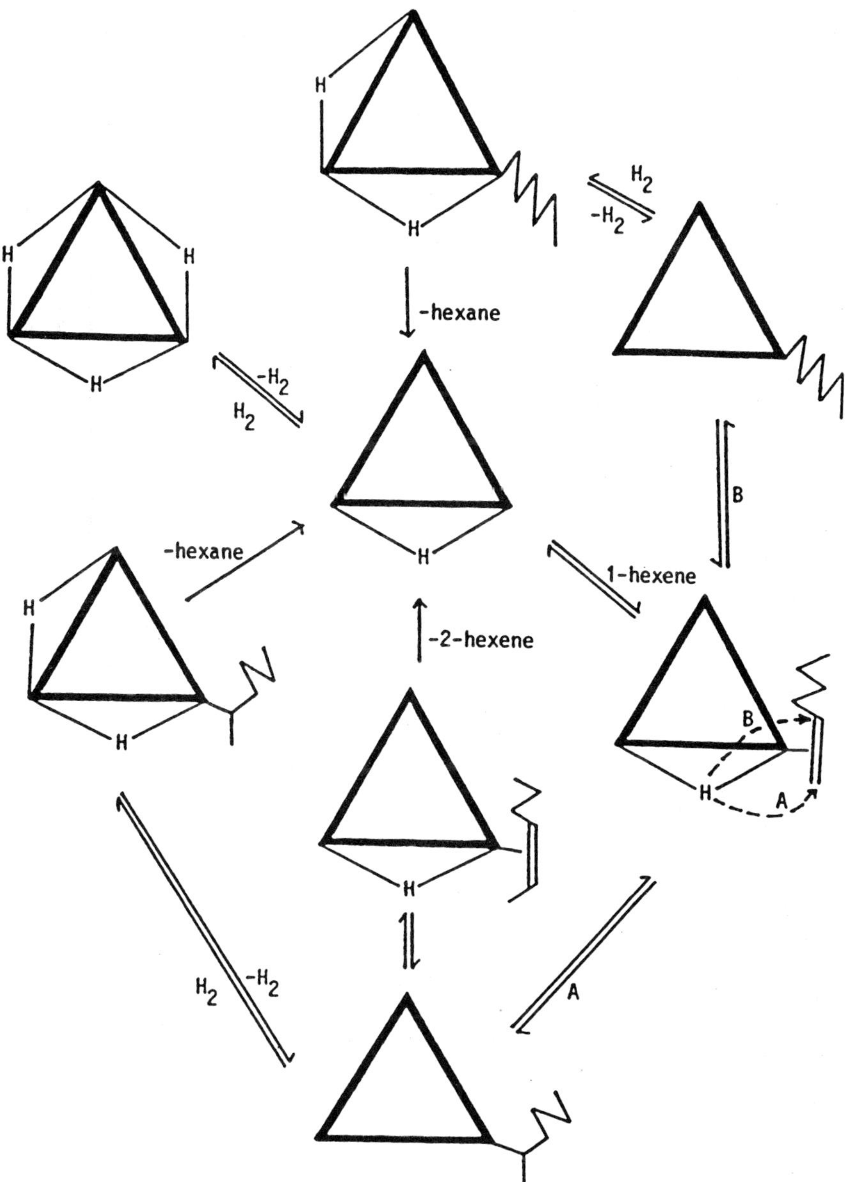

Fig. 4. Isomerisation and hydrogenation of 1-hexene catalysed by (4).

7. CONCLUSIONS

The hydrido cluster $[Ru_3(\mu\text{-}H)_3(CO)_3(\eta\text{-}C_5H_5)_3]$ evidently supports a substantial organo-triruthenium chemistry, and is characterised by a capacity for inducing organic transformations through hydrogen transfer to and from the triruthenium unit. Thermal reactions with alkenes and alkynes are observed to follow different pathways, involving essentially diruthenium chemistry, but a common alkylidyne species evolves and settles into the low energy 'sink' of μ_3-binding to the triruthenium site. This is the end of the road thermally, but activation of these complexes can be achieved photochemically, through CO substitution, when carbon-carbon bond formation processes follow. Early studies with triruthenium systems stabilised by $\eta\text{-}C_5Me_5$ indicate that the electron-donating power of this ligand promotes oxidation which can lead to the activation of co-ordinated hydrocarbons through their deprotonation. There are clearly a variety of paths available for organic transformations in a triruthenium complex and the parallels which already exist with a metal surface suggest that an insight into the chemistry of the latter is being gained.

ACKNOWLEDGEMENTS

The work described in the lecture was mainly performed by three talented post-graduate students, Michael J. Morris, Kirsty A. Macpherson, and Nigel J. Forrow, with substantial contributions coming from Dr. Melvyn C. Rendle and Professor Glenn S. Lewandos (on leave from Central Michigan University, U.S.A.). The collaborations with my colleagues Dr. Neil G. Connelly and Dr. A. Guy Orpen, on electrochemical and X-ray diffraction studies respectively, have been invaluable. The Science and Engineering Research Council and Johnson Matthey PLC have provided generous support. Acknowledgement is made to the Donors of the Petroleum Research Fund, administered by the American Chemical Society, for partial support of this research.

REFERENCES

1. A.F. Dyke, S.A.R. Knox, M.J. Morris, and P.J. Naish, J. Chem. Soc., Dalton Trans., 1983, 1417.
2. R.E. Colborn, D.L. Davies, A.F. Dyke, A. Endesfelder, S.A.R. Knox, A.G. Orpen, and D. Plaas, J. Chem. Soc., Dalton Trans., 1983, 2661.

3. D.L. Davies, S.A.R. Knox, K.A. Mead, M.J. Morris, and P. Woodward, J. Chem. Soc., Dalton Trans., 1984, 2293.
4. N.J. Forrow, S.A.R. Knox, M.J. Morris, and A.G. Orpen, J. Chem. Soc., Chem. Commun., 1983, 234.
5. A.J. Deeming, S. Hasso, and M. Underhill, J. Chem. Soc., Dalton Trans., 1975, 1614.
6. J.B. Keister, and J.R. Shapley, J. Am. Chem. Soc., 1976, 98, 1056.
7. J.B. Keister, and J.R. Shapley, J. Organomet. Chem., 1975, 85, C29.
8. C.P. Casey, S.R. Marder, and P.J. Fagan, J. Am. Chem. Soc., 1984, 106, 7197.
9. M. Green, A.G. Orpen, and C.J. Schaverein, J. Chem. Soc., Chem. Commun., 1984, 37.
10. N.J. Forrow, and S.A.R. Knox, J. Chem. Soc., Chem. Commun., 1984, 679.
11. L.L. Kesmodel, L.H. Dubois, and G.A. Somorjai, J. Chem. Phys., 1979, 70, 2180.
12. S.M. Davis, F. Zaera, B.E. Gordon, and G.A. Somorjai, J. Catal., 1985, 92, 240.
13. J.C. Jeffery, K.A. Mead, H. Razay, F.G.A. Stone, M.J. Went, and P. Woodward, J. Chem. Soc., Dalton Trans., 1984, 1383.
14. M.H. Chisholm, J.A. Heppert, and J.C. Huffmann, J. Am. Chem. Soc., 1984, 106, 1151.
15. L.R. Beanan, Z.A. Rahman, and J.B. Keister, Organometallics, 1983, 2, 1062.
16. K.A. Mahmoud, A.J. Rest, and H.G. Alt, J. Chem. Soc., Dalton Trans., 1985, 1365.
17. N.G. Connelly, N.J. Forrow, S.A.R. Knox, K.A. Macpherson, and A.G. Orpen, J. Chem. Soc., Chem. Commun., 1985, 16.
18. N.D. Feasey, N.J. Forrow, G. Hogarth, S.A.R. Knox, K.A. Macpherson, M.J. Morris, and A.G. Orpen, J. Organomet. Chem., 1984, 267, C41.

HYDROGEN ACTIVATION BY INORGANIC COMPLEXES AND SURFACES

Hung-Wen Jen and Alan Brenner*
Department of Chemistry, Wayne State University, Detroit, Michigan 48202, U.S.A.

ABSTRACT - H_2+D_2 exchange was studied over photolytically and thermally decomposed $Mo(CO)_6$ supported on Al_2O_3 and SiO_2. The parent complex was inactive and the activity of the catalysts increased with the removal of CO ligands. Decarbonylated Mo has a formal turnover frequency of 0.2 s^{-1} at 77 K. This is as active as some noble metals and is 96 fold higher than the turnover frequency for unsupported Mo. Some of the OH groups on the surface of the supports exchange with D_2. Unlike the case for bare Al_2O_3, the amount of exchangeable OH correlates with activity in a complex manner. CO_2 was found to be a selective poison for the exchange reaction over low valent supported Mo.

1. INTRODUCTION

Activation of hydrogen is an important step in catalytic hydrogenation. H_2+D_2 exchange, $H_2 + D_2 = 2HD$, is often used to measure the catalytic activation of hydrogen. Many transition metals are highly active for both olefin hydrogenation and H_2+D_2 exchange. The turnover frequency, N, is ~1 s^{-1} for evaporated films of Ti, V, Cr, and Ni [1], 23 s^{-1} for Pt wire [2], and 7 s^{-1} for Rh wire [3] at room temperature. Compared to transition metals the exchange reaction proceeds much slower over most homogeneous hydrogenation catalysts. For $RhCl(PPh_3)_3$, N has been reported to be $2x10^{-4}$ s^{-1} [4] and $7.4x10^{-3}$ s^{-1} [5], for trans-$IrCl(CO)(PPh_3)_2$ N = $3.5x10^{-7}$ s^{-1} [6], and for $Pd_2(Ph_3PCH_2PPh_3)_3$ N = $7.6x10^{-4}$ s^{-1} [7].

The Bonhoeffer-Farkas and Rideal-Eley mechanisms are most commonly used to explain the exchange reaction over solid catalysts.

```
H     D        H---D        H—D
|  +  |   ⇌    :   :    ⇌
S     S        S   S        S  S     Bonhoeffer-Farkas Mechanism

                 D
D-D   H        D'  'H       D   H-D
    + |   ⇌    :    :   ⇌   |
S     S        S    S       S   S    Rideal-Eley Mechanism
```

In the Bonhoeffer-Farkas Mechanism H_2 and D_2 are dissociatively adsorbed. Exchange then proceeds by the recombination of hydrogen and deuterium atoms held at two adjacent sites. In the Rideal-Eley Mechanism a chemisorbed hydrogen (deuterium) atom reacts with deuterium (hydrogen) which is molecularly adsorbed on a neighboring site [8]. In either mechanism two adjacent active sites are needed. The surface of a crystallite of a transition metal clearly has numerous chemisorption sites.

For transition metal oxides not only is the bonding of hydrogen to the transition metal cation an important factor [9], but also the involvement of neighboring oxygen ions. This has been demonstrated for the oxides of Cr^{+2} [10] and Cr^{+3} [11], ZnO [11,12], MgO [13,14], Al_2O_3 [15] and for Y-zeolites [16]. Thus, the low activity for the exchange over homogeneous hydrogenation catalysts is probably due to the lack of sufficient binding sites.

Supported carbonyl complexes, especially after at least partial loss of their ligands, can be highly active hydrogenation catalysts [17,18]. They can also be active for H_2+D_2 exchange and their activities can be compared to that of homogeneous and other heterogeneous catalysts. However, very little has been published on this. Howe [19] showed that the activity for exchange over $Ir_4(CO)_{12}$ supported on SiO_2 and Al_2O_3 increased with the loss of CO upon heating.

In this report the activity of a base metal catalyst, molybdenum hexacarbonyl supported on alumina and silica gel, is examined. The loss of ligands by thermal heating has been well studied [20-24]. Recently in this laboratory

›hotolysis of supported $Mo(CO)_6$ to remove CO ligands has ılso been studied. Photolysis should avoid oxidation and :educe sintering of Mo atoms which can result from thermal ıeating at temperature above 200 °C [20-24]. The activities ›f the decomposition products will allow the behavior of :he catalyst to be monitored as it changes from molecular :omplex to aggregated metal.

A few studies of H_2+D_2 exchange over Mo catalysts have ›een reported. Tanaka and Okuhara reported slow exchange ›ver a MoS_2 catalyst at -40 °C [25] and Hall et al. lescribed fast exchange at -78 °C over partially reduced Io/Al_2O_3 [26]. However, neither author reported specific ıctivities.

:. EXPERIMENTAL

:.1. Preparation of Catalysts

Gamma alumina (Conoco Catapal SB, S = 203 m^2/g standard, ; = 148 m^2/g dehydroxylated) and silica gel (Davison Grade i2, S = 340 m^2/g) were used as supports. Usually 0.25 g of ıl_2O_3 or 0.20 g of SiO_2 was used for a run. The support was ›retreated by calcination in O_2 flowing at 50 cc/min for 30 :in and then evacuation for 30 min at 600 °C (standard ›upport). Further evacuation at 1000 °C for 60 min was ›erformed if a dehydroxylated support (deh-Al_2O_3) was used. ully hydroxylated Al_2O_3 was prepared as suggested by renner [27]: (1) standard Al_2O_3 was heated at 100 °C in a low of He saturated with water vapor at room temperature ntil the water broke through the sample and then the emperature was raised to 300 °C, (2) after 30 min at 300 C the temperature was lowered to 100 °C, and (3) the ample was purged with dry He for 30 min and evacuated for 0 min.

The catalysts were prepared by subliming $Mo(CO)_6$ onto the upports in situ. The preparation and thermal decomposition f supported $Mo(CO)_6$ has been reported [23,24]. Photolysis f supported $Mo(CO)_6$ was accomplished by irradiating the catalyst inside a quartz reactor with He or H_2 sweeping through at 1000 cc/min. Al_2O_3 of 250-270 meshes and SiO_2 of 180-200 meshes were used to achieve fluidization. The light source was three or six Hanovia Hg lamps (450W, medium pressure). The reactor had a jacket of flowing water to remove heat (temperature of the catalyst was about 40 °C as measured with an internal thermocouple). Gases evolved by decomposition were trapped by silica gel at -196 °C and analyzed by gas chromatography (2' x 1/8" O.D. Spherocarb column at 40 °C) using a thermal conductivity detector.

2.2. Hydrogen-deuterium exchange

Catalytic activity was measured using a flow system. A flow of H_2+D_2+He mixture of 20 cc/min and a flow of pure He of 100 cc/min were passed through the catalyst bed. The H_2+D_2+He mixture contained 8% H_2, 8% D_2, and 84% He. The effluent from the reactor was analyzed by an UTI 100 Mass Analyzer controlled by an Apple II^+ computer and DAS System 5 A/D-D/A interface. Appropriate software was written in house. Thoria-coated iridium filaments were used (tungsten filaments gave a large background signal of mass 3 for the reaction gas).

The activity was calculated from the relative amounts of H_2 and HD: % conversion = 100x0.5[HD]/(0.5[HD]+[H_2]). The relative response of the mass analyzer for H_2 and HD was calibrated with known mixtures. The background of HD in the reaction gas was measured before each run. The background of HD corresponded to about 1.0% conversion as calculated by the above formula. The sensitivity of analysis was about 0.1% conversion and the accuracy was about 2% of the calculated conversion.

By using equilibrium constants calculated at different temperatures [8] values of X_e, the percent conversion at equilibrium, were calculated. X_e = 38.4%, 45.7%, and 47.5% at 77, 195, and 295 K respectively.

For catalysts containing Mo the activity is expressed as

formal turnover frequency, N_f = molecule D_2 reacted per atom Mo per s, regardless of the dispersion of the Mo. For catalysts with high activity, the exchange was run at the temperature of liquid nitrogen (77 K) or dry ice-acetone (195 K).

The amount of exchangeable OH on the surface of a catalyst was determined by the difference in the amounts of HD which were formed in two successive exchanges. In each exchange about 5 torr of pure D_2 was brought in contact with a catalyst inside a reactor (volume of about 40 mL) for 15 min at room temperature and the amount of HD in the gas phase was measured. After the first experiment the catalyst was evacuated for 15 min. Less than 3% of the gas phase was HD after exchange, so the back reaction was negligible and the amount of HD formed was equal to the amount of exchanged OH.

2.3. Gases

He was purified by passing through a purifier (Go Getter, General Electrics) to remove water and oxygen and then passing through molecular sieve 5A at liquid nitrogen temperature. H_2 was passed through Oxisorb (Messer Griesheim) and then molecular sieve 5A at liquid nitrogen temperature. The H_2+D_2+He mixture was purified by flowing through silica gel at liquid nitrogen temperature. The flow rate of gas was controlled by a Brooks Mass Flow Controller. The D_2 (Scott Specialty Gases) had a claimed chemical purity of 99.8 atom %, corresponding to 0.4 mole % HD.

2.4. Chemisorption

CO chemisorption was done at room temperature using an electronic pressure transducer (Setra, model 204). The amount of chemisorbed CO was determined by the difference between two successive isotherms. Details of the experimental procedures have been published [28].

CO_2 adsorption was done to poison catalysts for H_2+D_2 exchange. A known amount of CO_2 was introduced into a

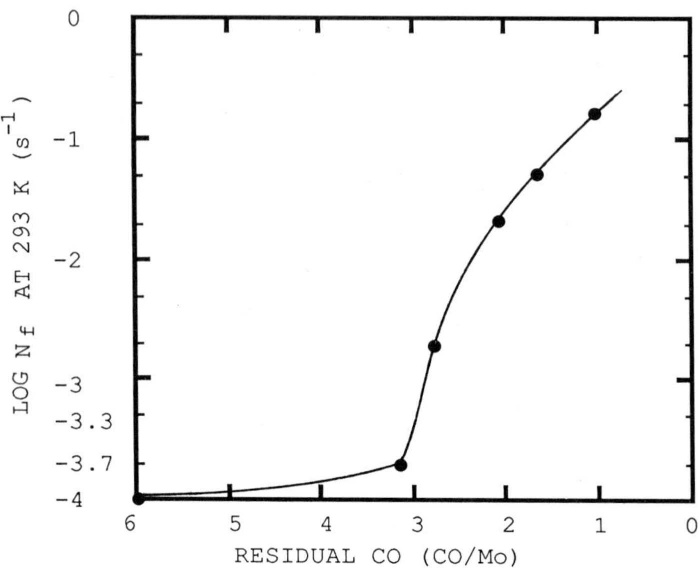

Figure 1. H_2+D_2 exchange over $Mo(CO)_6$/hydroxylated Al_2O_3 after photolysis (1.089% loading)

reactor at room temperature. The CO_2 was then desorbed by flowing He at 40 cc/min through the reactor at various temperatures. The amount of CO_2 evolved was determined by gas chromatography (2' x 1/8" O.D. Spherocarb column at 70 °C). If the activity remained high, the catalyst was again poisoned.

3. RESULTS AND DISCUSSION

3.1. H_2+D_2 Exchange over Mo/Al_2O_3

It is known that thermal heating of $Mo(CO)_6$ supported on Al_2O_3 results in the removal of CO. The evolution of more than three CO per Mo is accompanied by the oxidation of Mo by surface OH unless the Al_2O_3 is dehydroxylated above 900 °C. Photolysis at ambient temperature may avoid the oxidation of Mo since the OH groups are probably not mobile. Photolysis was found to be more efficient when hydroxylated Al_2O_3 was used. Extensive photolysis of

$Mo(CO)_6$ supported on hydroxylated Al_2O_3 led to the loss of about 5.2 CO per Mo. About 3.8 CO and 3.3 CO per Mo were removed by extended photolysis of $Mo(CO)_6$ supported on standard and dehydroxylated Al_2O_3, respectively.

Figure 1 shows that the activity at 293 K for H_2+D_2 exchange increases with the decrease of residual CO/Mo. Before photolysis $Mo(CO)_6$ is inactive. This demonstrates that Mo with more vacant bonding sites has higher activity and may explain the low activity of homogeneous (molecular) catalysts for H_2+D_2 exchange. It was unclear if sintering was also occurring during extended photolysis. Hydroxylated Al_2O_3 was used because it has no activity even after irradiation and photolysis proceeds more extensively than on standard or dehydroxylated Al_2O_3. Irradiation of partially or fully dehydroxylated Al_2O_3 yields high activity [29]. In Figure 1 the maximum activity is N_f = 0.15 s^{-1}. This is one to two orders of magnitude lower than the activities of metal wires [2,3]. However, since the limiting catalyst stoichoimetry is Mo(CO), the fraction of bare Mo may be low. Also, H_2O is known to poison H_2+D_2 exchange [13,30] so it is likely that the activity of Mo on hydroxylated Al_2O_3 is partially poisoned.

One disadvantage of photolysis is the difficulty of removing all CO ligands. To measure the activity for bare Mo/Al_2O_3, thermal heating of $Mo(CO)_6$ supported on

Table 1. H_2+D_2 exchange and exchangeable OH over Mo/deh-Al_2O_3

Loading (wt %)	Exch. OH (10^{16} groups)	Exchange at 77 K: Conversion (%)	Exchange at 77 K: N_f (s^{-1})
0.00	1.1	0.1	------
0.058	1.7	28.1	0.22
0.109	7.2	34.4	0.14
0.240	8.4	38.4	> 0.073
0.476	5.2	38.4	> 0.037
0.897	5.3	34.1	0.017
1.005	5.8	34.0	0.015

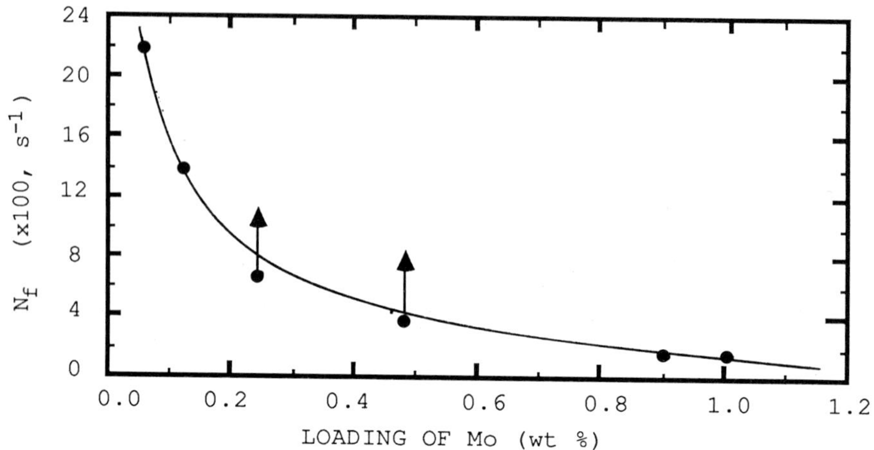

Figure 2. H_2+D_2 exchange at 77 K over Mo/deh-Al_2O_3

dehydroxylated Al_2O_3 was used. Table 1 shows the results for exchangeable OH and H_2+D_2 exchange over catalysts which were prepared by temperature programmed decomposition (TPDE) in He to 600 °C. The product catalysts contain Mo primarily in the zero valent state [17,24] and around 5.6 CO per Mo was removed during TPDE. These catalysts were very active so the exchange had to be run at 77 K since equilibrium conversion was obtained at 195 K.

Figure 2 shows N_f as a function of loading. For the catalysts with 0.240 and 0.476% Mo, N_f can only be estimated because equilibrium conversion was obtained (this is indicated by arrows). For the most active catalyst (0.058 % Mo), N_f = 0.22 s^{-1} which is about 96 fold higher than the value of N (true turnover frequency, not N_f) of 2.3×10^{-3} s^{-1} for unsupported Mo [29]. For the 1.005% Mo catalyst, N_f is 6.6 fold higher than the value of N for unsupported Mo. If the supported Mo is less than 100% dispersed, then the rate enhancement is even larger. This indicates that the support substantially modifies the activity of Mo. As loading increases, the average particle size may become larger and the special sites on the alumina necessary to activate Mo may become saturated. This would cause the value of N_f to

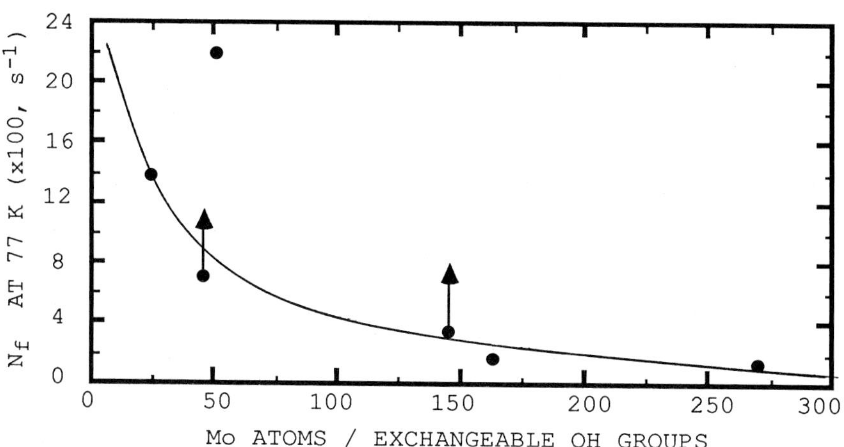

Figure 3. Relation between N_f and ratio of Mo atoms to exchangeable OH over Mo/deh-Al_2O_3

decrease and approach the value for bulk Mo.

It is known that OH groups on Al_2O_3 can exchange with D_2 [15]. The activity for H_2+D_2 exchange on bare Al_2O_3 strongly correlates with the number of exchangeable OH [29]. A site capable of stoichiometrically exchanging OH must also be a site for catalytic H_2+D_2 exchange according to the following equations (S = surface attachment).

(1) S-OH + D_2 = S-OD + HD

(2) S-OD + H_2 = S-OH + HD

H_2 + D_2 = 2HD

For bare dehydroxylated Al_2O_3 (about 0.6% residual OH), 4.4×10^{16} OH groups per gram were exchangeable, corresponding to about 0.18% of the total OH.

Table 1 shows that the activity and the amount of exchangeable OH increased up to a loading of ~0.3% Mo and then dropped. Unlike the case for bare Al_2O_3, the activity does not vary in a monotonic way with the amount of exchangeable OH. However, Figure 3 shows that N_f generally decreases as the ratio of Mo atoms to exchangeable OH increases (arrows again indicate equilibrium conversion). A

Table 2. H_2+D_2 exchange over Mo catalysts prepared from $Mo(CO)_6$/deh-Al_2O_3

Treatment	Loading (wt %)	rx. temp. (K)	conversion (%)	N_f (s^{-1})
TPDE in He to 600 °C.	0.0583	77	28.1	0.22
TPDE in He to 600 °C.	0.240	77	38.4	~ 0.1
O_2, 600 °C, 15 min.	0.236	295	0.0 [a]	< $2x10^{-4}$
O_2, 600 °C, 30 min; H_2, 600 °C, 30 min.	0.243	77	0.23	0.0004
	0.0534	77	0.15	0.0013
		195	16.4 [b]	0.14
O_2, 600 °C, 30 min; H_2, 1000 °C, 30 min.	0.243	77	4.5	0.008

a : Corrected for the background conversion of the support (41.1%).

b : Corrected for the background conversion of the support (24.4%).

larger ratio means less Mo atoms are interacting with exchangeable OH groups. Thus, it is likely that at higher loadings much of the exchange occurs on Mo without the participation of the Al_2O_3 surface and the activity should approach the lower value of bulk Mo. Although at a loading of 1.005% Mo the ratio of Mo to exchangeable OH is 270, the ratio of Mo to total OH is roughly 0.5. Thus, there are sufficient OH groups to interact with the Mo.

Table 2 compares the activities of Mo catalysts treated in different ways. MoO_3, the product of oxidation at 600 °C, is inactive and the activity increases with the increasing temperature of reduction of MoO_3. Temperature programmed reduction (TPR) of MoO_3/deh-Al_2O_3 indicated that the average oxidation state of Mo was about +5 after reduction at 600 °C

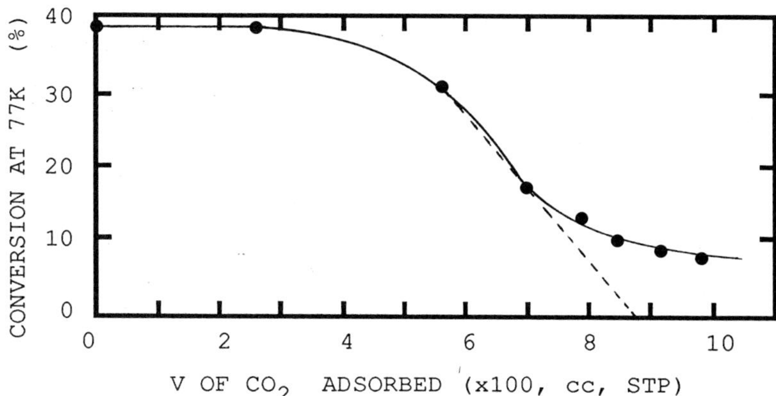

Figure 4. H_2+D_2 exchange over 0.236% Mo/deh-Al_2O_3 poisoned with CO_2 after heating at 100 °C

and MoO_3 was fully reduced after reduction at 1000 °C [29]. The 0.0534% Mo catalyst, after reduction at 600 °C, gave N_f = 0.14 s^{-1} at 195 K and at 77 K the activity was at the sensitivity level of the mass spectrometer. After reduction at 1000 °C, N_f was about 12 fold lower than for the catalyst of similar loading prepared by TPDE in He to 600 °C. This presumedly reflects sintering and less interaction with the alumina.

CO_2 poisoning was used to determine the number of active sites on Mo/deh-Al_2O_3 prepared by TPDE. Because deh-Al_2O_3 has no activity at 77 K, any decrease in activity must be due to poisoning of the Mo. A typical plot of activity versus the amount of CO_2 adsorbed (after desorption at 100 °C) is given in Figure 4. Various data are summarized in Table 3. Linear extrapolation of the data at smaller amounts of CO_2 added is used to obtain the amount of CO_2 necessary for complete poisoning since the CO_2 should be more selective for adsorption onto the active sites. Desorption at 250 °C and higher loading resulted in more selective poisoning. However, CO along with CO_2 was evolved when poisoned Mo/deh-Al_2O_3 was heated in flowing He at or above 250 °C. Therefore, a higher desorption temperature may not

Table 3. H_2+D_2 exchange at 77 K and CO_2-poisoning over Mo/deh-Al_2O_3

Loading (% Mo)	Original conversion (%)	V of CO_2 added (cc,STP)	Temp. of desorption (°C)	V of CO_2 remaining (cc,STP)	Final conversion (%)	CO_2/Mo for complete poisoning [a]	N [b] (s^{-1})
0.0508	26.9	0.0122	100	0.0122	21.4	2.0	0.24
	----	----	250 [c]	0.0088	13.3	0.59	0.41
0.0508	23.1	0.0539	250	0.0231	5.1	1.0	0.22
0.236	31.2 [d]	0.0149	100	0.0139	17.7	0.62	0.26

a: Linear extrapolation to zero % conversion is used (see text).
b: Number of active sites is measured by CO_2 poisoning; if $CO_2/Mo \geq 1$, then all Mo atoms are considered to be active.
c: Following the previous activation at 100 °C.
d: Catalyst had been partially poisoned with 0.0551 cc CO_2 to lower the conversion from 38.4% (equilibrium) to 31.2%.

be suitable.

The number of active sites measured by adsorbed CO_2 is an upper limit because Al_2O_3 also adsorbs CO_2. A blank experiment showed that 0.25 g of deh-Al_2O_3 retained 0.0252 and 0.0156 cc CO_2 after desorption at 100 and 250 °C, respectively. Table 3 shows that the adsorption of 0.01 cc CO_2 can significantly poison a catalyst. This and the ratio of CO_2/Mo for complete poisoning suggest that at least after desorption at 250 °C most of the CO_2 is adsorbed on the Mo and not on the Al_2O_3. Hall and Segawa [31] found that over reduced MoO_3/Al_2O_3, for which the average oxidation state of Mo was +5, CO_2 was only adsorbed on the Al_2O_3. For the essentially zero-valent Mo of this study, we find that CO_2 adsorbs on Mo and is a reasonably selective poison for H_2+D_2 exchange.

Howe [19] reported H_2+D_2 exchange at 77 K over Ir/Al_2O_3 prepared by heating $Ir_4(CO)_{12}/Al_2O_3$ in H_2 at 350 °C. His value of N = 0.2 s^{-1} is close to the value of N_f for some of the supported Mo catalysts. For Pt wire [2] at 77 K, N = 8.1×10^{-3} s^{-1}. Thus, Mo is as active as noble metals for catalyzing H_2+D_2 exchange.

3.2. H_2+D_2 Exchange over Mo/SiO_2

Photolysis of $Mo(CO)_6/SiO_2$ is more complete than for $Mo(CO)_6/Al_2O_3$. SiO_2 dehydroxylated at 600 °C is inactive for H_2+D_2 exchange even after irradiation, so the activity of irradiated $Mo(CO)_6/SiO_2$ is only due to the Mo. The activity after photolysis is shown in Table 4. The activation energy (E_a) of 1.1 kcal/mol was calculated from conversions at 77 K and 195 K for one catalyst. Thus, this value might not be highly accurate but is used here for comparison of data.

Figure 5 shows that the activity increases most rapidly when less than 1 CO per MO remains, indicating that completely decarbonylated Mo is the most active species. This suggests that one reason for the low activity of

Table 4. H_2+D_2 exchange over $Mo(CO)_6/SiO_2$ after photolysis

Residual CO/Mo	Temp. (K)	Conversion (%)	N_f (s^{-1})	N_f (s^{-1}) at 295K [a]
6.00 [b]	295	0.0	< $1.0x10^{-3}$	< $1.0x10^{-3}$
2.95	295	3.5	$1.7x10^{-2}$	$1.7x10^{-2}$
1.54	295	10.9	$5.2x10^{-2}$	$5.2x10^{-2}$
0.25	77	38.4	> $9.0x10^{-2}$	> 18
1.22 [c]	295	8.9	$2.2x10^{-2}$	$2.2x10^{-2}$
0.65	295	34.3	$8.5x10^{-2}$	$8.5x10^{-2}$
0.42	295	43.6	$1.1x10^{-1}$	$1.1x10^{-1}$
0.17	77	38.4	> $9.9x10^{-2}$	> 20

a : E_a = 1.1 kcal/mol
b : Loading of 0.253% Mo.
c : Loading of 0.230% Mo.

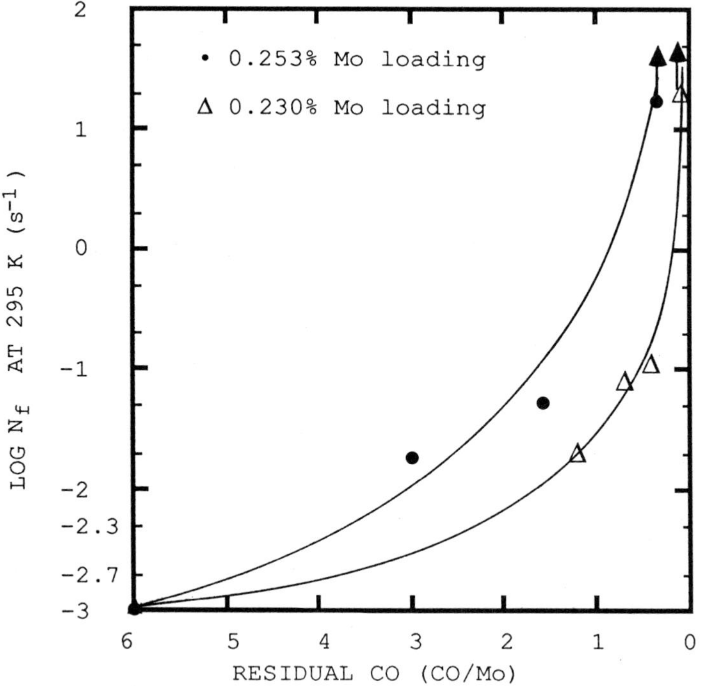

Figure 5. H_2+D_2 exchange over $Mo(CO)_6/SiO_2$ after photolysis

Table 5. H_2+D_2 exchange and exchangeable OH over 0.332% $Mo(CO)_6/SiO_2$

Catalyst	Treatment	Exch. OH (10^{16} groups)	Exchange Reaction: Conversion (%) [a]	Exchange Reaction: N_f (s^{-1})
SiO_2		0.0	0.0 (295)	
$Mo(CO)_{2.10}$	Irrad. in He 155 min.	2.3	4.9 (292)	0.009
$Mo(CO)_{0.42}$	Irrad. in H_2, 710 min.	5.6	19.7 (77)	0.034
Mo	H_2 at 600 °C, 30 min.	0.3	1.3 (77)	0.002
MoO_3	O_2 at 600 °C, 30 min.	--- [b]	0.0 (292)	< 0.001
Mo	H_2 at 1000 °C, 30 min.	0.6	6.6 (77)	0.011

a : Number in parenthesis are the temperatures of reaction (K).
b : Not measured.

photolytically decomposed $Mo(CO)_6$ on hydroxylated Al_2O_3 was due to the incomplete removal of CO. Long photolysis of $Mo(CO)_6$ supported on standard SiO_2 results in the removal of about 97% of the CO ligands and H_2+D_2 exchange reaches equilibrium conversion at 77 K. The N_f calculated from this conversion is therefore a lower limit and this is indicated by the arrows in Figure 5. The calculated N_f is at least 41 fold higher than the value of N for unsupported Mo [29] and similar to the values for Mo/deh-Al_2O_3 prepared by TPDE (Table 1). This again indicates the involvement of the support with Mo for the exchange reaction.

An attempt to correlate the amount of exchangeable OH with the exchange activity of Mo/SiO_2 is shown in Table 5. Exchangeable OH only exists upon the addition of Mo. For the last three active catalysts in Table 5, each of which is low valent and highly decarbonylated, the activity increases with the amount of exchangeable OH, but not in a linear fashion. The first Mo catalyst, however, has substantial exchangeable OH but low activity. This suggests that some exchangeable OH is associated with partially carbonylated Mo which is of low activity.

Table 6. Dispersion and H_2+D_2 exchange over 0.253% Mo/SiO_2

Catalyst	Treatment	Exchange at 77 K Conversion (%)	N_f (s^{-1})	N (s^{-1})	Dispersion (%)
$Mo(CO)_{0.25}$	Irrad. in He for 580 min	38.4	> 0.09	> 0.9	10
Mo	O_2 at 600 °C for 30 min, then in H_2 at 1000 °C for 30 min	12.2	0.029	0.88	3.3

Table 6 shows the dispersions and activities for two catalysts. Percent dispersion is calculated by 100(CO/Mo) where the amount of chemisorbed CO is determined as the difference between two isotherms at 293 K. The catalyst prepared by photolysis has higher dispersion than the catalyst prepared by reduction at 1000 °C, which presumedly reflects sintering. However, it is known that CO chemisorption on Mo underestimates the dispersion of Mo. On unsupported Mo(0) we found that the ratio of CO chemisorbed (at 195 K) to Mo atoms on the surface is 1/14 [29]. A study of CO chemisorption on Mo/deh-Al_2O_3 prepared by TPDE in He to 500 °C showed that the chemisorption of CO was weak and the measured dispersion (corrected for background) increased from 5% when measured at room temperature to a plateau of 35% when measured at -130 °C [29]. Thus, the calculated values of N in Table 6 are upper limits and may well be 7 to 10 fold high.

4. CONCLUSION

Organometallic compounds can be heterogenized on common supports and their activities can then be compared with conventional heterogeneous metal catalysts at similar reaction conditions. The H_2+D_2 exchange reaction has been used to compare several catalysts containing Mo. The activity increases with the extent of loss of CO from supported $Mo(CO)_6$. Both photolysis of $Mo(CO)_6/SiO_2$ and thermal decomposition of $Mo(CO)_6$/deh-Al_2O_3 yield highly active catalysts. However, the less exhaustive photolysis of $Mo(CO)_6/Al_2O_3$ leads to catalysts of lower activity and adsorbed H_2O can poison the catalyst.

The activity of supported Mo for H_2+D_2 exchange is much higher (per Mo atom) than for unsupported Mo. This suggests the direct involvement of the support in the active site ensemble. A small fraction of the support OH groups participate in the reaction, but the amount is only qualitatively related to the activity. CO_2 is a reasonably selective poison for H_2+D_2 exchange over Mo/deh-Al_2O_3.

ACKNOWLEDGEMENT. Support from the Department of Energy, Office of Basic Energy Science is gratefully acknowledged.

REFERENCES

1. Eley, D. D.; Shooter, D. J. Catal. 1963, 2, 259.
2. Breakspere, R. J.; Eley, D. D.; Norton, P. R. J. Catal. 1972, 27, 215.
3. Ross, P. N.; Stonehart, P. J. Catal. 1974, 35, 391.
4. Osborn, J. A.; Jardian, F. H.; Young, J. F.; Wilkinson, G. J. Chem. Soc., A 1966, 1711.
5. Strathdee, G.; Given, R. Can. J. Chem. 1974, 52, 2216.
6. Eberhardt, G. G.; Vaska, L. J. Catal. 1967, 8, 183.
7. Stern, E. W.; Maples, R. K. J. Catal. 1972, 27, 134.
8. Bond, G. C. "Catalysis by Metals", Academic Press, New york, 1962, p.149.
9. Dowden, D.A.; MacKenzie, N.; Trapnell, B. M. W. Pro. Roy. Soc., A 1956, 237, 245.
10. Indovina, V.; Cimino, A.; Valigi, M. "Proc. 6th Internat. Congr. Catalysis", The Chemical Society, London, Britian, 1976, p.216.
11. Conner, W. C.; Kokes, R. J. J. Phys. Chem. 1969, 73 (7), 2436.
12. Dent, A. L.; Kokes, R. J. J. Phys. Chem. 1969, 73 (11), 3772.
13. Boudart, M.; Delbouille, A.; Derouane, E. G.; Indovina, V.; Walters, A. B. J. Am. Chem. Soc. 1972, 94 (19), 6622.
14. Derouane, E. G.; Indovina, V. Chem. Phys. Lett. 1972, 16(2), 359.
15. van Cauwelaert, F. H.; Hall, W. K. Trans. Faraday Soc. 1970, 66, 454.
16. Heylen, C. F.; Jacobs, P. A.; Uytterhoeven, J. B. J. Catal. 1976, 43, 99.

17. Thomas, T. J.; Brenner, A. J. Mol. Catal. 1983, 18, 197.
18. Thomas, T. J.; Hucul, D. A.; Brenner, A. "Chemically Modified Surfaces in Catalysis and Electrocatalysis", ACS Symposium Ser. 192, 1982, p. 267.
19. Tanaka, K.; Watters, K. L.; Howe, R. F. J. Catal. 1983, 79, 251.
20. Brenner, A.; Burwell, R. L. J. Catal. 1978, 52, 353.
21. Brenner, A. J. Mol. Catal. 1979, 5, 157.
22. Brenner, A.; Hucul, D. A. J. Am. Chem. Soc. 1980, 102, 2484.
23. Brenner, A.; Hucul, D. A.; Hardwick, S. J. Inorg. Chem. 1979, 18, 1478.
24. Brenner, A.; Hucul, D. A. "Proc. Int. Conf. Chem. Uses Molybdenum, 3rd", Climax Molybdenum Company, Ann Arbor, 1979, p.194.
25. Tanaka, K.; Okuhara, T. J. Catal. 1982, 78, 155.
26. Lombardo, E. A.; Houalia, M.; Hall, W. K. J. Catal. 1978, 51, 256.
27. Brenner, A. Ph. D. Dissertation, Northwestern University, 1975.
28. Brenner, A.; Hucul, D. A. Inorg. Chem. 1979, 18, 2836.
29. Brenner, A., unpublished data.
30. Weller, S. W.; Hidin, S. G. J. Phys. Chem. 1956, 60, 1506.
31. Segawa, K.; Hall, W. K. J. Catal. 1982, 77, 221.

SITE CHARACTERISATION AND ACTIVITY IN METAL CLUSTER CATALYSTS

by P.B. Wells

Department of Chemistry, The University, Hull, HU6 7RX, United Kingdom.

ABSTRACT - The characterisation of active sites in supported metal catalysts is an important goal. This paper describes some first steps towards such a characterisation based on osmium cluster catalysts and precursors characterised by EXAFS spectroscopy. The nuclearity of the tetracapped octahedral structure of the ten metal-atom unit in $H_2Os_{10}C(CO)_{24}$ is retained on impregnation of $Os_{10}C(CO)_{24}^{2-}$ onto alumina and on thermal activation of the impregnated material. The activated material is an efficient adsorbent and an active catalyst for hydrogenation and hydrogenolysis. A catalyst having a similar activity and similar chemisorption properties is obtained by impregnating $Os_6(CO)_{18}$ onto alumina. In the pure compound the Os_6-unit has a bicapped tetrahedral structure; however, this cluster reacts with the support at room temperature, and EXAFS spectroscopy reveals (i) the existence of the cluster-support interaction, (ii) modifications that occur to the CO-ligands, and (iii) a change that occurs in the cluster configuration. Implications for future work and for progress toward the eventual goal are discussed.

1. INTRODUCTION

The geometrical arrangements of metal atoms that constitute sites at the surfaces of metal catalysts have, for a long time, attracted speculation. In the 1920s Taylor recognised that the fraction of a metal surface that exhibited catalytic activity might be quite small, and that a site for adsorption and catalysis might require a special local arrangement of metal atoms (1). This led Kobosev, in the next decade, to speculate about the likely arrangements of atoms in active ensembles (2). However, it was not until the 1960s and 1970s that various investigators, prompted by the predictions of

Poltorak and co-workers (3), showed that metal particle size in supported catalysts must be reduced below 5 nm before substantial effects on activity or selectivity can be expected. Subsequent experiments revealed that catalysed reactions may be divided into two classes, viz. those that are structure sensitive and those that are structure insensitive (4), and that reactions in the first category require sites comprising a group of metal atoms. Hydrogenolyses are an important class of reactions that show structure sensitivity; for example, claims have been made that twelve or more nickel atoms may be required to constitute a site for nickel-catalysed ethane hydrogenolysis (5). However, such studies simply add urgency to the question as to what minimum number of metal atoms is necessary to constitute a hydrogenolysis site, and whether the geometry of the arrangement is important.

The advent of new materials and techniques often provides fresh opportunities for the testing of concepts. Such has been the case in the development of our ideas concerning site characterisation. In the 1970s, advances in synthesis in organometallic chemistry enabled many new metal cluster compounds to be prepared and characterised (6). At the same time, the wider availability of synchrotron radiation and the high brightness of these sources in the X-ray region enabled the technique of extended X-ray absorption fine structure (EXAFS) spectroscopy to be developed and applied to the structural characterisation of non-crystalline systems including supported metal catalysts (7). These advances immediately permitted new investigations of site geometry to be undertaken, and this paper records some of the studies in this area that have been undertaken in the author's laboratory.

The metals ruthenium, rhenium, and osmium crystallise in the hexagonal close packed configuration and show high activities for hydrocarbon reactions, particularly the structure-sensitive reaction of ethane hydrogenolysis (8). They also have a rich cluster chemistry (6). In this paper, we concentrate attention on osmium clusters because, as is well known, the osmium-osmium bond is a particularly strong metal-metal bond and the osmium-oxygen bond which may participate in support-cluster interactions is also strong. Thus, if the clustered state of metal is to be retained in the turbulent environment of a catalytic reaction, then osmium chemistry is a likely context in which to find it.

In this work we were particularly attracted by the series of osmium cluster compounds $Os_3(CO)_{12}$, $H_4Os_4(CO)_{12}$, $Os_6(CO)_{18}$, and $H_2Os_{10}C(CO)_{24}$, in which the metal frameworks are, respectively, triangular, tetrahedral, bicapped-tetrahedral (I), and tetracapped-octahedral (II).

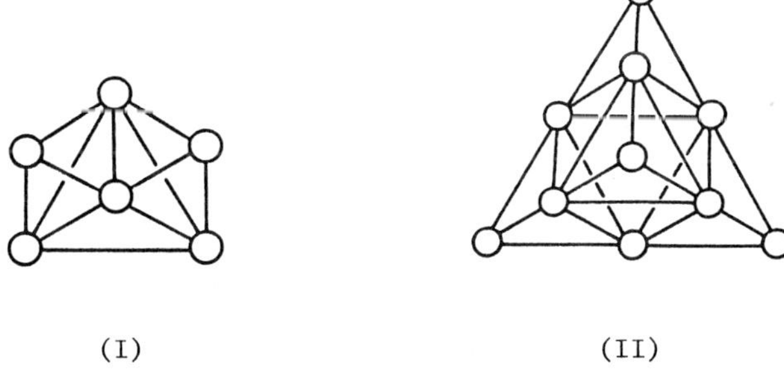

(I) (II)

Our studies concerning these cluster compounds are being published (9-15). In this paper the structural and catalytic properties of alumina-supported clusters derived from the Os_6- and Os_{10}-compounds are compared.

The Os_{10}-cluster (II) is particularly important from two standpoints. First, it is the smallest unit that possesses all the characteristics of close-packed structure and we may therefore regard it as the limiting case of a metal crystallite stabilised against crystal growth by its CO-ligands. In synthetic work, this configuration of osmium atoms is found to be particularly stable, and attempts to make higher clusters have encountered difficulties. We have therefore (i) supported $H_2Os_{10}C(CO)_{24}$ on alumina as the ion $Os_{10}C(CO)_{24}^{=}$ (which it forms in acetone solution), (ii) determined its structure by EXAFS spectroscopy both in the impregnated state and in the catalytically active state achieved after thermal activation, (iii) observed its chemisorption properties, and (iv) measured its catalytic activity for the structure insensitive reaction of ethene hydrogenation and for the structure sensitive reaction of ethane hydrogenolysis (13).

The Os_6-cluster (I), when supported on alumina and thermally activated, provides highly reproducible catalysts with chemisorption and catalytic properties closely similar to those obtained from the Os_{10}-system (11, 13). However, it is fundamentally a more labile cluster; for example, it converts to the octahedral configuration on reduction to the dianion $Os_6(CO)_{18}^{=}$ (16) and provides higher nuclearity clusters on pyrolysis (17). Thus, we suspected at the outset that the structural chemistry associated with its impregnation onto an alumina support and thermal activation would be complex and this, indeed, has proved to be the case (9, 14, 15).

2. EXPERIMENTAL

$Os_6(CO)_{18}$ was prepared by the conventional method (18). $[N(PPh_3)_2]_2[Os_{10}C(CO)_{24}]$ prepared by the literature method (19) was converted to $H_2Os_{10}C(CO)_{24}$ by dissolution in concentrated acetonitrile and acidification with H_2SO_4; the required material was formed as a precipitate in 95% yield. $H_2Os_{10}C(CO)_{24}$ does not dissolve as the neutral compound, but dissolves readily in acetone as the dianion $Os_{10}C(CO)_{24}^{=}$.

The alumina support was Aluminium Oxide C (Degussa) of surface area 97 m^2 g^{-1}. It was dried before use by heating to 773 K for 16 h in a stream of dry nitrogen.

$Os_6(CO)_{18}$/alumina was prepared by impregnation of the alumina with $Os_6(CO)_{18}$ in solution in 1,2-dichloroethane. Evaporation to dryness was carried out in a stream of dry nitrogen at 313 K, and all of the cluster was impregnated onto the support. $Os_{10}C(CO)_{24}^{=}$/alumina was similarly prepared by impregnation of the support with a solution of the dihydride in acetone; evaporation to dryness was carried out in a stream of dry nitrogen at 333 K, and again all of the cluster compound was impregnated onto the support. In each case the loading of the cluster compound on the support was such as to give 2% by weight Os.

Electron probe microanalysis showed $Os_6(CO)_{18}$ to be impregnated unevenly onto the support (9) and we have observed this behaviour in the impregnation of all neutral cluster compounds. However, the same technique showed that the dianion $Os_{10}C(CO)_{24}^{=}$ was impregnated evenly onto the alumina. In the freshly impregnated states, the supported materials gave infrared spectra in the CO-stretching region (2130 to 1900 cm^{-1}) closely similar to those of the pure cluster compounds in solution (9,13), indicating that impregnation had been achieved without substantial molecular change. However, $Os_6(CO)_{18}$/alumina underwent a substantial change on standing over a period of several days; its infrared spectrum changed from one containing 14-bands to one containing 3-bands (9) and electron probe microanalysis revealed that migration of osmium entities over the support had occurred to give an even distribution. This modified material is termed 'self-spread $Os_6(CO)_{18}$/alumina'.

Chemisorption properties and catalytic activity were induced in these impregnated materials by a process of thermal activation in which 0.3 g samples were placed in a vertical tube and heated in vacuum or in a flow of helium (23 ml(STP) min^{-1}) from 293 to 523 K at a steady rate of 10 to 12 K min^{-1}. As reported (9, 13), this activation resulted in carbon loss, mostly as CO_2; carbon losses were 14 to 15% in each case. Evidence has been presented (9) which suggests that the CO_2 was formed by the process: 2(ligand-CO) $\rightarrow$ ligand-C + CO_2(g).

The structures of the osmium clusters in the activated materials were determined by extended X-ray absorption fine structure (EXAFS) spectroscopy. EXAFS spectra were obtained at the Synchrotron Radiation Source (SRS) at the SERC Laboratory at Daresbury, England. The L_{III}-edge of osmium was examined. Spectra of Os_6-derived materials were obtained in the transmission mode, and those of Os_{10}-derived materials in the fluorescence mode using a set-up similar to that described by Hasnain et al. (20). Experimental spectra were fitted by use of the programme EXCURVE developed by Binsted (21) which generated a theoretical spectrum by use of single scattering curved wave theory and incorporates a least-squares fitting routine.

The spectra shown are k^2 weighted.

The apparatus used for chemisorption studies and the pulsed flow reactor for studies of catalytic activity have been described elsewhere (<u>10</u>, <u>11</u>).

3. NOMENCLATURE

Materials are described as $Os_6(CO)_{18}$/alumina and $Os_{10}C(CO)_{24}^{=}$/alumina. This nomenclature is used to describe freshly impregnated materials, self-spread materials (in the Os_6-system), and thermally activated materials. In each case it is intended to convey only the initial cluster used in the impregnation; it is <u>not</u> intended to describe the actual state of the clustered osmium in a given material.

4. RESULTS AND DISCUSSION

4.1 <u>Structural Chemistry in the Os_{10}-System</u>

High resolution transmission electron microscopy (JEOL 100C; Phillips EM 400) of freshly impregnated $Os_{10}C(CO)_{24}^{=}$/alumina gave micrographs which contained the expected image of the alumina together with small dark spots indicating the presence of osmium-containing species of apparent diameter 0.8 nm. These osmium entities were evenly distributed over the alumina surface, confirming the observation of even distribution also obtained by electron probe microanalysis. The preliminary conclusion from this, and from the infrared spectra of the samples (which resembled that of the pure starting material) was that individual Os_{10}-clusters had been imaged at the alumina surface. The EXAFS spectrum of the same material is shown in Figure 1, where it is compared with that of the pure compound $[N(PPh_3)_2]_2[Os_{10}C(CO)_{24}]$. The parameters used in the calculation are presented in Table 1. It is clear from the bonding and non-bonding osmium-osmium distances quoted in Table 1, and the coordination numbers shown in parenthesis, that the tetracapped

TABLE 1. Interatomic distances and coordination numbers for the Os_{10}-materials as derived from the osmium L_{III}-edge EXAFS spectra shown in Figure 1.

Shell	Atom type	Interatomic distances/Å (coordination numbers)		
		$Os_{10}C(CO)_{24}^{=}$ (a)	$Os_{10}C(CO)_{24}^{=}$/alumina	
		Pure compound	Freshly impregnated	Activated
1	Os — C in ligand-CO	1.91	1.88	1.89
	Os — C carbidic-C	2.04		
2	Os — Os	2.80 (4.8)	2.80 (4.8)	2.80 (4.8)
3	Os....O in ligand-CO	2.90	2.92	2.92
4	Os....Os	4.09 (0.6)	4.12 (0.6)	4.18 (0.6)
5	Os....Os	4.82 (2.4)	4.85 (2.4)	4.88 (2.4)
6	Os....O ? in support		5.18	5.17
7	Os....Os	5.75 (1.2)	5.84 (1.2)	5.90 (1.2)

(a) Mean values by X-ray diffraction for $[PPN]_2$-salt (17): Os — C, 1.90; Os — Os, 2.84; Os....O, 3.0; Os....Os, 4.01; 4.91, and 5.67 Å.

octahedral configuration of the Os_{10}-unit in the parent compound was preserved in the freshly impregnated $Os_{10}C(CO)_{24}^{=}$/alumina.

The same sample was activated in vacuum in the standard manner (see Experimental) in a controlled environment cell and exposed to CO at 293 K under conditions where saturation of active sites by adsorbed-CO is known to occur (13 and below). The EXAFS spectrum of the activated $Os_{10}C(CO)_{24}^{=}$/alumina is shown in Figure 1 and the

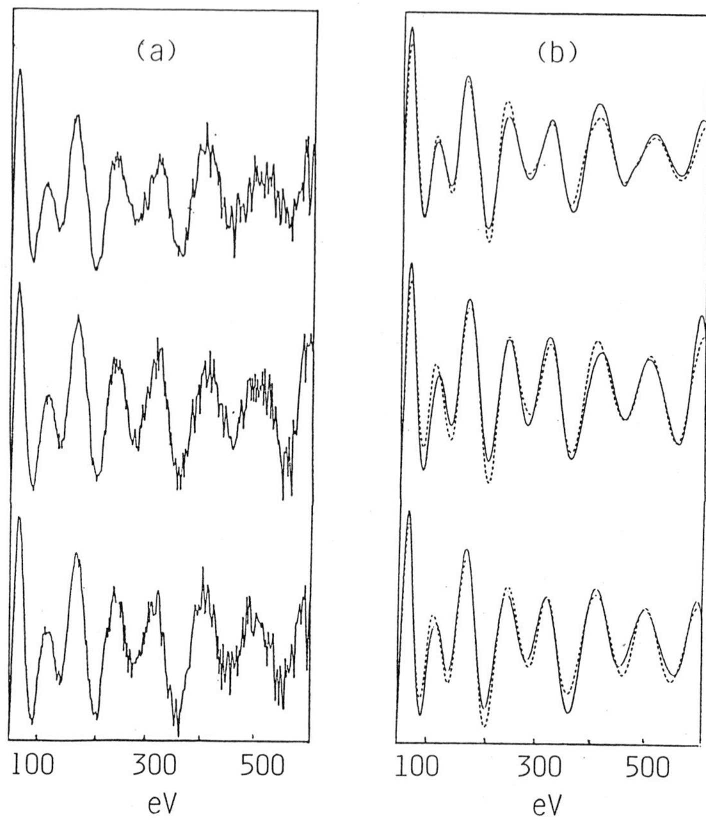

FIGURE 1. (a) EXAFS spectra as recorded of pure $[PPN]_2Os_{10}C(CO)_{24}$ (upper), freshly impregnated $Os_{10}C(CO)_{24}^{=}$/alumina derived from $H_2Os_{10}C(CO)_{24}$ (centre), and activated $Os_{10}C(CO)_{24}^{=}$/alumina + CO(g) lower. (b) Full curves represent smoothed versions of the spectra shown in (a), dashed curves represent theoretical spectra. All spectra are k^2 weighted.

parameters for the matching theoretical spectrum are shown in Table 1. The osmium-osmium interatomic distances and the coordination numbers of osmium atoms about osmium remain similar to those of the model compound and to those of the freshly impregnated material, indicating that the Os_{10}- cluster has retained its tetra-capped octahedral configuration during activation. The coordination number of carbon about osmium drops from 3.0 in the model compound to about 1.8 in the activated material; this is a lower value than that expected for a 14% loss of carbon reported in the Experimental section, and this suggests that the activation processes proceeded further in the EXAFS cell than in the conventional activation apparatus. However, the clear conclusion from these EXAFS experiments is that the original cluster structure of $Os_{10}C(CO)_{24}^{=}$ is retained on impregnation and activation, and hence a structurally defined material has been prepared for the study of chemisorption and cataysis.

Activated $Os_{10}C(CO)_{24}^{=}$/alumina gave an infrared spectrum consisting of three broad bands centred at 1960, 2063, and 2200 cm^{-1} (13). Such a spectrum has sometimes been interpreted as indicating cluster dissociation on the support (22). However, such spectra are always given, in our experience, by active catalysts prepared from $Os_3(CO)_{12}$, $H_4Os_4(CO)_{12}$, and $Os_6(CO)_{18}$ (9, 12) and an interpretation based on a carbonyl-protected-cluster model has been advanced (13). The present work demonstrates that activated clusters of known structure and visible by electron microscopy give the 3-band spectrum, and this greatly strengthens our interpretation.

4.2 Structural Chemistry of the Os_6-System

No clusters were visible in electron micrographs of self-spread $Os_6(CO)_{18}$/alumina; they are not resolved because of image degradation by electron scattering in the support. The EXAFS spectrum of self-spread $Os_6(CO)_{18}$/alumina is shown in Figure 2 and the parameters used to produce the theoretical spectrum are presented in Table 2. Comparison of the data with that obtained for the pure compound shows immediately that:

(i) chemical combination has occurred with the support as shown by the osmium-aluminium distance at 0.169 nm (this value, which appears low, is being re-examined) and the osmium-oxygen distance at 0.221(5) nm;

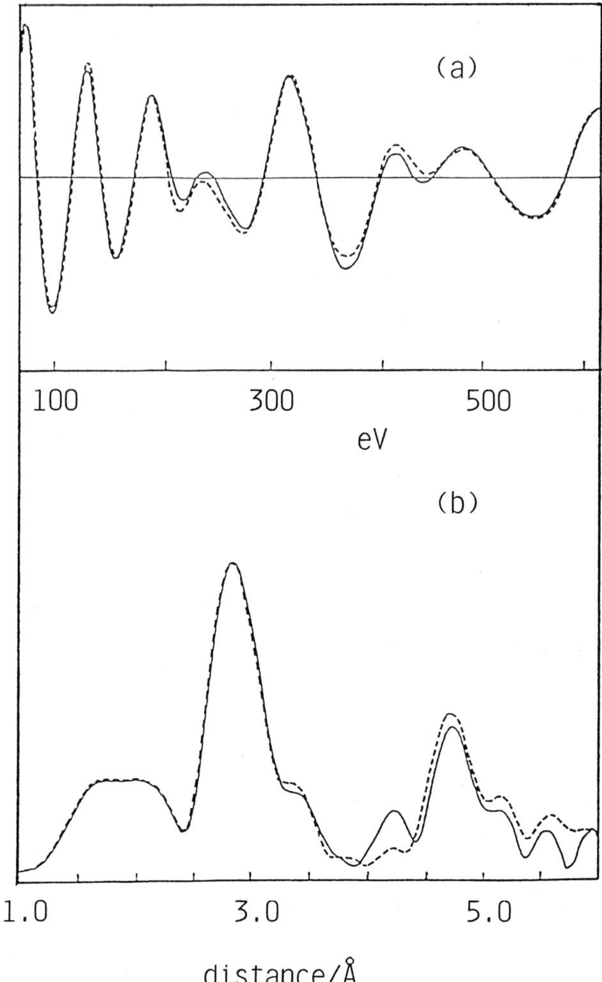

FIGURE 2. (a) Smoothed EXAFS spectrum of self-spread $Os_6(CO)_{18}$/alumina and (b) its Fourier transform. Full curves, experimental; dashed curves theoretical. All spectra are k^2-weighted.

TABLE 2. Interatomic distances and coordination numbers for the Os_6-materials as derived from the osmium L_{III}-edge EXAFS spectrum shown in Figure 2.

Shell	Atom type	Interatomic distances/Å (coordination numbers)	
		$Os_6(CO)_{18}$ (a)	$Os_6(CO)_{18}$/alumina
		Pure compound	Self-spread material
1	Os —— Al in support		1.69
2	Os —— C in linear CO	1.91	1.77
3	Os —— C in capped-CO		2.16
4	Os —— O ? in support		2.21
5	Os —— Os	2.76 (4.0)	2.83 (3.7)
6	Os....O in linear CO	2.90 (3.0)	2.95
7	Os....C in ligand-CO	3.53 (6.0)	3.65
8	Os....Os		3.87 (0.7)
9	Os....O in ligand-CO	4.47 (6.0)	4.50
10	Os....Os	4.58 (0.7)	4.61 (0.7)
11	Os....Os	4.76 (0.3)	5.05 (0.3)
12	Os....Os		5.54 (0.7)

(a) Mean values for $Os_6(CO)_{18}$ by X-ray diffraction (23): Os —— C, 1.89; Os —— Os, 2.78; Os....O, 3.03; Os....Os, 4.57 and 4.78 Å.

(ii) modification of the ligand-CO bonding has occurred, so that the osmium-carbon distance is retained 0.177 nm but a new distance at 0.216(5) nm is observed;

(iii) a modification of the osmium cluster structure has occurred as shown by the variation in the number and magnitudes of the observed non-bonded osmium-osmium distances.

Our provisional interpretations are as follows.
First, with respect to (i), evidence is present for a cluster-support interaction in which osmium atoms in the cluster interact with aluminium and oxygen centres in the support. This cluster-support interaction may be partially electrostatic in origin, involving (a) osmium atoms in a formally negative oxidation state (CO-ligands at which provide the band at 1960 cm^{-1} in the infrared spectrum (11)) interacting with positive aluminium centres and (b) osmium atoms in a formally positive oxidation state (CO-ligand at which provide the 2200 cm^{-1} band in the infrared spectrum) interacting with negatively charged oxygen centres. Second, with respect to (ii), the osmium-carbon distance of 0.216 nm suggests that some of the CO-ligands have undergone a change from the linear configuration to the capped configuration. Third, the osmium-osmium distances observed in self-spread $Os_6(CO)_{18}$/alumina are those that would be expected if the structure had partially unfolded as shown below.

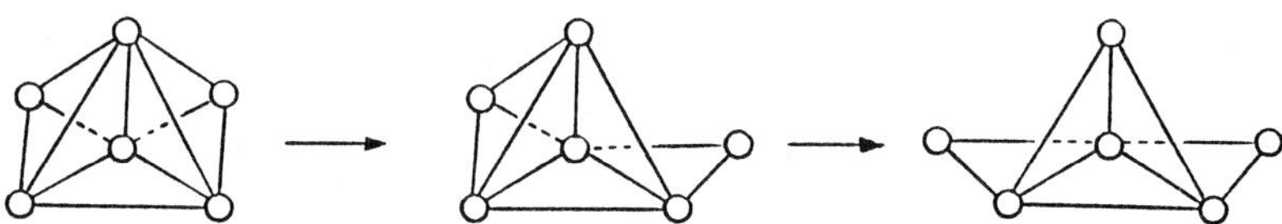

Progressive unfolding of the structure of the Os_6-cluster

These EXAFS spectra clearly demonstrate that a change in cluster structure occurs when $Os_6(CO)_{18}$ is impregnated onto alumina and that this Os_6-system is much more reactive when supported than the Os_{10}-system described above.

Self-spread $Os_6(CO)_{18}$/alumina when activated in the standard manner (9) gave catalytically active materials (11) with highly reproducible chemisorption properties (10). The EXAFS spectrum of activated $Os_6(CO)_{18}$/ alumina has been obtained, but not interpreted in terms of a structure. However, Os-Os bonding is retained in the activated material, and sintering to give polycrystalline osmium has not occurred (the EXAFS spectrum of polycrystalline osmium is strikingly different (24)); thus, a clustered state has been preserved. This is consistent with the absence of clearly visible particles in the electron micrographs, and the presence of a 3-band spectrum in the CO-stretching region in the infra-red (9).

4.3 Adsorption Properties of the Os_{10}- and Os_6-Derived Materials

Isotherms for the adsorption of CO were measured in the range 0.01 to 1 Torr at room temperature using a conventional volumetric apparatus. Little or no adsorption occurred on the freshly impregnated materials, or on the blank supports. Considerable adsorption occurred on the thermally activated materials as shown in Figure 3; the similarity of the extents of adsorption in the two systems is clear. Isotherms A and C represent the total adsorption recorded. When samples were evacuated at 10^{-4} Torr at room temperature and the experiment repeated curves B and D were obtained; these isotherms represent the extent of reversible adsorption. The difference between curves A and B and between C and D represent the extent of adsorption that is irreversible at room temperature and this strongly adsorbed CO was removed on heating at 10 K min^{-1} from 293 to 523 K and returning to 293 K, whereupon the system returned reproducibly to its initial state. These processes of adsorption at 293 K and thermal desorption could be repeated indefinitely without impairment of the chemisorption properties of the adsorbents. When $^{12}C^{18}O$ was adsorbed on $Os_6(CO)_{18}$/alumina only

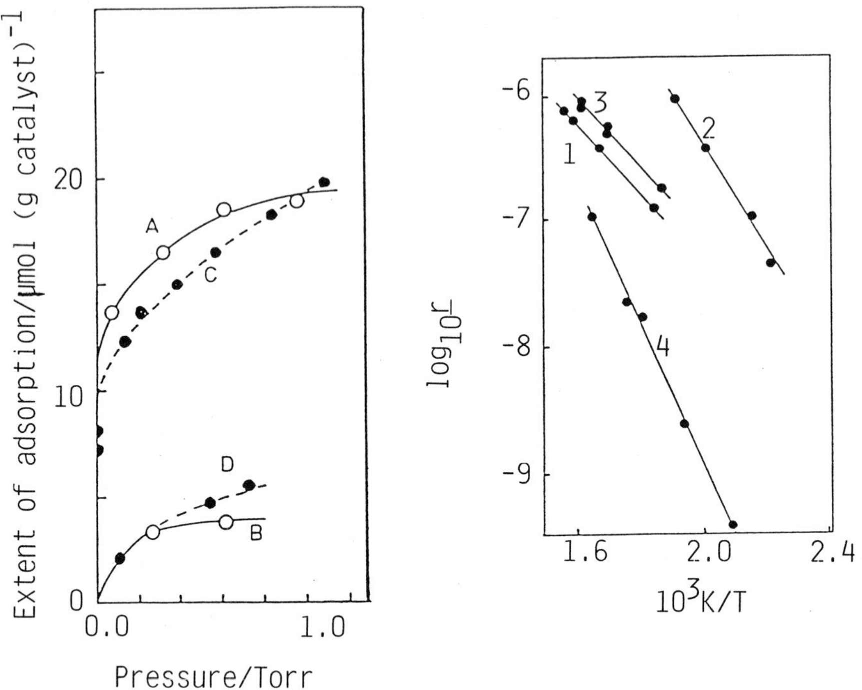

FIGURE 3 (above left) Isotherms for CO adsorption on activated $Os_{10}C(CO)_{24}^{=}$/alumina (curves A and B) and on activated $Os_6(CO)_{18}$/ alumina (curves C and D). Curves A and C represent the total extent of adsorption (reversible + irreversible); curves B and D represent the reversible component. Samples were evacuated at 293 K between A and B, and between C and D.

FIGURE 4 (above right) Variation of reaction rate, $\underline{r}$, with reciprocal temperature for ethane hydrogenolysis (ethane:hydrogen = 1:1). Line 1 = activated $Os_3(CO)_{12}$/alumina; line 2 = activated $Os_6(CO)_{18}$/alumina; line 3 = activated $Os_{10}C(CO)_{24}^{=}$/alumina; line 4 = conventional 2% Os/silica obtained by impregnation of $OsCl_3$ onto silica (Cab-O-Sil) from aqueous solution with subsequent reduction in H_2 at 723 K to highly dispersed metal. The units of $\underline{r}$ are: mol s^{-1} (g catalyst)$^{-1}$.

that isotopic variant was observed on thermal desorption, indicating that adsorbed-CO and ligand-CO are distinguishable (see reference 10 for suggested adsorbed states) and do not equilibrate at temperatures up to 523 K.

4.4 Activities of the Os_{10}- and Os_6-Derived Catalysts

Activated materials catalysed ethene hydrogenation and ethane hydrogenolysis. Temperatures of 473 K and above were required in order to achieve ethene hydrogenation, whereas conventional polycrystalline alumina-supported osmium catalyses ethene hydrogenation at room temperature (25). This indicates that these clusters, protected as they are against sintering by CO-ligands, behave as CO-poisoned metals in respect of ethene hydrogenation. The rates observed over the Os_{10}- and Os_6-derived catalysts at 500 K were 0.20 and 0.78 micromol s^{-1} (g catalyst)$^{-1}$ respectively, and the apparent activation energies were 48 and 53 kJ mol^{-1} respectively The activation energy over conventional osmium over a lower temperature range was 35 kJ mol^{-1} (25). Catalysts that had been used for ethene hydrogenation for long periods were re-examined; the three-band infrared spectrum was again observed but with a somewhat diminished intensity, and electron microscopy showed the presence of particles of 0.8 nm in diameter, no larger particles being visible. Thus cluster integrity was retained during ethene hydrogenation.

By contrast, the activities of the cluster-derived catalysts for ethane hydrogenolysis exceeded that of conventional (silica-supported) polycrystalline osmium (Figure 4), and the activation energies were always lower than that for conventional osmium. Thus;

	E_a/kJ mol^{-1}	
activated $Os_{10}C(CO)_{24}^{=}$/alumina	54	(535 - 625 K)
activated $Os_6(CO)_{18}$/alumina	99	(410 - 515 K)
activated $Os_3(CO)_{12}$/alumina	52	(540 - 665 K)
conventional 2% Os/silica (this work)	114	(475 - 605 K)
conventional 10% Os/silica (Sinfelt et al. (8))	146	(398 - 434 K)

The higher activities of the cluster-derived catalysts recorded in Figure 4, in comparison with that of the conventional metal, may be due in part to the higher dispersion of the metal in the case of the Os_6- and Os_{10}-derived catalysts. Curiously, the presence of CO ligands is not detrimental to activity in this reaction. The observed hydrogenolysis activity of $Os_6(CO)_{18}$/alumina was two orders of magnitude greater than that of the well-dispersed conventional metal, and it is therefore certain that the cluster catalyst exhibited the higher specific activity. The presence of the clustered state in $Os_{10}C(CO)_{24}^{=}$/alumina after use for ethane hydrogenolysis was confirmed by electron microscopy.

5. CONCLUSIONS AND A FORWARD LOOK

The clustered state of osmium is retained when $Os_6(CO)_{18}$ and $Os_{10}C(CO)_{24}^{=}$ are impregnated onto alumina and thermally activated. In the Os_{10}-system the tetracapped octahedral configuration is retained throughout; in the Os_6-system the bicapped tetrahedral configuration is modified on chemical interaction with the alumina. These results may assist the design of future experiments; since both retained and changed configurations of metal atoms can, in principle, be exploited to advantage.

This work demonstrates that hydrogenolysis activity can be shown by very small numbers of metal atoms. In one sense, these catalysts are more akin to catalytically active complexes in solution than to metal crystallites containing hundreds of atoms. Molecular definition of the site, which can be approached for some metal complexes in solution where the extent of solvolysis is known, is also approached in this work, and we hope in future studies to define more closely the actual atoms in clustered osmium systems that form the adsorption sites for ethene and for ethane.

In homogeneous systems, the active form of a catalyst is normally achieved when interactions between a metal complex and the solvent have achieved equilibrium. In heterogeneous systems, (for gas phase reactions where solvents are absent) the active form of a catalyst is often achieved when interactions have been established between the metal entity and the support. Solvent interactions and support interactions have each, in their own sphere, made the definition of catalytically active sites a difficult task. In the case of catalysis by supported clusters we are confident that EXAFS spectroscopy will, in carefully selected systems, provide definitive information about the metal-support interaction, and that the relationship between that interaction and site geometry will become clear.

Finally, in the context of heterogeneous catalysis, a site present in a metal surface is commonly considered as a static arrangement of metal atoms, and reaction is conceived to occur by movement of the molecules as they are converted from reactants through intermediates to products. But are the metal atoms at a multi atom site static? We know from surface science that reconstruction of surfaces on adsorption occurs. Is it possible to devise experiments from which we can deduce the movements of the metal atoms at a site during the catalytic act? It is certainly our hope that, by use of EXAFS spectroscopy, we shall be able to characterise any conformational changes that occur at cluster sites during adsorption, reaction, and desorption. Moreover, if such processes occur on an appropriate time-scale, it may be possible to examine the rates of such conformational changes by energy dispersive EXAFS spectroscopy. The study of site dynamics will then have begun and a new dimension will have been introduced into our understanding of catalysis.

ACKNOWLEDGEMENTS. This work was part of a Joint Research Scheme funded by Imperial Chemical Industries and carried out at the University of Hull; the EXAFS spectroscopy was part of an investigation funded by the Science and Engineering Research Council. I and my colleague Dr. R.B. Moyes wish to acknowledge that (i) the results for the Os_6-system were obtained by Dr. S.D. Jackson and Mr. D.J. Hunt, (ii) the Os_{10}-cluster was prepared by Dr. P.F. Jackson, (iii) the Os_{10}-system was investigated by Dr S.R. Morris, and (iv) the EXAFS spectroscopy was carried out by Mr. P. Worthington. My principal colleague at ICI New Science Group, Runcorn, was Dr. R. Whyman; Dr. S.D. Jackson is now with ICI Agricultural Division, Billingham.

REFERENCES

1. H.S. Taylor, Proc. Roy. Soc., A108, 105 (1925).

2. N.I. Kobosev, Zhur. fiz. Khim. 13, 1 (1939); idem, Russ. J. Phys. Chem., 19, 48, 142, (1945).
3. O.M. Poltorak and V.S. Boronin, Russ. J. Phys. Chem., 39, 1329 (1965); idem, ibid, 40, 1436 (1966); O.M. Poltorak, V.S. Boronin, and A.N. Mitrofanova, Paper No. 68, Fourth International Congress on Catalysis, Moscow, 1968.
4. P.B. Wells, in 'Surface and Defect Properties of Solids' (Specialist Periodical Report of the Chemical Society, M.W. Roberts and J.M. Thomas, eds.) 1, 236 (1972).
5. G.A. Martin, J. Catal., 60, 345 (1979).
6. B.F.G. Johnson (ed.) "Transition Metal Clusters", (Wiley Interscience, Chichester, 1980).
7. For example, J.H. Sinfelt, G.H. Via, and F.W. Lytle, J. Chem. Phys., 67, 3831 (1977); 68, 2009 (1978); 71, 690 (1979); 76, 2779 (1982).
8. J.H. Sinfelt, Catal. Rev., 3 175 (1970).
9. G. Collier, D.J. Hunt, S.D. Jackson, R.B. Moyes, I.A. Pickering, P.B. Wells, A.F. Simpson, and R. Whyman, J. Catal., 80, 154 (1983).
10. D.J. Hunt, S.D. Jackson, R.B. Moyes, P.B. Wells, and R. Whyman, J. Catal., 86, 333 (1984).
11. S.D. Jackson, R.B. Moyes, P.B. Wells, and R. Whyman, J. Catal., 86, 342 (1984).
12. D.J. Hunt, R.B. Moyes, P.B. Wells, S.D. Jackson, and R. Whyman, Proc. 8th Intern. Congr. Catal., Berlin, 1984, (Verlag Chemie, 1984), 5, p. 27.
13. S.D. Jackson, S.R. Morris, R.B. Moyes, P.B. Wells, R. Whyman, and P. Worthington, J. Catal., submitted.
14 B.S. Nicholls, S.D. Jackson, R.B. Moyes, A.F. Simpson, P.B. Wells, and R. Whyman, J. Catal., submitted.
15. G.N. Greaves, R.B. Moyes, P.B. Wells, and P. Worthington, J. Catal., submitted.
16. M. McPartlin, C.R. Eady, B.F.G. Johnson, and J. Lewis, J. Chem. Soc. Chem. Commun., 883 (1976).
17. P.F. Jackson, B.F.G. Johnson, J. Lewis, M. McPartlin, and W.J.H. Nelson, J. Chem. Soc. Chem. Commun., 224 (1980).
18. C.R. Eady, B.F.G. Johnson, and J. Lewis, J. Chem. Soc. Dalton, 2606 (1975).
19. P. F. Jackson, B.F.G. Johnson, J. Lewis, M. McPartlin, and W.J.H. Nelson, J. Chem. Soc. Chem. Commun., 224 (1980).
20. S.S. Hasnain, P.D. Quinn, G.P. Diakun, E.M. Wardell, and C.D. Garner, J. Phys. E, 17, 40 (1984).
21. N. Binsted, unpublished work.
22. For example, H. Knozinger, Y. Zhao, B. Tesche, R. Barth, R. Epstein, B.C. Gates, and J.P. Scott, Faraday Discuss. Chem. Soc. No. 72, 53 (1981).
23. R. Mason, K.M. Thomas, and D.M.P. Mingos, J. Am. Chem. Soc., 95, 3802 (1973).
24. D.J. Hunt, S.D. Jackson, R.B. Moyes, P.B. Wells, P. Worthington, and R. Whyman, J. Molec. Catal., 20, 289 (1983).
25. G.C. Bond, G. Webb, and P.B. Wells, Trans. Faraday Soc., 61, 999 (1965).

SURFACE INTERMEDIATES IN PREPARATION OF CLUSTER DERIVED BIMETALLIC CATALYSTS

László Guczi and Sándor Dobos
Institute of Isotopes of the Hungarian Academy of Sciences, P.O.Box 77, H-1525 Budapest, Hungary

ABSTRACT - The possible surface intermediates formed by supporting molecular mono- and bimetallic carbonyl clusters on Al_2O_3 and Cab-O-Sil carriers are studied by molecular isotope exchange (MIE) between ^{12}CO ligand of the cluster and gaseous ^{13}CO molecules, by retarded decomposition using i.r. technique as well as by considering the intermediate stability in different atmospheres and at various temperatures.

It has been established that depending on the ambient atmosphere and temperature the decomposition of the bimetallic clusters results in the formation of monoruthenium carbonyl species which become stabilized on the oxide carriers. These species may help the formation of bimetallic particles and partly participate in the creation of active sites in the CO + H_2 reaction.

1. INTRODUCTION

One of the most essential problems in heterogeneous catalysis is to prepare and stabilize very small metal particles supported on oxide carriers. Here an intermediate strength of bonding between metal particles and oxide support should be important: at too strong interaction the metallic state of the particles cannot be maintained whereas at too weak interaction the migration of the active component on the surface may cause coalescence of the metallic particles.

The application of molecular metal clusters may bear some advantages:

- small metal frameworks of the transition metals are

already preformed in the molecular cluster;

- it is comparatively easy to graft to oxide support, especially, if it contains surface hydroxyl groups;
- after grafting carbonyl clusters can be decomposed and stabilized at moderate temperature to obtain supported metal particles of high dispersion;
- under special conditions stabilization of small metallic framework leads to highly dispersed supported metal particles;
- in preparation of bimetallic catalysts molecular clusters containing heteronuclear bonds can be applied.

However, surface hydroxyl groups interact with the metal framework and as a result, metal atoms in zero valency state are easily oxidized even at low temperature. This effect is the most serious at the impregnation stage (Ref. 1) thus, in the further part of this paper we are mainly concerned with the mechanism of the formation of active metal catalysts from molecular carbonyl clusters by discussing it in sequence of the initial step (impregnation), decomposition of the supported cluster, and finally the formation of supported metal particles which are responsible for the catalytic activity.

2. INTERACTION OF MOLECULAR CLUSTER WITH THE SUPPORT DURING IMPREGNATION

In order to understand the behaviour of the supported molecular cluster, first its initial interaction with the support at impregnation should be discussed.

Since O^{2-} and OH^- groups are present on Al_2O_3 and on Cab-O-Sil surface, at impregnation different types of surface reactions between OH^-, O^{2-} and the cluster framework should be considered. The initial interaction between Cab-O-Sil and bimetallic molecular clusters such as $Fe_2Ru(CO)_{12}$ and $H_2FeRu_3(CO)_{13}$ is generally weak, the structure which is present in solution is maintained, i.e. no disintegration of the molecular cluster framework is observed by i.r. and Mössbauer spectroscopy (Ref. 2). Similar observation was reported by Choplin et al. (Ref. 3) for $H_2FeOs_3(CO)_{13}$ deposited on silica surface: the molecular cluster was first physisorbed at impregnation as shown by i.r. and Raman spectroscopy.

Alumina and magnesia supports behave in a completely different manner. As a first step, in the reaction between hydrated alumina and the molecular cluster $[HM_3(CO)_{11}]^-Al^+$ species are formed for e.g. $Fe_3(CO)_{12}$ (Ref. 4), $Ru_3(CO)_{12}$ (Ref. 5), $Os_3(CO)_{12}$ (Ref. 6) as shown by scheme 1:

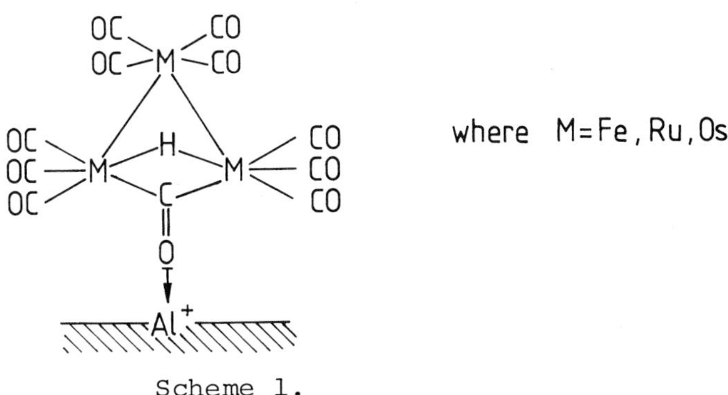

Scheme 1.

The Ru-Os bimetallic molecular cluster has also been proved to form this stable surface species. A complex of $Al^+[HRuOs_3(CO)_{13}]^-$ identified by i.r. spectroscopy was formed from $H_2RuOs_3(CO)_{13}$ (Ref. 7). By the treatment under an equimolar mixture of H_2 and CO at a temperature of 370-470 K it is transformed into $[H_3RuOs_3(CO)_{12}]^-$ on the surface and it can be extracted from the surface by ion exchange with $(AsPh_4)^+Cl^-$ and characterized in solution. Simultaneously, analogy was shown by i.r. method between this species and $Ph_4As^+[H_3Ru_4(CO)_{12}]^-$. This is the other relevant transformation on a bimetallic cluster which maintains an intact framework after a rather drastic treatment. However, these anionic species are stable only up to 373 K if heated under vacuum; at higher temperatures extensive decomposition occurs.

In addition to the i.r. evidences, recently these anions were prepared by extraction of a CH_2Cl_2 solution of bis-triphenyl phosphine imminium chloride (PPN^+Cl^-). Thus, for $H_2FeRu_3(CO)_{13}$ contacted with MgO, the surface complex could be extracted by PPN^+Cl^- solution in form of $[HFeRu_3(CO)_{13}]^-PPN^+$. It can be characterized by its i.r. and NMR spectra (Ref. 8).

2.1. <u>^{13}CO exchange.</u> Molecular isotope exchange (MIE) may serve as a useful tool for studying the initial interaction and a subsequent transformation of the surface species. MIE occurs when ^{12}CO ligands in the molecular carbonyl cluster or in the subcarbonyl species supported on an oxide surface are replaced by ^{13}CO molecules in the gas phase. MIE can be described by a first order equation provided that the metal-CO bonds are equivalent. Deviation has been observed particularly for partially decomposed rhodium clusters (Ref. 9).

In preliminary experiments the initial rate of exchange between ^{13}CO in the gas phase and ^{12}CO in supported $Fe_3(CO)_{12}$ was measured. It was established that in agreement with the CO evolution results (Ref. 2), the rate of isotope exchange on alumina is higher than that on Cab-O-Sil due to the stronger interaction with Al_2O_3 (Refs. 10,11). It was also determined that the amount of CO dissociated after decomposition increased with increasing pretreatment temperature of alumina.

Molecular isotope exchange between CO ligands in $Fe_3(CO)_{12}$, $Fe_2Ru(CO)_{12}$, $H_2FeRu_3(CO)_{13}$ and $Ru_3(CO)_{12}$ supported on alumina and on Cab-O-Sil and gaseous ^{13}CO in the temperature range 273 to 350 K has been studied (Ref. 12).

A typical experimental result for the change of the pressures of ^{13}CO and ^{12}CO for $H_2FeRu(CO)_{13}$ on Al_2O_3 measured at 298 K can be seen in Figure 1.

The quasi-equilibrium state indicated in Figure 1. after 100 min reaction time can be demonstrated by measuring the exchange in the temperature programmed mode (Figure 1b). Here initially the CO pressure is only slightly increased but the onset of a fast exchange indicates the presence of the strongly bonded CO ligands requiring higher temperature.

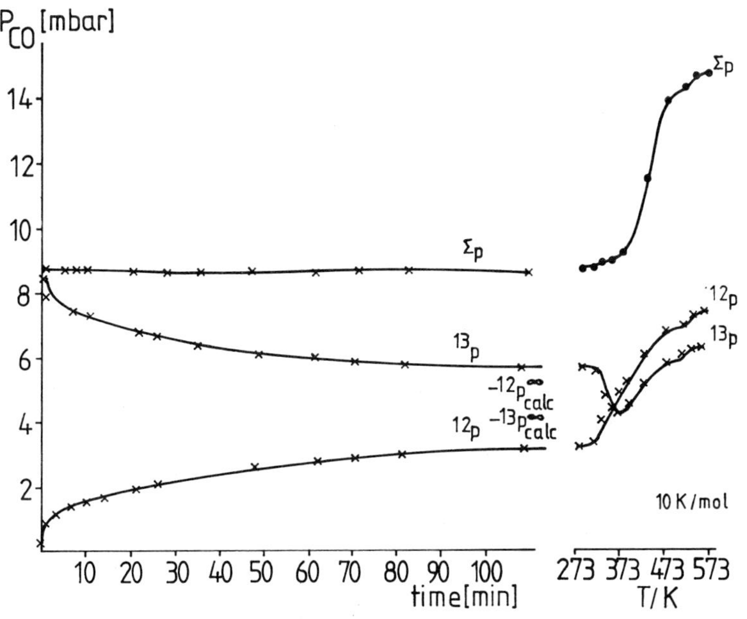

Figure 1. Plot of ^{12}CO, ^{13}CO and the total CO pressure for $H_2FeRu_3(CO)_{13}$ on Al_2O_3. a) vs time at 298 K; b) vs temperature [from (Ref. 12)].

These experiments suggest that at least two types of CO ligands exist on the surface. However, the simplest model has to be operated with three different types of

CO the third type being the non-removable form due to dissociation (Ref. 13). The resulting kinetic equation in integrated form is as follows:

$$^{12}G = {}^{12}G_1^{\infty}(1 - e^{-k_1' t}) + {}^{12}G_2^{\infty}(1 - e^{-k_2' t}) \qquad /1/$$

where ^{12}G, $^{12}G^{\infty}$ refer to the amount of ^{12}CO in the gas phase at time t, and at equilibrium, respectively. k_1' and k_2' are the initial rates of fast and slow exchange, respectively.

The rate represented by k_3' is obviously zero. /CO cannot be removed without dissociation./ From these rates the rate constant can be calculated by equation $k_1' = k_i(G + C_i)$ where G and C_i are the amounts of CO in the gas phase and in the supported cluster, respectively.

From the results detailed in (Ref. 12) it can be established that generally on the stronger electron donor alumina the rate constant for exchange is higher than those for Cab-O-Sil-supported clusters in agreement with the CO evolution results. In addition, the number of CO ligands involved in the three types of exchange for $H_2FeRu_3(CO)_{13}$ is 3.5, 4.5, 5 and 1.5, 11.5, 0 for alumina and Cab-O-Sil, respectively, i.e. on the Cab-O--Sil support practically one type of exchange is prevailing.

Another factor which influences the interaction is the carbonyl cluster itself. Generally, higher exchange rates, i.e. stronger interaction, are experienced for clusters with higher iron content.

The sequence of the decreasing exchange rate coincides with the order of interatomic CO scrambling in these carbonyl clusters:

$$Fe_3 \sim Fe_2Ru > H_2FeRu_3 > Ru_3$$

revealing the importance of intramolecular CO mobility

in the exchange (Refs. 14-17).

Both the effect of the support and the structure of the molecular cluster on the isotope exchange are well supported by Mössbauer spectroscopy (Refs. 2,11). On Cab-O-Sil the $H_2FeRu_3(CO)_{13}$ (Ref. 2) and the $H_2FeOs_3(CO)_{13}$ clusters (Ref. 3) retain their structure but on the same support $Fe_3(CO)_{12}$ disintegrates (Ref. 11) to form high spin Fe^{3+} and Fe^{2+} which is a clear indication of the oxidation of iron atoms.

The most plausible explanation for the CO ligands not being equivalent in the supported state unlike the unsupported case (Ref. 14) is the reaction between the support and the carbonyl cluster producing metal subcarbonyl species with different metal-CO bond strength. The deviation from simple exchange is certainly indicative of the CO ligand being in different environments.

The validity of this assumption also seems to be supported by the experimental fact that the extent of the surface process increases in time. Overnight standing shows that most of the clusters are to be found in a strongly bonded environment indicated by the lack of exchange at room temperature when using an alumina support.

2.2. Retarded decomposition on the surface. Isotope exchange may give only phenomenological description on the interaction. Further studies are, therefore, needed to elucidate the mechanism of these surface reactions which ultimately lead to the creation of the catalytically active formulation of the metal species. Since the decomposition on alumina starts immediately after contacting the cluster solution with the support, the intermediate steps are difficult to monitor even by i.r. spectroscopy. The characteristic bands observed after impregnation on Al_2O_3 are 2138, 2075, 2053, 2001 and 1980 cm^{-1} for $Ru_3(CO)_{12}$, $H_2FeRu_3(CO)_{13}$ and $RuFe_2(CO)_{12}$

which completely differ from the spectra shown in solution (Refs. 2,18,19).

In order to understand the mechanism of the decomposition, immediately after impregnation nujol was added to the supported cluster. The high molecular weight hydrocarbon may effectively compete with the molecular carbonyl cluster for the active sites. Using this technique the instantaneous cleavage of the cluster metal framework can be avoided and the decomposition can be kinetically followed.

I.r. spectra of the alumina supported $Fe_2Ru(CO)_{12}$ in nujol are presented in Figure 2 (Ref. 18). The difference spectra indicate first the formation of $Fe(CO)_5$ characterized by the bands at 2022 and 2000 cm^{-1}. These bands disappear after a long contact time resulting in the formation of the oxidized iron species. The rest of the spectra very much resembles that originating from $Ru_3(CO)_{12}$.

However, in all other systems containing iron, there is also a rise of the $Fe(CO)_5$ species which must be considered as an "endproduct" of the first stage of the process rather than an intermediate. On the other hand, $Ru_3(CO)_{12}$ also seems to be an "endproduct" for the systems of Fe_2Ru/Al_2O_3/nujol and H_2FeRu_3/Al_2O_3/nujol (Ref. 18).

Although it is quite obvious that formation of intermediate species must be suggested between the starting clusters and the above mentioned "endproducts", no sign of new, well defined molecular structures has been observed in the spectra. Instead, there have always been broad overlapping bands present in the stretching region of terminal carbonyls offering assignment to hypothetical intermediates. This assignment would mean that here the "intermediates" are a series of subcarbonyls in the adsorbed layer having different numbers of CO ligands and occupying very different sites. The broad band system that belongs to these species decreases and its shape is changed with the time, and after a series of surface rearrangements the 5 band spectra is developed.

In order to obtain more information about these processes the data points were analysed by fitting Gaussians.

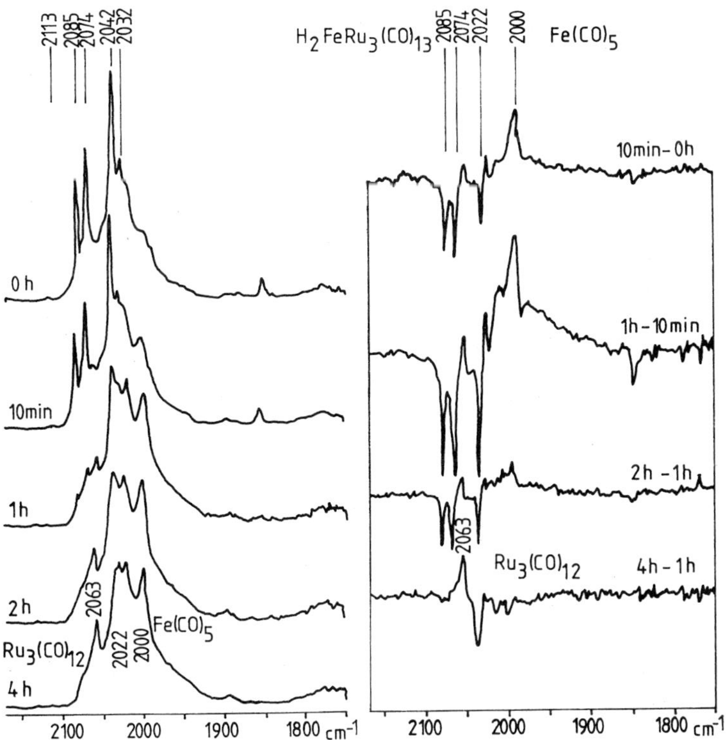

Figure 2. Infrared spectra of Al_2O_3 supported $H_2FeRu_3(CO)_{13}$ at different times. Left side: the original spectra of the nujol covered sample; Right side: difference spectra [from (Ref. 18)].

In Figure 3 the result of Gaussian approximation is presented for $Fe_2Ru(CO)_{12}/Al_2O_3$. The calculated bands of the intermediates are very broad in contrast to those of physisorbed molecular species and they drastically change with time. At t = 70 h the bands 2128, 2063 and 2005 cm^{-1} clearly show the presence of "incipient" anchored species of $Ru^{0}(CO)_2$ and $Ru^{II}(CO)_2$ (Ref. 20).

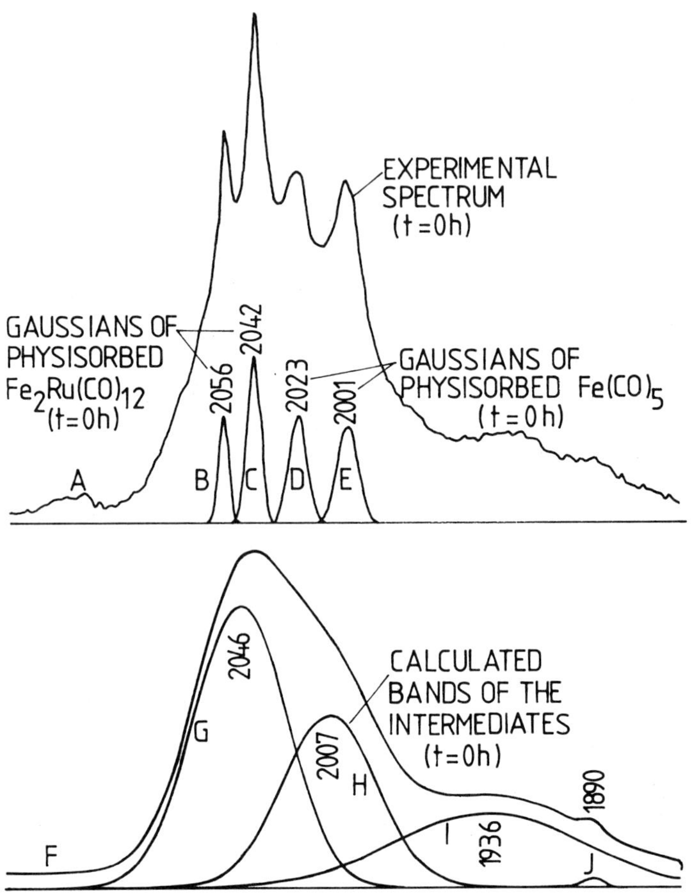

Figure 3. A: Experimental spectrum of Fe_2Ru/Al_2O_3/nujol system at t = 0; B-C and D-E: Gaussian components of physisorbed $Fe_2Ru(CO)_{12}$ and $Fe(CO)_5$, respectively, at t = 0; F is the computed spectra of the intermediate subcarbonyls at t = 0; G-J are the Gaussian components of the intermediate subcarbonyls at t = 0 [from (Ref. 18)].

From the MIE and the retarded surface decomposition data the following conclusions can be drawn:

Bimetallic clusters $Fe_2Ru(CO)_{12}$ and $H_2FeRu_3(CO)_{13}$ follow similar pathway. The first step is physisorption where the molecules are slightly perturbed. The surface layer may be regarded as a two-dimensional liquid with highly mobile molecules in it. When the cluster molecule comes to an active site of the alumina e.g. Lewis sites of Al^+, the metal-metal bonds of the clusters must quickly split /i.e. neither highly distorted original nor intermediate polynuclear carbonyl cluster molecules have been detected!/ producing a large variety of mobile and supposedly monometallic subcarbonyls containing different numbers of CO groups. These intermediates serve as precursors for molecular $Fe(CO)_5$ and $Ru_3(CO)_{12}$ and for incipient anchored species $Ru^0(CO)_2$, $Ru^{II}(CO)_2$ and $Ru^{III}(CO)_2$.

For $H_2FeRu_3(CO)_{13}$ the direct decomposition into $Fe(CO)_x$ /x = 0-5/ and $Ru_3(CO)_{12}$ might have also been assumed /similarly as it was shown for $H_2RuOs_3(CO)_{13}/Al_2O_3$(Ref. 21)/, but first we ruled it out because of the presence of very high amount of intermediate subcarbonyls. Furthermore, the $Fe_2Ru(CO)_{12}$ ends up in the identical molecular products though for this molecule no direct path leading to $Ru_3(CO)_{12}$ can be formulated.

Similar statements concern the monometallic cluster framework ($Ru_3(CO)_{12}/Al_2O_3$/nujol and $Fe_3(CO)_{12}/Al_2O_3$/nujol). The complete suggested mechanism for these surface processes is summarized in Scheme 2.

3. STABILITY OF THE IMPREGNATED CLUSTER AT CHANGING AMBIENT ATMOSPHERE AND AT DIFFERENT TEMPERATURES

The interaction between Al_2O_3 and molecular cluster which proceeds in the chemisorbed layer, is largely affected by the conditions, e.g. by the ambient atmosphere. This effect is demonstrated for $Ru_3(CO)_{12}$ deposited on Al_2O_3 and Cab-O-Sil in Figures 4 and 5 by i.r. spectra.

Under static conditions both in He and in CO the

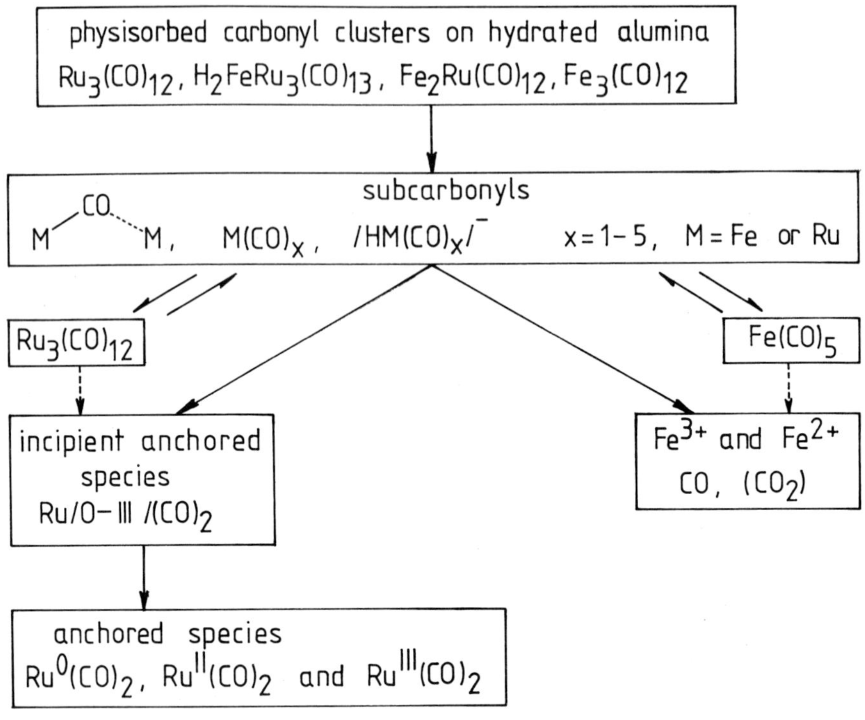

Scheme 2.

structure of the Al_2O_3- and Cab-O-Sil-supported clusters is similar. On the other hand, under dynamic conditions in flowing He or in vacuum, $Ru(CO)_n$ subcarbonyls characterized by bands at 2138, 2070, 2050, 2000 and 1980 cm^{-1} are formed (Ref. 19) (Figure 4).

When the spectra of the $Ru_3(CO)_{12}$ having been in contact for 12 h with the support, are compared with those supported on Cab-O-Sil, the band broadening points to the coexistence of various subcarbonyl fragments and the original cluster framework. If the sample is evacuated, the original $Ru_3(CO)_{12}$ is further decomposed into a structure similar to that obtained under dynamic conditions. This is shown in Figure 5.

For the preparation of supported, highly dispersed

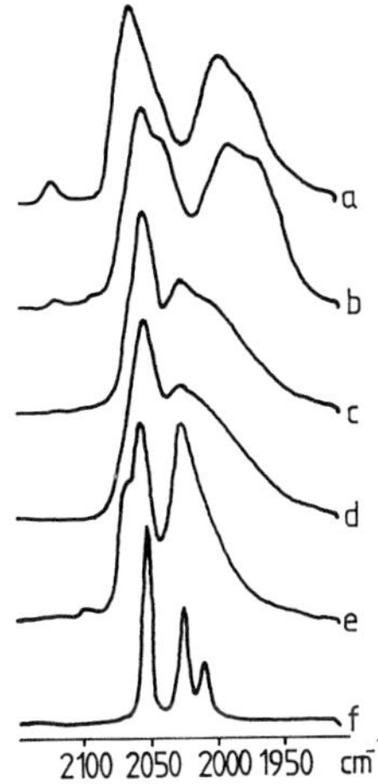

Figure 4. I.r. spectra of $Ru_3(CO)_{12}/Al_2O_3$ stored for 12 h after impregnation from n-hexane in: a/ vacuum; b/ He flow; c/ static He; d/ static CO; e/ vacuum but Cab--O-Sil supported; f/ unsupported $Ru_3(CO)_{12}$ in n-hexane.

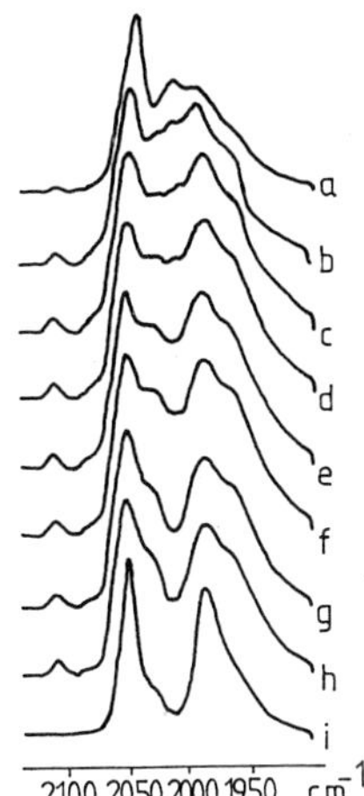

Figure 5. I.r. spectra of $Ru_3(CO)_{12}/Al_2O_3$ stored for 12 h in static He and evacuated for a/ 0.05 h; b/ 0.5 h; c/ 1 h; d/ 2 h; e/ 5 h; f/ 12 h; g/ 24 h; h/ 28 h; i/ stored in air for 168 h.

metal catalysts the impregnated, partially decomposed carbonyl cluster should be thermally decomposed. The i.r. spectra for $H_2FeRu_3(CO)_{13}$ are presented in Figure 6. The bimetallic system shows five bands which are identical with the main spectral features of the supported $Ru_3(CO)_{12}$. By this fact our prediction that at the end of an interaction with hydrated alumina surface iron-ruthenium bimetallic clusters decompose into identical ruthenium carbonylic surface species, like pure ruthenium clusters, is verified.

The main features of the decomposition on Al_2O_3 are as follows:

i/ The five bands, under the experimental circumstances used in our investigations, proved to be very stable in their frequencies and band widths. The bands are rather narrow and are in the same range as those calcula-

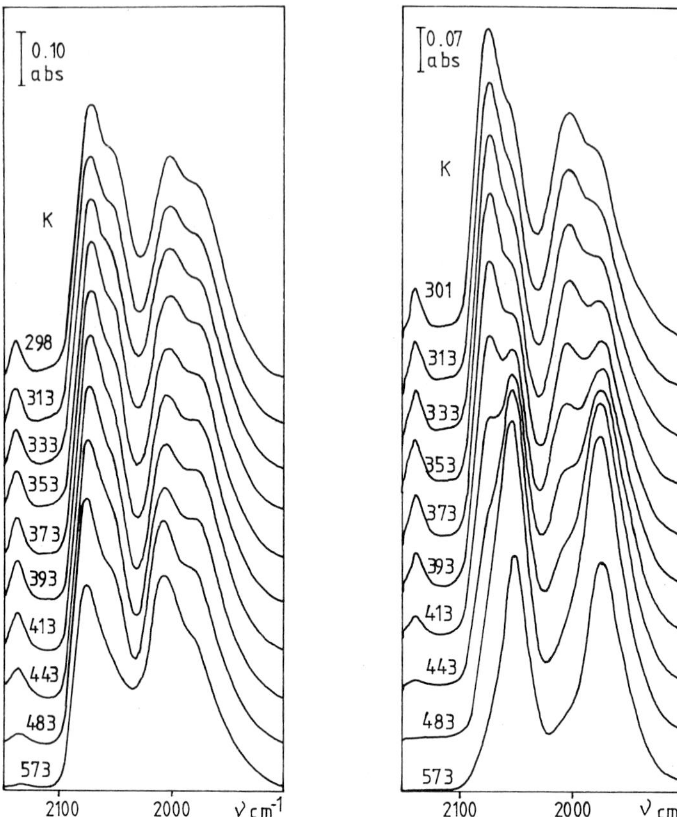

Figure 6. Baseline-corrected spectra of systems H_2FeRu_3/Al_2O_3/vacuum and $H_2FeRu_3/Al_2O_3/H_2$ [from (Ref. 20)]

ted for chemisorbed or physisorbed species. This means that these bands should be assigned to well defined, molecule-like species;

ii/ Taking into account the relative intensities of the five bands and the qualitative parallelisms, the bands can be arranged in three pairs /A: 2138 - 2075; B: 2075 - - 2008 and C: 2055 - 1980 cm^{-1}/ for all systems, as already suggested in the literature for Ru_3/Al_2O_3 systems (Refs. 22-26). Consequently, we extend the assignment introduced by Zecchina et al. (Ref. 26) to the main surface species found in the bimetallic systems, too, as follows: $Ru_A = Ru^{III}(CO)_2$, $Ru_B = Ru^{II}(CO)_2$ and $Ru_C = Ru^{O}(CO)_2$ [from (Ref. 10)].

iii/ The anchored species Ru_B and Ru_C are stable in a very wide temperature range /300-500 K/ and they can be interconverted by oxidation and reduction. Ru_A is the less stable /300-400 K/.

iv/ Beside the anchored species Ru_A, Ru_B and Ru_C, there are analogous surface species present producing very weak /1-5%/ and very broad bands centred near to the main strong bands. These species are mobile /not anchored/, they supposedly occupy a large variety of different sites.

The decomposition scheme for Cab-O-Sil supported bimetallic systems is somewhat different. As shown in Figure 7 $H_2FeRu_3(CO)_{13}$ is transformed into a stable molecular cluster with different compositions and this is stable in a wide range of temperatures (Ref. 31). Serious decomposition starts at around 400 K. Similar features have been demonstrated for $H_2FeOs_3(CO)_{13}$ on silica (Ref. 3) on $(\mu_2\text{-H})AuOs_3(CO)_{10}PPh_2(CH_2)_2SIL$ /with closed tetrahedral structure/ or $(\mu_2\text{-Cl})AuOs_3(CO)_{10}PPh_2(CH_2)_2SIL$ /with open-butterfly structure/ (Ref. 27).

It appears to be a general rule that bimetallic clusters - even if they are stable at room or moderate temperature - decompose at higher temperature with the formation of separate monometallic carbonyl species and metallic particles. On highly dehydroxylated alumina the metal /iron/ dispersion may be quite high (Ref. 28). As it was demonstrated, high temperature reduction may lead to the formation of supported bimetallic catalysts. It is, therefore, plausible to assume that the use of two monometallic clusters may give the similar results.

A mixture of $Fe_3(CO)_{12}$ and $Ru_3(CO)_{12}$ was first applied to prepare iron-ruthenium bimetallic catalysts deposited on Cab-O-Sil at different Ru/Fe ratios. Broadening in the temperature programmed decomposition peak at 510 K

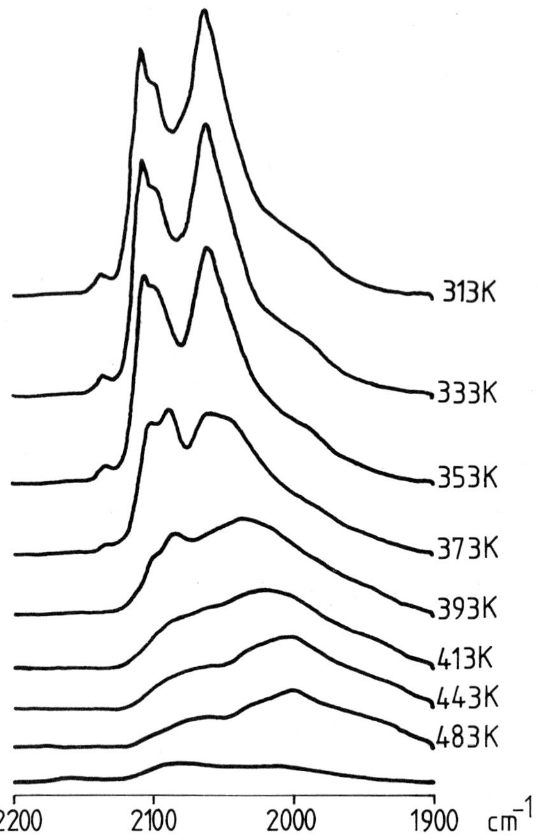

Figure 7. Decomposition of $H_2FeRu_3(CO)_{13}$ deposited on Cab-O-Sil under vacuum [from (Ref. 31)].

for various Fe-Ru samples may suggest the formation of bimetallic catalysts. The catalytic activity and selectivity in carbon monoxide hydrogenation provided further indication for the influence of ruthenium on iron (Ref. 13).

Although the kinetic and chemisorption data strongly suggested that at higher temperature interaction exists between $Fe_3(CO)_{12}$ and $Ru_3(CO)_{12}$, evidence is given by i.r. and Mössbauer spectroscopy (Refs. 29,30). With increasing temperature the i.r. spectra show that first $Fe_3(CO)_{12}$ is decomposed and then the decarbonylation of

$Ru_3(CO)_{12}$ started. At this stage decomposition of $Ru_3(CO)_{12}$ was accelerated by the presence of iron both in hydrogen and in He. The latter data are presented in Figure 8.

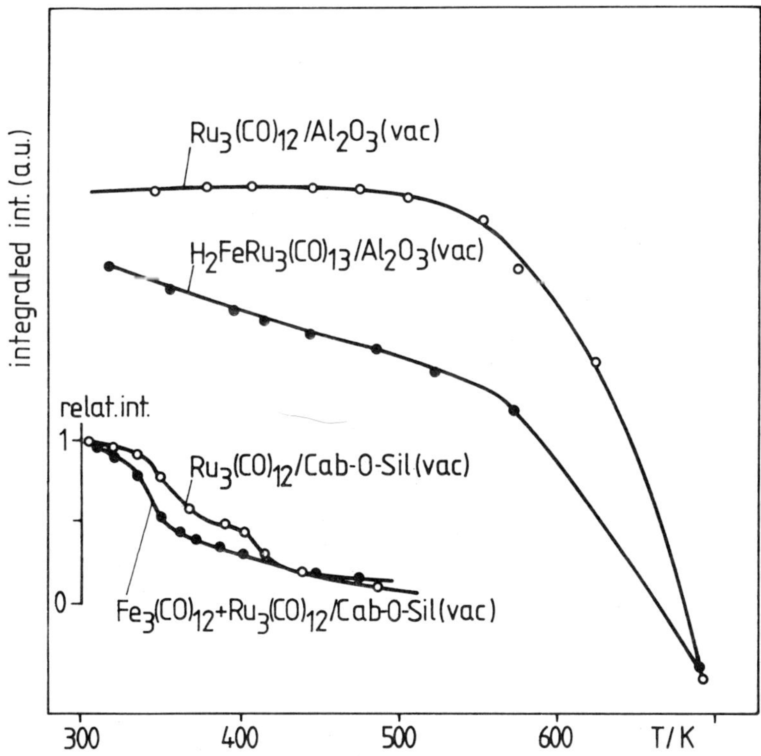

Figure 8. The change of the integrated intensities of the CO stretching modes, calculated by numerical integration over the frequency range of 2170-1900 cm^{-1}. Bottom left figure represents the change of intensity of the 2060 cm^{-1} band for Cab-O-Sil supported clusters [from (Refs. 20 and 29)].

It is obvious from the figure that subcarbonyl species are more stable on Al_2O_3 than on Cab-O-Sil. Furthermore, both for bimetallic clusters and for a mixture of two monometallic clusters decomposition is accelerated by iron, i.e. in both cases interaction exists between iron and ruthenium. The nature of this interaction at this stage is still not clear, what we know is that iron is present in ionic form.

It is obvious that the amount of molecular type sub-carbonyl species considerably differs for Al_2O_3 and Cab-O-Sil, but their structure is similar on Cab-O-Sil regardless whether they are prepared from bimetallic or a mixture of two monometallic clusters as indicated in Figure 9. It is not unreasonable to support that in both cases the surface species leading to the formation of bimetallic catalysts are nearly identical.

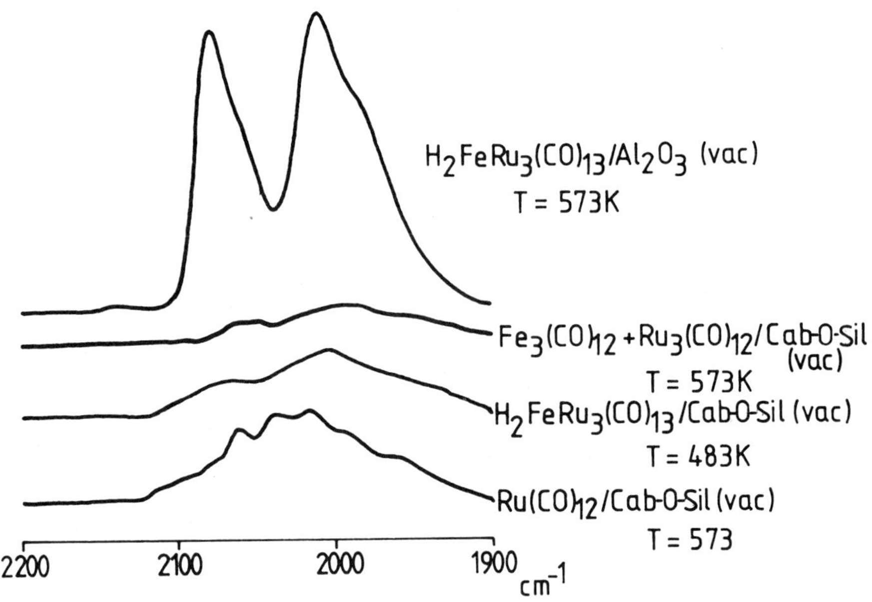

Figure 9. I.r. spectra of the surface species on different clusters supported on Cab-O-Sil and on Al_2O_3.

Decomposition at higher temperature in hydrogen leads to the formation of bimetallic metal particles as indicated by low temperature /1.6 K/ and high field /up to 6 T/ Mössbauer spectroscopy (Ref. 32) regardless of the source of clusters used. Similar results have been received by Kaminsky et al. (Ref. 33) using amorphous carbon black as a support.

4. NATURE OF THE ACTIVE SITES: ARE THEY METAL SITES OR SUPPORTED SUBCARBONYL SPECIES?

It is not easy to answer this question. It was first suggested for osmium supported on Al_2O_3 (Ref. 23) that the active sites for the $CO + H_2$ reaction are the supported monoosmium dicarbonyl species as no metal particle was found in the system. Later, large Os particles were found by TEM (Ref. 34) but the activity was still ascribed to the surface carbonyl species.

Without making a final decision there is some evidence that the above statement is - at least - partly valid For iron-ruthenium system it was found earlier (Refs. 13,30,35-41) that generally

i/ metal dispersion is higher when the molecular cluster is decomposed in He and

ii/ the catalytic activity revealed in $CO + H_2$ reaction is also higher as compared to the sample decomposed in H_2.

A plausible explanation was given by suggesting the presence of carbon which originates from CO-ligand dissociation as a result of hydrogen depletion on the surface (Ref. 29). The encapsulated carbon may prevent coalescence of the small metal particles thereby leading to high dispersion and high catalytic activity.

However, in the light of the present study it is clearly indicated that a part of the CO ligands is indeed stabilized on the surface, especially, when decomposition occurs only up to 573 K /normally reduction is carried out up to 770 K/ which temperature is already higher than that normally used for $CO + H_2$ reaction. The proportion of the CO molecules retained by the surface depends on the support, but at one type of support it is a function of the ambient atmosphere in which decomposition took place. That is, Al_2O_3 retains more CO than Cab-O-Sil, and more CO is left on the surface when decomposition occured in helium or in vacuum.

We do not have evidence that the monoruthenium dicarbonyl species are directly involved in developing the catalytically active sites, but it is experimentally proven that in its presence the resulting catalysts reveal higher activity. The difference between the samples decomposed up to 570 and 770 K, however, may be ascribed to the enhanced dissociation of the CO ligands at higher temperature, leading to the metal surface desactivation (Ref. 39). The iron oxide matrix formed both on Cab-O-Sil and on Al_2O_3 makes the picture even more complex. I.r. studies on CO adsorption at room temperature (Ref. 10) clearly indicate that the prevailing band resembles that observed on conventional Ru/SiO_2 catalysts only if $Ru_3(CO)_{12}$ is used on Cab-O-Sil and on Al_2O_3. In any other cases the observed bands are shifted to higher wavenumber. It means that beyond the carbon formation the oxide environment also influences the catalytic activity. On the other hand, if decomposition ceased at lower temperature metallic ensembles are stabilized by the CO ligands.

Without decreasing the importance of the mono-ruthenium carbonyl species in the reaction, the selectivity values in CO + H_2 reaction unambiguously point to the presence of bimetallic particles. (When $Ru_3(CO)_{13}$ is replaced by $Fe_2Ru(CO)_{12}$, CH_4 selectivity is considerably shifted for olefin formation.)

We believe that the presence of subcarbonyl species helps the bimetallic formation which is not the case for the supported metal catalysts prepared from inorganic salt solution at very low metal loading. This may be the main advantage of supporting molecular carbonyl clusters.

REFERENCES

1. L. Guczi, Catal. Rev. Sci. Eng., 23, 329 (1981)
2. I. Böszörményi, S. Dobos, L. Guczi, L. Markó, K. Lázár, W.M. Reiff, Z. Schay, L. Takács, A. Vizi-Orosz, Proc. 8th Int. Congress on Catalysis, Vol. 5, p. 183, Verlag Chemie, Weinheim, 1984.
3. A. Choplin, M. Leconte, J.M. Basset, S.G. Shore, W.-l Hsu, J. Mol. Catal., 21, 389 (1983)
4. F. Hugues, J.M. Basset, Y. Ben Taarit, A. Choplin, D. Rojas, J. Am. Chem. Soc., 104, 7020 (1982)
5. A. Theolier, A. Choplin, L.D'Ornelas, J.M. Basset, G. Zanderighi, R. Ugo, R. Psaro, C. Sourisseau, Polyhedron, 2, 119 (1983)
6. R. Psaro, R. Ugo, G.M. Zanderighi, B. Besson, A.K. Smith, J.M. Basset, J. Organomet. Chem., 213, 215 (1981)
7. J.R. Budge, J.P. Scott, B.C. Gates, J.C.S. Chem. Comm., 1983, 342.
8. A. Choplin, A. Theolier, L.D'Ornclas, P. Defour, L. Huang, J.M. Basset, J. Organomet. Chem., submitted.
9. L.I. Bulgakova, V.L. Kuznetsov, Yu.I. Yermakov, L.A. Savonov, React. Kin. Catal. Lett., 22, 259 (1983)
10. I. Böszörményi, S. Dobos, K. Lázár, Z. Schay, L. Guczi, Surf. Sci., 156, 995 (1985)
11. K. Lázár, K. Matusek, J. Mink, S. Dobos, L. Guczi, A. Vizi-Orosz, L. Markó, W.M. Reiff, J. Catal., 87, 163 (1984)
12. I. Böszörményi, L. Guczi, Inorg. Chim. Acta, 112, 5 (1986)
13. L. Guczi, Z. Schay, K. Matusek, I. Bogyay, G. Stefler, Proc. 7th Int. Cong. Cat., Part A., Elsevier Amsterdam, 1981, p. 211.
14. A. Forster, B.F.G. Johnson, J. Lewis, T.-W. Matheson, B.H. Robinson, W.G. Jackson, J. Chem. Soc. Chem. Comm., 1042 (1974)
15. W.L. Gladfelter, G.L. Geoffroy, Inorg. Chem., 19, 2579 (1980)
16. H. Dorn, E. Motell, Inorg. Chim. Acta. Lett., 54, 71 (1981)
17. A.K. Baer, J.A. Connor, N.I. El-Saied, H.A. Skinner, J. Organomet. Chem., 213, 151 (1981)
18. S. Dobos, I. Böszörményi, J. Mink, L. Guczi, Inorg. Chim. Acta, Part 3., in press.
19. S. Dobos, I. Böszörményi, V. Silberer, L. Guczi, J. Mink, Inorg. Chim. Acta, 96, L13 (1985)
20. S. Dobos, I. Böszörményi, J. Mink, L. Guczi, Inorg. Chim. Acta, Part 2, in press.
21. J.R. Budge, B.F. Lücke, J.R. Scott, B.C. Gates, Proc. 8th Int. Cong. Cat., Vol. 5, p. 89, Verlag Chemie, Weinheim, 1984.
22. V.L. Kuznetsov, A. Bell, Y.I. Yermakov, J. Catal., 65, 374 (1980)
23. H. Knözinger, Y. Zhao, B. Tesche, R. Barth, R. Epstein,

B.C. Gates, J.P. Scott, Faraday Discussion, 72, 53 (1981)
24. A. Zecchina, E. Guglielminotti, A. Bossi, M. Camia, J. Catal., 74, 225 (1982)
25. E. Guglielminotti, A. Zecchina, A. Bossi, M. Camia, J. Catal., 74, 240 (1982)
26. E. Guglielminotti, A. Zecchina, A. Bossi, M. Camia, J. Catal., 74, 252 (1982)
27. H. Knözinger, B. Tesche, M. Wolf, J. Mol. Catal., 25, 273 (1984)
28. A. Brenner, J.C.S. Chem. Comm., p. 251 (1979)
29. Z. Schay, K. Lázár, J. Mink, L. Guczi, J. Catal., 87, 179 (1984)
30. L. Guczi, Z. Schay, K. Lázár, A. Vizi, L. Markó, Surf. Sci., 106, 516 (1981)
31. S. Dobos, I. Böszörményi, J. Mink, L. Guczi, to be published.
32. K. Lázár, W.M. Reiff, L. Guczi, J. Catal., submitted for publication.
33. M. Kaminsky, K.J. Yoon, G.L. Geoffroy, M.A. Vannice, J. Catal., 91, 338 (1985)
34. E.O. Odebunmi, B.A. Matrana, A.K. Datye, L.F. Allard, Jr., J. Schwank, W.H. Manague, A. Hayman, J.H. Onuferk H. Knözinger, B.C. Gates, J. Catal., 95, 320 (1985)
35. Z. Schay, L. Guczi, React. Kinet. Catal. Lett., 14, 207 (1980)
36. Z. Schay, L. Guczi, Acta Chim. Hung., 111, 607 (1983)
37. L. Guczi, Z. Schay, I. Bogyay, Preparation of Catalysts III. Elsevier, Amsterdam, 1983, p. 451.
38. K. Lázár, Z. Schay, L. Guczi, J. Mol. Catal., 17, 205 (1982)
39. L. Guczi, J. Mol. Catal., 25, 13 (1984)
40. L. Guczi, I. Böszörményi, S. Dobos, K. Lázár, L. Takács, Z. Schay, Reactivity of Solids, p. 1087, Elsevier/Amsterdam, 1985.
41. Z. Schay, I. Bogyay, L. Guczi, Acta Chim. Hung., 119, 105 (1985)

CATALYTIC FEATURES OF Pd AND Pd/Pt ALLOY CATALYSTS PREPARED FROM ORGANOMETALLIC PRECURSORS

Giovanni Carturan and Renzo Campostrini

Facoltà di Ingegneria, Università di Trento, ITALY

ABSTRACT - The features of Pd and Pd/Pt alloy catalysts obtained from organometallic precursors are reported and discussed. The selectivity of these systems in hydrogenation reactions has been studied in function of catalyst and support properties.

INTRODUCTION

The achievement of heterogeneous catalysts with selectivity features constitutes one of the current challenges attracting researchers from both academic and industrial institutions. The attention devoted to this topic comes from reports of the open literature, involvement of leading research groups, and the interest aroused on occasion of national and international meetings. In this area, it may be considered as a sort of aphorism that industrial experience on specific cases far exceeds academic knowledge.

The aims and working methods of academic organizations pay paramount attention to fundamental aspects, so that academic research furnishes stimulating contributions but leaves prospective exploitation in other hands.

Within these guide-lines, we focused on the selectivity of Pd catalysts in the partial hydrogenation of poly-unsaturated hydrocarbons.

Some facts facilitated relevant researches, such as the exhaustively studied properties of Pd catalysts in hydrogenation reactions |1,2| , the already-made claim that form-factor is involved in determining selectivity |3,4| , and knowledge on Pd reactivity *vs.* hydrogen |5-8|.

GENERAL REMARKS

Selectivity in the partial hydrogenation of poly-unsaturated substrates with supported metallic catalysts is determined by two main factors: A, metal surface chemistry and B, transport phenomena due to physical support features.

A. Chemical Aspects of Hydrogenation Reactions on Metallic Surfaces.

Most hydrogenation kinetics of olefins and alkynes with VIII group metallic catalysts are linearly dependent on H_2 pressure and zerothorder on unsaturated substrate concentration: the reaction is considered a well-established example of a structure-insensitive process |9,10|.

Scheme 1 represents the possible surface chemistry of a two-fold unsaturated substrate hydrogenation.

The metal surface activated hydrogen H_a, the nature of the metallic surface (metal properties and relevant morphology), and the chemical features of S constitute the catalytic ensemble A, manifesting its specific reactivity in the partial hydrogenation of S to SH_2 (reaction 1). This reactivity is different than that of ensemble B due to the hydrogenation of SH_2 to SH_4 (reaction 2).

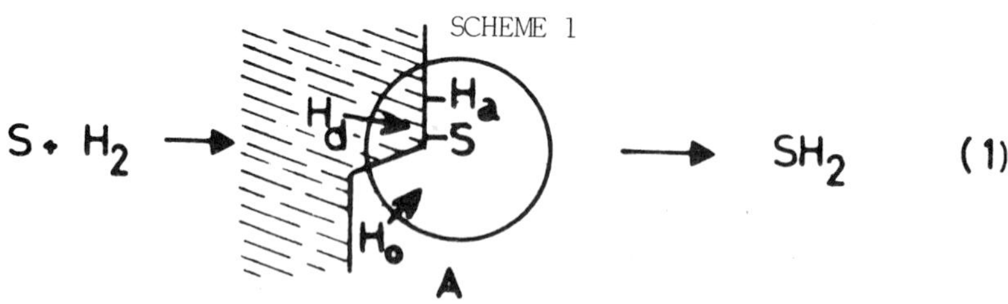

$$SH_2 + H_2 \rightarrow \quad H_d \quad H_a \quad SH_2 \quad H_o \quad B \quad \longrightarrow SH_4 \qquad (2)$$

According to this generalization, in the absence of factors disrupting the original properties of ensemble A, the selectivity of each system may be treated as a physical-chemical constant dictated by the ratio between the thermodynamics of A and B and the energy balance required for their occurrence. Thus, the selectivity of each system may only be lowered by external factors: occluded hydrogen H_o, and dissolved hydrogen H_a, may furnish additional hydrogen to ensemble A, favouring the total hydrogenation of S to SH_4 and ultimately leading to selectivity loss.

For both reactions (1) and (2), H_2 activation is rate-determining so that selectivity, defined as the ratio $|SH_2|/(|SH_2|+|SH_4|)$ at any time during the catalytic run, arises from the increased coordination ability of S <u>vs</u>. SH_2; in other words, from the different strenghts of "metallic surface" - S" and "metallic surface - SH_2" bonds |9-10|. This fact was often invoked to account for the impressive selectivity of some catalysts: for example, the bonding mode of alkynes <u>vs</u>. a metal center is very different than that of olefins |11,12|. More recent studies reject this simple interpretation, rate-determining being ascribed to hydrogen transfer from "metallic surface" - C_xH_y" rather than from "metallic surface - H". Selectivity is therefore determined by the stability of "metallic surface - C_xH_y" species in comparison with that of "metallic surface - C_xH_{y+2}". While it maintains the difference between olefin and alkyne bonding modes, this interpretation also considers the dynamics of the catalytic process: different inter-

mediates are expected owing to the different nature of the original components of the metallic surface species |13,14|.
On the other hand, mechanisms in organometallic chemistry and homogeneous catalysis offer quite a number of examples supporting the occurrence of different intermediates depending on the original metal - substrate interactions |15-17|.
In any case, whatever the intimate mechanism for H_2 activation may be, in order to observe selectivity the overall energy content of intermediate A must favourably compete with ensemble B with preferential occurrence of reaction (1).

With these premises, the original selectivity of the catalytic system is maintained as the contribution of external factors is suppressed; we focused our attention on the effects of dissolved hydrogen H_a and occluded hydrogen H_o on the selctivity of Pd catalysts.

B. Physical Aspects and Support Features.

Metal support interactions |18,19| and other phenomena such as the support-determined stability of metallic dispersions |20, 21| are examples of support involvement in catalyst performance.
Moreover, even simple physical features may affect the selectivity of hydrogenation processes.

Ordinary supports display a multimodal pore distribution function; the ratios between different pore sizes are due to preparation procedures and the physical state of the support.
Whatever the distribution of the supported metallic phase, metallic particles display different degrees of exposure to the reacting substrate, as shown in scheme 2.

SCHEME 2

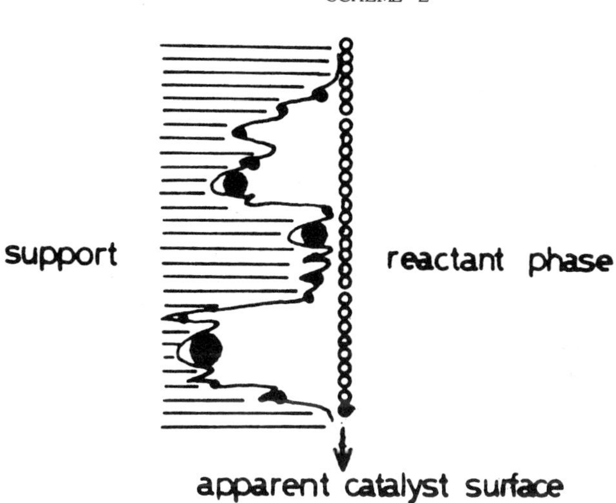

It is evident that, in the case of two-step hydrogenation, diffusion-controlled phenomena inside the pores may affect reactant concentration: this should not affect selectivity since hydrogenation reactions are of zeroth-order in substrate concentrations. However, for very active catalysts, the original poly-unsaturated substrate may be completely half-hydrogenated by the more exposed metallic particles so that the catalyst on the bottom of the pores is only involved in reaction (2). This phenomenon has been studied in detail |22-24|: it appears that real improvement of selectivity may be achieved by reducing the complexity of surface porosity. We tested this possibility using as supports glass beads covered by a thin porous layer of inorganic amorphous materials.

CATALYST PREPARATION AND CHARACTERIZATION

The negative effects of H_o and H_a on the selectivity of ensemble A may be reduced by increasing the dispersion of the metallic phase. Occlusion zones and "cavity" effects are suppressed by lowering the metal particle diameter |3,4|.

Palcewska et al. |6-8| have pointed out that, under hydrogenation conditions, the occurrence of ß-PdH (where the H_d content reaches a ratio of H/Pd=1)is operative for metallic particles ≥ 50 Å in diameter whereas small metallic particles maintain the α-Pd H state, having practically no dissolved hydrogen.

Synthesis of highly dispersed metallic catalysts can be prepared in various ways. In our case, the availability of an amorphus support with a high surface area and the notable presence of surface -OH groups suggested using the Yermakov preparation procedure |27,28|. Our supports were xerogels obtained by hydrolysis of Si, Al and Na alkoxides giving gels which, under controlled heating, furnished the amorphous powders used as supports. OH group content and surface area may be predicted from experimental conditions adopted during preparation |27,28|.

The -OH groups were used for anchoring purely organometallic precursors which may be reduced under mild conditions. The general preparation procedure is presented in scheme 3.

SCHEME 3

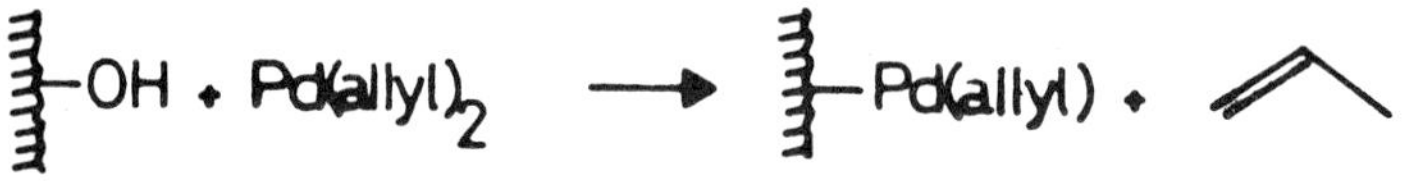

—OH + Pd(allyl)$_2$ ⟶ —Pd(allyl) + propene (3)

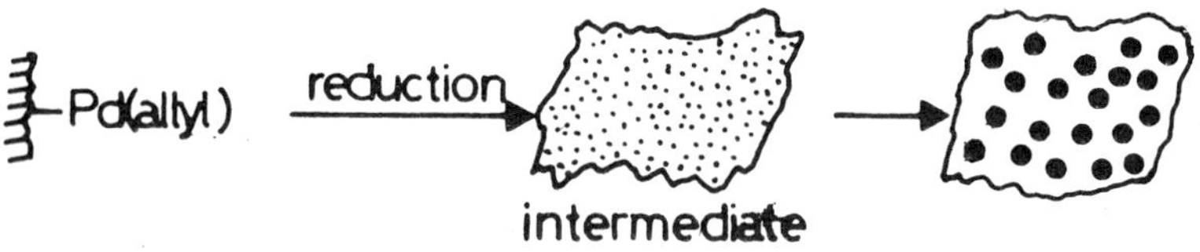

—Pd(allyl) —reduction⟶ intermediate ⟶ (4)

The advantage of this process is that Pd metal particles (equation 4) are generated by reduction of single molecular species, so that particles are built up with single atoms formed as transient intermediates from the support-anchored molecular species.

The small size of the support pores and their amorphous nature suggested that metallic particles are small and uniformly distributed. In the case of supports with extensive surface -OH's, so that the entire $Pd(allyl)_2$ loading remains anchored to the support (reaction 3), the resulting catalysts display a monomodal distribution function of Pd particles, as shown in figure 1 |29|.

On the other hand, as $Pd(allyl)_2$ loading exceeds the -OH group availability, only a fraction of the molecules are bonded to the support and the rest are simply sorbed; in this case, the catalyst is composed of a fraction of small metallic particles equivalent to the amount of $Pd(allyl)_2$ reacted according to (3) and a fraction of bigger particles equivalent to sorbed $Pd(C_3H_5)_2$ |30,31|. Typical distribution of metallic particles is shown in figure 2.

This leads to the conclusion that the ratio between small and large metallic particles at constant Pd loading is determined by the availability of -OH groups on the support. Therefore the type of metallic dispersion may be predicted from the -OH content of the support.

The main features of some prepared catalysts are reported in table 1.

The preparation procedure of scheme 3 may by extended to any metallic catalyst for wich precursors suitable for reaction with surface -OH's and for reduction under mild conditions are available. We used this method to achieve high dispersion of bimetallic alloy catalysts |32|.

For this aim the crucial point is that reduction of different chemical species anchored to the support generates alloy component atoms for the growing metallic particles so as to avoid segregation of single metals or non-uniform metal distribution into the alloy

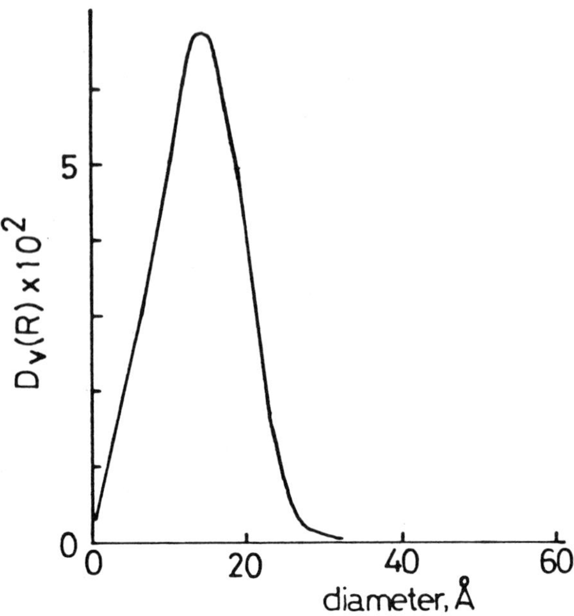

Figure 1. Particle size distribution of 0.67% Pd on amorphous xerogel with high -OH group availability. Determined by SAXS | 29 |.

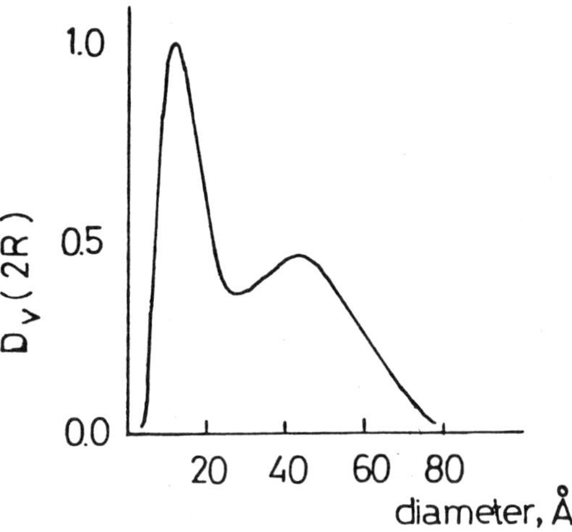

Figure 2. Particle size distribution of 0.3% Pt on amorphous xero gel with low -OH group availability. Determined by SAXS | 28 |.

TABLE 1. - Physical and Chemical Features of Some Pd Catalysts Used for PhC≡CH Hydrogenation.

Sample	-OH [a] (ml H_2, g^{-1})	Surface area [b] (m^2 g^{-1})	Pd%	Diameter [c] (Å)
A (support)	16.2	330	-	-
A_1	-	-	0.67	13
A_2	-	-	1.61	16
A_3	-	-	2.27	19
B (support)	4.9	520	-	-
B_1	-	-	0.85	21
B_2	-	-	1.42	27
C (charcoal)	-	860	-	-
C_1	-	-	1.72	54
C_2	-	-	5.0	106

a. - Determined by reaction support |-OH + LiH → support |-OLi + H_2; data report H_2 evolved in 1 hr.

b. - Determined by Small Angle X-Ray Scattering.

c. - Porod's diameter of Pd particles.

particle.

We managed to prepar real Pd/Pt alloy catalysts with various Pd/Pt rations by using the procedure shown in scheme 4.

SCHEME 4

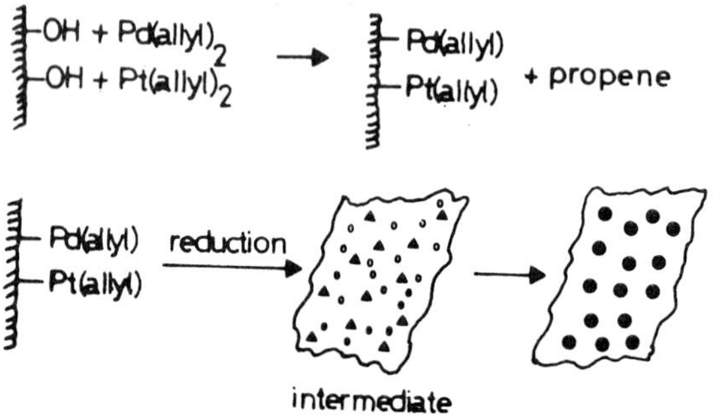

$$-OH + Pd(allyl)_2, \; -OH + Pt(allyl)_2 \rightarrow -Pd(allyl), \; -Pt(allyl) + \text{propene} \quad (5)$$

$$-Pd(allyl), \; -Pt(allyl) \xrightarrow{\text{reduction}} \text{intermediate} \rightarrow \quad (6)$$

The prepared catalysts and their physical and chemical features are reported in table 2 |32|. Real alloy occurrence was demostrated by X-ray analysis in the presence of H_2 and by RDF experiments.

Beside xerogels as support, we also used new materials composed of glass beads coated with a thin porous layer of amorphous inorganic oxides.

The preparation procedure involves dipping the glass beads into solution of alkoxides suitable for producing amorphous oxides by hydrolysis and heating. The new supports displayed the physical and chemical features reported in table 3 |33,34|. Microscopy observations indicated that the layer was 100-500 Å deep. The support surface maintained high porosity affording high specific surface area, but the reduced indentation prevented the diffusion phenomena wich negatively affect catalyst selectivity. On the other hand, surface -OH group availability assured metal deposition according to reactions (3) and (4) or (5) and (6), generating high metal dispersions.

TABLE 2. - Main Parameters for Pd/Pt Catalyst Characterization

Catalyst	Metal content[a] (wt.%)	Pt (at.%)	S_{sp}[b] (m^2 g^{-1}_{metal})
Al 00	1.45	0.0	217
Al 10	1.44	9.9	142
Al 18	1.13	18.4	146
Al 31	1.30	31.1	124
Al 56	1.19	56.2	119
Al 72	1.26	71.8	143
Al 100	1.20	100.0	132
M1[a]	1.42	10.0	199
M2[a]	1.4	20.1	175
M3[a]	1.32	50.0	163
M4[a]	1.25	80.0	132

a. - Mechanical mixture of Al 00 + Al 100 catalysts; relevant data are calculated.

b. - From Small Angle X-Ray Scattering analysis.

TABLE 3. - Physical and Chemical Features of Some Glass/Oxide Coated Supports.

Oxide	Oxide Deposited (%)	Surface area ($m^2\ g^{-1}$) a	Surface area ($m^2\ g^{-1}$) b	Surface OH group content[c]
Al_2O_3	1.20	3.3***	270	4.8
TiO_2	0.44	0.98*	220	5.0
ZrO_2	0.32	0.6**	190	3.7
MgO	0.25	0.6**	240	6.8
SiO_2	0.89	2.2***	250	4.9
Fe_2O_3	0.12	0.097*	80	3.1

a. - Total surface area of coated material determined: * BET (krypton), ** BET(nitrogen), *** Small Angle X-Ray Scattering.

b. - Value calculated from total surface area (column a)x100 divided by % of deposited oxide.

c. - Expressed in ml of H_2 evolved per g of support treated with LiH in tetrahydrofurane suspension.

CATALYTIC ACTIVITY AND SELECTIVITY

The catalyctic activity and selectivity of our Pd and Pd/Pt catalysts was tested in the case of phenyl-acetylene, acetylene and poly-olefin hydrogenations.

A. Effect of Metal Dispersion

Pd catalysts with different degrees of metal dispersion were used in the hydrogenation of phenyl-acetylene to ethylbenzene accordirg to equations (7) and (8):

$$C_6H_5{-}C{\equiv}CH + H_2 \rightarrow C_6H_5{-}CH{=}CH_2 \quad (7)$$

$$C_6H_5{-}CH{=}CH_2 + H_2 \rightarrow C_6H_5{-}C_2H_5 \quad (8)$$

Reactions (7) and (8) are independent of substrate concentration; catalytic activity is expressed as pseudo-zeroth order rate constant, $\underline{k}_{obs}$, ($\underline{M}$ sec^{-1} mg$_{Pd}^{-1}$) measured at constant H_2 pressure and monitoring the progress of the hydrogenation from H_2 and substrate consumptions (see table 4).

X-ray analysis of catalysts before and after catalysis indicate (see table 1) that highly dispersed catalysts preserve their original αPdH structure, being unaffected by the catalytic process |29|. On the contrary, catalysts with large metallic particles are converted to βPdH, having a Pd/H ratio $\simeq$ 1.

In the case of catalysts with a bimodal distribution function of Pd particles, both α PdH and ß PdH were detected. For well-dispersed catalysts, selectivity turns out to be $\simeq$ 100%, so that complete consumption of the acetylenic species (reaction 7) is needed before starting reaction (8). Other catalysts involved with ß PdH occurrence displayed lower selectivities.

These results agree with the expectation that H_d and H_o negative effects would be avoided by using high dispersion of the metallic phase.

TABLE 4. – Hydrogenation of $PhC{\equiv}CH$[a].

Catalyst	Physical state of Pd		$10^4 \cdot k_{obs}$ reaction(7)	$10^4 . k_{obs}$ reaction(8)	Selectivity[b]
	Before catalysis	After catalysis	($M\ sec^{-1}\ mg_{Pd}^{-1}$)	($M\ sec^{-1}\ mg_{Pd}^{-1}$)	($\pm$ 2%)
A_1	α	α	3.99	5.14	100
A_2	α	α	4.02	5.01	100
A_3	α	$\alpha+\beta$ (trans?)	3.84	4.95	100
B_1	α	$\alpha+\beta$	2.05	2.19	94
B_2	α	$\alpha+\beta$	1.91	2.00	91
C_1	α	β	1.55	–	78
C_2	α	β	1.25	–	71

a. – Rate constants per mg of Pd, t = 25°C, $|PhC{\equiv}CH|$ = 0,26 M, solvent n-heptane 10 ml, atmospheric H_2 pressure.

b. – Selectivity = 100 x % $PhCH{=}CH_2$ / (% $PhCH{=}CH_2$ + % $PhCH_2{-}CH_3$), % $PhCH{=}CH_2$ kept at the maximum value during each kinetic run.

The catalytic activity data deserve some comments, particularly as regards catalysts with the highest degree of dispersion. We found that catalytic activity for the same catalyst was lower in the hydrogenation of acetylenic species than for olefins. On the other hand, the rate law is identical for both reactions (7) and (8). Since H_2 activation is rate-determining, the observed difference of catalytic activity between hydrogenations of olefinic and acetylenic species implies more difficult H_2 activation for the latter. This agrees with the results of Somorjai showing that hydrogen trasfer from a dissociatively adsorbed substrate is more difficult as unsaturation increases |35,36|. On the other hand, this indicates that unsaturated substrate coordination to the metallic surface precedes hydrogen activation and that the metal - alkyne bond is stronger than the metal - olefin one.

The resulting mechanism can be presented as shown in scheme 5,

SCHEME 5

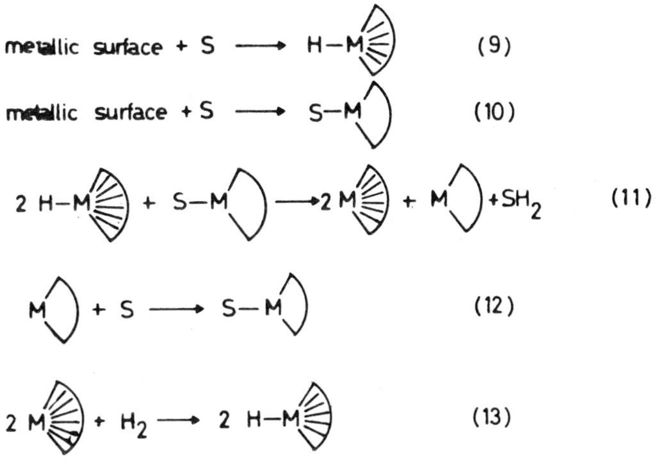

metallic surface + S ⟶ H–M (9)

metallic surface + S ⟶ S–M (10)

2 H–M + S–M ⟶ 2 M + M + SH_2 (11)

M + S ⟶ S–M (12)

2 M + H_2 ⟶ 2 H–M (13)

where both (11) and (13) may be rate-determing and display different rates as the coordination ensemble around the metallic surface is composed of olefin or acetylenic species. Selectivity originates from the favourable competition of the acetylenic substrate with respect of the olefin during reactions (9) and (10) or (12).

For catalysts composed of large metallic particles, H_o and H_d contributions become important; selectivity is lowered and rates for reactions (7) and (8) become equivalent. This means that the catalytic ensembles involved in reactions (11,13) are no longer able to discriminate effectively between olefinic or acetylenic compounds. Using the same arguments invoked to account for the selectivity of highly dispersed Pd catalysts, the catalytic ensembles on the metallic surface when engaged in H_o and H_d participation becomes equivalent whether the cocordination sphere is composed of either olefinic or acetylenic species.

The reactions of scheme 5 parallel analogous reactions in the homogeneous phase. In particular, the hydrogenation process on Pd catalysts with low metallic dispersion resembles the homogeneous phase hydrogenation in complexes capable of expanding their coordination spheres, ultimately having two hydrides and one π-bonded olefin |37|.

B. Effect of Surface Chemical Composition

Pd/Pt alloy catalysts obtained from reactions (5) and (6) of scheme 4 display definite catalytic activity in hydrogenation reactions (7) and (8). Catalysts are composed of metal particles 25 - 45 Å in diameter, so that the Pd/Pt surface composition may be assumed to be identical to nominal bulk composition |38,39|. Selectivity for phenylacetylene hydrogenation to ethylbenzene is $\geq$ 95% for all catalysts, in agreement with the good dispersion of the metallic phase. Moreover, for Pd/Pt ratios $\leq$ 80/20, hydrogen dissolution is completely suppressed |32,40|; this absence of H_d improves the selectivity of the system.

Catalytic activity is affected by the Pd/Pt ratio, as reported in figures 3 and 4 which shows the behaviours of specific catalytic activity *vs*. Pt content for reactions (7) and (8) respectively |41|. The results of figures 3 and 4 confirm the already discussed different reactivity between catalytic ensembles composed of olefins and acetylenic species; the effect of changing the Pd/Pt ratio exalts this difference. The identical kinetic behaviour of pure Pd, pure Pt and Pd/Pt catalysts for both reactions (7) and (8) suggests that the reaction mechanism of scheme 5 is valid for all these systems. This implies that the observed non-linear dependence of catalytic activity on Pd/Pt may be related to the effect of surface chemical composition on steps determining hydrogenation rate, *i.e.* hydrogen transfer (reaction 11) and hydrogen activation (reaction 13).

C. Effect of Support Porosity

Pd metal catalysts on supports composed of glass beads covered with a porous microlayer of amorphous oxide were used in the selective hydrogenation of fats, acetylene and phenylacetylene |34,42|.

The achievement of a porous surface with a low degree of identation effectively improves catalyst selectivity. Table 5 shows the catalytic parameters for the hydrogenation of C_{18} fats with Pd on amorphous TiO_2 (catalyst A) and on glass beads covered with a microlayer of amorphous TiO_2 (catalyst B).

The resulting catalytic behaviours indicate greater selectivity for catalyst B, with selectivity = 98% for the hydrogenation of metyl oleate to stearate. This improvement means that the approach we used to suppress the negative effects of inside pore diffusion was

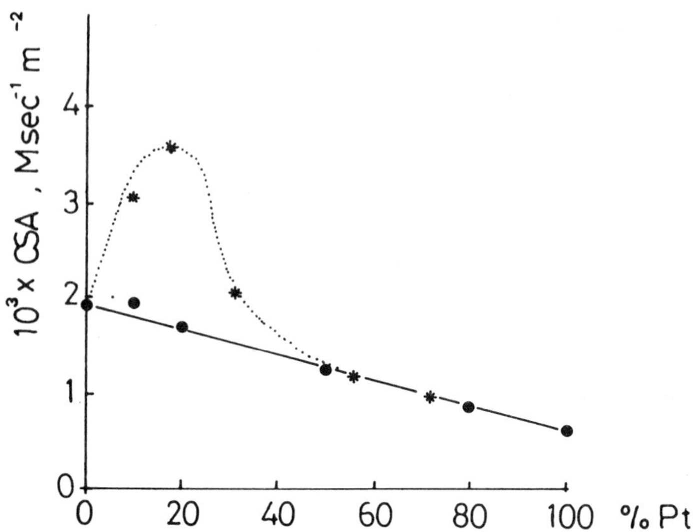

Figure 3. Specific catalytic activity (CSA) of Pd/Pt catalysts on amorphous xerogel as a function of %Pt. PhC=CH hydrogenation to $PhCH=CH_2$ alloy catalysts mechanical mixtures. |41| .

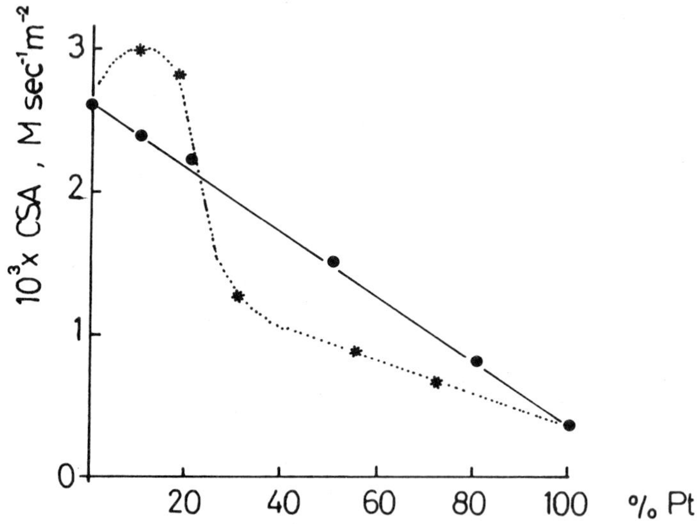

Figure 4. Specific catalytic activity (CSA) of Pd/Pt catalysts on amorphous xerogel as a function of % Pt. $PdCH=CH_2$ hydrogenation to $PhCH_2$-CH_3. alloy catalysts; mechanical mixtures |41| .

TABLE 5. - Hydrogenation of Fat Methyl Esters[a]

Catalyst	$10^5.k_1$ [bc]	$10^5.k_2$ [bc]	k_1/k_2	Selectivity|1|[d]	Selectivity|2|[e]
C	2.0	0.12	16.7	70	93
	(2.1)	(0.12)		(3)	(6)
D	5.0	0.25	21	73	98
	(5.1)	(0.27)		(0.6)	(2)

a. Solvent = THF (10-20 ml), |substrate| = 0.1-0.3 M, catalyst = 200-500 mg, t = 25°C, pressure (P_{H_2} + solvent vapour pressure) = 760 ± 8 mm Hg.

b. k_1 and k_2 refer to pseudo - zero order rate constants, M s^{-1} mg_{Pd}^{-1} for hydrogenations of methyl linolate |1| and metyl oleate |2| respectively.

c. Values in brackets calculated from H_2 volumetric comsumption.

d. Selectivity |1| = 100 x |% linolate/(% linolate + % oleate + % stearate)| kept at maximum values for each kinetic run. Values in brackets give relevant % stearate.

e. Selectivity |2| = 100 x |% oleate / (%oleate + % stearate | kept at maximum value during each kinetic run. Values in brackets give relevant % stearate.

successful.

CONCLUSIONS

The selectivity of Pd catalysts in the partial hydrogenation of poly-unsaturated hydrocarbons may be improved by increasing the degree of metal dispersion and lowering the macroporosity of the support.

The complexity of surface reactions does not allow definite comparisons with specific reactions of homogeneous catalysis. On the other hand, metallic surface intermediates are affected by the same parameters wich determine the thermodynamic stability and reactivity of organometallic species. Thus, discussion of heterogeneous catalysis results may involve concepts of homogeneous phase process. This approach was used when dealing with the concept of catalytic ensembles for wich some of the features proper to the transition metal ions may be usefully applied.

REFERENCES

1. - C.C.Bond, "Catalysis by Metals", Academic Press, New York (1962).
2. - P.N.Rylander, "Catalytic Hydrogenation over Pt Metals", Academic Press, New York (1967).
3. - A.G.Burden, J.Grant, J.Martos, R.B.Moyes and P.B. Wells, Faraday Disc. Chem. Soc. 72, 109 (1981).
4. - P.B.Wells, J. Catal. 52, 498 (1978).
5. - F.A.Lewis, "The Palladium-Hydrogen System", Academic Press, New York (1967).
6. - W.Palaczewska, Adv. Catal. 24, 245 (1975).
7. - A.Borodzinski, R.Dus, R.Frank, A.Janko and W.Palczewsca, Proc. Int. Conf. Catal. 5th, A7, London (1976).
8. - A.Janko, W.Palczewska and I.Szmerska, J. Catal. 61, 264 (1980).
9. - G.C.Bond and P.B.Wells, Adv. Catal. 15, 91 (1964).
10. - E.G.Allison and C.G.Bond, Catal. Rev. 7, 233 (1973).
11. - J.P.Collman and J.W.Krang, J. Am. Chem. Soc. 89, 844 (1967).
12. - S.Otsuka and A.Nakamura, Adv. Organomet. Chem. 14, 245 (1976).
13. - S.J.Thomson and G. Webb, J. S. C. Chem. Comm., 526 (1976).
14. - F.Zaera and G.A.Somorjai, J. Am. Chem. Soc. 106, 2288 (1984).
15. - "Advances in Organometallic Chemistry" 17, Catalysis and Organic Syntheses, F.G.A.Stone and R.West Eds Academic Press 1979.
16. - R.F.Heck, "Organotransition Metal Chemistry" Academic Press, New York (1974).
17. - Y.K.Kochi, "Organometallic Mechanisms and Catalysis", Academic Press, New York 1978.
18. - S.J.Tauster, S.C.Fung and R.L.Garten, J. Am. Chem. Soc. 100, 170 (1978).
19. - "Metal-Support and Metal-Additive Effects in Catalysis", B.Imelik et al. Eds;Elsevier,Amsterdam (1982).
20. - E.Ruckenstein and B. Pulvermacher, J.Catal. 29, 224 (1973).
21. - P.Ehrburger and P.L.Walker,Jr., J.Catal. 55, 63 (1978).
22. - G.J.Acres and B.J.Cooper, Appl.Chem.Biotechnol. 22,769 (1972).
23. - G.W.Roberts, "Catalysis in Organic Syntheses", Academic Press, New York (1976).
24. - A.J.Bird and D.T.Thompson, "Catalysis in Organic Syntheses", p. 61, Academic Press, New York (1980).
25. - Yu.I.Yermakov, Catal.Rev.13, 77 (1976).
26. - Yu.I.Yermakov, B.N.Kuznetsov and V.A.Zakharov, "Catalysis by Supported Complexes", Elsevier,Amsterdam (1981).
27. - G.Carturan and V.Gottardi, J.Mol.Ctal.4, 349 (1979).
28. - G.Carturan, G.Cocco, L. Schiffini and G.Strukul, J.Catal.65, 359 (1980).

29. - G.Carturan,G.Facchin,G.Cocco,S.Enzo and G.Navazio, J.Catal.76, 405 (1982).
30. - G.Cocco,L. Schiffini,G.Strukul and G.Carturan, J.Catal.65, 348 (1980).
31. - G.Cocco,G.Fagherazzi,G.Carturan and V.Gottardi, J.C.S. Chem. Comm., 979 (1978).
32. - G.Cocco, G.Carturan,S.Enzo and L.Schiffini, J.Catal.85, 405 (1984).
33. - G.Carturan,G.Facchin,G.Navazio,V.Gottardi and G.Cocco, "Ultra structure Processing of Ceramics, Glasses and Composites", Chapter 16, J.Wiley and Sons, NewYork (1984).
34. - G.Carturan,G.Facchin, V.Gottardi and G.Navazio, J.Non-Crystalline Solids 63, 273 (1984).
35. - S.M.Davis, F.Zaera,B.F.Gordon, and G.A.Somorjai, J.Catal.92, 240 (1985).
36. - S.M.Davis,F.Zaera and G.A.Somorjai, J.Catal.77, 439 (1982).
37. - B.R.James, "Homogeneous Hydrogenation", Chapter II, Wiley-Interscience, New York (1973).
38. - F.L.Williams and D.Nason, Surf Sci.45, 377 (1974).
39. - Z.Karpinski and T.Koscielski, J.Catal.56, 430 (1979) and J.Catal.63, 313 (1980).
40. - I.T.Caga and J.M.Winterbottom, J.Catal.57, 494 (1979).
41. - G.Carturan,G.Cocco,G.Facchin and G.Navazio, J.Mol.Catal.26, 375 (1984).
42. - G.Carturan,G.Facchin and G.Navazio, La Chimica e l'Industria 65, 688 (1983).

METAL ALKYL COMPOUNDS IN PREPARATION AND MODIFICATION OF SUPPORTED MONO AND BIMETALLIC CATALYSTS

Jozsef L.Margitfalvi

Central Research Institute for Chemistry of the Hungarian Academy of Sciences, Budapest, 1525 POB 17, Hungary

ABSTRACT - The use of metal alkyl compounds of transition metals to prepare supported mono and bimetallic catalysts is reviewed. The preparation of highly dispersed monometallic platinum and different types of bimetallic catalysts with direct metal-metal interactions is based on different types of Controlled Surface Reactions. Examples are given on the surface chemistry involved in preparation of these catalysts and an emphasis is placed on the modes and ways of control of the selectivity of these surface reactions. Evidences obtained in n-hexane conversion and in selective hydrogenation of acrylonitrile have clearly demonstrated that the activity and selectivity of catalysts prepared by this method strongly depends on the type of Controlled Surface Reactions (CSRs) used during their preparation.

1. INTRODUCTION

One of the major challenges in catalysis science is the application of scientific methods in the preparation of catalysts with the required catalytic properties. We have been witnessing continuous changes in the contemplation of various problems connected with the preparation of heterogeneous catalysts. The idea of molecular level of understanding, mentioned first by surface scientists and people working in the field of homogeneous catalysis, has been attracting more and more followers indluding those who are applying scientific methods in the catalyst preparation.

There are different approaches in catalyst preparation. Our scientific interest is focused on the modes and ways of control of surface reactions responsible for the exclusive formation of a given, catalyticaly active, surface species and the methods of surface organometallic chemistry is used for the preparation of different types of supported mono and bimetallic catalysts.

Recently the preparation of supported metal and metal -complex catalysts using organometallic compounds has obtained great scientific interest. Three approaches have become especially of great importance: (i) the use of allylic complexes of different metals (Ref. 1), (ii) the application of carbonyl clusters (Ref. 2) and (iii) anchoring metal halides into lithiated supports (Ref. 3). The supported or anchored organometallic compounds can be used directly as heterogenized metal complex catalysts or can be decomposed with the aim to prepare highly dispersed supported mono or bimetallic catalysts.

However, there are only scarce data in the literature on the use of metal alkyl complexes for the preparation of supported mono and bimetallic catalysts.

This contribution can be considered as an attempt to summarize results obtained in our laboratories upon using $[(CH_3)_3PtX]_4$ type complexes for the preparation of alumina supported platinum catalysts and tin or lead alklys for the modification of alumina supported Pt and Ni catalysts in order to obtain surface species with direct metal-metal interaction.

2. BASIC PRINCIPLES OF OUR APPROACH

With respect to the preparation of supported monometallic catalysts the reactivity of the surface hydroxyl groups is often used to control the direction of surface reactions with the formation of well dispersed ionic surface species of known structure (Ref. 1), which upon subsequent reduction may result in metallic particles with high dispersion. This approach has been followed in our preparation technique too, i.e. in the use of $[(CH_3)_3PtX]_4$ complexes to prepare alumina supported platinum catalysts.

However, in the preparation of supported bimetallic catalysts the control of surface reactions by using surface OH involved reactions will not ensure the exclusive formation of bimetallic surface species.

In the course of the preparation of supported bimetallic catalysts by conventional methods the formation of bimetallic surface species takes place during the high temperature decomposition of the precursor compounds adsorbed on inorganic supports. Not much is known about these high temperaure surface reactions involved in the preparation of supported bimetallic catalysts via conventional preparation techniques and these reactions can not be considered to be controlled.

In our approach a supported monometallic catalyst is used to prepare bimetallic catalysts and hydrogen preadsorbed on the metals provides the driving force for the primary interaction, i.e. for the exclusive formation of bimetallic surface entities with direct metal-metal interaction.

Upon using another type of controlled surface reaction, based on the reactivity of the support OH group, the second metal can be introduced exclusively into the support in its ionic form.

In this way upon starting from a monometallic catalyst, applying different types of CSRs we can control not only the forms (metallic or ionic) but the environment (metal or support) of the second metal introduced.

3. PREPARATION OF Pt/Al_2O_3 CATALYSTS FROM $[(CH_3)_3PtX]_4$ COMPLEXES

3.1. General aspects

It is known that alkyl platinum halides (APH) with general formula of $[(CH_3)_3PtX]_4$ (where X = Cl, Br, I) are air stable well characterized complexes (Ref. 4). However, there are only scarce data on their thermal stability and decomposition.

In our recent studies it has been demonstrated that the thermal stability of these complexes increases in the series I < Br < Cl (Ref. 5). The main decomposition products observed in pyrolysis and DTG experiments (Refs. 5,6) were methane, ethane, ethylene, C_3

hydrocarbons, methyl halides and hydrogen halides. It has been observed that in the presence of hydrogen or in the presence of surface hydroxyls or traces of water the selectivity for methane formation increases. These experimental data initiated us to study the decomposition of silica and alumina supported APH complexes with the aim to prepare highly dispersed Pt/Al_2O_3 catalysts.

3.2. Thermal decomposition of silica and alumina supported APH complexes

The first attempt to use APH complexes was done in the late seventies (Refs. 7,8). Upon studying the decomposition of silica supported and unsupported $[(CH_3)PtCl]_4$ complexes in vacuum we have clearly demonstrated that the proton of the surface hydroxyl group is directly involved in the decomposition reaction. We have observed that the temperature of decomposition of the supported complex has decreased from 245^oC to 185^oC. This fact, as well as the shift of the product distribution to methane formation, pointed towards strong interaction between Si-OH and $Pt-CH_3$ fragments.

The decomposition of the alumina supported APH complexes has been investigated in more details (Refs. 9-11). In these works the decomposition of the Al_2O_3 supported APH complexes has been investigated. The studied experimental variables are (i) the type of the APH complexes, (ii) the temperature of dehydroxylation of the alumina, and (iii) the atmosphere used during the decomposition.

In these studies Thermal Programmed Reaction (TPR) technique was used to follow the decomposition. Typical TPR curves obtained during thermal decomposition of different APH complexes supported on alumina are given in Figure 1. The decomposition of the cloride complex in nitrogen atmosphere is shown in Figure 1/A using alumina dehydroxylated at 300 and 500^oC, respectively. The decomposition was not selective. The main reaction product was methane with parallel formation of C_2 and C_3 hydrocarbons and methyl halides (not shown). The ratio of methane to higher hydrocarbons (C_2+C_3) was significantly influenced by the temperature of dehydroxylation of

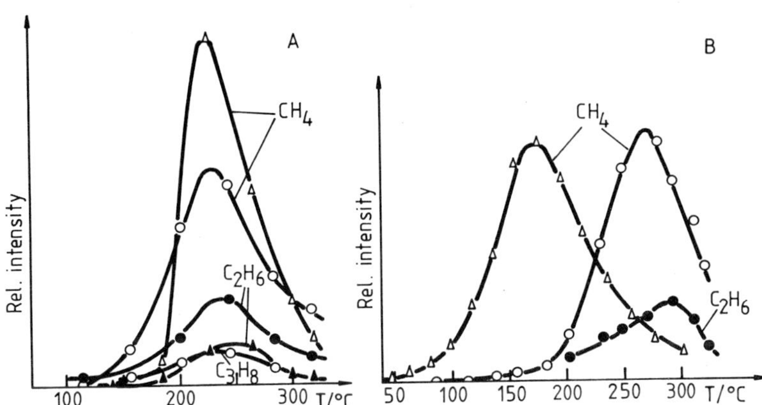

Figure 1. Temperature programmed decomposition of $[(CH_3)_3PtX]_4/Al_2O_3$. A. The influence of the temperature of dehydroxylation: o, Δ-300°C; ●, ▲ -500°C; X = Cl; decomposition in N_2. B. The influence of the atmosphere of decomposition: o, ● - in N_2; Δ - in H_2; temperature of dehydroxylation: 500°; X = Br.

the support. After thermal treatment at 300°C the concentration of surface hydroxyls is relatively high, i.e., conditions are favourable for an intermolecular reaction between the surface OH group and the methyl group of the APH complex. Contrary to that on highly dehydroxylated alumina the routes of decomposition via intermolecular reactions, i.e. with the formation of ethane, ethylene and C_3 hydrocarbons, have become of importance.

In hydrogen atmosphere, the formation of higher hydrocarbons and methyl halides was almost entirely suppressed and the decomposition of these complexes took place at much lower temperatures, i.e. the maxima of the methane peaks were shifted from 260-270°C to 160-170°C as shown in Figure 1/B. Only trace of hydrogen halide could be detected, probably, due to its adsorption on the alumina.

Both in nitrogen and hydrogen atmosphere the position of the maxima of the methane peak was independent of the temperature of dehydroxylation, however, the initial temperature of decomposition of the APH complexes was always lower if the alumina was dehydroxylated at 500°C.

These data clearly demonstrated that the reactivity of the "free" hydroxyl groups formed at 500°C is higher than that of the "hydrogen bridged" hydroxyls which prevail after dehydroxylation at 300°C

From results presented above it can be concluded that in the decomposition of the alumina supported APH complexes hydrogen and proton induced reactions are preferentialy involved, and intermolecular reactions, i.e. formation of ethane and methyl halide, are sufficiently suppressed.

4. PREPARATION OF SUPPORTED BIMETALLIC CATALYSTS USING TIN AND LEAD ALKYLS

4.1. Surface chemistry

We have recently observed that tin and lead alkyls with general formula of $ER_xCl_{(4-x)}$ (where E=Sn or Pb, and R=$-CH_3$, $-C_2H_5$, $-C_3H_7$ and $-C_4H_9$) react with hydrogen adsorbed on nickel or platinum surfaces with the selective formation of corresponding alkane (Ref. 12-17). The reaction with lead or tin tetraethyl can be written as follows:

$$xPtH + E(C_2H_5)_4 \longrightarrow Pt\text{-}E(C_2H_5)_{(4-x)}\ (I) + x\ C_2H_6 \qquad (1)$$

In the absence of preadsorbed hydrogen catalytic decomposition of the alkyl compounds was observed on these metals with the parallel formation of ethane and ethylene.

Special care was taken to avoid additional side reactions between the alkyl compounds and the surface OH group of the support. In order to avoid this undesired side reaction the concentration of the metal alkyl compounds was low in their solution. Reaction (1) was monitored by gas volumetry followed by GC analysis of the gases evolved (Refs. 12, 16).

Surface complex (I) formed in reaction (1) was decomposed in hydrogen or inert gas atmospheres. Temperature Programmed Reaction Technique (TPR) was used to follow the decomposition. In hydrogen atmosphere this reaction can be written as follows:

$$(I) \xrightarrow{H_2} Pt\text{-}E + (4\text{-}x)\ C_2H_6 \qquad (2)$$

In reactions (1) and (2) the selective formation of ethane was the indication for the control of surface reactions used.

4.2. Reactivity of hydrogen chemisorbed on alumina supported platinum

In these series of experiments a Pt/Al_2O_3 catalyst with Pt content 0.5 % and platinum dispersion around 10 % was used (Ref. 12). Reaction (1) was carried out at 35 and 50^0C. In the presence of adsorbed hydrogen and if the initial concentration of the tin tetraethyl was not higher than 0.0025 mol/dM^3 selective ethane formation was observed (Ref. 12,16). Ethylene formation was observed only in the absence of adsorbed hydrogen or at reaction temperatures and concentrations higher than mentioned above. Ethylene formation was an indirect indication to the loss of control in reaction (1).

On pure alumina, under identical experimental conditions the formation of ethane and ethylene was observed only in trace amounts and the amount of the adsorbed tin alkyl was also neglegible. Considerable adsorption of tin tetraalkyls onto the support,i.e.the creation of tin- alumina surface species, was observed upon increasing (i) the concentration of the tin alkyl solutions and (ii) the time of reaction (1) (Refs. 15,16).

The activity of different types of tin alkyls as methyl, ethyl and buthyl with general formula of $SnR_{(4-x)}Cl_x$ were studied under identical experimental condition. The activity of the tin alkyls increased in the following order: $Sn(C_4H_9)_4 < Sn(C_2H_5)_4 < Sn(CH_3)_4$ (Refs. 14-16). $Sn(C_2H_5)_2Cl_2$ appeared to be completely inactive in reaction (1). However, the adsorption of this compound onto pure or cloride containing alumina was much more pronounced than that of the $Sn(C_2H_5)_4$ (Ref. 17).

The decomposition of the primary formed surface complex (I) was studied in a series of experiment using tin tetraethyl. In these

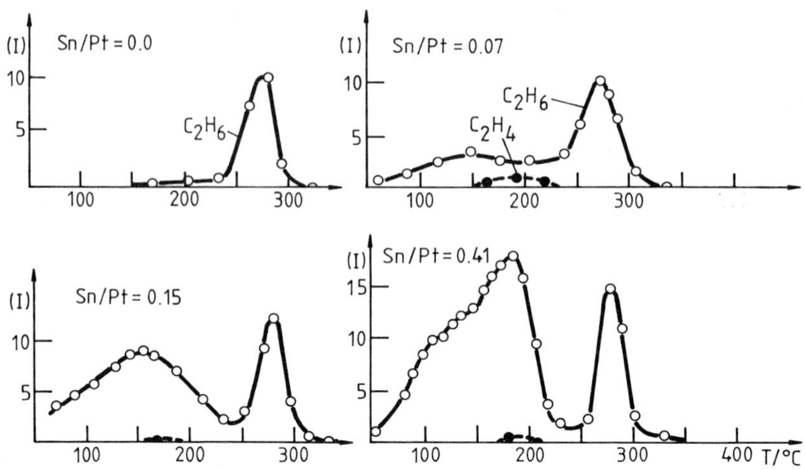

Figure 2. Preparation of Sn-Pt/Al_2O_3 catalysts (controlled reaction condition, Sn-Pt interaction). Decomposition of the primary formed surface complex (I) with different Sn/Pt ratios by TPR. Tin precursor in reaction (1): $[Sn(C_2H_5)_4]_0$ = $2.5.10^{-3}$ mol.dM^{-3}

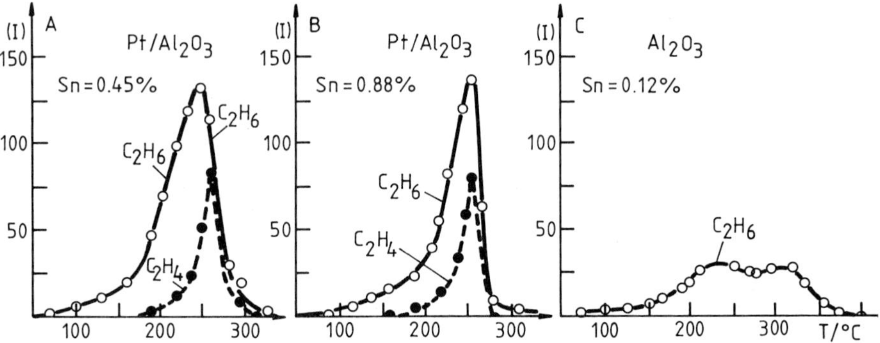

Figure 3. Preparation of Sn-Pt/Al_2O_3 catalysts (loss of control, Sn-Al_2O_3 interaction). Decomposition of the primary formed surface species by TPR. A and B. Pt/Al_2O_3, C-pure Al_2O_3. Tin precursor: A and C: $[Sn(C_2H_4)_4]_0$ = 5×10^{-3}mol.dM^{-3}, B: $[SnCl_2(C_2H_5)_2]_0$=$5.0.10^{-3}$ mol.dM^{-3}

series of experiments samples with different Sn/Pt ratios were prepared creating conditions favourable for both tin-platinum and tin-alumina intractions (Ref. 12).

Typical TPR curves are shown in Figure 2. At low Sn/Pt ratios (i.e. under conditions favourable for tin-platinum interactions) ethane was the main reaction product, and formation of ethylene and methane were observed only in trace amounts. From a blank experiment it could be established that both the high temperature ethane peak around 240-260^{0}C and the small amount of methane evolved in the temperature interval of 200-300^{0}C could be attributed to the hydrogenolysis of the trace of the solvent (usually pentane) used in the final step during the preparation procedures. It should be added that TPR peaks shown in Figure 2. were not observed on pure alumina (see Figure 3). Good agreement was obtained between the integrated peak areas and the tin content of the bimetallic catalysts. No loss of tin was observed in TPR experiments. The calculated values of x in reactions (1) and (2) were 1.0 ± 0.12. From these experimental data it has been concluded that reactions involved in the formation of bimetallic Sn-Pt surface species can be really controlled.

Parallel to the creation of controlled reaction condition it was also interesting to study reaction conditions in which the control was lost partly or entirely.

Two samples of the base Pt/Al_2O_3 catalyst were treated with excess $Sn(C_2H_5)_2Cl_2$ and $Sn(C_2H_5)_4$. The TPR curves of these samples given in Figure 3. are completely different from those shown in Figure 2.. Due to the increased amount of the tin tetraalkyl or to inactivity of the $Sn(C_2H_5)_2Cl_2$ in reaction (1), the conditions for controlled surface reactions has been lost, i.e. the main part of the tin alkyls has adsorbed on the support and has reacted with the surface hydroxyl groups only at high temperatures. This side reaction resulted in a significant change in the forms of the TPR curves. The position of the peak maxima was shifted to high temperatures close to those obtained on pure alumina. This observed shift of the ethane peaks as well as the parallel formation of ethylene can be considered as indirect evidences for the tin-alumina interaction, while the small shoulder bellow 180^{0}C can be related to the platinum-tin interaction.

The differences in the position of the TPR peaks shown on Figures 2. and 3. can be attributed to the differences in the amount and type of the hydrogen source consumed in the decomposition of the primary surface species formed either on the platinum or on the alumina sites. We suggest that under conditions of our TPR experiments hydrogen can adsorb on the platinum sites which are not covered by $SnR_{(4-x)}$ moieties. This process can take place at relatively low temperatures. This hydrogen on the platinum sites increases the probability of the decomposition of the $SnR_{(4-x)}$ moieties resulting in (i) exclusive formation of ethane and (ii) the low temperature peak on the TPR curves.

Contrary to that, this type of hydrogen is not available at the alumina sites. Due to the absence of atomic hydrogen on the alumina sites either surface OH involved reactions or the thermal decomposition of the $Sn(R)_{(4-x)}$ moieties can take place resulting in (i) the observed shift of the maxima of the TPR curves into the high temperature range and (ii) the appearence of the ethylene in the reaction products.

4.3. Reactivity of hydrogen adsorbed on alumina supported nickel

In these experiments a semiindustrial Ni/Al_2O_3 catalyst with 40% Ni content was used (Refs. 13 14). The high nickel content resulted in a large amount of hydrogen adsorbed. This fact gave us additional possibilities to study the reaction taking place between the adsorbed hydrogen and lead or tin alkyls. In this section results, obtained by using lead tetraethyl will be discussed (Refs. 13-16). Upon studying the desorption of hydrogen from this catalyst by TPD technique (Ref. 18) we have observed that the amount of hydrogen desorbed after reduction at 150°C strongly depended on the procedure used when cooling the catalyst to toom temperature. These results initiated us to study the impact of the cooling procedures on the rate and selectivity of the reaction (1).

Three different cooling procedures were applied (Ref. 14):

(1) cooling in hydrogen up to 25°C;

(2) Cooling in hydrogen up to 25°C, and purging with N_2

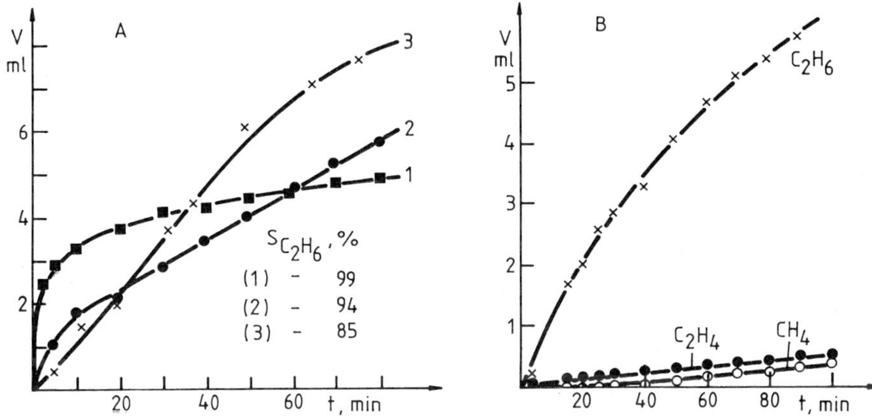

Figure 4. Reactivity of hydrogen adsorbed on Ni/Al_2O_3. Catalysts reduced in hydrogen at 150^0C. A. The influence of the cooling procedures; (1) -cooling in hydrogen up to 25^0C; (2) - as (1) followed by purging with N_2; (3) - purging in nitrogen at 150^0C, followed by colling in N_2 up to 25^0C. Lead precursor: $[Pb(C_2H_5)_4]_0$: 0.014 mol.dM^{-3}, reaction temperature: 27^0C. B. Kinetic curves of product formation: cooling procedure (2).

(3) purging in nitrogen at 150^0C, followed by cooling in nitrogen up to 25^0C.

Typical kinetic curves of the total gas formation are shown in Figure 4A. (Ref. 14). The selectivity of ethane formation after 30 minutes were 99, 94 and 85 %, on catalysts (1), (2) and (3), respectively. Good correlation was found between the ethane selectivity values and the amount of hydrogen desorbed from these catalysts after corresponding cooling procedures. The highest ethane selectivity was observed on catalyst (1) which resulted in the highest value of the desorbed hydrogen. It can be seen that any changes in the overall hydrogen pool result in a strong influence on the initial rates and product selectivities.

The characteristic feature of the Ni/Al_2O_3 catalyst is that parallel to the formation of ethane in reaction (1) ethylene and methane were also detected in the reaction prducts (Refs. 13-16). The formation of ethylene can be attributed to the catalytic decomposition of the lead tetraethyl, while the appearence of methane may indicate that the nickel itself or the Ni-H surface species can hydrogenolyze the C-C bond in the $Ni-Pb(C_2H_5)_{(4-x)}$ moieties. The rate of methane and ethylene formation strongly depended on the experimental conditions used, i.e. on the temperature of the reaction and on the concentration of the lead alkyl used. The rates of formation of these by products was relatively low at room temperature and at low concentration of lead alkyl as shown in Figure 4B. Details on the nature of the side reactions are given and discussed elsewhere (Ref. 15).

In this system the formation of ethylene and methane can be considered as indirect evidences for the loss of control of surface reactions used to prepare bimetallic surface species with direct metal-metal interaction. From these studies it can be concluded that special care should be taken to find the optimal reaction condition for the selective introduction of the second metal by using reaction (1). More details in this respect can be found elsewhere (Refs. 14-17).

5. CATALYTIC PROPERTIES OF THE PREPARED MONO AND BIMETALLIC CATALYSTS

5.1. Catalytic activity of Pt/Al_2O_3 catalysts prepared from APH complexes.

Catalysts prepared as described in Section 3.2. were tested in a continuous flow reactor at 500^oC (Ref. 9). Catalysts prepared from the iodine complex appeared to be quite inactive, probably, due to the poisoning effect of the iodine (Refs. 9-11). The dispersion, (i.e. the H/Pt ratios) the activity and selectivity of these catalysts measured in n-hexane conversion were compared with that of a catalyst prepared from H_2PtCl_6 These data are given in Table 1.

In these studies no correlation was found between the platinum dispersion and the activity and selectivity of these catalysts. The dispersion, the activity and the selectivity of the catalyst prepared from H_2PtCl_6 were strongly altered by the oxygen treatment

Table 1. Catalytic properties of different Pt/Al_2O_3 catalysts prepared form APH complexes. Conversion of n-hexane in continuous flow reactor

N°	Catalysts dehydroxylation (°C) Atmosphere of decomposition	O_2[a] treatment	H/Pt[b] (%)	X[c] (%)	Selectivities (%)					
					Hy	I	MCP	B	T	Ol
1	Br,500,N_2	no	79.6	69.6	24.7	26.7	4.0	29.9	2.0	7.2
2	Br,500,H_2	no	77.8	69.3	26.8	24.8	4.3	29.9	2.2	9.1
3	Br,500,H_2	yes	81.6	67.9	28.0	25.0	4.5	23.0	2.3	10.8
4	Cl,300,N_2	no	-	44.3	19.5	25.4	6.9	30.8	1.5	15.9
5	Cl,300,N_2	yes	34.0	43.1	22.5	24.9	7.9	30.0	1.7	13.5
6	H_2PtCl_6	no	8.0	41.2	63.9	3.1	6.2	25.1	1.7	-
7	H_2PtCl_6	yes	41.0	66.6	49.1	10.6	8.1	30.3	1.9	-

a) at 400°C, final reduction at 500°C; b) measured by pulse hydrogen chemisorption at 25°C; c) conversion after 4 hours on stream. Reaction temperature: 500°C, LHSV: 1.5 $g/g.h^{-1}$, CH/H = 1/5; Hy-hydrogenolysis products, I - isohexanes, MCP - methlylcyclopentane - B - benzene, T - toluene, Ol - C_6 unsaturated.

applied before final reduction. Contrary to that the properties of catalysts prepared from APH complexes were only slightly changed by this treatment procedure. Catalysts prepared from the bromide complex had the highest activity in n-hexane conversion.

Catalysts prepared from APH complexes had lower selectivities for hydrogenolysis and considerable higher selectivities for isohexanes than catalyst prepared from H_2PtCl_6. Significant increase in the formation of C_6 olefins was also observed on catalysts prepared from APH complexes. In this respect let us discuss results obtained on catalysts prepared in different way from the chloride complex. In this connection product selectivities measured at the same extent of conversion (~41%) are compared and shown in Figure 5.

It has already been mentioned in Section 3.2. that the route of decomposition of the APH complexes was strongly affected by the atmosphere applied. In nitrogen the chloride content of the complex appeared in the form of methylchloride. While in hydrogen atmosphere no methylchloride was detected, only methane formation was observed. Under this experimental condition the formation of hydrogen chloride and its strong adsorption on the alumina was suggested. This adsorbed chloride ion might be responsible for the differences in the selectivity for isohexane and C_6 olefins observed on these catalysts. On the chloride containing catalyst isomerisation of the primary formed C_6 olefins can take place at higher rates resulting in lower selectivities for the olefins and higher selectivities for the isohexanes. Just the opposite effect has been observed on the chlorine free catalyst prepared upon decomposition of the APH complex in nitrogen atmosphere (see catalyst N^o1. and 2. v.s. N° 3. and 4 on Figure 5.).

5.2. Catalytic properties of Sn-Pt/Al_2O_3 catalysts prepared using tin alkyl compounds.

Catalysts prepared as described in Section 3.3. were tested in n-hexane conversion at atmospheric pressure using a microcatalytical continuous flow reactor operated in slug pulse mode (Ref. 19). Experimental techniques and conditions are described elsewere (Ref. 12). In these studies following types of catalysts were tested:

(A) - base Pt/Al_2O_3 catalyst;

(B) - catalysts with direct Sn-Pt interaction;

(C) - catalyst without Sn-Pt interaction containing tin in its ionic form anchored into alumina by another type of CSRs (Refs. 12,20);

(D) - catalyst prepared by impregnation with $SnCl_4$ from aceton solution (Ref. 21).

Before the test reaction all of the catalysts were finally reduced in hydrogen at 550^oC without calcination. The H/Pt ratios measured after this treatment procedure can be found elsewhere (Ref. 12.) The H/Pt ratios of the Sn-Pt/Al_2O_3 catalysts were much lower than that of the base Pt/Al_2O_3 catalyst,

The initial rates of n-hexane conversion and the initial selectivities of the main reaction products obtained on catalysts type (B) with different Sn/Pt ratios are shown in Figure 6.

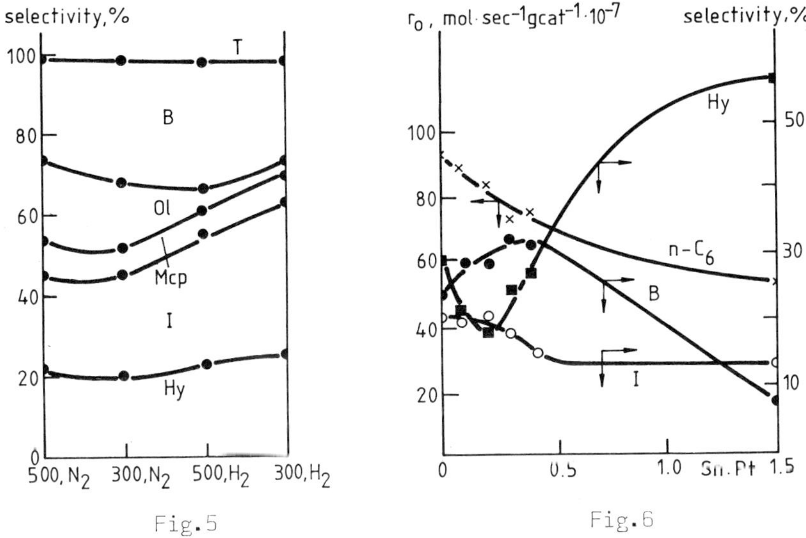

Fig.5 Fig.6

Figure 5. The influence of the atmosphere of decomposition of the chloride complex on the selectivity of the catalysts prepared. Experimental condition: See Table 1.

Figure 6. Sn-Pt/Al_2O_3 catalysts with direct Sn-Pt interaction. Reaction rates and selectivity changes as a function of tin content. Reaction temperature: 520^{0}C. Long pulse technique (Ref. 19).

Upon increasing the Sn/Pt ratio a constant decrease in the n-hexane conversion was observed. This could be attributed to the side blocking effect of tin introduced. A constant increase in the benzene selectivities was observed up to Sn/Pt = 0.4, i.e. up to the highest tin concentration which resulted in direct tin-platinum interaction. This observation can be explained by ligand and electronic effect of tin introduced. It is suggested that after the high temperature hydrogen treatment used prior to the reaction the tin is alloyed with the platinum. The primary ESCA (Ref.17) and X-ray diffraction studies (Ref. 14) on platium black supports this suggestion. It was evidenced earlier (see Section 3.3.) that in a Sn-Pt catalyst with Sn/Pt = 1.4 part of the tin was introduced into the alumina. The unusual catalytic properties of this catalyst can be attributed to this fact.

It was interesting to compare the discussed modes of tin

introduction with other conventional or controlled modes. In these series of experiment an other charge of base platinum catalyst was used both in the preparation of the Sn-Pt/Al_2O_3 catalyst and in n-hexane conversion. The product selectivities of hexane conversion were measured on this base Pt/Al_2O_3 catalyst and were compared with the results obtained on one of the (B), (C) and (D) types of Sn-Pt/Al_2O_3 catalysts containing the same amount of tin. In all of these catalysts the Sn/Pt ratio was 0.4. These results are shown in Figure 7. (Refs. 13, 15, 22).

Upon comparing selectivities for benzene formation (see Figure 7.A) it can be seen that at low extent of conversion catalysts type (A), (C) and (D) have almost the same actual selectivity values. This fact can be interpreated that the introduced form of tin did not alter the platinum sites responsible for the benzene formation. We suggest that in all of these catalysts tin-alumina interaction prevails. However on catalyst type (B), containing bimetallic tin-platinum surface entities, the benzene formation has been more pronounced than on the monometallic platinum catalyst. This result is in a good agreement with literature data (Ref. 23). Further analysis with respect to the differences between these catalysts in the selectivities for benzene formation observed at higher extent of conversion can be found elsewhere (Ref. 24).

Upon comparing the selectivity for hydrogenolysis or cracking products (C_1-C_5 hydrocarbons, S_{Hy} values) (see Figure 7.B) it can be seen that catalyst type (B) strongly resembles the base catalyst. It has been shown that in catalysts type (C) and (D) the acidity of the support has strongly been altered by the form of tin introduced (Refs. 20,22). The increase of the acidity resulted in an increase (catalyst type D)), the decrease of the acidity (catalyst type (C)) a decrease in the selectivity for these products (Refs. 12, 20, 22).

Let us compare the selectivity for isohexanes formed on these catalysts (see Figure 7.C). With respect to the base catalyst the selectivity of isohexanes decreased on all catalysts studied, however, this decrease can be attributed to different phenomena. In catalyst type (B) it can be explained by the increase in the benzene selectivity. In catalyst type (D) this fact can be connected with

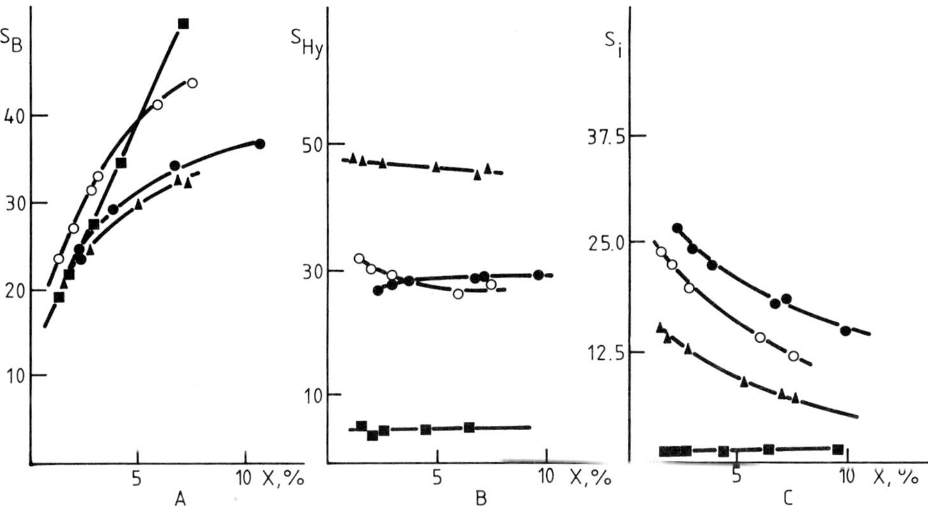

Figure 7. Conversion (x) - selectivity (S_i) dependences. ● - base Pt/Al_2O_3 (catalyst A), o - $Sn-Pt/Al_2O_3$ (catalyst B), ■ - $Sn-Pt/Al_2O_3$ (catalyst C). Sn/Pt = 0.4 in all catalysts. Final reduction at 550^{0}C, reaction at 520^{0}C. Long pulse technique (Ref. 19) S_B, S_{Hy} and S_i, selectivities for benzene, hydrogenolysis and isomerization, respectively.

the high selectivity for hydrogenolysis or hydrocracking products. With respect to catalyst type (C) it has already been mentioned that the introduced form of tin is ionic by its nature (see Section 3.3). No metallic tin could be detected by ESCA even after reduction at 550^{0}C (Ref. 24). This form of tin strongly reduced the acidity or the bifunctional character of the catalyst (Refs. 12,13,20). This fact resulted in strong hindering of the isomerization of the primary formed olefins (1-hexene or 2-hexene). On this type of Sn-Pt catalyst a drastic increase in the selectivity of C_6 olefins was observed (compare 15 % on the base catalyst (A) v.s. 70 % on catalyst type (C) at 2 % conversion (Refs. 12, 22)).

These data clearly demonstrate that in the case of $Sn-Pt/Al_2O_3$ catalysts the overall catalytic properties depends both on the form of tin (metallic or ionic) and its environment (platium or alumina), however, the exclusive formation of these surface species cannot be provided in another way as using Controlled Surface Reactions.

Table 2. Hydrogenation of acrylonitrile on Ni/Al_2O_3 and $Pb-Ni/Al_2O_3$ catalysts[a]

Catasysts	Lead content (%)	Temperature (^oC)	Conversion (%)	Selectivities (%)[b]				
				NPA	NPI	DNPA	TNPA	PN
Ni/Al_2O_3	-	100	100.0	13.1	10.0	3.9	-	73.0
Ni/Al_2O_3	-	104	99.6	23.9	10.2	9.9	0.3	53.7
Ni/Al_2O_3	-	135	100.0	51.2	-	41.4	6.1	1.3
A4	4.0	106	100.0	25.5	15.5	15.5	0.9	41.7
A5	6.0	104	72.8	-	1.1	-	-	98.9
B4	3.1	108	100.0	29.8	14.1	26.4	4.2	25.4
B6	6.1	106	98.9	2.5	4.0	1.3	0.3	91.9
B7	9.9	102	81.0	-	-	-	-	100.0
C1	5.1	100	100.0	1.4	3.6	1.1	-	93.9

a) Trickle-bed reactor: liquid flow-rate: 0.3 $kg.kg^{-1}.h^{-1}$ acrylonitrile in n-hexane solution; gas flow rate: 6.6 $l.h^{-1}$; p = 15 bar.
b) NPA = n-propylamine, NPI = n-propylimine, DNPA = di-n-propylamine, TNPA = tri-n-propylamine, PN = propionitrile

5.3. Catalytic properties of lead modified nickel catalysts in selective hydrogenation of acrylonitrile

In these experiments a Ni/Al_2O_3 catalyst was used. Its modification with lead was described in Section 3. 4.. The hydrogenation reaction was carried out in a conventional trickle bed reactor operated in the pressure range of 10-20 bar. Details on the experimental conditions and methods can be found elsewhere (Ref. 25).

In order to demonstrate the importance of CSRs three types of catalysts were prepared as follows (Refs. 13, 25):

Catalyst type (A): adsorption of $Pb(C_2H_5)_4$ at 25^oC on catalyst without thermal treatment:

Catalyst type (B): adsoprtion of $Pb(C_2H_5)_4$ at 25^oC treated in nitrogen at 150^oC:

Catalyst type (C): CSRs at 25^oC between $Pb(C_2H_5)_4$ and hydrogen adsorbed on the catalyst treated in hydrogen at 150^oC.

On the Ni/Al_2O_3 catalyst (with Ni content 40 %), parallel to the selective hydrogenation of acrylonitrile into propionitril the formation of propylamine, di- and tri-n-porpylamine and n-propylamine were observed.

The aim of this work was to maintain the high conversion of acrylonitrile (up to 90-95 %) with strong suppression of the undesired reaction products. The experimental data are given in Table 2. (Refs. 13, 25). The best results, i.e. the highest selectivity values for the propionitrile formation were obtained on catalyst type (C) prepared by CSRs.

These results clearly demonstrate and give further evidence that good selectivity control can be achieved only on catalysts with direct lead - nickel interaction obtained by Controlled Surface Reactions.

CONCLUSION

In this review an attempt has been made to demonstrate that different metal alkyl compounds can be used for the preparation of supported platinum catalysts and for the modification of different monometallic catalysts with the aim to obtain supported bimetallic catalysts with direct metal-metal interaction. It was demonstrated that the main point in the preparation of these catalysts was to find the driving force which could guarantee the exclusive formation of the required surface species. This way unique catalysts can be prepared the activity and selectivity of which can be tailored depending on the type of reactions in which they will be used.

ACKNOWLEDGEMENTS - The author wishes to express thanks to all colleagues of his research group: S. Göbölös, E.Kern-Tálas, M.Hegedüs and P. Szedlacsek, whose results were used in this review. Additional thanks to those who encouraged this work (prof. F.Nagy) or helped to characterize the surface species formed (Dr.Z.Schay and Dr.I.Bertoti).

REFERENCES

1. Y.I. Yermakov, B.N. Kuznecov and V.A.Zakharov, Catalysis by Supported Complexes, Elsevier, Amsterdam, 1981.
2. D.C.Baily and S.H.Langer, Chem. Rev., 81, (1981) 110.
3. G.V.Lisichkin, A.Ya. Yuffa and G.V. Kudryavcev, 8th International Congress on Catalysis, Proceedings, Vol. IV., p. 669.,

Verlag Chemie, 1984
4. D.E. Clegg, I.R.Hall, Inorg. Synth. 10, (1967) 71.
D.E.Clegg, I.R.Hall, J.Organometal.Chem., 22 (1970) 491.
5. J.Margitfalvi and L.Koltai, Therm. Anal., 15 (1979) 251.
6. J.Margitfalvi, L.Koltai and Z.Poko, to be published.
7. L. Guczi, K.Matusek and J.Margitfalvi, React. Kinet. Catal. Lett. 8, (1978) 309.
8. L.Guczi, K. Matusek, J. Margitfalvi, M. Eszterle, F.Till, Acta Chimica Acad. Sci. Hung., 101, (1979) 107.
9. J. Margitfalvi, S.Göbölös, M.Hegedüs and F.Nagy, Reaction Kinet. Catal.Lett., 21, (1982) 541.
10. J.Margitfalvi, S.Göbölös, M.Hegedüs and F.Nagy, First Czechoslovak Conference on Preparation and Properties of Heterogeneous Catalysts, Bechyne, 1982.
11. J.Margitfalvi, S.Göbölös, M.Hegedüs and F.Nagy, J. Mol. Cat., (submitted for publication)
12. J.Margitfalvi, M.Hegedüs, S.Göbölös, E.Kern-Tálas, P.Szedlacsek, S.Szabo and F. Nagy, 8th International Congress on Catalysis, Proceedings, Vol. IV., p. 903, Verlag Chemie, 1984.
13. J.Margifalvi, E.Kern-Tálas, S.Göbölös and M.Hegedüs, 9th North American Meeting of the Catalysis Society, Houston, March 1985, Abstract, p. P-24.
14. J.Margitfalvi, E.Kern-Tálas and S.Göbölös, XIIth International Conference on Organometallic Chemistry, Vienna, Sept. 1985., Abstract p. 518.
15. S.Göbölös, E.Kern-Tálas, M.Hegedüs, P.Szedlacsek, J.Margitfalvi, IVth International Symposium: Scientific Bases for the Preparation of Heterogeneous Catalysts, Leuven-la-Neuve,Sept. 1986 (in preparation).
16. J.Margitfalvi, S.Göbölös and E.Kern-Tálas, J.Organomet. Chem., (submitted for publication).
17. J.Margitfalvi, S.Göbölös, E.Kern-Tálas and I.Bertoti IIIrd European Conference on Solid State Chemistry,Regensburg,May 1986 (in preparation)
18. M.Hegedüs and J.Margitfalvi, unpublished results.
19. J.Margitfalvi., M.Hegedüs, P.Szedlacsek and F.Nagy, Applied Catal., 15, (1985) 69.
20. J.Margitfalvi, E.Kern-Tálas, S.Göbölös and G.Resofszky, to be published.
21. R.Burch, J.Catal., 71, (1981) 348.
22. J.Margitfalvi, P.Szedlacsek, M.Hegedüs, E.Kern-Tàlas and F. Nagy, J.Catal. (submitted for publication)
23. B.Coq and Figueras, J. Catal. 85, (1984) 197.
24. E.Kern-Tálas, Z.Schay, S.Göbölös and J.Margitfalvi, to be published
25. J.Margitfalvi and S.Göbölös, Appl. Catal. (submitted for publication).

MODIFIED NOBLE METAL CATALYSTS FOR THE METHANOL SYNTHESIS FROM SYNGAS : Pt-SiO_2+CrOx*, Pt-SiO_2+MoOx

P. Mériaudeau, K. Albano, H.P. Nguyen, M. Primet and C. Naccache
Institut de Recherches sur la Catalyse, CNRS, conventionné à l'Université Claude Bernard Lyon 1, 2 avenue Albert Einstein 69626 Villeurbanne cédex

SUMMARY

Pt/SiO_2 catalysts, modified by adducts of oxides MeOx (Me=Cr or Mo) are active and selective for the syngas conversion into methanol.

Characterization of the catalyst by different methods indicate the adduct is deposited in the vicinity of the metal.

It is suggested that the resulting catalyst acts as a bifunctional system for the formation of methanol.

Since Poutsma et al (1) have discovered that supported noble metals are active and selective for alcohols synthesis from syngas, it is known that the nature of the support greatly influences the performances of the metal :
Ichikawa (2) reported that basic supports favor the alcohol formation. For Pd, a series of reports (3) (4) (5) (6) (7) indicates that rare earth oxides or adducts (Li, Mg etc...) enhanced the activity and the selectivity of the catalyst.

Similarly, the same observations were reported for Pt (2) (8). Although the experimental results are in agreement there is still a matter of contreversy to explain this benifit effect of the support or of the adduct : 3 more or less contradictory conclusions have been reached at :

* A part of this work will appear in J. Appl. Catal.; 21, 337 1986

i) The activity and the selectivity depend on the surface structure of the metal, Pd(100) plane being more active than (111) plane (4) ; in addition an electronic effect, charge transfert from the support to the metal which is benefit for the alcohol formation, is suggested (4).

ii) The support (or the adduct) favors the formation of ions on the metals particles (or at the interface), these ions being the active centers of the alcohols synthesis (7) (9).

iii) The formation of alcohol is a bi-site reaction, thus supports or adducts which are able to activate one of the two reactants are promotors of the alcohol activity, the other reactant being activated on the metal (8) (10).

The aim of this report is to discuss in view of recent results obtained on $Pt-SiO_2+Cr$ or Mo adducts, the possible role of the support or of the adduct.

EXPERIMENTAL

Material

Pt supported catalysts were prepared by the incipient wetness impregnation technique of SiO_2 Degussa (200 m^2g^{-1}) with H_2PtCl_6 aqueous solution. The samples were dried in air at 473K and reduced in flowing H_2 for 12 hours at 723K. Reduced $Pt-SiO_2$ was used a basic material to prepare $Pt-SiO_2+CrO_x$ or $Pt-SiO_2+MoO_x$ catalysts.

$Pt-SiO_2+CrO_x$: $Pt-SiO_2$ was impregnated with aqueous chromic acid solution and dried in air at 473K.

$Pt-SiO_2+MoO_x$: $Pt-SiO_2$ was impregnated with aqueous ammonium paramolybdate and dried in air a 473K.

Characterization

Pt dispersion was measured by hydrogen chemisorption at 298K (volumetric apparatus equipped with a Texas Instrument pressure gauge) ; the hydrogen uptake was estimated by extrapolating the isotherm to zero pressure. In some cases, transmission electron microscopy (T.E.M.) was used to estimate the metal particle diameters.

For Pt-SiO_2+CrO_x catalysts, the rate of toluene hydrogenation was considered to be proportional to the number of surface platinum atoms and measured in a differential flow reactor at 288K. This indirect method, to estimate the surface area of Pt, was not used for Pt-SiO_2+MoO_x since it is known in this particular case that the hydrogenation rate of the aromatic is not proportional to the number of Pt surface atoms (11) (12).

For I.R. measurements, the samples were pressed in disks, reduced under H_2 at the required temperature, evacuated under vacuum and cooled down to room temperature ; then they were contacted with CO or CO+H_2 and the spectrum registred, in presence of the gas phase or after evacuation.

For Pt-SiO_2+CrO_2 catalysts spectra were recorded with a Perkin Elmer I.R. spectrometer 580 and for Pt-SiO_2+MoO_x with a Brucker I.F. 110 Fourier transform spectrometer ; on this apparatus the spectrum of adsorbed CO was computed from accumulated interferograms recorded before and after adsorption thus cancelling out the contribution of the Si-O bands in the final spectrum. S.T.E.M. spectra were recorded with a V.G. HB5 system.

Methanol synthesis

The catalytic activity for alcohol synthesis from CO+H_2 was measured in a differential flow reactor at 30 atmospheres of total pressure with H_2/CO=2. Effluents were analyzed after 18-20 hours on stream by using gas chromatographs. The catalysts were reduced in situ prior reaction.

RESULTS

1) Pt-SiO_2+CrO_x

Pt dispersion : H/Pt was calculated by assuming the stoechiometry H/Pt_s=1. In our experimental conditions CrO_x-SiO_2 (or MoO_x-SiO_2) did not chemisorb H_2. Results are summarized on table 1.

TABLE 1

3.4% $Pt-SiO_2$				6.4% $Pt-SiO_2$*			
%Cr	0	0.3	1	%Cr	0	0.5	1
H/Pt	0.27	0.17	0.13		0.53	0.29	0.17

* Euro Pt.1 SiO_2 support : Sorbosil AQ U30

Toluene hydrogenation

The toluene hydrogenation rate was measured at atmospheric pressure, 288K. Under experimental conditions used, CrO_x-SiO_2 was inactive. Results are given in Table 2.

TABLE 2

Catalyst	3.4% $Pt-SiO_2$					6.4% $Pt-SiO_2$				
% Cr	0	0.15	0.3	1	3	0	0.15	0.5	1	3
V(a)	35	30	22	17	7	105	83	52	23	16

V(a) rate of hydrogenation - Arbitrary units

Infrared study of adsorbed CO

The infrared spectrum of carbon monoxide adsorbed on $Pt-SiO_2$ showed an I.R. band at 2075 cm^{-1} due to the strenching frequency of CO linearly bonded to surface Pt atoms. When $Pt-SiO_2$ was promoted with Cr, CO adsorption produced an I.R. band which intensity decreased with the increase of Cr content (see table 3). The frequency of this I.R. band remained virtually inchanged (2075-2070) when Cr is added to Pt. On table 3 are reported the adsorbances of

CO adsorbed on Pt, together with the H/P_t values and the rate of toluene hydrogenation. All the values are normalized to 1 for Cr free sample.

TABLE 3

Catalyst 3.4% $Pt-SiO_2$	% Cr 0	0.3	1
Adsorbance of CO adsorbed on Pt[(a)]	1	0.6	0.42
H/Pt	1	0.62	0.54
Toluene hydrognation rate	1	0.63	0.48

a) Sample was reduced under H_2 at 673K, then H_2 evacuated at 673K, cooled down to room temperature. CO was introduced (10 torr) over the sample and then evacuated at R.T. and the spectrum registred.

T.E.M. and STEM analysis

In order to have a more precise description of the catalyst TEM and STEM analysis were performed on 4 samples :

TABLE 4

Sample	% Cr	TEM	STEM
6.4% $Pt-SiO_2$	0	X	
"	0.1		X
"	0.5	X	X
"	1		X

Table 4 summarized the samples analyzed.

On fig.1, are given TEM of a Pt free sample and the corresponding with 0.5%Cr: the particle diameters do not increase significantly when 0.5% Cr is added to the 6.4% Pt-SiO_2.

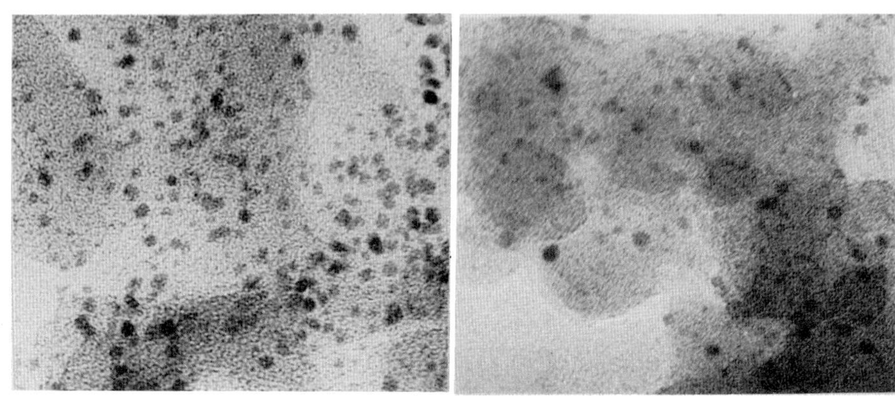

Fig.1a. TEM of 6.4% Pt-SiO_2 1b. TEM of 6.4% Pt-SiO_2 + 0.5% Cr

By analyzing the different spectra obtained with STEM when focusing on :

- large ensemble of particles
- individual particles (20-30 Å)
- parts of the support SiO_2 with no apparent Pt particle

One can summarize as follow

a) For low Cr loading : (0.1 and 0.5)

- The $|\frac{Pt}{Cr}|_{at}$ ratios vary greatly from one particle to the other indicating that the repartition of Cr is very inhomogeneous
- The $|\frac{Pt}{Cr}|_{at}$ for large ensemble of particles is in good agreement with that measured by chemical analysis.
- No Cr was detected on the support in absence of any visible Pt particle.

b) 1% Cr loading

As previously the two first observations were obtained for a) and b) but in addition, chromium was detected onto the support in areas of the support having no Pt particle.

Methanol synthesis

In our experimental conditions, the pseudo stationnary state was reached after 16-20 hours on stream. On fig.2 and fig.3 are reported the activities and the selectivities for CH_3OH production. The apparent activation energies of CH_3OH formation measured in the temperature range 518-563K were 16 $\pm$ 1.5 kcal $mole^{-1}$ for $Pt\text{-}SiO_2$ and 13.5 $\pm$ 1.5 kcal $mole^{-1}$ for $Pt\text{-}SiO_2$+CrO_x respectively.

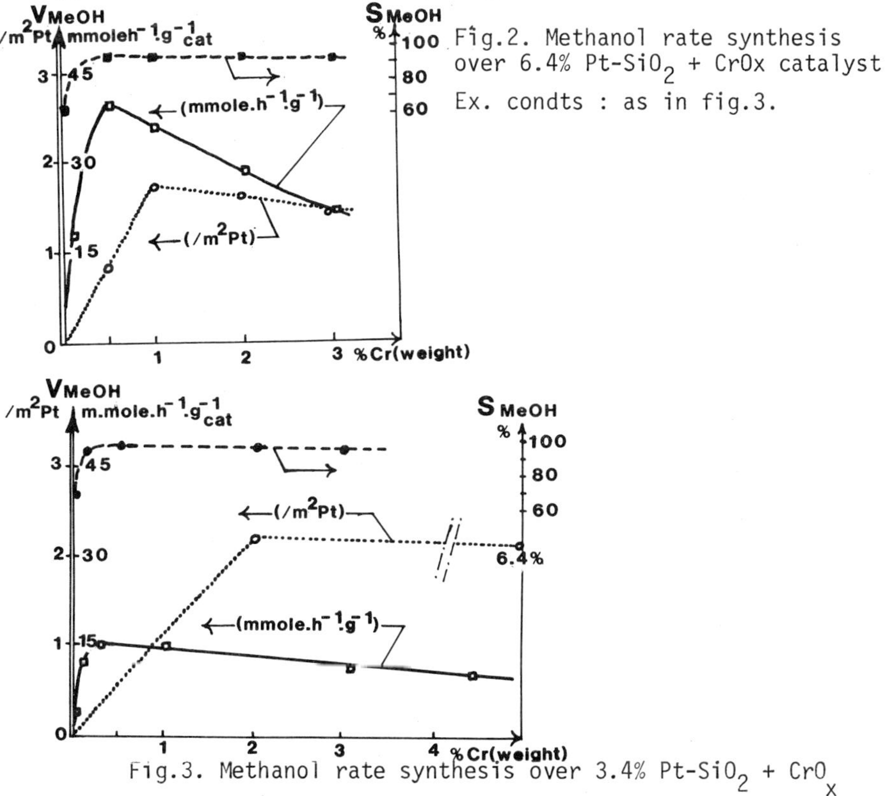

Fig.2. Methanol rate synthesis over 6.4% $Pt\text{-}SiO_2$ + CrOx catalyst Ex. condts : as in fig.3.

Fig.3. Methanol rate synthesis over 3.4% $Pt\text{-}SiO_2$ + CrO_x catalyst. P = 30 bar T = 563K H_2/CO = 2 wgt 30 mg Flow rate 3.61 h^{-1}

To better describe the effect of Cr adduct, we tested, in the same experimental conditions a 6.4% $Pt-SiO_2$ catalyst, a 3% CrO_x-SiO_2 catalyst and a mechanical mixture of 6.4% $Pt-SiO_2$+3% CrO_x-SiO_2 catalyst (the mechanical mixture was prepared by grinding in an agathe mortar the two components (50% by weight).

Results are given in TABLE 5

Catalyst	V_{MeOH} mmole^{-1}g^{-1}		S_{MeOH} %
	measured	calculated*	
6.4% Pt-SiO2	0.47	--	72
3% CrOx-SiO2	0.01	--	97
Mechanical mixture	2.3	0.24	97

* by adding the activities of 6.4% $Pt-SiO_2$+3% CrO_x-SiO_2

2) $Pt-SiO_2+MoO_x$

Pt dispersion : As for chromium adduct, it was observed that the addition of molybdenum decrease the initial dispersity of $Pt-SiO_2$ as indicated in Table 6.

TABLE 6

Sample	1.7% $Pt-SiO_2$	1.7% $Pt-SiO_2$ + 2 % MoO_x
H/Pt	0.60	0.20

Transmission electron microscopy (see fig.4) indicate that the addition of Mo to Pt-SiO_2 (at least for percentage 2%) do not increase greatly particle diameters.

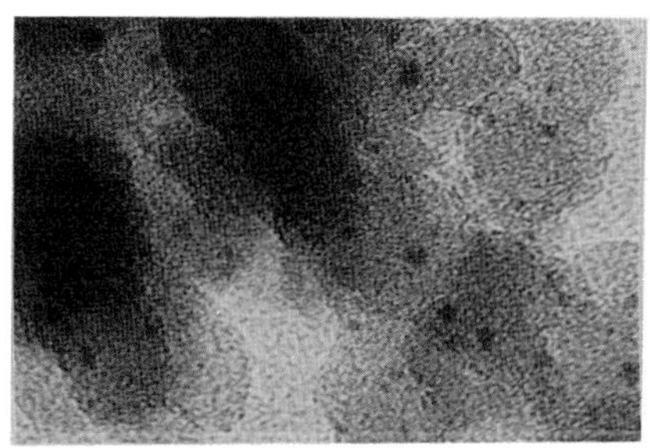

Fig.4a. TEM of 1.7% Pt-SiO_2

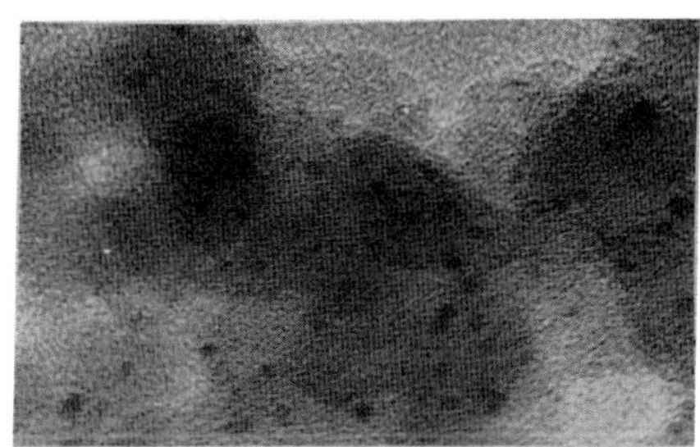

Fig.4b. TEM of 1.7% Pt-SiO_2 + 2 %MoO_x

STEM analysis

One sample (1.7% Pt-SiO_2+2% Mo) was analyzed by STEM. As for Cr adduct, it appears that the composition of individual particles is inhomogeneous, some particles having a high Mo/Pt atomic ratio, some others having a low one (compared to the atomic ratio expected from chemical analysis).

For the 2% Mo loading, molybdenum was detected onto the support in areas having no Pt particle.

Infrared study of CO adsorption

The adsorption of CO on pure Pt-SiO_2 (not represented on fig.5) gave 2 bands : one corresponding to a linear CO on Pt, the other to a bridged CO.

After Mo addition, the magnitude of these two bands decrease (that of the bridged CO more than that of the linear CO) and a typical spectrum is represented on fig.5a (in presence of 20 torr CO) and on fig.5b after evacuation of CO at R.T.

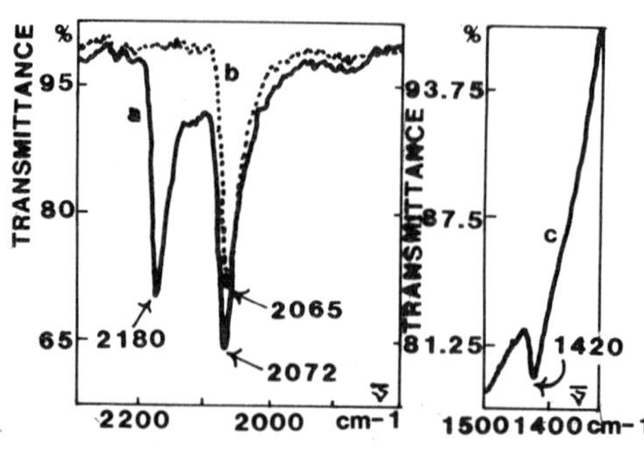

Fig. 5. Infrared spectra of CO on 1.7%Pt-SiO_2+2%Mo

a) in presence of 20 torr of CO
b) after evacuation at 298K

c) in presence of H_2 after reduction at 673K under H_2

It appears on fig.5b that the band at 2180 cm^{-1} disappears when CO is evacuated indicating that the bond between CO and the Mo adsorption center is weak ; if CO is reintroduced again the 2180 cm^{-1} band appears again, but if the mixture CO+3H_2 (20 torr CO + 60 torr H_2) is contacted the sample the 2180 cm^{-1} band is no more formed and in that frequency region only gas phase ν_{CO} is detected (ν_{CO}=2175 cm^{-1} , (see discussion).

As for Cr adduct, it has been observed that the frequency shift of the linear ν_{CO} on Pt decrease very slightly with the addition of Mo (ν_{CO}=2065 cm^{-1} for 1.7% Pt-SiO_2 + 2% Mo).

Methanol synthesis

On fig.6 is represented the rate of MeOH formation as a function of the added Mo.

A mechanical mixture of 1.7% Pt-SiO_2 + 1% MoO_x-SiO_2 was also tested, and as for chromium, an synergic effect was observed : (see table 7).

TABLE 7

Catalyst	V_{MeOH} mmole $h^{-1}g^{-1}$ cata.	S_{MeOH}
1.7% Pt-SiO_2	0.32	75
1% MoOx-SiO2	0.18	59
Mechanical mixture	1.68	86

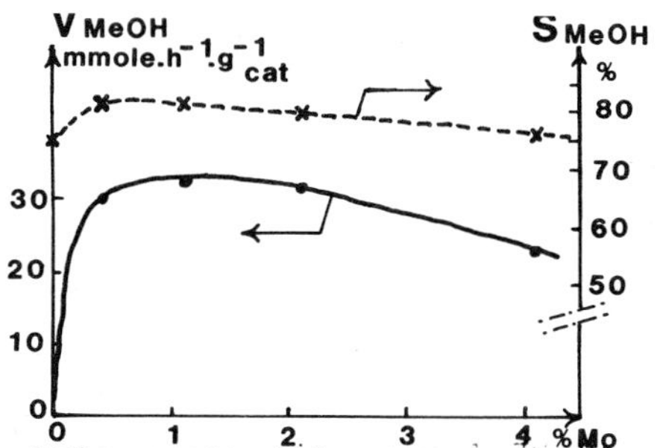

Fig.6. Methanol rate synthesis over 1.7% Pt-SiO_2 + MoO_x as a function of the molybdenum content
30 bar, T= 443K, H_2/CO = 2 Catalyst weight 30 mg,
Flow rate : 3,6 lh^{-1}

Dicussion

1) Morphology of Pt-SiO_2+CrO_x or Pt-SiO_2+MoO_x catalysts

The results listed in Table 3 indicate a strong influence of the addition of Cr on H_2 or CO adsorption and on the overall rate of toluene hydrogenation suggesting a coverage of the Pt particles with the chromium adduct. TEM indicated that there is no large increase of particle size diameters with Cr adduct (at least for low percentage) ; in addition STEM analysis showed that the coverage of Pt with Cr (or Mo) is inhomogeneous. Thus, one can suggest the next picture (fig.7) for the catalyst.

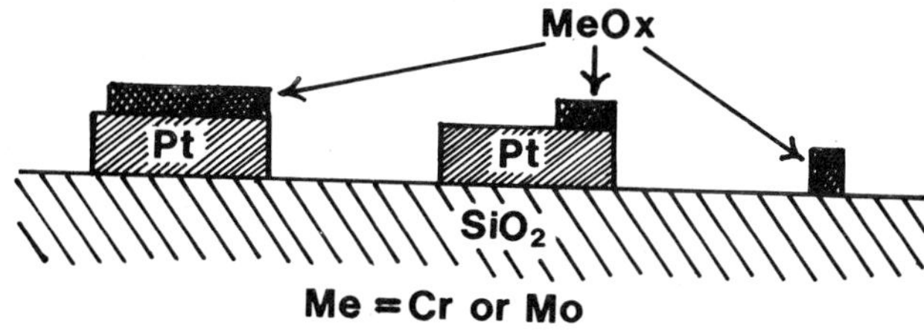

Fig.7. Schematic picture of Pt-SiO_2 + CrO_x (or Pt-SiO_2 + MoO_x)

with such a picture, one now can understand the experimental results obtained with I.R. spectroscopy i.e. the virtually unchanged ν_{CO} frequency of the Pt linear carbonyl, results which are in contrast with those reported for Pt-Cu (13) or PtPb (14) alloys were a large downward frequency shift was observed, this shift being attributed to a geometric effect (dilution of Pt atoms with Cu) or to an electronic effect. In the case of Pt-SiO_2+CrO_x, the adduct decorate the Pt particles : would be the decoration process homogeneous, one would observe a frequency shift due to the dilution effect (13) but in the case of an inhomogeneous coverage, the contribution to the I.R. CO band of the Pt particles having few Cr (in that case ν_{CO} = ν_{CO} of bare Pt) is much more important than the contribution of the particles highly covered with Cr (for which the expected ν_{CO} is lower than ν_{CO} on bare Pt). Thus, the picture of fig.6 explains well our experimental results. This model seems to be adequat to describe the Pt-SiO_2+MoO_x system since we observed similar experimental facts : decrease in H_2 or CO chemisorption, little change in frequency of the linear Pt-CO vibration, inhomogeneous distribution of Mo on Pt.

Concerning the oxidation states of Cr as studied by I.R. and E.S.R. spectroscopies by using CO and NO as probes, results will be reported elsewhere (15). These results suggest that Cr exists at least as Cr^{+2} and Cr^{+3} states, this observation being in agreement with those reported earlier for CrO_x-SiO_2 (16).

For Pt-SiO_2+MoO_x it appears that after the reduction under H_2 at 673K, at least a part of the molybdenum is in+4 or + 5 state since the I.R. band (2180 cm^{-1}) is attributed to a CO adsorbed on ionic molydenum (17) ; in addition, ESR results (not reported here) showed the presence of Mo^{5+} in low concentration.

To summarize up, the major effect of the impregnation of Pt-SiO_2 with chromic acid or ammonium paramolybdate is to cover Pt particles with patches of CrO_x or MoO_x. In the present work, no Pt ions were detected by I.R. spectroscopy (in contrast with Pd-SiO_2+MgO_x catalyst as reported in (18) suggesting that Pt ions (if they exist) have a negligible concentration compared to that of Pt atoms.

Catalytic properties

From a catalytic viewpoint, the effect of Cr or Mo adducts is quite spectacular since the activity towards methanol formation is increased by an important factor ; the MeOH rate increased with increasing Cr or Mo content up to 1-2% despite the fact that in parallel the Pt surface areas decreased. Before to discuss these results, one must recall that similar effects were observed on Pd-SiO_2 with Mg, Li, Sr, Mo or Ba adducts. By using La_2O_3 or ThO_2 supports instead of SiO_2, the same effects (large increase in MeOH yields) were observed for Pd(5) Pt(2), Rh(2)(10). At present time, it has been proposed that :

i) The adduct (or the nature of the support) favors the stabilization of metallic ions which are active sites for MeOH formation (9)
ii) The support favors the formation of Pd(100) plane which is the most active plane for MeOH formation ; in addition, the La_2O_3 support creates a "metal support interaction" giving an additionnal increase in MeOH formation (5)

iii) The methanol formation occurs via a bifunctional mechanism ; thus the performances of a given metal for MeoH synthesis are strongly dependent on the properties of the support (8)(10)(19). We will now examine the three proposed explanations in view of our experimental results.

i) I.R. results do not indicate that ionic platinum is present in our Pt-SiO_2+MeO_x (Me=Cr or Mo) system as the CO frequency does not vary significantly with the amount of the adduct and since this ν_{CO} is close to that observed for bare Pt-SiO_2.

The maximum values of V_{MeOH}/m^2 Pt (see fig.2 and 3) are not too different for catalysts having different Pt loadings and dispersions, which is not in favor of i).

ii) Due to the procedure that we used to prepare our catalysts (formation of Pt-SiO_2 and then addition of MeO_x) it seems improbable that during the activation process the presence of the added oxide induces a change in crystal surface planes of Pt. Concerning the metal support interaction as pointed out by Bell et al.(5), it appears that if this interaction exists on Pt-SiO_2+MeO_x the magnitude is small, the shift of $^{\nu}CO$ on Pt is very low for Pt-SiO_2+CrO_x and small (10 cm^{-1} maximum) for Pt-SiO_2+MoO_x.

Morover one can expect that this interaction would not be created by a single mechanical mixing of Pt-SiO_2 and MeO_x, nevertheless, an enhancent of the activity of each component on the other is observed (table 5, 7). For these reasons the proposal ii) seems unlikely in our system.

iii) It is known that chromium oxide has an intrinsic activity for MeOH synthesis (table V, (19) ; a low activity is also detected for MoO_x-SiO_2 (table VII).

If one accept the assumption that the mechanical mixtures of Pt-SiO_2+MeO_x do not change morphology or metallic properties of Pt-SiO_2, the large enhancement of the activity of Pt-SiO_2 by MeO_x adduct can be understood through a bi site mechanism, the CO component being activated on one component, H_2 on the other.

It has been proposed for Ir-Mo catalysts that Mo could be the active site for CO activation (20) or in the case of Rh-SiO_2+Mo_x ($M=Mn^{+2}$, Ti^{+3} etc) that the CO molecule could be bonded to 2Rh atoms, and to the promotor metal ion (21)(22).

For our systems, at this stage, it is not possible to indicate which component activates the CO, since we observe that for Pt-SiO_2+MoO_x, CO is adsorbed onto Pt and onto Mo and since the CO is displacedby hydrogen on Mo ; in addition hydrogen is adsorbed onto $Mo^{+?}$ (1430 cm^{-1})(23).

Further experiments are required to clear this important question and I.R. studies under pressure are now in progress and will probably help to have a better picture of the elementary steps of the reactions.

ACKNOWLEDGMENTS are due to :

G.WICKER (TEM), H.URBAIN (Chemical Analysis), C.LECLERCQ and her assistants (STEM).

REFERENCES

1. M.L.Poutsma, F.Elek, P.A.Ibarbia and J.A.Rabo, J. Catal., 52 (1978) 157.
2. M.Ichikawa and K.Shikakura "New Horizon in Catalysis", Proceedings 7th. Int. Cong. on Catal., Tokyo (1980). Seiyama, Tanabé Eds, Part B p.925, Elsevier, Amsterdam (1981)
3. S.W.Kaiser ; U.S. Patent 4,289,709. Sept.15. 1981 Union Carbide
4. F.Fajula, R.G.Antony and J.H.Lunsford, J. Catal., 73 (1982) 237
5. R.F.Hicks and A.T.Bell, J. Catal., 90 (1984) 205 and ref. therein
6. Y.Kikuzono, S.Kagami, S.Naito, T.Oniski and K.Tamaru. Farad. Disc. Chem. Soc., 72 (1981) 135
7. J.M.Driessen, E.K.Poels, J.P.Hindermann and V.Ponec, J. Catal., 82 (1983) 26
8. P.Mériaudeau, M.Dufaux and C.Naccache, VIII Int. Cong. on Catal., Berlin, 1984, Vol II, p.185, Verlag Chemie, Berlin, 1984
9. F.G.A.Van den Berg, J.H.E.Olezer and W.M.H.Sachtler, J. Catal., 93 (1985) 340
10. V.L.Kuznetsov, A.V.Romanenko, I.L.Mudrakoskii, V.N.Natikhin, V.A.Shmachkov and Yu.I.Yermakov, VIII Int. Cong. on Catal., Berlin, 1984, Vol.V, p.3, Verlag Chemie. Berlin 1984
11. Yu.I.Yermakov, B.N.Kuznetsov and Yu.A.Ryndin, React. Kinetics and Catalysis Letters Vol.2 (1975) 151
12. G.Leclercq, T.Romero, S.Pietrzyk, J.Grimblot and L.Leclercq, J. of Mol. Catal., 25 (1984) 67
13. F.J.Toolenaar, F.Stoop and V.Ponec, J. Catal., 82 (1983) 1
14. (a) A.Palazov, Ch. Bonev, G.Kadinov and D.Shopov and G.Lietz and J.Völter, J. Catal., 71 (1981) 1
(b) A.G.Bastein, F.J.Toolenaar and V.Ponec, J. Catal., 90 (1984) 88
15. H.P.Nguyen (to be published)

16. a) A.Zecchina, E.Garrone, G.Ghiotti, G.Morterra and E.Borella, J. Phys. Chem., 79 (1975) 966
b) C.Groeneveld, P.P.M.M.Wittgen, A.M.Van Kersbergen, P.L.M.Mestrom, C.E.Nuijten and G.C.A.Schuit, J. Catal., 59 (1979) 153
c) D.D.Beck and J.H.Lunsford, J. Catal., 68 (1981) 121 and ref. therein
17. E.Delgado, G.A.Fuentes, C.Hermann, G.Kuntzmann and H.Knözinger, Bull. Soc. Chim. Belg., 93 (1984) 735
18. E.K.Poels, R.Koolstra, J.W.Gens and V.Ponec. Studies in Surface Science and Catalysis 11, B.Imelik et al Edts, p.233, Elsevier 1982
19. E.Druet. Thesis Paris (1982)
20. Y.Kuwahara, H.Hamada, Y.Kintaichi, T.Ito and K.Wakabayaski, Chem. Letters, (1985) 205
21. Schatler, W.M.H., Proc. Int. Congr. Catal., 8th 1984, 1, 151
22. M.Ichikawa and T.Fukushima, J. Phys. Chem., 89 (1985) 1564
23. Y.I.Yermakov, B.N.Kuznetsov, V.A.Zakharoy, Studies in Surface Science and Catalysis, 8 (1981) 59.

STRUCTURE ANALOGIES AND TRANSFORMATIONS BETWEEN METAL AGGREGATES AND ORGANOMETALLIC CLUSTERS

P. GALLEZOT
Institut de Recherches sur la Catalyse
Laboratoire Propre du C.N.R.S.,
conventionné à l'Université Claude Bernard LYON 1
2, avenue Albert Einstein - 69626 - Villeurbanne
FRANCE

ABSTRACT - This article reviews briefly the literature showing that the structure of metal surfaces is flexible in response to adsorbates. Especially the structure of metal aggregates encaged in zeolites is compared to that of organometallic clusters of similar nuclearity. It is concluded that in both species the structure of the metal framework is modified to an extent depending upon the nature and number of surface ligands.

INTRODUCTION

From a survey of the large body of literature covering metal aggregates (naked clusters) on the one hand, organometallic clusters (clusters) on the other hand, there seems to be little relationships between the structure of these species. Although the cohesive forces in both cases involve metal-metal bondings, clusters exhibit an almost infinite variety of structures whereas metal aggregates are often viewed as fixed models where the metal framework does not change whatever its environment (support, adsorbates). It is fair to say that the structure of the clusters can be determined with a high accuracy from crystal structure analyses whereas little is known on the actual structure of metal aggregates. The static picture of the metal aggregate structure is typically illustrated by the past debates on the relative merit of the cubooctahedron and of the icosahedron as the most probable structure for very small metal particles. It was overlooked that the difference in stabilization energy between a 13 atoms cubooctahedron and a 13 atoms icosahedron is only 0.5 kcal $mole^{-1}$ (1), a negligible value with respect to the heat of chemisorption of most adsorbates which might easily induce the transformation from an arrangement to the other. Interestingly, the enthalpies of metal-adsorbate bonds are generally larger than those of metal-metal bonds. Therefore from a mere consideration of bond energy, it is expected that the structure of metal particles must be perturbed by adsorbates.

The aim of the present paper is to review very briefly some examples of metal structure flexibility in single crystal surface and in metal particles. Then, we will concentrate our attention on metal aggregates containing 6 to 50 atoms which can be prepared in zeolite cages. The flexibility of their structure will be discussed along with that of clusters, especially the large metal carbonyl clusters (2), having a comparable nuclearity. Finally, possible transformations of metal aggregates into clusters will be examined shortly.

1. FLEXIBILITY OF THE STRUCTURE AND MORPHOLOGY OF METALS

The cases of single crystal faces, particles larger than 20-25 Å, particles smaller than 15-20 A and metal molecules containing a few atoms are discussed separately because they are prepared and characterized in very different ways.

1.1. Single crystal faces

LEED studies of single crystal faces provide several examples of surface reconstruction (3, 4) whereby the atoms in the topmost layer rearrange into a surface structure incommensurate with the underlying bulk layers. There are more subtle rearrangement taking place on crystal faces. Thus, because the binding forces exerting toward the inner side of the surface are not compensated, a contraction of the first interplanar spacing of clean metal surface is generally observed (3, 4). On the other hand, a small expansion has been reported for contaminated surfaces (5) because the coordination of surface atoms is improved. These relaxations perpendicular to the surface are evidenced by quantitative analysis of LEED data. Unfortunately, the accuracy in bond length between atoms in the surface plane is only ± 0.2 Å. Furthermore LEED is a diffraction technique mainly sensitive to the long range order of atoms so that many types of local distortions possibly produced by adsorbates can be overlooked. Indeed other surface science techniques point to metal atom displacement in the surface plane. Thus, sulfur atoms adsorbed on Ni(100) planes are bonded preferentially with two Ni atoms which distorts the initial square arrangement of Ni atoms in this plane (6, 7). Another example of the flexibility of metal surfaces is the adsorbate-induced facetting of crystals whereby a given surface plane can be recrystallized into planes of different orientation because the surface free energy of these planes is lowered by the presence of adsorbates (8).

1.2. Metal particles larger than 20-25 Å

The morphology of particles larger than 20 Å can be studied by transmission electron microscopy (TEM). The weak-beam imaging technique is especially useful since the three dimensional shape of the particles can be reconstructed from the weak beam fringes spacing (9).

Changes in particle morphology have been reported at high temperatures in oxygen atmosphere, possibly leading to metal redispersion into smaller particles (10). Marked changes in the morphology of polycrystalline catalysts (etching and pitting of wires or foils) were also observed during catalytic reactions such as the NH_3 oxidation on platinum (11) or the HCN synthesis from CH_4 and NH_3 on rhodium (12).

Adsorbates-induced modifications of particle morphology occuring at room temperature are fascinating examples of metal flexibility. These transformations must be studied by *in situ* TEM under controlled temperature and atmosphere (13). Thus it has been shown (13) that 15-30 Å Pd particles deposited on MgO (111) under high vacuum coalesce in a few minutes when exposed to oxygen or air at room temperature. The particles spread and move over the support forming rafts. These and other transformations were attributed to the change in the interfacial (metal/support, metal/adsorbate) free-surface energies, as well as to the change in the surface diffusion energy favouring atom mobility.

1.3. Metal aggregates smaller than 15-20 Å

Informations on the structure of particles smaller than 15-20 Å can be obtained by Extended X-ray Absorption Fine Structure (EXAFS) or by Radial Electron Distribution (RED). The respective advantages and limitations of these techniques have been discussed (14). Recently, the latter has been improved by using anomalous X-ray scattering data to calculate partial distribution functions giving the local structure around each element separately (15).
Although these techniques have been applied to various metal supported catalysts, only the results obtained on metal aggregates supported in zeolites will be examined since the dispersion of metal is very homogeneous and the nuclearity of metal aggregates is comparable to that of large metal carbonyl clusters. The structure of Pt, Rh and Ir aggregates in Y-type zeolite is discussed in section II.

1.4. Isolated clusters (metal molecules)

Vaporized metal atoms can be trapped and embedded in a solid matrix of rare gas (16). The isolated naked clusters can be studied by spectroscopic techniques but they are inaccessible to reactants and unstable upon warming up. Isolated metal clusters containing 2 to 100 metal atoms can also be obtained from metal vapour via an adiabatic expansion resulting in a supersonic beam (17) or more recently by laser-induced vaporization of metal which is subsequently coexpanded with an inert gas (18, 19). The nuclearity of the clusters is characterized by mass-resolved photoionization technique.

There are still very few structural data on these materials. Very short Fe-Fe distances (2.02 Å) compared to bulk iron (2.48 Å) have been observed by EXAFS on Fe_2 molecules isolated in cold matrix (20). Cluster beams can be analysed after reaction with molecular jets of H_2, O_2, CO, hydrocarbons (19, 21). It appears that n-hexane can be aromatized by metal molecules as small as Pt_2 but the ease of dehydrogenation increases from Pt_2 to Pt_8 (21).

Further comparison of metal molecules with metal aggregates and with clusters must await additional characterization. Also their reactivity and catalytic properties could not be properly evaluated until they are stabilized in the normal temperature conditions.

2. STRUCTURE ANALOGIES BETWEEN METAL AGGREGATES AND CLUSTERS

The structure of platinum, iridium and rhodium aggregates encaged in Y-zeolites have been studied extensively during the past few years (22-27, 30) by Radial Electron Distribution. This technique gives the interatomic distances present in the samples. The modifications of the aggregate structure induced by adsorbates are compared with the structure of organometallic clusters known from crystal structure analysis.

2.1. Nuclearity of Pt aggregates

It was found by TEM (28) and by small angles scattering (29) that the Pt aggregates are 9 ± 3 Å large. The number of atoms per aggregate is therefore in the range 6 to 40. More precise information on the number of atoms and on the morphology of the Pt aggregates have been obtained by comparing the experimental RED with RED calculated from model aggregates. Figure 1 shows the comparison of the experimental distribution (curve b) with distributions calculated for 19 and 43 atoms spherical aggregates. Clearly the experimental distribution falls between the two calculated ones so that the aggregates should have a near spherical shape and contain between 20-40 atoms. A better fit was obtained with a 40-atoms, truncated tetrahedron (27) (figure 1). However, the first peak corresponding to the nearest neighbours distance is higher in the experimental distribution (curve 1) than in the model (curve 2).

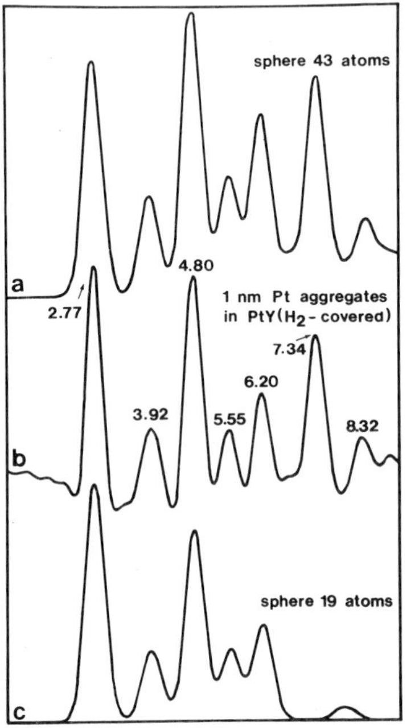

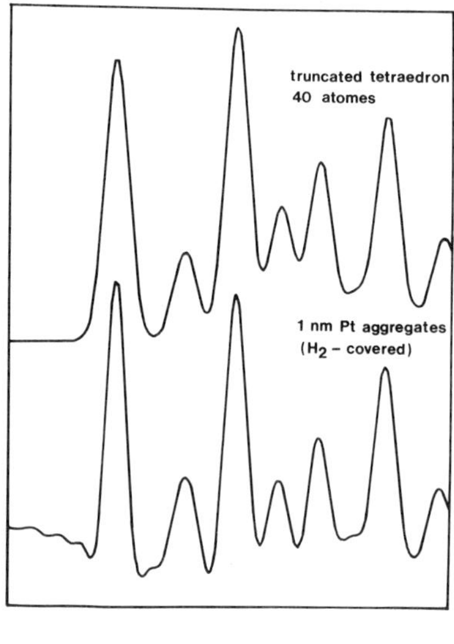

Figure 1. Comparison between the experimental distribution function of Pt aggregates in H_2 atmosphere (curve b) with the distribution functions calculated from model aggregates of spherical shape and f.c.c. structure containing 43 and 19 atoms (curves a and c respectively), and from a 40 atoms truncated tetrahedron.

This indicates that in addition to the 40-atoms aggregates there are a significant number of much smaller aggregates contributing mainly to the first peak. Model calculations have been done with a bimodal distribution of 6 atoms octahedra and 40 atoms truncated tetrahedra. There is a good fit with the experimental distribution for a proportion of 70 % of 6 atoms and 30 % of 40 atoms aggregates (Note that the latter encompass 75 % of the total number of Pt atoms) (30).

Similar experiments performed on IrY and RhY zeolites activated and reduced under proper conditions show that the number of atoms per aggregate is rather in the range 6 to 20 (30).

It turns out that Pt or Rh aggregates encaged in zeolites have a nuclearity comparable to that of large metal carbonyl clusters (2) such as $|Pt_{38}(CO)_{44}|^{4-}$ $|Pt_{26}(CO)_{32}|^{2-}$ $|Pt_{19}(CO)_{22}|^{4-}$ $|Pt_{15}(CO)_{30}|^{2-}$ $|Rh_{15}(CO)_{27}|^{3-}$ $|Rh_{22}(CO)_{37}|^{4-}$. These clusters have also a near spherical morphology except $|Pt_{15}(CO)_{30}|^{2-}$.

2.2. Structure of naked aggregates (23, 24)

The RED study of Pt aggregates outgassed in vacuum at high temperatures, i.e. without hydrogen left on the surface, shows that the f.c.c. structure is contracted and distorted with respect to bulk platinum (23, 24). Table 1 gives the interatomic distances in the naked aggregates. The contraction of the distances is expected from various theoretical calculations including the simple model of compression due to surface tension in small spherical particles. The distortion could correspond to an intermediate step between f.c.c. and non-periodic structures, indeed icosahedra and decahedra can be described as multiply-twinned, distorted f.c.c. tetrahedra (31).

The interatomic distances in naked aggregates are smaller than in large Pt carbonyls. Thus, the average nearest neighbours distances in $|Pt_{19}(CO)_{22}|^{4-}$ and $|Pt_{38}(CO)_{44}|^{4-}$ are 2.79 Å (32) and 2.85 Å (2) respectively compared to 2.77 Å for the bulk Pt metal and 2.70 Å for the naked Pt aggregates (table 1). This is due to the presence of CO ligands in the carbonyl clusters which fulfill the coordination of the Pt atoms whereas the coordination of the Pt atoms in the aggregates is incomplete. However, the excess of negative charges in the cluster anions and the electron deficiency of the Pt aggregates (33, 34) might also account for the difference in distances.

TABLE 1 : Interatomic distances (± 0.2 Å) in Pt aggregates.

Bulk Pt	H_2-covered aggregates	bare aggregates	% of contraction
2.774	2.77	2.70	2.7
3.923	3.93	3.90	0.6
4.805	4.80	4.72	1.8
5.548	5.52	5.32	4.1
6.203	6.20	6.12	1.3
7.331	7.32	7.22	1.6

2.3. Structure of hydrogen-covered aggregates (22, 24)

The adsorption of hydrogen on the Pt aggregates produces a relaxation of the distances towards those of bulk platinum (22, 24) (table 1). This is because dissociated hydrogen complete the coordination of the Pt atoms. The fact that the relaxed distances are the same as in bulk Pt is coincidental. Indeed hydrogen adsorption on Ir aggregates also produces a relaxation of the Ir-Ir distances, but they remain slightly contracted with respect to those of bulk Ir (30).

Theoretical study of Pt clusters hydrogenation (35) indicates that hydrogen atom should adsorb first in bridging position along the cluster edges. The elongation of the metal-metal distances observed upon hydrogen adsorption on platinum or iridium aggregates could well be due to an hydrogen addition in bridging position. If so, there would be a strong analogy with the oxidative addition of hydrogen in clusters which also leads to a lengthening of metal-metal bonds between H-bridged metal atoms.

2.4. Structure of CO-covered aggregates (24)

The RED distribution obtained after CO adsorption on naked Pt aggregates is given in figure 2, curve 2, in comparison with the distributions of the naked and H_2-covered aggregates (figure 2, curve 1 and 4 respectively). Except the second peak, all the distances are similar to those on H_2-covered aggregates i.e. they are relaxed with respect to those of naked aggregates. However, the heights of the peaks are smaller on curve 2 than on curve 1 or 4, thus the intensities of the four main peaks at 2.76, 4.81, 6.20 and 7.32 Å have decreased by 46, 57, 52 and 51 % respectively. This

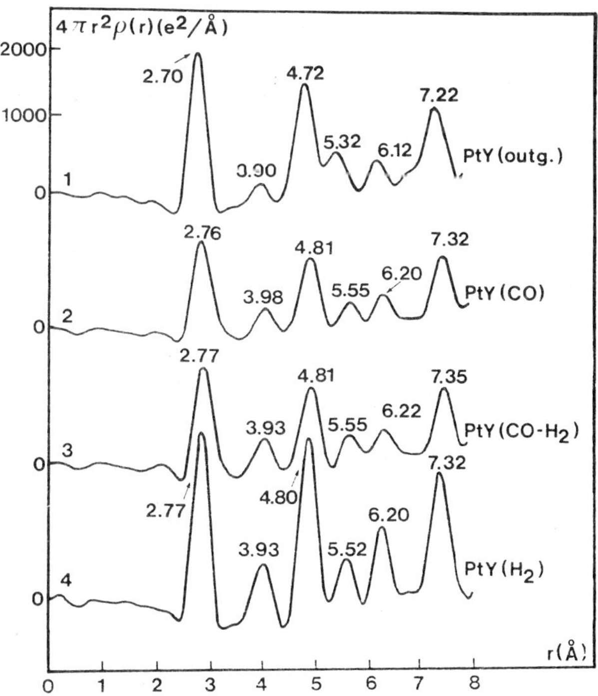

Figure 2. Distribution function of Pt aggregates under different atmosphere (1) naked aggregates (2) aggregates covered with CO (3) hydrogen adsorption on CO-precovered aggregates (4) H_2-covered aggregates.

effect of CO adsorption was interpreted by a model where the platinum atoms bonded to adsorbed CO molecules are shifted from their equilibrium positions. There is a displacement disorder involving elongation and contraction of the bond lengths which statistically cancel each other since the distances in the "average lattice" are similar to those of the regular f.c.c. lattice or to those of the H_2-covered aggregates.

There is a close analogy between the structure of the Pt aggregates covered with CO and the structure of large metal carbonyl clusters. Their nuclearity and morphology are comparable and the distribution of metal-metal distances is also similar. Indeed, in a given cluster, the metal-metal distances are very different from each other. They depend upon the local coordination of the metal atoms with other metal atoms and with the CO ligands which are terminal, bridging or capping. This induces a large spreading of distances around an average value. Thus in $|Pt_{19}(CO)_{22}|^{4-}$ the Pt-Pt distances are in the range 2.52-2.95 Å around an average distance of 2.79 Å (32). This structure, as well as that of $|Rh_{15}(CO)_{27}|^{3-}$ (36) where the nearest neighbours distances range in a continuum from 2.60 to 3.91 Å provide good models for the structure of the Pt aggregates covered with CO. On the other hand, the structure of $|Pt_{15}(CO)_{30}|^{2-}$ (37), where the Pt-Pt distances fall in two separate groups at 2.66 and 3.08 Å (corresponding to inlayer and interlayer distances respectively), is very unlikely for the Pt aggregates since the RED would exhibit two peaks instead of one at 2.76 Å.

2.5. Structure of hydrocarbon-covered aggregates (23, 25, 27)

The adsorption of hydrocarbon on bare Pt aggregates induces different changes in the structure according to the type of bonding. The associative chemisorption of benzene *via* a π bond produces only a little modification of the RED (figure 3). The structure is hardly affected because the binding energy per platinum atom is weak. On the other hand, the dissociative adsorption of methane or ethylene and the cracking of benzene lead to a displacement disorder because of the formation of strong σ bonds between the hydrocarbon fragments and the Pt atoms. Even so, the displacement disorder is weaker than that produced by CO adsorption. This is due to the difference in the stoichiometry of adsorption : at least one Pt out of two is directly involved in Pt-CO bonding (CO : Pt = 0.5) whereas an hydrocarbon fragment is likely to cap three metal atoms as suggested by the molecular structure of $H_3M(CO)_9CCH_3$ (M = Os, Ru) (38). The binding energy is then distributed over several Pt atoms which attenuates the magnitude of the Pt atom displacement .

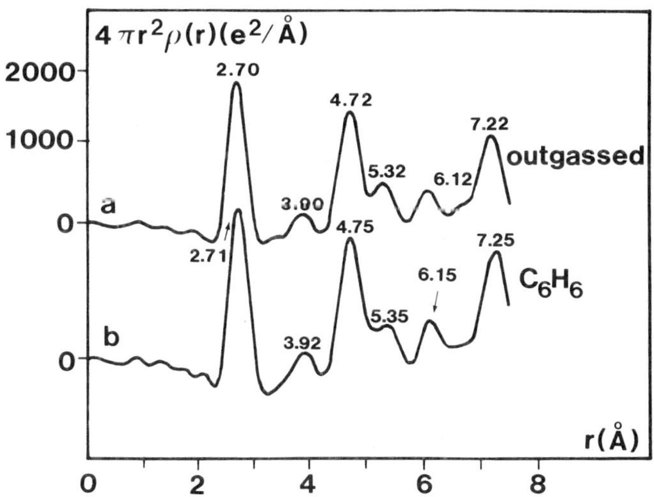

Figure 3. Distribution function of Pt aggregates (a) naked aggregates (b) benzene covered aggregates.

There is an interesting analogy between the structure of hydrocarbon-covered aggregates and that of organometallic clusters involving hydrocarbon ligands since in both species the effect of an hydrocarbon ligand on the metal framework depends upon the type of bonding. π-bonded ligands do not distort cluster frameworks. Thus in $|Rh_6(CO)_{14}C_3H_5|^-$ (39) the allyl ligand does not produce any asymmetry in the cluster since the four Rh-Rh(C_3H_5) bonds have similar lengths (2.75 ± 0.01 Å) which are also very close to the average distance of the 12 Rh-Rh bonds of the octahedra (2.753 Å). On the other hand, σ-bonded hydrocarbon ligand or carbide carbon atom could lead to marked changes in cluster structures. Thus, once

the nearly perfect octahedra Os_6CO_{18} reacts with ethylene, a complete structure rearrangement occurs : two CO ligands are replaced by two $C\text{-}CH_3$ ligands capping a triangular and a square face respectively, the Os cluster framework bears no relationship with the initial octahedron (40).
The structure of the Pt aggregates has been studied during benzene hydrogenation at 300 K (23) and n-butane conversion at 573 K (27). From the change in the distribution of distances it was concluded in both cases that more than 50 % of the Pt atoms are covered by hydrogen in the course of these reactions. Even in excess of hydrocarbon, only a small fraction of the Pt atoms are bonded with the adsorbed hydrocarbon fragment. Similarly, in the proposed mechanism of ethylene hydrogenation on Os_3CO_{12} grafted on silica (41), only one Os atom out of three, accomodates the adsorbed ethylene or the hydrogenated intermediate.

2.6. Effect of oxygen adsorption (22, 24)

The adsorbates examined previously (H, CO, hydrocarbon) are all "corrosive" in so far as they induce Pt atom displacement. However the nuclearity of the aggregate is maintained and the cohesion is still assured by metal-metal bonds (this is not true for CO adsorbed on rhodium aggregates, see section 3). In contrast, oxygen adsorption on Pt aggregates leads to the disappearance of the metal bonds between Pt atoms exposed on surface. Although the structure is then highly disordered, the peaks at 2.04 and 3.32 Å attributed to Pt-O and Pt $\overset{O}{}$ Pt bonds respectively prove the formation of an oxide-like structure. Heating above room temperature results in the development of the 3.32 Å peak <u>i.e</u> to an ordered oxide structure.
Similarly the structure of many clusters is unstable in oxygen. However because the metal framework of metal carbonyls is shielded by CO ligands, they are less prone to oxidation than metal aggregates. Nevertheless they are disrupted and oxidized at moderate temperatures.

3. TRANSFORMATION OF RHODIUM AGGREGATES INTO CLUSTERS

The transformations of clusters into metal aggregates are well documented. Once the ligands are removed - e.g. by thermal decomposition in reducing atmospheres - the naked metal frameworks could either keep the same nuclearity or agglomerate into larger particles. These transformations could be useful to prepare bimetallic aggregates of controlled composition from heteronuclear clusters (42).

On the other hand, clusters are not prepared from metal aggregates but synthesized by organometallic chemistry. However, we have recently found that <10 Å Rh aggregates supported in Y zeolites are transformed into Rh carbonyls clusters under mild conditions (43).

Figure 4, curve a, shows the RED of Rh aggregates under H_2 atmosphere. The peaks correspond to the f.c.c. structure. After contacting the zeolite with CO at 300 K, all the distances characteristic of the metal disappear (curve b). A parallel IR study shows that $Rh^{I}(CO)_2$ species are formed. Therefore the metal aggregates have been completely disintegrated by a process involving oxydation, complexation and extraction of the Rh atoms by CO molecules.

Cluster fragmentation into reactive monomeric species are common in cluster chemistry. In contrast, few other examples of complete fragmentation of a metal particle under the effect of adsorbates at low temperature have been reported. These transformation should be energetically favourable since a large number of M-L bonds can be formed but they could be limited kinetically and thus occur only on very small aggregates.

Upon further treatment with CO : H_2O mixture the $Rh^{I}(CO)_2$ species migrate and rearrange into polynuclear carbonyls. Indeed, the RED (figure 4, curve c) shows a large peak at 2.77 Å which clearly corresponds to a metal-metal distance different from that of Rh metal. Rather, it is typical of a rhodium polynuclear carbonyl, especially $Rh_6(CO)_{16}$ where the average Rh-Rh distance is 2.78 Å. The

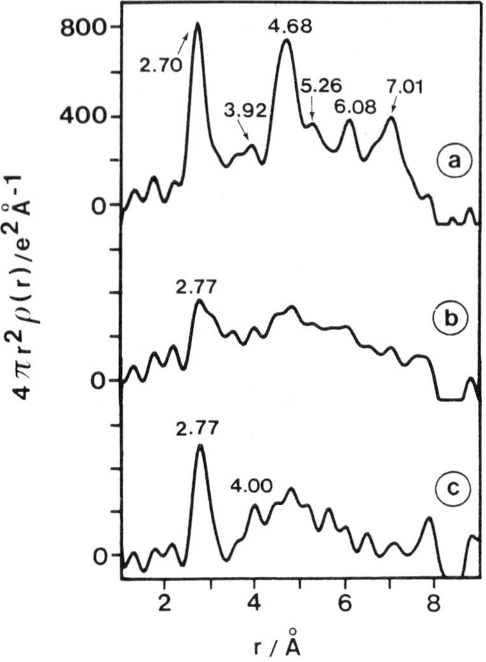

Figure 4. (a) Distribution of <10 Å Rh aggregates under H_2 atmosphere (b) effect of CO at 300 K, (c) effect of CO : H_2O mixture at 300 K.

formation of $Rh_6(CO)_{16}$ is also supported by IR spectroscopy (44). Therefore, the $Rh^I(CO)_2$ species issued from the Rh aggregate disintegration migrate along the zeolite pores and rearrange into Rh clusters inside zeolite cages. A similar process has been described previously on alumina support (45).

CONCLUSION

Theoretical calculations of cohesion energy in platinum aggregates (46) as well as experimental data on atomization and adsorption enthalpies (47) indicate that surface metal atom may be bonded more strongly to ligands like H or CO than to each of their metal atom neighbours. Accordingly, the arrangement of surface atoms should be perturbed by adsorbates. The flexibility of the structure of metal aggregates should even be larger because of the low coordination of the atoms favouring their accessibility and their affinity to surface ligands. Since the strengths of the metal-metal and metal-ligand bonds are comparable to those of metal carbonyl clusters one should expect for metal aggregates and clusters of comparable nuclearity the same diversity of configurations and the same wide range of metal-metal bond lengths depending upon the local coordination of each atom.

The experimental structure data obtained on metal aggregates encaged in zeolites support these expectations. Although their structure cannot be described with the same details as that of clusters, the set of interatomic distances obtained by RED proves unambiguously that the structure is as flexible as that of organometallic clusters. Depending upon the nature and coverage of adsorbates the f.c.c. packing might undergo all the modifications from a slight distortion to a complete disintegration with all the intermediate states of disorder. These modifications have been also observed during catalytic reactions which means that the structure of the fresh catalysts should not be viewed as an immobile set of atoms in the course of surface reactions. The ultimate disorder, namely the disintegration of metal aggregates into monomeric species could depend upon the oxidability of the metal. Indeed structure disintegration has been observed for CO and NO adsorption on Rh and Pd aggregates respectively (43, 48) giving Rh^{I} and Pd^{II} carbonyl and nitrosyl complexes respectively, but it has not been detected on Pt aggregates at room temperature.

The structure of metal aggregates on other supports has probably the same flexibility provided they are as small as those encaged in zeolites. The structure and morphology modifications mentioned in this paper show that the long lasting picture of static metal structure is fading out. It appears that the structure of metal aggregates is similar in many respects to that of organometallic clusters. Thus the active species involved in heterogeneous and homogeneous catalysis might be less different than expected previously.

REFERENCES

1. M.B. Gordon, F. Cyrot-Lackmann, M.C. Desjonquères, Surf. Sci., 80, 159 (1979).
2. P. Chini, J. Organometal. Chem. 200, 37 (1980).
3. Adsorption on Metal Surfaces (J. Benard Ed.), Elsevier, Amsterdam 1983, pp. 15.
4. M.A. Van Hove in The Nature of the Surface Chemical Bond (T.N. Rhodin and G. Ertl eds.) North Holland, Amsterdam 1979, pp. 277.
5. W.C. Turkenburg, R.G. Smeenk and F.A. Saris, Surface Sci., 72, 562 (1978).
6. H.D. Hagstrum and G.F. Becker, J. Vac. Sci. Technol., 14, 369 (1977).
7. G.B. Fisher, Surf. Sci., 62, 31 (1977).
8. G.A. Somorjai, J. Catal., 27, 453 (1972).
9. M.J. Yacaman and T. Ocana, Appl. Phys. Lett., 30, 359 (1977).
10. T. Wang and L.D. Schmidt, J. Catal. 66, 301 (1980).
11. R.W. Mc Cabe, T. Dignet and L.D. Schmidt, J. Catal., 32, 114 (1974).
12. D. Hasenberg and L.D. Schmidt, J. Catal., 91, 116 (1985).
13. H. Poppa, Vacuum, 34, 1081 (1984).
14. P. Gallezot, X-ray techniques in catalysis in "Catalysis, Science and Technology" (J.R. Anderson and M. Boudard eds) Springer Berlin 1984, Vol. 5, Chapter 4.
15. M. Saman, G. Bergeret, P. Gallezot, M. Boudart, to be published.
16. G.A. Ozin, Faraday Symp. Chem. Soc., 14, 7 (1980).
17. A. Herrmann, E. Schumacher and L. Wöste, J. Chem. Phys., 68, 2327 (1978).
18. D.F. Powers, S.G. Hansen, M.E. Gensic, A.C. Puiv, J.B. Hopkins, T.G. Dietz, M.A. Duncan and R.E. Smalley, J. Phys. Chem., 86, 2556 (1982).
19. R.L. Whetten, D.M. Cox, D.J. Trevor and A. Kaldor, J. Phys. Chem., 89, 566 (1985).
20. H. Purdum, P.A. Montano, G.K. Shenoy and T. Morrison, Phys. Rev. B, 25, 4412 (1982).
21. D.J. Trevor, R.L. Whetten, D.M. Cox and A. Kaldor, J. Am. Chem. Soc., 107, 518 (1985).
22. P. Gallezot, A. Bienenstock and M. Boudart, Nouv. J. Chim., 2, 263 (1978).
23. P. Gallezot and G. Bergeret, J. Catal., 72, 294 (1981).
24. P. Gallezot, Zeolites, 2, 103 (1982).
25. P. Gallezot, J. Chim. Phys., 78, 881 (1981).
26. G. Bergeret and P. Gallezot, J. Catal. 87, 86 (1984).
27. G. Bergeret and P. Gallezot, Proc. 7th Int. Congr. Catalysis, Berlin, Dechema Frankfurt 1984, Vol. 5, pp. 659.

28. P. Gallezot, I. Mutin, G. Dalmai-Imelik and B. Imelik, J. Microsc. Spectr. Electron, 1, 1 (1976).
29. P. Gallezot, A. Alarcon-Diaz, J.A. Dalmon, A.J. Renouprez and B. Imelik, J. Catal., 39, 334 (1975).
30. G. Bergeret and P. Gallezot, to be published.
31. C.Y. Yang, J. Crystal Growth, 47, 274 (1979).
32. D. Washecheck, F.J. Wucherer, L.F. Dahl, A. Ceriotti, G. Longoni, M. Manassero, M. Sansoni and P. Chini, J. Amer. Chem. Soc., 101, 6110 (1979).
33. P. Gallezot, J. Datka, J. Massardier, M. Primet and B. Imelik in Proc. 6th Int. Congr. on Catalysis (G.C. Bond et al. eds) The Chemical Society, London, 1977, p. 144.
34. P. Gallezot, R.S. Weber, R.A. Dalla Betta and M. Boudart, Z. Naturforsch, 34A, 40 (1979).
35. C. Minot, B. Bigot and A. Hariti, J. Amer. Chem. Soc. Accepted.
36. S. Martinengo, G. Ciani, A. Sironi and P. Chini, J. Amer. Chem. Soc., 100, 7096 (1978).
37. J.C. Calabrese, L.F. Dahl, P. Chini, G. Longoni and S. Martinengo, J. Amer. Chem. Soc. 96, 2614 (1974).
38. G.M. Sheldrick and J.P. Yesinowski, J. Chem. Soc., Dalton Trans. 873, (1975).
39. G. Ciani, A. Sironi, P. Chini, A. Ceriotti and S. Martinengo, J. Organometal. Chem., C39, 192 (1980).
40. C.R. Eady, J.M. Fernandez, B.F.G. Johnson, J. Lewis, P.R. Raithby and G.M. Sheldrick, Chem. Commun., 421, 1978.
41. B. Besson, A. Choplin, L. d'Ornelas and J.M. Basset, J.C.S. Chem. Comm. 842 (1982).
42. A. Choplin, T. Huang, A. Theolier, J.M. Basset, S.G. Shore, R. Mathieu, to be published.
43. G. Bergeret, P. Gallezot, F. Lefebvre, C. Naccache and Y. Ben Taarit, to be published.
44. F. Lefebvre, P. Gelin, C. Naccache and Y. Ben Taarit, in Proc. 6th Int. Conf. on Zeolites (D. Olson and A. Bisio eds.) Butterworths, Guildford UK, 1984, p. 435.
45. A.K. Smith, F. Hugues, A. Theolier, J.M. Basset, R. Ugo, G.M. Zanderighi, J.L. Bilhou-Bouanol and W.F. Graydon, Inorg. Chem., 18, 3105 (1979).
46. B. Bigot and C. Minot, J. Amer. Chem. Soc., 106, 6601 (1984).
47. C. Hansecraft, M.E. O'Neill and K. Wade, J. Organomet. Chem., 213, 35 (1981).
48. M. Che, J.F. Dutel, P. Gallezot and M. Primet, J. Phys. Chem., 80, 2371 (1976).

CATALYSIS BY MONO AND POLYNUCLEAR COMPOUNDS SUPPORTED ON ZEOLITES

by

M.Lenarda and R.Ganzerla

Dipartimento di Chimica, Universita' di Venezia (Italy)

J.Kaspar and M.Graziani

Dipartimento di Scienze Chimiche, Universita' di Trieste (Italy)

ABSTRACT - Osmium, iridium and ruthenium complexes supported on zeolites are used as catalysts for water gas shift, hydrogenation and hydroformylation reactions. The reactions are carried out in batch or in a flow reactor. In all cases the catalytic activity depends on the thermal treatment and on the conditions of the heterogeneization process. The catalyst precursors undergo several modifications during the reaction, and some of the species which are formed were identified by i.r. spectroscopy.

1. INTRODUCTION

Over the last decade there has been growing interest in the use of catalytic systems formed supporting transition metal complexes and cluster carbonyls on inorganic oxides (Ref. 1).

Zeolites are quite a unique family of inorganic supports owing to their particular cage and pore structure. The peculiar zeolitic framework can furnish, first of all, the adsorbed catalytic species with the same shape selective properties rarely detected on amorphous supports. Secondly the thermal stability of zeolite entrapped clusters and complexes results much higher, compared with silica or alumina; the zeolitic framework in fact allows only limited aggregation of carbonyl species, because of its particular structure.

Besides that, if metallic catalysts are generated inside the zeolitic framework by reduction of the precursor complexes, the zeolite can control the nuclearity of the metallic clusters, preventing the formation of large metal atom aggregates and providing stabilization of small particles.

It has been proved, that adsorbtion of certain transition metal carbonyls and particularly carbonyl clusters into the zeolitic framework, followed by thermal desorption, leads to carbonyl ions or to small metal particles (Ref. 1a). These species are active catalysts for the Fischer-Tropsch synthesis and have a good selectivity for light olefins (Ref. 2).

We have used a zeolite as support for carbonyl clusters or transition metal complexes as catalyst precursors, to obtain a better understanding of the factors governing formation, trapping and stabilization of transition metal complexes on zeolites and their catalytic activity in water gas shift, hydrogenation and hydroformylation reactions.

2. RESULTS and DISCUSION

2.1. Osmium system. $Os_3(CO)_{12}$ was adsorbed on acidic zeolite (HX), prepared by conventional methods from 13X type zeolite (Ref. 3), to give $[Os_3(CO)_{12}.HX]$. This compound was heated to 180°C to give a carbonyl compound having an i.r. spectrum with bands at 2120, 2080, 2068, 2015, 1980 cm^{-1} which is very similar to that of osmium species already reported (Ref. 4-7). Comparison of the spectrum with that of the product of the reaction between $Os_3(CO)_{12}$ and Ph_3SiOH (Ref. 4) indicated that the compound could be formulated as $[HOs_3(CO)_{10}(O\text{-}Surf)]$, herein after indicated as Os-1,(spectrum 1, figure 1). This compound catalyzes the water gas shift reaction at 140°C under 1 atm of CO and in the presence of an excess of water, but its activity increases with time as reported in table 1. Turnover number calculated as indicated in table 1, increases from 0.4 to a value of 3 in run 1. The i.r. spectrum of the osmium containing zeolite at the end of the run 1 shows carbonyl bands at 2130, 2050 cm^{-1} (figure 1, spectrum 2). This compound, herein after indicated as Os-2, was reused and the catalytic activity still increased with time to reach a maximum. At the end of the run 2 the i.r. spectrum was checked and it was found changed, the catalytic species infact shows bands at 2120, 2035 and 1960 cm^{-1} (figure 1, spectrum 3.). This sample, herein after indicated as Os-3, was tested as catalyst after addition of a large excess of water. At the beginning, the catalytic activity was as expected, but it decreased and at the end of the run 3 (table 1) it became rather low. The i.r. spectrum of the catalyst at this stage was that of $Os_3(CO)_{12}$ adsorbed on zeolite.

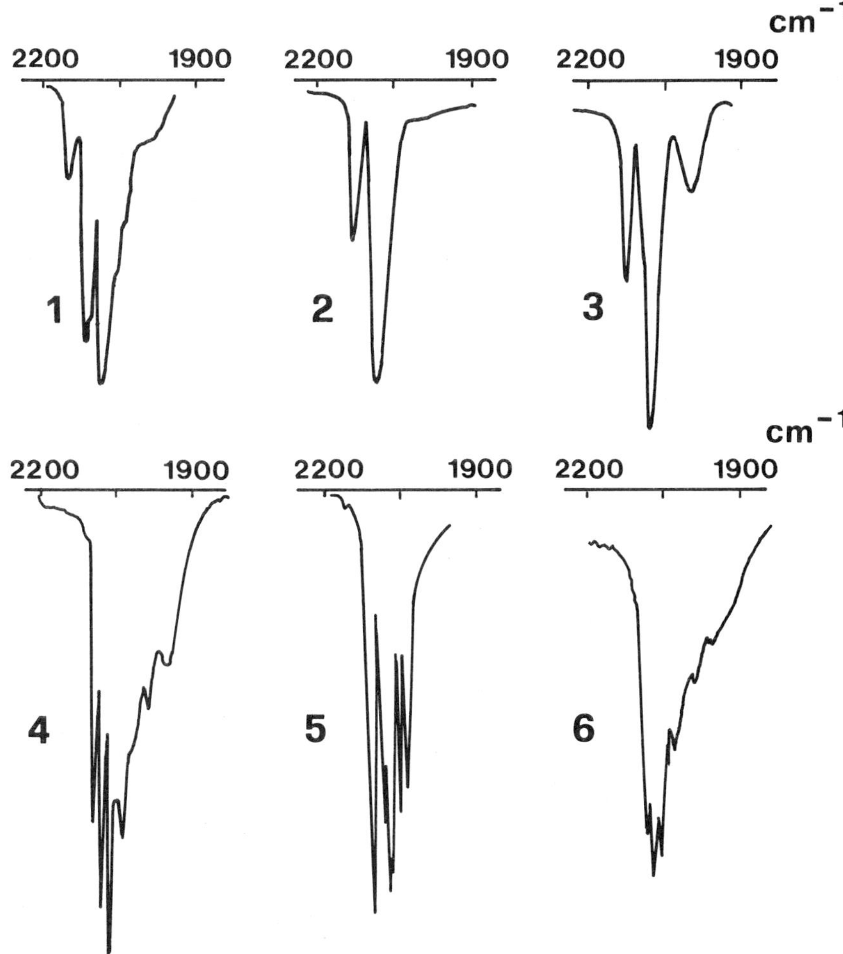

Figure 1. 1. Os-1; 2. Os-2; 3. Os-3; 4. $[H_3Os_4(CO)_{12}]^-$ in KBr pellet; 5. $Os_3(CO)_{12}$ adsorbed on zeolite; 6. $[H_3Os_4(CO)_{12}]^-$ trapped in the zeolite framework.

Table 1. Water gas shift reaction catalyzed by $Os_3(CO)_{12}$ on zeolite as catalyst precursor.[a]

Run	Catalyst	Turnover[b]	
1	Os-1	0.4	calculated over first 25 hrs of reaction.
		3	calculated over last 20 hrs of reaction.Overall reaction 140 hrs.
2	Os-2	4	calculated over first 25 hrs of reaction.
		12	calculated over last 20 hrs of reaction.Overall reaction 285 hrs.
3	Os-3[c]	12	calculated over first 25 hrs of reaction.
		0.5	calculated over the last 20 hrs of reaction.Overall reaction 100 hrs.

a. Reaction conditions : temperature 140°C; p_{CO} 1 atm; water added in excess.
b. Turnover number calculated as moles of H_2 formed x (gr.atom of metal)$^{-1}$ x day^{-1}.
c. Excess of water was added to the catalyst obtained from run 2.

A species with the spectrum very similar to Os-3 has been reported (Ref. 4,10,11) in literature, studying the thermal behaviour of osmium clusters on various oxides, and it has been formulated as a mononuclear oxidized Os(II) carbonyl bound to the surface. It has been also reported (Ref. 8,9) that when osmium clusters as $Os_3(CO)_{12}$, $Os_6(CO)_{18}$, $H_4Os_4(CO)_{12}$, impregnated on silica, alumina or titania, are heated up to 250 °C, the surface product obtained has similar spectral properties.

The very simple i.r. spectrum (it shows three bands in the carbonyl region) is attributed (Ref. 9) in this case to a family of high nuclearity clusters. In spite of several spectroscopic, physico-chemical techniques employed up to now, the nature of this species is still uncertain. Indeed there are quite strong suggestions that we are in the presence of the same species produced in different conditions on various supports (Ref. 1d), even if it is still impossible to completely wipe out the suspicion that the close similarity of the i.r. spectra could be possibly due to a coincidence.

The Os-1 species has also been tested in a flow reactor operating at atmospheric pressure in differential conditions. Also in these conditions an evolution of the catalyst was observed. Infact, after 4 hours on stream (W/F_{tot} = 0.21 g_{cat} x min x ml^{-1}; H_2O:CO:Ar = 1:1:5; at 180 °C), the species Os-2 is formed from the grafted cluster (Os-1). This species is stable for weeks in the flow conditions as long as CO is present. In the absence of CO, the species Os-2 evolves into the species Os-3, which is also catalytically active for the WGSR, but its activity decreases from an initial reaction rate of 8 umoles x g_{cat}^{-1} x h^{-1} to a value of 4 umoles x g_{cat}^{-1} x h^{-1} (W/F_{tot} = 0.43 g_{cat} x min x ml^{-1}; p_{H_2O} = 66 torr; p_{CO} = 87 torr; 180 °C) after 4 days on stream. A similar deactivation of the catalyst was reported in the Fischer-Tropsch synthesis for an analogue osmium species supported on silica and it was attributed to the deposition of carbon on the catalyst surface (Ref. 10).

The catalytic behaviour of Os-2 has been studied in the temperature range of 206-265°C at various H_2O and CO pressures. The rate of the reaction measured at 206°C at constant pCO of 85 torr depends on pH_2O up to a value of 200 torr, but it becomes independent on the water content at higher values of pH_2O. Higher reaction temperatures need higher value of pH_2O to reach the zero order dependence. The apparent activation energy resulted 17.8 kcal/mole. These observations probably show that the activation of water molecules proceeds through a strong chemisorbtion by active sites of the support, before to react with supported osmium carbonyl. The chemisorbtion process can be dissociative, non dissociative or both (Ref. 12-14) and it competes with the thermal dehydration process of the zeolite. The reaction rate is consequently controlled by the number of water molecules present on the support. When water pressure is sufficiently high, for a given temperature, to saturate the zeolite active sites, the reaction appears independent from pH_2O. The dependence on pCO at constant pH_2O (85 torr) shows a rather different behaviour. The reaction infact appears independent from pCO, but at pCO > 400 torr a decrease in the reaction rate is observed. The accepted mechanism for WGSR proceeds via reaction of water molecule to CO coordinated to transition metal with formation of coordinated formiato anion which decomposes to produce CO_2. If the metal carbonyl species is already present, as Os-2 seems to be, pCO could not have any influence in the formation of active species, but assuring its stability. At higher pressure CO can compete for the coordination to the metal center depressing the overall reaction rate. These results seems to confirm that the reaction, as was previously observed, occurs bifunctionally with the metal activating the carbon monoxide molecule by coordination and the surface activating the water molecule by chemisorbtion (Ref. 12-14).

The Os-3 species, prepared by various methods in zeolite, regenerates $Os_3(CO)_{12}$, when heated with CO, H_2 and H_2O in 1:1:20 molar ratio at 140°C. The presence of H_2, has been proved not necessary, because it is produced via WGSR from CO and H_2O; when present, it shortens the reaction time. Regeneration of the starting cluster species from surface carbonyls in the presence of CO and excess of water has been observed in other cases. In the case of rhodium, the surface carbonyls "$Rh(CO)_2$" on silica (Ref. 15,16) regenerate the starting $Rh_4(CO)_{12}$ and $Rh_6(CO)_{16}$. The reaction of the osmium covalently grafted cluster $Os_3(CO)_{10}(H)(O\text{-}Si\lessgtr)$ on silica, with water and carbon monoxide is outlined by Psaro et al. (Ref. 4b) and the starting cluster $Os_3(CO)_{12}$ is obtained. Hydrolysis of the metal-O-surface bond and successive carbonylation, are invoked to tentatively explain this unusual reaction that seems not have homogeneous counterpart. We have confirmed these results recovering $Os_3(CO)_{12}$.

It is worth to mention that if the isopropanol is used in place of water, the reaction occurs at 140°C in the presence of CO but the product $Os_3(CO)_{10}(OR)(H)$ (Ref. 17) is obtained, probably via the easier reaction of ROH with $Os_3(CO)_{12}$.

A comparison of the results obtained by other research groups (Ref. 4,5,8-11) suggests that reaction of the osmium carbonyl clusters with the surface hydroxils of the various kind of inorganic supports follows different reaction paths. Each path strongly depends on the previous thermal treatment of the support, on the temperature and the gaseous environment and finally it gives different catalytic species.

The catalytic activity of the species Os-1 has been studied also in the presence of a base.

Table 2. Water gas shift reaction catalyzed by Os-1 in the presence of added base[a]

Run	Added base	TN[b]
1	ethylenediamine	330
2	NaOH	616
3	n-propylamine	288
4[c]	ethylenediamine	1125

a. Reaction conditions: all experiments were carried out in diglyme at 130°C in the presence of $3.5x10^{-4}$ moles of added base; p_{CO}= 1 atm.
b. TN: turnover number after 24 hours of reaction as moles H_2 x (gr.atom of metal)$^{-1}$ x day^{-1}.
c. p_{CO}= 3 atm.

A suspension in diglyme, in the presence of base, of Os-1 (0.6% in osmium) obtained as previously reported (Ref. 7), catalyzes the water gas shift reaction and the results are reported in table 2. Temperatures higher than 100°C seems to be necessary for a good activity of the system, but at 150°C after a good initial activity there is a fast deactivation, and 130°C is the best temperature for the study. The system is rather insensitive to the amount of the base added, but it appears rather dependent on the CO pressure (table 2, run 4).

After 24 hours of reaction, the catalysts were filtered off and dried in vacuum (sample A). The filtrate was evaporated to dryness (sample B). Both solids were examinated by i.r. spectroscopy. The patterns of the spectra of samples A and B depend on the nature of the base used. Type A sample recovered from the experiment described in run 1 of table 2 shows two broad bands in the carbonyl region at 2040 and 1960 cm^{-1}. The solid recovered from the reaction solution (B type sample) shows the same pattern with the maxima slightly shifted to lower frequencies (2000 and 1920 cm^{-1}). We can conclude that during the reaction, the original Os-1 sample is modified to form $Os^{II}(CO)_n$ (n=2,3) species bonded to the zeolites and that about 50% of the catalyst is released into the solution. A similar i.r. band pattern (ν_{CO}: 2050, 1960 cm^{-1}) was attributed (Ref. 18) to $Ru^{II}(CO)_2$-zeolite. Osmium carbonyl supported on various oxides gives analogous species at temperature higher than 120°C (Ref. 19). The species $Os^{II}(CO)_2$-zeolite generated in the first part of the reaction in the presence of ethylenediamine was reused in the absence of base. This species catalyzed the WGSR in minor extent (about 50 cycles per day), but it shows a constant activity for at least 10 days. Occasional addition of CO to the system is necessary to keep the system runnig. The system obtained with NaOH as added base behaves quite differently. After 24 hours of reaction both the modified zeolite (sample A) and the solution were analyzed by i.r. spectroscopy. Type A samples show strong bands ay 2040, 2020, 2000 and 1970 cm^{-1} together with broad absorption of medium intensity in the range 1920-1950 cm^{-1}. Type B samples show bands at 2040, 2020, 2000, 1970 and 1950 cm^{-1} (see figure 1, spetrum 4). Samples of the type A obtained in run 2 and 3 of table 2 were reused as catalysts in the WGSR under the same reaction conditions but without added base. In these conditions the modified zeolite, herein after indicated as Os-1-H, catalyzed the WGSR with a constant activity for at least 160 hours. Analysis of the solution shows that osmium is no longer released from the zeolite.

To obtain a better understanding of the nature of the compounds formed in the zeolite during the reaction we have examined the reactivity of $Os_3(CO)_{12}$ as catalyst of WGSR in homogeneous phase under the conditions used for Os-1. Some activity data were already available in the literature (Ref. 20). The cluster catalyzes WGSR in the presence of base with an activity of few tens of turnover per day. The i.r. spectra of the solution evaporated to dryness (type B sample) shows pattern dependent on the base used. The spectrum of the compound obtained in the reaction with ethylenediamine as added base, shows bands at 2000 and 1920 cm^{-1}, similar to those obtained from run 1 of table 2, and attributed to mononuclear osmium carbonyl. On the other hand in the presence of NaOH the compound separated from the solution shows a spectrum with bands at 2040, 2020, 2000, 1970, 1950 and 1920 cm^{-1} which is identical to the spectrum of the B type sample separated from run 2 of table 2 and very similar to that of the zeolitic species obtained in the same run. $Os_3(CO)_{12}$ reacts in iso-butanol in the presence of KOH to give $[H_3Os_4(CO)_{12}]^-$ as major product. The spectrum of this species shows bands at 2048, 2022, 2000, 1976 cm^{-1} (Ref. 21)(figure 1, spectrum 4), which is identical to that of the compound we obtained from the reaction of $Os_3(CO)_{12}$ in homogeneus phase. Thus it is probable that the active species in the homogeneous reaction and the species lost from the zeolite in run 2 and 3 reported in table 2 is the anion $[H_3Os_4(CO)_{12}]^-$. Recently Gates et al. (Ref. 22) prepared the anion $[H_3RuOs_3(CO)_{12}]^-$ supported on γ-alumina and the i.r. spectra is also similar to that of Os-1-H we formed in zeolite. We treated our compound with a $NEt_4^+Cl^-$ solution, but observed no extraction of the anion which is probably trapped in the cages of the zeolite. We prepared $Os_3(CO)_{12}$ supported on γ-alumina as described in the literature, to obtain the alumina analogue of Os-1. This species was subjected to the reaction conditions reported in run 2 of table 2. The hydridocarbonyl anion $[H_3Os_4(CO)_{12}]^-$ is formed, but nothing remains on the support, and the compound is found only in solution. This proves that the above anion can be prepared from the supported $Os_3(CO)_{12}$, but it remains on the support only when a zeolite is used.

It appears that under our experimental conditions a carbonyl anion is formed during the reaction and some of it, presumably situated on the external surface is lost in the solution and some is trapped in the cages. The latter catalyzes the WGSR for a long period of time at constant activity. To have a better proof of this assumption we used this compound (Os-1-H) as hydrogenation catalyst at 3 atm of H_2 pressure and 70°C. It readily catalyzes the hydrogenation of propene, but not that of the bulkier norbornadiene, following a typical zeolite shape selectivity pattern. By reaction of suitable reagents with complexes supported on zeolites it is possible to prepare catalysts for various reactions (for example for WGSR or olefin hydrogenation), which are stable in the reaction conditions and exhibit shape selctivity due to the structure of the zeolite. Such a catalyst, suspended in a suitable solvent from which it can be easily separated, mantains its identity and selectivity in the above reactions. Furthermore we have always observed enhanced reactivity in the supported catalyst, compared with the homogeneous counterpart in the WGSR. This result can be tentatively attributed to a possible activating interaction of the coordinated carbonyls with the acid centers of the zeolite structure.

2.2. <u>Iridium system.</u> $[Ir(NH_3)_6]^+$ supported on 13-X zelite by conventional methods (Ref. 23,24) was treated with CO at 1-4 atm and in the range 100-300°C with H_2O, keeping the H_2O/CO ratio at a constant value of 2. Under these conditions the WGSR proceeds and the results are shown in figure 2 and in table 3. The significant aspect of the figure (which refers to CO/Ir ratio of 1700; cfr. table 3) is that three different regions can be identified (A,B and C). This behaviour can be attributed to the formation of three different species having different catalytic activities. The species A,B and C were isolated by keeping the supported iridium hexamino complexes under the reaction conditions for suitable times. The infrared spectra of the species A,B and C are shown in figure 3. Carbonyl compounds are formed, but substantial changes in the form of the spectra occur during the reaction in which the catalyst changes from A to B to C and at the same time the catalytic activities decreases.

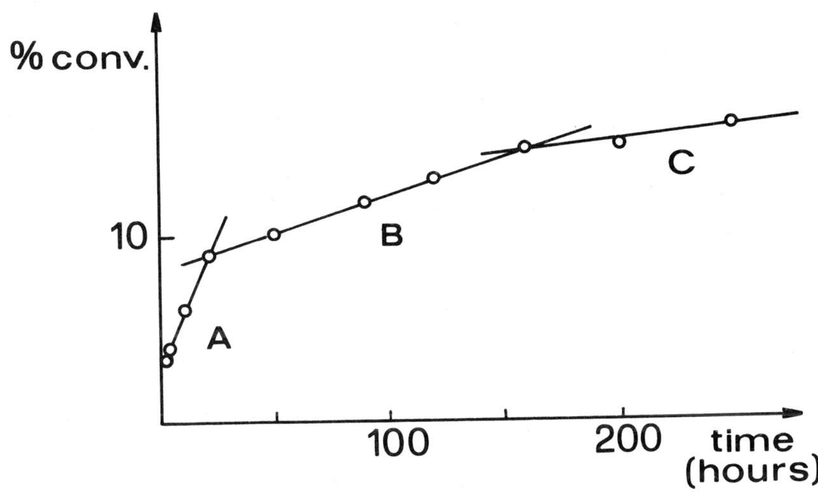

Figure 2. Percentage conversion vs.time for the WGSR catalyzed by [Ir(NH_3)$_6$-zeolite]

The interesting feature is the presence of a band at 1330 cm^{-1} which can be attributed to coordinated NH_3. The species A,B and C appear to be bonded to the support by reaction with the hydroxo groups present on the surface, so that washing the zeolite - supported products with various solvents leaves the i.r. spectrum unchanged. The species A which we have formulated as an iridium carbonylamino compound, appears to be the most active. This species seems to be stabilized by the zeolite framework, probably via some chemical interaction with the reactive groups present on the surface. After 5 hours of reaction the catalyst became completely inactive and gave iridium metal (Ref. 25). This catalyst was also tested at 220°C (the temperature was selected as high as possible, but below 250°C to avoid formation of metal). The results are similar to those obtained at 140°C, and the final spectrum has the same pattern as C in figure 3. At this temperature however we were not able to detect any "aminocarbonyl" intermediate. The reaction probably follows the same path as at 140°C, but the higher temperature prevents the trapping of the intermediate.

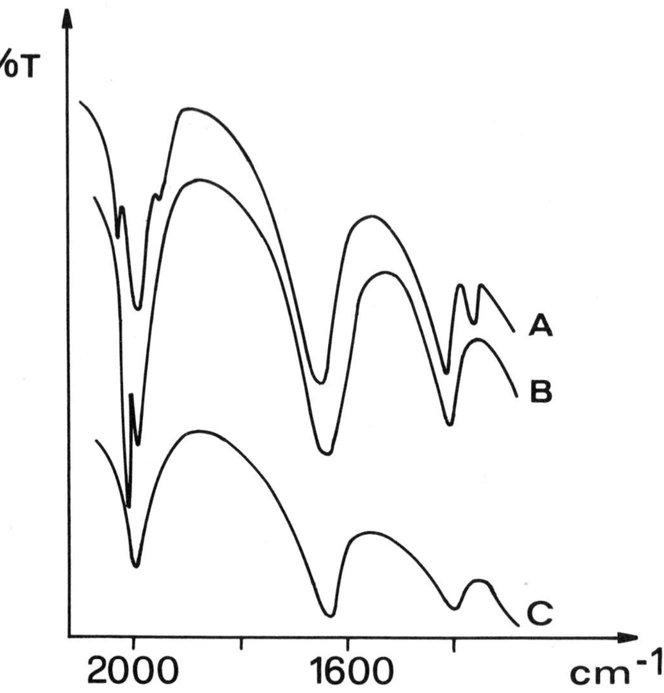

Figure 3. Infrared spectra of the recovered [Ir(NH_3)$_6$-zeolite] after treatment with CO and H_2O under the conditions corresponding to the ranges for A,B, and C respectively.

As mentioned before, we feel that the first species formed (A) could be an "iridiumaminocarbonyl" compound somehow trapped in the zeolite cages, probably without homogeneous counterpart. A similar compound has been proposed for WGSR catalyzed by ruthenium complexes supported on zeolite (Ref. 18,26). To obtain a better evidence in supporting this hypothesis we carried out some experiments both in homogeneous and heterogeneous phases. [Ir(NH_3)$_6$]$^+$ was used as homogeneous catalyst in water for the water gas shift reaction or in water/methanol mixture, but in neither cases any activity was observed, and at the end of the experiment the starting material was recovered unchanged. The same compound on zeolite was also used as catalyst for WGSR in a flow reactor at 1 atm and in the range 140-220°C. The catalytic activity was constant for about two hours, but it was followed by a fast deactivation. The deactivation process is much faster under flow conditions, and we think that it could be attributed to an easier displacement of NH_3 under such conditions leading to a completely carbonylated iridium complex having the same i.r. spectrum as C in figure 3.

Table 3. WGSR catalyzed by iridium compounds supported on zeolite[a]

Run	Catalyst	Reaction time(h)	% Conversion	TN[b]	T (°C)	pCO (atm)	H_2O/CO	CO/Ir
1	[Ir(NH_3)$_6$-X]	2	3.5	30	140	1	2	1700
		4	4	17				
		22	9	7				
		90	11.5	2.2				
		120	13	1.8				
2	[Ir(NH_3)$_6$-X]	1	7	110	220	2	2	1600
		4	12	50				
		9	19	35				
		24	42	28				
		30	51	27				
		35	55.5	26				
		50	63	25				
3	[Ir(en)$_2$(NH_3)$_2$-X]	1	2	30	220	2	2	1600
		3	29	150				
		5	53	170				
		7	70	160				
		9	78	140				
4	[Ir(en)$_3$-X]	1	0.8	15	220	2	2	1700
		3	2.3	13				
		4	2.8	12				
		7	4.5	11				
		10	6	10				
		22	11	9				
		30	16	9				

a. X = 13-X faujasite type zeolite.
b. Turnover calculated as (moles of product) x hour^{-1} x (gr.atom of metal)$^{-1}$

To hinder or prevent the easy displacement of the ligand at the working temperature the zeolite supported procatalyst was treated at 90°C and 200°C with an excess of ethylenediamine to give products we designated as $(Ir-en)_1$ and $(Ir-en)_2$ respectively. Comparison of the i.r. spectra of the two products with those of authentic samples indicates that they can be formulated as $[Ir(en)_2(NH_3)_2$-zeolite$]$ and $[Ir(en)_3$-zeolite$]$ respectively. The catalyst prepared by supporting the homogeneous analogue of $(Ir-en)_1$ showed the same reactivity as $(Ir-en)_1$ generated on the support. The i.r. spectra of used catalyst obtained from $(Ir-en)_1$ shows significant bands at 2060, 1480 and 1400 cm^{-1}, attributable to coordinated CO, ethylenediamine and NH_4^+ bonded to the zeolite, but no bands attributable to coordinated NH_3 were found.

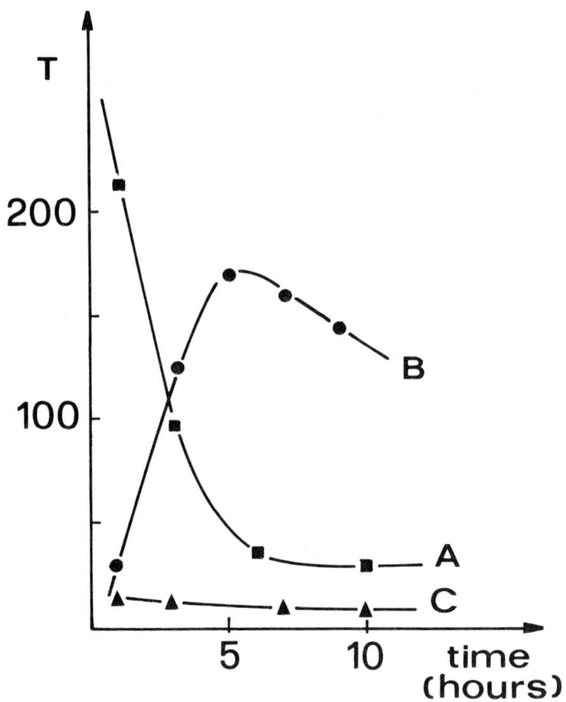

Figure 4. Turnover (molecules of products x h^{-1} x metal atom^{-1} vs. time. A= $[Ir(NH_3)_6$-zeolite$]$; B= $[Ir(en)_2(NH_3)_2$- zeolite$]$; C= $[Ir(en)_3$-zeolite$]$; en= ethylene diamine

In the case of (Ir-en)$_2$ only the bands attributable to the starting complex were found.

(Ir-en)$_1$ was also tested as catalyst in flow reactor. The reaction was run for 72 hours in the range of temperature 180-220°C, without loss of activity. The i.r. spectrum of the final product is the same as was found when (Ir-en)$_1$ was used for the WGSR in autoclave. These results are clearly rappresented in figure 4. Curve A describes the behaviour of [Ir(NH_3)$_6$-zeolite]. The activity is rather high at first, but it decreases rapidly as NH_3 ligands are replaced by CO to give fully carbonylated product. Curve B refers to [Ir(en)$_2$(NH_3)$_2$-zeolite]; the activity initially increases due to the replacement of NH_3 ligands by CO. After 5 hours the catalytic activity decreases, probably because of the considerable drop of CO pressure, and upon restoring the original pressure, the catalytic activity increases again, and it remains constant until the next pressure drop is reached. Curve C shows the behaviour of a sample with all the ammonia molecules replaced by ethylenediamine. The activity is very low since CO hardly replaces the chelating ligand.

In conclusion the active species seems to be an iridium-ethylenediamine carbonyl compound which requires high temperatures and longer times for its formation than the corresponding amino complex. The catalyst however lasts longer than the amino analog under our reaction conditions. We attribute this behaviour to the difficulty to replace ethylenediamine by CO at these temperatures and pressures, so that the inactive, totally substituted iridium carbonyl cannot be formed.

2.3. Ruthenium system. Faujasite type zeolite (NaX) was exchanged with [Ru(NH_3)$_6$]$^{+2}$ to form a wine coloured ruthenium(III) containing zeolite, for which the species [Ru(NH_3)$_5$OH]$^{+2}$ has been proposed (Ref. 27). This complex (Ru-X) has proved to be an active catalyst for water gas shift reaction and the resulting carbonyl containing zeolite is equivalent to that produced from the similarly active [Ru(NH_3)$_6$]$^{+3}$-X (Ref. 18). Two different batches of catalyst were prepared and were of 3% (sample A) and 5% (sample B) ruthenium content respectively. In both cases no appreciable loss of crystallinity was detected by X-ray powder spectra but they exhibited a rather different behaviour in catalitic activity.

Sample A was subjected to 100 atms at 200°C of $H_2/CO/C_2H_4$ (2.5:2.1:1.5) for 46 hours. The pressure fell appreciably and gas-phase mass spectroscopy indicated that most of the ethylene had been consumed, ethane but not methane had been produced. The liquid phase present in the autoclave was analyzed by g.l.c.-mass spectroscopy. The catalyst turnover was calculated to be at least 20 molecules of C_2H_4 per Ru atom per hour and approximately 60% of the products comprised CH_3CH_2CHO, $CH_3CH_2CH_2OH$ and $CH_3-CH(CHO)CH_2CH_3$ (in the ratio 1:3:1). Other organic products were present but they were not completely identified. The formation of aldehydes by hydroformylation was obviously accompanied by side reactions and the main product appears to be propan-1-ol. Evidence was also observed for the dimerisation of ethylene and subsequent hydroformylation to C_5 products. Higher oligomers were also detected. Sample A was then subjected to 100 atm at 200°C of (1:1) CO/H_2 for 46 hours but no Fischer-Tropsch reaction products were observed and the resulting complex also failed to dimerise ethylene at 200°C and 25 atm pressure. During the hydroformylation reaction, ruthenium was lost from the zeolite as carbonyl compound C (ν_{CO} in CH_2Cl_2, 2014, 1950 cm^{-1}) which in the mass spectrometer gave a parent peak at 406 m/e and the typical monoruthenium isotope pattern. The fragmentation pattern also suggested the presence of a hydrocarbon chain. I.r. spectra of the recovered zeolite D revealed the presence of supported ruthenium carbonyl (ν_{CO} 2047, 1991, 1958 cm^{-1}) and compound C, however, the latter was extracted with CH_2Cl_2.

Both the zeolite D and a diglyme solution of C were reused under the conditions described above. Both were active but although D produced a similar product distribution C was more selective and gave CH_3CH_2CHO and $CH_3CH_2CH_2OH$ (about 1:1) with only very small amounts of the other products. Interestingly only negligible amounts of metal were lost from D on reuse.

Catalyst B was tested under those conditions used for A but although an appreciable fall in pressure was again registered, gas analysis revealed considerable amounts of methane as well as ethane indicating Fischer-Tropsch activity. The liquid phase consisted of several constituents and although no propanaldehyde was detected, a large amount of the hydrogenation product, propan-1-ol, was present. The remaining products included hydrocarbons and compounds arising from crossed Fischer-Tropsch and hydroformylation reactions. The resulting black zeolite appeared to contain only metal thus accounting for the formation of CH_4 and higher hydrocarbons (Ref. 28). In the liquid phase only very small amounts of ruthenium carbonyls were present.

These results indicate that among many other factors the loading of metal complex plays a vital role in such reactions. In the reducing envinroment, the reduced ruthenium species of the "concentrated" (B) zeolite appear to aggregate to form metal particles. At lower loading, metal carbonyls of varying nuclearity are produced, but while the mononuclear species are lost from the zeolite, compounds of higher nuclearity may be trapped in the zeolite cages.

ACKNOWLEDGMENTS. The authors thank the University of Trieste and Ministry of Education (Roma) for financial support.

References

1. a) P.Gallezot, Catal.Rev., 20,121, 1979; b)D.C.Bailey and S.H.Langer, Chem.Rev., 81, 109, 1981; c) J.M.Basset and A.Choplin, J.Mol.Catal., 21, 95, 1983; d) J.Zwart and R.Snel, J.Mol.Catal., 30, 305, 1985
2. H.H.Nijs, P.A.Jacobs, J.J.Verdonck, and J.B.Uytterhoeven, in "Growth and Properties of Metal Clusters", J.Bourdon (ed.), p.479, Elsevier, Amsterdam, 1980
3. D.W.Breck, in "Zeolite Molecular Sieves", J.Wiley, New York, N.Y., 1974
4. a) A.K.Smith, B.Besson, J.M.Basset, R.Psaro, A.Fusi, and R.Ugo, J.Organometal.Chem., 192, C31, 1980; b) R.Psaro, R.Ugo, G.M.Zanderighi, B.Besson, A.K.Smith, and J.M.Basset,

J.Organometal.Chem., 213, 215, 1981; c) R.Psaro, C.Dossi, and R.Ugo, J.Mol.Catal., 21, 331, 1983

5. B.Besson, B.Moraweck, A.K.Smith, J.M.Basset, R.Psaro, A.Fusi, and R.Ugo, J.Chem.Soc.Chem.Commun., 569, 1980
6. P.L.Watson and G.L.Schrander, J.Mol.Catal., 9, 129, 1980
7. R.Ganzerla, M.Lenarda, F.Pinna, and M.Graziani, J.Organometal.Chem., 208, C43, 1980
8. G.Collier, D.J.Hunt, S.D.Jackson, R.B.Moyes, I.A.Pickering, P.B.Wells, A.F.Simpson, and R.Whyman, J.Catal., 80, 154, 1983
9. D.J.Hunt, R.B.Moyes, P.B.Wells, S.D.Jackson, and R.Whyman, "Proceedings VIII International Conference on Catalysis", Paper V-27, Berlin 1984
10. H.Knozinger, Y.Zhao, B.Tesche, R.Barth, R.Epstein, B.C.Gates, and J.P.Scott, Faraday.Disc.Chem.Soc., 72, 53, 1981
11. a) M.Deeba and B.C.Gates, J.Catal., 67, 303, 1981; b) M.Deeba, J.P.Scott, R.Barth, and B.C.Gates, J.Catal., 71, 373, 1981
12. M.Iwamoto, T.Hasuwa, H.Furukawa, and S.Kagawa, J.Catal., 79, 291, 1983
13. D.C.Grenoble, M.M.Estadt, and D.F.Ollis, J.Catal., 67, 90, 1981
14. B.L.Gustafson and J.Lunsford, J.Catal., 74, 393, 1982
15. A.Theolier, A.K.Smith, M.Leconte, J.M.Basset, G.M.Zanderighi, R.Psaro, and R.Ugo, J.Organometal.Chem., 191, 415, 1980
16. J.M.Basset, B.Besson, A.Choplin, F.Hugues, M.Leconte, D.Rojas, A.K.Smith, A.Theolier, Y.Chauvin, D.Commereuc, R.Psaro, R.Ugo, and G.M.Zanderighi, in "Fundamental Research in Homogeneous Catalysis", M.Graziani and M.Giongo (eds.), p.19, Plenum Press, (NY), 1984
17. K.A.Azam, A.J.Deeming, R.E.Kimber, and P.R.Shukla, J.Chem.Soc., 1853, 1976
18. J.J.Verdonck, P.A.Jacobs, and J.B.Uytterhoeven, J.Chem. Soc.Chem.Commun., 181, 1979
19. X.-Li, J.H.Onuferko, and B.C.Gates, J.Catal., 85, 176, 1984
20. P.C.Ford, Acc.Chem.Res., 14, 32, 1981
21. B.F.G.Johnson, J.Lewis, P.R.Raithby, G.M.Sheldrick, K.Wong, and M.McPartlin, J.Chem.Soc.Dalton Trans., 673, 1978

22. J.R.Budge, J.P.Scott, and B.C.Gates, J.Chem.Soc.Chem. Commun., 342, 1983

23. E.Mantovani, N.Palladino, and A.Zenobi, J.Mol.Catal., 3, 285, 1977

24. P.Jackson, B.F.G.Johnson, J.Lewis, R.Ganzerla, M.Lenarda, and M.Graziani, J.Organometal.Chem., 190, Cl, 1980

25. P.Gelin, G.Condurier, Y.Ben Taarit, and C.Naccache, J.Catal., 70, 32, 1980

26. H.H.Nijs, P.Jacobs, and J.B.Uytterhoeven, J.Chem.Soc.Chem. Commun., 180, 1979

27. K.R.Laing, R.L.Leubner, and J.H.Lunsford, Inorg.Chem., 14, 1400, 1975

28. M.J.Doyle, A.P.Kouwenhoven, C.A.Schapp, and B.Van Oort, J.Organometal.Chem., 174, C55, 1979

THE EFFECTS OF STRONG METAL-SUPPORT INTERACTION IN CATALYSTS PREPARED WITH THE USE OF ORGANOMETALLIC COMPOUNDS

Yu.I. Yermakov and Yu.A. Ryndin
Institute of Catalysis, Novosibirsk 630090, USSR

ABSTRACT - Systematic studies of the effect of anchored on silica surface ions of Group IV B elements (Ti, Zr, Hf) and of lanthanides on the proportios of dioporood particlos of Ni, Pd and Pt have been carried out. Catalysts were prepared from organometallic precursors. The following effects of anchored ions have been observed: a) increased stability of metal particles to their sintering in H_2 atmosphere; b) increased catalytic activity in CO hydrogenation and c) decreased activity in benzene hydrogenation and ethane hydrogenolysis over Pd and Pt catalysts. The data obtained have been discussed in terms of various hypotheses concerning the reasons for the strong metal-support interaction.

1. INTRODUCTION

Application of oxides of transition elements as supports offers additional possibilities to control catalytic properties of supported Group VIII metals owing to the observed effects of strong metal-support interaction (SMSI). After the first studies (Refs. 1-3) describing these effects, the number of publications on SMSI catalysts has impetuously been growing and this problem was even the subject of the International Conference (Ref.4).

According to the experimental data of many authors (Refs. 1-7), when using oxides of transition elements as supports, the typical SMSI effects are:

1. Changes in the specific catalytic activity of metals. Activity increase is observed in CO hydrogenation, whereas in hydrogenation and hydrogenolysis of hydrocarbons it decreases.

2. Decrease in H_2 and CO chemisorption on metal.

These peculiarities are more distinct with increasing the reduction temperature of catalysts and disappear upon the oxidative treatment with subsequent reduction under mild conditions.

Supporting Group VIII metals on lanthanide (Ln) oxides also changes catalytic properties of these metals compared to the SiO_2-supported catalysts. In this case their activity and selectivity in the synthesis of lower alcohols from CO and H_2 increases, whereas in reactions of hydrocarbon conversions the activity decreases (Refs. 8, 9,10).

Note that the effects similar to SMSI have been observed previously for the catalysts obtained by anchoring highly dispersed particles of Group VIII metals on the surface of low-valent ions of molybdenum, tungsten or rhenium (Refs. 11-15). Thus, Pt particles on SiO_2 that contains tungsten ions possess the following properties (Ref. 12): a) decreased chemisorption of H_2 and CO and b) altered (increased) catalytic activity in benzene hydrogenation and ethane hydrogenolysis. Besides, Pt particles in these catalysts are more stable to sintering compared to Pt/SiO_2.

A previously formulated and supported in experiment concept (refs. 11-15) implies that the probable reason for the effect of addition of promoting elements of hardly to reduce oxides on the properties of supported Pt and Pd particles is the interaction of dispersed particles with surface ions of promoters. This interaction increases the stability of dispersed metal particles to sintering and changes electronic properties of these particles, which leads to the change of their chemisorption and catalytic properties. A question arises whether it is also possible to apply this concept to explain the specific properties of metals when they are supported on hardly reducible oxides of transition metals.

To test the above hypothesis, in our recent studies we have investigated SiO_2-supported systems containing ions of promoting elements and Group VIII metals. Group IV B elements (Ti, Zr and Hf hereafter designated as E_{IV}) and lanthanides (Ln in what follows) were used as promoting additives. Ni, Pd and Pt were applied as catalytically active Group VIII elements (M_{VIII}).

To support promoters and metals, we have used organometallic precursors.

Here we discuss data that have been partially published (Ref. 16) or are to be published.

2. PECULIARITIES IN PREPARATION OF SUPPORTED CATALYSTS FROM ORGANOMETALLIC PRECURSORS

To prepare catalysts, we have used an approach described previously in several publications (Refs. 11-15). Synthesis of supported catalysts on SiO_2 (BET surface 250 m^2/g) involved the following steps:

1. Treatment of SiO_2 with a pentane solution of organometallic compounds of promoting elements. In the case of Group IV elements, tetrabenzyltitanium, tetraallylcirconium or tetrabenzylhafnium were used for supporting. These compounds react with SiO_2 dehydrogenated at 870 K by the scheme (Refs. 11, 15, 17):

$$\begin{matrix}-OH \\ -OH\end{matrix} + E_{IV}R_4 \longrightarrow \begin{matrix}-O \\ -O\end{matrix} > E_{IV}R_2 + 2RH\uparrow$$

(⫽ is the SiO_2 surface, R is an organic ligand).

To modify SiO_2 surface by lanthanide ions, tris-isopropylcyclopentadienyllanthanides were supported. These compounds are rapidly and irreversibly adsorbed by the SiO_2 surface from a solution in pentane.

2. Treatment of the obtained samples with hydrogen at 573-873 K or oxygen (in the case of E_{IV} and Ln compounds at 673 and 873 K, respectively) to remove organic ligands and to obtain promoter ions in different oxidation states.

3. Adsorption of allyl Ni, Pd and Pt complexes on SiO_2 modified either by E_{IV} or by Ln ions.

4. Treatment of the samples with hydrogen. As usual, all procedures to prepare and examine catalysts were carried out in glass apparatus without contacting with air.

For comparison with SiO_2-supported catalysts, metallic catalysts supported on massive oxides: TiO_2 (S=62 m^2/g, anatase in the main) and ZrO_2 (S=80 m^2/g) dehydrated in vacuum at 873 K were prepared.

3. DATA ON THE STATE OF SURFACE IONS OF PROMOTING ELEMENTS

3.1. Before Supporting Group VIII Metals

Hydrogen treatment of supported organometallic Ti and Zr complexes yields either surface hydride complexes or low-valent ions not bound to hydrogen. The composition of surface compounds depends on the treatment conditions (primarily on temperature). Data on the composition of surface compounds for the 1.1% Ti/SiO_2 catalyst that was used primarily to examine the effect of Ti ions on the properties of Group VIII metals, are listed in Table 1.

Surface hydride Ti and Zr complexes possess several interesting catalytic properties. Samples containing Ti hydrides have high catalytic activity in the hydrogenation of benzene and olefins (commensurable, e.g. with that of supported nickel catalysts) (Ref. 16 b). SiO_2, containing anchored Ti hydrides, adsorbs $Co_2(CO)_8$ (Ref. 18). Hydrogen treatment of such samples provides super-dispersed Co particles that are highly selective towards the formation of alcohols in CO hydrogenation.

TABLE 1. Surface titanium compounds in 1.1% Ti/SiO_2 catalyst obtained by treatment of supported complexes *

Treatment conditions	Surface compounds **
Without treatment	$(\text{–O})_2\text{Ti}^{4+}\text{R}_2$
Hydrogen treatment	
at 423 K	$(\text{–O})_2\text{Ti}^{4+}\text{H}_2$ $\text{–O–Ti}^{3+}\text{H}_2$
at 573 K	$(\text{–O})_2\text{Ti}^{3+}\text{–H}$
at 873 K	$(\text{–O})_2\text{Ti}^{2+}$ and $(\text{–O})_3\text{Ti}^{3+}$
Oxygen treatment at 673 K	$(\text{–O})_2\text{Ti}^{4+}\text{=O}$

* According to the data of measuring the amount of O_2 adsorbed by the samples (Ref. 16 b).

** Formulas represented account for the oxidation states of titanium and the stoichiometry of the compounds but not for the coordination number of Ti ions.

Bond energy of Ti $2p_{3/2}$ for the Ti/SiO_2 samples treated with hydrogen at 823 K is E_b=459.5 eV that practically coincides with $E_b Ti^{3+}$ in $TiCl_3$ (Ref. 19). According to the high-resolution electron microscopic data, no morphological changes or formation of any new structures take place in the samples containing E_{IV} or Ln on SiO_2 supported by the aforesaid methods. It indicates the absence of dispersed oxide phases or large two-dimensional clusters of E_{IV} or Ln compounds.

In several of the revized studies (Ref. 16) the effect of the initial state of promoting additives (after either reductive or oxidative treatments of supported organometallic complexes) on the properties of supported metals has been tested. It has been established that the difference in the catalytic activities of the samples obtained by supporting M_{VIII} on either prereduced or preoxidized M_{IV}/SiO_2 (or Ln/SiO_2) samples is low. Apparently, the composition of surface species in promoted $(M_{VIII}+E_{IV})/SiO_2$ and $(M_{VIII}+Ln)/SiO_2$ catalysts is largely determined by the temperature of their treatment in hydrogen after supporting M_{VIII} on E_{IV}/SiO_2 samples.

3.2. After Supporting Group VIII Metals

Data on the oxidation state of promoting element ions in the presence of M_{VIII} can be obtained from those concerning the amount of O_2 consumed for the oxidation of samples at various temperatures. Samples were subjected to the following treatment:

a) preoxidation by oxygen at 673 K (to remove the carbon formed by the decomposition of organic ligands) and

b) hydrogen reduction at 823 K.

After this treatment the amount of O_2 consumed for the catalyst oxidation at temperatures ranging within 300-673 K was measured.

In the case of $(Pt + E_{IV})/SiO_2$ catalysts the amount of O_2 consumed for the oxidation was higher than that corresponding to the stoichiometry O:Pt=1 that is typical for Pt/SiO_2 (Table 2). These data indicate that the promoter ions are in the lower oxidation state (average oxidation number ~ 3.5) after the reduction of catalysts at 823 K.

TABLE 2. Amount of O_2 consumed for the oxidation of $(M_{VIII}+E_{IV})/SiO_2$ catalysts at 673 K.

Nos. of samples	Catalyst composition (wt%)	O/M_{VIII} Atomic ratio	O/E_{IV} *	n **
1	2.5% Pt	0.97	-	-
2	0.94% Ni+1.1% Ti	2.44	1.0	2.7
3	3.2% Pt +1.1% Ti	1.52	0.37	3.2
4	2.7% Pt +3.5% Zr	1.31	0.11	3.8
5	2.0% Pt + 4.0% Hf	1.26	0.13	3.7

* O_2 amount consumed for E_{IV} oxidation (atomic ratio) calculated assuming that $O/M_{VIII}=1$ like for sample 1;

** average oxidation number of E_{IV} calculated from $n=4-2\ O/E_{IV}+H/E_{IV}$, where $H/E_{IV}=0.7$ for sample 2 (Ref. 16 b) and $H/E_{IV}=0$ for other samples.

Similar data for $(Ni+Ti)/SiO_2$ indicate (Ref. 16) that the average oxidation number of Ti ions in catalysts reduced at 673 K is ~2.7.

XPS data have not provided a possibility to observe variations in the bond energy at Ti $2p_{3/2}$ and Pr 3d upon supporting platinum on Ti/SiO_2 and Pr/SiO_2 samples (Table 3). But the supporting of platinum leads to the 20% and 30% decrease of intensity in the spectrum of Ti and Pr promoting elements, respectively. These data can indicate that the ions of promoting elements are covered by metal particles.

TABLE 3. XPS data for the state of Ti, Pr and Pt in supported catalysts*

Data on Ti or Pr state			Data on Pt state		
	E_b(Ti $2p_{3/2}$) (eV)	Relative intensity (arbitrary units)		E_b(Pt $4f_{7/2}$) (eV)	Relative intensity (arbitrary units)
Ti/SiO_2	459.5	1.0	Pt/SiO_2	71.7	1.0
(Pt+Ti)/SiO_2	459.3	0.8	(Pt+Ti)/SiO_2	71.9	1.5
	E_b(Pr $3d_{5/2}$)				
Pr/SiO_2	933.7	1.0	Pt/SiO_2	71.7	1.0
(Pt+Pr)/SiO_2	934.0	0.7	(Pt+Pr)/SiO_2	72.3	1.8

* Ti = 1.1%, Pr = 5.0%, Pt = 3-4%.

4. DISPERSION OF METAL PARTICLES

In all cases the introduction of E_{IV} and Ln additives into the composition of Ni, Pd and Pt/SiO_2 catalysts leads to the decrease of the size of metal particles and to their narrower size distribution. At the same time the stability of metal particles to sintering increases. As an example, Fig. 1 represents comparative data for Pt/SiO_2 and (Pt+Ti)/SiO_2 samples. The effect of Pr ions on the stability of disperse Pt is still more pronounced compared to Ti ones. In the presence of Pr ions the size of Pt particles (8 Å) remains unchanged at reduction temperatures as high as 873 K. It should be noted that the stability increase of Pd and Pt particles has been observed previously in the presence of Mo, W and Re ions (Ref. 15).

Increased dispersion of supported Pt is confirmed by the XPS data. In the case of promoted Ti and Pr samples, the intensity of XPS signals increases approximately 1.5-2 times (Table 3).

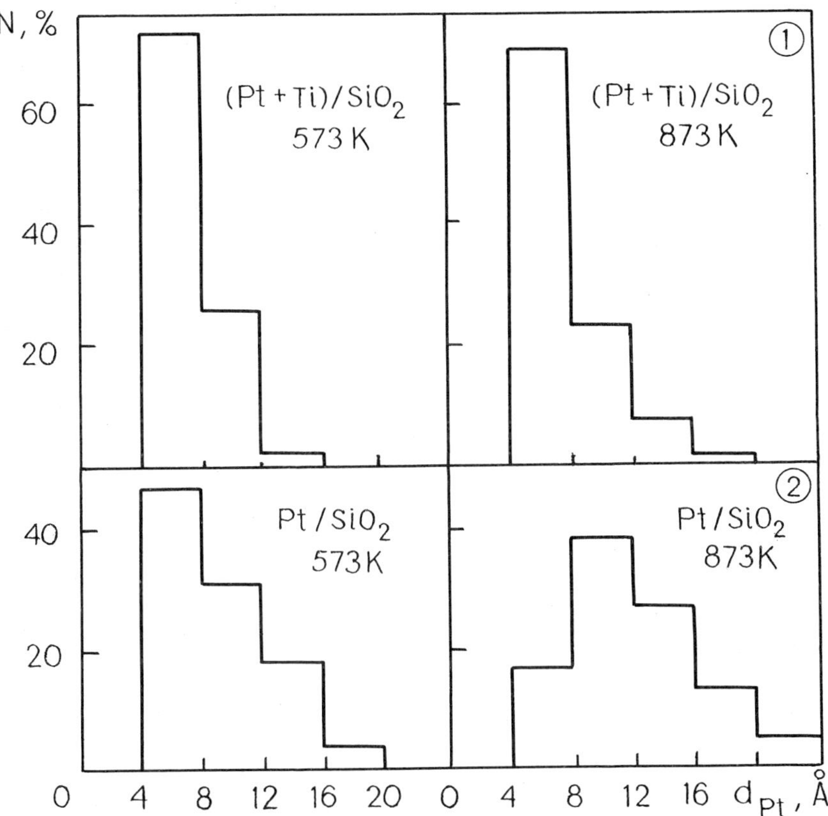

Fig. 1. Size distribution of metal particles vs. the reduction temperature of catalysts.

5. CHEMISORPTION PROPERTIES

Chemisorption properties of Pt and Pd particles prepared from SiO_2-supported organometallic complexes and by conventional methods (Refs. 11, 15) do not differ. Chemisorption data for the dispersion of metal particles in the samples of "organometallic" origin agree well with electron microscopic data.

Pt/TiO_2 catalysts obtained from allyl complexes, contain dispersed Pt particles (of ~15 Å in size after reduction at 823 K). These catalysts adsorb almost no hydrogen (H:Pt~0.01) and (CO:Pt~0.05) and in this respect they do not differ from the samples obtained by conventional methods (Refs. 1-7).

In the presence of Ti ions on SiO_2, platinum is characterized by the decreased chemisorption of hydrogen and CO(H:Pt~0.2, CO:Pt~0.2). Similar decrease of H_2 and CO adsorption is typical for Pd supported on Pr/SiO_2 Thus the introduction of titanium or lanthanide ions onto the SiO_2 surface can yield qualitatively the same effects on the chemisorption on Group VIII metals as the supporting of these metals on TiO_2 or lanthanide oxides.

H_2 and CO chemisorption, however, does not decrease on platinum supported on Zr/SiO_2 and Hf/SiO_2. At the same time Zr and Hf ions exert a more pronounced effect on the catalytic properties of disperse metal (see below than titanium ions. Probable reasons for this contradic tion will be treated in what follows.

6. CATALYTIC PROPERTIES

6.1. CO Hydrogenation

In all cases the presence of E_{IV} or Ln ions in Ni, Pd or Pt catalysts increases the activity. In this case selectivity also changes: on nickel catalysts the yield of C_2 and C_{2+} hydrocarbons increases, on Pd and Pt samples their yield decreases, whereas that of methanol is as high as 95-99% at 523 K and 10 atm. These effects are enhanced in the sequence: Ti < Zr < Hf (Fig. 2 illustrates activity of $(M_{VIII}+E_{IV})/SiO_2$ catalysts).

Variation of lanthanide ions provides changes in the activity (Fig. 3); similar dependences with extrema in the region of Pr and Ho are also observed in reactions on bulk lanthanide oxides (Refs. 20-22). Activity increase in CO hydrogenation upon supporting Pd on bulk lanthanide oxides has been observed previously (Refs.8,9).

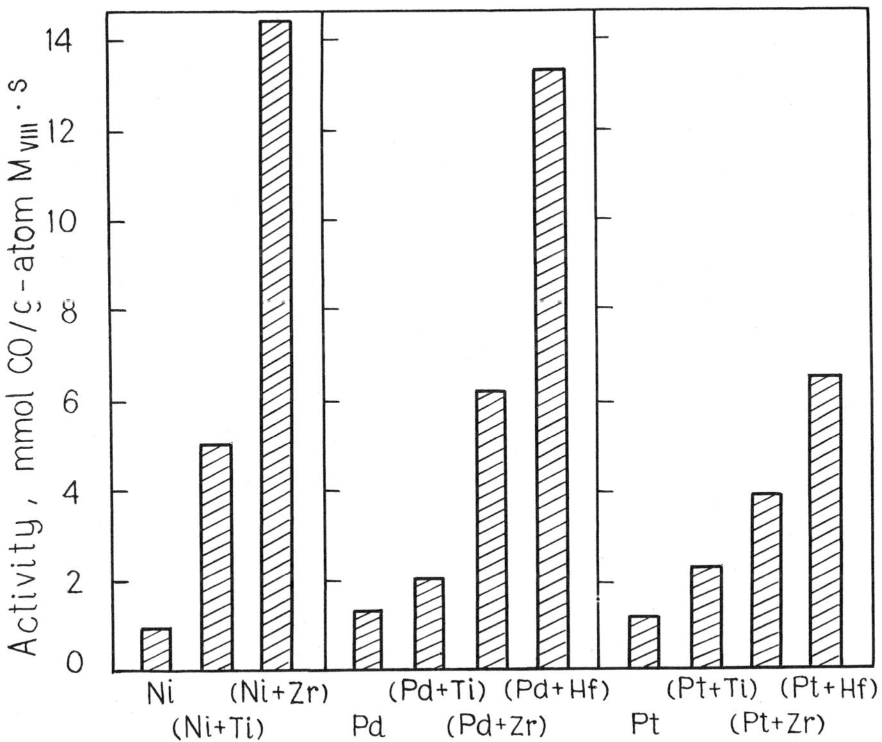

Fig. 2. Influence of Group IV B elements on the activity of Group VIII metals in CO hydrogenation at 523 K

6.2. Benzene Hydrogenation

Platinum activity in Pt/SiO_2 catalysts slightly decreases with increasing reduction temperature, apparently, due to its sintering (Table 4). The activity of Pt/TiO_2 catalyst drops sharply with increasing reduction temperature, which agrees with the literature data (Ref.

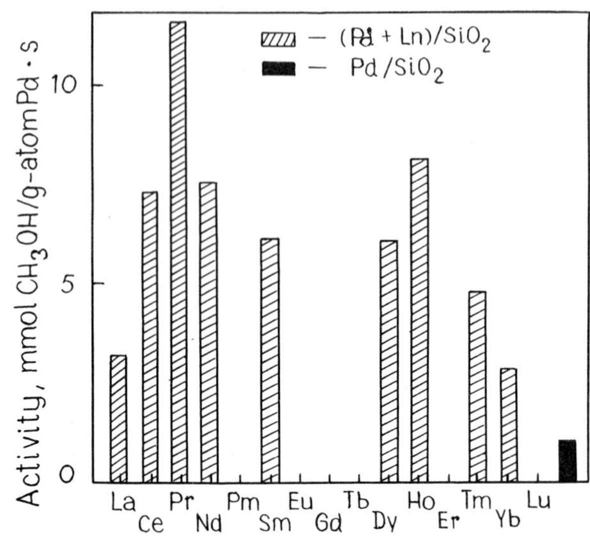

Fig 3. Influence of the lanthanides on the activity of $(Pd+Ln)/SiO_2$ catalysts in CO hydrogenation at 523 K.

TABLE 4. Influence of the reduction temperature on the activity of Pt catalysts of various composition in the hydrogenation of benzene.

Nos.	Catalyst composition	Activity at various reduction temperatures ($mol\ C_6H_6/g\ Pt \cdot h$)				
		373 K	573 K	773 K	873 K	873 K*
1	$2.5\%Pt/SiO_2$	0.230	0.205	0.220	0.180	0.171
2	$(3.2\%Pt+1.1\%Ti)/SiO_2$	0.040	0.079	0.095	0.052	-
3	$(2.7\%Pt+5.3\%Zr)/SiO_2$	0.006	0.045	0.060	0.073	0.152
4	$(2.0\%Pt+4.0\%Hf)/SiO_2$	0.026	0.039	0.065	0.065	0.131
5	$(1.0\%Pt/TiO_2$	0.108	0.047	0.009		-

* Samples were pretreated with oxygen at 673 K

23). SiO_2 modification by Group IV ions diminishes the catalyst activity compared to Pt/SiO_2, but upon increasing reduction temperature, its activity slightly rises. High temperature oxidation accompanied by reduction eliminates almost completely the effect of Group IV ions. Similar data have also been obtained previously for the Pt catalysts modified by lanthanides (Ref. 16).

6.3. Ethane Hydrogenolysis

The activity of $(Pt+E_{IV})/SiO_2$ catalysts in this reaction is decreased compared to Pt/SiO_2 (Fig. 4). The similar activity decrease is also observed for $(Pd+Ln)/SiO_2$ catalysts (Fig. 5), and the maximum change (decrease) in the activity is observed for the same lanthanides as for CO hydrogenation where the activity rises.

7. PROBABLE REASONS FOR THE EFFECT OF GROUP IV ELEMENTS AND LANTHANIDES ON THE PROPERTIES OF SUPPORTED METAL PARTICLES

Several reasons were proposed in literature to explain unusual properties of dispersed metals on TiO_2 and some other oxides:

1. Electron transfer to metal particles from bulk TiO_2(or from other oxides of transition elements)(Ref.5).
2. Direct interaction of metal atoms with low-valent E_{IV} ions (Ref. 2).
3. Localization of flat metal particles (with "pill--box" morphology) on the reduced sites of TiO_2(Ref. 3).
4. Formation of intermetallic compounds of a supported Group VIII metal with a transition element of oxide supports (Ref. 1).
5. Diffusion of TiO_2 (or some other carrier) reduction products to the surface of metal particles (Refs. 7,24).

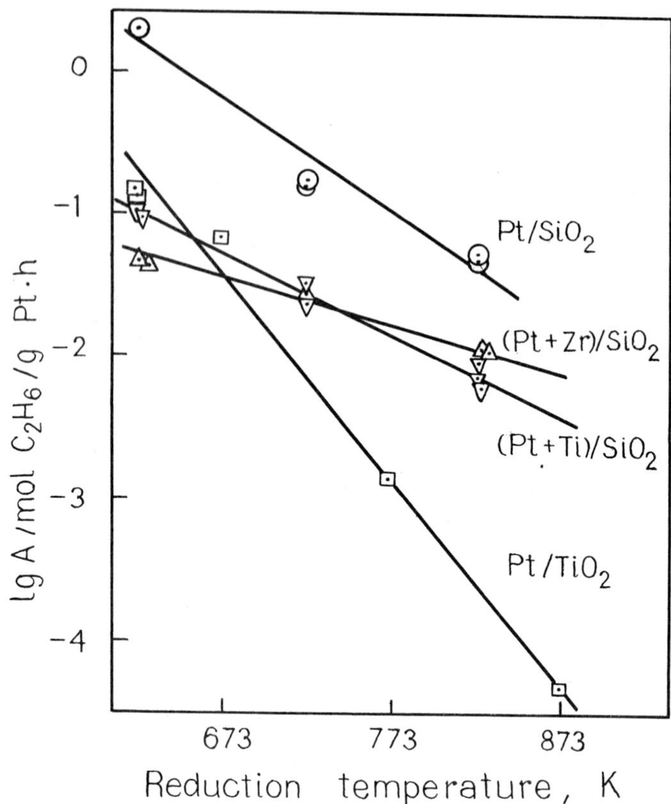

Fig. 4. Activity of Pt catalysts in ethane hydrogenolysis vs. their reduction temperature.

6. Activation of reactants (in particular, of CO) on the metal particle-carrier interface (Ref. 6).

None of the hypotheses put forward to treat SMSI has not, however, been provided with a decisive experimental support. It is possible that for the catalysts containing metal particles on a bulk TiO_2 (and the other transition metal oxides), the effects responsible for SMSI cannot be attributed to a single reason, and various peculiarities in the properties of such metals are due to different factors.

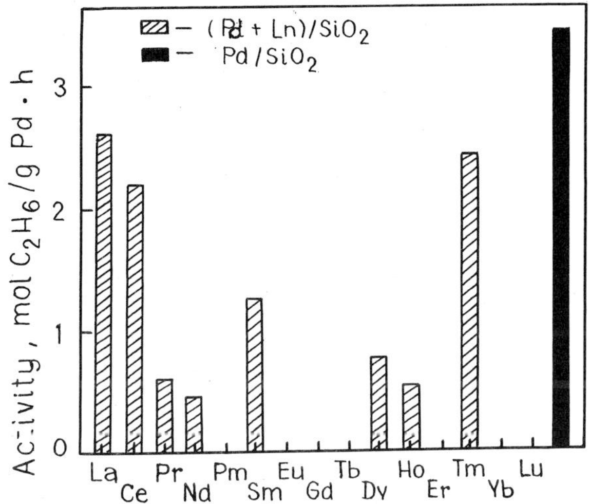

Fig. 5. Influence of lanthanides on the properties of $(Pt+Ln)/SiO_2$ catalysts in ethane hydrogenolysis at 623 K.

Let us consider the results obtained in our studies of supported metal catalysts on SiO_2 promoted by M_{IV} and Ln ions. These catalysts have the following peculiarities:

1. Increased stability of superdispersed Ni, Pd and Pt particles to sintering. It has been confirmed by electron microscopic, chemisorption and XPS data. Similar effects have been observed previously upon adding Mo, W and Re ions onto the SiO_2 surface.

2. Altered catalytic properties: a) increased activity in CO hydrogenation and b) decreased activity in benzene hydrogenation and ethane hydrogenolysis.

Note that this effect (a 2-10 fold activity decrease) is considerably lower than those observed for metals on bulk TiO_2 where the activity can decrease by a factor of several orders of magnitude (Ref. 4).

3. The effect of activity decrease is observed even at low (573-773 K) reduction temperatures. In this

respect the systems examined differ from metals on bulk oxides where SMSI effects are more distinct after high temperature reduction.

4. The effects of altered activity disappear after high temperature oxidation accompanied by reduction. This behaviour of the catalysts is similar to that observed in the case of metals on bulk oxides.

5. According to the chemisorption and XPS data, the reduced catalysts contain low-valent ions of transition promoting element.

6. Intensity decrease of the signals in X-ray photoelectron spectra due to promoting ions upon supporting Group VIII metals.

7. For several catalysts (Pd and Pt in the presence of titanium and lanthanide ions) an evident decrease in H_2 and CO chemisorption is observed. This effect is similar to that observed previously for Pt in the presence of Mo and W ions (Refs. 11, 12). It is less pronounced than in the case of supporting Pd and Pt on bulk oxides (e.g., TiO_2), where after the high temperature reduction the adsorption of H_2 and CO can be inhibited almost completely (Ref. 24). But for the catalysts obtained by supporting Pt on Zr/SiO_2 and Hf/SiO_2, no adsorption decrease of H_2 and CO is observed.

Let us interpret the above data from the viewpoint of probable reasons for the appearance of SMSI in supported catalysts.

Before metal introduction the initial catalysts contain ions that do not form a phase or superdisperse particles of oxides. In this case one cannot rule out a neighbour localization of several ions (the so-called "island" distribution). According to the recent literature data the ions on the SiO_2 surface supported from halide (Ref. 25) or organometallic (Refs. 17, 26) compounds are localized on the surface with formation of "islands" distribution.

Schematic representation of the initial state of SiO_2 surface with introduced ions of promoting elements is illustrated in Fig. 6.

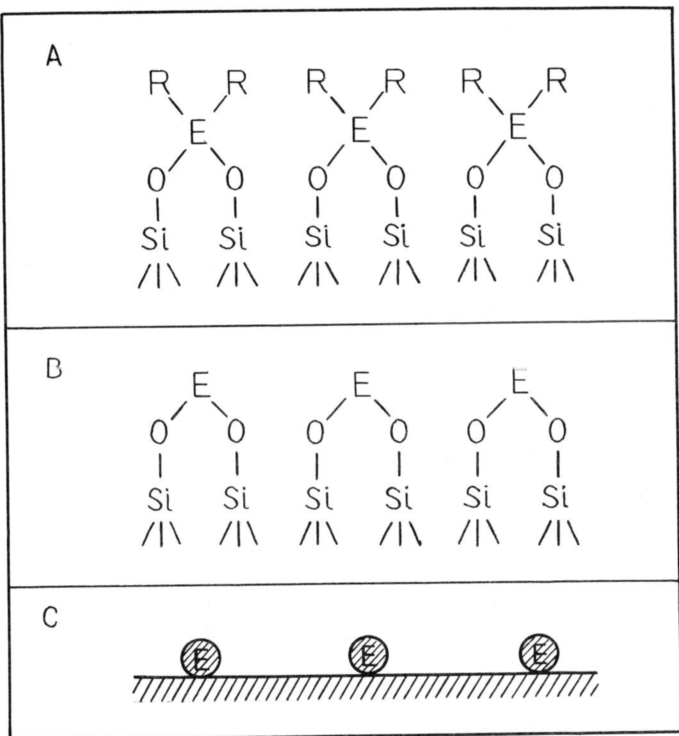

Fig. 6. Schematic representation of SiO_2 surface promoted by element E ions.
A - Upon supporting organometallic compound of element E.
B - After treatment of surface organometallic compounds by hydrogen. The scheme does not account for the coordination state of element E in the oxygen environment whose coordination number seems to be not lower than 4.
C - Schematic representation of SiO_2 surface used in Fig. 7 to represent models of supported metal particles.

Absence of a phase or very dispersed three-dimensional particles of a promoting element oxide indicates that for the SMSI effects to take place, the interaction of metal particles with a bulk oxide phase is not obligatory. Apparently, SMSI effects can be treated in terms of the local interaction between metal clusters and surface transition metal ions.

Let us consider the probable models of surface compounds on SiO_2-supported catalysts (Fig. 7).

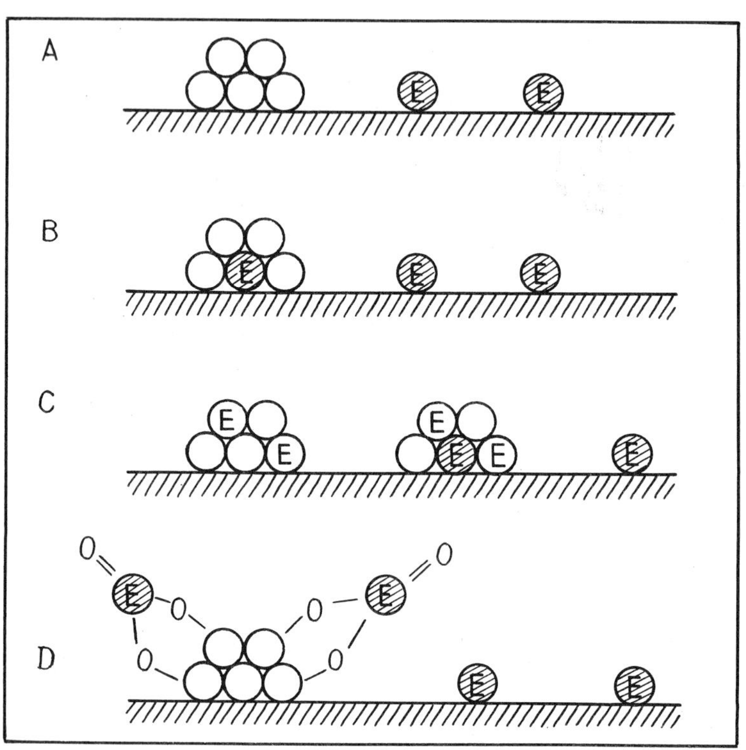

Fig. 7. Probable models for metal particles on SiO_2 promoted by element M.
○ - Group VIII metal atoms; Ⓔ (hatched) - surface ions of promoting elements; Ⓔ - zero-valent atoms of element E.

<u>Model A</u>. The surface of SiO_2 contains metal particles that are not bound to E. This segregated localization of M and E accounts neither for the increase in dispersion nor for the decrease in catalytic activity in conversion of hydrocarbons. However, one cannot rule out tha part of the metal particles, particularly, in the case of

Ni, are localized on the SiO_2 surface without interacting with E.

Model B. In this case metal particles are suggested to interact directly with surface ions of transition elements like in the previously suggested model (Ref. 14). This model has been considered in Ref. 2 to explain reasons for SMSI. In a recent study (Ref. 27) it is stated that the EXAFS data for the Pt/TiO_2 catalyst reveal the existence of Pt-Ti bonds.

Type B model can account for the following data: a) dispersion increase of metal particles in the presence of E; b) change of catalytic properties due to the variations in the electron state of metal clusters; c) intensity decrease of XPS signals of ions E due to their screening by clusters M; and d) influence of preoxidation. High temperature treatment of the samples with oxygen leads to the oxidation of low-valent ions E and the seggregation between E and M with metal sintering.

This model permits some other additional (besides variations in the electron properties of metal clusters M) factors responsible for the change in catalytic activity. Thus, the activity increase in CO hydrogenation can be due to the CO activation on the cluster boundary according to the scheme suggested previously (Refs. 6,28,29):

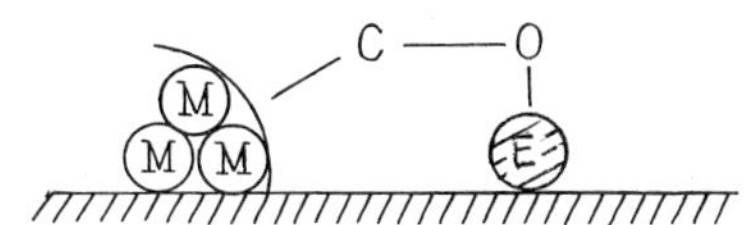

Model C. This model suggests the formation of alloy particles between promoter element and metal. Such hypotheses have also been put forward to treat the properties of M/TiO_2 systems (Ref. 1). For the supported catalysts examined in the present study, however, this model cannot be applied, since these catalysts have ions in the intermediate oxidation state.

Model D. The surface of metal clusters M adsorbs oxid E. But this model would be contradicted to by the decrease in the surface concentration of E, which is supported by the XPS data. For the catalysts on bulk oxides however, the screening of metal surface by the support oxide is possible (Ref. 7).

7. CONCLUSION

The results of studying the properties of super dispersed metal particles supported on SiO_2 have revealed that Ti, Zr, Hf and lanthanide ions exert the same effects on Ni, Pd and Pt that have been observed previously for the other low-valent ions (Refs. 11,15) and referred to as strong interaction effects when they were established for metal catalysts on bulk oxides.

These effects can be interpreted in terms of the interaction of clusters with low-valent ions (model B, Fig. 7).

Compared to metal catalysts on bulk oxides of transition elements, the SiO_2-supported systems have several peculiarities: a) dispersion increase without change of metal particles morphology. One can observe round spot--line particles whose size is largely of 5-15 Å. In catalysts on bulk oxides obtained by conventional impregnation methods the formation of structure (with morphology of "pill-box" type (Ref. 3)) is observed; b) catalytic effects in reactions of hydrocarbons (hydrogenatio and hydrogenolysis) are not so distinct as for the metal on bulk oxides; c) with increasing the reduction temperature of catalysts, no sharp increase in the SMSI effects is observed; d) not always the presence of promoting ions decreases H_2 and CO chemisorption on metals.

A probable reason for the above differences is the different structures of the surface sites interacting with metal clusters in these two cases. In the case of promoted SiO_2, the cluster is likely to interact with one or two surface ions of the transition element. Changes in the electron properties of M are relatively insignificant. In the case of a bulk oxide of element E, the cluster can interact with the extended reduced parts of the surface; the fraction of metal M ions directly bound with E is high, which accounts for a more sharp change in the electron properties of metal.

ACKNOWLEDGEMENT. The authors are grateful to their colleagues who have carried out experimental studies of the system examined: O.S.Alekseev (studies of the systems containing Ti, Zr and Hf), Yu.N.Nogin (studies of the systems containing lanthanides), T.M.Lozovaya (catalytic activity measurements), V.I.Zaikovskii and A.L. Chuvilin (electron microscopy) and A.V.Pashis (XPS studies).

REFERENCES

1. S.F. Tauster, S.C. Fung, R.L. Garten, JACS, 100, 1, 170 (1978).
2. J.A. Horsley, JACS, 101, 10, 2870 (1979).
3. R.T.K. Baker, E.P. Prestridge, R.L. Garten, J.Catal., 56, 3, 390 (1979).
4. B. Imelik, C. Naccache, G. Coudurier, H. Praliaud, P. Meriaudeau, P. Gallezot, G.A. Martin and J.C. Vedrine, Studies in Surface Science and Catalysis, Elsevier, 11 (1982).
5. P. Meriaudeau, O.H. Ellestad, M. Dufaux, C. Naccache, J.Catal., 75, 2, 243 (1982).
6. R. Burch, A.R. Flambard, J.Catal., 78, 2, 389 (1982).
7. D.E. Resasko, G.L. Haller, J.Catal., 82, 2,279 (1983).
8. M. Ichikawa, Shokubai (Catalyst), 21, 253 (1979).
9. Yu.A. Ryndin, R.F. Hicks, A.T. Bell, Yu.I. Yermakov, J.Catal., 70, 2, 287 (1981).
10. P. Meriaudeau, J.F. Dutel, M. Dufaux, C. Naccache, Studies in Surface Science and Catalysis, Elsevier, 11, 95 (1982).
11. Yu.I. Yermakov, Catal.Rev.- Sci.Eng., 12,1,93 (1977).
12. M.S. Ioffe, B.N. Kuznetsov, Yu.A. Ryndin, Yu.I. Yermakov, The 6th Intern.Congr.Catal., London, A5 (1976).
13. Yu.I. Yermakov, The 7th Intern.Congr.Catal., Tokyo, P5 (1980).
14. Yu.I. Yermakov, B.N. Kuznetsov, Yu.A. Ryndin, V.K. Duplyakin, Kinet.Katal., 15, 4, 1093 (1974).
15. Yu.I. Yermakov, B.N. Kuznetsov, V.A. Zakharov, Catalysis by Supported Complexes, Elsevier Scientific Publishing Company, Amsterdam-Oxford-New York, 1981.

16. a) Yu.I. Yermakov, Yu.A. Ryndin, O.S. Alekseev, M.N. Vassilieva, Chem.Commun., 21-24, 1480 (1984).
b) Yu.I. Yermakov, O.S. Alekseev, V.A. Shmachkov, V.I. Sobolev, Yu.A. Ryndin, Kinet.Katal.,26,5, 1270 (1985).
c) Yu.A. Ryndin, Yu.N. Nogin, V.I. Zaikovskii, K. Jakob, K.-H. Thiele, Yu.I. Yermakov, React.Kinet.Catal. Lett., 29 (1985).
d) Yu.A. Ryndin, Yu.N. Nogin, Yu.B. Zverev, N.P. Chernyaev, F.G. Abdikova, Yu.I. Yermakov, React.Kinet. Catal.Lett., (1985).
17. N.S. Maksimov, G.A. Nesterov, V.A. Zakharov, V.F. Stchastnev, V.F. Anufrienko, Yu.I. Yermakov, J.Molec. Catal., 4, 2, 167 (1978).
18. A.S. Lisitsyn, A.V. Golovin, V.L. Kuznetsov, Yu.I. Yermakov, J.Catal., 95, 3, 527 (1985).
19. V.I. Nefedov, X-Ray Electron Spectroscopy of Chemical Compounds, Khimiya, Moskva (1984).
20. M.D. Mitchel, M.A. Vannice, Ind.Eng.Chem.Fundam.,23 1, 88 (1984).
21. C. Sudhakar, M.A. Vannice, Appl.Catal., 14, 1-3, 47 (1985).
22. C. Sudhakar, M.A. Vannice, J.Catal., 95, 1,227 (1985)
23. M.A. Vannice, P. Chon, The 8th Intern.Congr.Catal., 5, 99 (1984).
24. M.S. Spencer, J.Catal., 93, 2, 216 (1985).
25. A.Yu. Stakheev, A.Ya. Yuffa, E.S. Shpiro, G.V. Antoshin, A.A. Slinkin, E.A. Fedorovskaya, Kh.M. Minachev Kinet.Katal., 25, 3, 750 (1984).
26. C.H. Lochmüller, A.S. Colborn, M.L. Hunnicntt, Analyt.Chem., 55, 8, 1344 (1983).
27. G.L. Haller, M. McMillan, D.E. Resasco, H.R. Sadeghi, S. Sakellson, The 8th Intern.Congr.Catal., 5, 1355 (1984).
28. W.M.H. Sachtler, D.F. Shriver, W.B. Hollenberg, A.F. Lang, J.Catal., 92, 2, 429 (1985).
29. M. Ichikawa, T. Fukushima, J.Phys.Chem., 89, 9, 150 (1985).

SUPPORTED RHENIUM SUBCARBONYL CATALYSTS FOR HIGH-PRESSURE CO HYDROGENATION

Masaharu Komiyama, Jin Sato, Kagehisa Yamamoto and Yoshisada Ogino

Department of Chemical Engineering, Faculty of Engineering,
Tohoku University, Aramaki Aoba, Sendai 980 JAPAN

ABSTRACT - A high-pressure infrared spectroscopy (IR) found the formation of surface rhenium subcarbonyl complex during high-pressure CO hydrogenation over supported rhenium catalysts. The complex formation was support specific, titania and zirconia being the two support materials that gave the surface subcarbonyl. Catalyst preparation using dirhenium decacarbonyl, and the close examination of its process by IR, identified the complex to be rhenium tricarbonyl. High-pressure CO hydrogenation over Re/titania indicated that the complex may be the active species for dimethylether formation.

INTRODUCTION

Rhenium metal and its compounds, such as oxides and sulfides, are known to catalyse a wide range of reactions from hydrogenation to oxidation (Refs. 1 & 2). Its ability to catalyse hydrogenation has been extensively explored in connection with the commercial-scale reforming process using a Re-Pt/alumina catalyst, in which rhenium component apparently plays a role of hydrogenating coke precursors (Ref. 3). It has also been reported recently that rhenium is also active for ammonia synthesis (Ref. 4).

Despite these facts, rhenium has been scarcely studied as a CO hydrogenation catalyst (Ref. 5), and little is known on its CO hydrogenation characteristics or its catalytically active species for the reaction. This has prompted us to initiate the study of supported rhenium as a CO hydrogenation catalyst, their characterization and their reaction characteristics (Refs. 6-12).

Viewed from the CO hydrogenation product distribution, supported rhenium is very much like a Fischer-Tropsch catalyst. Primary products are hydrocarbons, with relatively small amount of oxygenates, mainly primary alcohols (Refs. 6 & 7). The carbon number distribution among hydrocarbons produced is well represented by the Schultz-Flory distribution.

On catalyst characterization, much are still to be learned of supported rhenium. We have examined its X-ray photoelectron spectra (XPS) (Ref. 8), its reduction characteristics by ion bombardment (Ref. 8), and electron paramagnetic resonance (EPR) (Ref. 9). The effect of combining rhenium with iron oxide has also been studied by means of soft X-ray spectroscopy (SXS) and Mössbauer spectroscopy (Ref. 10), and with ion bombardment used in conjunction with Auger electron spectroscopy (AES) or XPS (Refs. 8 & 11).

Recently we have found, by means of infrared spectroscopy (IR), that on certain supports a rhenium subcarbonyl species is formed on the catalyst surface during high-pressure CO hydrogenation (Ref. 12). The present work focuses on studying the formation, characterization and CO hydrogenation characteristics of this supported rhenium subcarbonyl species.

Organometallic complex-derived supported rhenium catalysts reported so far have been limited mainly to the ones that uses allyl or alkoxy rhenium complexes (Ref. 13). Although there has been few attempts to prepare supported rhenium carbonyls (Refs. 14 & 15), their characteristics as heterogeneous catalysts have not been reported. The present paper will be the first to report the catalytic activity of supported rhenium subcarbonyl catalysts.

EXPERIMENTAL

Catalyst Preparation

Two series, A and B, of supported rhenium catalysts were prepared. Series A catalysts were obtained by impregnating a commercial titania powder (Nippon Aerosil Co., P-25) with aqueous solutions of ammonia perrhenate, heating the resulting slurry to dryness, and further drying them in an air oven at ca. 300 K overnight. Series B catalysts were prepared by impregnating the same titania powder with tetrahydrofuran (THF) solutions of dirhenium decacarbonyl until the adsorption of the carbonyl on the titania surface reaches equilibrium, then separating the carbonyl-adsorbed titania by filtration followed by air-drying them. Support materials other than titania are also used when necessary, with similar rhenium supporting procedures.

Infrared Spectroscopy

Two types of infrared spectrophotometer, wave-dispersive Hitachi model 260-50 and Fourier-transform JEOL model 100, were employed. Depending on the purposes, three types of infrared cells were used. One of them is a liquid cell with NaCl windows, and used for monitoring the carbonyl catalyst (series B) preparation process. Another is a diffusion-reflectance cell used to examine the prepared or spent catalyst powder surfaces. The third is a transmission-type *in situ* high-pressure cell with 8 mm-thick NaCl windows. The design of this high-pressure cell follows that of Hicks *et al.* (Ref. 16). The cell is capable of withstanding up to 570 K and 6 MPa.

CO Hydrogenation Tests

Measurement of the catalytic activities of the above catalysts follow the previous description (Refs. 10 & 11), *viz.*, a conventional high-pressure flow reactor was used with typically a 2 ml of pressed-and-crushed catalyst (1 to 2 mm size) charge. Prior to the activity tests series A catalysts were reduced with hydrogen at 573 K and 0.1 MPa for 2 h. The CO hydrogenation activities were measured at 2.1 MPa and 523 K with reaction gas of 3/7 CO/H_2 mixture. Reaction products were analysed by on-line gas chromatography.

RESULTS AND DISCUSSION

Formation of Rhenium Subcarbonyl under CO Hydrogenation Conditions

Figure 1 shows *in situ* IR spectra in the carbonyl stretching region of series A Re/titania (1.0 wt % Re) exposed to 1.0 MPa of CO/H_2 at various temperatures for 15 min and then evacuated at room temperature.

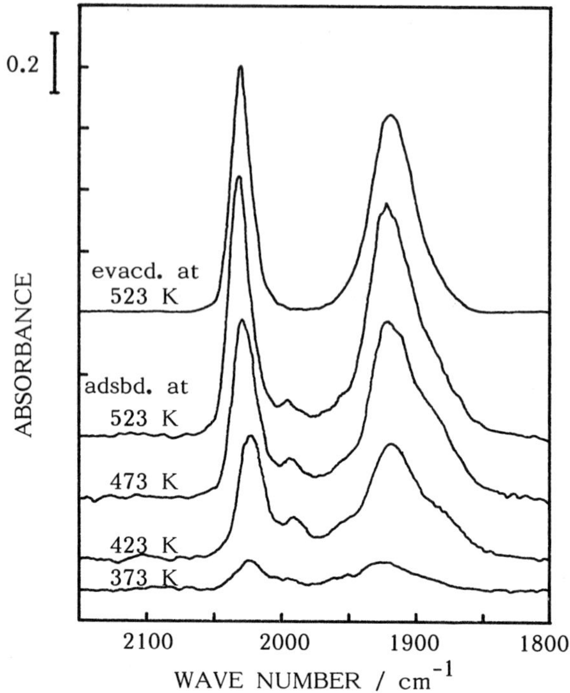

Figure 1. IR spectra of series A Re/titania (1.0 wt % Re) showing stable and evacuatable CO species.

A spectrum from the one that was evacuated at 523 K is also shown. Four CO stretching absorption bands are obvious in the figure. It is also apparent in the figure that, through the responce to high-temperature evacuation, those four CO peaks differentiate themselves into two sets of peaks, in accordance with the previous observation (Ref. 12): a stable set with absorption peaks at the wave numbers of 2030 and 1920 and an unstable set at 1995 and 1885. The constant intensity ratio between the two peaks in the stable set (Ref. 12) indicates that the CO species which gives rise to the stable absorption bands is a stoichiometric entity, most likely a subcarbonyl species. As has been reported (Ref. 12), the formation of this subcarbonyl species are observed only on rhenium supported on titania or zirconia: supports such as silica or alumina did not show the formation of this stable surface species.

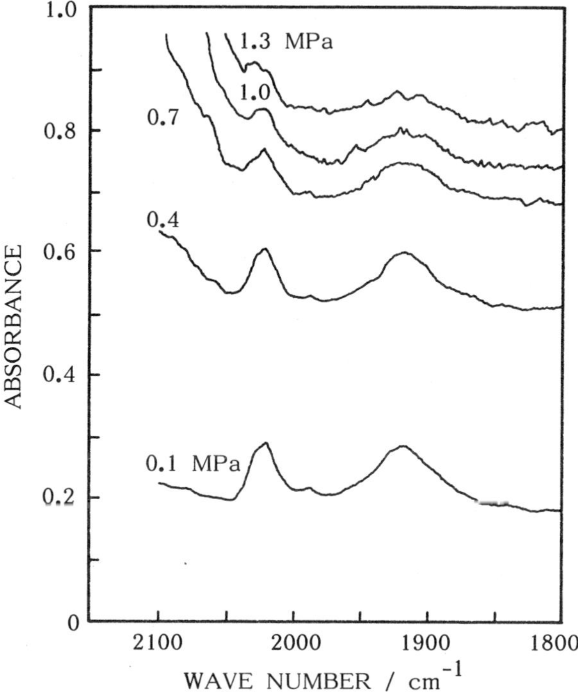

Figure 2. IR spectra of series A Re/titania (1.0 wt % Re) at 423 K under various CO/H_2 pressures.

Catalyst surfaces under reaction conditions may be more closely simulated with the presence of reactant in the gas phase. Figure 2 shows the IR spectra from the same series A catalyst at 423 K under various CO/H_2 pressures. Two features are observed in the spectra. Firstly, as gas phase pressure increases, the overall optical density of the sample increases, i.e., the sample becomes "dark," as seen by the rising baseline with gas phase pressure. This is a phenomenon that occures reversibly with gas phase hydrogen pressure, and has been attributed to the electron density increase in titania support due to the electron donation from the adsorbed and possibly spilled-over hydrogen (Ref. 17). The second feature of the IR spectra in Figure 2 is that, while no apparent shift in the energetic positions of the CO peaks is observed, the CO absorption peak intensities seemingly decrease with this "hydrogen pressure-induced" optical density increase.

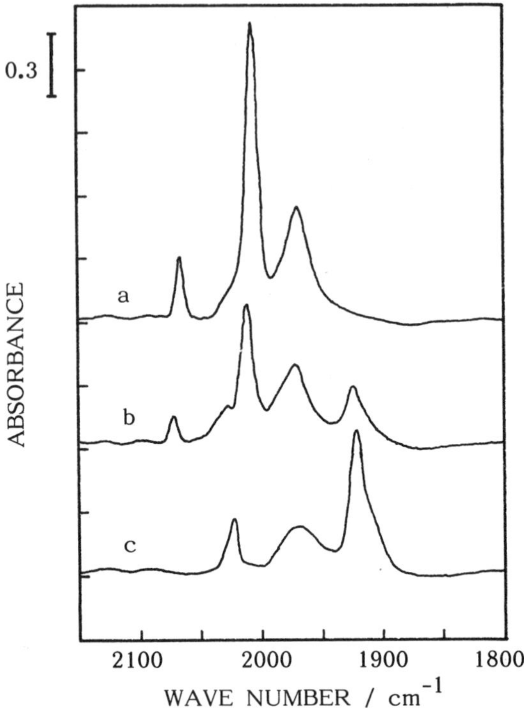

Figure 3. IR spectrum of dirhenium decacarbonyl THF solution (spectrum a) and its changes with sunlight irradiation (spectra b and c).

Identification of the Surface Rhenium Subcarbonyl

Figure 3 shows IR spectra of a THF solution of dirhenium decacarbonyl (spectrum a) and its changes as it is irradiated with sunlight (spectra b and c). While spectrum a shows only features attributable to decacarbonyl species (wave numbers at 2070, 2010 and 1970), sunlight irradiation gives rise to the features attributable to rhenium tricarbonyl species (2020 and 1920) (Refs. 14, 18-20). An absorption band at the wave number of 1970 (overlapping with one of the decacarbonyl peaks) is due to the solvent THF.

In Figure 4 diffusion-reflectance FT-IR spectra of fresh and spent catalysts prepared from the decacarbonyl solution (series B catalyst, 0.3 wt % Re) is shown. On the fresh catalyst rhenium is mainly

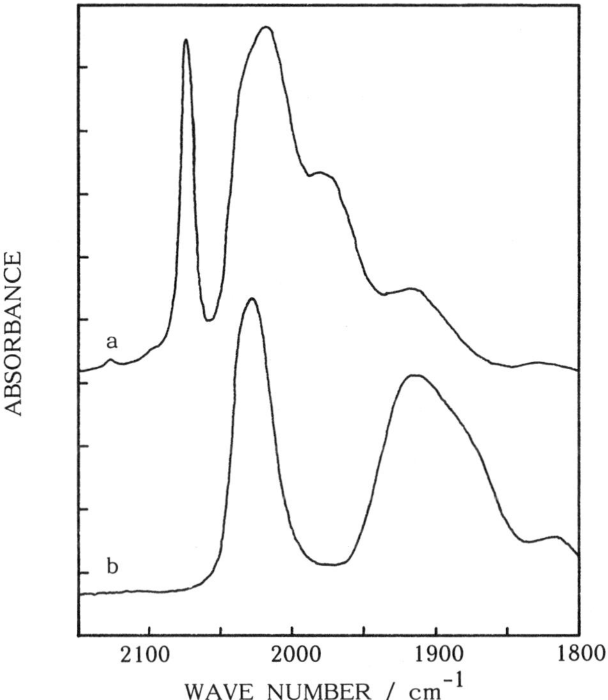

Figure 4. IR spectra of a series B catalyst (0.3 wt % Re) as prepared (spectrum a) and after high-pressure CO hydrogenation (spectrum b).

supported in the form of decacarbonyl while minor part of it is supported as tricarbonyl. However, after high pressure CO hydrogenation for 12 h the surface species observable by IR is entirely converted to tricarbonyl. The spectrum is very similar to that of species found on the series A catalyst evacuated at 523 K after CO adsorption (cf. Figure 1).

CO Hydrogenation Characteristics of Supported Rhenium Subcarbonyl

CO hydrogenation products from supported rhenium, both series A and B catalysts, include hydrocarbons, alcohols, dimethylether and carbon dioxide. In Figure 5 the rate of dimethyether formation over series A catalysts of various rhenium content are plotted along with the IR

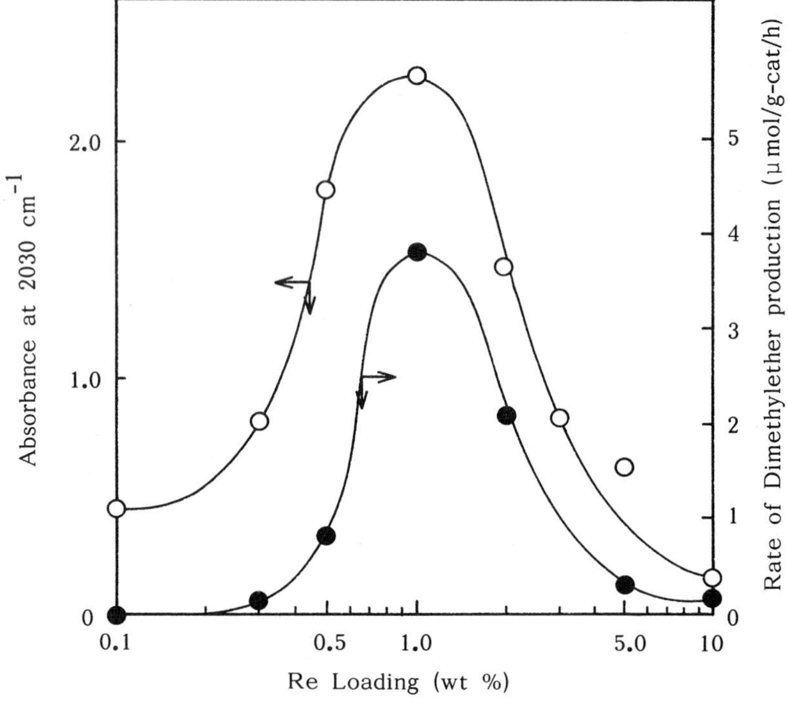

Figure 5. Dimethylether yield over series A catalysts and the intensity of the tricarbonyl species on their surfaces.

intensity of the tricarbonyl species formed on their surfaces, measured separately by using an in situ diffusion-reflectance IR cell. The parallelism between the dimethylether production rate and the overall tricarbonyl intensity on the surface is striking.

Table compares the CO hydrogenation characteristics between series A and B catalysts. At 0.5 wt % Re loading, major hydrogenation product from the series A catalyst is hydrocarbons. In contrast, over the series B catalyst it is alcohols, giving more than one order of magnitude higher alcohols yield. Apparently dirhenium decacarbonyl as a starting material increases the number and/or activity of the alcohols formation site on the finished catalyst.

A rough estimate of the amount of surface tricarbonyl on the spent series A and B catalysts using the extinction coefficient obtained

Table. Comparison of CO hydrogenation characteristics between series A and series B catalysts (0.5 wt % Re loading) after 2 h on stream.

	Conversion to (C %)			
	Hydrocarbons	Alcohols	Dimethylether	CO_2
Catalyst A	0.30	0.02	0.03	0.51
Catalyst B	0.22	0.29	0.03	0.58

from Figure 3 indicated that less than half of the rhenium on the surface exists as tricarbonyl. The rest of the surface rhenium is likely to exist in the form of metallic rhenium crystallites. Diffusion-reflectance IR on the spent catalysts found that the amount of surface tricarbonyl on catalyst A and B are similar to each other, corresponding to the similar dimethylether yield between these catalysts.

REFERENCES

1. R.H. Blom, V. Kollonitsch and C.H. Kline, Ind. Eng. Chem., 54(4), 16-22 (1962)
2. W.H. Davenport, V. Kollonitsch and C.H. Kline, Ind. Eng. Chem., 60(11), 10-19 (1968)
3. R.J. Bertolacini and R.J. Pellet, in Catalyst Deactivation, B. Delmon and G.F. Fromet (Eds.), pp. 73-77, Elsevier, Amsterdam (1980)
4. N.D. Spencer and G.A. Somorjai, J. Catal., 78, 142-146 (1982)
5. R.A. Dalla Betta, A.G. Piken and M. Shelef, J. Catal., 40, 173-183 (1975)
6. T. Tsunoda, H. Ogasawara, M. Komiyama, S. Ozawa and Y. Ogino, Chem. Lett., 819-822 (1981)
7. T. Tsunoda, H. Ogasawara, M. Komiyama, S. Ozawa, Y. Ogino, T. Nakamura and M. Osugi, Nippon Kagaku Kaishi, 305-309 (1982)
8. M. Komiyama, Y. Ogino, Y. Akai and M. Goto, J. Chem. Soc., Faraday Trans. 2, 79, 1719-1728 (1983)
9. T. Komatsu, M. Komiyama, Y. Ogino and M. Iwamoto, J. Chem. Soc., Faraday Trans. 1, in press (1986)
10. M. Komiyama, T. Tsunoda and Y. Ogino, Ind. Eng. Chem., Prod. Res. Develop., in press (1986)
11. M. Komiyama, T. Tsunoda and Y. Ogino, Carbon, 23, 613-617 (1985)
12. M. Komiyama, T. Okamoto and Y. Ogino, J. Chem. Soc., Chem. Commun., 618-619 (1984)
13. Yu. I. Yermakov, B.N. Kuznetsov and V.A. Zakharov, Catalysis by Supported Complexes, pp. 99-103, Elsevier, Amsterdam (1981)
14. W.P. McKenna, B.E. Higgins and E.M. Eyring, J. Mol. Catal., 31, 199-206 (1985)
15. P. Gallezot, G. Coudurier, M. Primet and B. Imelik, in Molecular Sieves-II, J.R. Katzer (Ed.), pp. 144-155, American Chemical Society, Washington (1977)

16. R.F. Hicks, C.S. Kellner, B.J. Savatsky, W.C. Hecker and A.T. Bell, J. Catal., 71, 216-218 (1981)
17. M. Komiyama, M. Izumisawa, T. Okamoto and Y. Ogino, Chem. Lett., 823-826 (1985)
18. H.D. Kaesz, B. Fontal, R. Bau, S.W. Kirtley and M.R. Churchill, J. Am. Chem. Soc., 91, 1021-1023 (1969)
19. M. Herberhold and G. Süss, Angew. Chem., Int. Ed. Engl., 14, 700 (1975)
20. A. Bertolucci, M. Freni, P. Romiti, G. Ciani, A. Sironi and V.G. Albano, J. Organomet. Chem., 113, C61-C64 (1976)

$HOs_3(CO)_{10}OH$ AND $HOs_3(CO)_9(OH)PPh_3$ CLUSTERS SUPPORTED ON SILICA: THEIR STATE AND ACTIVITY IN CO HYDROGENATION

B.N. Kuznetsov, V.I. Kovalchuk*, L.K. Kedrova, A.A. Efremov and L.F. Dmitrieva

Institute of Chemistry and Chemical Engineering, Krasnoyarsk 660097;

* Krasnoyarsk State University, Krasnoyarsk 660062, USSR

ABSTRACT - Properties of $HOs_3(CO)_{10}OH$ and $HOs_3(CO)_9(OH)PPh_3$ clusters anchored on SiO_2 via reactive OH-groups have been studied. IR spectroscopy and temperature programmable decomposition were employed to examine the behaviour of supported osmium clusters upon thermal treatments. Catalytic properties of samples were tested in the synthesis of hydrocarbons from CO and H_2. A mechanism of CO hydrogenation on the supported osmium clusters is considered.

1. INTRODUCTION

The main products of the $CO+H_2$ reaction on conventional catalysts are known to be a broad class of organic compounds. The use of metal clusters for catalyst preparation gives the hope to create homogeneous active sites on the surface of the support and to obtain individual organic products from synthesis-gas with a high selectivity.

This work is concerned with a study of the properties of catalysts obtained by anchoring osmium clusters on the surface of SiO_2. In previous works (Refs. 1-4), the $Os_3(CO)_{12}$ was used. In this case the surface active sites are formed predominantly at the step of catalyst heating. During the thermal treatment the supported cluster can decompose to produce dispersed species whose catalytic properties are essentially similar to those of metallic osmium.

To prepare the catalyst, we choose osmium compounds containing a reactive hydroxyl group. The presence of this group allows the clusters to be chemically attached to the surface of the oxide support. Data on the interaction of $HOs_3(CO)_{10}OH$ and $HOs_3(CO)_9(OH)PPh_3$ with the surface of SiO_2, thermal stability of supported clusters and specific catalytic properties in synthesis of hydrocarbons from CO and H_2 are reported.

2. EXPERIMENTAL

Silica with a BET area of 300 m^2/g was used as a support. Sodium and iron content was less than or equal to 0.01% by weight. Prior to experiments, the support was calcined at 873 K for five to six hours.

Catalysts were prepared by supporting $HOs_3(CO)_{10}OH$ and $HOs_3(CO)_9(OH)PPh_3$ from benzene or ether solutions. The initial compounds were synthesized as described in (Refs. 5-6) and identified on the basis of their IR-spectra.

IR-spectra were recorded using a Specord-75 IR in the region of stretching vibrations of carbonyl groups (1600-2200 cm^{-1}) and in the region of vibrations of surface hydroxyl groups of the support (3300-3800 cm^{-1}). A special vacuum cell having sodium chloride windows was used to ensure high temperature treatment and to avoid contact with air during spectra registration.

For an IR-study of the process of cluster chemisorption, silica had been pressed into thin wafers (20 mg/cm^2 which were placed into a Pyrex cell-reactor and then treated in vacuum (10^{-5}Torr) at 873 K for an hour to remove water. The SiO_2 wafer was cooled down to 300 K, and a solution of the initial cluster in cyclohexane was added avoiding contact with air. Once added to SiO_2, the solution was almost decolored; then the solvent was frozen out in a trap filled with liquid nitrogen and evacuated at 300 K to achieve a pressure of 10^{-5} Torr.

The catalytic activity of CO hydrogenation was measured in a flow reactor with a catalyst fixed bed; the content of C_1-C_7 hydrocarbons was determined chromatographically. The reaction mixture with the composition $CO:H_2 = 1:1$ was used at a pressure of 0.1 MPa.
Conversion with respect to CO did not exceed 5%.

3. RESULTS AND DISCUSSION

3.1. Study of the Interaction of $HOs_3(CO)_{10}OH$ and $HOs_3(CO)_9(OH)PPh_3$ with SiO_2

Treatment of silica with a $HOs_3(CO)_{10}OH$ solution leads to a gradual decoloring of the latter. The substitution of one carbonyl group by triphenylphosphine ligand enhances the $HOs_3(CO)_9(OH)PPh_3$ cluster basicity and its reactivity in the reaction with the surface of silica.

The initial IR-spectrum of the $HOs_3(CO)_{10}OH$ cluster has the following absorption bands in the ν_{CO} region: 1987, 2001, 2024, 2071 and 2113 cm^{-1} (Fig. 1 A, curve 7). After chemisorption of the cluster on SiO_2, absorbance above 3400 cm^{-1} is observed (no such effect occurs when a pure solvent is added). Probably, this is evidence that water is produced during the interaction of the OH-groups of the supported cluster with those of the support surface.

The IR-spectrum of the silica-supported cluster shows the absorption bands at 2000, 2020, 2035, 2070, 2080, 2120 and 2135 cm^{-1} (Fig. 1 A, curve 2) in the ν_{CO} re-

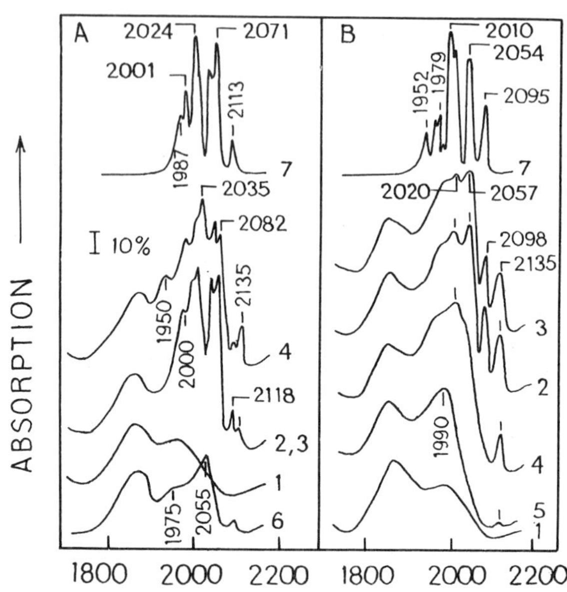

Fig. 1. IR spectra of $HOs_3(CO)_{10}OH$ (A) and $HOs_3(CO)_9(OH)PPh_3$ (B) supported on silica
1 - initial SiO_2; 2 - the SiO_2-supported cluster treated in vacuum at 298 K; 3 - sample 2 after treatment in H_2 at 373 K; 4 - sample 3 after treatment in H_2 at 473 K; 5 - sample 4 after treatment in H_2 at 573 K; 6 - CO adsorption at 378 K on the cluster decomposed at 573 K; 7 - cluster solution in cyclohexane.

gion. The following scheme of $HOs_3(CO)_{10}OH$ attachment to SiO_2 is proposed:

$$2HOs_3(CO)_{10}OH + \text{Si(OH)–O–Si (surface)} \rightarrow (CO)_3Os(\mu\text{-}H)(\mu\text{-}O\text{–Si})Os(CO)_3\,[Os(CO)_4] + (CO)_3Os(\mu\text{-}H)Os(CO)_3\,[Os(CO)_4](O\text{–Si})(O\text{–Si})$$

A B

The existence of surface structures of type A was reported by some authors (Refs. 7-8) who, however, obtained these structures by a less obvious way. Evidences for the formation of surface structures of type B may be (i) the existence of their molecular analog (Ref. 9), (ii) a blue-shift of the carbonyl frequencies of the chemisorbed cluster, and (iii) the appearance of a new band at 2135 cm^{-1}.

Structure A has the same local symmetry as the initial cluster. Structure B is characterized by a lower symmetry; thus, the corresponding spectrum must have a greater number of bands in the carbonyl region. In addition, electron density on the osmium triangle must be lower than that of structure A. The appearance of a new band at 2135 cm^{-1} characteristic of the charged Os_3 clusters indicates formation of structure B whose relative amount increases as the temperature is raised (see Fig. 1 A, curves 3 and 4).

A similar behaviour is observed for the $HOs_3(CO)_9(OH)PPh_3$ cluster attached to SiO_2, with the only difference that structures of type A (scheme 1) are formed to a considerably lesser extent. This is suggested by the appearance of an intensive band at 2135 cm^{-1} after supporting the cluster at 378 K and by the constancy of this band upon temperature variations within the cluster stability.

3.2. Thermal Stability of Supported Osmium Cluster

The $HOs_3(CO)_{10}OH$ and $HOs_3(CO)_9(OH)PPh_3$ clusters supported on SiO_2 are stable in air at ambient temperature; however their decarbonylation occurs at elevated temperatures in helium, hydrogen or vacuum. The IR-spectrum of the $HOs_3(CO)_{10}OH$ cluster, supported on SiO_2 and treated with H_2 at 373 K, does not differ significantly from that of the initial cluster. When increasing temperature up to 473 K, the intensities of the bands at 2000, 2035 and 2080 cm^{-1} decrease noticeably. In addition, a new band appears at 1950 cm^{-1}, the intensity ratio of the high frequency absorption bands changes, and resolution of the bands at 2035 and 2080 cm^{-1} diminishes (Fig. 1 A, curve 4).

All of the above indicates that over the temperature range from 423 K to 473 K the supported cluster is either partly or completely reconstructed. Note that complete decarbonylation of the supported cluster occurs only at 573 K. In the case of its treatment with H_2 methane is the only reaction product. Catalysts obtained by supporting $Os_3(CO)_{12}$ on SiO_2 are completely decarbonylated even at 473 K (Ref. 7).

With a view to get more detailed information about the stability of the supported $HOs_3(CO)_{10}OH$ cluster, its temperature programmable decomposition (TPD) in helium has been examined. As is seen in Fig. 2 A (curve 1), the curves of TPD have two peaks at 435 and 580 K. The decomposition of the cluster occurs more intensively at 473 K, which is in good agreement with our IR spectroscopic data.

Experiments performed under isothermal conditions (temperature rise in helium up to the definite temperature and keeping samples at this temperature for 1-2 h) have shown that the supported cluster loses one CO-group at 433 K and six CO-groups at 533 K. The main product of the thermal decomposition of the supported cluster is CO.

After CO adsorption at 278 K on the $HOs_3(CO)_{10}OH/SiO_2$ sample decomposed in helium at about 573 K, the TPD curve shows the appearance of more strongly bound forms of adsorbed CO than those observed in the initial cluster.

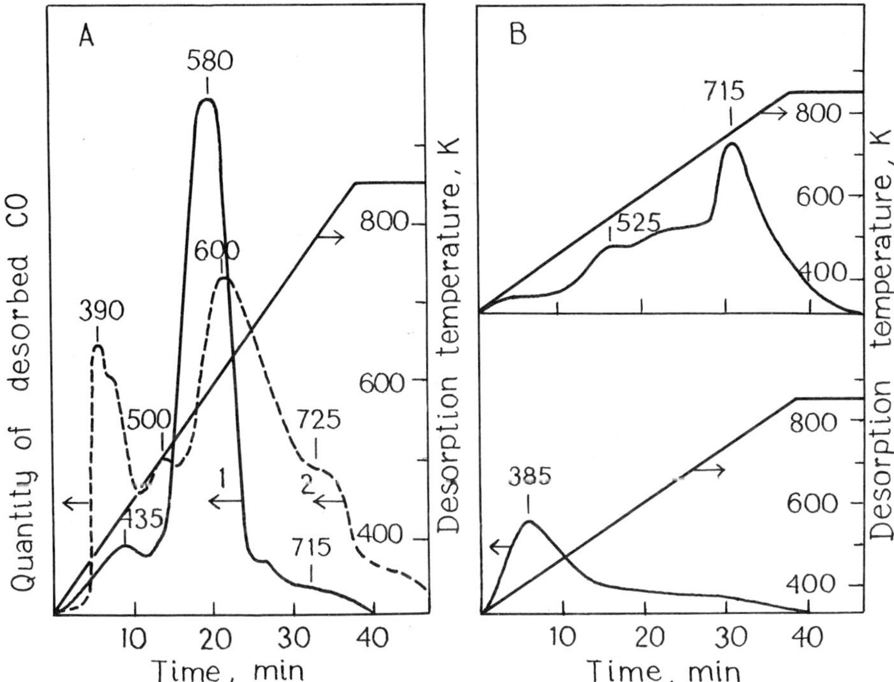

Fig. 2. Temperature-programmable decomposition curves of $HOs_3(CO)_{10}OH/SiO_2$ (1) and $HOs_3(CO)_9(OH)PPh_3/SiO_2$ (2) - A; thermal desorption of CO - B and H_2 - C obtained for the completely carbonylated $HOs_3(CO)_{10}OH/SiO_2$ catalyst. CO and H_2 were adsorbed at room temperature and 573 K, respectively.

CO adsorption on the supported $HOs_3(CO)_{10}OH$ cluster allowed to complete thermal decomposition in hydrogen or helium gives rise to ν_{CO} bands at 2135, 2055 and 1980 cm^{-1} (Fig. 1, curve 6). As suggested in (Ref. 10), these bands correspond to vibrations of carbonyl groups in compounds with the structure $[Os^{II}(CO)_2X_2]_n$ or $[Os^{II}(CO)_3X_2]_n$. Consequently, whatever the method of thermal decomposition of the supported cluster, a subsequent adsorption of CO leads to the formation of surface subcarbonyl structures containing ions of divalent osmium.

A similar behaviour at different thermotreatment procedures is shown by the $HOs_3(CO)_9(OH)PPh_3$ supported on SiO_2.

Thus, when $HOs_3(CO)_{10}OH$ and $HOs_3(CO)_9(OH)PPh_3$ are used, the interaction of the clusters with the surface of silica follows the route different from that of $Os_3(CO)_{12}$ interaction with silica. Besides, the thermal stability of surface osmium compounds increases considerably. Note that the supported $HOs_3(CO)_{10}OPh$ cluster, which cannot bind chemically with the surface of SiO_2, completely decomposes in H_2 atmosphere even at 373 K.

3.3. Catalytic Properties in CO Hydrogenation

The dependence of the yield of organic products and the activity of the supported clusters in the $CO+H_2$ reaction on temperature is plotted in Figs. 3 and 4. By inspecting Fig. 3 one can see that the content of methane in the reaction products is maximum at low temperatures (323-363 K) and tends to the minimum with the temperature increase. For $HOs_3(CO)_{10}OH/SiO_2$, the dependence of the total reaction rate on temperature in Arrhenius coordinates can be divided into three regions (Fig. 4, curve 1). The first region is characterized by methane formation which is most probably a stoichiometric reaction. This is suggested by the fact that after the catalyst treatment with hydrogen at this temperature during 20 h the reaction products have almost no methane. Starting from 373 K the supported clusters exhibit the true catalytic activity and the high selectivity with respect to individual hydrocarbons. The high selectivity seems to be caused by the homogeneity of surface carbonyl clusters which produce active reaction sites.

The region from 373 to 500 K is characterized by a low activation energy of about 38 kJ/mol. In the temperature region from 500 to 623 K, the activation energy increases up to approximately 88 kJ/mol.

The decomposition products of the supported $HOs_3(CO)_{10}OH$ clusters catalyze primarily methane forma-

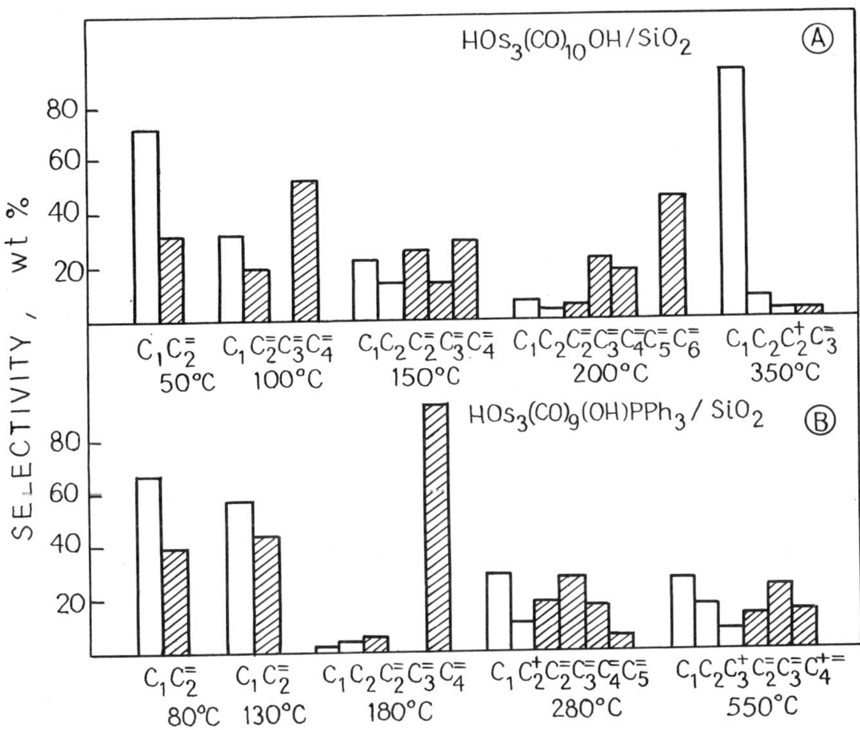

Fig. 3. Selectivity of $HOs_3(CO)_{10}OH$ (A) and $HOs_3(CO)_9(OH)PPh_3$ (B) clusters supported on SiO_2 in the CO + H reaction as a function of temperature.

tion. A similar composition of the products was observed (Ref. 1) on the catalyst prepared by the decomposition of $Os_3(CO)_{12}$ supported on SiO_2.

As distinct from this, the decomposition products of the supported $HOs_3(CO)_9(OH)PPh_3$ cluster catalyze both methanation and synthesis of olefins. In this case modification of the catalytic properties of osmium is caused by poisoning of methanation active centers by phosphorus. In our speculations on the mechanism of CO hydrogenation on supported osmium clusters, we consider the existence of organometallic compounds containing one of the groups: $-CHO$, $-CH_2OH$, $=CH_2$, $-CH_3$ as ligands.

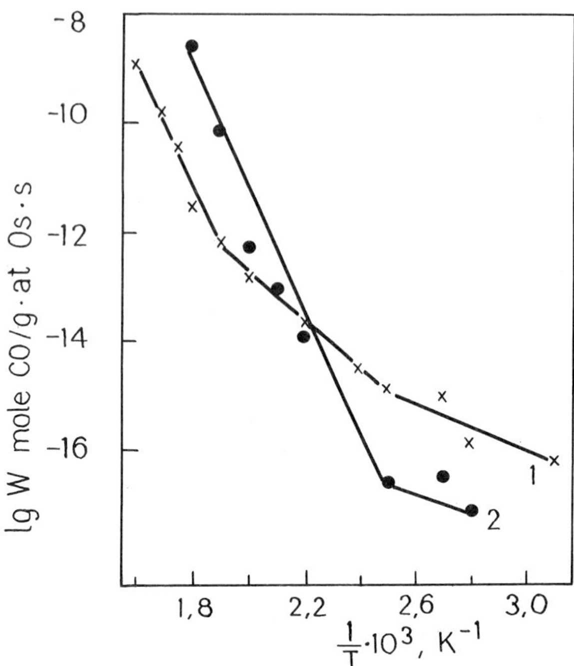

Fig. 4. Arrhenius plot for SiO_2-supported $HOs_3(CO)_{10}OH$ (1) and $HOs_3(CO)_9(OH)PPh_3$ (2). The catalytic activty was determined after the catalyst had been kept at the reaction temperature during 3-10 h.

Stable adducts having -CHO, $-CH_2OH$ and CH_3-groups as ligands were successfully isolated by the stoichiometric reduction of $[(\eta^5-C_5H_5)Re(CO)_2][BF_4]$ with sodium borohydride (Ref. 11). Thus, for the first time the sequence of CO to $-CH_2$ transformations postulated for most of the known mechanisms of CO hydrogenation, has been proven on one metal atom. The hydrocarbon chain propagation in the conventional "organometallic" mechanisms involves CO insertion into the $Me-CH_3$ bond (or migration of $-CH_3$ to the CO ligand) resulting in the formation of the $-CO-CH_3$-group that is hydrated further to $-CH_2CH_3$, etc.

The deviation of the product composition from Schulz--Flory distribution cannot be interpreted in terms of the "organometallic" mechanism of CO hydrogenation, suggested by Olive for active centers of the subcarbonyl structures.

In this work we have made an attempt to explain formation of methane, ethylene and propylene by a concerted mechanism of CO hydrogenation on two osmium atoms in the carbonyl cluster.

Methane is formed by hydrogenation of the carbene ligand, $=CH_2$. Ethylene is produced by dimerization of the following ligands formed on neighbouring atoms of osmium:

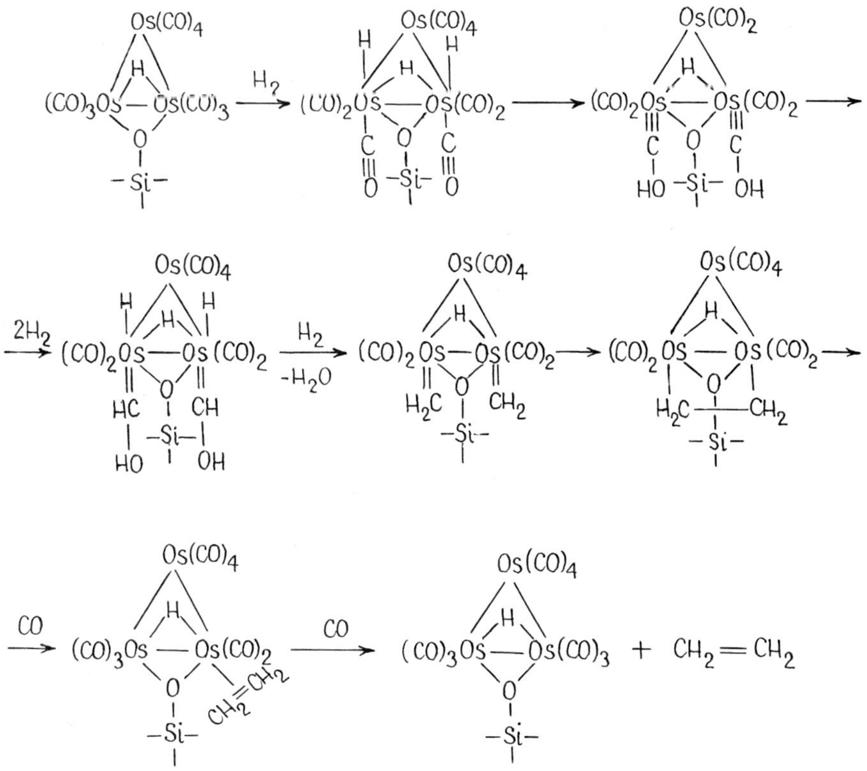

Propylene is produced by dimerization of $=CH_2$ and $=CHCH_3$ ligands:

$$(CO)_2Os(H)(\mu\text{-}O\text{-}Si)(=CH_2)-Os(CO)_2(=CH_2)[Os(CO)_4] \xrightarrow{H_2} (CO)_2Os(H)(CH_3)-Os(CO)_2(=CH_2) \xrightarrow{CO} (CO)_2Os(=C(OH)CH_3)-Os(CO)_2(=CH_2) \xrightarrow[-H_2O]{H_2}$$

$$\longrightarrow (CO)_2Os(=CHCH_3)-Os(CO)_2(=CH_2) \longrightarrow (CO)_2Os(H_2C(CH_3))-Os(CO)_2(CH_2) \xrightarrow{2CO} (CO)_3Os-Os(CO)_3(\mu\text{-}O\text{-}Si)[Os(CO)_4] + CH_3CH{=}CH_2$$

Finally, butene results from dimerization of $=CH-CH_3$ ligands that belong to neighbouring osmium atoms. The scheme of olefin formation postulated here may be supported by the results of Ref.12, in which thermal decomposition of the $L_nR_n(=R'_n)-L_nR_n(=R)$ cluster was shown to give the R'=R olefins. Formation of hexenes at elevated temperatures seems to be associated with the participation of the third osmium atom in the catalytic reaction. This atom is known to be inactive in low temperature substitution reactions.

REFERENCES

1. H. Knözinger, Yaping Zhao, K.B. Gates, J.P. Scott and B. Tesche, in Mechanisms of Catalysis, part 2, pp.-167-180, Nauka, Novosibirsk (1984).
2. J.M. Basset and A. Choplin, J. Mol. Catal., 21, 95 (1983).

3. J.-M. Basset, B. Besson, A. Choplin and A. Theolier, ibid, p. 154.
4. M. Deeba, J.P. Scott, R. Barth and R.C. Gates, J. Catal., 71, 373 (1981).
5. S.P. Gubin, V.A. Maksakov, L.K. Kedrova and E.D. Korniets, Dokl. Akad. Nauk SSSR, 271, 354 (1983).
6. N.V. Podberezskaya, V.A. Maksakov, L.K. Kedrova, E.D. Korniets and S.P. Gubin, Koord. Khim., 10, 919 (1984).
7. A.K. Smith, B. Besson, J.M. Basset, et al., J. Organometal. Chem., 192, C31 (1980).
8. B. Besson, B. Noraweck, A.K. Smith et al., J. Chem. Soc., Chem. Commun., No 12, 569 (1980).
9. E.D. Korniets, V.A. Maksakov, L.K. Kedrova, N.I. Shakotko and S.P. Gubin, Izv. Akad. Nauk SSSR, ser. Khim., 2, 435 (1983).
10. L.A.W. Hales and R. Irwing, J. Chem. Soc. (A), No. 12, 1932 (1967).
11. J.R. Sweet and W.A.G. Graham, J. Organometal. Chem., 173, 09 (1979).
12. R.E. Colborn, A.F. Dyke, S.A.R. Knox, K.A. Macpherson, K.A. Mead, A.G. Orpen, J. Roné and P. Woodward, Phil. Trans. Royal Soc., London, A308, 67 (1982).

PECULIARITIES OF ACTIVATION AND HYDROGENATION OF CARBON MONOXIDE ON SUPERDISPERSED COBALT PARTICLES OBTAINED BY DECOMPOSITION OF $Co_2(CO)_8$ ON THE SURFACE OF SILICA

V.L. Kuznetsov, A.S. Lisitsyn, A.V. Golovin,
M.N. Aleksandrov and Yu.I. Yermakov
Institute of Catalysis, Novosibirsk 630090, USSR

ABSTRACT - The interaction of CO with cobalt catalysts of different dispersion (d_{Co}=1-12 nm) was examined by IR spectroscopy. The catalytic properties with regard to CO hydrogenation were compared. Highly dispersed catalysts (d_{Co}=1-1.5 nm), obtained by thermal decomposition of $Co_2(CO)_8$ on silica modified by titanium hydride compounds, showed a low activity in CO conversion but a high selectivity in the synthesis of low alcohols (primarily ethanol). IR data evidence that while the dissociation of CO on highly dispersed cobalt particles is rather difficult to occur, carbon deposition on low dispersed catalysts in the presence of CO proceeds even at ambient temperature. The difference in the character of CO activation on large and fine cobalt particles is thought to be a reason of their different catalytic properties.

1. INTRODUCTION

In recent years, supported catalysts prepared with the use of transition metal carbonyl complexes have occupied great deal of attention (Refs. 1,2). In a number of cases these catalysts reveal the properties essentially different from those of the catalysts prepared by conventional techniques. For example, thermal decomposition of Co carbonyl complexes at quite low temperatures allowed one to obtaine highly dispersed metal particles on the surface of oxide supports (Refs. 3-6). As found (Ref. 7), modification of the surface of silica by Ti, Zr or Al hydride compounds leads to the formation of superdispersed (< 1.5 nm average size) cobalt particles.

Catalysts obtained with the use of Co carbonyls were most widely studied in CO hydrogenation (Refs. 3-10) - a promising reaction of the production of organic compounds and liquid fuel from nonoil raw materials. The diminuation of the size of Co particles was found (Refs. 6,10) to decrease the catalytic activity of Co and increase the relative yield of low hydrocarbons. In addition, in the case of cobalt particles of less than 2.0 nm in size the reaction products had noticeable amounts of low alcohols (mainly ethanol) even at the atmospheric pressure of a $CO+H_2$ mixture (Ref. 7). The dependence of the catalytic properties on the size of metal crystallites was also reported for Ni (Refs. 11-20), Ru (Refs. 13-15), Fe (Refs. 16,17), Co (Refs. 18,19) and Os (Ref. 20).

The objective of this work was to elucidate the nature of the effect of Co particle size on the properties of Co catalysts in the $CO+H_2$ reaction.

2. EXPERIMENTAL

2.1. CATALYST PREPARATION

Silica (200 m^2/g) preheated under vacuum (10^{-2}Torr) at 670 K and silica modified by titanium hydride complexes (denoted as TiH_x-SiO_2 in the following discussion) were used as supports. TiH_x-SiO_2 was prepared by treating silica with a pentane solution of $Ti(CH_2C_6H_5)_4$ followed by decomposition of the anchored benzyl complexes under H_2 at 430 K (Ref. 21).

Catalysts were prepared using cobalt nitrate or $Co_2(CO)_8$ (denoted in the following discussion as catalysts of (n) and (c) series, respectively). The content of cobalt (wt.%) will be indicated by numerals before letter symbols of the catalysts.

Pentane, used as solvent, was dried above the molecular sieves and outgassed under vacuum. Procedures of silica modification and $Co_2(CO)_8$ supporting were carried out in all-glass system with thin glass partitions (see, for example, Ref. 22).

The catalysts $5Co/SiO_2$ (n) and $10Co/SiO_2$(n) were prepared by impregnating the silica with a water solution of $Co(NO_3)_2$ followed by drying in air at 390 K and reducing with hydrogen at 670 K.

The catalyst $1.1Co/TiH_x-SiO_2$ was obtained by the adsorption of $Co_2(CO)_8$ from a pentane solution on TiH_x-SiO_2. The sample was then washed several times by a pure pentane, evacuated and treated at 520 K (20 m).

$4.7Co/TiH_x-SiO_2$ was prepared by the impregnation of TiH_x-SiO_2 with a $Co_2(CO)_8$ solution with a subsequent outgassing of the solvent in vacuum and thermal treatment of the sample at 520 K.

2.2. STUDY OF THE CATALYSTS

The catalytic properties of the samples were examined, like in (Refs. 6,7), in a flow differential reactor (atmospheric pressure, $CO:H_2$=1:2).

Magnetization of the samples was determined using a magnetic balance by the Faraday method at a magnetic field of 2.1-15.6 kOe and temperatures 77 and 293 K.

ESR spectra were registered on a ER 200D (Bruker) spectrometer at 293 K and at a frequency of 9.6-9.7 GHz.

IR spectra were recorded on a "Specord 75 IR" instrument. Prior to experimentation, the CO/SiO_2 (n) catalyst, passivated in air, was pressed into wafers (10-15 mg/cm^2) and placed into an IR cell. The cell allowed the further operations to be performed without contact of the sample with air. In the case of Co/TiH_x-SiO_2, a SiO_2 wafer was placed in the cell, and all steps of catalyst preparation were performed there without contact with air.

In some cases, we registered the spectra of the samples suspended in Nujol. To prevent the oxidation of the supported component, the catalysts were impregnated with Nujol under vacuum, then were rapidly ground in the atmosphere of argon and placed between the NaCl discs.

3. RESULTS

3.1. STATE OF Co IN THE CATALYSTS

All $\underline{Co/SiO_2 (n)}$ catalysts had Co particles of 5-15 nm in size (average diameter 10-12 nm), as ascertained by electron microscopy data. The catalysts were characterized by a broad FMR signal (Fig. 1) and by a relatively high, under the experimental conditions, magnetization, which was independent of temperature and slightly dependent on the strength of magnetic field. These observations correspond well to the behaviour of large particles of metallic cobalt (Ref. 6).

The $\underline{1.1Co/TiH_x-SiO_2}$ sample showed the same properties as an analogous sample, $1.0Co/TiH_x-SiO_2$, examined in (Ref. 7). Our sample, after the thermal treatment, produced no ferromagnetic resonance (FMR) signal in the ESR spectrum; its extremely low magnetization depended strongly on temperature and on the strength of magnetic field. This is evidence that particles of metallic cobalt of less than 1 nm in size were formed in the catalyst. The catalyst operated in the $Co+H_2$ reaction gave a FMR signal in the ESR spectrum. The magnetization of the sample increased, but it still remained strongly dependent on temperature and on the strength of magnetic field. This suggests the appearance of larger cobalt particles; however, their average size, calculated by the Langevin equation, remained small (ca. 1.2 nm).

Data on magnetic properties of $4.7\ Co/TiH_x-SiO_2$ are

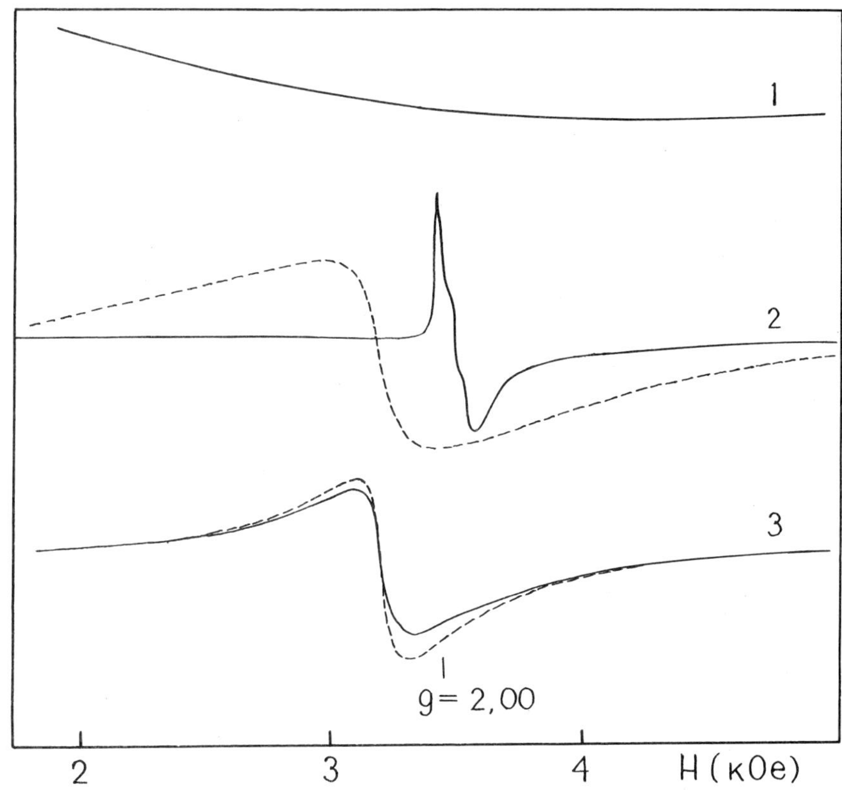

FIGURE 1. ESR spectra of $5Co/SiO_2$ (N) (1), $1.0Co/TiH_x-SiO_2$ (2) and $4.7Co/TiH_x-SiO_2$ (3). The spectrum of the samples used in hydrogenation of CO (470 K) is shown by dotted line.

illustrated in Figs. 1 and 2. In contrast to $1.1Co/TiH_x-SiO_2$, the ESR spectrum of $4.7Co/TiH_x-SiO_2$ had the FMR signal immediately after the thermal decomposition of the supported carbonyl (Fig. 1); its magnetization being higher than that of the former sample. These facts indicate the formation of larger metal particles in the $4.7Co/TiH_x-SiO_2$ catalyst. But the average size of these particles is still only ~ 1 nm, as follows from magnetic measurements. After the reaction, the catalyst magnetization slightly increased as a result

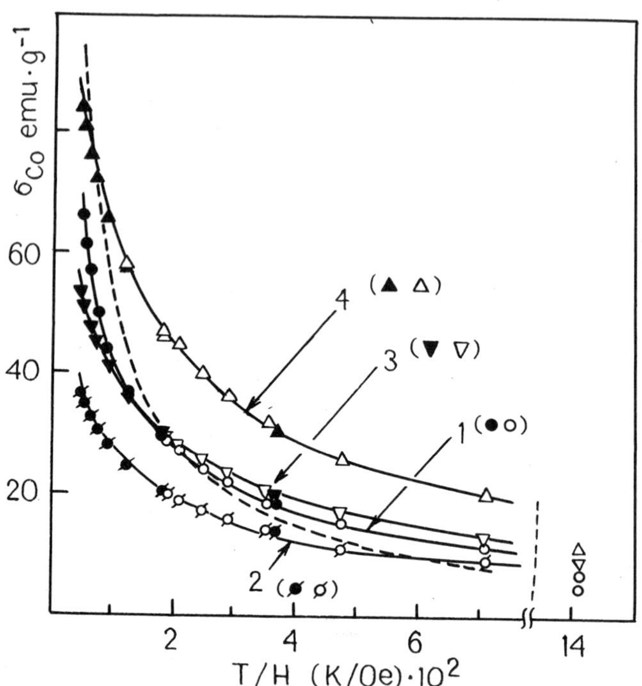

FIGURE 2. Specific magnetization of cobalt in the 4.7Co/TiH_x-SiO_2 catalyst after a series of successive treatments:
1 - initial sample (after thermal decomposition of supported $Co_2(CO)_8$ in vacuum at 520 K);
2 - treated with a CO+H_2 mixture at 295 K and outgassed at the same temperature;
3 - used in the CO+H_2 reaction at 473 K, cooled in the CO+H_2 flow and outgassed at 295 K;
4 - additionally outgassed (5 min) at 450 K.

of a slight increase (up to 1.5 nm) in the average size of Co particles.

3.2. CATALYTIC PROPERTIES OF THE SAMPLES

The results of the study of the catalytic properties of cobalt catalysts are compiled in Table 1. Here are

TABLE 1. Properties of cobalt catalysts in the $CO+H_2$ reaction (453 K)

No	Catalyst	Size of cobalt particles, nm	Activity, µmole CO / g Co×s	Selectivity [a]						
				HC_1	HC_2	HC_3	HC_{1-7}	α [e]	Alcohols	ethanol/methanol [f]
1	$5.0Co/SiO_2$(N)	10[c]	2.5	29	4	9	58	0.85	1	-
2[b]	$0.9Co/SiO_2$(C)	2[d]	4	30	6	16	43	0.72	5	1.2
3	$4.7Co/TiH_x$-SiO_2(C)	1.5[d]	4.5	30	11	16	28	0.6	15	3
4[b]	$1.0Co/TiH_x$-SiO_2(C)	1[e]	3	-	-	-	-	0.5	25	4
5	$1.1Co/TiH_x$-SiO_2(C)	1[e]	4	32	11	14	23	0.55	25	3

a) [(mole CO, converted to a product)/(mole CO, converted to all products)] $\times 10^2$;

b) after (Ref.7); c) from electron microscopy data;

d) from magnetic measurements for the catalysts after reaction;

e) probability of the hydrocarbon chain propagation (Ref. 24) in the region of C_{4-7} hydrocarbons;

f) mole/mole.

given the activities and selectivities of the catalysts in the synthesis of C_1-C_7 hydrocarbons, methanol and ethanol, which were the main organic products of the interaction between CO and H_2 at 453 K on highly dispersed catalysts. The results were obtained at high space velocities (10^4 cm^3 $(g.cat\ h)^{-1}$) of the reaction mixture flow. In this case, the change in the flow velocity had little effect on the distribution of the products. The relative yield of the oxygen-containing products had a tendency to decrease at the low space velocities and at rising temperature of the reaction.

3.3. IR STUDY OF CO ADSORPTION ON COBALT CATALYSTS

Fig. 3 shows the time variations of IR spectra of CO adsorbed on $5Co/SiO_2$ (n) and Co/TiH_x-SiO_2 (prepared directly in the cell), at room temperature. The adsorption of CO on the $5Co/SiO_2$ (n) catalyst containing large cobalt particles brought about weak bands at 1900-2075cm^{-1} (~1930 and 2050-2075 cm^{-1}), whose positions and intensities changed as functions of the pressure of CO in the gas phase (Fig. 3A). Analogous bands (15 Torr CO) were also registered for the $10Co/SiO_2$ (n) sample. However, in this latter case, the low wave numbers shift (2035 cm^{-1}) of the most intensive peak was observed. The spectrum also had two shoulders at 2060 and 2000 cm^{-1} and a weak band at 1930 cm^{-1}.

The exposure of $5Co/SiO_2$ (n) and $10Co/SiO_2$ (n) in the atmosphere of CO for 15-30 hrs resulted in that the absorption bands at 1900-2070 cm^{-1} were almost completely removed (Fig. 3). The treatment of the samples with H_2 at 670 K and the subsequent adsorption of CO at room temperature did not recover the bands at 1900-2070 cm^{-1}, but gave rise to a weak band at 2175 cm^{-1}, which could be removed by outgassing under vacuum.

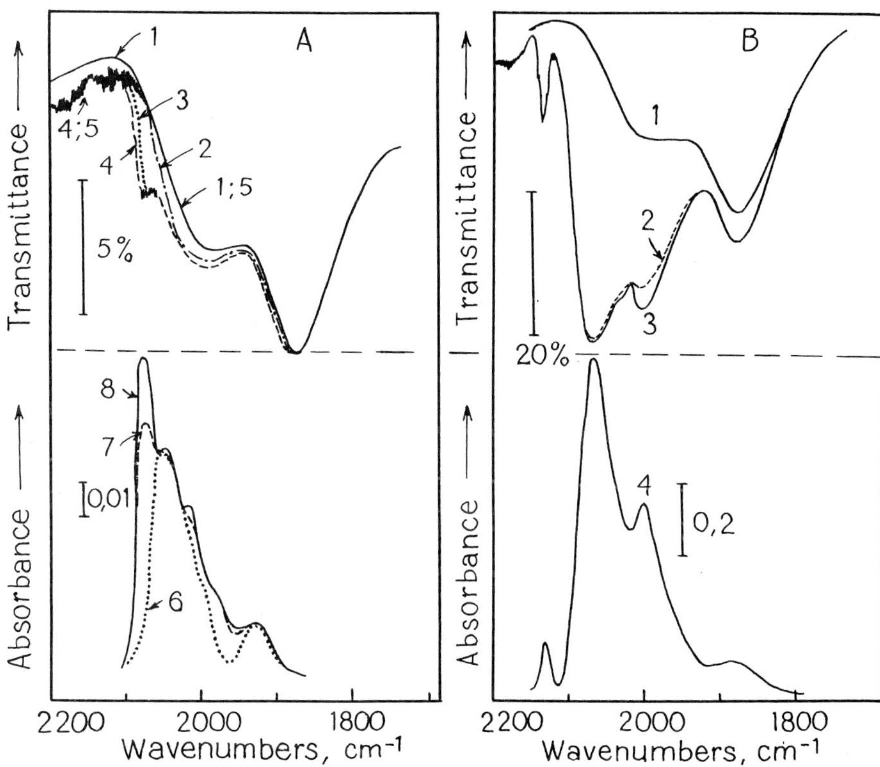

FIGURE 3. IR spectra of CO adsorbed on cobalt catalysts at 293 K.
A. $5Co/SiO_2$ (n). 1 - after the treatment of the catalyst wafer with H_2 (thrice at 100 Torr H_2, evacuation at 670 K); 2 - 0.12 Torr CO, 10 min; 3 - 1.6 Torr CO; 4 - 10 Torr CO, 10 min; 5 - after the exposure of the wafer in the atmosphere of CO for 30 hrs; 6-8 - difference spectra (optical density scale); 6 - (2)-(1); 7 - (3)-(1); 8 - (4)-(1).
B. Co/TiH_x-SiO_2 (prepared directly in the IR cell). 1 - after thermal decomposition of adsorbed $Co_2(CO)_8$; 2 - subsequent adsorption of CO at 15 Torr, 10 min; 3 - the same as in (2) after 16 hrs; 4 - difference spectrum (3)-(1).

These were only successive treatments with oxygen at 620 K and hydrogen at 670 K that restored the initial spectrum of CO on freshly reduced catalysts.

The adsorption of CO on Co/TiH_x-SiO_2 after the thermal decomposition of the adsorbed cobalt carbonyl led

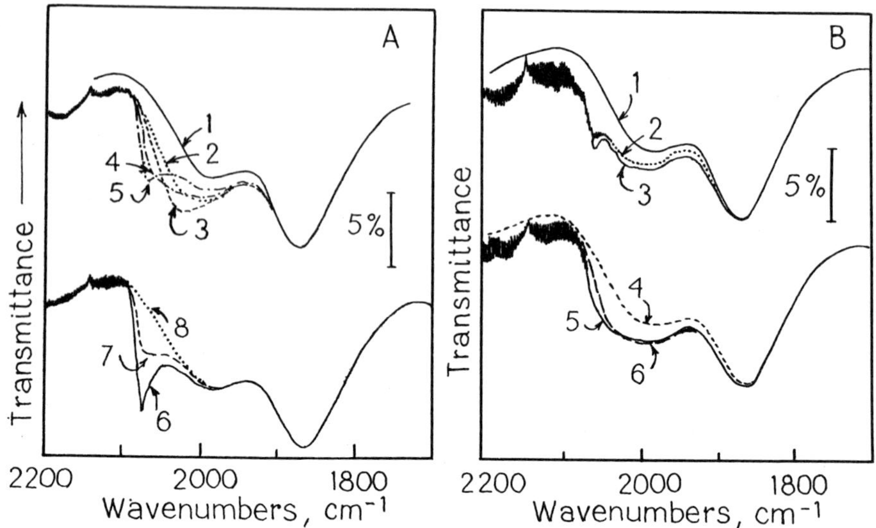

FIGURE 4. A. IR spectra of CO adsorbed on $10Co/SiO_2$ (n) at various temperatures. 1 - after treatment of the catalyst wafer with H_2 (3 portious at 100 Torr) at 670 K; 2 - 15 Torr CO, 293 K, 10 min; 3 - heating to 400 K, 5 min; 4 - 450 K, 5 min; 5 - 460 K, 30 min; 6 - 460 K, 3 hrs; 7 - 460 K, 7 hrs; 8 - 460 K, 14 hrs.
B. IR spectra of CO, adsorbed (30 Torr) on $10Co/SiO_2$(n) in the presence of H_2 (60 Torr). 1 - after treatment of the catalyst wafer with H_2 (3 portious at 100 Torr) at 670 K; 2 - adsorption of the $CO+H_2$ mixture at 293 K, 5 min; 3 - 25 min; 4 - catalyst spectrum at 460 K without a $CO+H_2$ mixture; 5 - adsorption of the mixture at 460 K, 10 min; 6 - 20 hrs.

to the absorption bands with the maxima at 2132, 2070, 2000 and 1930 cm^{-1}. These bands were much more intensive than those of CO adsorbed on the low dispersed Co/ $/SiO_2$ catalyst.

On the further exposure of Co/TiH_x-SiO_2 in the atmosphere of CO, the band intensities even tended to increase (Fig. 3B), in contrast to the Co/SiO_2 (n) sample.

Heating the $10Co/SiO_2$ (n) catalyst in CO (Fig. 4A) first enhanced the intensity of the 2025 cm^{-1} band. However, with increasing temperature, the intensities of this and other low-frequency bands rapidly decreased, while absorbance at 2065-2070 cm^{-1} increased, but became weaker with time.

On heating the wafer of $10Co/SiO_2$ (n) in a $CO+H_2$ mixture (Fig. 4C) at 460 K for several tens of hours, the spectrum showed broad bands with the maxima at 2020-2040 and 1930 cm^{-1}. The intensities of the bands were decreasing during 2-3 hrs and then remained almost constant.

We failed to register CO absorbance for the Co/SiO_2 (n) catalysts used in the $CO+H_2$ reaction (450 K, 1 atm) which had been cooled in the reaction mixture, suspended in Nujol and placed, avoiding contact with air, between the NaCl discs. This failure seems to be due to the low intensity of these bands. Contrary to this, for the Co/TiH_x-SiO_2 catalysts under the same conditions we have observed a sharp band at 2040 cm^{-1} (Fig. 5).

In principle the decrease in the absorption band intensity from the adsorbed CO may be caused by the carbon deposition on the surface of metallic particles, as well as by their oxidation and sintering. However, there was no evidence that the catalyst oxidation occurs in the experimental conditions employed (neither the decrease in magnetization of the samples, nor the appearance of absorption bands, which could be assigned to CO adsorbed on Co ions, were detected).

A slight increase in the size of cobalt particles under the reaction conditions was found for only highly dispersed Co/TiH_x-SiO_2 catalysts.

Thus, the main reason for the fall of the absorbance band intensities ascribed to adsorbed CO seems to be carbon deposition on the catalyst surface. As found from temperature programmable desorption (TPD) data (Ref. 23), the carbon deposition on the surfaces of Co monocrystals (1012) noticeably suppresses the metal ability to adsorb CO. On carbonized metal surfaces the amount of adsorbed CO did not exceed 20% of the amount

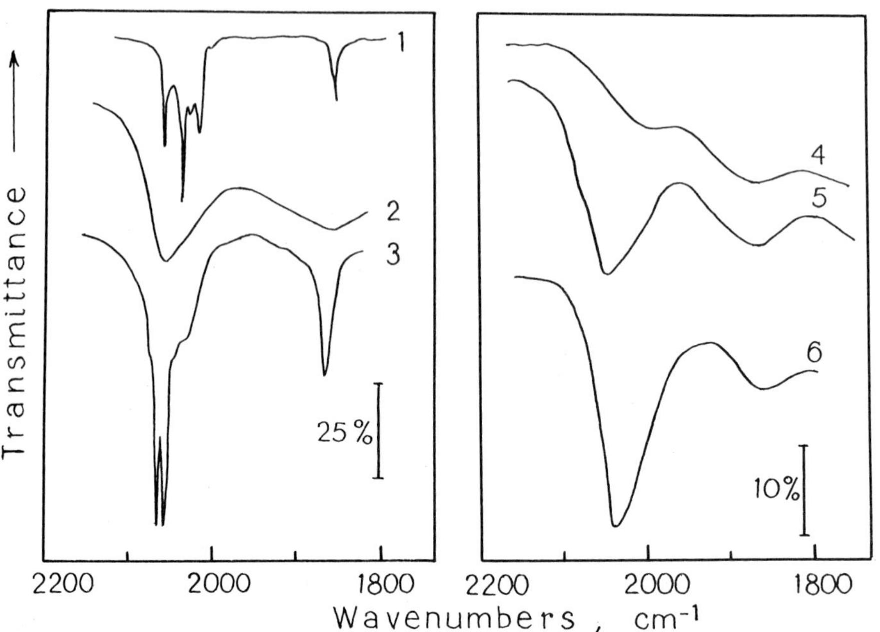

FIGURE 5. IR spectra of $1.0Co/TiH_x-SiO_2$ (2,4,5) and $4.7Co/TiH_x-SiO_2$ (3,6) catalysts after supporting $Co_2(CO)_8$ (2,3), after thermal decomposition of supported carbonyl (4) and after their use in the $CO+H_2$ reaction at 430 K (cooling the catalysts in the reaction mixture) (5,6). 1 - a starting solution of $Co_2(CO)_8$ in pentane. Spectra (2-6) were recorded for catalyst suspensions in Nujol.

of CO adsorbed on a clean surface.

From all of the above, we may conclude that

(1). Carbon deposition on low dispersed cobalt catalysts, Co/SiO_2 (n), can occur slowly in the atmosphere of CO (15-20 Torr) even at room temperature. The surface carbon cannot be removed completely by hydrogen treatment at 673 K;

(2). Carbon deposition on low dispersed cobalt catalysts at elevated temperatures removes, first, low-frequency bands characteristic of bridged CO. Simultaneously, a high-frequency band with a maximum at 2065-2075 cm^{-1}, associated with terminal CO groups appears in the spectrum, but tends to disappear with time. This might be evidence for the initial carbon deposition on the metal sites, in which the molecule of CO is bound to several Co atoms;

(3). Carbon deposition on the Co/SiO_2 catalysts in a $CO+H_2$ mixture occurs far more slowly, most probably, due to the hydrogenation of the surface carbon to hydrocarbons. This latter fact is supported by the appearance of absorbance bands in the region of C-H stretching bands (3018 cm^{-1});

(4). Highly dispersed cobalt particles are stable to carbon deposition in the atmosphere of CO at room temperature. Even when used in the reaction of CO with H_2, despite some sintering of cobalt particles, they produced a sharp band with a maximum at 2040 cm^{-1} in the IR spectra (Fig. 5).

4. DISCUSSION

4.1. NATURE OF THE EFFECT OF THE SIZE OF COBALT PARTICLES ON THE CATALYTIC PROPERTIES OF THE SAMPLES

The results listed in Table 1 confirm the effect of the size of cobalt particles on their catalytic properties in the hydrogenation of CO, which was previously found in (Refs. 6,7,10, 18,19). For example, on varying the average size of cobalt particles from 1 to 12 nm, the catalytic activity per 1 g of the metal did not change significantly. This indicates that the specific catalytic activity increases with increasing the size of the metal particles. The decrease in the average size of cobalt particles leads also to the increase in the yield of low hydrocarbons, as can be judged from a considerable decrease of parameter α in the Schulz-Flory equation which characterizes the rate ratio of the propagation and termination of the growing hydrocarbon chain (Ref. 24).

Note also that with decreasing the size of Co particles, the yield of alcohols is increased. Since this phenomenon was observed for many highly dispersed cobalt catalysts, prepared using the various supports (Al_2O_3, WgO (Ref. 4), TiO_2-SiO_2, TiH_x-SiO_2 (Ref. 7) and ZrH_x-SiO_2 (Ref. 25)), it seems reasonable to relate it to the structural effect rather than to the specific effect of the support.

Fu and Bartholomew (Ref. 19) have suggested that the increase in the yield of higher hydrocarbons and in the specific catalytic activity with increasing the size of cobalt particles is caused by the strengthening of the binding of CO and other carbon-containing intermediates to the metal surface. In their opinion, this causes the increase in the rate ratio of the propagation and termination of the growing hydrocarbon chains and the decrease in the yield of low hydrocarbons.

It has been assumed in (Ref. 7) that the change in the size of Co particles affects the ratio between the surface concentrations of carbon-containing intermediates such as (I) CO adsorbed in the molecular form, $(CO)_s$;
(II) $(CH_x)_s$ groups formed on hydrogenation of the carbon from dissociatively adsorbed CO; and
(III) growing hydrocarbon chains, $(C_nH_{2n\pm x})_s$.

The ratio between the surface concentrations of these species should determine the probability of their interaction and, consequently, the reaction activities (scheme I in Ref. 7). The change in the character of the activation of CO may have a crucial effect on the ratio of surface concentrations of intermediate hydrocarbon groups.

IR data suggest that the dissociative adsorption of CO easily occurs on large cobalt particles, but is not facile on fine particles. The latter fact may be explained by structural peculiarities of the small particles. These small particles have a great number of edge and angular metal atoms, which must lead to the increase in the CO adsorption stoichiometry; it was really observed in (Ref.26) (up to 2.3 CO molecules per one Co atom). The high intensity of the IR bands of adsorbed CO on the highly dispersed Co/TiH_x-SiO_2 catalysts may also be regarded as indirect evidence for the multiplied adsorption of CO. In this case, the gain in energy during the coordination of several CO molecules may be more than that during the formation of M-C and M-O bonds in the case of the dissociative adsorption of CO.

There is an opinion (Refs. 11,12) that the dissociation of CO is provided by the ensembles containing several atoms located on the planes of crystallites. The probability of the formation of these ensembles will be the lower the smaller the size of metal particles. In particular, the authors (Refs. 11,12) used this idea to explain the drop in the specific catalytic activity of Ni-Cu and Ru-Cu alloys in the Co hydrogenation upon decreasing the concentration of the active component (Ni, Ru) in them.

Along with the structural effects on the activation of CO, one should also consider possible variations in the electronic state of Co. For example, the dissociation of CO on dispersed metal species may slow down as they incorporate small amounts of carbon. Some carbon deposition on highly dispersed cobalt particles can occur at the step of thermal decomposition of the carbonyl (Ref. 6) (when the catalyst is prepared with the use of carbonyl complexes)or in the beginning of the CO+H_2 reaction (when dispersed catalysts are prepared by reduction of Co^{II}). For example, partial carbon deposition on the surface of Co monocrystal (1012) suppressed completely its ability to provide the dissociative adsorption of CO (Ref. 23). This effect seems to be most pronounced for dispersed metal particles. The change in the state of Co due to the interaction of fine particles with the support surface is not excluded either.

The lowered ability of highly dispersed particles of Co (< 2.0 nm) to adsorb CO dissociatively and the related with it decrease in the concentration of CH_x fragments appear to be responsible for the fall of their specific catalytic activity and for the increase in the relative yield of alcohols and low hydrocarbons. This situation is favoured also by the high surface concentrations of CO groups. The high yields of alcohols on larger Co particles (~4.0-5.0 Å) can also be obtained, but it needs the high pressure of the reaction mixture. This condition is fullfilled in the synthesis of C_1-C_6 alcohols on the Co-Cu-ZnO-Cr_2O_3 catalysts developed by French researchers (Ref. 27).

Note that hydrogenation of CO on homogeneous cobalt catalysts occurs without dissociative adsorption of CO (Ref. 28), the main products being methanol, methylformate and ethyleneglycol. The hydrocarbons are formed only upon the decomposition of carbonyl complexes, for example, at a low pressure of CO (Ref. 28).

The difference in composition of the products formed on the homogeneous catalysts and on the highly dispersed cobalt catalysts obtained by us may evidence for the difference in the mechanisms of their action. It is not excluded, however, that on increasing the pressure of CO, the products, similar to those formed in the presence of homogeneous catalysts, can be obtained on supported small cobalt particles as well.

REFERENCES

1. D.S. Bailey and S.H. Langer, Chem. Rev., 81, 109 (1982 (1982).
2. Yu.I. Yermakov, J. Mol. Catal., 21, 35 (1983).
3. D. Vanhove, D. Makambo and M. Blanchard, J. Chem. Soc. Chem. Commun., 605 (1979).

4. A.S. Lisitsyn, V.L. Kuznetsov and Yu.I. Yermakov, React. Kinet. Catal. Lett., 14, 445 (1980).
5. A.S. Lisitsyn, A.V. Golovin, V.L. Kuznetsov and Yu.I. Yermakov, React. Kinet. Catal. Lett., 19, 187 (1982).
6. A.S. Lisitsyn, A.V. Golovin, V.L. Kuznetsov and Yu.I. Yermakov, C_1 Mol. Chem., 1, 115 (1984).
7. A.S. Lisitsyn, A.V. Golovin, V.L. Kuznetsov and Yu.I. Yermakov, J. Catal., 95, 527 (1985).
8. D. Vanhove, Z. Zhuyong, P. Makambo and M. Blanchard, Appl. Catal., 9, 327 (1984).
9. A.L. Lapidus, A.Yu. Krylova and L.T. Kondratiev, Neftekhimiya, 21, 397 (1981).
10. A.S. Lisitsyn, V.L. Kuznetsov and Yu.I. Yermakov, Kinet. Katal., 23, 919, 926 (1982).
11. W.A.A. van Barneveld and V. Ponec, J. Catal., 51, 426 (1978).
12. J.A. Dalmon and G.A. Martin, in: New Horizons in Catalysis, Proc. of the 7th Int. Congr. on Catalysis, Tokyo 1980, T. Seiyama and K. Tanabe (Eds.), Kodansha, Ltd., Tokyo 1981, p. 402.
13. D.L. King, J. Catal., 51, 386 (1978).
14. C.S. Kellner and A.T. Bell, J. Catal., 75, 251 (1982).
15. Y.W. Chen, H.T. Wang and J.G. Goodwin Jr., J. Catal., 83, 415 (1983).
16. D. Storm and M. Boudart, Abstracts of the 6th North American Meeting of the Catalysis Society, Chicago, R. Bertolacini (Ed.), Chicago Catalysis Club, G5, 1979.
17. H.J. Jung, P.L. Walker and M.A. Vannice, J. Catal., 75, 416 (1982).
18. R.C. Reuel and C.H. Bartholomew, J. Catal., 85, 78 (1984).
19. L. Fu and C.H. Bartholomew, J. Catal., 92, 376 (1985).
20. E.O. Odebunmi, B.A. Batrana, A.K. Datye, L.F. Allard Jr., J. Schwank, W.H. Manoque, A. Hayman, J.H. Onnferko, H. Knözinger and B.C. Gates, J. Catal., 95, 370 (1985).
21. G.A. Nesterov, V.A. Zakharov, E.A. Paukshtis, E.N. Yurtchenko and Yu.I. Yermakov, J. Mol. Catal., 3, 163 (1978).
22. Yu.I. Yermakov, B.N. Kuznetsov, Yu.P. Grabovski, A.N. Startsev, A.M. Lazutkin, V.A. Zakharov and A.I. Lazutkina, J. Mol. Catal., 1, 93 (1976).
23. K.A. Prior, K. Schwaha and R.M. Lambert, Surf. Sci., 77, 193 (1978).
24. G. Henrici-Olive and S. Olive, Angew. Chem. Intern. Ed. Engl., 15, 136 (1976).
25. A.S. Lisitsyn, V.L. Kuznetsov and Yu.I. Yermakov, Kinet. Katal., 24, 762 (1983).
26. R.C. Reuel and C.H. Bartholomew, J. Catal., 85, 63 (1984).
27. P. Courty, D. Durand, E. Freund and A. Sugier, J. Mol. Catal., 17, 241 (1982).
28. L.C. Costa, Catal. Rev.-Sci. Eng., 25, 325 (1983).

SURFACE ORGANOMETALLIC CHEMISTRY AND CATALYSIS BY GRAFTED CLUSTERS : $Ru_3(CO)_{12}$ SUPPORTED ON SILICA.

L. d'Ornelas, A. Théolier, A. Choplin, J.M. Basset

Institut de Recherches sur la Catalyse, Laboratoire Propre du C.N.R.S., conventionné à l'Université Claude Bernard, LYON I, 2 avenue Albert Einstein -69626 Villeurbanne Cédex - France

ABSTRACT

The interaction of $M_3(CO)_{12}$ (M=Ru,Os) with the surface of silica leads to (μ-H)(μ-OSi)$M_3(CO)_{10}$ anchored to the silica surface via a three electrons donor oxygen. We report here that this type of bonding is responsible for catalytic activity (hydrogenation of ethylene, isomerization of butene-1) by opening of a Ru-O bond. Evidence is given for a cluster catalyzed reaction. Two mechanisms, based on IR data, involving either a monohydride or a dihydride intermediate are discussed.

The formation of the grafted clusters $HOs_3(CO)_{10}(OSi\equiv)$ (1) and $HRu_3(CO)_{10}(OSi\equiv)$ (2) by interaction of respectively $Os_3(CO)_{12}$ and $Ru_3(CO)_{12}$ with the OH surface groups of silica is now well established. The recent synthesis of the molecular analog (μ-H)(μ-OSiEt$_3$)$Os_3(CO)_{10}$ and its full characterization by X-ray structure (3) has unambiguously confirmed that these clusters are bound to the surface of silica via a three electrons donor oxygen : Os$\searrow$O$\nearrow$Os ; the same Os-Os bond is also bridged by an hydride ligand. This type of bonding suggests the possibility of a catalytic behavior, involving the reversible opening of an Os-O bond, without breakage of the M_3 cluster frame or dissociation of a CO ligand. Indeed, $HOs_3(CO)_{10}(OSi\equiv)$ has been suggested to be the catalytic active species for hydrogenation of ethylene between 70° and 90°C

(4) and isomerisation of butene-1 at 50-90°C (5).

We have now investigated the catalytic properties of $HRu_3(CO)_{10}(OSi\equiv)$ in the isomerization of butene-1 and the hydrogenation of ethylene ; the results are compared with those obtained for Ru metal particles supported on silica.

I. EXPERIMENTAL PART

$(\mu\text{-H})(\mu\text{-OSi}\equiv)\ Ru_3(CO)_{10}$ has been prepared as previously described (2) and transferred, under strict exclusion of air, by Schlenk type techniques, to a microreactor. The catalytic experiments have been carried out under differential conditions, at atmospheric pressure. Helium has been used as diluting gas in order to vary the partial pressure of the reactants. Temperatures were regulated within ± 1°C. The catalyst contained 0.9 % wt Ru. IR experiments have been carried out as already described (1).

II. ISOMERIZATION OF BUTENE-1

When $(\mu\text{-H})(\mu\text{-OSi}\equiv)\ Ru_3(CO)_{10}$ (I) is brought in contact with 1-butene at temperatures between 30° and 50°C, under dynamic conditions, instantaneous formation of 2-butene (cis + trans) is observed. From kinetic data, it appears that the reaction is first order in 1-butene ; the apparent activation energy is respectively 15 kcal/mole for $1\text{-}C_4 \longrightarrow \text{cis-2-}C_4$ and 19 kcal/mole for $1\text{-}C_4 \longrightarrow \text{trans-2-}C_4$ over the 30-50°C range.

The reaction rate is strongly enhanced in the presence of hydrogen (figure 1), but at no time (even after 48 h) could any butane be detected.

As (I) decomposes to metal particles under fairly mild conditions (2), we have studied the isomerization activity of a Ru/SiO_2 catalyst obtained by reduction at 350°C under H_2 of (I). Under similar conditions of temperature and pressure, we could not detect any isomerization product in the absence of hydrogen. In the presence of hydrogen, isomerization is occuring as well as hydrogenation.

These experiments are strongly in favor of a cluster catalysed reaction by (I).

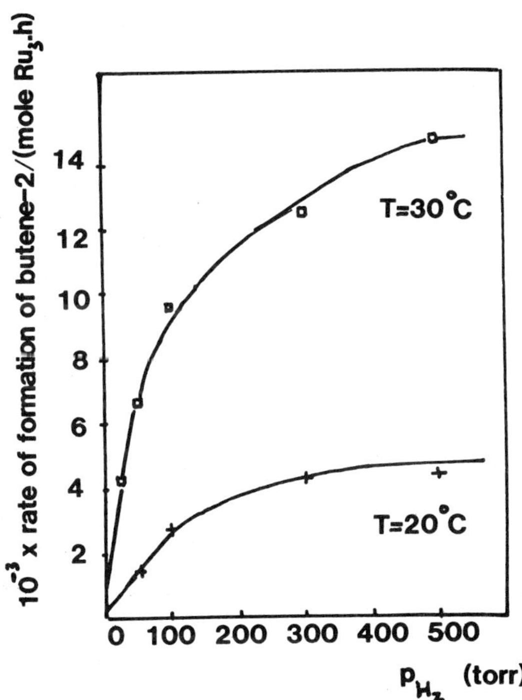

Figure 1 : Isomerization of 1-butene in the presence of (μ-H)(μ-OSi$\leqslant$) $Ru_3(CO)_{10}$. Influence of P_{H_2} on the reaction rate under the following conditions : m_{cata} = 18.3 mg, $p_{butene-1}$ = 100 torr, feed flow rate = 150 ml/min.

III. HYDROGENATION OF ETHYLENE

When a freshly prepared sample of (I) is brought in contact with (C_2H_4 + H_2) ethane is observed immediately, but the conversion is time dependent (fig. 2). A stationnary state is reached only after few hours.

We have determined the kinetic parameters once the system is stabilized. Partial pressures have been varied in the following ranges : $100 \leqslant p_{H_2} \leqslant 700$ torr and $20 \leqslant P_{C_2H_4} \leqslant 80$ torr. The reaction is found to be first order in hydrogen and zero order in ethylene over the range 30-50°C. The apparent activation energy is E_a = 10.9 kcal/mole.

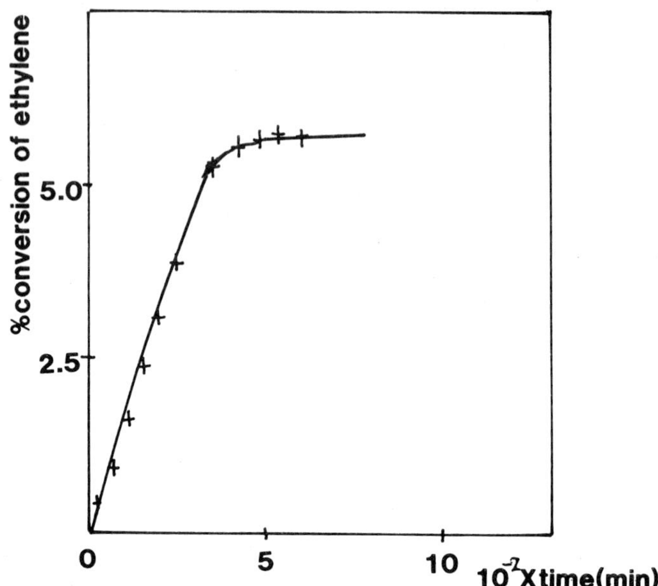

Figure 2 : Hydrogenation of ethylene in the presence of (μ-H)(μ-OSi$\equiv$) $Ru_3(CO)_{10}$: time dependence of the conversion. Experimental conditions : $p_{ethylene}$ = 63 torr, $p_{hydrogen}$ = 500 torr, feed flow rate = 50 ml/min. T = 50°C, m_{cata} = 98.7 mg.

IV. STUDY OF THE REACTIVITY OF (μ-H)(μ-OSi$\equiv$)$Ru_3(CO)_{10}$ WITH OLEFIN AND HYDROGEN BY INFRA-RED SPECTROSCOPY

1) <u>(μ-H)(μ-OSi$\equiv$)$Ru_3(CO)_{10}$ plus C_2H_4</u>

When a disc of (μ-H)(μ-OSi$\equiv$)$Ru_3(CO)_{10}$ is contacted with C_2H_4 (55 torr) at room temperature, the immediat changes in the ν(CO) region of the IR spectrum are small : the weak bands at 2113 cm^{-1} and 1956 cm^{-1} are totally replaced by small bands at 2105 cm^{-1} and 1980 cm^{-1} respectively ; simultaneously the intensity of the 2079 cm^{-1} and 2035 cm^{-1} bands diminish. (Fig. 3b). These changes are completely reversible by simple evacuation at room temperature.

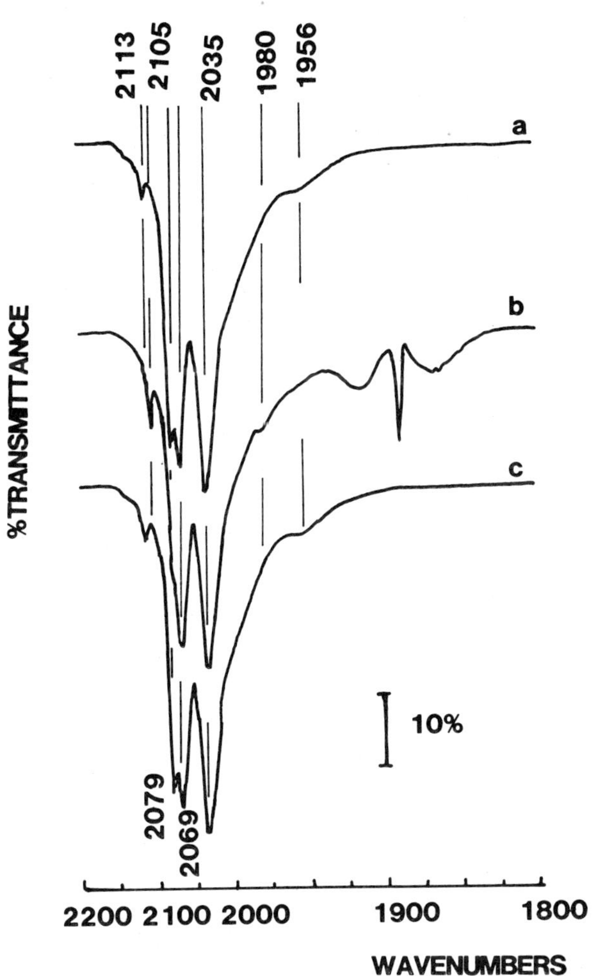

Figure 3 : Infrared spectra in the ν(CO) region of - a $(\mu\text{-H})(\mu\text{-OSi}\equiv)Ru_3(CO)_{10}$ - b - a, after treatment during 15 min. with ethylene (55 torr) at 25°C, - c - b, after evacuation (10^{-5} torr) during 5 min. at 25°C.

2) <u>$(\mu\text{-H})(\mu\text{-OSi}\equiv)Ru_3(CO)_{10}$ plus pentene-1</u>

When pentene-1 (95 Torr) is introduced in an IR cell on a disc of (<u>I</u>), changes in the ν(CO) region are comparable to those observed for C_2H_4, even at room temperature. At 50°C, under H_2 (510 torr) and pentene-1 (95 torr), pentene-1 disappears in the gas phase (the sharp band at 3080 cm^{-1}, characteristic of terminal olefinic ν(C-H), disappears) and pentene-2 appears. Simple evacuation under vacuum at 50°C (16 h) restores the ν(CO) spectrum of (I).

3) $(\mu\text{-H})(\mu\text{-OSi}\equiv)Ru_3(CO)_{10}$ plus H_2

In presence of H_2 (220 torr), the spectrum of (I) in the (CO) region does not change. It is only after 48 h at 25°C, or a few hours at 50°C (fig. 4b, c), that minor changes are noticed :

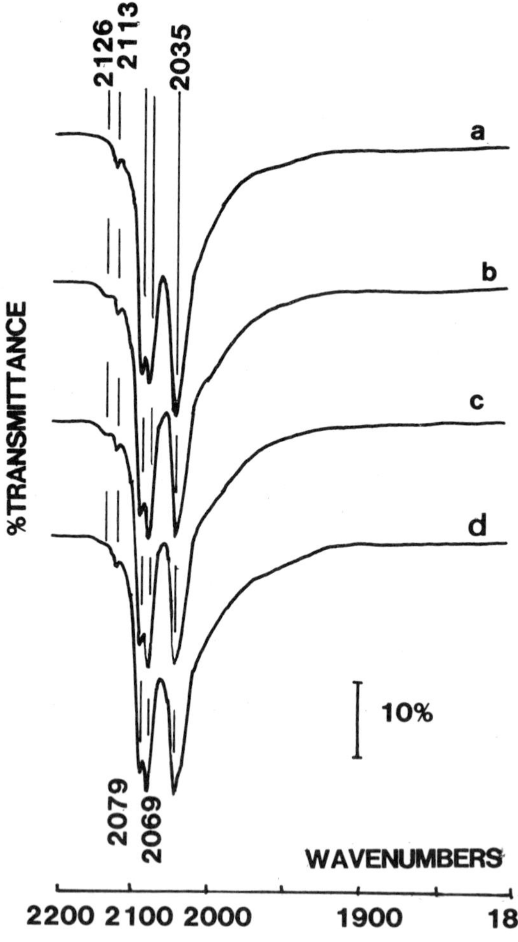

Figure 4 : Infrared spectra in the ν(CO) region of - a - $(\mu\text{-H})(\mu\text{-OSi}\equiv)Ru_3(CO)_{10}$ after treatment during 15 min, at 25°C, by hydrogen (220 torr) - b - same as - a - after 48 h - c - same as - b - after 16 h at 40°C - d - c, after evacuation (10^{-5} torr) at 50°C during 3 h.

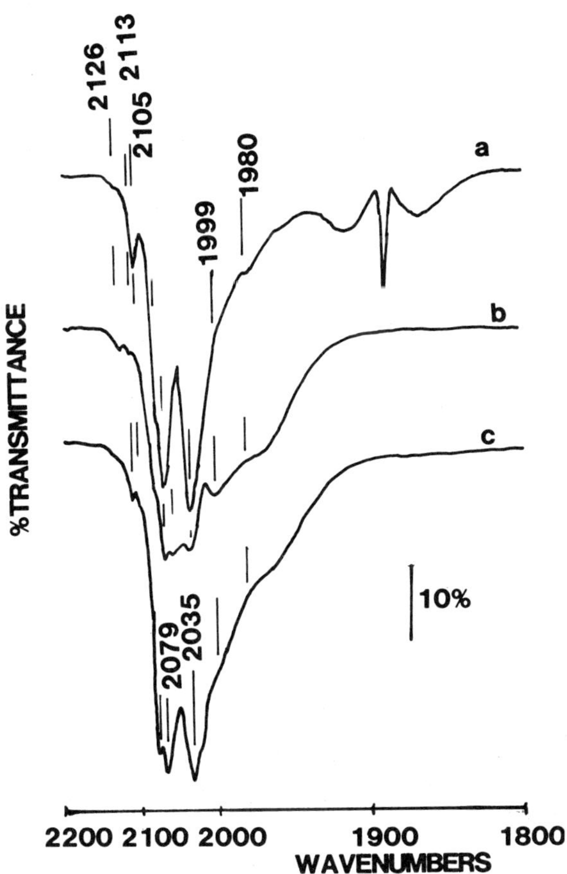

Figure 5 : Infrared spectra in the ν(CO) region of a -(μ-H)(μ-OSi$\equiv$)$Ru_3(CO)_{10}$ in the presence of ethylene (55 torr) and hydrogen (430 torr), after 15 min. at 25°C. - b - same as - a -, after 16 h at 50°C - c ; - b -, after evacuation (10^{-5} torr) at 25°C during 30 min.- d - same as - c -, after 4 h at 50°C.

a small band appears at 2126 cm^{-1}, while the intensity of the bands at 2113, 2079 and 2035 cm^{-1} diminishes slightly. These changes are reversible by simple evacuation at 50°C, under vacuum (fig. 4d).

4) $(\mu\text{-H})(\mu\text{-OSi}\equiv)Ru_3(CO)_{10}$ plus C_2H_4 and H_2

When C_2H_4 (55 torr) and H_2 (430 torr) are introduced simultaneously on (I) at room temperature, the spectrum observed (fig. 5a) is identical to the spectrum observed when (I) is contacted with C_2H_4 (fig. 3), even after 2h30 at 50°C. After 16h at 50°C, C_2H_4 is totally replaced by C_2H_6 in the gas phase and in the ν(CO) spectrum, new bands appear at 2126 (vw) 2060 (s) and 1999 (m). The spectrum of (I) is completely restored after evacuation at 50°C, for 4hrs (fig.5 d).

DISCUSSION

All the reported results suggest a catalysis by the grafted cluster $(\mu\text{-H})(\mu\text{-OSi}\equiv)Ru_3(CO)_{10}$ (I) ; indeed, the possibility of a catalysis by metal particles can be reasonably rejected on the following basis :

- A Ru/SiO_2 catalyst does not isomerize 1-butene in the absence of H_2, whereas (I) does this kind of isomerization.

- A Ru/SiO_2 catalyst hydrogenates and isomerizes 1-butene in the presence of H_2, whereas (I) in the same conditions isomerizes, but does not hydrogenate 1-butene.

Spectroscopic studies indicate that (I) coordinates reversibly ethylene or 1-butene at reaction temperature, probably by opening of the $(\mu\text{-OSi}\equiv)$ bridging ligand. (I) also coordinates hydrogen at a much lower rate to give likely $(\mu\text{-H})(H)(HOSi\equiv)$-$Ru_3(CO)_{10}$.

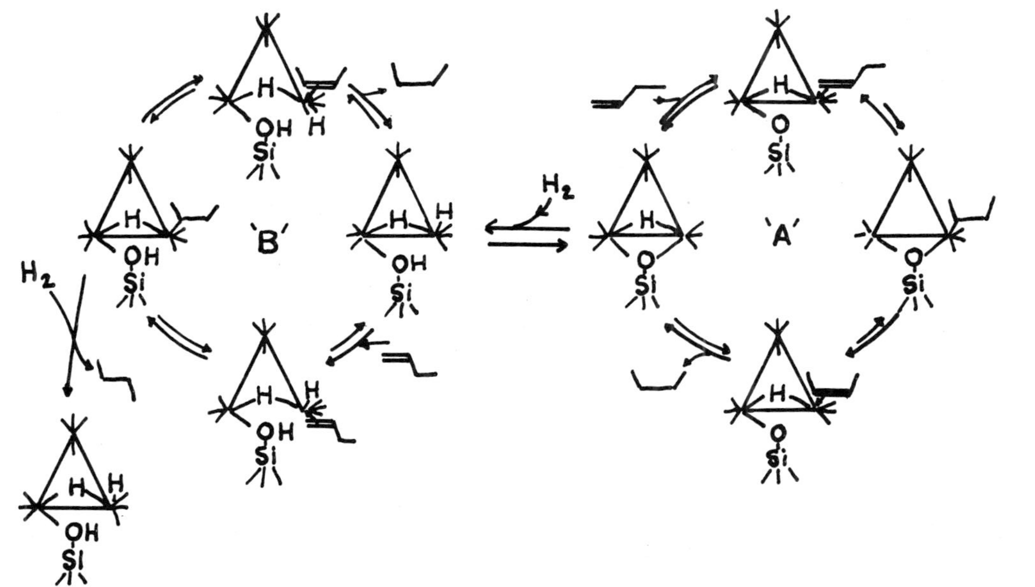

Scheme : Catalytic cycle for isomerization and hydrogenation of α-olefins catalyzed by $(\mu\text{-H})(\mu\text{-OSi}\equiv)$ $Ru_3(CO)_{10}$. The carbonyl ligands are omitted for simplicity.

In the absence of hydrogen, the first step of the catalytic cycle is probably coordination of the olefin, after opening of a Ru — O bond (scheme, cycle A). The next step is then the insertion of the olefin in a Ru-H bond, followed by β-hydrogen elimination and decoordination of the isomerized olefin (6).

In the presence of hydrogen, there is a probable competition between hydrogen and olefin coordination to the cluster (I). Once hydrogen is coordinated, the olefin can insert into a Ru-H bond. This second cycle (scheme, cycle B) is much faster than the cycle A, as suggested by the isomerization rate enhancement in the presence of hydrogen.

Further studies on molecular analogs are currently in progress to try to get more insight into the mechanism of catalytic isomerization of olefins by grafted clusters.

REFERENCES

1. R. Psaro, R. Ugo, G.M. Zanderhighi, B. Besson, A.K. Smith and J.M. Basset, J. Organomet. Chem., 213, 215 (1981).
2. A. Théolier, A. Choplin, L. d'Ornelas, J.M. Basset, R. Ugo, R. Psaro and C. Sourisseau, Polyhedron, 2, 119 (1983).
3. L. d'Ornelas, A. Choplin, J.M. Basset, L.Y. Hsu and S.G. Shore, Nouv. J. Chim., 9, 155 (1985).
4. B. Besson, A. Choplin, L. d'Ornelas and J.M. Basset, J. Chem. Soc. Chem. Commun. 843 (1982).
5. R. Barth, B.C. Gates, Y. Zhao, H. Knözinger and J. Hulse, J. Catal. 82, 147 (1983).
6. A.J. Deeming, S. Hasso and M. Hunderhill, J. Chem. Soc., Dalton Trans., 1614 (1975).

SURFACE SUPPORTED METAL CARBONYL CLUSTERS : CATALYSIS FOR 1-BUTENE ISOMERIZATION.

R.Psaro[1], C.Dossi, A.Fusi, R.Ugo and G.M.Zanderighi

Centro C.N.R.[1] and Dipartimento di Chimica Inorganica e Metallorganica, Università di Milano.

P.Doldi and V.Ragaini

Dipartimento di Chimica Fisica ed Elettrochimica, Università di Milano

R.Zanoni

Dipartimento di Chimica, Università "La Sapienza", Roma

ABSTRACT - The title reaction has been studied in a flow reactor in the range 343-389 K. The evolution of the grafted cluster $HOs_3(CO)_{10}(OSi\equiv)$ has been characterized under catalytic conditions and by infrared and XP spectroscopy. The activation energies are reported.

1. INTRODUCTION

The heterogeneous catalytic reactions with metal carbonyl clusters supported either on inorganic oxides or on zeolites (Ref. 1) can be divided into : (i) "homogeneous type" reactions carried out with molecular organometallic species anchored to the surface by chemical bonds (Refs. 2-4); (ii) reactions which involve as catalysts metal aggregates originated from the decomposition of the supported carbonyl metal cluster (Ref. 5). The separation between the two classes remains often largely untested, primarily because of the difficulty of controlling the cluster structure during catalysis. In this respect, supported metal catalysts prepared from triosmium carbonyl clusters may be the ideal systems to be investigated, because it is known that well characterized organometallic species derived from $Os_3(CO)_{12}$ are grafted to the surface of inorganic oxides and remain anchored intact in a well defined range of temperatures and reaction conditions (Ref. 6).

Few reports have described the ethylene hydrogenation at room temperature (Ref. 2) and the 1-butene isomerization between 323 and 363 K (Ref. 3), catalyzed by the covalently grafted cluster $HOs_3(CO)_{10}(OSi\equiv)$, which seems to remain intact under catalytic conditions (Ref. 7). However, above 373 K, the surface hydrido clusters are broken down into mononuclear oxidized carbonyl species (Ref. 6), organized in isolated ensembles of three atoms (Ref. 8). These new species are less active as isomerization catalysts (Ref. 9) although it is reported that they hydrogenate olefins (Ref. 10) and even CO (Ref. 11). In this work, we have investigated the isomerization of 1-butene catalyzed by species derived from $Os_3(CO)_{12}$ supported on silica, with the aim of defining the role of the support and of producing some evidence for catalytically active surface organometallic species.

2. EXPERIMENTAL

2.1. Catalyst preparation. Porasil 80-100, a porous silica, has been chosen because suitable to perform the activity tests even in differential conditions. The supported osmium catalyst was prepared by refluxing a n-octane solution of $Os_3(CO)_{12}$ with the silica under Argon (Ref. 3). The highly dispersed Os/SiO_2 catalysts were obtained starting either from $Os_3(CO)_{12}$ or $(NH_4)_2OsCl_6$, by thermal reduction with H_2. The catalytic runs were performed in the range 343-389 K (±0.5 K) using a glass flow reactor on line with a gaschromatograph. The feeding mixture was 1-butene in nitrogen with a total flux of 4.5 $l_{(NTP)}/h$ (contact time $\tau = 20$ g(catalyst) / $|(mol/h)_{1\text{-butene}}|$). Gaschromatographic analyses were performed on a 3mm x 2m s.s. column of 0.19% picric acid on Carbopack C.

2.2 Spectroscopic characterization. The infrared spectra were recorded with a Nicolet MX-1E infrared spectrometer. The samples were handled as pressed wafers in a pyrex cell interfaced to a vacuum and gas line.

The X-ray photoelectron spectra of fresh and used catalysts (as pressed wafers) were recorded with a VG ESCA-3 spectrometer using Al Kα radiation. Binding energies of 285.0 and 103.6 eV for the C1s and Si2p levels, respectively, were used as internal standards.

3. RESULTS AND DISCUSSION

3.1. Catalytic activities. Whilst $Os_3(CO)_{12}$ supported on glass beads was completely inactive at 343 K for 1-butene isomerization, the grafted cluster $HOs_3(CO)_{10}(OSi\equiv)$ exhibited a very low catalytic activity (0.1-0.3% conversion at 343-371 K under our experimental conditions). By raising the temperature up to 389 K, an increase in conversion with time was observed, with a parallel change in selectivity (Fig. 1) and a steady condition was maintained for a long reaction time (over 160 hours).

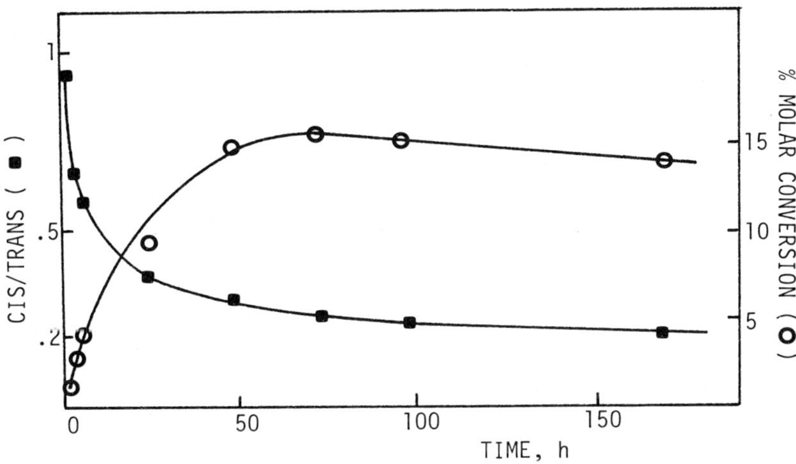

Fig.1 - Activity and selectivity of $HOs_3(CO)_{10}(OSi\equiv)$ under catalytic conditions.

Once the steady condition was attained, the catalytic bed was cooled down to r.t. in a flux of nitrogen and left for 16 h in contact with different gases (nitrogen, dry or wet air, CO) (Fig. 2). Then the

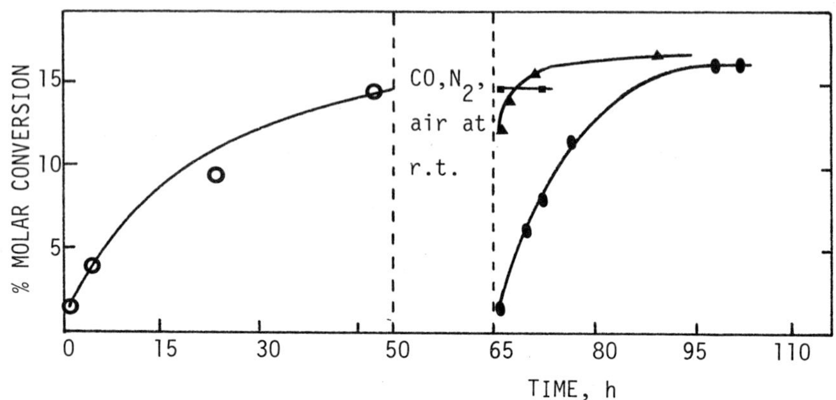

Fig.2 - Effects of N_2 (▪), air (▲) and CO (●) on the catalytic activity.

temperature was raised again up to 389 K in a flux of the feeding mixture. As expected, the contact with nitrogen did not affect both activity and selectivity, whilst the action of air resulted in an initial decrease of activity, followed by the attainment of the original steady condition after 4 h. On the contrary, CO induced a strong, but reversible inhibition. The starting activity was nihil and about 35 hours were required to restore the original steady state activity. These results indicate that both oxygen and CO interact with the catalyst formed in situ, but without relevant modifications of its structure, so that the active species is regenerated reversibly by treatment with 1-butene. The inhibiting properties of carbon monoxide were further evidenced by flowing at 389 K a mixture of 1-butene and CO (13/87 v/v) over the fresh catalytic bed. Under these conditions, the rate of isomerization was nihil, confirming thus that the decarbonylation of $HOs_3(CO)_{10}(OSi\equiv)$ is a necessary step for the formation in situ of the new catalytic species. This is supported also by the different values obtained for the activation energies in two characteristic ranges of temperatures : 60.7 KJ/mole in the range 341-361 K (where it is known that $HOs_3(CO)_{10}(OSi\equiv)$ is thermally stable), but 100.8 KJ/mole in the range 389-397 K.

This new catalyst formed in situ is not highly dispersed metallic osmium because the two samples prepared exhibited constant activity without any induction period and a higher selectivity (Table 1). The catalytic properties of Os(II) surface carbonyl species were also investigated (Table 1).

TABLE 1. Activity and selectivity of various Os catalysts at 389 K.

Catalyst	CIS/TRANS (a) ; % MOLAR CONVERSION	
	INITIAL	STEADY STATE
$HOs_3(CO)_{10}(OSi\equiv)$	0.93 ; 1.22	0.25(b) ; 15.3
$Os(CO)_x(OSi\equiv)_2$	0.48 ; 7.60	0.20(c) ; 13.0
Os/SiO_2(d)	1.12 ; 20.0	
Os/SiO_2(e)	1.14 ; 23.6	

a) thermodynamic value = 0.397 at 389 K
b) induction period 48 h; c) induction period 25 h
d) from $Os_3(CO)_{12}$; e) from $(NH_4)_2OsCl_6$

An initial conversion of about 8% without any induction period and a selectivity near to the thermodynamic value were observed. Moreover, after about 25 h both the activity and selectivity changed towards the values of the steady condition typical of the species generated in situ from the grafted hydride. This suggests the formation of the same active species, independently of the nature of the catalyst precursor.

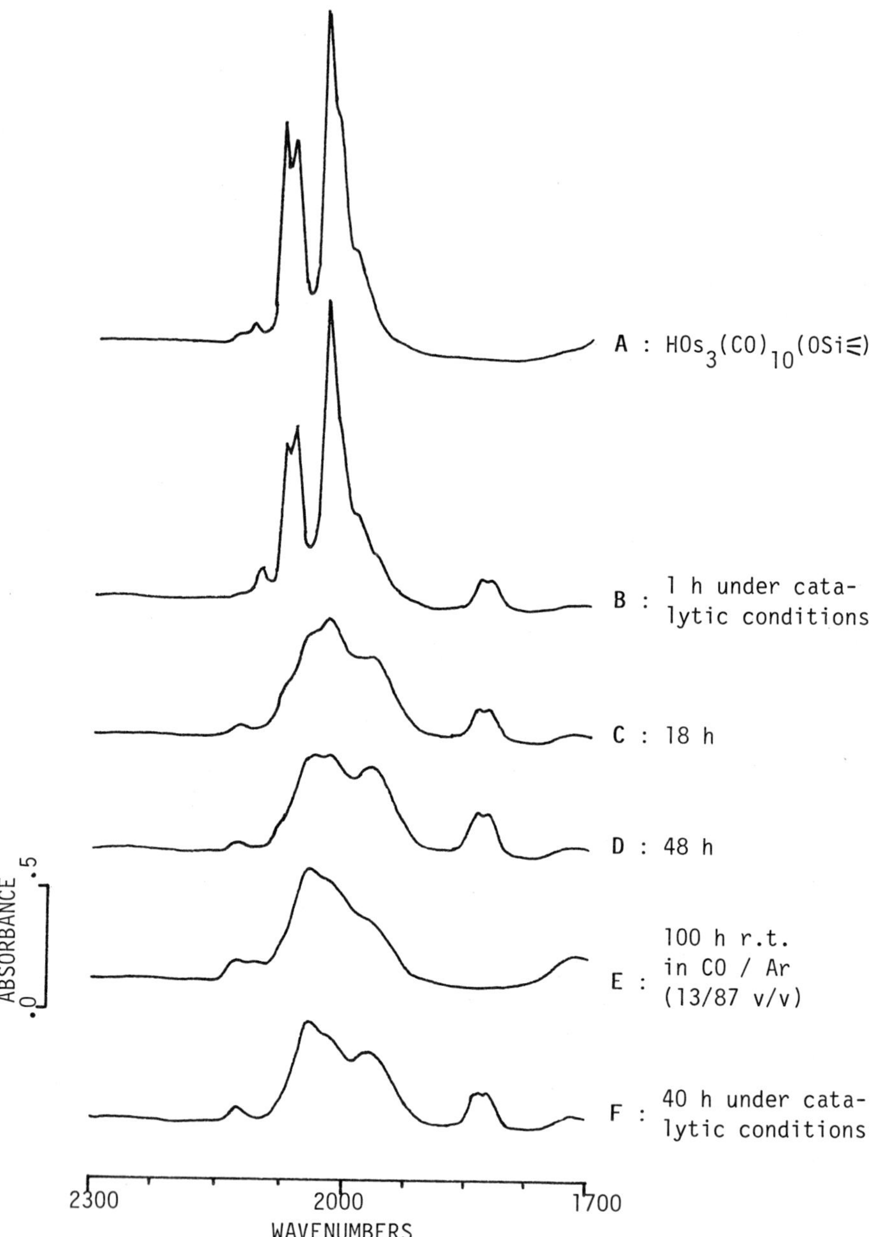

Fig.3 - Evolution of the grafted hydride and effect of CO.

3.2. Spectroscopic investigations

3.2.1. Infrared spectra. The evolution of the grafted cluster (Fig. 3-A) has been followed in the infrared cell under catalytic conditions at 389 K. The small band at 2115 cm^{-1} disappears in about one hour, while two new bands at 2105 and 1969 cm^{-1} (Fig. 3-B) are formed. A further modification performs slowly (about 18 h), characterized by bands at 2126w, 2077m,br, 2044s,br, 2022vs and 1972vs,br cm^{-1} (Fig. 3-C). After 48 h a final spectrum is achieved (Fig. 3-D), similar to that of the catalyst recovered after the catalytic runs, but markedly different from that of oxidized osmium(II) (Ref. 6). Therefore, the interaction of the grafted osmium cluster with 1-butene is confirmed (Ref. 3) and leads to the formation of a new and stable surface species. This is probably the real catalytic entity, which corresponds to the steady condition of the catalytic investigations. The reversible poisoning action of CO is evidenced in Fig. 3-E and F.

3.2.2. XP spectra. The relevant spectral data are reported in Table 2.

Table 2. XPS data of the osmium catalysts.

Catalyst	Os (wt%)[a]	Os $4f_{7/2}$ b.e. (eV) ; FWHM (eV) FRESH	USED
$HOs_3(CO)_{10}(OSi\equiv)$	1.0	52.5 ; 5.6	52.5 ; 5.6
$HOs_3(CO)_{10}(OSi\equiv)$	2.5	52.6 ; 5.4	52.4 ; 5.9
$Os(CO)_x(OSi\equiv)_2$	2.5	53.1 ; 5.6	52.7 ; 5.7
$Os_3(CO)_{12}/SiO_2$	1.0	51.2 ; 4.8	
Os/SiO_2(b)	2.5	51.2 ; 6.6	
Os powder	-	50.7 ; 4.8	

a) determined by X-ray fluorescence; b) from $Os_3(CO)_{12}$

The Os $4f_{7/2}$ b.e. value of the grafted cluster is between that of physisorbed $Os_3(CO)_{12}$ and Os(II) oxidized carbonyl species, but quite far away from that of the surface metallic particles of osmium. This b.e. trend reflects the oxidation state of the metal, in agreement with the reported XPS study on selected Os clusters (Ref. 12). Noteworthy is also the constancy of Os $4f_{7/2}$ b.e. values of the species under catalytic conditions, regardless of the osmium loading (Table 2). The full width at half maximum (FWHM) values are nearly constant in all the molecular samples and their range (5.4-5.9 eV) is smaller than that reported by Knözinger et al. (7.2-7.9 eV) (Ref. 13). These results confirm the constancy of the surface species formed during the catalytic runs, as suggested by the infra-red studies. Finally, the Os $4f_{7/2}$ b.e. values of the oxidized Os(II) surface species lower after the catalytic runs and are not too different from those of the species generated in situ from the grafted cluster. This observation is in agreement with the catalytic results and suggests that some rearrangments of the Os(II) entities incapsulated into the silica surface occur.

4. CONCLUSIONS

Our catalytic investigations indicates that the grafted hydride has a very low catalytic activity for the isomerization of 1-butene, whilst Os(II) carbonyl species have a higher activity. Both species transform slowly under catalytic conditions into a more active species, which is quite stable. This species is poisoned either by air or carbon monoxide, but the original activity (and probably the original species) is slowly restored under catalytic conditions.

The nature of this new species is unknown, but both infrared and XP spectra, together with the catalytic properties, would suggest a molecular structure different from either metallic particles or oxidized surface ions.

It seems that the low oxidation state and probably the cluster nature are maintained. The attribution of the catalytic activity to supported clusters or ions should be handled with great care, because under catalytic conditions the support itself behaves as a ligand (Ref. 14) and promotes chemical rearrangments leading to new organometallic surface species.

REFERENCES

1. M.Ichikawa, in Tailored Metal Catalysts, Y. Iwasawa (Ed.), pp. 183-259, Reidel Publishing Company, Dordrecht (1985) and references therein; J.Zwart and R.Snell, J.Mol.Catal., 30, 305-352 (1985) and references therein.

2. B.Besson, A.Choplin, L.D'Ornelas and J.M.Basset, J.Chem.Soc. Chem.Comm., 843-845 (1982).

3. R.Barth, B.C.Gates, Y.Zhao, H.Knözinger and J.Hulse, J.Catal., 82,147-159 (1983).

4. T.R.Krause, M.E.Davies, J.Lieto and B.C.Gates, J.Catal., 94, 195-202 (1985).

5. J.Phillips and J.A.Dumesic, Appl.Catal., 9, 1-30 (1984).

6. R.Psaro, R.Ugo, G.M.Zanderighi, B.Besson, A.K.Smith and J.M. Basset, J.Organomet.Chem., 213, 215-247 (1981).

7. The 1-butene isomerization has been previously studied in the presence of ungrafted osmium compounds or dispersed metallic osmium : P.B.Wells and G.R.Wilson, J.Catal., 9, 70-75 (1967); S.D.Mellor and P.B.Wells, Trans.Faraday Soc., 65, 1873-1890, (1969); V.Ragaini and M.G.Cattania-Sabbadini, J.Catal., 93, 161-175 (1985).

8. J.Schwank, L.F.Allard, M.Deeba and B.C.Gates, J.Catal., 84, 27-37 (1983).

9. X.J.Li, B.C.Gates, H.Knözinger and E.Alizo-Delgado, J.Catal., 88, 355-361 (1984).

10. M.Deeba and B.C.Gates, J.Catal., 67, 303-307 (1981).

11. J.M.Basset, B.Besson, A.Choplin, F.Hugues, M.Leconte, D.Rojas, A.K.Smith, A.Theolier, Y.Chauvin, D.Commereuc, R.Psaro, R.Ugo and G.M.Zanderighi, in Fundamental Research in Homogeneous Catalysis, M.Giongo and M.Graziani (Eds.), Vol. 4, pp. 19-53, Plenum Press, New York (1984).

12. R.Zanoni, V.Carinci, Raja H.Abu-Samn, R.Psaro and C.Dossi, J.Mol.Struct., 131, 363-369 (1985).

13. H.Knözinger, Y.Zhao, B.Tesche, R.Barth, R.Epstein, B.C.Gates and J.P.Scott, Faraday Discuss. Chem. Soc., 74, 53-71 (1981).

14. B.F.G.Johnson and R.E.Benfield, in Transition Metal Clusters, B.F.G.Johnson (Ed.), pp. 471-499, John Wiley & Sons, New York, (1980); H.P.Boehm and H.Knözinger, in Catalysis Science and Technology, J.R.Anderson and M.Boudart (Eds.), Vol. 4, pp.39-189, Springer-Verlag, Berlin (1983).

PREPARATION OF Os/SiO_2 CATALYSTS FROM $Os_3(CO)_{12}$: IR, EPR AND XPS STUDY

D.K. Chakrabarty, Ajay A. Desai and V. Vasumathi
Solid State Laboratory, Chemistry Department,
Indian Institute of Technology, Bombay 400 076,
India

ABSTRACT - Interaction of $Os_3(CO)_{12}$ with silica has been studied by IR spectroscopy. Physically adsorbed $Os_3(CO)_{12}$ changes to $Os_3H(CO)_{10}(-OSi)$ between 50 - 150 °C. Further heating leads to the formation of mononuclear di- and tri-carbonyl species and complete decarbonylation is achieved at 300 °C. EPR study of the decomposition products shows the presence of Os(I) and Os(III) as minor products. Reduction in H_2 at 400 °C yields mostly zero-valent osmium.

1. INTRODUCTION

Metal carbonyl clusters are good homogeneous catalysts for many reactions. It is hoped that interaction with the support will heterogenise them retaining at the same time the high selectivity of the homogeneous catalysts. Besides, partial decarbonylation of the supported clusters will lead to coordinative unsaturation that is so necessary for catalysis. Complete decarbonylation of the supported clusters under milder conditions may lead to highly dispersed supported metal catalysts leading to unique catalytic properties (Ref. 1).

Interaction of several osmium cluster compounds with silica has been described by Besson et al (Ref. 2) and Watson and Schrader (Ref. 3). Psaro et al (Ref. 4) concluded that physically adsorbed $Os_3(CO)_{12}$ on silica changes to a hydrido-triosmium carbonyl anchored on the surface that changes to mononuclear carbonyls at 200 °C. This was confirmed by Knözinger, Zhao and coworkers (Ref. 5) and also by Deeba and Gates (Ref 6). Collier et al (Ref. 7), on the other hand, suggested a further growth of the cluster on heating forming $Os_n(CO)_{xn}C_{yn}$ where n is as high as 12.

In the present work, we have used several techniques to characterise the samples prepared by the interaction of $Os_3(CO)_{12}$ with silica. Infrared spectra have been recorded in a high vacuum infrared cell as well as under atmospheric conditions since the catalysts will be normally exposed to atmosphere while loading on a conventional reactor. EPR spectroscopy has been used to detect minor species other than zero and divalent osmium. To check the correctness of the contradictory views on cluster breaking, UV reflectance spectra were studied. X-ray photoelectron spectroscopy (XPS) has been used to identify the osmium species and estimate their dispersion in the completely decarbonylated and the reduced catalysts.

2. EXPERIMENTAL

The compound $Os_3(CO)_{12}$ was obtained from Strem Chemicals. The support was Davison grade 979 silica (specific surface 280 m^2/g). It was heated at 500 °C for 4 h and stored in a dessicator. $Os_3(CO)_{12}$ was dissolved in dry HPLC grade hexane and added to the support kept in a flask through which dry helium was passed. The mixture was stirred by a magnetic stirrer for 4 h. The supernetant liquid was colourless. The solvent was then evaporated by slow heating to obtain a dry mass and was kept under helium. The metal loading was 1% by weight.

A small amount of this was quickly transferred into another tube and heated under vacuum 10^{-4} torr for 2 h at various fixed temperatures and infrared spectra were recorded in nujol after each heating using a Fourier transform spectrometer Niolet 170 SX at a resolution 1 cm^{-1}. Spectra were also recorded in a vacuum cell using a thin wafer.

The ESR tube was fitted with a HV stopcock and ground joint so that the sample once transferred into it can be subjected to vacuum, heating or various gases, and ESR spectra can be recorded without exposing it to the atmosphere. A X-band spectrometer (Varian E-112) was employed. Unless otherwise stated all spectra were recorded at 77 K. Reflectance spectra were recorded on UV-visible spectrophotometer (Shimadzu UV-260) using pure support as the reference.

XPS spectra were recorded on a VG Scientific ESCA 3 Mark II. The spectrometer was calibrated using a gold foil. Binding energy correction due to charging was done by using Si(2s) signal at 154 eV as internal calibration.

Supported metal catalysts were prepared by vacuum heating of $Os_3(CO)_{12}/SiO_2$ (1% loading) at 400 °C (catalyst D) and also by reducing it in flowing H_2 at 400 °C (catalyst A). Another catalyst (B) with same metal loading was prepared by impregnating the support with $OsCl_3$ followed by reduction in hydrogen at 400 °C.

3. RESULTS AND DISCUSSIONS

The infrared spectrum in solution of $Os_3(CO)_{12}$ showed four infrared bands (2068, 2036, 2016 and 2004 cm^{-1}) in agreement with the literature (Ref. 8). The solid state spectrum in nujol shows some extra bands (see Fig. 1a) which is very similar to that reported by Collier et al (Ref. 7). The spectrum of freshly prepared $Os_3(CO)_{12}/SiO_2$ (Fig. 1b) shows that $Os_3(CO)_{12}$ molecule has remained intact, possibly physisorbed on silica.

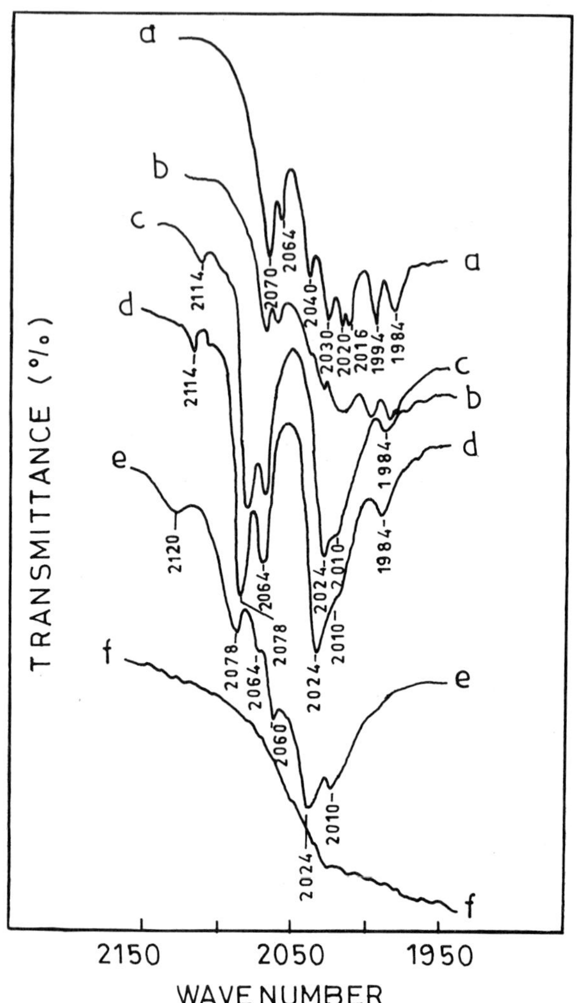

Figure 1. Infrared spectra in nujol (a) $Os_3(CO)_{12}$, (b) $Os_3(CO)_{12}/SiO_2$ (1%), (c), (d), (e) and (f) are sample (b) thermoevacuated at 50, 125, 200 and 300 °C, respectively for 2 h.

Spectra c and d are obtained after heating the supported carbonyl at 50 and 125 °C respectively under vacuum. These are identical with that reported for $Os_3(H)(CO)_{10}O\text{-}Si$

(to be called C hereafter). Unlike Psaro et al (Ref.4) we did not observe any band at 2127 cm^{-1} for pure or physisorbed $Os_3(CO)_{12}$. In fact, this band is not expected for $Os_3(CO)_{12}$ as well as C if one compares with the spectrum of the molecular compound $Os_3(H)(CO)_{10}OSiPh_3$ (Ref. 4). An IR band in this region is generally related to cluster breaking (Refs. 4-6). Since the support was heated to 500 °C prior to impregnation, it had lesser number of hydroxyl groups and this must be responsible for higher stability of the cluster. Our results show that C is formed even at 50 °C and not above 100 °C as reported by Psaro et al (Ref. 4) and also that it is stable in the atmosphere and hence can be used as a heterogenised cluster catalyst. At 200 °C a new band appears at 2060 cm^{-1} and the band at 2114 cm^{-1} is shifted to 2120 cm^{-1}. Complete decarbonylation occurs at 300 °C. This shows that the mononuclear Os(II) di- and tri-carbonyl species reported to be stable even at 400 °C under vacuum or helium (Ref. 6) are not stable under atmospheric conditions. The three-band IR spectrum is, however, observed when a thin wafer of the sample is heated at 250 °C in a vacuum IR cell.

Next we shall take up the question of cluster breaking on interaction of $Os_3(CO)_{12}$ with silica. The reflectance spectra are presented in Figure 2. Based on the work of Tyler et al (Ref. 9), the two bands at 324 and 394 nm for fresh $Os_3(CO)_{12}/SiO_2$ may be assigned to transitions due to triangular osmium frame. On heating at 125 °C under vacuum, an additional band appears at 358 nm. One can conclude that the osmium cluster is still intact. On heating to 250 °C, these bands disappear indicating breaking of the cluster.

Freshly prepared $Os_3(CO)_{12}/SiO_2$ shows an intense EPR signal at g = 4.12 (peak to peak width ΔH = 80 G) at 77 K. At 300 K the intensity reduces to a sixth. The spectrum remains unchanged on heating at 50 °C in vacuum.

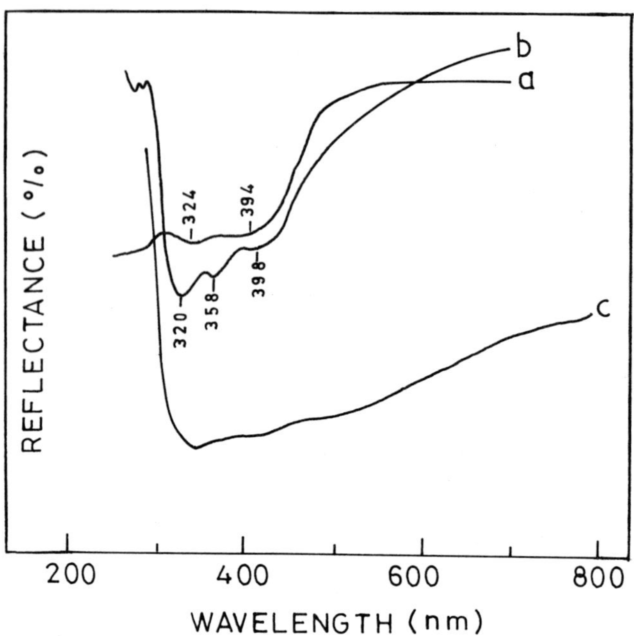

Figure 2. Reflectance spectra; (a) freshly prepared $Os_3(CO)_{12}/SiO_2$. (b) and (c) are sample (a) heated at 125 and 250 °C, respectively for 2 h.

At 100 °C a second sharp ($\Delta H'$ = 15 G) asymmetric signal appears with $g'_\perp$ = 2.00 and $g'_\parallel$ = 1.97 (Fig. 3) the intensity of which increases when the sample is prepared at 150 °C. Unlike the low field signal, they show little temperature dependence. On treatment with H_2 at 250 °C, the signal with $g \sim 4.2$ reduces to half its original intensity but the signal with $g' \sim 2$ remains unchanged. On subsequent treatment with air at 50 °C, the signal g remains unchanged but g' reduces in intensity (Fig. 4).

Both Os(I) and Os(III) with an effective electron spin S' = 1/2 are expected to show EPR signals (osmium in higher than +3 state is not considered under these mild treatments). In the literature, there are a few reports (Refs. 10 & 11) on the EPR study of Os(III) complexes that show a wide variation of g values de-

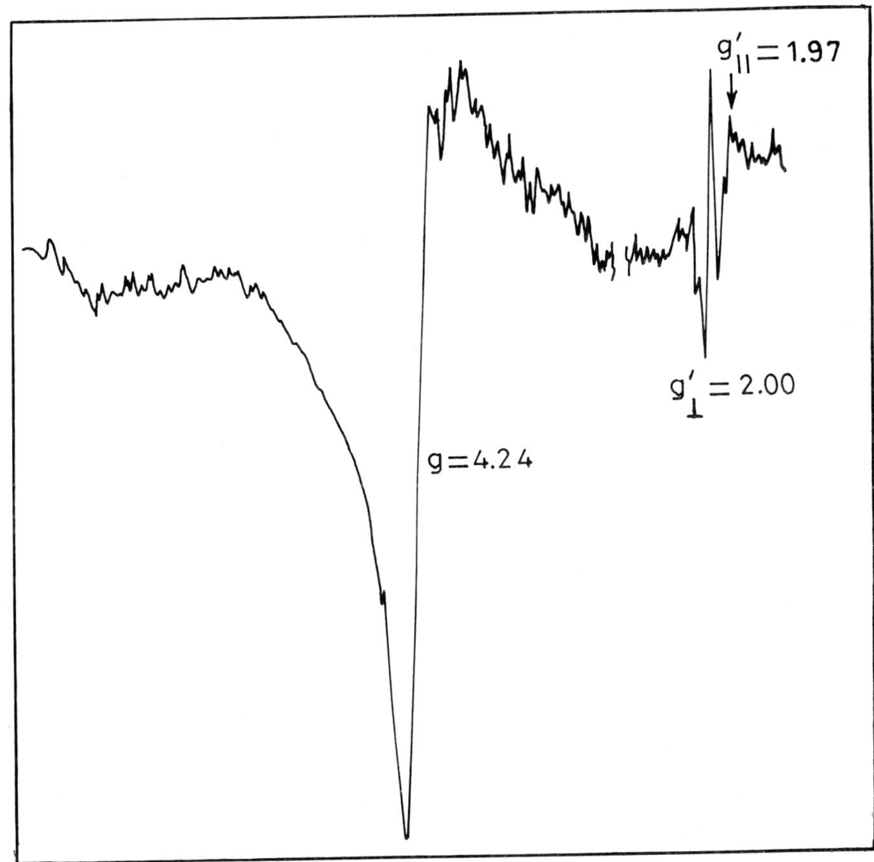

Figure 3. EPR spectrum of $Os_3(CO)_{12}/SiO_2$ thermoevacuated at 100 °C (recorded at 77 K)

pending on the ligand and the symmetry of the ligand field. Based on these meagre data, it is not possible to make definite assignment for the surface species of unknown structure. The report by Shvets et al (Ref. 12) on EPR spectra of $Os_3(CO)_{12}/Al_2O_3$ does not mention the signal with $g \approx 4.2$. However, the spectra near $g'_{\perp} = 2.00$ is very much similar to that observed by them. We assign the signal g to Os(III) and the signals g' to

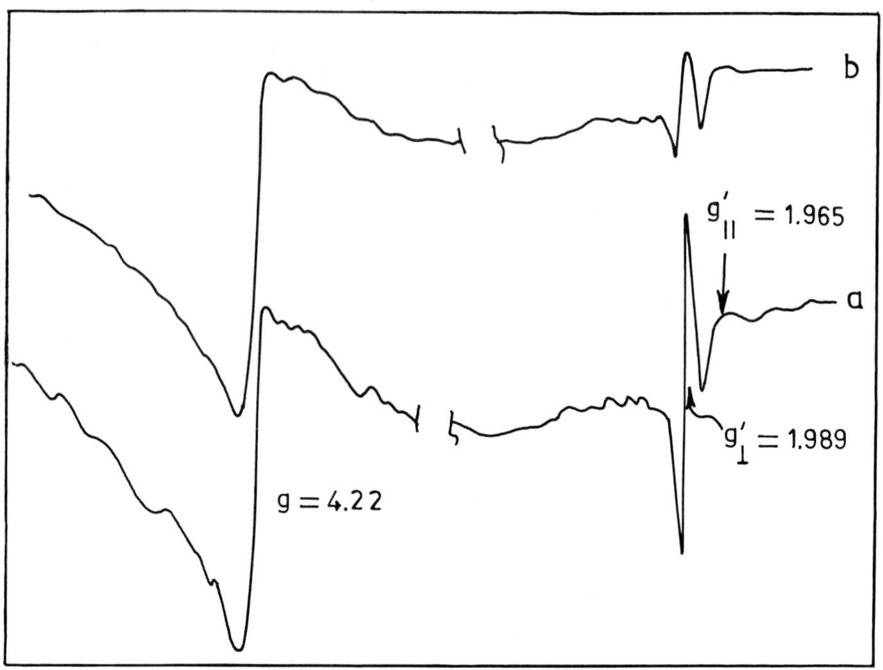

Figure 4. EPR spectra: (a) $Os_3(CO)_{12}/SiO_2$ evacuated at 100 °C followed by treatment with H_2 at 250 °C, (b) sample (a) after treatment in air at 50 °C.

Os(I). This looks justified as the reduction in H_2 at 250 °C is likely to reduce Os(III) thus leading to a lower intensity signal. Since the intensity of g' does not simultaneously increase, it may be concluded that Os(III) is reduced to diamagnetic Os(II). Subsequent treatment in air at 50 °C oxidises Os(I) to Os(II) that is known to be more stable on the support (Ref. 5). This leads to a decrease in the intensity of the signal g' without any increase in the signal g. It can also be concluded that small amount of Os(III) is formed right at the beginning, but Os(I) appears in the process of thermal degradation. ESR spectroscopy, thus, has established the presence of Os(III) and Os(I) as minor species along with the major zero-valent Os.

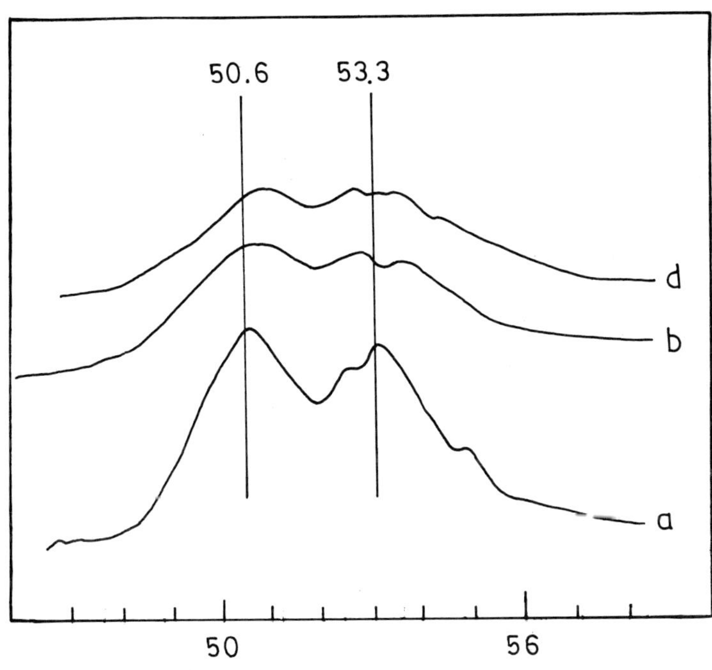

Figure 5. Os(4f) XPS spectra: (a) $OsCl_3/SiO_2$ reduced at 400 °C, (b) $Os_3(CO)_{12}/SiO_2$ reduced at 400 °C in H_2, (d) $Os_3(CO)_{12}/SiO_2$ evacuated at 400 °C.

Figure 5 shows the Os(4f) XPS spectra of the reduced catalysts prepared from $OsCl_3$ (b) and $Os_3(CO)_{12}$ (a) and also that of the thermoevacuated $Os_3(CO)_{12}/SiO_2$ at 400 °C (d). From the intensity of the spectra, it can be seen that the sample B has a much higher dispersion of osmium on the surface. Unlike the Os(4f) spectra reported by Knözinger et al (Ref. 13), sample b shows clear resolution into $4f_{7/2}$ (50.6 eV) and $4f_{5/2}$ (53.3 eV) peaks that can be assigned to metallic osmium. The shape of the spectrum is suggestive of the presence of a second osmium species at a higher oxidation state, although in very small concentration. Higher dispersion of the reduced catalyst obtained from $OsCl_3$ is compatible with the results of Prestridge and Sinfelt (Ref. 14) who reported the formation of raft-like structure on the support when the catalyst was prepared by impregnating the support with H_2OsCl_6 followed by high temperature reduction. Close similarity of the spectra A and D indicates that thermoevacuation or reduction in H_2 of $Os_3(CO)_{12}/SiO_2$ lead to the formation of the same osmium species consisting mainly of zero-valent osmium.

ACKNOWLEDGEMENTS. A research grant from the Department of Science and Technology is gratefully acknowledged. One of the authors (AAD) is grateful to C.S.I.R. for research fellowship.

REFERENCES

1. F. Hugues, P. Bussiere, J.M. Basset, D. Commereuc, J. Chauvin, L. Bonneviot and D. Olivier, Intern. Congr. Catal., 7th, Part A, p. 418, Tokyo (1980); Kodansha, Tokyo/Elsevier, Amsterdam (1980).
2. B. Besson, B. Moraweck, A.K. Smith, J.M. Basset, R. Psaro, A. Fusi and R. Ugo, J. Chem. Soc., Chem. Commun., 569 (1980).
3. P.L. Watson and G.L. Schrader, J. Catal., 9, 129 (1980).
4. R. Psaro, R. Ugo, G.M. Zanderighi, B. Besson, A.K. Smith and J.M. Basset, J. Organometal. Chem., 213, 215 (1981).
5. H. Knözinger and Y. Zhao, J. Catal., 71, 337 (1981); R. Barth, B.C. Gates, Y. Zhao, H. Knözinger and J. Hulse, J. Catal., 82, 147 (1983).
6. M. Deeba and B.C. Gates, J. Catal., 67, 303 (1981).
7. G. Collier, D.J. Hunt, S.D. Jackson, R.B. Moyes, I.A. Pickering, P.B. Wells, A.F. Simpson and R. Whyman, J. Catal., 80, 154 (1983).
8. M.A. El-Sayed and H.D. Kasez, J. Mol. Spectrosc., 9, 310 (1962).
9. D.R. Tyler, R.A. Levenson and H.B. Groz, J. Am. Chem. Soc., 100, 7888 (1979).
10. S. Sakaki, N. Hagiwara, Y. Yanase and A. Ohyoshi, J. Phys. Chem., 82, 1917 (1978).
11. R.E. Desimone, J. Am. Chem. Soc., 95, 6238 (1973).
12. V.A. Shvets, A.L. Tarasov, V.B. Kazansky and H. Knözinger, J. Catal., 86, 223 (1984).

13. H. Knözinger, Y. Zhao, B. Tesche, R. Barth, R. Epstein, B.C. Gates and J.P. Scott, Faraday Discuss. Chem. Soc., 53 (1981).
14. E.B. Prestridge, G.H. Via and J.H. Sinfelt, J. Catal., 50, 115 (1977).

ANIONIC CARBONYL CLUSTERS AS HOMOGENEOUS AND SUPPORTED CATALYSTS IN HYDROGENATION AND TRANSFER HYDROGENATION REACTIONS

A. Basu, S. Bhaduri* and K.R. Sharma
Alchemie Research Centre, CAFI Site, P.O.B. 155, Thane-Belapur Road, Thane 400 601, Maharashtra, India

ABSTRACT - Catalytic hydrogenation reactions based on redox chemistry of $[Pt_3(CO)_6]_n^{2-}$ (n=5,4) in solution are reported. Homogeneous transfer hydrogenation and isomerisation reactions with $[H_3Ru_4(CO)_{12}]^-$ as catalyst are also described. Both anionic clusters anchored on anion exchangers have been characterized and used as supported catalysts. The platinum cluster is considerably mobile on the support, while the ruthenium cluster slowly reacts with the surface halide.

1. INTRODUCTION

Homogeneous catalysis by carbonyl cluster complexes has attracted considerable attention in the past ten years (Ref. 1). It has been proposed that such reactions could give a valuable insight into the mechanism of catalytic reactions involving metal surfaces (Ref. 2). The presence of more than one metal atom in a cluster molecule also gives rise to the possibility of achieving catalytic activities in reactions where multicentre interactions might be important. Carbonyl clusters supported on inorganic and organic polymers have also been extensively used as catalysts for a variety of reactions with the expectations that anchored carbonyl clusters under suitable conditions could be ideal precursors for bare metal clusters or very small metal particles of unique catalytic properties (Ref. 1,3). Investigations on supported cluster complexes may therefore provide valuable information regarding the relationships between homogeneous and heterogeneous catalysis, particularly in systems where the chemistry of the cluster complex in solution has been explored in some detail. Required is a comparative study between the chemistry of solution and of anchored species and their relevance to the observed catalytic activities. Special attention has to be paid to questions, such as: does the cluster retain its molecular identity after anchoring? Is there a possibility of a reaction between the anchored cluster and the polymer leading to the formation of other clusters or mononuclear species? How mobile are the anchored clusters and can they migrate on the support producing metal aggregates or clusters of different nuclearity? In this paper we describe the use of two anionic cluster carbonyls as catalysts in homogeneous hydrogenation and transfer hydrogenation reactions. The same clusters supported on anion exchange resins have also been used as catalysts for the same set of reactions. Since the surface hydroxyl groups of inorganic oxides like alumina and silica react with carbonyl clusters, in most cases the supported species on such support materials are different from the parent clusters and are difficult to characterize. Although there are a few examples of use of ion exchange resins in cluster catalyzed reactions, further exploration of this as a general method for anchoring a variety of anionic carbonyl clusters is desirable (Ref. 3,4). This is so since there exist a large number of fully characterized anionic carbonyl clusters, a few of which exhibit unique catalytic properties at relatively low temperatures (Ref. 5,6).

2. RESULTS AND DISCUSSION

2.1. Synthesis and Characterization

Two anionic carbonyl clusters $[Pt_{15}(CO)_{30}]^{2-}$ and $[H_3Ru_4(CO)_{12}]^-$ recently reported as having unusual catalytic properties (Ref. 5), have been used as homogeneous and anion exchanger supported catalysts in a few hydrogenation and transfer hydrogenation reactions. The platinum cluster originally reported by Longoni and Chini is one of the well characterized classes of anions of the general formula $[Pt_3(CO)_6]_n^{2-}$ (n=6,5,4 and 3) (Ref. 7). The ruthenium cluster, first synthesized and reported in 1975, has recently been shown to play an important role in the $[Ru_3(CO)_{12}]$ pre-catalyst-based Water Gas Shift Reaction and in transfer hydrogenation reactions. Infra-red spectra, C^{13} and H^1 n.m.r. studies and single crystal X-ray diffraction work have been employed to characterize the two cluster anions. In addition, $[Pt_3(CO)_6]_n^{2-}$ anions have been shown to posess highly characteristic electronic spectra (Ref. 5). X-ray photoelectron spectroscopic studies have also been reported on them (Ref. 8).
The synthesis reported for $[Pt_{15}(CO)_{30}]^{2-}$ involves reaction of $[Na_2PtCl_6]$ with CO in methanol in the presence of appropriate amount of the base. The ruthenium cluster is synthesized by treating $[H_4Ru_4(CO)_{12}]$ with a stoichiometric amount of the base in ethanol (Ref. 9). The anion exchange reactions have been effected in methanol at ambient temperatures under inert atmosphere with perfectly dry IRA 401 in the chloride form. The amounts of metal incorporation have been determined by atomic absorption or spectrophotometric analysis and found to be 1-2% for both polymersupported derivatives. Both supported complexes have i.r. spectra identical with those of the original clusters (Ref. 3).

$[Pt_{15}(CO)_{30}]^{2-}$ is one of the few carbonyl clusters for which X-ray photoemission studies have been reported (Ref. 8). The anion exchanger supported complex has a platinum $4f_{7/2}$ signal of binding energy 72.6 eV, which is consistent with the value of 72.8 eV for the free complex. A weak signal corresponding to Pt^{4+} is also observed and is thought to arise from small quantities of unreacted $[Na_2PtCl_6]$, the starting material for the synthesis of $[Pt_{15}(CO)_{30}]^{2-}$. The supported complex like free $[Pt_{15}(CO)_{30}]^{2-}$ also has a characteristic u.v. - visible band at 700 nm.
X-ray photoelectron spectra of the free complex [PPN] $[H_3Ru_4(CO)_{12}]$ (PPN = Bis (triphenylphosphoranylidiene) ammonium) and the anion exchanger supported $[H_3Ru_4(CO)_{12}]$ have been recorded (Ref. 5). The close similarity in the two ruthenium $3d_{5/2}$ binding energies (281.1 and 282.0) confirms that on anchoring the cluster does not undergo any change.

2.2 Catalysis by $[Pt_3(CO)_6]_n^{2-}$ (n=5,4) in solution and on the support

The platinum cluster anions are known to react with hydrogen according to reaction (1)

$$(n-1)\ [Pt_3(CO)_6]_n^{2-} + H_2 \rightleftharpoons 2H^+ + n[Pt_3(CO)_6]_{(n-1)}^{2-} \quad (1)$$

$$n = 6,5 \text{ and } 4$$

It is also known that the increase in hydrogen pressure only, at a moderate temperature (of about 40 °C) cannot reduce $[Pt_9(CO)_{18}]^{2-}$ any further (Ref. 10). When tested as precursor materials for homogeneous hydrogenation of a few unsaturated functionalities, $[Pt_3(CO)_6]_n^{2-}$ (n=3,4,5 and 6) did not show any activity except in the conversion of benzaldehyde to benzylalcohol. In benzaldehyde hydrogenation, all the parent clusters are irreversibly reduced to some other common but uncharacterized cluster(s) considered to be the catalytically active species. Very little or no activity is observed until this common intermediate with characteristic i.r. bands is generated in situ. Spectrophotometric studies show that $[Pt_9(CO)_{18}]^{2-}$ can be completely oxidised back to $[Pt_{12}(CO)_{24}]^{2-}$ by several inorganic, organic and biochemical oxidants, according to reaction (2)

$$4\,[Pt_9(CO)_{18}]^{2-} + 2H^+ + Q \rightleftharpoons 3\,[Pt_{12}(CO)_{24}]^{2-} + Q\,H_2 \qquad (2)$$

(Q = oxidant)

Contribution of reactions (1) and (2) suggests that for the hydrogenation of these oxidants $[Pt_{12}(CO)_{24}]^{2-}$ and $[Pt_9(CO)_{18}]^{2-}$ should be able to act as active catalytic intermediates without undergoing any further reduction. The hydrogenation should proceed according to the catalytic cycle shown in Scheme 1.

$$[Pt_{12}(CO)_{24}]^{2-} \underset{Q \to QH_2}{\overset{H_2}{\rightleftharpoons}} [Pt_9(CO)_{18}]^{2-} + 2H^+$$

The organic and biochemical oxidants that are capable of oxidising $[Pt_9(CO)_{18}]^{2-}$ to $[Pt_{12}(CO)_{24}]^{2-}$ are: β-benzoquinone, 1-chloro,2,4 dinitrophenol, flavinadenine dinucleotide (FAD) and $CytC_{ox}$. Thus β-benzoquinone and 1-chloro,2,4-dinitrobenzene could be quantitatively hydrogenated to β-dihydroxybenzene and 1-chloro,2,4-diaminobenzene (acetonitrile solvent, 40 lbf in^{-2} H_2, 30 °C and 1 h) in the presence of $[Pt_{12}(CO)_{24}]^{2-}$ as catalyst. During the course of the reaction the i.r. and u.v.-visible spectra of the reactant solution show the presence of only $[Pt_{12}(CO)_{24}]^{2-}$ and $[Pt_9(CO)_{18}]^{2-}$.

We were not able to use homogeneous catalytic hydrogenation of FAD to $FADH_2$ and reduction of $CytC_{ox}$ with $[Pt_{12}(CO)_{24}]^{2-}$ or $[Pt_9(CO)_{18}]^{2-}$ as catalysts, because of the absence of a suitable solvent. However, using the supported clusters as catalysts these transformations could be achieved (see below).

As already mentioned, $[Pt_{15}(CO)_{30}]^{2-}$ in solution undergoes stepwise reductions which can be monitored by i.r. or more effectively by u.v.-visible spectroscopy. The anion exchanger supported $[Pt_{15}(CO)_{30}]^{2-}$ that has reacted with hydrogen shows characteristic changes in the u.v.-visible spectra suggesting a stepwise formation of $[Pt_{12}(CO)_{24}]^{2-}$ and $[Pt_9(CO)_{18}]^{2-}$. This indicates considerable mobility of the cluster molecules on the polymer surface. A prolonged ($\sim$ 48 h) treatment with hydrogen leads to complete decarbonylation of the supported $[Pt_{15}(CO)_{30}]^{2-}$ as is evident from the total loss of i.r. carbonyl bands. Although the decarbonylated material could be recarbonylated under atmospheric pressure of CO, on the basis of photoelectron and i.r. spectroscopic data the resultant species is considered to be different.

Resin supported $[Pt_{15}(CO)_{30}]^{2-}$ is convenient to use for the reduction of FAD to $FADH_2$ and $CytC_{red}$ with hydrogen in the aqueous medium. Infra-red and u.v.-visible spectra of the supported species after the reaction indicate the presence of $[Pt_{12}(CO)_{24}]^{2-}$ and $[Pt_9(CO)_{18}]^{2-}$ on the resin surface. Presumably, while the conversion of FAD to $FADH_2$ involves a catalytic cycle as shown in Scheme 1, $CytC_{ox}$ reduction does not require protons. It is worth mentioning that catalytic hydrogenation of flavoquinones with conventional hydrogenation catalysts such as platinum on carbon leads to "over-reduction", i.e. formation of 6,7,8,9,9a,10-Octahydro flavin (Ref. 11). Formation of the platinum metal from the supported $[Pt_{15}(CO)_{30}]^{2-}$ under the reaction conditions is therefore unlikely. Resin supported $[Pt_{15}(CO)_{30}]^{2-}$, its thermally decarbonylated derivative and the decarbonylated material obtained by treating resin supported $[Pt_{15}(CO)_{30}]^{2-}$ with hydrogen for a long time, exhibit noticeable differences in cyclohexene hydrogenation activities. Decarbonylation and the associated mobility of the $[Pt_{15}(CO)_{30}]^{2-}$ ions are expected to lead to the formation of highly dispersed metal crystallites and consequently to the high catalytic activity. This is the case for the thermally decarbonylated derivative, but, surprisingly, predecarbonylation with hydrogen results in a lowering of activity. A possible explanation could be that metal crystallites formed differ in size and shape depending on the mode of decarbonylation.

2.3. Catalysis by $[H_3Ru_4(CO)_{12}]^-$ in solution and on the support

The anionic ruthenium cluster has been found to catalyze conjugation of unconjugated olefins in solution. However, this reaction is complicated by a competing transfer hydrogenation reaction in protic solvents. In fact, when tested as a homogeneous transfer hydrogenation catalyst, the cluster shows high activity (Ref. 5). Since conjugated vegetable oils are industrially important because of their improved drying property, conjugation of methyl linoleate catalyzed by $[H_3Ru_4(CO)_{12}]^-$ has been studied. A mixture of conjugated dienes consisting of cis-cis, cis-trans and trans-trans isomers, with a turnover of about 4 h^{-1}, is formed at 65 °C, in tetrahydrofuran, the amount of each isomer being about 33%. However, the same reaction, when carried out in a protic solvent such as isopropyl alcohol, gives small quantities of methyl oleate (turnover 0.5 h^{-1}). The metal hydride addition-elimination mechanism involving 6-alkyl intermediates has been previously demonstrated for isomerization of pentenes with $[H_4Ru_4(CO)_{12}]$[12]. A similar mechanism may be involved in the isomerization of methyl linoleate with $[H_3Ru_4(CO)_{12}]^-$ as the catalyst.

The activity of $[H_3Ru_4(CO)_{12}]^-$ as a transfer hydrogenation catalyst has been tested for conjugated double bonds, aldehydes, ketones and α, β-unsaturated carbonyl compounds. With isopropyl alcohol as the hydrogen donor and solvent, high turnovers of hydrogenated ketones and acetone are obtained. The catalyst shows low activity for aldehydes or α, β-unsaturated aldehydes. The following observations are relevant in proposing a general mechanism for the transfer hydrogenation reactions.

First, the i.r. spectra of the solutions containing the ketones, isopropyl alcohol and $[H_3Ru_4(CO)_{12}]^-$ do not change the carbonyl region during a catalytic run. Second, this is not the case with the aldehydes, and definite changes in the i.r. spectra observed indicate conversion of $[H_3Ru_4(CO)_{12}]^-$ to some other carbonyl species. Third, the hydrogenation of cyclohexenone proceeds through a consecutive formation of cyclohexanone and cyclohexanol. Finally, the anionic cluster is far

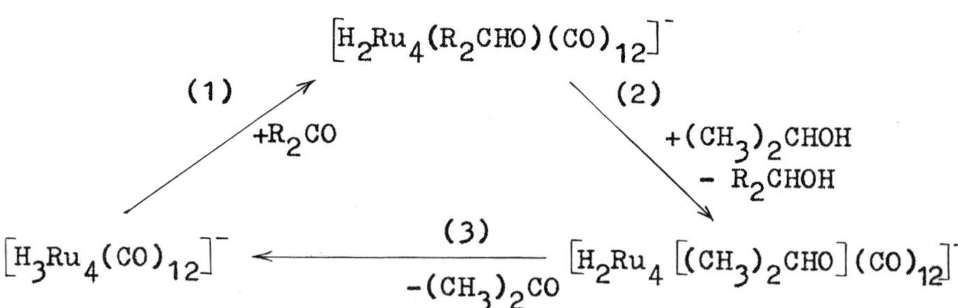

more active than the parent cluster $[H_4Ru_4(CO)_{12}]$ or its phosphine substituted derivatives. A mechanism shown in

Scheme 2 may be proposed for the transfer hydrogenation of ketones and α, β-unsaturated ketones. In step (1) an alkoxo complex is formed by the nucleophilic attack of the anionic hydride on the ketonic carbonyl group. In an α, β-unsaturated system the attack could take place at the 4-position. The alkoxo species undergoes exchange reaction with isopropyl alcohol (Step (2)), which is followed by regeneration of the anionic cluster through β-elimination and formation of acetone. Indeed, formation of the alkoxo complex in step (1) is consistent with preliminary kinetic data.

Resin supported $[H_3Ru_4(CO)_{12}]^-$ has been found to be totally inactive in the isomerization and transfer hydrogenation reactions of methyl linoleate. This seems to be due to the steric restrictions imposed by the resin around the catalytic centers. Diffusion of the substrate into the polymer porous system, which depends on the size of the substrate, is known to play an important role in these reactions (Ref. 13). Methyl linoleate with its large size probably has a large diffusion barrier and is consequently unable to approach the catalytic sites on the resin.

With other substrates such as cyclohexenone, cyclohexanone and acetophenone by using the resin supported catalyst good conversions to the hydrogenated products are obtained.

Although metal leaching to an observable extent does not take place over a period of 50 h, conversion and consequent deactivation of the original anionic cluster to some other carbonyl species is observed at the end of this period. Fragmentation and redistribution of other resin anchored mixed metal anionic cluster to give mononuclear and polynuclear species have recently been reported (Ref. 4). The residual chloride on the resins play an important part in cluster degradation. Since i.r. spectra of the deactivated catalysts from cyclohexenone, cyclohexanone or acetophenone transfer hydrogenation reactions are identical, deactivation through reaction with residual chloride is suggested. This conclusion is further supported by the observation that the deactivated material, characterized by its i.r. spectra, could be generated by taking a freshly prepared supported catalyst in a solvent such as isopropyl alcohol or tetrahydrofuran and keeping it at 65 °C for a long period of time (~ 60 h).

REFERENCES

1. R. Whyman, in Transition Metal Clusters, B.F.G. Johnson (Ed.), John Wiley, N.Y. (1980).
   E.L. Mutterties and M.J. Kranse, Angew Chem. Int. Ed. Eng., 22, 135-148 (1983).
2. E.L. Mutterties, T.N. Rhodin, E. Band, C.F. Brucker and W.R. Pretzer, Chem. Rev., 79, 91 (1979).
3. S. Bhaduri and K.R. Sharma, J. Chem. Soc. Dalton Trans., 2309-2313 (1984) and references therein.
4. R.C. Ryan, G.M. Wilemon, M.P. Dalsanto and C.U. Pittman, Jr., J. Molec. Catal., 5, 319 (1979);
   Y. Doi, A. Yokota, H. Miyake and K. Soga, J. Chem. Soc. Chem. Commun., 394 (1984);
   R.A. Dubois, P.E. Garrow, G.E. Hartwell and D.L. Hunter, Organometallics, 3, 95-101 (1984).
5. S. Bhaduri and K.R. Sharma, J. Chem. Soc. Chem. Commun., 1412-1414 (1983);
   A. Basu, S. Bhaduri and K.R. Sharma, J. Chem. Soc. Dalton Trans. 2315-2318 (1984);
   A. Basu, S. Bhaduri and K.R. Sharma, to be published.
6. J.C. Bricker, C.C. Nagel, A.A. Bhattacharya and S.G. Shore, J. Am. Chem. Soc., 107, 377-384 (1985);
   G. Suss-Fink and J. Reiner, J. Molec. Catal., 16, 231-242 (1982).
7. G. Longoni and P. Chini, J. Am. Chem. Soc., 98, 7225 (1976).
   J.C. Calabrese, L.F. Dahl, A. Cavalieri, P. Chini, G. Longoni and S. Martinengo, J. Am. Chem. Soc., 96, 2616 (1974).
8. G. Apai, S.T. Lee, M.G. Mason, L.J. Grenser and S.A. Gardner, J. Am. Chem. Soc., 101, 6880 (1979).
9. S.A.R. Knox, J.W. Koepke, M.A. Andrews and H.D. Kaesz, J. Am. Chem. Soc., 97, 3942-3947 (1975).
10. S. Bhaduri and K.R. Sharma, J. Chem. Soc. Dalton Trans., 727 (1982).

11. Flavins and Flavoproteins, 3-d Intern. Symp., H.Kamin (Ed.), pp. 83-90, University Park Press, Baltimore, Butterworths, London (1971).
12. M. Valle, D. Osella and G.A. Vagilo, Inorg. Chim. Acta, 20, 213-216 (1976).
13. S. Bhaduri and H. Khwaja, J. Chem. Soc. Dalton Trans., 419-423 (1983) and references therein.

ANCHORED HETERONUCLEAR Pd-Co CARBONYL COMPLEXES AS CATALYSTS FOR HYDROFORMYLATION OF OLEFINS

B.L. Moroz, O.N. Shumilo, E.A. Paukshtis,
V.A. Likholobov, N.N. Bulgakov, E.N. Yurchenko
and Yu.I. Yermakov
Institute of Catalysis, Siberian Branch, USSR
Academy of Sciences, Novosibirsk 630090, USSR

ABSTRACT - Homo- and heteronuclear Pd- and Co-containing complexes of the composition P_2PdCl_2, P_3Pd^0, $P_nCo_2(CO)_{6-7}$ and $P_nPdCo_2(CO)_{6-8}$, where P_n= Ⓢi-[$(CH_2)_3P(C_6H_{11})_2$]$_n$, Ⓢi is the surface of SiO_2, n = 2 or 3, anchored on phosphinated silica have been synthesized. IR spectroscopy was employed to study the structure of these compounds and their transformations in CO atmosphere. Catalytic properties of anchored complexes have been examined in gas-phase hydroformylation of propylene. According to the relative activity at 100 °C, the catalysts in question can be arranged as sequence:

$P_nPd\text{-}Co_2(8\text{-}10) > Pn_nCo\text{-}Co\ (1) \gg P_2PdCl_2,\ P_3Pd^{\circ}$ (inactive). In the presence of Pd-Co catalysts, products were produced at temperatures as low as 40-70 °C. Interacting bonds method was used to estimate enthalpies of the steps of ethylene hydroformylation involving homo- and heteronuclear Pd- and Co-containing complexes. As shown, oxo-reaction on Pd-Co complexes is characterized by lowest enthalpies of the steps of CO insertion into a metal-carbon bond and hydrogenolysis of acylcarbonyl. A possible reason for synergetic effect observed is the simultaneous formation of reactive Pd-H and Co-C bonds at activating reactants on heteronuclear centers.

1. INTRODUCTION

Anchoring of metal complexes on supports is considered to be a promising method for purposeful synthesis of active centers of heterogeneous catalysts (Ref. 1). An urgent trend in this area is synthesis and study of catalytic properties of anchored polynuclear complexes (clusters) (Ref. 2). General approaches to preparation of anchored polynuclear metal complexes are:

- interaction of dissolved polynuclear complexes with ligand-containing supports;
- synthesis of clusters directly on the support surface proceeding from anchored mononuclear complexes (the so-called "assembling" of clusters on the support surface).

Note that by using the "assembling" technique, anchored clusters that have no analogs among soluble compounds can be obtained. For example, this method was used previously (Ref. 3) to prepare polynuclear Pd complexes anchored on phosphinated silica.

Heteronuclear clusters present great interest as potential catalysts for reactions following complicated mechanisms with reagent activation at several sites of different nature. In some reactions these heteronuclear complexes show the catalytic properties essentially different from those of homonuclear complexes (Refs. 4-8). However, heteronuclear clusters are difficult to investigate because a number of problems associated with their synthesis and isolation, as well as with their low solubility or instability in solutions are to be solved.

Using the "assembling" technique, we have synthesized heteronuclear Pd-Co carbonyl complexes anchored on phosphinated silica. Catalytic properties of the compounds obtained were studied in the hydroformylation of olefins and compared with those of homonuclear complexes of Co and Pd.

Phosphine complexes of cobalt corbonyl are known as catalysts for the hydroformylation of olefins. This process is carried out effectively under rather severe conditions, that is, at increased (5-150 atm) pressure of an olefin + H_2 + CO gas mixture and at temperatures above 100 °C. The rate-determining reaction step is probably hydrogenolysis of acylcobalt carbonyl (Ref. 9).

On the other hand, Pd complexes are known as effective catalysts for hydrogenation of both various organic unsaturated groups and metal-carbon bonds (Ref. 10). Therefore, we have suggested that the step of Co-(COR) bond hydrogenolysis may be facilitated by introducing palladium into the carbonylcobalt catalyst.

2. RESULTS AND DISCUSSION

2.1 Synthesis and Identification of Anchored Palladium and Cobalt Complexes

For the catalyst preparation, silica (BET surface 110 m^2/g) heated in vacuum at 600 °C was used. Phosphine ligands were anchored by the reaction of surface hydroxyl groups of silica with $(EtO)_{4-n}Si[(CH_2)_3P(C_6H_{11})_2]_n$ as described in (Ref. 11), to form Ⓢi$-[(CH_2)_3P(C_6H_{11})_2]_n$ which we denoted by Ⓢi$\sim P_n$. The catalyst preparation and sample treatment under further study were performed without contact with air.

Elemental analysis data are presented in Table 1.

Anchored mononuclear palladium complexes Ⓢi$\sim P_3Pd^0$ and Ⓢi$\sim P_2PdCl_2$ were prepared by interaction of phosphinated silica with solutions of $Pd_2[(PhCH{=}CH)_2\ C{=}O]_4$ and $(PhCN)_2PdCl_2$ (Ph - phenyl), respectively. Preparation procedure and identification of similar anchored palladium compounds were as described in (Refs. 11,12).

Anchored cobaltcarbonyl complexes Ⓢi$\sim P_n + Co_2(CO)_8$ were prepared by treating phosphinated silica with pentane solution of $Co_2(CO)_8$ at room temperature for 24 h. An excess carbonyl complex was thoroughly washed off by pentane, then the sample was dried at 25 °C, at 0.01 Torr.

IR spectra of the catalysts containing anchored cobalt carbonyl complexes are given in Fig. 1A. IR spectrum

TABLE 1. Elemental analysis[a)] data for the samples containing anchored palladium and cobalt complexes

Catalyst[b)]	Content of elements			Atomic ratios		
	P	Co	Pd	P/Co	P/Pd	Co/Pd
1 $P_2 + Co_2(CO)_8$	1.3	1.2	-	1.1	-	-
2 $P_3 + Co_2(CO)_8$	1.8	1.2	-	1.5	-	-
3 $P_3 + Pd_2L_4$	1.8	-	0.7	-	2.7	-
4 $P_3Pd^o + Co_2(CO)_8$	1.8	1.2	0.7	1.5	2.7	1.8
5 $P_2 + L'_2PdCl_2$	1.3	-	0.7	-	1.9	-
6 $P_2PdCl_2 + Na[Co(CO)_4]$	1.3	1.1	0.7	1.2	1.9	1.6

a) Elemental analysis of P was performed by iodometry and the analysis of Pd and Co was carried out by atomic absorption;

b) Designations: P_n = (Si) $[(CH_2)_3P(C_6H_{11})_2]_n$, n=2 or 3; L = $(PhCH = CH)_2$ C=O, L' = PhCN

of the sample (Si)$\sim P_2 + Co_2(CO)_8$ after storing under vacuum (spectrum 1) is essentially similar to the spectrum of solution of monophosphine substituted complex $[(C_6H_{11})_3P]Co_2(CO)_7$ (Ref. 13).

Small shift and broadening of CO absorption bands in the IR spectra of (Si)$\sim P_2 + Co_2(CO)_8$ can be due to some nonuniformity of surface ligands (Si)$\sim P_n$. The most distinct difference between the spectra is that the intensity of the band at 1945 cm^{-1} relative to other strong bands at 2073,2015,1982 cm^{-1} in spectrum 1 is much higher than in the spectrum of individual $[(C_6H_{11})_3P]Co_2(CO)_7$. The evacuating of the sample under vacuum at 60 °C resulted in disappearance of all the strong CO absorption bands, except for the band at 1945 cm^{-1} (spectrum 2). After a subsequent treatment of the evacuated sample with CO the initial pattern of the bands are restored. In this case the relative intensity of the band at 1945 cm^{-1} in the spectrum of (Si)$\sim P_2$ +

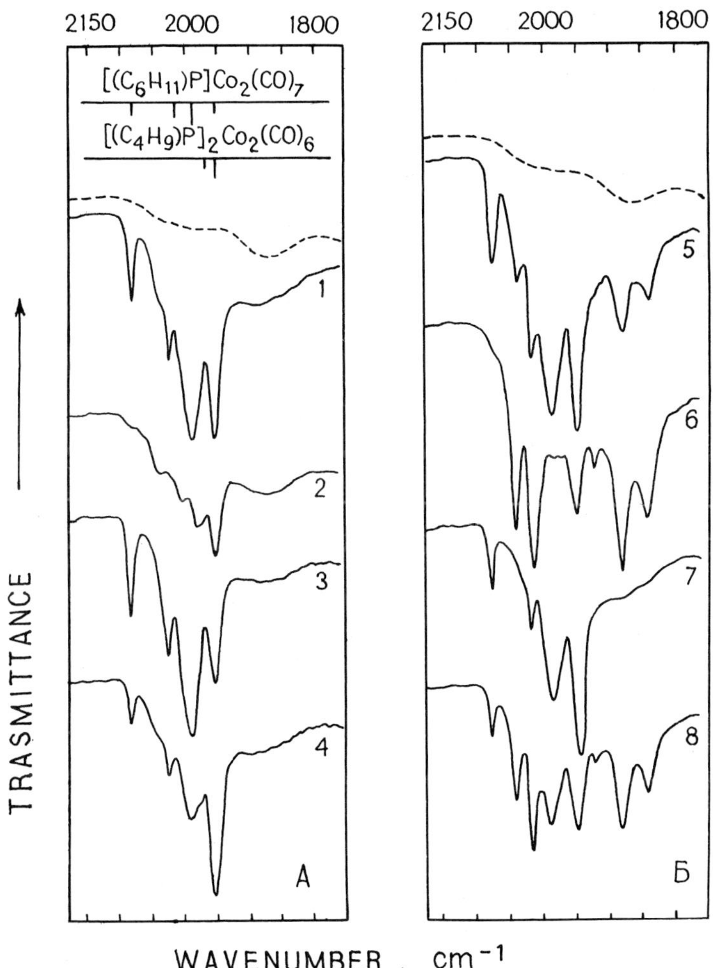

Figure 1. IR spectra of samples containing anchored Pd and Co complexes (for spectra recording the samples were impregnated with outgassed Nujol and placed between CaF_2 disks).

A - (Si) $\sim P_2$ + $Co_2(CO)_8$ sample: 1 - after storing in outgassed cell at 20 °C; 2 - after 30 min evacuation at 60 °C; 3 - under CO after 24 h at 20 °C; 4 - under CO after 1 h at 150 °C.

B - (Si) $\sim P_3Pd^0$ + $Co_2(CO)_8$ (spectra 5-7) and (Si) $\sim P_2PdCl_2$ + + Na $Co(CO)_4$ (spectrum 8): 5-8 - after storing in the evacuated cell at 20 °C; 6 - after 30 min evacuation at 60 °C; 7 - under CO after 1 h at 100 °C

$Co_2(CO)_8$ decreases noticeably after storing the samples in CO atmosphere (spectrum 3). These data indicate that the band at 1945 cm^{-1} in the spectra of (Si) ~ P_2+ $Co_2(CO)_8$ is not a singlet; it can be assigned to some unresolved CO stretching bands arising from different carbonyl complexes. The band at 1945-1953 cm^{-1} is typical not only for mono- but also for diphosphine substituted derivatives of cobalt carbonyl, and is the only intensive band in the IR spectra of $(R_3P)_2Co_2(CO)_6$ (Ref. 14). A comparison made permits us to attribute the bands at 2073, 2015 and 1982 cm^{-1} and one of the bands at 1945 cm^{-1} in the spectra of (Si) ~ P_n + $Co_2(CO)_8$ to vibrations of terminal CO groups in anchored complexes (Si) ~ P-Co_2(CO the second band at 1945 cm^{-1} can be assigned to anchored complexes (Si) ~ $P_2Co_2(CO)_6$.

Thus the interaction of $Co_2(CO)_8$ with phosphinated silica (Si) ~ P_2 leads to the consecutive formation of mono- and diphosphine substituted complexes 1 and 2 on the support surface.

(Si)<(P)(P) + $Co_2(CO)_8$ ⇌ [-CO(1a) / +CO(-1a)] (Si)<(P—Co(CO)(CO)(CO)—Co(CO)(CO)(CO)—CO)(P) ⇌ [Δ; - CO(1b) / +CO(-1b)]

⇌ [Δ; - CO(1b) / +CO(-1b)] (Si)<(P—Co(CO)(CO)—CO)(P—Co(CO)(CO)—CO)

During the first step of interaction $Co_2(CO)_8$ molecules are coordinated by anchored ligands to form complexes 1. According to the elemental analysis data (Table 1 approximately half of the surface phosphine groups do not enter into the reaction here. Preparation and storage of catalysts under vacuum resulted in the detachment of a CO molecule from part of complexes 1 and their transformation to complexes 2. A slight increase in evacuation temperature shifts almost completely equilibrium (1b) towards the formation of complexes 2. But the treatment of the evacuated samples with CO at room temperature permits us to regenerate the initial complexes 1. Note that upon increasing temperature of the treatment with CO up to 100-150 °C, complexes 2 reappear (spectrum 4).

IR spectra of Ⓢi $\sim P_3$ + $Co_2(CO)_8$ catalysts are adequate to those represented in Fig. 1A, and assignment of CO stretching bands in these spectra is similar to that given above.

The results suggest that the composition of surface compounds formed by the $Co_2(CO)_8$ interaction with Ⓢi $\sim P_n$ is independent of the number of donor groups in a molecule of the surface ligand. In all cases monophosphine substituted derivatives of cobalt carbonyl are primarily formed. Thus Ⓢi $\sim P_n$ + $Co_2(CO)_8$ catalysts contain anchored complexes that can be treated as heterogenized analogs of soluble phosphine-modified cobalt carbonyls. These complexes are well-known as catalysts for liquid-phase oxo-reaction.

<u>Anchored heteronuclear palladium-cobalt carbonyl complexes.</u> General methods to prepare heteronuclear carbonyl clusters have been described in detail previously (Refs. 8,15). By analogy with the chemistry of the homogeneous systems, "surface assembling" of Pd-Co complexes anchored on phosphinated silica was carried out by the following methods:

- interaction of anchored palladium (0) with dicobaltoctacarbonyl (Method A)

$$\text{Ⓢi} \sim P_3Pd^{\circ} \xrightarrow[\text{pentane}]{Co_2(CO)_8} \text{Ⓢi} \sim P_3PdCo_2(CO)_x\ , \qquad (2a)$$

- interaction of anchored palladium (II) chloride complexes with anion carbonyl complex $Na^+[Co(CO)_4]^-$ (Method B)

$$\text{(Si)} \sim P_2PdCl_2 \xrightarrow[-NaCl]{Na[Co(CO)_4];\ ether} \text{(Si)} \sim P_2PdCo_2(CO)_x . \quad (2b)$$

Reactions (2a) and (2b) were performed at 20 °C for 24 hours. An excess Co carbonyl compound was washed off by pentane and ether, respectively. The samples were dried under vacuum.

Catalysts prepared by Method A. When a solution of $Co_2(CO)_8$ is added to the sample of (Si) $\sim P_3Pd^0$ cobalt absorption from solution is observed. After this treatment the content of palladium in the sample remains almost unchanged (Table 1). The Co/Pd ratio for the samples prepared by Method A is close to 2.0.

The IR spectrum of the (Si) $\sim P_3Pd^0 + Co_2(CO)_8$ catalyst (spectrum 5 in Fig. 1B) has as many as 8-10 CO stretching bands. Such a great number of bands in IR spectra may indicate the nonuniformity in composition of anchored compounds. To divide CO stretching bands observed in spectrum 5 into patterns of bands arising from different surface carbonyl complexes, we examined intensity variations of these bands after thermal treatment of (Si) $\sim P_3Pd^0 + Co_2(CO)_8$ samples under vacuum and in CO atmosphere (similarly to assignment of (Si) $\sim P_n + Co_2(CO)_8$). The data obtained are listed in Table 2.

On the basis of the data obtained, all bands observed in the IR spectra of (Si) $\sim P_3Pd + Co_2(CO)_8$ can be divided into patterns I and II (Table 2). Pattern I is very similar to the IR spectra of (Si) $\sim P_n + Co_2(CO)_8$ samples in the number of bands and their positions (compare to spectrum 1, Fig. 1A). Under various treatments of (Si) $\sim P_3Pd^0 + Co_2(CO)_8$ the intensities of the bands of I change in the same way as the intensities of the (Si) $\sim P_n + Co_2(CO)_8$ bands.

Pattern II comprises the bands at 2035, 2010, 1917, 1878 and 1837 cm^{-1}, which are observed only in the spectra of Pd-Co catalysts. The behaviour of those bands under

TABLE 2. Positions of CO stretching bands in IR spectra of Ⓢi$\sim P_3Pd^o + Co_2(CO)_8$ after treatment under vacuum and in CO atmosphere (spectra 5-7, Fig. 1B). I and II are patterns of bands, whose intensities are changed under various treatments of samples

Nos of spectra	Treatment conditions	Wavenumber, cm^{-1}
5.	Storage under vacuum at 20 °C (initial spectra)	I: 2073,2015[b),1982, 1945 II: 2035,2010[b)],1917, 1878, 1837
6.	Evacuation at 60 °C for 30 min[a)]	I: 1945 (weak) II: 2035,2010,1917, 1878,1837
7.	Heating in a CO atmosphere at 100 °C for 1 h	I: 2073, 2015, 1982, 1945 II: no bands

a) Treatment of evacuated samples under CO at 20 °C restores their initial spectra.

b) As can be seen from spectra in Fig. 1B, the band at 2010-2015 cm^{-1} may be included into both patterns. Therefore, we have suggested that this band be assigned to two unresolved bands, one belonging to pattern I and the other to II.

various treatments of Ⓢi$\sim P_3Pd^o + Co_2(CO)_8$ is opposite to that of the bands I. In fact, when the Ⓢi$\sim P_3Pd^o$ + $Co_2(CO)_8$ sample was evacuated at 60 °C, the bands assigned to pattern II remained unchanged (spectrum 6). On the other hand, these bands, simultaneously disappeared on heating this sample under CO (spectrum 7). Note that after the Pd-Co catalyst treatment by CO, its IR spectrum is very similar to that of Ⓢi$\sim P_n + Co_2(CO)_8$ (compare spectra 4 and 7 in Fig. 1).

The results suggest that the patterns of bands I and II arise from different types of surface carbonyl compounds on Pd-Co catalysts. Apparently, I is due to the anchored complexes which involve (without a change of their structure) either Ⓢⓘ$\sim$P-$Co_2(CO)_7$ or Ⓢⓘ$\sim P_2Co_2(CO)_6$ fragments. When interacting with Ⓢⓘ$\sim P_3Pd^o$, $Co_2(CO)_8$ molecules can displace phosphine ligands from the coordination sphere of Pd to form anchored complexes of the composition

Ⓢⓘ< P–$Co_2(CO)_7$; P, P > Pd (complexes 3)

Ⓢⓘ< P–Co$(CO)_3$; P–Co$(CO)_3$ (Co–Co bonded) ; P–Pd (complexes 4)

that have neither Pd-Co nor Pd-CO-Co bonds.

We have failed to assign bands II because of the lack of literature spectroscopic and structural information for Pd-Co and Pd carbonyl clusters. Note that pattern II includes strong bands at 1878 and 1837 cm^{-1} in the region typical of bridged CO ligands. Consequently, one can assume that besides complexes 3 and 4 the $Co_2(CO)_8$ interaction with Ⓢⓘ$\sim P_3Pd^o$ produces anchored clusters which include Pd-CO-Co, Co-CO-Co or Pd-CO-Pd fragments. However, these surface species with bridged CO ligands seem to be unstable under CO.

<u>Catalysts prepared by Method B.</u> In preliminary tests it has been established that the anion carbonyl complex $Na^+[Co(CO)_4]^-$ does not interact with a phosphinated silica without attached metal ions. However, the silica containing complexes of Ⓢⓘ$\sim P_2PdCl_2$ is capable of irreversibly adsorbing $Na[Co(CO)_4]$ from solution (Table 1). IR spectra of Ⓢⓘ$\sim P_2PdCl_2$ + $Na[Co(CO)_4]$ are very similar to those of Ⓢⓘ$\sim P_3Pd^o$ + $Co_2(CO)_8$ (compare, e.g. spectra 5 and 8 in Fig. 1B). It is important here that the

spectrum of Ⓢⓘ$\sim P_2PdCl_2$ + $Na[Co(CO)_4]$ also contains carbonyl bands arising from the $Co_2(CO)_8$ complexes with a phosphinated silica. If one takes this fact and data of (Ref. 16) into account, it may be suggested that the interaction of Ⓢⓘ$\sim P_2PdCl_2$ with $Na[Co(CO)_4]$ is accompanied by redox transformations of the initial compounds. The reaction scheme is therefore:

$$\text{(Si)}\sim P_2PdCl_2 \xrightarrow[-NaCl]{Na[Co(CO)_4]} \text{(Si)}\sim P_2Pd^o + Co_2(CO)_8 \qquad (3a)$$

$$\text{(Si)}\sim P_2Pd^o + Co_2(CO)_8 \longrightarrow \text{anchored complexes similar to those prepared by Method A} \qquad (3b)$$

Thus, the two different methods for preparing Pd-Co catalysts lead to samples with nearly the same composition of their anchored compounds. Additional investigations are needed, however, for a detailed description of the composition of anchored complexes in the Ⓢⓘ$\sim P_3Pd^o$ + + $Co_2(CO)_8$ and Ⓢⓘ$\sim P_2PdCl_2$ + $Na[Co(CO)_4]$ samples.

2.2. Catalytic Properties of Anchored Palladium-Cobalt and Cobalt Complexes

Data on catalytic properties of prepared samples in the gas-phase hydroformylation of propylene are compiled in Table 3 (Ⓢⓘ$\sim P_3Pd^o$ and Ⓢⓘ$\sim P_2PdCl_2$ samples appeared to be inactive in this reaction at 40-130 °C).

Under the conditions examined, the main products of oxo-reaction were n- and iso-butyraldehydes, the yield of butyl alcohols being not more than 10-15% of the overall yield of oxo-products.

Cobalt catalysts Ⓢⓘ$\sim P_n$ + $Co_2(CO)_8$ are active in hydroformylation of propylene only at temperatures above 100 °C (Table 3). After the catalyst attains the steady state, its activity remains unchanged throughout the

TABLE 3. Catalytic properties of anchored Pd-Co and Co complexes in the hydroformylation of propylene. Reaction conditions: a flow differential reactor, total pressure of reaction mixture ($H_2:CO:C_3H_6$ = 1:1:1) 3 ata, space velocity 1 s^{-1}

Catalyst[a)]	Reaction temperature, °C	Activity[b)] μmol (C_4H_8O+ C_4H_9OH/ g-atom Co/s	Selectivity for straight-chain products, %
1 $P_2+Co_2(CO)_8$	60	-	-
	80	10	-
	100	50	87
2 $P_3+Co_2(CO)_8$	100	60	85
	130	640	84
3 $P_3Pd^0+Co_2(CO)_8$	60	270	79
	80	400	78
	100	560	78
4 P_2PdCl_2 + +Na[Co(CO)$_4$]	60	230	86
	100	500	86

Designation:
a) P_n = (Si)-[$(CH_2)_3P(C_6H_{11})_2]_n$ (n=2 or 3);
b) For Co catalysts steady state activities are represented; for Pd-Co catalysts they were determined after a 30 min catalytic run at the temperature indicated.

whole course of the reaction (8-10 h). Product distribution here is similar to that observed for the liquid-phase reaction in the presence of soluble phosphine complexes of cobalt carbonyl (Ref. 9). It should also be noted that the catalytic properties of (Si)~P_n + + $Co_2(CO)_8$ samples are slightly affected by the number of phosphine groups in the molecules of anchored ligand.

Palladium-cobalt catalysts (Si)~P_3Pd^0 + $Co_2(CO)_8$ and (Si)~P_2PdCl_2 + Na[Co(CO)$_4$] are considerably more active

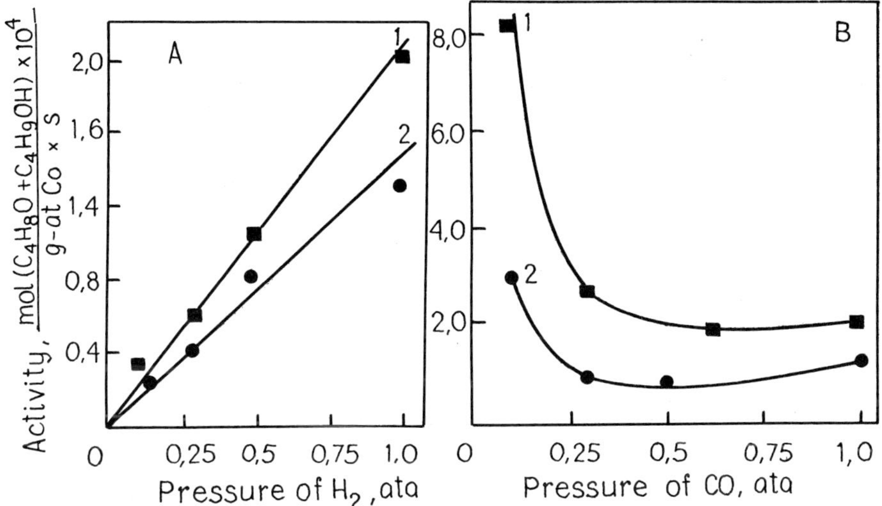

Figure 2. Effect of H_2 (A) and CO (B) partial pressure on the activity of (Si)$\sim P_3$ + $Co_2(CO)_8$ (1) and (Si)$\sim P_3Pd^0$ + $Co_2(CO)_8$ (2). Reaction mixtures $H_2:CO:C_3H_6:Ar = P_{H_2}:1:1:(1-P_{H_2})$ (A) and $H_2:CO:C_3H_6:Ar = 1:P_{CO}:1:(1-P_{CO})$ (B), total pressure 3 ata, space velocity 1 s^{-1}, temperature 120 (1) or 50 °C (2).

in propylene hydroformylation than (Si)$\sim P_n$ + $Co_2(CO)_8$ samples (Table 3). In the presence of palladium-cobalt catalysts, oxo products are produced at temperatures as low as 40-70 °C, while samples containing anchored homonuclear Co complexes are inactive at these temperatures. During the reaction the activity of Pd-Co catalysts gradually decreases. At elevated temperatures (above 100 °C) deactivation is faster.

(Si)$\sim P_3Pd^0$ + $Co_2(CO)_8$ and (Si)$\sim P_2PdCl_2$ + $Na[Co(CO)_4]$ catalysts have demonstrated similar catalytic properties. This fact agrees with the above conclusion about the identical composition of Pd-Co compounds obtained by the two different methods.

Effects of CO and H_2 partial pressures on the rate

of propylene hydroformylation in the presence of Ⓢⓘ$\sim P_3 + Co_2(CO)_8$ and Ⓢⓘ$\sim P_3Pd^0 + Co_2(CO)_8$ catalysts are shown in Fig. 2. The rate of reaction catalyzed by both Co and Pd-Co anchored complexes decreases with decreasing partial pressure of H_2 (Fig. 2A) and is at P_{CO} = 0.25 ÷ 1.0 ata almost independent of CO pressure (Fig. 2B). At very low pressures of CO ($P_{CO} < 0.25$ ata) the reaction rate sharply increases (under these conditions an undesirable hydrogenation of olefins also occurs). The decrease in CO and, partially, in H_2 pressure slows down deactivation of the Pd-Co catalyst.

From the Arrhenius plots obtained the following activation energies for propylene hydroformylation were determined: ~20 kcal/mol (at temperatures ranging within 100-130 °C) for the reaction catalyzed by Ⓢⓘ$\sim P_3 + Co_2(CO)_8$ (this value is close to that for the propylene hydroformylation in the presence of soluble cobalt carbonyls (Ref. 17)); and 10 kcal/mol (at 40-80 °C) for the reaction catalyzed by Ⓢⓘ$\sim P_3Pd^0 + Co_2(CO)_8$. Activation energies were determined within different temperature ranges, since at T $\geqslant$ 100 °C Pd-Co catalysts are rapidly deactivated.

Thus the results indicate that the introduction of palladium into the cobaltcarbonyl catalyst leads to a synergetic increase in the rate of hydroformylation reaction. Note that Pd-Co catalysts are characterized by the same dependences of their activity on partial pressures of reactants that were observed for homonuclear cobaltcarbonyl complexes.

Consequently, it seems reasonable to suggest that the kinetic schemes of oxo process catalyzed by Pd-Co complexes and by Co complexes are similar. However, when the Pd-Co complexes are used as catalysts, the activation energy of the rate-determining step of this process is significantly lower than that of the reaction catalyzed by Co complexes.

2.3. Estimation of Enthalpies of the Steps of Olefin Hydroformylation and Characterization of the Reactivity of Anchored Complexes

In order to interprete the above synergetic effect, we have carried out a comparative estimate of enthalpies of the steps of olefin hydroformylation catalyzed by homo- and heteronuclear Pd and Co complexes. Estimation of the enthalpies were carried out by interacting bonds method (IBM) (Ref. 18). Enthalpy calculations in terms of the IBM are based on the concepts about the structure of reacting complexes and on the data for the formation heats of clue compounds. As a model of the catalysts we have chosen binuclear compounds of tho composition L'Pd-PdL', L''Co-CoL'' and L'Pd-CoL'', where $L' = (PF_3)_2$ $L'' = (CO)_3(PF_3)$. We have assumed that metal atoms in the structure of these homo- and heteronuclear complexes are bonded either directly by metal-metal bonds or by bridged CO ligands.

By analogy with the well-known mechanism for the hydroformylation of olefins, a catalytic cycle of the process catalyzed by model complexes was suggested to involve H_2, alkene and CO coordination to the initial complexes followed by CO insertion into the metal-carbon bond to form acylmetalcarbonyl complex, and decomposition of this intermediate to form product-aldehyde. Thus, to estimate reactivity of L'Pd-Co-L'', we have used the following scheme of hydroformylation (to simplify calculations, it was ethylene hydroformylation):

$$L'\text{-}Pd\text{-}Co\text{-}L'' + H_2 \longrightarrow L'(H)Pd\text{-}Co(H)L'' \quad (4a)$$

$$L'(H)Pd\text{-}Co(H)L'' + C_2H_4 \longrightarrow L'(H)Pd\text{-}Co(C_2H_5)L'' \quad (4b)$$

$$L'(H)Pd\text{-}Co(C_2H_5)L'' + CO \longrightarrow L'(H)Pd\text{-}Co(CO)(C_2H_5)L'' \quad (4c)$$

$L'(H)Pd\text{-}Co(CO)(C_2H_5)L'' \longrightarrow L'(H)Pd\text{-}Co(COC_2H_5)L''$ (4d)

$L'(H)Pd\text{-}Co(COC_2H_5)L'' \longrightarrow L'\text{-}Pd\text{-}Co\text{-}L'' + CH_3CH_2COH$. (4e)

Enthalpy for each step (4a)-(4e) was determined as a difference in sums of atomization energies of products and initial compounds taking part in this step. Atomization energies of model binuclear complexes and products of their interaction with reactants were calculated by the equation:

$$H_{at} = \Sigma V_i (2 - V_i)E_i - \sum_{i>k} \sum V_i V_k \Delta_{ik},$$

where E and Δ_{ik} are the semiempirical parameters for the strength of the i-th bond and the interaction between the i-th and the k-th bonds meeting on one atom, respectively. Bond coefficients V_i are determined from the condition of the maximum of H_{at} and are ranging within 0-1. E and Δ parameters were calculated by Eq. (5). For the calculations of these parameters IBM approach was applied to individual compounds containing the same bonds as in complexes under study. These individual compounds were chosen between these for which the structures were established and formation enthalpies were measured. In our case the calculated values of E and Δ are: E_{Co-Co}= 60.7, E_{Pd-Co}= 56.8, E_{Pd-Pd}=52.8, E_{Co-C}=92.7, E_{Pd-C}=102.8, E_{Co-H}=93.7, E_{Pd-H}=67.5, E_{C-O}=238, 220 and 193 (for linear, bridged and aldehyde CO groups, respectively), E_{Co-P}=79.0, E_{Pd-O}=65.3, E_{C-H}=136.4, E_{C-C}= 145.4, Δ_{Pd}=12.3, Δ_{Co}=14.1, Δ_{C}=33.8 and Δ_{P}=13.5. All values of E and Δ are given in kcal/mol.

Enthalpies of the steps of ethylene hydroformylation (4a)-(4e) catalyzed by Pd-Pd, Co-Co and Pd-Co complexes are listed in Table 4 (data for complexes having metal-metal bonds and no bridged CO ligands).

A comparison of the enthalpies of hydroformylation

TABLE 4. Enthalpies (ΔH) of the steps of ethylene hydroformylation catalyzed by homo- and heteronuclear Pd and Co complexes

Nos of steps	Step	ΔH, kcal/mol		
		Pd-Pd	Co-Co	Pd-Co
4a	Coordination of H_2	10.2	-13.1	-2.0
4b	Coordination of C_2H_4	-55.7	-28.5	-29.1
4c	Coordination of CO	-14.6	-13.0	-12.2
4d	Insertion of CO	25.2	8.2	7.5
4e	Decomposition of acylcarbonyl complex	2.9	14.4	3.7

steps calculated for various catalysts (Table 4) indicates that when this reaction is carried out with participation of Pd-Co carbonyl complexes, a significant decrease in the endothermic effect of the steps of H_2 coordination to the initial complex and CO insertion into a metal carbon bond to form an acyl derivative (compared to Pd complexes) as well as of the step of this derivative decomposition (compared to Co-carbonyl complexes) should be observed. According to the Polanyi-Semenov principle, this decrease has to diminish the step energy barrier. Consequently, the enthalpies, calculated in terms of our model for the Pd-Pd, Co-Co and Pd-Co complexes, indicate that if oxo reaction is carried out on heteronuclear active sites, activation barriers of the steps, which are high for the reaction catalyzed by homonuclear Pd or Co complexes, can be diminished. Similar estimations were performed also for binuclear complexes with bridged CO ligands M'-CO-M'' (where M' = M'' = Pd or Co or M' = Pd, M'' = Co) whose structures have no metal-metal bonds. In terms of such model the data, qualitatively the same as for the complexes that do contain metal-metal bonds, have been obtained.

3. CONCLUSIONS

By using the "assembling" technique, we have synthesized Pd-Co carbonyl compounds anchored on phosphinated silica, that have unusual catalytic properties in propylene hydroformylation. The results obtained provide support for the existence of a synergetic effect in propylene hydroformylation when palladium is introduced in the carbonylcobalt catalyst. This effect is pronounced at low temperatures (40-70 °C) when homonuclear complexes are inactive, whereas the rate of the reaction catalyzed by Pd-Co catalyst is noticeably high.

Kinetic data and calculated enthalpies of hydroformylation steps obtained by the interacting bonds method suggest that if the reaction is carried out on heteronuclear active sites, the activation energy of the most demanding steps is decreased. In fact, when substrates are activated by Pd-Co active sites both Pd-H and Co-C bonds can form at the same site. These bonds are more reactive compared to their respective Co-H (at Co-Co sites) and Pd-C (at Pd-Pd sites). We believe it is the formation of the optimal combination of metal-substrate bonds at Pd-Co sites that is responsible for the decrease in activation barriers and, consequently, for synergetic effect observed. Apparently, in this case the type of binding between Pd and Co atoms in the structure of heteronuclear sites is not very important. For synergetic effect, however, it is important that the distance between metal atoms be sufficiently close to transfer hydride from Pd to acyl ligand coordinated to Co. This situation should be realized when Pd and Co atoms are attached to the neighbouring ligand groups anchored on the support surface.

REFERENCES

1. Yu.I. Yermakov, B.N. Kuznetsov and V.A. Zakharov, Catalysis by Supported Complexes, Elsevier, N.Y. (1981).
2. Yu.I. Yermakov, L.N. Arzamaskova and V.L. Kuznetsov, Koord. Khim., 10, 87 (1984).
3. V.A. Semikolenov, V.A. Likholobov and Yu.I. Yermakov, Kinet. Katal., 22, 1026 (1981).
4. G. Doyle, J. Mol. Catal., 48, 251 (1983).
5. P.C. Ford, R.G. Rinker, C. Ungermann, R.M. Laine, V. Landis and S.A. Moya, J. Am. Chem. Soc., 100, 4595 (1978).
6. M. Hidai, A. Fukuoka, Yu. Koyasu and Ya. Uchida, J. Chem. Soc., Chem. Commun., 516 (1984).
7. M. Hidai, M. Orisaki, M. Ue and Ya. Uchida, Chem. Lett., 143 (1981).
8. G.L. Geoffrey, Acc. Chem. Res., 13, 469 (1980).
9. F.E. Paulic, Catal. Rev., 6, 49 (1972).
10. P.M. Maitlis, in The Organic Chemistry of Palladium, v. 1, p. 316, v. 2, p. 216, Acad. Press, N.Y. (1971).
11. V.A. Semikolenov, D.K. Mikhailova, Ya.I. Sobchak, V.A. Likholobov and Yu.I. Yermakov, React. Kinet. Catal. Lett., 10, 105 (1979).
12. V.A. Semikolenov, V.A. Likholobov, P.A. Zhdan, A.P. Shepelin and Yu.I. Yermakov, Kinet. Katal., 21, 429 (1980).
13. P. Szabo, L. Fekete, G. Bor, Z. Nagy-Magos and L. Marco, J. Organometal. Chem., 12, 245 (1968).
14. A.R. Manning, J. Chem. Soc., A, 1135 (1968).
15. E.V. Leonova, V.Kh. Syundyukova, F.S. Denisov, A.A. Koridze, V.S. Khandkarova and A.Z. Rubezhov, in Cobalt, Nickel, Platinum Metals (Ser. Methods of Elemento-Organic Chemistry), Nauka, Moskva (1978).
16. P. Braunstein, J. Dehand, and J.P. Neuning, J. Organometal.,Chem., 92, 117 (1975).
17. V.Yu. Gankin and G.S. Gurevich, in Technology of Oxosynthesis (in Russian), p. 75, Khimia, Leningrad (1981).
18. N.N. Bulgakov, Yu.A. Borisov and V.V. Popovskii, Kinet. Katal., 14, 468 (1973).

HOMOMOLECULAR EXCHANGE OF METHANE OVER SUPPORTED CATALYSTS OBTAINED BY DECOMPOSITION OF SURFACE METAL COMPLEXES

A.A. Shestov, Yu.A. Ryndin, V.S. Muzykantov and Yu.I. Yermakov
Institute of Catalysis, Novosibirsk 630090, USSR

ABSTRACT - Homomolecular exchange of methane on dispersed Pt/SiO_2, $(Pt+W)/SiO_2$ and $(Pt+Sn)/SiO_2$ catalysts obtained by decomposition of anchored surface metal complexes, has been studied by a new isotopic-kinetic method. It has been established that the introduction of tin decreases sharply the activity and alters significantly the exchange mechanism. The effect of tungsten is less pronounced. Modifying effect of additives on the exchange mechanism is similar to that in catalytic reactions of hydrocarbons.

1. INTRODUCTION

Supported platinum catalysts having unique properties in hydrocarbon reactions have been extensively studied. The most effective method to control their catalytic properties is modification by other elements (Refs. 1-8), when activity and selectivity can be changed by varying the chemical nature of these modifying elements. Application of organometallic compounds as precursors for catalyst preparation makes it possible to obtain dispersed catalysts in which practically all atoms of active metal are accessible to reacting molecules (Refs. 1-10). These catalysts are convenient objects to study the so-called "ligand" effects of modifying elements on the properties of supported metal clusters. Evidence for these effects comes from the change in the mechanism of formation of intermediate hydrocarbon species on metal surface. Valuable information about these mechanisms can be provided by isotopic methods that have not been applied to these catalysts previously.

Kinetics of the isotopic transfer between different components (heteromolecular exchange or heteroexchange) and redistribution of isotope molecules of the same component (homomolecular exchange or homoexchange) reflects such important features of the reaction mechanism as the atomicity of intermediate species, the rearrangement character of chemical bonds and the rate ratio of the steps whose reversible course leads to an exchange. Exact theory of isotope redistribution (Refs. 11-14) permits us to relate kinetic constants of the exchange to the parameters of reaction mechanisms. It seemed reasonable that this theory be first applied in experiments with a homomolecular exchange of the simplest organic molecule, namely, methane, so that to elucidate primarily the effect of catalyst nature on the mode of activation of hydrogen-carbon bonds in surface species. Note that in numerous isotopic studies of metal catalysts prepared by "conventional" methods (see, e.g., (Refs. 15-18)) an empirical Kemball approach (Ref. 15) is used, that provides only a qualitative interpretation of experimental data in heteroexchange of hydrocarbon with dihydrogen and is, in principle, unapplicable to the homoexchange of hydrocarbons. Therefore, in most cases the behaviour of hydrocarbons solely remains unrevealed in the presence of H_2. Homoexchange studies are very few in number (Refs.16-17) ~~and provide~~ practically only primary kinetic data.

2. CATALYSTS

For our studies we have chosen three well characterized catalysts, namely 3.1 Pt, 3.0 Pt + 6.0 W and 2.5 Pt+ + 2.2 Sn (numerals before chemical symbols are weight per cents of the respective metals), obtained by decomposition of metal complexes supported on a carrier (SiO_2 dehydrated in vacuum at 873 K with BET surface area ~250 m^2/g). It should be noted that the carrier was inert both in catalytic reactions and in exchange processes examined. SiO_2 was modified in two steps (Refs.1--6): 1). Treatment by tetrakis-π-metallyltungsten in pentane at room temperature or by Sn(II) diacetate vapours at 423 K; 2). Hydrogen treatment of W/SiO_2 at 873 K and of Sn/SiO_2 in vacuum at 523 K to remove organic ligands and to obtain surface low-valent tungsten and tin ions. Highly dispersed platinum on the surface of both modified and nonmodified carriers was obtained by the interaction of bis-π-metallylplatinum with SiO_2 in pentane with subsequent removal of the unreacted complex by the solvent and drying in vacuum (Refs. 1-6). Then the catalysts were reduced with hydrogen at 873 K. Prior to reaction the catalysts were reoxidized by O_2 at 673 K to ensure a more complete removal of the solvent and then reduced with hydrogen at 873 K (Refs. 4-6).

According to the hydrogen titration and electron microscopic data (Ref. 3), the average size of metal particles for Pt catalyst and modified catalysts was 12 and 9-10 Å, respectively. Average oxidation number of tungsten and tin, as ascertained by the chemisorption measurements and XPS and Mössbauer spectroscopic data (Refs.1,2), is close to two. Changes in the electronic properties of Pt due to modification (Refs. 1-2) indicate the direct interaction of Pt clusters with the ions of a second element.

Catalytic properties of modified Pt catalysts have been examined in several reactions of different types: benzene hydrogenation (Ref. 8), ethane hydrogenolysis (Refs. 7-8) and hexane conversions (Refs. 4-5). The modifying effect of transition (Mo,W,Re) (Refs. 5,8) and non-transition (Ga, Sn, Pb) (Refs. 4,8) elements was qualitatively different, which is clearly seen in Table 1 representing the rates of various reactions (the number of reacted molecules per unit time with respect to one surface atom on Pt) for the catalysts examined. For comparison with the isotopic exchange on the same samples, we have chosen reactions where in the formation or the rupture of C-H bonds must play a significant role. Therefore Table 1 does not list data on n-hexane conversions that occur simultaneously with hydrogenolysis, i.e. C_6- and C_5-cyclizations and skeleton isomerization (Refs.4, 5), since the determining role is played by the transformation of C-C bonds. Note, however, that besides a sharp drop in activity, modification by tin leads to a significant change in the selectivity (the yield of benzene considerably increases (Ref. 4)). Due to the great difference in the activity of modified catalysts, the rates of n-hexane hydrogenolysis are given in Table 1 at two different temperatures (773 and 823 K). Besides the reaction rate (the turn-over number) for ethane hydrogenolysis, Table 1 lists its kinetic parameters (activation energy and reaction orders with respect to C_2H_6 and H_2).

TABLE 1. Hydrogenation of C_6H_6 (6.7 kPa P_{H_2}= 40 kPa) and hydrogenolysis of n-C_6H_{14} (3.3 kPa, P_{H_2}=10^2 kPa) and of C_2H_6 (6.7 kPa, P_{H_2}= 6.7 kPa)

Catalyst	C_6H_6	n-C_6H_{14}		C_2H_6			
	(Turn-over number) x 10^2 (s^{-1})				E	$n_{C_2H_6}$	n_{H_2}
	300 K	773 K	823 K	623 K	(kJ/mol)		
Pt	1.5	2.5	4.0	6.7	230	0.9	-2.5
Pt+W	4.5	4.5	-	65	117	0.9	-0.5
Pt+Sn	<0.05	-	0.5	≪0.5	-	-	-

3. EXPERIMENTAL

Isotopic exchange of methane was examined in a static circulation system (of 300 cm^3 in volume) with a minireactor (of 1-2% of the whole volume) that could be isolated from a circulation system, which permitted us to replace gas mixtures before run without distorting the catalyst state. Pretreatment of catalyst samples (0.5-0.6 g) was similar to the above procedure, i.e. a 2 h treatment in vacuum (10^{-5} Pa) at 450 K, a 3 h oxidation in oxygen (1.3 kPa) at 650 K, a 3 h reduction in hydrogen (1.3 kPa) at 850 K and then a 2 to 24 h treatment in vacuum at the same temperature. During tho gas treatment a trap in the circulation system was cooled by liquid nitrogen to remove desorption products. In the system high-vacuum metallic valves and connections were used.

To study the homomolecular exchange of methane, we have prepared a gas mixture of CH_4 (of natural isotopic composition) and CD_4 (98 at%D) in the molar ratio of 5:2 that contains 28 at%D with the nonequilibrium distribution of isotope molecules of $CH_{4-i}D_i$: 71.4; 0; 0.1; 2.1 and 26.4 mol%, respectively for i = 0,1,2,3 and 4. Most experiments were carried out at a pressure of methane of 1.3 kPa. The reaction order was determined upon several-fold pressure variations. For the isotopic exchange of methane with hydrogen, H_2 (natural isotopic composition) plus certain amounts of either the above mixture or CD_4 were used. In these experiments we determined the rate of heteroexchange, the effect of hydrogen on homo-exchange and the dependence of its rate on methane and hydrogen pressures. In the course of experiments no noticeable chemical conversions were observed, that is, gas pressure remained almost unchanged, and changes in the mass spectra corresponded to the redistribution of isotope molecules.

An isotopic analysis was made upon continuous inlet of gas from a reaction volume to a mass-spectrometer (MI-1330) connected to a computer (on the base of an Elektronika-60 processor) to treat mass-spectra and to perform primary kinetic calculations during experiment. When treating mass-spectra (with mass numbers ranging within 14-21), we made a correction for the natural concentration of ^{13}C isotope and took into account the molecule fragmentation (ionization potential is 22 eV). Complete calculation of the exchange rates and the mechanism parameters (Refer to the following Section) was carried out on a BESM-6 computer.

4. ISOTOPIC KINETICS OF HOMOMOLECULAR EXCHANGE OF METHANE

Let us designate the fractions of isotope molecules (IM) of $CH_{4-i}D_i$ (i = 0,1,2,3 and 4) as x_i and the atomic fraction of D as α. It is evident that

$$\alpha = \frac{1}{4} \Sigma i x_i \quad \text{and} \quad \Sigma x_i = 1 . \tag{1}$$

During homoexchange α = const; consequently, if one considers Eqs.(1), IM distribution is determined by three independent isotopic concentrations (e.g., x_2, x_3 and x_4), and the redistribution kinetics is described by three independent equations. In terms of the general theory (Refs. 11-14), a strict set of isotopic-kinetic equations (without taking into account isotopic effects) can be obtained for any homoexchange mechanism of methane. In terms of x_i variables, these sets of differential equations (nonlinear in the general case) are cumbersome and practically inapplicable to discriminate mechanisms. But if one substitutes concentrations x_i (i = 2, 3 and 4) by new isotopic variables Z_m (m = 2,3 and 4) that are determined by the linear relationships:

$$Z_2 = -\alpha^2 + \frac{1}{6}x_2 + \frac{1}{2}x_3 + x_4 \qquad (2\text{-a})$$

$$Z_3 = 2\alpha^3 - \frac{1}{2}\alpha x_2 + \frac{1}{4}(1-6\alpha)x_3 + (1-3\alpha)x_4 \qquad (2\text{-b})$$

$$Z_4 = -3\alpha^4 + \alpha^2 x_2 - \alpha(1-3\alpha)x_3 + (1-4\alpha + 6\alpha^2)x_4 \qquad (2\text{-c})$$

all sets of isotopic-kinetic equations for methane homo-exchange will be reduced to a very simple general form:

$$(N/N_S)\dot{Z}_2 = -A_2 K Z_2 \qquad (3\text{-a})$$

$$(N/N_S)\dot{Z}_3 = -A_3 K Z_3 \qquad (3\text{-b})$$

$$(N/N_S)\dot{Z}_4 = -A_4 K Z_4 + A_{22} K Z_2^2 \qquad (3\text{-c})$$

Here the point above variables means time (τ) differentiation, N is the number of methane molecules, N_S is the number of active sites (surface Pt atoms in our case), and K is the exchange rate per one active site (turn-over number).

We do not intend to discuss here the physical sense of new isotopic variables, Z_m; we note only that their equilibrium values (at $\tau \rightarrow \infty$) are equal to zero. Besides, if in the initial mixture the isotope molecules of CH_4 and CD_4 are taken in the equimolar ratio (1:1), then $Z_3^o = 0$, as follows from Eq.(2-b). Here according to Eq. 3-b, this equality ($Z_3 = 0$) preserves throughout the whole experiment. Thus using the generally accepted initial composition of $CH_4:CD_4 = 1:1$ (see, e.g., (Ref. 17)), one of the three independent variables vanishes, and as a result, the informational value of experiment is lost. It is for this reason that in the present study we have used the initial composition of $CH_4:CD_4 = 5:2$ (Refer to the previous Section).

A concrete form of the right-hand part in the system of equations (3) is specified by the parameters A_k (k= = 2,3,4 and 22) that are determined by mechanisms of the reactions leading to homoexchange, to be more precise, by the character of reaction steps and their rate ratios. Altogether there are four linearly independent sets of the <u>mechanism parameters</u> A_k naturally related to the four types of methane dissociative adsorption:

	A_2	A_3	A_4	A_{22}
1) $CH_4 + 4* \rightleftharpoons *CH + 3*H$	1	1	1	0
2) $CH_4 + 3* \rightleftharpoons *CH_2 + 2*H$	5/6	1	1	0
3) $CH_4 + 2* \rightleftharpoons *CH_3 + *H$	1/2	3/4	1	0
4) $CH_4 + 2* \rightleftharpoons *CH_2 + *H_2$	2/3	1	1	1

Here the symbol * identifies adsorption sites (active centers).

If the number of adsorbed species is very small compared to that of molecules in a gas phase, the kinetics of IM redistribution due to the reversible course of these processes is described by the system of isotopic-kinetic homoexchange equations (3) with the sets of right-hand mechanism parameters. These four sets account for the four <u>types of homoexchange</u> of methane. In case several exchange types (or any other multi-step mechanisms) are realized, the observed parameters are the linear combinations of the above sets with contributions that are proportional to the respective types of homoexchange rates:

$$A_K = \sum_{j=1}^{4} \varkappa_j A_K(j), \qquad (4)$$

where $\varkappa_j = K_j/K$ (K_j is the j-type homoexchange rate).

Thus from the experimental values of A_k which are readily determined using Eqs.(3), one can easily calculate homoexchange rates K and contributions of various types $\varkappa_j$.

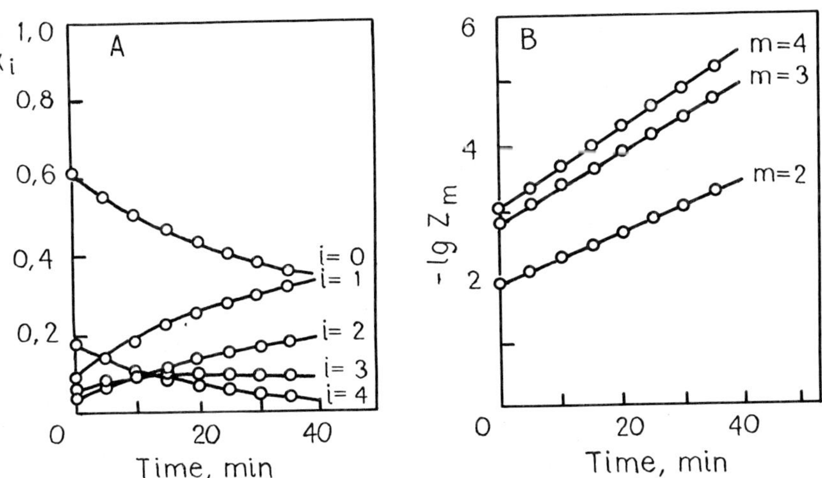

Fig. 1. Time dependence (τ, min) of: A) molar fractions of $CH_{4-i}D_i$ isotope molecules (x_i) and B) isotopic variables Z_m in methane homoexchange on $(Pt+W)/SiO_2$catalyst (T= 387 K, P= 1.3 kPa).

5. RESULTS AND DISCUSSION

It has been established that in the systems examined A_{22}= 0 (i.e. the 4th type of homoexchange is absent). In this case all three differential equations (3) are linear and can be readily integrated:

$$\ln(Z_m/Z_m^o) = -A_m K \frac{N_S}{N} \tau \quad (m = 2,3,4), \qquad (5)$$

where Z_m^o is the initial value of Z_m (at $\tau = 0$). Experimental data for one of the runs (dots) and theoretical dependences obtained are represented in Fig. 1. As is seen, experimental data are very well described by the derived isotopic-kinetic equations for CH_4 homoexchange. Since $A_4=1$ (always) and $A_{22}=0$ (in our case), the data of each experimental run were used to calculate three values homoexchange rate K and its mechanism parameters A_2 and A_3 (Table 2 and Fig. 2).

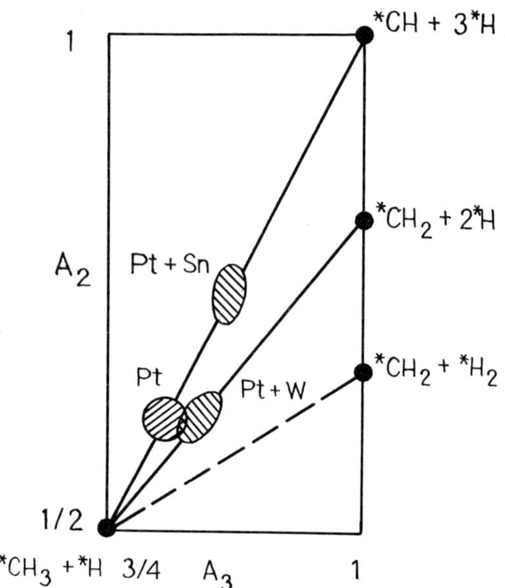

Fig. 2. Field of parameters for methane homoexchange mechanisms. Dark circles in triangle vertices correspond to the four types of dissociative adsorption; cross-hatched are the regions of averaged experimental data.

TABLE 2. Homomolecular exchange of CH_4(1.3 kPa) without H_2(*) and in its presence (P_{H_2} = 1.3 kPa).

Cata-lyst	Tempera-ture range(K)	K x $10^3(s^{-1})$		E (kJ/mol)	n_{CH_4}	n_{H_2}	R/K
		300 K	450 K				
Pt	250-450	8.1*	13.5	15* 115	1	-1	0.2-0.25
Pt+W	300-650	1.5*	>3	10÷40*	~1	-	0.1-0.2
Pt+Sn	600-700	–	~0.03	40	1	~0	0.3-0.4

In the presence of hydrogen, besides the homoexchange rate and parameters, we have also measured the rate of CH_4-H_2 heteroexchange (R) and reaction orders for homo-exchange with respect to CH_4 and H_2. Hydrogen affects in essentially different way the state of the catalysts. In the case of Pt/SiO_2,H_2 addition drastically decreases the homoexchange rate, the reaction order being negative (of -1). On the contrary, in the case of (Pt+Sn)/SiO_2 the homoexchange rate in the presence of H_2 slightly in-creases, though the reaction order is close to zero (to be more precise, it is slightly negative). For (Pt+W)/SiO_2 we failed in observing the effect of hydrogen because of the catalyst instability. Homoexchange order with respect to CH_4 on all catalysts is close to unity both with and without hydrogen. No noticeable effect of hydrogen on the homoexchange mechanism parameters has been found.

On Pt/SiO_2 and (Pt+W)/SiO_2 catalysts the homoexchange of methane is activated already at room temperature, whereas on (Pt+Sn)/SiO_2 samples the exchange is observed only at elevated temperatures. Modifying effect of tin in methane homoexchange is similar to that observed in catalytic reactions (compare Tables 1 and 2). Besides a drastic activity decrease upon adding tin, substantial changes in the homoexchange mechanism are also observed (Fig. 2), which are similar to the selectivity variations in catalysis. In both cases the effect of tungsten is less pronounced, but here a complete analogy is not observed; while in catalytic hydrocarbon conversion a noticeable increase in activity is observed, in homoexchange the activity changes only slightly.

The values observed for the A_2 and A_3 parameters (Fig. 2) indicate that in all cases the exchange mechanisms are complex (multi-step) and differ both in atomicity of intermediate species and in step rates. The main features of these mechanisms can be explained by a scheme that involves consecutive processes of the surface dissociation of hydrocarbon species:

$$CH_4 \underset{1}{\rightleftharpoons} {}^{*}CH_3 \underset{2}{\rightleftharpoons} {}^{*}CH_2 \underset{3}{\rightleftharpoons} {}^{*}CH.$$

In principle, it is possible to relate quantitatively the parameters of homoexchange mechanisms to the ratios of step rates, but here we will restrict ourselves to a qualitative consideration. Note, first, that the homoexchange rate is determined by the first step. On Pt/SiO_2 and $(Pt+W)/SiO_2$ catalysts this step is predominant and its rates are approximately the same. Upon adding Sn, this rate sharply drops. If the second step is much slower than the third one, the superposition of second- and third-type exchanges will be observed, as is the case on a platinum catalyst and on a sample modified by tin, whose addition decreases essentially the relative rate of step 1. The increase in the relative rate of step 2 leads to the appearance of the second-type exchange, which is observed on a tungsten-modified catalyst. At least some of these steps (or of steps that are similar to them) must also take place in ethane hydrogenolysis (and in the other reactions studied), which may account for a certain aforesaid analogy in the modifying effect of additives.

Thus the application of isotopic kinetic methods to the exchange of simple organic molecules indicates the significant discriminating possibilities of these methods to analyze the mechanism of intermediate species formation.

ACKNOWLEDGEMENT. The authors are grateful to S.G. Shubnikov for his valuable assistance in experiments and for the development of the minireactor.

REFERENCES

1. M.S. Ioffe, Yu.M. Shulga, Yu.A. Ryndin, B.N. Kuznetsov, A.N. Startsev, Yu.G. Borodko and Yu.I. Yermakov, React.Kinet.Catal.Lett.,4, 2, 229-234 (1976).
2. A.P. Shepelin, A.P. Chernyshev, V.I. Kovalchuk, P.A. Zhdan, E.N. Yurchenko, B.N. Kuznetsov and Yu.I. Yermakov, Kinet.Katal.,22, 716-721 (1981).
3. V.I. Zaikovskii, Yu.A. Ryndin, V.I. Kovalchuk, L.M. Plyasova, B.N. Kuznetsov and Yu.I. Yermakov, Kinet. Katal., 22, 443-451 (1981).
4. Yu.A. Ryndin, J. Laurent, B.N. Kuznetsov, V.I.Kovalchuk, A. Pentenero and Yu.I. Yermakov, Kinet. Katal., 22, 675-679 (1981).
5. J. Laurent, Yu.A. Ryndin, B.N. Kuznetsov, A.N. Startsev, A. Pentenero and Yu.I. Yermakov, Kinet. Katal., 22, 680-683 (1981).
6. B.N. Kuznetsov, Yu.I. Yermakov, Yu.A. Ryndin, J. Laurent, V.K. Duplyakin and A. Pentenero, Kinet. Katal., 22, 1200-1206 (1981).
7. Yu.I. Yermakov, V.A. Zakharov and B.N. Kuznetsov,in Supported Complexes on Oxide Carriers in Catalysis (in Russian), Nauka, Novosibirsk(1980).
8. M.S. Ioffe, B.N. Kuznetsov, Yu.A. Ryndin and Yu.I. Yermakov, 6th Intern.Congr.Catal., London 1976, Proc. v.1, pp. 131-137.
9. Yu.I. Yermakov, 7th Intern.Congr.Catal., Tokyo, 1980 Pure Appl.Chem., 52, 9, 2075-2089 (1980).
10. Yu.I. Yermakov, B.N. Kuznetsov and V.A. Zakharov, in Catalysis by Supported Complexes, Studies in Surface Science and Catalysis, v. 8, Elsevier (1981), 522 p.
11. V.S. Muzykantov, React.Kinet.Catal.Lett., 13, 4, 419-424 (1980).

12. V.S. Muzykantov, React.Kinet.Catal.Lett.,14,1, 113-118 (1980).
13. V.S. Muzykantov, React.Kinet.Catal.Lett.,14, 2, 161-167 (1980).
14. V.S. Muzykantov, React.Kinet.Catal.Lett., 14, 3, 311-316 (1980).
15. C. Kemball, Adv.Catal., 11, 223-262 (1959);
16. A. Frennet, Cat.Rev., 10(1), 37-68 (1974).
17. J.G. Larson, and W.K. Hall, J.Phys.Chem., 69, 9, 3080-3089 (1965).
18. L. Guczi, B.S. Gudkov and P. Tetenyi, J.Catal., 24, 2, 187-196 (1972).

LIST OF CONTRIBUTORS

AKHREM, I.S. Institute of Organoelement Compounds, USSR Academy of Sciences, Moscow 117813, USSR

ALBANO, K. Institut de Recherches sur la Catalyse, CNRS, convention l'Université Claude Bernard Lyon 1, 2 avenue Albert Einstein 6, Villeurbanne cédex

ALEKSANDROV, M.N. Institute of Catalysis, Novosibirsk 630090, USSR

ALT, L.Ya. Institute of Catalysis, Omsk 664029, USSR

ASAKURA, K. Department of Chemistry, Faculty of Science, The University of Tokyo, Hongo, Bunkyo-ku, Tokyo 113, Japan

BABA, T. Department of Industrial Chemistry, Kochi College of Technology, Monobe, Nankoku-shi, Kochi 783, Japan

BASSET, J.M. Institut de Recherches sur la Catalyse, Laboratoire Propr. C.N.R.S., conventionné à l'Université Claude Bernard, Lyon 2, avenue Albert Einstein - 69626 Villeurbanne Cédex - France

BASU, A. Alchemie Research Centre, CAFI Site, P.O.B. 155, Thane-Belapur Road, Thane 400 601, Maharashtra, India

BHADURI, S.* Alchemie Research Centre, CAFI Site, P.O.B. 155, Thane-Belapur Road, Thane 400 601, Maharashtra, India

BEIJNUM, Van, J. Department of Inorganic Chemistry, University of Utrecht, Croesestraat 77a;

BERENTSWEIG, V.V. State University, Tyumen 625003; State University, Moscow 117234, USSR

BOAG, N.M. Department of Chemistry & Biochemistry, University of California, Los Angeles, CA 90024, USA

BRACA, G. Dipartimento di Chimica e Chimica Industriale, Università di Pisa, 56100 Pisa, Italy

BRENNER, A.* Department of Chemistry, Wayne State University, Detroit Michigan 48202, U.S.A.

BROUWER, W.M. Laboratory of Polymer Chemistry, Eindhoven University of Technology, P.O. Box 513, 5600 MB EINDHOVEN, The Netherlands

BROVKO, V.S. Lensovet Technological Institute, Leningrad 198013, USSR

BULGAKOV, N.N. Institute of Catalysis, Novosibirsk 630090, USSR

BUTKOVA, O.L. A.V. Topchiev Institute of Petrochemical Synthesis of the Academy of Sciences of the USSR, Moscow 117912, USSR

CAI, F.X. Laboratoire de Réactivité de Surface et Structure, UA 1106, CNRS, 4, Place Jussieu, Tour 54-55, 2ème étage, 75252 Paris Cédex 05, France

CAMPOSTRINI, R. Facoltà di Ingegneria, Università di Trento, Italy

ČAPKA, M.* Institute of Chemical Process Fundamentals, Czechoslovak Academy of Sciences, 165 02 Prague 6-Suchdol, Czechoslovakia

CARLINI, C.* Dipartimento di Chimica e Chimica Industriale, Università di Pisa, Centro del C.N.R. per le Macromolecole Stereordinate ed Otticamente Attive, Via Risorgimento 35, 56100 Pisa, Italy

CARTURAN, G.* Facoltà di Ingegneria, Università di Trento, Italy

CECCHI, A. Dipartimento di Chimica e Chimica Industriale, Università di Pisa, 56100 Pisa, Italy

CHAKRABARTY, D.K. Solid State Laboratory, Chemistry Department, Indian Institute of Technology, Bombay 400 076, India

CHERKASHIN, G.M. A.V. Topchiev Institute of Petrochemical Synthesis of the USSR Academy of Sciences. Moscow 117912, USSR

CHOPLIN, A. Institut de Recherches sur la Catalyse, Laboratoire Propr. C.N.R.S., conventionné à l'Université Claude Bernard, LYON 2 avenue Albert Einstein - 69626 Villeurbanne Cédex - France

CIARDELLI, F. Dipartimento di Chimica e Chimica Industriale, Università di Pisa, Centro del C.N.R. per le Macromolecole Stereordinate ed Otticamente Attive, Via Risorgimento 35, 56100 Pisa, Italy

CONESA, J.C. [2]Ins. Catálisis y Petroleoquimica, C.S.I.C., Madrid, Spain

DESAI, A.A. Solid State Laboratory, Chemistry Department, Indian Institute of Technology, Bombay 400 076, India

Di BUGNO, C. Dipartimento di Chimica e Chimica Industriale, Università di Pisa, 56100 Pisa, Italy

DILLEN, Van, A.J. Department of Inorganic Chemistry, University of Utrecht, Croesestraat 77a

DMITRIEVA,L.F. Institute of Chemistry and Chemical Engineering, Krasnoyarsk 660097; USSR

DOBOS, S. Institute of Isotopes of the Hungarian Academy of Sciences, P.O.Box 77, H-1525 Budapest, Hungary

DOLDI, P. Dipartimento di Chimica Fisica ed Elettrochimica, Università di Milano, Italy

DOLGOPLOSK, B.A.* A.V. Topchiev Institute of Petrochemical Synthesis of the USSR Academy of Sciences, Moscow, 117912, USSR

DORONIN, V.P. Institute of Catalysis, Omsk 664029, USSR

DOSSI, C. Centro C.N.R.[1] and Dipartimento di Chimica Inorganica e Metallorganica, Università di Milano, Italy

DRENTH, W. Department of Organic Chemistry, University of Utrecht, Transitorium 3, Padualaan 8, 3584 CH, Utrecht, The Netherlands
DUPLYAKIN, V.K. Institute of Catalysis, Omsk 664029, USSR
DYACHKOVSKII, F.S.* Institute of Chemical Physics of the USSR Academy of Sciences, Moscow Region, Chernogolovka, 142432, USSR
EFENDIEV, A.A. Institute of Theoretical Problems of Chemical Technology, Baku 370143, USSR
EFREMOV, A.A. Institute of Chemistry and Chemical Engineering, Krasnoyarsk 660097, USSR
ELEV, I.V. N.D. Zelinsky Institute of Organic Chemistry, Academy of Sciences of the USSR, Moscow 117913, USSR
ELNATANOVA, A.I. A.V. Topchiev Institute of Petrochemical Synthesis of the USSR Academy of Sciences, Moscow 117912, USSR
FARAFONOV, V.V. I.M. Gubkin Institute of Petrochemical and Gas Industry, 117917 Moscow, USSR
FEDOTOV, A.V. Institute of Catalysis, Novosibirsk 630090, USSR
FENELONOV, V.B. Institute of Catalysis, Novosibirsk 630090, USSR
FROLOV, V.M. A.V. Topchiev Institute of Petrochemical Synthesis of the USSR Academy of Sciences, Moscow 117912, USSR
FUJIMOTO, K. Department of Synthetic Chemistry Faculty of Engineering, The University of Tokyo Hongo 7-3-1, Bunkyo-ku, Tokyo 113, Japan
FUSI, A. Centro C.N.R.[1] and Dipartimento di Chimica Inorganica e Metallorganica, Università di Milano, Italy
GALLEZOT, P.* Institut de Recherches sur la Catalyse, Laboratoire Propre du C.N.R.S., conventionné à l'Université Claude Bernard LYON 1, 2, avenue Albert Einstein - 69626 - Villeurbanne, France
GANZERLA, R. Dipartimento di Chimica, Università di Venezia, Italy
GARNETT, J.L. University of New South Wales, Kensington, N.S.W. Australia 2033
GERASIMOV, O.V. Institute of Catalysis, Novosibirsk 630090, USSR
GERMAN, A.L.* Laboratory of Polymer Chemistry, Eindhoven University of Technology, P.O. Box 513, 5600 MB EINDHOVEN, The Netherlands
GEUS, J.W. Department of Inorganic Chemistry, University of Utrecht, Groesestraat 77a
GONZÁLEZ-ELIPE, A.R. [1]Dpto. Quimica General, Universidad de Sevilla, Seville, Spain
GOLOVIN, A.V. Institute of Catalysis, Novosibirsk 630090, USSR

GRAZIANI, M.* Dipartimento di Scienze Chimiche, Universita' di Trieste, Italy

GRYAZNOV, V.M. Institute of Petrochemical Synthesis, Moscow 117912, USSR

GUCZI, L.* Institute of Isotopes of the Hungarian Academy of Sciences, P.O.Box 77, H-1525 Budapest, Hungary

HJORTKJÆR, J. Technical University of Denmark, Building 227, 2800 Lyngby, Denmark.

HSU, Chao-Yang Applied Research and Development Department, Sun Refining and Marketing Company, P.O. Box 1135, Marcus Hook, PA 19061 USA

ICHIKAWA, M.* Research Institute for Catalysis, Hokkaido University, Sapporo, 060 Japan

ITTEL, S.D.* Central Research & Development Department* E.I. du Pont de Nemours & Company, Experimental Station, Wilmington, DE 19898 USA

IWASAWA, Y.* Department of Chemistry, Faculty of Science, The University of Tokyo, Hongo, Bunkyo-ku, Tokyo 113, Japan

JANKIEWICZ, S.V. University of New South Wales, Kensington, N.S.W. Australia 2033

JEN, Hung-Wen Department of Chemistry, Wayne State University, Detroit, Michigan 48202, USA

KABANOV, V.A.* Institute of Petrochemical Synthesis, USSR Academy of Sciences, Moscow, USSR

KAESZ, H.D.* Department of Chemistry & Biochemistry, University of California, Los Angeles, CA 90024, USA

KAGAN, Yu.B. A.V. Topchiev Institute of Petrochemical Synthesis of the Academy of Sciences of the USSR, Moscow 117912, USSR

KAMINSKY, W.* Institute of Technical and Macromolecular Chemistry, University of Hamburg, F.R. of Germany

KARAKHANOV,R.A. I.M. Gubkin Institute of Petrochemical and Gas Industry, 117917 Moscow, USSR

KASPAR, J. Dipartimento di Scienze Chimiche, Universita' di Trieste, Italy

KATO, A. Hitachi Research Laboratory of Hitachi Ltd, Hitachi, Ibaraki, Japan 319-12

KAZANSKY, V.B.* N.D. Zelinsky Institute of Organic Chemistry, Moscow 117913, USSR

KEDROVA, L.K. Institute of Chemistry and Chemical Engineering, Krasnoyarsk 660097, USSR

KEIM, W.* Institut für Technische Chemie und Petrolchemie der Rheinisch-Westfälischen Technischen Hochschule Aachen, Worringer Weg 1, D-5100 Aachen, F.R.G.

KERMAREC, M. Laboratoire de Réactivité de Surface et Structure, UA 1106, CNRS,4, Place Jussieu, Tour 54-55, 2ème étage, 75252 Paris Cédex 05, France

KLABUNDE, U. Central Research & Development Department* E.I. du Pont de Nemours & Company, Experimental Station, Wilmington, DE 19898 USA

KLIGER, E.G. A.V. Topchiev Institute of Petrochemical Synthesis of the USSR Academy of Sciences, Moscow 117912, USSR

KNOBLER, C.B. Department of Chemistry & Biochemistry, University of California, Los Angeles, CA 90024, USA

KNOX, S.A.R.* Department of Inorganic Chemistry, The University, Bristol, BS8 1TS, U.K.

KNÖZINGER H.* Institut für Physikalische Chemie, Universität München, Sophienstrasse 11, 8000 München 2, W.-Germany

KOMIYAMA, M. Department of Chemical Engineering, Faculty of Engineering, Tohoku University, Aramaki Aoba, Sendai 980 Japan

KORNEEVA, G.A. A.V. Topchiev Institute of Petrochemical Synthesis of the Academy of Sciences of the USSR, Moscow 117912, USSR

KOVALCHUK, V.I. Krasnoyarsk State University, Krasnoyarsk 660062, USSR

KRONE-SCHMIDT, W. Department of Chemistry & Biochemistry, University of California, Los Angeles, CA 90024, USA

KUDRYAVTSEV, G.V. Chemical Department, State University, Moscow 119899, USSR

KUZNETSOVA, L.I. Institute of Catalysis, Novosibirsk 630090, USSR

KUZNETSOV, B.N. Institute of Chemistry and Chemical Engineering, Krasnoyarsk 660097, USSR

KUZNETSOV, V.L. Institute of Catalysis, Novosibirsk 630090, USSR

LAPIDUS, A.L. I.M. Gubkin Institute of Petrochemical and Gas Industry, 117917, Moscow, USSR

LENARDA, M. Dipartimento di Chimica, Universita' di Venezia, Italy

LEPETIT, C. Laboratoire de Réactivité de Surface et Structure, UA 1106, CNRS, 4, Place Jussieu, Tour 54-55, 2ème étage, 75252 Paris Cédex 05, France

LEVOT, R. University of New South Wales, Kensington, N.S.W. Australia 2033

LIKHOLOBOV, V.A. Institute of Catalysis, Novosibirsk 630090, USSR

LISICHKIN, G.V.* Chemical Department, State University, Moscow 119899, USSR

LISITSYN, A.S. Institute of Catalysis, Novosibirsk 630090, USSR

LOKTEV, S.M. A.V. Topchiev Institute of Petrochemical Synthesis of the Academy of Sciences of the USSR, Moscow 117912, USSR

LONG, M.A. University of New South Wales, Kensington, N.S.W. Australia 2033
LUNIN, A.F. I.M. Gubkin Institute of Petrochemical and Gas Industry, 117917 Moscow, USSR
LYMAR, S.V. Institute of Catalysis, Novosibirsk 630090, USSR
LYONS, E.* Applied Research and Development Department, Sun Refining and Marketing Company, P.O. Box 1135, Marcus Hook, PA 19061 USA
MARGITFALVI, L.* Central Research Institute for Chemistry of the Hungarian Academy of Sciences, Budapest, 1525 POB 17, Hungary
MATSENKO, G.P. State University, Tyumen 625003; State University, Moscow 117234, USSR
MATSUDA, S. Hitachi Research Laboratory of Hitachi Ltd, Hitachi, Ibaraki, Japan 319-12
MATVEEV, K.I. Institute of Catalysis, Novosibirsk 630090, USSR
MAYER, L.I. Institute of Catalysis, Novosibirsk 630090, USSR
MÉRIAUDEAU, P.* Institut de Recherches sur la Catalyse, CNRS, conventioné à l'Université Claude Bernard Lyon 1, 2 avenue Albert Einstein 69626, Villeurbanne cédex, France
MICHALSKA, Z.M. Institute of Polymers, Technical University, ul. Zwirki 36, 90-924 Lódź, Poland
MIRI, M.H. Institute of Technical and Macromolecular Chemistry, University of Hamburg, F.R. of Germany
MIRSKAYA, E.Ya. A.V. Topchiev Institute of Petrochemical Synthesis of the USSR Academy of Sciences, Moscow 117912, USSR
MKRTYCHAN, V.R. I.M. Gubkin Institute of Petrochemical and Gas Industry, 117917 Moscow, USSR
MOROZ, B.L. Institute of Catalysis, Novosibirsk 630090, USSR
MULHAUPT, R. Central Research & Development Department* E.I. du Pont de Nemours & Company, Experimental Station, Wilmington, DE 19898 USA
MUNUERA, G. [1]Dpto. Quimica General, Universidad de Sevilla, Seville, Spain
MUZYKANTOV, V.S. Institute of Catalysis, Novosibirsk 630090, USSR
NACCACHE, C. Institut de Recherches sur la Catalyse, CNRS, conventionné à l'Université Claude Bernard Lyon 1, 2 avenue Albert Einstein 69626, Villeurbanne cédex, France
NGUYEN, H.P. Institut de Recherches sur la Catalyse, CNRS, conventionné à l'Université Claude Bernard Lyon 1, 2 avenue Albert Einstein 69626, Villeurbanne cédex, France

NESTEROV, G.A. Institute of Catalysis, Novosibirsk 630090, USSR

NOVIKOVA, A.V. A.V. Topchiev Institute of Petrochemical Synthesis of the USSR Academy of Sciences, Moscow 117912, USSR

OGINO, Y. Department of Chemical Engineering, Faculty of Engineering, Tohoku University, Aramaki Aoba, Sendai 980 Japan

OLIVIER, D. Laboratoire de Réactivité de Surface et Structure, UA 1106, CNRS, 4, Place Jussieu, Tour 54-55, 2ème étage, 75252 Paris Cédex 05, France

ONO, Y.* Department of Chemical Engineering, Tokyo Institute of Technology, Ookayama, Meguro-ku, Tokyo 152, Japan

OMATA, K. Department of Synthetic Chemistry Faculty of Engineering, The University of Tokyo Hongo 7-3-1, Bunkyo-ku, Tokyo 113, Japan

ORESHKIN, I.A. A.V. Topchiev Institute of Petrochemical Synthesis of the USSR Academy of Sciences, Moscow, 117912, USSR

D'ORNELAS, L. Institut de Recherches sur la Catalyse, Laboratoire Propre du C.N.R.S., conventionné à l'Université Claude Bernard, Lyon I, 2 avenue Albert Einstein -69626 Villeurbanne Cedex - France

ORUJEV, D.D. Institute of Theoretical Problems of Chemical Technology, Baku 370143, USSR

PARENAGO, O.P. A.V. Topchiev Institute of Petrochemical Synthesis of the USSR Academy of Sciences, Moscow 117912, USSR

PARMON, V.N. Institute of Catalysis, Novosibirsk 630090, USSR

PASHIS, A.V. Institute of Catalysis, Novosibirsk 630090, USSR

PAUKSHTIS, E.A. Institute of Catalysis, Novosibirsk 630090, USSR

PAVLOVA, S.N. Institute of Catalysis, Novosibirsk 630090, USSR

PIET, P. Laboratory of Polymer Chemistry, Eindhoven University of Technology, P.O.Box 513, 5600 MB Eindhoven, The Netherlands

POMOGAILO, A.D. Institute of Chemical Physics of the USSR Academy of Sciences, Moscow Region, Chernogolovka, 142432, USSR

POPOVSKII, V.V. Institute of Catalysis, Novosibirsk 630090, USSR

PRIMET, M. Institut de Recherches sur la Catalyse, CNRS conventionné à l'Université Claude Bernard Lyon 1, 2 avenue Albert Einstein 69626, Villeurbanne cedex

PRITULA, N.A. A.V. Topchiev Institute of Petrochemical Synthesis of the Academy of Sciences of the USSR, Moscow 117912, USSR

PSARO, R. Centro C.N.R. and Dipartimento di Chimica Inorganica e Metallorganica, Università di Milano, Italy

RAGAINI, V. Dipartimento di Chimica Fisica ed Elettrochimica, Università di Milano, Italy

RAKOVSKY, S.K. Institute of Kinetics and Catalysis, Bulgarian Academy of Sciences, Sofia 1113, Bulgaria

REIKHSFELD, V.O. Lensovet Technological Institute, Leningrad 198013, USSR

RIVES-ARNAU, V.* [1]Dpto. Quimica General, Universidad de Sevilla, Seville, Spain

ROBEERST, A.J.M.S. Laboratory of Polymer Chemistry Eindhoven University of Technology, P.O.Box 513 5600 MB Eindhoven, The Netherlands

SKUPINSKI, W. Institute of Organic Technology, Faculty of Chemistry, Technical University (Politechnika), 00-662, Warszawa, Poland

SKVORTSOV, N.K. Lensovet Technological Institute, Leningrad, 198013, USSR

SMETANYUK, V.I. Institute of Petrochemical Synthesis, USSR Academy of Sciences, Moscow, USSR

SORIA, J., Dpto. Química General, Universidad de Sevilla, Seville, Spain

STEINGAUER, L.G. Institute of Catalysis, Novosibirsk 630090, USSR

SULD, G. Applied Research & Development Depertment, Sun Refining & Marketing Company, P.O. Box 1135, Marcus Hook, PA 19061, USA

TACHI, T. Hitachi Research Laboratory of Hitachi Ltd., Hitachi, Ibaraki, Japan 319-12

TANAKA, Ken-ichi * The Institute for Solid State Physics, The University of Tokyo, Roppongi, Minato-ku, Tokyo 106, Japan

TANAKA, K. Research Institute for Catalysis, Hokkaido University, Sapporo 060, Japan

THÉOLIER, A. Institut de Recherches sur la Catalyse, Laboratoire Propr. C.N.R.S., conventionné à l'Université Claude Bernard, Lyon 1, 2 avenue Albert Einstein-69626 Villeurbanne Cédex, France

TOMINAGA, H. Department of Synthetic Chemistry Faculty of Engineering, The University of Tokyo, Hongo 7-3-1, Bunkyo-ku, Tokyo 113, Japan

TOROMANOVA-PETROVA, P. Higher Institute of Chemical Technology, Burgas 8010, Bulgaria

TSVETKOV, I.M. Institute of Catalysis, Novosibirsk 630090, USSR

UGO, R. Centro C.N.R. and Dipartimento di Chimica Inorganica e Metallorganica, Università di Milano, Italy

VALENTINI, G. Dipartimento di Chimica e Chimica Industriale, Università di Pisa, 56100 Pisa, Italy
VASNETSOV, S.A. Institite of Catalysis, Novosibirsk 630090, USSR
VASUMATHI, V. Solid State Laboratory, Chemistry Department, Indian Institute of Technology, Bombay 400 076, India
VOLPIN, M.E.* Institute of Organoelement Compounds, USSR Academy of Sciences, Moscow 117813, USSR
VORONTSOVA, N.V. State University, Tyumen 625003; State University, Moscow 117234, USSR
WELLS, P.B.* Department of Chemistry, The University, Hull, HU6 7RX, United Kingdom
YAMAMOTO, K. Department of Chemical Engineering, Faculty of Engineering, Tohoku University, Aramaki Aoba, Sendai 980 Japan
YAMASHITA, H. Hitachi Research Laboratory of Hitachi Ltd, Hitachi, Ibaraki, Japan 319-12
YERMAKOV, Yu.I. Institute of Catalysis, Novosibirsk 630090, USSR
YUFFA, A.Ya. State University, Tyumen 625003, USSR
YURCHENKO, E.N. Institute of Catalysis, Novosibirsk 630090, USSR
ZAKHAROV, V.A. Institute of Catalysis, Novosibirsk 630090, USSR
ZANDERIGHI, G.M. Centro C.N.R. and Dipartimento di Chimica Inorganica e Metallorganica, Università di Milano, Italy
ZANONI, R. Dipartimento di Chimica, Università "La Sapienza", Roma, Italy
ZAMARAEV, K.I. * Institute of Catalysis, Novosibirsk 630090, USSR
ZHIDOMIROV, G.M. Institute of Catalysis, Novosibirsk 630090, USSR
ZHIZHINA, E.G. Institute of Catalysis, Novosibirsk 630090, USSR
ZVEZDKINA, L.I. A.V. Topchiev Institute of Petrochemical Synthesis of the Academy of Sciences of the USSR, Moscow 117912, USSR
ZYMBAL, T.V. Institute of Catalysis, Omsk 664029, USSR

* Invited speakers